T0092044

Introduction to Nanophotonics

Introduction to Nanophotonics

Henri Benisty, Jean-Jacques Greffet

Institut d'Optique Graduate School, Université Paris-Saclay

Philippe Lalanne

Institut d'Optique Graduate School, Université de Bordeaux, CNRS

OXFORD
UNIVERSITY PRESS
Great Clarendon Street, Oxford, OX2 6DP,

United Kingdom

Oxford University Press is a department of the University of Oxford.
It furthers the University's objective of excellence in research, scholarship,
and education by publishing worldwide. Oxford is a registered trade mark of
Oxford University Press in the UK and in certain other countries

Impression: 1

Published in the United States of America by Oxford University Press
198 Madison Avenue, New York, NY 10016, United States of America

British Library Cataloguing in Publication Data
Data available

Library of Congress Control Number: 2021945081

ISBN 978–0–19–878613–9

DOI: 10.1093/oso/9780198786139.001.0001

Printed and bound in the UK by
TJ Books Limited

Cover image: Early-day optical metasurfaces composed of arrayed
TiO2 nanowaveguides, courtesy of Edmond Cambril (C2N, Palaiseau, France)

To

Anne and Georges

Marie-Madeleine, Damien, Matthieu, and Laetitia

Dominique, Inès, Pierre, and Elsa

Foreword

The purpose of this book is to cover the scope of nanophotonics, a discipline of interest for physicists, chemists, and electrical engineers. Our feeling when we started the project was that a book covering most of its aspects all together was lacking. A couple of years later, we hope that its 'centripetal' motivation remains welcome in a blossoming field.

Nanophotonics results from the merging of different communities that took place when a strong emphasis was put on nanoscience at the turn of the millennium. These groups had followed rather separate paths until then as regards nanoscience. Near-field microscopy, which focused on developing subwavelength resolution imaging started in 1984.[1] The study of photonic crystals became a very active topic starting in the late 1990s.[2] Propagating surface plasmons had been a very active topic in the 1980s. With the advent of imaging techniques, chemical synthesis of nanoparticles, and progress of nanofabrication around 2000, it became a new field called plasmonics.[3] On the photonics side, confined electrons played a key role, either from optoelectronic devices based on quantum wells or quantum dots[4] or from chemically synthesized colloidal quantum dots.[5] Finally, a major topic in the field of nanophotonics is metamaterials, a diverse playground where all these communities are very active.

It is very clear that nanophotonics is a field where people with very different backgrounds (optics, semiconductor physics, quantum physics, electrical engineering, chemistry) are contributing. The purpose of *Introduction to Nanophotonics* is to provide a self-consistent introduction to all the aspects of the topic. The purpose of the book is to be used as a textbook by instructors at the master's and PhD levels and as a reference by researchers working in the field.

The current book is a collaborative project between three researchers that have been actively involved in the field from different communities. Henri Benisty has a background in semiconductor physics and optoelectronics, Jean-Jacques Greffet has a background in near-field optics and light scattering, and Philippe Lalanne has a background in diffractive optics and photonic crystals. All of them have made significant contributions to the advancement of the field. The material herein is based on lectures that have been given by them at the Institut d'Optique Graduate School (Palaiseau, Bordeaux, and Saint-Etienne). The book is organized into 21 chapters with complements containing advanced material and 146 exercises.

[1] The first Near-Field Optics conference took place in 1992 in Arc-et-Senans, France.

[2] The first Photonic and Electromagnetic Crystal Structure conference took place in Laguna Beach, USA in 1999.

[3] The first Surface Plasmon Photonics conference took place in Obernai, France, in 2001.

[4] The conference on Optics and Excitons in Confined Systems started in 1987 in Rome, Italy.

[5] The conference on Quantum Dots started in 2000 in Munich, Germany.

The first part of the book is mainly devoted to an introduction to electromagnetism with an emphasis on optics applications. The purpose of this section is to give a unified presentation of material that is usually scattered among electrical engineering books (waveguides, radiation) and optics-oriented ones (propagation, reflection, refraction, diffraction). It also includes linear response theory with a discussion of the main models of absorption and temporal and spatial dispersion. It is concluded by a chapter giving a vision as universal as can be of resonators and cavities, relying on quasinormal-mode and coupled-mode theories, both of which deserve increasing attention in nanophotonics nowadays.

The second part of the book also covers introductory basic material on electronic and optical properties of semiconductors, and particularly their confined version (quantum wells, wires, dots, graphene).

The third part of the book is devoted to a detailed introduction to optics at the nanoscale, also known as near-field optics. Its purpose is to clearly define what near-field optics is and to emphasize the differences from far-field optics and the similarities with electrical engineering. This is also the section where super-resolution imaging is discussed and where the theoretical tools for computing near fields are introduced.

The fourth part of the book covers surface plasmons, the emblematic use of metals in nanophotonics. Two chapters deal with extended and localized surface plasmons.

The fifth part of the book (the longest after Part I) covers artificial media. Here, we introduce propagation in periodic media and Bloch modes, in various dimensionalities. We then move to periodic waveguides, slow light, and homogenization and also articulate the nanophotonics knowledge to the expanding fields of metamaterials and metasurfaces.

The sixth part deals with interaction of confined photons with material media, and the further device implication. One chapter covers the nanoantenna concept and analyses in detail how a nanostructure can be used to tailor spontaneous emission, Raman scattering, etc. The other deals with nanophotonics in devices of chosen areas such as sensors and lasers, whereby the significance of the nanophotonics concepts of the book make most sense.

Finally, the seventh part is devoted to fluctuational electrodynamics. We analyse thermal emission and radiative heat transfer at the nanoscale.

ForewordForeword ForewordForeword

Acknowledgements

Science is a collective construction. We are deeply indebted to all the students and colleagues with whom we have collaborated over the last 30 years. We wish to express our thanks for the scientific input we received from the master's students attending the courses, our PhD students, post docs and all our coauthors for the inspiring interactions. We also warmly remember our former colleagues who, early in our careers, thoughtfully paved our way towards areas that would only later be named as nanophotonics: Pierre Chavel, Jean-Michel Gérard, Romuald Houdré, Jean-Pierre Huignard, Mike Morris, Dieter Pohl, Ross Stanley, Claude Weisbuch, and the late Emil Wolf. Many colleagues are gratefully acknowledged for having agreed to read parts of the book and for providing us with invaluable critical remarks: Elise Bailly, Pierre

Chavel, Alexandre Delga, Yannick de Wilde, Flore Hentinger, Thomas Krauss, Yves Lassailly, Aurelian Loirette-Pelous, Anne Nguyen, Emmanuel Rousseau, Kevin Vynck, Léo Wojszvzyk, and Tong Wu.

The book reflects many exchanges with close colleagues, which have contributed to deepening our understanding of the field through repeated valuable discussions. We particularly thank Mark Brongersma, Rémi Carminati, Carsten Henkel, Jean-Paul Hugonin, Jean-Pierre Huignard, Karl Joulain, Marine Laroche, Haitao Liu, François Marquier, Lukas Novotny, Anne Sentenac, Christophe Sauvan, Benjamin Vest, Kevin Vynck, and Wei Yan.

We are particularly indebted to Jean-Paul Hugonin and Mondher Besbes for providing many figures for the book and to Alexandre Gras for providing Fig. 7.2. We also thank Giulia Lipparini and Sonke Adlung from Oxford University Press for their assistance and support.

Contents

Part I

Basics of electromagnetic optics

1 Basics of electrodynamics of continuous media

The purpose of this chapter is to briefly summarize the basics of electrodynamics. We assume that the reader knows Maxwell equations and start with Maxwell equations with classical and microscopic descriptions of matter. We then introduce the potentials and their connection to the longitudinal and transverse fields. In section 1.3, we introduce the macroscopic electromagnetic fields which result from spatial averaging of the microscopic fields. It is shown that these averaged fields obey macroscopic Maxwell equations where additional terms (polarization and magnetization) appear as a result of the spatial average of the microscopic charge and current densities. We then introduce key ideas about energy conservation of electromagnetic fields, derive the reciprocity theorem, and discuss its physical content. We close the chapter by classifying a few generic topics: radiation (treated in Chapter 2); propagation and diffraction (treated in Chapter 4); reflection and transmission (treated in Chapter 5); scattering (treated in Chapter 12).

1.1 Microscopic Maxwell equations

1.1.1 Fundamental equations

In classical electrodynamics, the interaction between charged particles is described by introducing electromagnetic fields. The direct interaction between particles is then replaced by a set of equations connecting the force to the fields and the fields to the charges. The force $\mathbf{F}(t)$ on a system with charge density $\rho(\mathbf{r},t)$ and current density $\mathbf{j}(\mathbf{r},t)$, known as the Coulomb–Lorentz force,[1] is given by

$$\mathbf{F}(t) = \int \mathrm{d}^3\mathbf{r}[\rho(\mathbf{r},t)\mathbf{E}(\mathbf{r},t) + \mathbf{j}(\mathbf{r},t) \times \mathbf{B}(\mathbf{r},t)]. \tag{1.1}$$

The equations connecting the fields to the charges are Maxwell equations. In the SI units system, they are given by ref. (Jackson, 1975; Zangwill, 2013):

$$\nabla \cdot \mathbf{E} = \frac{\rho}{\varepsilon_0} \qquad \nabla \cdot \mathbf{B} = 0. \tag{1.2}$$

$$\nabla \times \mathbf{E} = -\frac{\partial \mathbf{B}}{\partial t} \qquad \nabla \times \mathbf{B} = \mu_0 \mathbf{j} + \varepsilon_0\mu_0\frac{\partial \mathbf{E}}{\partial t}. \tag{1.3}$$

[1] Named after the French physicist Charles Coulomb and the Dutch physicist Anton Hendrik Lorentz.

Introduction to Nanophotonics. Henri Benisty, Jean-Jacques Greffet, and Philippe Lalanne, Oxford University Press.
 DOI: 10.1093/oso/9780198786139.003.0001

We consider that the material medium is made of point-like charges with mass m_α, position $\mathbf{r}_\alpha(t)$, velocity $\mathbf{v}_\alpha(t)$, and charge q_α. The microscopic expressions of the charge density and current density are:

$$\rho(\mathbf{r},t) = \sum_\alpha q_\alpha \delta[\mathbf{r} - \mathbf{r}_\alpha(t)], \tag{1.4}$$

$$\mathbf{j}(\mathbf{r},t) = \sum_\alpha q_\alpha \mathbf{v}_\alpha(t) \delta[\mathbf{r} - \mathbf{r}_\alpha(t)]. \tag{1.5}$$

1.1.2 Helmholtz theorem

In this section, we first address a simple question. Are Maxwell equations sufficient to determine the fields? The answer is positive because it can be shown that the knowledge of the divergence and the curl of a field $\mathbf{X}(\mathbf{r})$ is sufficient to fully specify the field, provided that the field is twice continuously differentiable and vanishes faster than $1/r$ as $r \to \infty$. The explicit form of the field is known as the Helmholtz theorem. It can be shown that the field can be cast in the form of the sum of a field with zero curl and a field with zero divergence:

$$\mathbf{X}(\mathbf{r}) = \nabla \times \mathbf{A}(\mathbf{r}) - \nabla\varphi(\mathbf{r}), \tag{1.6}$$

where the potentials $\mathbf{A}(\mathbf{r})$ and $\varphi(\mathbf{r})$ are given by:

$$\mathbf{A}(\mathbf{r}) = \frac{1}{4\pi}\int \mathrm{d}^3\mathbf{r}' \frac{\nabla_{\mathbf{r}'} \times \mathbf{X}(\mathbf{r}')}{|\mathbf{r} - \mathbf{r}'|} \quad \text{and} \quad \varphi(\mathbf{r}) = \frac{1}{4\pi}\int \mathrm{d}^3\mathbf{r}' \frac{\nabla_{\mathbf{r}'}\mathbf{X}(\mathbf{r}')}{|\mathbf{r} - \mathbf{r}'|}. \tag{1.7}$$

From this theorem, we can infer two properties that will be used when introducing the potentials. A field with zero divergence everywhere has a zero scalar potential, so that it can be written as a curl of a vector potential. Similarly, a field with a zero curl everywhere has a zero vector potential, so that it can be written as the gradient of a scalar potential.

1.1.3 Longitudinal and transverse fields

We now revisit this decomposition in Fourier space and introduce the concept of longitudinal and transverse fields. We start by introducing the Fourier transform of the field:

$$\mathbf{X}(\mathbf{r}) = \int \frac{\mathrm{d}^3\mathbf{k}}{(2\pi)^3} \mathbf{X}(\mathbf{k}) \exp(i\mathbf{k}.\mathbf{r}). \tag{1.8}$$

The curl and the divergence of the field can be easily derived:

$$\nabla \times \mathbf{X}(\mathbf{r}) = \int \frac{\mathrm{d}^3\mathbf{k}}{(2\pi)^3} [i\mathbf{k} \times \mathbf{X}(\mathbf{k})] \exp(i\mathbf{k}.\mathbf{r}), \tag{1.9}$$

$$\nabla \cdot \mathbf{X}(\mathbf{r}) = \int \frac{\mathrm{d}^3\mathbf{k}}{(2\pi)^3} [i\mathbf{k} \cdot \mathbf{X}(\mathbf{k})] \exp(i\mathbf{k}.\mathbf{r}). \tag{1.10}$$

It is thus seen that $i\mathbf{k} \times \mathbf{X}(\mathbf{k})$ is the Fourier transform of $\nabla \times \mathbf{X}(\mathbf{r})$ and that $i\mathbf{k} \cdot \mathbf{X}(\mathbf{k})$ is the Fourier transform of $\nabla \cdot \mathbf{X}(\mathbf{r})$. We now introduce the natural definitions:

- a transverse field $\mathbf{X}(\mathbf{r})$ is such that $\mathbf{X}(\mathbf{k}) \cdot \mathbf{k} = 0$,
- a longitudinal field $\mathbf{X}(\mathbf{r})$ is such that $\mathbf{X}(\mathbf{k}) \times \mathbf{k} = 0$.

This amounts to writing the Fourier transform of the field as the sum of two terms:

$$\mathbf{X}(\mathbf{k}) = \mathbf{X}_{\parallel}(\mathbf{k}) + \mathbf{X}_{\perp}(\mathbf{k}), \tag{1.11}$$

where $\mathbf{X}_{\parallel} = [\mathbf{X} \cdot \mathbf{k}]\mathbf{k}/\mathbf{k}^2$ and $\mathbf{X}_{\perp} = \mathbf{X} - \mathbf{X}_{\parallel} = -\mathbf{k} \times \mathbf{k} \times \mathbf{X}/k^2$. It turns out that this geometrical decomposition in Fourier space is identical to the Helmholtz decomposition of the field in a curl term (the transverse field) and a gradient term (the longitudinal field).

1.2 Potentials

We now introduce the vector and scalar potentials. Using these potentials, the two Maxwell equations without source terms are automatically satisfied. We start with the Maxwell–Gauss equation $\nabla \cdot \mathbf{B} = 0$. According to the Helmholtz theorem, $\mathbf{B}$ can be written as:

$$\mathbf{B} = \nabla \times \mathbf{A}. \tag{1.12}$$

By inserting this form in Eq. (1.3), we get:

$$\nabla \times (\mathbf{E} + \frac{\partial \mathbf{A}}{\partial t}) = 0. \tag{1.13}$$

Thus, the electric field can be cast in the form:

$$\mathbf{E} = -\frac{\partial \mathbf{A}}{\partial t} - \nabla \varphi. \tag{1.14}$$

Note that this decomposition is not a decomposition in transverse and longitudinal fields. Here, $-\nabla \varphi$ is longitudinal, but there is no constraint on $\mathbf{A}$. Working with $\mathbf{A}$ and φ ensures that the two Maxwell equations without sources are automatically satisfied. By inserting these forms in the two Maxwell equations with source terms, we obtain:

$$\begin{aligned} \nabla^2 \mathbf{A} - \varepsilon_0 \mu_0 \frac{\partial^2 \mathbf{A}}{\partial t^2} &= -\mu_0 \mathbf{j} + \mathbf{grad} \left[\nabla \cdot \mathbf{A} + \varepsilon_0 \mu_0 \frac{\partial \varphi}{\partial t} \right] \\ \nabla^2 \varphi &= -\frac{\rho}{\varepsilon_0} - \nabla \cdot \left(\frac{\partial \mathbf{A}}{\partial t} \right). \end{aligned} \tag{1.15}$$

It is seen that Eqs (1.15) form a coupled system relating potentials $\mathbf{A}$ and φ to the sources ρ and $\mathbf{j}$. An important remark is in order. The vector potential is not fully determined, as only its curl was specified by Eq. (1.12). Any choice is possible for $\nabla \cdot \mathbf{A}$. In other words, it is possible to choose different potentials producing the same fields. This is called gauge invariance or gauge symmetry. The choice of $\nabla \cdot \mathbf{A}$ is a gauge choice. There are two choices which are used in practice. The Lorenz gauge[2] is given by:

[2] Named after the Danish physicist Ludwig Lorenz.

$$\nabla \cdot \mathbf{A}_L + \varepsilon_0 \mu_0 \frac{\partial \varphi_L}{\partial t} = 0. \tag{1.16}$$

It has the convenient property of decoupling the two equations. Another possible choice that is often used is the Coulomb gauge,

$$\nabla \cdot \mathbf{A}_C = 0. \tag{1.17}$$

With this choice, the transverse component of the electric field is derived from the vector potential, whereas the longitudinal component is derived from the scalar potential.

1.2.1 Gauge invariance

So far, we have given two examples of gauge choices, but many others are possible. In this section, we seek the general forms of $\mathbf{A}$ and φ, which do not affect the values of the fields $\mathbf{E}$ and $\mathbf{B}$. Since the magnetic field does not depend on the longitudinal part of $\mathbf{A}$, we need to ensure that the electric field remains invariant when changing $\nabla \cdot \mathbf{A}$. This analysis is more easily performed in Fourier space.[3] We introduce the following representation of the electric field:

$$\mathbf{E}(\mathbf{r}, t) = \int \frac{\mathrm{d}^3\mathbf{k}}{(2\pi)^3} \int \frac{\mathrm{d}\omega}{2\pi} \mathbf{E}(\mathbf{k}, \omega)] \exp(i\mathbf{k}.\mathbf{r} - i\omega t). \tag{1.18}$$

Note that the sign convention is not the same for time and space variables. With this choice, a plane wave with wavevector $\mathbf{k}$ propagates in the direction of $\mathbf{k}$. Using the same get:

$$\begin{aligned} \mathbf{E}(\mathbf{k}, \omega) &= i\omega \mathbf{A}(\mathbf{k}, \omega) - i\mathbf{k}\varphi \\ &= i\omega \mathbf{A}_\perp(\mathbf{k}, \omega) + i\left[\omega \mathbf{A}_\parallel(\mathbf{k}, \omega) - \mathbf{k}\varphi\right], \end{aligned} \tag{1.19}$$

where we have separated the transverse and longitudinal components. This form shows very clearly what a gauge choice is. A given longitudinal component of the electric field can be obtained by different couples, $(\mathbf{A}, \varphi)$. As $\mathrm{div}\mathbf{A}$ is not specified, the longitudinal part of the vector potential can be changed at will. It is clearly seen that if the couple $(\mathbf{A}_\parallel, \varphi)$ is a solution, then any couple of the form $(\mathbf{A}_\parallel + i\mathbf{k}X(\mathbf{k}, \omega), \varphi + i\omega X(\mathbf{k}, \omega))$ is also a solution. In space and time domain, this amounts to using a solution of the form $\mathbf{A} + \nabla X(\mathbf{r}, t), \varphi - \frac{\partial X(\mathbf{r}, t)}{\partial t}$. Hence, moving from the Lorenz gauge $\mathbf{A}_L$ to the Coulomb gauge $\mathbf{A}_C$ amounts to adding a longitudinal component ∇X to the vector potential $\mathbf{A}_L$ such that $\mathbf{A}_{L,\parallel} + \nabla X = 0$ in order to ensure that the Coulomb vector potential is purely transverse (see Exercise 1.2).

[3] The Fourier transform can be defined for square integrable deterministic functions. When dealing with light, a statistical description is needed. See Complement 1.A for a discussion of the spectral analysis of a stationary random process.

1.2.2 Potential propagation equation in Lorenz gauge

By inserting Eq. (1.17) in Eq. (1.15), we find:

$$\nabla^2 \mathbf{A}_L - \frac{1}{c^2}\frac{\partial^2 \mathbf{A}_L}{\partial t^2} = -\mu_0 \mathbf{j}, \tag{1.20}$$

$$\nabla^2 \varphi_L - \frac{1}{c^2}\frac{\partial^2 \varphi_L}{\partial t^2} = -\frac{\rho}{\varepsilon_0}. \tag{1.21}$$

It is seen that the two equations are decoupled and that both potentials satisfy a propagation equation whose solutions are the retarded potentials, as will be seen in the radiation chapter.

1.2.3 Potential propagation equation in the Coulomb gauge

By inserting Eq. (1.17) in Eq. (1.15), we find:

$$\nabla^2 \mathbf{A}_C - \frac{1}{c^2}\frac{\partial^2 \mathbf{A}_C}{\partial t^2} = -\mu_0 \mathbf{j} + \varepsilon_0\mu_0 \nabla\left[\frac{\partial \varphi_C}{\partial t}\right], \tag{1.22}$$

$$\nabla^2 \varphi_C = -\frac{\rho}{\varepsilon_0}. \tag{1.23}$$

At first sight, the Coulomb gauge appears to be awkward, as i) the scalar and vector potentials seem to be coupled, ii) the scalar potential appears to be the non-retarded Coulomb potential, suggesting that instantaneous action may be possible in contradiction with special relativity. In fact, the equations are not coupled. Furthermore, the fields are retarded, as expected from the fact that the fields are gauge invariant. To make this point clear, we move to Fourier space again. The system becomes:

$$\left(-k^2 + \tfrac{\omega^2}{c^2}\right)\mathbf{A}_C = -\mu_0 \mathbf{j} + \tfrac{\omega \mathbf{k}}{c^2}\varphi_C \tag{1.24}$$

$$\varphi_C = \tfrac{\rho}{\varepsilon_0 k^2}. \tag{1.25}$$

The charge conservation can be written in direct space or in Fourier space:

$$\nabla\cdot\mathbf{j} + \frac{\partial \rho}{\partial t} = 0 \quad ; \quad \mathbf{k}\cdot\mathbf{j} = \omega\rho(\mathbf{k},\omega). \tag{1.26}$$

Inserting Eqs (1.25) and (1.26) in Eq. (1.24) gives:

$$\left(-k^2 + \frac{\omega^2}{c^2}\right)\mathbf{A}_C = -\mu_0\left[\mathbf{j} - \frac{\mathbf{j}\cdot\mathbf{k}}{k^2}\mathbf{k}\right] = \mu_0 \mathbf{j}_\perp. \tag{1.27}$$

Returning to direct space, this equation can be cast in the form:

$$\nabla^2 \mathbf{A}_C - \frac{1}{c^2}\frac{\partial^2 \mathbf{A}_C}{\partial t^2} = -\mu_0 \mathbf{j}_\perp, \tag{1.28}$$

where it is seen that in the Coulomb gauge, the vector potential depends only on the transverse current density and is not coupled to the scalar potential.

We now check that the electric field is retarded, although the scalar potential appears to be non-retarded in Eq. (1.23). It turns out that in the Coulomb gauge, the vector potential also contains a non-retarded contribution. Both contributions cancel exactly when they are added so that the total electric field is retarded, whereas the longitudinal and the transverse electric fields have an unphysical non-retarded contribution. This can be checked directly by inspection (see Exercise 1.3).

1.2.4 Propagation equations for the fields

The propagation equation satisfied by the electric field $\mathbf{E}$ is derived by eliminating the magnetic field in Maxwell equations. To proceed, we take the curl of the Maxwell–Faraday equation in order to introduce $\nabla \times \mathbf{B}$. This quantity can be eliminated using the Maxwell–Ampere equation,

$$\nabla \times \nabla \times \mathbf{E} = -\frac{\partial}{\partial t}(\nabla \times \mathbf{B}).$$

Since $\nabla \times \mathbf{B} = \mu_0 \mathbf{j} + \varepsilon_0 \mu_0 \frac{\partial \mathbf{E}}{\partial t}$, we obtain

$$\nabla \times \nabla \times \mathbf{E} + \varepsilon_0 \mu_0 \frac{\partial^2 \mathbf{E}}{\partial t^2} = -\mu_0 \frac{\partial \mathbf{j}}{\partial t}. \tag{1.29}$$

Similarly, we get:

$$\nabla^2 \mathbf{B} - \varepsilon_0 \mu_0 \frac{\partial^2 \mathbf{B}}{\partial t^2} = -\mu_0 \nabla \times \mathbf{j}. \tag{1.30}$$

1.3 Macroscopic Maxwell equations

In this section, we summarize the key elements that account for propagation in material media. We first remind the reader that the fields are spatial averages of the microscopic fields generated by a distribution of point-like charges. We then introduce Maxwell equations satisfied by the spatially averaged electromagnetic fields.

1.3.1 Microscopic form of the source terms

The field produced by the microscopic sources in Eq. (1.4) contains a very large number of singularities. Each point-like charge produces a diverging electric field which is given by Coulomb law. As there are typically N_A electrons, where N_A is the Avogadro number in a 1 cm^3 of condensed matter, it is clear that accounting for all these singularities is intractable. Hence, when dealing with electromagnetic fields in material media, there is an absolute need to regularize the field. We need to define a smoothing procedure. This raises the question of the nature of the equations satisfied by the fields after smoothing. Are Maxwell equations still applicable? The short answer to these questions is that the fields are spatially averaged to remove spatial singularities. After this spatial averaging procedure, the general form of Maxwell equations is preserved but new terms appear in the equations. These terms account for the fact that the fields produced by charges are not the same in vacuum, in a metal, or in a dielectric. Hence, some information about the materials needs to be included in the equations relating the fields to the sources. The new terms appear naturally when performing the spatial average procedure.

1.3.2 Spatial average of the microscopic fields

The basic idea of macroscopic electrodynamics is to spatially average the electromagnetic fields on an intermediate length scale L that is large enough to remove singularities and small enough to keep the relevant information on the system, in particular the wavelength. In practice, a spatial average is performed using a normalized function $f_L(\mathbf{r})$:

$$< \mathbf{E}(\mathbf{r}) >= \int \mathrm{d}^3\mathbf{r}' f_L(\mathbf{r}-\mathbf{r}')\, \mathbf{E}(\mathbf{r}'). \tag{1.31}$$

To illustrate the averaging procedure, let us consider the case of a charge density. If we use this definition to average a point-like charge distribution $\rho_{micro}(\mathbf{r}) = q\delta(\mathbf{r}-\mathbf{r}_0)$, we obtain $< \rho_{micro}(\mathbf{r}) >= qf_L(\mathbf{r}-\mathbf{r}_0)$. It is seen that the average procedure amounts to replacing a point-like charge by a smooth continuous charge distribution with typical radius L. There are some cases where this classical reasoning cannot be used. If one deals with X-rays, the wavelength is on the order of the atomic size so that there is no intermediate length scale. Similarly, when studying the electromagnetic field at a subnanometre scale in nanophotonics, the same issue appears. So it seems that our procedure fails. In fact, when averaging over the position of a point-like charge, a quantum procedure should be used. The nature of the spatial average is the quantum average over the wave function of the relevant observable, be it a field, a current density, or a charge density. In summary, all electromagnetic fields in material media are spatially averaged in order to regularize these quantities. This raises an important question: what are the equations satisfied by the averaged fields? These are known as macroscopic Maxwell equations. To derive them, we average the microscopic Maxwell equations. This involves averaging the fields and the current and charge densities.

1.3.3 Spatial average of Maxwell equations

We start by averaging the fields. Let us consider, for example, a component of the electric field. We shall show that the average procedure commutes with a spatial derivative, namely:

$$\frac{\partial < E_{micro}(\mathbf{r}) >}{\partial x} =< \frac{\partial E_{micro}(\mathbf{r})}{\partial x} > . \tag{1.32}$$

Let us check this property for a 1D variable:

$$\begin{aligned}
\frac{d < E(x) >}{dx} &= \frac{d}{dx}\int f(x-x')\, E(x')\, \mathrm{d}x' = \int \frac{df(x-x')}{dx} E(x')\, \mathrm{d}x' \\
&= -\int \frac{df(x-x')}{dx'} E(x')\, \mathrm{d}x' = \int f(x-x')\frac{dE(x')}{dx'}\, \mathrm{d}x' \\
&= \langle \frac{\mathrm{d}E(x')}{\mathrm{d}x'} \rangle.
\end{aligned} \tag{1.33}$$

The last equality follows from an integration by parts where we have used the fact that the function f used for spatial averaging has a finite support so that it vanishes at infinity. It follows that we can commute nabla and spatial averaging.

Let us now consider the spatial averaging of the charge density and the current density. To proceed, let us consider an atom A with a nucleus located at $\mathbf{r}_A$ and electrons located at $\mathbf{r}_A + \mathbf{r}_i$. Given that the distance between the electron and the nucleus is much smaller than the averaging length scale L, we can expand the charge density as follows: $< \rho(\mathbf{r}) >= \sum_i q_i f_L(\mathbf{r} - \mathbf{r}_A - \mathbf{r}_i) = \sum_i q_i f_L(\mathbf{r} - \mathbf{r}_A) - \sum_i q_i \mathbf{r}_i \cdot \nabla f_L(\mathbf{r} - \mathbf{r}_A)$. Introducing the net charge of the atom q_A and its dipole moment $\mathbf{p}_A$, we get $< \rho(\mathbf{r}) >= q_A f_L(\mathbf{r} - \mathbf{r}_A) - \mathbf{p}_A \cdot \nabla f_L(\mathbf{r} - \mathbf{r}_A) = q_A f_L(\mathbf{r} - \mathbf{r}_A) - \nabla \cdot [f_L(\mathbf{r} - \mathbf{r}_A)\mathbf{p}_A]$. Summing now over many atoms yields the net charge density $\rho_{net}(\mathbf{r}) = \sum_A q_A f_L(\mathbf{r} - \mathbf{r}_A)$ and an additional contribution due to the polarization density $\mathbf{P}(\mathbf{r}) = \sum_A \mathbf{p}_A f_L(\mathbf{r} - \mathbf{r}_A)$. Finally, we obtain:

$$< \rho(\mathbf{r}) >= \rho_{net}(\mathbf{r}) + \rho_{pol}(\mathbf{r}) = \rho_{net}(\mathbf{r}) - \nabla \cdot \mathbf{P}(\mathbf{r}), \tag{1.34}$$

where we recognize two contributions. The first contribution $\rho_{net}(\mathbf{r})$ is due to the net charge of the system and is zero for a neutral system. This term is often called 'free charge' because electrons, which are particles that can move freely, can be added to a conductor. However, this name is misleading, as a localized ionized impurity in a semiconductor will also produce a contribution to the net charge. The second contribution is a polarization charge density $\rho_{pol}(\mathbf{r})$ and is given by the divergence of the polarization density in the medium. Here, the polarization density is the dipole moment per unit volume.

A similar average procedure of the microscopic current density would yield three terms:

$$< \mathbf{j}_{micro} >= \mathbf{j}_{net} + \frac{\partial \mathbf{P}}{\partial t} + \nabla \times \mathbf{M}, \tag{1.35}$$

where $\mathbf{j}_{net}$ is the contribution of the displacement of net charges over distances much larger than the atomic size. It can be due to free electrons but also to ions in liquids or plasmas, or to holes in semiconductors. This term is also often named 'free current density'. The second term is the current density associated with the time dependence of the polarization density $\mathbf{P}$. This is, for instance, the current density in a dielectric inside a capacitor. The displacement of polarization charges contributes to the net current in a time-dependent regime, whereas it does not in the direct current (DC) regime. Finally, the last term is due to the magnetization $\mathbf{M}$. Indeed, a uniform magnetization in an area S is equivalent to a current loop around that area.

1.3.4 Macroscopic Maxwell equations

We start with Maxwell equations without source terms.

$$\begin{aligned} \nabla\times < \mathbf{E}_{micro} > &= -\frac{\partial < \mathbf{B}_{micro} >}{\partial t} \\ \nabla\cdot < \mathbf{B}_{micro} > &= 0. \end{aligned} \tag{1.36}$$

Hence, it is seen that Maxwell equations without source terms are the same for microscopic and macroscopic fields. We now consider the Maxwell–Gauss equation,

$$\nabla \cdot \mathbf{E}_{micro} = \frac{\rho_{micro}}{\varepsilon_0}, \tag{1.37}$$

which, after averaging, becomes:

$$\nabla\cdot < \mathbf{E}_{micro} >= \frac{\rho_{net} - \nabla \cdot \mathbf{P}}{\varepsilon_0}. \tag{1.38}$$

It is seen that Eq. (1.38) for $< \mathbf{E}_{micro} >$ requires a knowledge of $\mathbf{P}$. In other words, the electric field in the material depends on the neighbouring charges. It is possible to cast Maxwell–Gauss equation in a form similar to the vacuum equation by introducing an auxiliary vector:

$$\mathbf{D} = \varepsilon_0 < \mathbf{E}_{micro} > +\mathbf{P}. \tag{1.39}$$

With this definition of $\mathbf{D}$, we have

$$\nabla \cdot \mathbf{D} = \rho_{net}. \tag{1.40}$$

However, we have only hidden our ignorance of $\mathbf{P}$ by introducing an unknown auxiliary field, $\mathbf{D}$[4]. In any case, we need to establish a connection between the field $\mathbf{E}$ and either $\mathbf{P}$ or $\mathbf{D}$. The Maxwell–Ampere equation can be written as:

$$\nabla \times \mathbf{B}_{micro} = \mu_0 \mathbf{j}_{micro} + \varepsilon_0 \mu_0 \frac{\partial \mathbf{E}_{micro}}{\partial t}. \tag{1.41}$$

Averaging this equation yields the macroscopic Maxwell–Ampere equation:

$$\begin{aligned}\nabla\times < \mathbf{B}_{micro} > &= \mu_0 < \mathbf{j}_{micro} > +\varepsilon_0\mu_0 \frac{\partial < \mathbf{E}_{micro} >}{\partial t} \\ &= \mu_0 \left[\mathbf{j}_{net} + \frac{\partial \mathbf{P}}{\partial t} + \nabla \times \mathbf{M}\right] + \varepsilon_0\mu_0 \frac{\partial < \mathbf{E}_{micro} >}{\partial t}.\end{aligned} \tag{1.42}$$

Again, this equation can be formally simplified by removing the unknown magnetization from Maxwell equations and introducing an unknown auxiliary field $\mathbf{H}$[5]:

$$\nabla \times \mathbf{H} = \mathbf{j}_{net} + \frac{\partial \mathbf{D}}{\partial t}, \tag{1.43}$$

where we have defined

$$\mathbf{H} = \frac{< \mathbf{B}_{micro} >}{\mu_0} - \mathbf{M}. \tag{1.44}$$

To summarize this section, we write the set of four Maxwell equations in terms of vectors $\mathbf{D}$ and $\mathbf{H}$. Hereafter, electric and magnetic fields are assumed to be averaged quantities, so that we drop the notations $<>$.

[4] This vector was called electric displacement in the early literature on electrodynamics.

[5] **H** and **B** were called the magnetic field and the magnetic induction, respectively, in the early literature on electrodynamics.

$$\nabla \cdot \mathbf{D} = \rho_{net} \qquad \nabla \cdot \mathbf{B} = 0, \tag{1.45}$$

$$\nabla \times \mathbf{E} = -\frac{\partial \mathbf{B}}{\partial t} \qquad \nabla \times \mathbf{H} = \mathbf{j}_{net} + \frac{\partial \mathbf{D}}{\partial t}. \tag{1.46}$$

$$\nabla \times \mathbf{E} = -\frac{\partial \mathbf{B}}{\partial t} \qquad \nabla \cdot \mathbf{D} = \rho_{net}, \tag{1.47}$$

$$\nabla \times \mathbf{H} = \mathbf{j}_{net} + \frac{\partial \mathbf{D}}{\partial t} \qquad \nabla \cdot \mathbf{B} = 0. \tag{1.48}$$

An alternative form involves the current density and the polarization density:

$$\nabla \times \mathbf{E} = -\frac{\partial \mathbf{B}}{\partial t}, \quad \nabla \cdot \mathbf{E} = \frac{\rho_{net} - \nabla \cdot \mathbf{P}}{\varepsilon_0}, \tag{1.49}$$

$$\nabla \times \mathbf{B} = \mu_0 \left[\mathbf{j}_{net} + \frac{\partial \mathbf{P}}{\partial t} + \nabla \times \mathbf{M} \right] + \varepsilon_0 \mu_0 \frac{\partial \mathbf{E}}{\partial t}, \qquad \nabla \cdot \mathbf{B} = 0. \tag{1.50}$$

The set of equations involving **D** and **H** is more compact and retains the structure of Maxwell equations in vacuum. The set of equations involving the electric field **E** and the magnetic field **B** connects these fields to their sources.

1.3.5 Physical discussion of the polarization charges

In the previous discussion, a polarization charge density was introduced. This result is surprising at first glance, as it seems that a neutral medium can become charged. The key idea is that while it remains globally neutral, a medium can be locally charged. The aim of this section is to clarify this point by illustrating the concept of polarization charge density $\rho_{pol} = -\nabla \cdot \mathbf{P}$ using a few examples.

Example 1: We consider the situation depicted in Fig. 1.1a. It consists of a medium where a positive charge q is located at point M. It can be, for example, a phosphorus

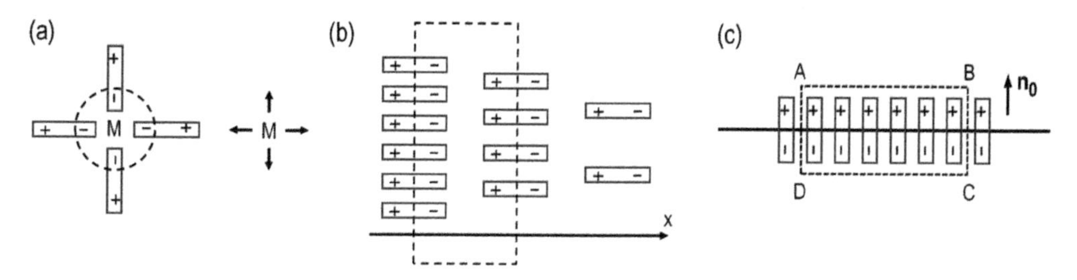

Fig. 1.1 Polarization charge density. (a) An external positive charge q introduced in the medium at point M attracts the negative charges, thus generating a radial distribution of polarization. It is seen that the polarization charges contribute to a negative charge inside the sphere surrounding M, so that they can screen the positive charge. (b) We consider a gas with a density gradient. It is seen that the rectangle contains a negative charge. (c) An interface separating two homogeneous materials is a particular case of the gradient seen in (b). The polarization discontinuity generates a surface charge.

impurity in n-doped silicon (see Complement 8.C) or an H_3O^+ ion in water. This net charge generates an electrostatic field that polarizes the surrounding molecules: electrons are attracted. It follows that a negative cloud of charges is formed around point M in the sphere shown in Fig. 1.1a. This negative charge tends to screen the positive charge q. As shown in Fig. 1.1a, the divergence of the polarization at the centre is positive, so the quantity $-\nabla \cdot \mathbf{P}$ indeed corresponds to a negative charge.

Example 2: We now consider a different situation shown in Fig. 1.1b. We consider an atomic gas with a gradient of density along the x-axis. An electrostatic field is applied along the x-axis. The field polarizes the atoms, as depicted in Fig. 1.2b. It is clearly seen in the figure that there is a net charge in the volume, indicated by the rectangle as a consequence of the density gradient. In turn, the presence of the gradient implies that $\nabla \cdot \mathbf{P} = \frac{\partial P_x}{\partial x}$ is non-zero.

Example 3: We now consider an interface between a dielectric and vacuum. This can be viewed as a limiting case of the density gradient case. By inspection of Fig. 1.1c, it is readily seen that a surface charge density σ_S appears at the interface. It can be derived from $\rho_{pol} = -\nabla \cdot \mathbf{P}$ by applying the Green–Ostrogradski theorem to the contour ABCD and letting BC and AD tend to zero. We assume that AB is in vacuum and CD is in the dielectric so that $\mathbf{P}$ is zero along AB and perpendicular to the outgoing normal along BC and DA:

$$\begin{aligned}\int_{ABCD} -\nabla \cdot \mathbf{P}\, \mathrm{d}^3\mathbf{r} &= -\int_{AB} \mathbf{P}.\mathbf{n}_{ext}\, \mathrm{d}S - \int_{CD} \mathbf{P}.\mathbf{n}_{ext}\, \mathrm{d}S \\ &= \int_{CD} -\mathbf{P}.\mathbf{n}_{ext}\, \mathrm{d}S \\ &= \int_{CD} \sigma_S\, \mathrm{d}S,\end{aligned} \tag{1.51}$$

where $\sigma_S = -\mathbf{P}.\mathbf{n}_{ext}$ and $\mathbf{n}_{ext}$ is the outgoing normal vector so that $\mathbf{n}_{ext}(CD) = -\mathbf{n}_0$. Finally, we find that the charge per unit surface is a particular case of $-\nabla \cdot \mathbf{P}$:

$$\sigma_S = \mathbf{P}.\mathbf{n}_0. \tag{1.52}$$

The concept of surface charge is a simplified model. In practice, the polarization is due to the electronic deformation of the electron orbitals for a metal and to the displacement of ions in an ionic crystal. Hence, the 'surface' charge is in reality a volume charge distribution that exists over a transition layer with a typical length scale on the order of 0.1 nm.

1.3.6 Screening

We have seen that a positive charge embedded in a medium generates a polarization field. The divergence of this field corresponds to a local polarization charge density of opposite sign so that the apparent charge is reduced. This phenomenon is called screening. It can be measured quantitatively by the dielectric relative permittivity ε_r, which is defined for a linear and isotropic medium as $\mathbf{D} = \varepsilon_0\varepsilon_r\mathbf{E}$. Inserting this definition in Eq. (1.46), we get $\nabla \cdot \mathbf{E} = \rho_{net}/\varepsilon_0\varepsilon_r$. It is thus seen that the electric field generated by a net charge is reduced by ε_r. In electrostatics, this explains the increase of the capacitor's capacity C when introducing a dielectric material with large permittivity. A larger charge is needed to produce the same field as in vacuum.

1.4 Energy conservation

We first simply remark that because of finite light speed, it is required to assign energy to the electromagnetic field. Indeed, when the sun emits light, it takes 8 minutes and 19 seconds to reach the earth. During this time, the energy is lost by the sun and not yet absorbed (or scattered) by the earth. Energy conservation requires this energy to be assigned to the electromagnetic field.

1.4.1 Power exchanged between matter and the electromagnetic field

In this section, we are going to discuss the rate of energy exchanged between the field and matter. We will then study this quantity using Maxwell equations, and it will be seen that, indeed, an energy term corresponding to the field energy is found.

Considering a single charge, q, and an electromagnetic force, we can derive the power $P_{\text{EMfield}\rightarrow\text{matter}}$ transferred from the field to the charge using the mechanical definition of power:

$$P_{\text{EMfield}\rightarrow\text{matter}} = \mathbf{F}\cdot\mathbf{v} = q\mathbf{E}\cdot\mathbf{v}. \tag{1.53}$$

The magnetic field does not contribute to the work, as the magnetic force is perpendicular to the velocity. If we now consider a distribution of N charges, we find:

$$P = \sum_{i=1}^{N} q_i\mathbf{v}_i\cdot\mathbf{E}(\mathbf{r}_i) = \int \mathbf{j}_{\text{micro}}(\mathbf{r})\cdot\mathbf{E}_{\text{micro}}(\mathbf{r})d^3\mathbf{r}. \tag{1.54}$$

We now assume that the field is produced by charges which are in a source, away from the ensemble of N charges. In that case, the field varies slowly so that it is equal to its spatial average, $\mathbf{E}_{\text{micro}}(\mathbf{r}) = \mathbf{E}(\mathbf{r})$. Taking the spatial average yields:

$$\frac{\partial P}{\partial V} = \mathbf{j}\cdot\mathbf{E}. \tag{1.55}$$

This term can be either positive or negative. A positive term corresponds to absorption of radiation in a medium. This energy transferred is then converted into heat. This is called the Joule effect for metals and dielectric loss in dielectrics. A negative term corresponds to energy transferred from the currents to the field, as happens in an antenna.

1.4.2 Energy conservation equation: Poynting's theorem

We now combine the form of the absorbed power with Maxwell equations. We start from the equation:

$$\nabla\times\mathbf{H} = \mathbf{j}_{ext} + \frac{\partial\mathbf{D}}{\partial t}. \tag{1.56}$$

Here, we consider that the current density $\mathbf{j}_{ext}$ is the current density in an external source. For instance, it may be the current flowing in an antenna imposed by a current source. The term $\frac{\partial\mathbf{D}}{\partial t} = \varepsilon_0\frac{\partial\mathbf{E}}{\partial t} + \frac{\partial\mathbf{P}}{\partial t}$ accounts for all the induced currents in the material media, due to either polarization charges or net charges. We now form the scalar product $\mathbf{j} \cdot \mathbf{E}$:

$$\mathbf{E} \cdot \nabla \times \mathbf{H} = \mathbf{E} \cdot \mathbf{j}_{ext} + \mathbf{E} \cdot \frac{\partial \mathbf{D}}{\partial t}. \tag{1.57}$$

Using the identity $\nabla \cdot (\mathbf{E} \times \mathbf{H}) = \mathbf{H} \cdot \nabla \times \mathbf{E} - \mathbf{E} \cdot \nabla \times \mathbf{H}$, we find:

$$\mathbf{H} \cdot \nabla \times \mathbf{E} - \nabla \cdot (\mathbf{E} \times \mathbf{H}) = \mathbf{E} \cdot \mathbf{j}_{ext} + \mathbf{E} \cdot \frac{\partial \mathbf{D}}{\partial t}. \tag{1.58}$$

Finally, replacing $\nabla \times \mathbf{E}$ by $\partial \mathbf{B}/\partial t$, we obtain:

$$\mathbf{E} \cdot \frac{\partial \mathbf{D}}{\partial t} + \mathbf{H} \cdot \frac{\partial \mathbf{B}}{\partial t} = -\mathbf{j}_{ext} \cdot \mathbf{E} - \nabla \cdot (\mathbf{E} \times \mathbf{H}). \tag{1.59}$$

This form of energy conservation can be analysed as follows. The term $-\mathbf{j}_{ext} \cdot \mathbf{E}$ corresponds to the power provided to the fields by the external current, or, in other words, by the source. The term $-\nabla \cdot (\mathbf{E} \times \mathbf{H})$ corresponds to the energy radiated given by the flux of the Poynting vector $\mathbf{E} \times \mathbf{H}$. The terms $-\mathbf{E} \cdot \frac{\partial \mathbf{P}}{\partial t} + \mathbf{M} \cdot \frac{\partial \mathbf{B}}{\partial t}$ correspond to energy transferred from the field to matter so that the term $\mathbf{E} \cdot \frac{\partial \mathbf{D}}{\partial t} + \mathbf{H} \cdot \frac{\partial \mathbf{B}}{\partial t}$ corresponds to time variation of the energy in the material. This includes both the dissipation and the variation of electromagnetic energy stored in the medium.

It is tempting to try to view this energy as the sum of field energy, matter energy stored in potential and kinetic energy, and energy dissipated in the medium and transformed into heat. It is, indeed, possible to identify the losses. This will be done in Chapter 3. By contrast, it is generally not possible to split the energy stored in the medium as the sum of field energy and matter energy. The reason is that the field is coupled to the degrees of freedom of the charges.

Let us consider a particular case to illustrate this point. We consider a finite-sized crystal illuminated by a narrow-band infrared radiation on resonance with the crystal vibration. We assume that the currents in the external source are turned on at time $t = 0$. The crystal phonons are excited, so energy is stored as kinetic energy and potential energy in the crystal. An electromagnetic field also penetrates the medium so that there is electromagnetic energy within. At first glance, we can compute the kinetic energy and potential energy of each atom and thus separate these terms. Note, however, that while it is of course possible to compute the kinetic energy of any given atom at any time, it is not possible to assign an energy to each atom. Indeed, the atoms are *coupled*, so that they continuously exchange energy between them. It can be shown that the energy can be defined for modes which are linear combinations of the atoms' movement. Neglecting interactions between modes, the crystal energy is the sum of the modes' energy. In the presence of an electromagnetic field *coupled* to the atoms, the same phenomenon appears. The coupling of the field to the vibration gives rise to a new mode called a polariton (Hopfield, 1958). Hence, an energy can be assigned to this coupled mode which involves both electromagnetic energy and mechanical energy. This particular example is given to illustrate the difficulty in assigning energy to field and matter unambiguously. We can cast the energy conservation equation in the form:

$$\frac{\partial}{\partial t}\left(\varepsilon_0\frac{\mathbf{E}^2}{2}+\frac{\mathbf{B}^2}{2\mu_0}\right)=-\mathbf{j}_{ext}\cdot\mathbf{E}-\mathbf{E}\cdot\frac{\partial \mathbf{P}}{\partial t}+\mathbf{M}\cdot\frac{\partial \mathbf{B}}{\partial t}-\mathrm{div}(\mathbf{E}\times\mathbf{H}), \tag{1.60}$$

and we stress that it is not possible to assign the left-hand-side term to field energy, because that would ignore the coupling of the fields with the polarization and magnetization. A further discussion of the form of the energy in a material media is given in section 3.4.

1.5 Reciprocity theorem

1.5.1 Derivation of the reciprocity theorem

In this section, we derive the general form of the reciprocity theorem for electromagnetic waves (Landau et al., 1984), often called Helmholtz reciprocity. Reciprocity is a property related to the exchange between source and detector in a given environment. We must therefore compare two different situations obtained by exchanging source and detector without modifying the medium, namely without modifying the permittivity. In other words, the source and detector are immaterial. In a real experiment, a source is a material object such as an antenna. In what follows, inserting a source means turning on and off the currents in an antenna. These currents are sources for the field equations: they can be introduced even though there are no physical objects. We will often use fictitious dipolar sources which have zero polarizability.

We consider that there are sources with a current density $\mathbf{j}_1$ in a volume V_1 in a first situation radiating at frequency ω. Let us call $\mathbf{E}_1$ and $\mathbf{H}_1$ the fields created by this source in the presence of some scatterer described by its constitutive tensors $\overleftrightarrow{\varepsilon}(\mathbf{r},\omega)$ and $\overleftrightarrow{\mu}(\mathbf{r},\omega)$. We also consider a second situation, where the scattering medium is left unchanged but the sources are now due to a different distribution of currents $\mathbf{j}_2$ in a volume V_2. Let us call the fields created by this second source $\mathbf{E}_2$ and $\mathbf{H}_2$. The fields in each situation satisfy Maxwell equations, with k = 1, 2:

$$\nabla\times\mathbf{E}_k=i\omega\mathbf{B}_k, \qquad \nabla\times\mathbf{H}_k=\mathbf{j}_k-i\omega\mathbf{D}_k, \tag{1.61}$$

and the constitutive relations yield:

$$\mathbf{D}_k(\mathbf{r})=\varepsilon_0\,\overleftrightarrow{\varepsilon_r}(\mathbf{r},\omega)\cdot\mathbf{E}_k(\mathbf{r}) \quad , \quad \mathbf{B}_k(\mathbf{r})=\mu_0\,\overleftrightarrow{\mu_r}(\mathbf{r},\omega)\cdot\mathbf{H}_k(\mathbf{r}).$$

Then, provided that $\varepsilon_{ij}=\varepsilon_{ji}$ and $\mu_{ij}=\mu_{ji}$, it can be shown[6] that:

[6] This equality is obtained by combining Maxwell equations (1.61) for the two situations:

$$(\mathbf{H}_2\cdot\nabla\times\mathbf{E}_1-\mathbf{E}_1\cdot\nabla\times\mathbf{H}_2)+(\mathbf{E}_2\cdot\nabla\times\mathbf{H}_1-\mathbf{H}_1\cdot\nabla\times\mathbf{E}_2)=i\omega(\mathbf{B}_1\cdot\mathbf{H}_2-\mathbf{H}_1\cdot\mathbf{B}_2)$$
$$-i\omega(\mathbf{D}_1\cdot\mathbf{E}_2-\mathbf{E}_1\cdot\mathbf{D}_2)+(\mathbf{j}_1\cdot\mathbf{E}_2-\mathbf{j}_2\cdot\mathbf{E}_1). \tag{1.62}$$

The left-hand side can be rewritten as $\nabla\cdot(\mathbf{E}_1\times\mathbf{H}_2-\mathbf{E}_2\times\mathbf{H}_1)$. Provided that $\varepsilon_{ij}=\varepsilon_{ji}$ and $\mu_{ij}=\mu_{ji}$, the first two terms of the right-hand side cancel. This shows that the reciprocity theorem is valid for media with symmetric constitutive tensors. It is not valid for gyrotropic media or in the presence of the Faraday effect, for instance.

$$\nabla \cdot (\mathbf{E}_1 \times \mathbf{H}_2 - \mathbf{E}_2 \times \mathbf{H}_1) = \mathbf{j}_1 \cdot \mathbf{E}_2 - \mathbf{j}_2 \cdot \mathbf{E}_1. \tag{1.63}$$

This is the local form of the reciprocity theorem. When integrated over a volume without sources, it yields:

$$\int (\mathbf{E}_1 \times \mathbf{H}_2 - \mathbf{E}_2 \times \mathbf{H}_1) \cdot d\mathbf{A} = 0.$$

This is the form most used for antenna theory. It allows us to deal with propagation of waves through media without sources. Using this form, one can demonstrate that the emission pattern of an antenna is identical to its detection pattern. We now consider a system with sources inside the integration volume, and we let the integration surface tend to infinity. It can be shown that the surface integral vanishes[7] so that we obtain:

$$\int \mathbf{j}_1 \cdot \mathbf{E}_2 \, d^3\mathbf{r} = \int \mathbf{j}_2 \cdot \mathbf{E}_1 \, d^3\mathbf{r}.$$

For the case of a point-like dipolar source located at $\mathbf{r}_k$, since $\mathbf{j}_k = -i\omega \mathbf{p}_k \delta(\mathbf{r} - \mathbf{r}_k)$, the theorem can be cast in the form:

$$\mathbf{p}_1 \cdot \mathbf{E}_2(\mathbf{r}_1) = \mathbf{p}_2 \cdot \mathbf{E}_1(\mathbf{r}_2). \tag{1.64}$$

This last form of the reciprocity theorem has a simple interpretation: the amplitude of the signal produced by a dipolar source on a dipolar antenna does not change when interchanging their roles. Note that this is valid for polarized fields. Namely, the orientation of the antenna does not change—only their roles are exchanged.

1.5.2 Reciprocity and time reversal

It is important to realize that reciprocity is different from time-reversal invariance. At first glance, it seems that exchanging source and detector amounts to changing the direction of propagation. And, of course, reversing the sign of time changes the direction of propagation. This is why these two properties can be confused. Reciprocity is equivalent to time-reversal symmetry for lossless media and for propagating waves. However, reciprocity is not equivalent to time reversal for evanescent waves, as shown in ref. (Carminati et al. 2000).

Let us first point out that time-reversal invariance is not valid if there are losses. A simple example suffices to illustrate this point. Upon time reversal of a beam propagating in a homogeneous lossy medium, the amplitude decay becomes an exponential growth. In contrast, it is possible to exchange source and detector even if losses are present and Eq. (1.62) is still valid. This is a crystal clear difference between reciprocity and time reversal.

A second example will show the strong difference between the two cases. Let us consider an opaque screen with a small hole illuminated by an incident wave coming

[7] This can be seen by using the far-field asymptotic form of the fields derived using the stationary phase approximation. It is found that the fields at $\mathbf{r}$ are proportional to the Fourier transform of the field for the wavevector $\mathbf{k} = k_0\mathbf{r}/r$, where $k_0 = \omega/c$. Hence, we find that the leading term of $\mathbf{E}_1 \times \mathbf{H}_2 - \mathbf{E}_2 \times \mathbf{H}_1$ is proportional to $\mathbf{E}_1 \times \mathbf{k} \times \mathbf{E}_2 - \mathbf{E}_2 \times \mathbf{k} \times \mathbf{E}_1$ so that it vanishes.

from a point-like source located at point A (see Fig. 1.2). The light impinging on the aperture is diffracted, and part of it reaches a detector located at point B. However, most of the incident light is reflected in the left half-space. If we now exchange the position of source and detector, the source located at B illuminates the screen and part of the light is diffracted and reaches the point A, whereas most of the light is reflected into the right half-space. It is seen that the distribution of light is completely different in both cases and cannot be described by using a time-reversal symmetry. However, reciprocity tells us that the amplitude that is detected by the detector is the same in both cases.

We now give a few examples of possible applications. We have mentioned already the fact that reciprocity is the basis of the proof of the equality between the emission and the detection pattern of an antenna. Similarly, reciprocity is a key ingredient of the proof of Kirchhoff's Law, which states that thermal emission of a body at a given temperature for a given direction, frequency, and polarization is proportional to its absorptivity for the same conditions. We give some other examples of application in the next section. We will use reciprocity in Chapter 11 to model the signal of a near-field microscope and in Chapter 12 to derive some symmetry properties of a scattering matrix.

1.5.3 Application 1: Green tensor symmetry

The field produced by a point-like electric dipole is linearly related to the dipole moment $\mathbf{p}_1$. It can be cast in the form:

$$\mathbf{E}(\mathbf{r}) = \mu_0\omega^2 \overleftrightarrow{G}(\mathbf{r}, \mathbf{r}_1, \omega) \cdot \mathbf{p}_1,$$

where $\overleftrightarrow{G}(\mathbf{r}, \mathbf{r}_1, \omega)$ is called the Green tensor. Using the form of the theorem established for dipoles, we can immediately derive a very important symmetry property of the Green tensor. Since the reciprocity theorem is valid for any dipole, Eq. (1.64) entails

$$G_{nm}(\mathbf{r}_1, \mathbf{r}_2, \omega) = G_{mn}(\mathbf{r}_2, \mathbf{r}_1, \omega).$$

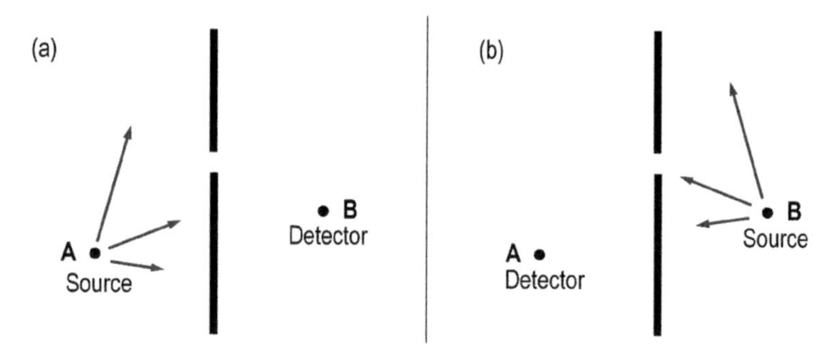

Fig. 1.2 Illustration of the difference between time-reversal symmetry and the reciprocity theorem. We consider an opaque screen with a tiny aperture. The source is on the left in case 1 (panel (a)) and on the right in case 2 (panel (b)). In both cases, a tiny fraction of the emitted light is diffracted through the screen. Reciprocity imposes that the signal delivered by the detector in cases 1 and 2 is the same. It is clear that case (2) cannot be derived from case (1) by time-reversal symmetry.

1.5.4 Application 2: Finding where to place a source

A very practical application of the reciprocity theorem is to find the location of a dipole source such that it radiates maximally in a given direction. Alternatively, when looking in the far field of a source, one may wonder what the parts of the source that contribute most to the signal are. Since the reciprocity theorem states that the connection between the source and the detector is invariant upon interchanging source and detector, let us place a source at the detector location. The distribution of fields in the system that is obtained in this reciprocal situation immediately gives the answer: one has to place the dipole source in a bright spot parallel to the electric field (see Exercise 1.4).

1.5.5 Application 3: Computing fields generated by a distribution of sources

We consider a current distribution, $\mathbf{j}(\mathbf{r}_1)$, in a volume V generating an electric field $\mathbf{E}(\mathbf{r}_2)$ at point 2. We aim to compute that field, accounting for all the interferences between the fields generated by all the volume elements in V. A given volume element $\mathrm{d}\mathbf{r}_1$ has a dipole moment $\mathbf{j}_1\mathrm{d}\mathbf{r}_1/(-i\omega)$, so that the field can be written as

$$\mathbf{E}(\mathbf{r}_2) = i\mu_0\omega \int_V G(\mathbf{r}_2, \mathbf{r}_1)\mathbf{j}_1(\mathbf{r}_1)\mathrm{d}\mathbf{r}_1. \tag{1.65}$$

The m-component of the field can be written using the unit vector $\mathbf{e}_m$:

$$E_m(\mathbf{r}_2) = \mathbf{e}_m \cdot \mathbf{E}(\mathbf{r}_2) = i\mu_0\omega \int_V G_{mn}(\mathbf{r}_2, \mathbf{r}_1) j_{1,n}(\mathbf{r}_1)\mathrm{d}\mathbf{r}_1. \tag{1.66}$$

We now make use of the reciprocity property of the Green tensor:

$$E_m(\mathbf{r}_2) = \mathbf{e}_m \cdot \mathbf{E}(\mathbf{r}_2) = i\mu_0\omega \int_V j_{1,n}(\mathbf{r}_1) G_{nm}(\mathbf{r}_1, \mathbf{r}_2)\mathrm{d}\mathbf{r}_1. \tag{1.67}$$

We recognize in the last expression the form of the field $E_n(\mathbf{r}_1)$ emitted by a current density $\mathbf{j}_2 = \mathbf{e}_m\delta(\mathbf{r} - \mathbf{r}_2)$ with unit amplitude and located at the detector position:

$$E_{2,n}(\mathbf{r}_1) = i\mu_0\omega \int_V G_{nm}(\mathbf{r}_1, \mathbf{r})\delta(\mathbf{r} - \mathbf{r}_2)\mathrm{d}\mathbf{r}, \tag{1.68}$$

so that, finally, the field emitted by a current distribution $E_m(\mathbf{r}_2)$ is given by the overlap integral of that current distribution $j_{1,n}$ by the field $E_{2,n}(\mathbf{r}_1)$ that would be emitted by a dipole located at the detector position:

$$E_m(\mathbf{r}_2) = \int_V j_n(\mathbf{r}_1) E_{2,n}(\mathbf{r}_1)\mathrm{d}\mathbf{r}_1. \tag{1.69}$$

This technique is extremely powerful for deriving fields generated by complex systems. It is also extremely powerful for analysing the signal received by a detector.

1.5.6 Application 4: Checking the accuracy of a solution

The reciprocity theorem can be very useful for checking the accuracy of numerical or approximate solutions to scattering problems. It is always necessary to check the consistency of the solution. To this aim, it is useful to check whether the solution satisfies energy conservation and the reciprocity theorem.

1.6 Radiation, propagation, scattering, and diffraction

In this brief section, we classify the different types of problem generally encountered (see Fig. 1.3). Such a classification is to some extent arbitrary. It is, nevertheless, useful to try to define some classes of problems and to introduce a terminology.

A radiation problem consists in finding the electromagnetic fields when the source currents are given. This is the typical problem of finding the fields radiated by an antenna knowing the intensity distribution in the wires of the antenna. Owing to the linearity of Maxwell equations, the solution is a linear operator connecting currents at $\mathbf{r}'$ (the sources) to the fields at $\mathbf{r}$. Such a linear operator connecting two vectors is a tensor. It is called a Green tensor, often denoted $\overleftrightarrow{G}(\mathbf{r}, \mathbf{r}')$.

A propagation problem consists in finding the electromagnetic fields at a point $z > 0$, provided that the field is known in the plane $z = 0$ in a homogeneous medium. The solution is thus a linear operator called a propagator, connecting the fields in the plane $z = z_0$ to the fields in the plane $z = 0$.

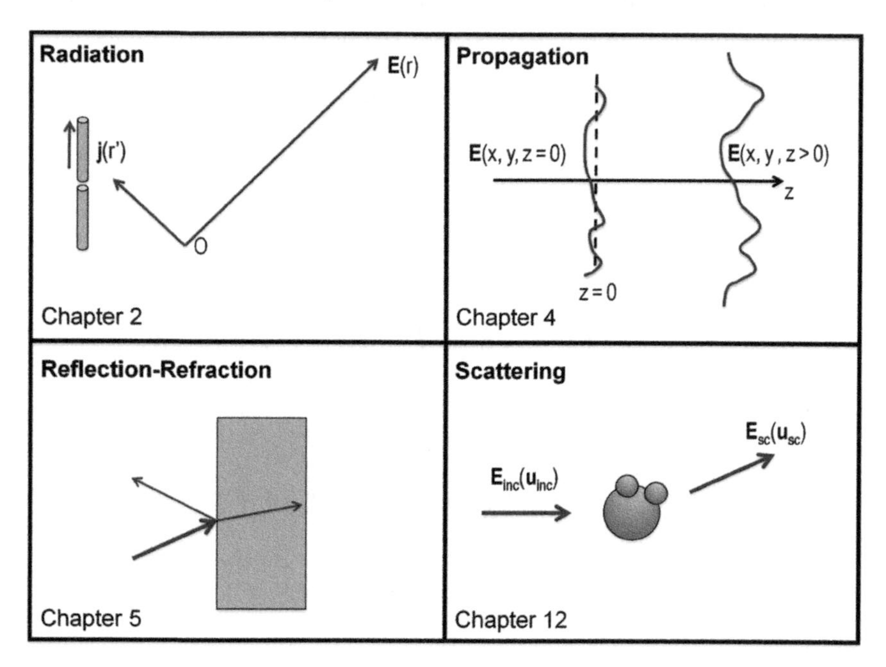

Fig. 1.3 Sketch of different classes of problems. Radiation: finding the field generated by known sources. The linear operator connecting the sources to the fields is the Green tensor. Propagation: finding a field in the plane $z > 0$ from the knowledge of the field in the plane $z = 0$. The linear operator connecting the field at z to the field at $z = 0$ is the propagator. Reflection/Refraction: finding the amplitude of the fields reflected and refracted at a flat interface. The linear operators connecting the reflected and transmitted fields to the incident field are the reflection and transmission operators. Scattering: finding the far-field amplitude of the field scattered by a known arbitrary object illuminated by a plane wave. The linear operator relating the incident and scattered fields is the scattering matrix.

A scattering problem consists in finding the electromagnetic field generated when a finite size object is illuminated by an incident field. This is given by the scattering matrix. A particular case of a scattering problem occurs when the object has a perfectly smooth flat surface and a size much larger than the wavelength. In that case, the terms of reflection and transmission are used. They are characterized by the Fresnel factors.

The term diffraction has been introduced historically to describe the deviations from geometrical optics when electromagnetic radiation passes through a small aperture or is close to a sharp edge. It is essentially a manifestation of wave behaviour when the field is spatially confined. The screen with an aperture can be viewed as a particular type of scatterer, so that the problem is a particular case of a scattering problem. When assuming that the field in the plane of the aperture is known (i.e. equal to the incident field in the aperture and null outside the aperture), the diffraction problem appears to be a subclass of a propagation problem. Note that the same terminology is used for objects with a regular shape and geometrical sizes on the order of or smaller than the wavelength, such as gratings or sharp edges.

In all these canonical problems, it is assumed that the incident field or the source currents are given and are independent of the environment. In other words, the feedback of the environment on the sources is neglected. This assumption may not be valid, particularly when the sources and environment are separated by subwavelength distances. This will be extensively discussed in Chapter 19.

In the next chapter, we will introduce the basic concepts of radiation. Chapter 4 is devoted to the discussion of propagation in vacuum and in material media, Chapter 5 deals with reflection and refraction, and Chapter 12 discusses scattering.

1.A Complement: Temporal coherence and spectral analysis of light (stationary random electromagnetic fields)

1.A.1 How to define the light spectrum

In this section, we address the issue of the spectral analysis of light. Let us first consider the case of radiation emitted by deterministic currents flowing in antennas. The fields are described by functions which are square-integrable over time. It is possible to define their Fourier transform. The energy spectrum can be easily defined using the Fourier transform of the field and the Parseval identity. For instance, for the electric energy, we have:

$$\int_{-\infty}^{\infty} \varepsilon_0 \frac{\mathbf{E}^2(t)}{2} \mathrm{d}t = \int_{-\infty}^{\infty} \varepsilon_0 \frac{|\mathbf{E}(\omega)|^2}{2} \frac{\mathrm{d}\omega}{2\pi}. \tag{1.70}$$

When studying light emitted by the sun or an incandescent lamp, by a light-emitting diode or a continuous laser, we know that the spectrum can be measured using spectrometers. We now address the issue of its mathematical representation. We first need to account for the random character of light. This is due to the fact that light emission results from the addition of a very large number of independent spontaneous emission events, such as radiative relaxation of atoms or molecules in a gas or a plasma, electron-hole recombination in a solid, etc. Even in the case of a laser, despite the fact that most of the energy is emitted by stimulated emission, spontaneous emission still takes place and is responsible for the finite spectral width of the emission. Hence, we have to represent the field as a random process; this is known as statistical optics (Goodman, 2015). We often assume that the electromagnetic field is a stationary random process. The stationarity assumption is made if the light source does not change over time during the duration of an experiment. An immediate consequence of the stationarity assumption is that the field is not square-integrable over time. Hence, Eq. (1.70) cannot be used. Thus, we need to address the question of the definition of the spectrum of a stationary random process.

1.A.2 Power spectral density

To deal with this issue, it is possible to introduce the restriction of the function to a limited domain in time. Let us consider a random function f of time t corresponding to a stationary random process. This means that $f(t)$ takes random values with a probability density function which does not depend on time. The statistical average value is denoted $< f >$. We introduce the restriction f_T of the function f to a finite time window with width T:

$$\begin{aligned} f_T(t) &= f(t) \quad \text{if} \quad 0 < t < T \\ &= 0 \quad t \leq 0 \quad \text{or} \quad t \geq T. \end{aligned} \tag{1.71}$$

It is now possible to compute the Fourier transform of f_T:

$$\tilde{f_T}(\omega) = \int_{-\infty}^{\infty} f_T(t)\exp(i\omega t)dt = \int_0^T f_T(t)\exp(i\omega t)dt. \tag{1.72}$$

There is no limit for f_T, as T tends to infinity. However, it may be possible to define the limit:

$$\lim_{T\to\infty} \frac{1}{T} < |\tilde{f_T}(\omega)|^2 >, \tag{1.73}$$

for a class of random processes. This is typically the case for white noise, for instance. This quantity is called 'power' spectral density and denoted $W_f(\omega)$. Here, the word power refers to the fact that the square of the function is associated with energy for electrical quantities. Dividing energy by time yields power. In other words, as T goes to infinity, the energy diverges but the power remains constant.

1.A.3 The Wiener–Khinchin theorem and temporal coherence

A key property of spectral analysis of stationary random processes is the Wiener–Khinchin theorem, which states that the correlation function is the Fourier transform of the power spectral density (Goodman, 2015):

$$< f(t+\tau)f(t) >= C_f(\tau) = \int_{-\infty}^{\infty} W_f(\omega)\exp(-i\omega\tau)\frac{d\omega}{2\pi}. \tag{1.74}$$

This is a major result. It shows that the broadening of the emission spectrum leads to a correlation function that takes significant values on smaller time intervals. This is often called reduced temporal coherence. It has a direct consequence on the ability to observe interferences, as we now discuss. The time correlation of an electromagnetic field is given by an interference measurement with, e.g., a Michelson interferometer, as the signal is proportional to $< (E(t) + E(t+\tau))^2 >=< E^2(t) > + < E^2(t+\tau) > +2 < E(t)E(t+\tau) >= 2I_0[1+ < E(t)E(t+\tau) > /I_0]$, where we have introduced the notation $I_0 =< E^2(t)>$ for the time-independent intensity. It is seen that the interference term is given by the time correlation of the field. From the Wiener–Khinchin theorem, we see immediately the connection with the field spectrum. According to the Fourier properties, a broad spectrum entails a short correlation time and a narrow spectrum gives rise to a long time correlation, known as long coherence time in the jargon of optics.

We finally note that the Wiener–Khinchin theorem is the basis of the Fourier Transform Infra Red (FTIR) spectrometers. These instruments are Michelson interferometers that measure the time correlation function and then yield the spectrum from a numerical calculation of the Fourier transform.

1.A.4 Formal definition of the Fourier transform of the fields

Although we have noted that the Fourier transform cannot be defined in the sense of functions for stationary processes, it is possible to use it formally as follows. We start by writing:

$$f(t) = \int_{-\infty}^{\infty} f(\omega) \exp(-i\omega t) dt, \tag{1.75}$$

so that:

$$< f(t)f(t') >= \int_{-\infty}^{\infty} \frac{\mathrm{d}\omega}{2\pi} \int_{-\infty}^{\infty} \frac{\mathrm{d}\omega'}{2\pi} < f(\omega)f(\omega') > \exp(-i\omega t)\exp(-i\omega' t'). \tag{1.76}$$

We now use the Wiener–Khinchin theorem for a stationary process:

$$< f(t)f(t') >= \int_{-\infty}^{\infty} W_f(\omega) \exp[-i\omega(t-t')] \frac{\mathrm{d}\omega}{2\pi}. \tag{1.77}$$

Comparing both forms, we obtain:

$$< f(\omega)f(\omega') >= 2\pi\delta(\omega+\omega')W_f(\omega). \tag{1.78}$$

This simple relation allows us to formally manipulate the Fourier transform of stationary random processes. Since measurable quantities are quadratic, they can be computed using (1.78) and thus the well-defined power spectral density.

Exercises

Exercise 1.1 Hertz potential. The Hertz potential $\mathbf{\Pi}_e$ is defined by the equation:

$$\mathbf{A}(\mathbf{r}, t) = \frac{1}{c^2}\frac{\partial \mathbf{\Pi}_e}{\partial t}.$$

(1) Derive the scalar potential in terms of the Hertz potential in the Lorenz gauge.

(2) We consider a medium with currents only due to the polarization density $\mathbf{P}(\mathbf{r}, t)$. Derive the propagation equation:

$$\nabla^2 \mathbf{\Pi}_e - \frac{1}{c^2}\frac{\partial^2 \mathbf{\Pi}_e}{\partial t^2} = -\frac{\mathbf{P}}{\varepsilon_0}.$$

(3) Derive the form of the fields:

$$\mathbf{E} = \nabla \times \nabla \times \mathbf{\Pi}_e - \frac{\mathbf{P}}{\varepsilon_0}$$
$$\mathbf{B} = \frac{1}{c^2}\nabla \times \frac{\partial \mathbf{\Pi}_e}{\partial t}.$$

Exercise 1.2 Transformation from the Lorenz gauge to the Coulomb gauge. Show that the function X such that $\varphi_L(\mathbf{k}, \omega) = \varphi_C(\mathbf{k}, \omega) + i\omega X(\mathbf{k}, \omega)$ is given by $-i\frac{\omega\rho}{\varepsilon_0 c^2}\frac{1}{k^2(\mathbf{k}^2 - k_0^2)}$, where $k_0 = \omega/c$.

Exercise 1.3 Retardation in the Coulomb gauge. We saw in section 1.2.3 that the Coulomb gauge has non-retarded potentials. This seems to generate non-retarded fields. On the other hand, we know that fields are gauge invariant, so the fields given by both gauges are equal. Here, we directly check their equality. We work in Fourier space. We use the notation $k_0 = \omega/c$.

(1) Derive the equation

$$\mathbf{E}_L = \frac{\mu_0 i\omega \mathbf{j}}{k_0^2 - k^2} + \frac{i\rho \mathbf{k}}{\varepsilon_0(k_0^2 - k^2)}.$$

In these formulas, the denominator $k_0^2 - k^2$ comes from the propagation equation and is the signature of the retardation.

(2) Derive the equation

$$\mathbf{E}_C = \frac{\mu_0 i\omega \mathbf{j}_\perp}{k_0^2 - k^2} + \frac{i\rho \mathbf{k}}{\varepsilon_0 k^2}.$$

Here, the denominator k^2 is typical of the non-retarded Coulomb potential.

(3) In the equation obtained in question (2), replace the quantity $\mathbf{j}_\perp$ by $\mathbf{j} - \mathbf{k}(\mathbf{k}\cdot\mathbf{j})/k^2$ and use the charge conservation to express $\mathbf{j}_\parallel$ in terms of ρ. Combining the terms proportional to the charge density, retrieve the equation derived in question 1. This direct calculation clearly illustrates that the transfer of a longitudinal component between the scalar potential and the vector potential does not change the field but changes the potentials. This is a direct illustration of the gauge invariance.

Exercise 1.4 Enhancing light emission. We consider a metallic optical antenna consisting of two metallic wires, as shown in Fig. 1.4 with white dotted lines. The modulus of the three components of the electric field is shown in the figure when the antenna is illuminated with a plane wave propagating along the y-axis and polarized along the x-axis. This antenna can be used to enhance the field emitted by a molecule, which can be viewed as a monochromatic electric dipole. We assume that the dipole is along the y-axis. Use the reciprocity theorem to find the best position and orientation to maximize the emitted field in the y-direction.

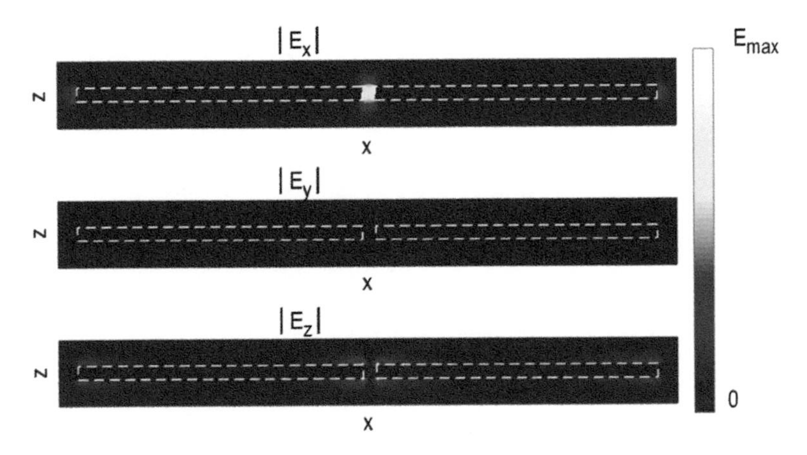

Fig. 1.4 Maps of the modulus of the Cartesian components of the electric field.

(3) In the equation obtained in question (2), replace the quantity [illegible] use the charge conservation equation [illegible] combining the [illegible] to the charge density [illegible] the equation derived in question 1. [illegible] [illegible] [illegible] [illegible] [illegible]

[illegible]

2
Radiation

The aim of this chapter is to derive the electromagnetic fields generated by time-dependent sources. Here, we assume that the sources are known and that they radiate in vacuum. This typically corresponds to the problem of a wire antenna in vacuum. The procedure for computing the fields is based on the fact that Maxwell equations are linear. It follows that a key role is played by the field radiated by a point-like source, as any source can be decomposed into a linear superposition of point-like sources. In this chapter, we will also introduce a concept that plays a fundamental role in electrodynamics: the electric dipole moment.

Radiation plays a key role in classical optics, as many elementary phenomena involving light–matter interaction can be viewed as specific cases of radiation. Scattering by an object is nothing but radiation by currents induced in this object. The term 'current' refers to oscillating charges, be it polarization charges in a dielectric or free electrons in a conductor. Reflection or transmission by a slab is also radiation due to the induced currents in the slab. Diffraction by a grating is also radiation by the induced currents in the grating. While scattering is, from a physics point of view, a radiation problem, solving a scattering problem involves finding the induced currents in the scatterer. This question will be addressed in Chapter 12.

2.0.1 Qualitative discussion

In this short paragraph, we briefly give some physical insight into radiated fields. We aim to show that radiated fields are the consequence of the acceleration of a charge. We also aim to stress that there are two major ingredients: Coulomb force and special relativity. A more detailed discussion can be found in ref. (Crawford, 1968). The electrostatic force is given by the Coulomb expression for a point-like charge at rest at point O. The field lines are radial and centred on O. Let us now imagine that this charge is accelerated between time $t = 0$ and time Δt, reaching a velocity $v \ll c$, where c is the speed of light in vacuum and then moves at constant velocity. In the frame moving at constant velocity, the line fields are radial and constant. Let us now consider the structure of the field in space at a given time $t \gg \Delta t$. For an observer at a distance $r > ct$, the electric field is given by the lines centred on O because the interaction is not instantaneous and the information on the change of the source position has not yet reached the observer. For an observer at a distance $r < c(t - \Delta t)$, the electric field is given by the electrostatic radial lines centred on O' with coordinate $x = vt$. In the region of space between $r = ct$ and $r = c(t - \Delta t)$, there is a transition between radial line fields centred on O and radial line fields centred on O'. In this thin region, the

Introduction to Nanophotonics. Henri Benisty, Jean-Jacques Greffet, and Philippe Lalanne, Oxford University Press.
 DOI: 10.1093/oso/9780198786139.003.0002

electric field connects both field lines and is squeezed between the two spherical shells, resulting in a transverse field. This is the radiated field.

2.0.2 Retarded potentials

From a mathematical point of view, computing radiated fields amounts to solving a propagation equation accounting for the source terms. Source terms in Maxwell equations are the current density $\mathbf{j}(\mathbf{r},t)$ and the charge density $\rho(\mathbf{r},t)$. The equations to be solved can be written using the potentials

$$\begin{aligned} \nabla^2 V - \frac{1}{c^2}\frac{\partial^2 V}{\partial t^2} &= -\frac{\rho(\mathbf{r},t)}{\varepsilon_0} \\ \nabla^2 \mathbf{A} - \frac{1}{c^2}\frac{\partial^2 \mathbf{A}}{\partial t^2} &= -\mu_0\, \mathbf{j}(\mathbf{r},t), \end{aligned} \tag{2.1}$$

where the source terms are considered to be known. For instance, a source drives an antenna, and the current flowing in the antenna is known. In order to deal with this problem, we will introduce a general technique for solving linear equations with source terms. This can be applied to radiation problems but also to the diffusion equation, for instance. The basic idea is to find the response of the system to a point-like excitation. This is called the *Green function* of the problem. Then, as the system amounts to a superposition of elementary sources, the solution is obtained by a linear superposition of the solutions. To illustrate the procedure, we start with the elementary example of electrostatics. We aim to compute the potential $V(\mathbf{r})$ produced by a static distribution of charges $\rho(\mathbf{r})$.

2.1 Green function: Electrostatic potential produced by a point-like charge

2.1.1 Introduction

Finding the electrostatic potential produced by the charge distribution $\rho(\mathbf{r})$ amounts to solving Poisson's partial differential equation

$$\nabla^2 V = -\frac{\rho(\mathbf{r})}{\varepsilon_0}. \tag{2.2}$$

The basic idea is based on two remarks:

- Poisson equation is linear.
- The charge distribution can be viewed as a superposition of point-like charges:

$$\rho(\mathbf{r}) = \int \rho(\mathbf{r}')\, \mathrm{d}\mathbf{r}'\, \delta(\mathbf{r}-\mathbf{r}'). \tag{2.3}$$

2.1.2 Point-like charge located at $\mathbf{r} = \mathbf{r}_0$

The charge density corresponding to a single point-like charge is given by

$$\rho(\mathbf{r}) = q\, \delta(\mathbf{r}-\mathbf{r}_0). \tag{2.4}$$

It can be checked that upon integration of this charge density over the volume, we find a total charge q. We now use the mathematical identity:

$$\nabla^2 \frac{1}{\mid \mathbf{r} - \mathbf{r}_0 \mid} = -4\pi\, \delta(\mathbf{r} - \mathbf{r}_0). \tag{2.5}$$

It follows that Poisson's equation solution for a point-like charge can be written as:

$$V(\mathbf{r}) = \frac{1}{4\pi\varepsilon_0} \frac{q}{\mid \mathbf{r} - \mathbf{r}_0 \mid}. \tag{2.6}$$

Of course, this elementary result could be derived more directly by recognizing that the Coulomb force derives from a scalar potential. Yet, this procedure serves as an introduction to the general procedure for solving linear equations. We search the solution of the partial differential equation with a Dirac source. The corresponding solution is called the *Green function* and is given here by $\frac{-1}{4\pi|\mathbf{r}-\mathbf{r}_0|}$. The second step implies adding all the solutions weighted by their amplitudes, as we will now illustrate.

2.1.3 Distribution of charges

We will consider two cases in parallel: a *discrete* and a *continuous* distribution of charges. For a discrete charge distribution, $\rho(\mathbf{r})$ can be cast in the form:

$$\rho(\mathbf{r}) = \sum_i q_i\, \delta(\mathbf{r} - \mathbf{r}_i). \tag{2.7}$$

For a continuous charge distribution, we can use the analogue form:

$$\rho(\mathbf{r}) = \int \rho(\mathbf{r}')\, \mathrm{d}\mathbf{r}'\, \delta(\mathbf{r} - \mathbf{r}'). \tag{2.8}$$

The correspondence between the two cases is clear: the charge q_i located at $\mathbf{r}_i$ corresponds to the charge $\rho(\mathbf{r}')\, d\mathbf{r}'$ located in the volume element $d\mathbf{r}'$ centred on $\mathbf{r}'$. The potential is then obtained by simply linearly adding the contributions. For the discrete case, we have

$$V(\mathbf{r}) = \sum_i \frac{q_i}{4\pi\varepsilon_0|\mathbf{r} - \mathbf{r}_i|}. \tag{2.9}$$

Similarly, for the continuous distribution, we have:

$$V(\mathbf{r}) = \int \frac{\rho(\mathbf{r}')\, \mathrm{d}^3\mathbf{r}'}{4\pi\varepsilon_0|\mathbf{r} - \mathbf{r}'|}. \tag{2.10}$$

It can be checked that these formulas are solutions of Poisson's equation by inserting them in Eq. (2.2). It is important to note that the linear superposition is given by the *convolution product of the Green function* $\frac{-1}{4\pi|\mathbf{r}-\mathbf{r}_0|}$ *by the source term* $(-\rho/\varepsilon_0)$. We note that when the term $1/|\mathbf{r}-\mathbf{r}'|$ does not diverge, the field can be computed without difficulty. Note, in particular, that the ordering of derivation and integration can be exchanged. Computing the electric field from the potential is thus possible without difficulty. Conversely, when we are interested in the field inside the charge distribution, the order of integration and derivation cannot be exchanged. However, the derivation can be done in the sense of generalized functions. In particular, by noting that there is a convolution product, we have:

$$\nabla^2 \left[\rho \otimes \frac{1}{r}\right] = \rho \otimes \nabla^2 \frac{1}{r}. \tag{2.11}$$

Using Eq. (2.5), it can be checked that Eq. (2.10) is indeed a solution of Eq. (2.2).

2.1.4 Generalization

This procedure can be generalized to any linear operator

$$L\, f = s, \tag{2.12}$$

where L is a linear operator such as ∇^2 or $(\nabla^2 + k^2)$, and where f is the unknown and s is the known source term. In order to find a solution, we seek the Green function G that obeys

$$L\, G(\mathbf{r} - \mathbf{r}') = \delta(\mathbf{r} - \mathbf{r}'), \tag{2.13}$$

which satisfies the boundary conditions. The second step consists in superposing the contributions of all sources. As noted before, this amounts to performing a convolution product:

$$f(\mathbf{r}) = \int G(\mathbf{r} - \mathbf{r}')\, s(\mathbf{r}')\, \mathrm{d}^3\mathbf{r}'. \tag{2.14}$$

It can be checked that this function is a solution of the partial differential equation using Eq. (2.13).

2.2 Retarded potentials

2.2.1 Green function

We return now to the solution of the propagation equations with source terms. We first specify the equations for the Green function. Here, we have four variables: three variables for the position, and time. Hence, the source is point-like in space along the three coordinates and also in time. We thus have to define G to obey

$$\nabla^2 G - \frac{1}{c^2}\frac{\partial^2 G}{\partial t^2} = -\delta(\mathbf{r} - \mathbf{r}')\delta(t - t') \tag{2.15}$$

and a similar equation for the vector potential $\mathbf{A}$.[1] The potential solution will be obtained in two steps. First, we need a derivation of the Green function. Second, the potential is derived by linear superposition. In order to simplify the notations, we use:

$$\mathbf{r} - \mathbf{r}' = \mathbf{R} \quad \text{and} \quad t - t' = \tau. \tag{2.16}$$

[1] Note that the minus sign is conventional. It is used here to obtain a Green function without a minus sign.

2.2.2 Derivation of the Green function

Finding a Green function amounts to solving a partial differential equation. This can be converted into an algebraic equation by introducing its Fourier transform. The procedure is summarized below:

$$\begin{array}{c}\text{Partial differential equation satisfied by } G(\mathbf{R},\tau)\\ \Big\downarrow {\scriptstyle FT}\\ \text{algebraic equation satisfied by } G(\mathbf{k},\omega)\\ \Big\downarrow {\scriptstyle \text{algebraic solution}}\\ G(\mathbf{k},\omega)\\ \Big\downarrow {\scriptstyle FT^{-1}}\\ G(\mathbf{R},\tau)\end{array}$$

The calculation steps are thus to set the equation obeyed by $G(\mathbf{R},\omega)$, and from it the equation obeyed by $G(\mathbf{k},\omega)$, to solve for $G(\mathbf{k},\omega)$, and then to calculate the inverse Fourier transform in space to get an explicit form of $G(\mathbf{R},\omega)$. A last time domain Fourier-transform yields $G(\mathbf{R},\tau)$; see details in Complement 2.A. The final result is the following:

$$G(\mathbf{R},\tau) = \frac{1}{4\pi|\mathbf{R}|}\delta\left(\tau - \frac{|\mathbf{R}|}{c}\right). \tag{2.17}$$

This solution shows that the wave propagates at speed c. It decays as $1/r$, as required to preserve energy conservation. A simple and familiar 2D version of this solution is the wave generated by a stone thrown into a pond of water at a given point and a given time.

2.2.3 Retarded potentials

The solution $V(\mathbf{r},t)$ is obtained by taking the convolution product of ρ/ε_0 and the Green function:

$$V(\mathbf{r},t) = \int \left[-\frac{\rho(\mathbf{r}',t')}{\varepsilon_0}\right][-G(\mathbf{r}-\mathbf{r}',t-t')]\,\mathrm{d}^3\mathbf{r}'\,\mathrm{d}t'. \tag{2.18}$$

Upon inserting G into the integral, we get:

$$V(\mathbf{r},t) = \int \frac{\rho(\mathbf{r}',t')}{4\pi\varepsilon_0}\frac{1}{|\mathbf{r}-\mathbf{r}'|}\,\delta\left(t-t'-\frac{|\mathbf{r}-\mathbf{r}'|}{c}\right)\mathrm{d}^3\mathbf{r}'\,\mathrm{d}t'. \tag{2.19}$$

After integration over t', we obtain the retarded potential solution:

$$V(\mathbf{r},t) = \int \frac{\rho(\mathbf{r}',t-\frac{|\mathbf{r}-\mathbf{r}'|}{c})}{4\pi\varepsilon_0}\frac{1}{|\mathbf{r}-\mathbf{r}'|}\,\mathrm{d}^3\mathbf{r}'. \tag{2.20}$$

A similar calculation yields the vector potential $\mathbf{A}$:

$$\mathbf{A}(\mathbf{r},t) = \int \frac{\mu_0}{4\pi}\frac{\mathbf{j}(\mathbf{r}',t-\frac{|\mathbf{r}-\mathbf{r}'|}{c})}{|\mathbf{r}-\mathbf{r}'|}\,\mathrm{d}^3\mathbf{r}'. \tag{2.21}$$

In these formulas, the time dependence introduces a time retardation $\frac{R}{c}$ corresponding to the travel time between the source point and the observation point. This is the origin of the name 'retarded potential'. A simple example of this retardation is the fact that light coming from a star was emitted many years before we detect it.

It is very useful to revisit this interpretation of the retardation time. The distance R is the length of an optical path. Eq. (2.21) shows that the vector potential is the sum of many contributions of the source. When adding the amplitudes of these contributions, we account for the phase variations due to the optical path.

This is better seen if we consider a monochromatic source. Let us define our notations. We consider a monochromatic field $\mathbf{X}(\mathbf{r},t)$ characterized by a complex amplitude denoted $\mathbf{X}(\mathbf{r})$ so that $\mathbf{X}(\mathbf{r},t)$ =Re$\left[\mathbf{X}(\mathbf{r})e^{-i\omega t}\right]$. Hence, the complex amplitude of the vector potential is given by:

$$\mathbf{A}(\mathbf{r}) = \frac{\mu_0}{4\pi} \int \mathbf{j}(\mathbf{r}') \frac{\exp\left(i\frac{\omega}{c}|\mathbf{r}-\mathbf{r}'|\right)}{|\mathbf{r}-\mathbf{r}'|} \, \mathrm{d}^3\mathbf{r}'. \tag{2.22}$$

It is seen that the retardation term appears as an optical path difference between different points of the source and the observation point. In other words, the fields produced at the observation point $\mathbf{r}$ by different parts of the source *interfere.*

2.3 Far-field radiation

2.3.1 Far-field approximation

In this section, we restrict our attention to the case of an observation point located far away from the source. This corresponds to situations where waves propagate away from the source. In order to define precisely the concept of far field, we first introduce a characteristic size (length) L of the source. We denote by $\mathbf{r}'$ a point in the source and by $\mathbf{r}$ the observation point (see Fig. 2.1). We will consider only points such as

$$|\mathbf{r}-\mathbf{r}'| \gg L.$$

This condition is a necessary one to be in the far field, but as we will see, it is not sufficient. We now proceed to simplify the optical path difference, assuming that we choose the origin within the source so that $|\mathbf{r}'| \ll |\mathbf{r}|$:

$$R^2 = |\mathbf{r}-\mathbf{r}'|^2 = \mathbf{r}^2 + \mathbf{r}'^2 - 2\mathbf{r}.\mathbf{r}' = r^2[1 - 2\mathbf{r}\cdot\frac{\mathbf{r}'}{r^2} + \frac{r'^2}{r^2}].$$

By neglecting the term of second order in $\frac{r'}{r}$, we obtain:

$$R \approx r - \frac{\mathbf{r}\cdot\mathbf{r}'}{r} = r - \mathbf{u}_r\cdot\mathbf{r}'. \tag{2.23}$$

It is now possible to simplify the form of the vector potential in Eq. (2.22). The denominator $1/|\mathbf{r}-\mathbf{r}'|$ is simply replaced by $1/r$. The approximation of the phase

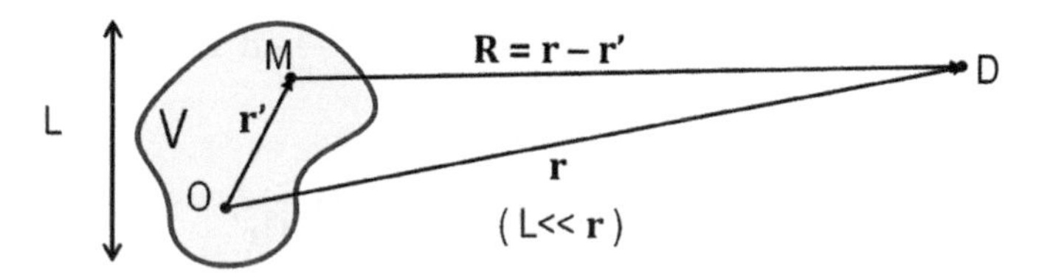

Fig. 2.1 The currents are confined in volume V. Observation point position is given by $\mathbf{r}$.

term needs more care. As the phase term is given by $2\pi R/\lambda$, the accuracy required to approximate R is given by the wavelength. We thus need to keep the term $\mathbf{r}' \cdot \mathbf{u}_r$, whose typical value is given by L and does not vanish as r increases. The second-order term in the expansion is on the order of L^2/r. It can be neglected at large distances, provided that

$$\frac{L^2}{\lambda} \ll r.$$

Finally, we obtain:

$$\mathbf{A}(\mathbf{r}) = \frac{\mu_0}{4\pi} \frac{e^{i\frac{\omega}{c}r}}{r} \int \mathbf{j}(\mathbf{r}')\, e^{-i\frac{\omega}{c}\mathbf{u}_r.\mathbf{r}'}\, \mathrm{d}^3\mathbf{r}'. \tag{2.24}$$

This result is the far-field form of the vector potential. We recognize from left to right: (i) a constant which depends on the unit system, (ii) a spherical wave, (iii) the Fourier transform of the current density in the source. Here, the wavevector is given by $\mathbf{k} = \frac{\omega}{c}\,\mathbf{u}_r$. It is very remarkable to note that this term contains the dependence on the emission angle through the unit vector $\mathbf{u}_r$. It follows that any emitting system has an angular pattern given by the Fourier transform of the current density distribution. This is a very general result, as we did not make any specific assumption. It can be applied to any electromagnetic system, from an atom to an antenna.

We already mentioned that the field produced by different points of the source interfere. This is particularly clear in Eq. (2.24). Each point M of the source generates a spherical wave whose amplitude is proportional to the current density at point M. The integral corresponds to the addition of the fields produced by all source points. The phase term

$$\frac{\omega}{c}\mathbf{u}_r.\mathbf{r}' = \frac{2\pi}{\lambda}\mathbf{u}_r.\mathbf{r}'$$

accounts exactly for the path difference $[OD]$–$[MD]$ between paths OD and MD, where D stands for the detector located at point $\mathbf{r}$.

We now add a word of caution regarding the far-field approximation, as its domain of validity is not always intuitive. The condition $|r| \gg L$ is simple and ensures that $1/R \approx 1/r$. However, the term neglected in Eq. (2.23) corresponds to a path difference that must be negligible as compared to the wavelength. The neglected term is r'^2/r and its order of magnitude is given by L^2/r. The validity condition

$$\frac{L^2}{\lambda} \ll r$$

is exactly the same condition that is used in diffraction theory and is known as the Fraunhofer condition. Let us consider a few examples. At a distance of 5 m from a 5 mm aperture illuminated by visible light ($\lambda \approx 0.5\,\mu$m), the far-field approximation is not valid! Similarly, a terrestrial observer of light emitted by the sun is not in the far field! Indeed, the earth–sun distance is on the order of $1.5\,10^{11}$ m and the diameter of the sun is approximately $1.4\,10^9$ m. By contrast, at a distance of 50 m from the Eiffel tower in Paris, an observer is in the far field of a 5 m radio antenna emitting at a wavelength of $\lambda \approx 1800$ m. In summary, the condition depends quadratically on the source size. Its physical meaning is illustrated in Fig. 2.2. When the far-field validity condition is satisfied, the curvature of the phase fronts can be neglected.

2.3.2 Dipole approximation in the far field

In this section, we still consider that $|\mathbf{r} - \mathbf{r}'| \gg L$. In addition, we consider sources whose size L is smaller than the wavelength $L \ll \lambda$ so that the phase differences of the fields produced by different parts of the source can be neglected. In that case, the source is equivalent to a single electric dipole, as we will see now. Since the argument is much smaller than 2π, we can expand the exponential $\exp[-i\frac{\omega}{c}\mathbf{u}_r.\mathbf{r}']$ as follows:

$$\mathbf{A}(\mathbf{r}) = \frac{\mu_0}{4\pi}\frac{e^{-i\frac{\omega}{c}r}}{r}\int \mathbf{j}(\mathbf{r}')[1 - i\frac{\omega}{c}\mathbf{u}_r.\mathbf{r}' + \ldots]\mathrm{d}^3\mathbf{r}'. \tag{2.25}$$

The expansion of the different terms corresponds to a multipolar expansion. The first term in Eq. (2.25) corresponds to the dipolar electric contribution. The second term corresponds to the dipolar magnetic term and the quadrupolar electric term, etc. In order to justify this terminology, we now cast the vector potential in a different form, explicitly introducing the electric dipole moment. Let us first define the dipole moment by the formula:

$$\mathbf{p}(t) = \int \rho(\mathbf{r}, t)\mathbf{r}\,\mathrm{d}^3\mathbf{r}, \tag{2.26}$$

where $\rho(\mathbf{r}, t) = Re\left[\rho(\mathbf{r})e^{-i\omega t}\right]$. We introduce the notation $\mathbf{p}(t) = Re\left[\mathbf{p}_0 e^{-i\omega t}\right]$. We will now establish the identity

$$\frac{d\mathbf{p}(t)}{dt} = \int \mathbf{j}(\mathbf{r}, t)\,\mathrm{d}^3\mathbf{r}, \tag{2.27}$$

which allows us to show that the lowest-order contribution in the expansion in Eq. (2.25) is due to the dipole moment.

The simplest approach is to start with the discrete form of the charge density Eq. (1.4)[2]. Let us consider a volume element δV containing N particles with speed $\mathbf{v}_i$ and located at $\mathbf{r}_i$. We now integrate over the volume δV the current density Eq. (1.5) $\mathbf{j}(\mathbf{r}, t) = \sum_i q_i\mathbf{v}_i(t)\delta[\mathbf{r} - \mathbf{r}_i(t)]$ and the polarization density $\mathbf{P}(\mathbf{r}, t) = \sum_i q_i\mathbf{r}_i(t)\delta[\mathbf{r} - \mathbf{r}_i(t)]$:

$$\mathbf{p}(t) = \sum_{i\in\delta V} q_i\,\mathbf{r}_i \quad \text{and} \quad \int_{\delta V} \mathbf{j}(\mathbf{r}, t)\,d\mathbf{r} = \sum_{i\in\delta V} q_i\,\mathbf{v}_i, \tag{2.28}$$

[2] The equality in Eq. (2.27) can also be established without invoking a discrete model of the charge density: let us simply derive the equality for a single Cartesian component along the axis Ox. We start with the identity $\nabla\cdot(x\mathbf{j}) = x\nabla\cdot\mathbf{j} + \nabla x\cdot\mathbf{j}$ (that is, the vector analogue of the identity $[uv]' = u'v + v'u$). By integrating this identity by parts, we obtain:

$$\int_{\delta V} \nabla\cdot(x\mathbf{j})\mathrm{d}^3\mathbf{r} = \int_{\delta V} [x\,\nabla\cdot\mathbf{j} + \nabla x\,\mathbf{j}]\mathrm{d}^3\mathbf{r}.$$

Using the Green–Ostrogradski theorem, we transform the first integral in a surface integral. By choosing a surface far from the source, the current density $\mathbf{j}(\mathbf{r})$ vanishes on the surface. We then get:

$$\int_{\delta V} x\,\nabla\cdot\mathbf{j}\,\mathrm{d}^3\mathbf{r} = -\int_{\delta V} \nabla x.\mathbf{j}\,\mathrm{d}^3\mathbf{r}.$$

Using $\nabla\cdot\mathbf{j} + \frac{\partial\rho}{\partial t} = 0$ and $\nabla x = \mathbf{u}_x$, we get:

$$\int_{\delta V} x\,\frac{\partial\rho}{\partial t}\mathrm{d}^3\mathbf{r} = \frac{d}{dt}\int_{\delta V} \rho(\mathbf{r}, t)x\,\mathrm{d}^3\mathbf{r} = \int_{\delta V} j_x(\mathbf{r}, t)\mathrm{d}^3\mathbf{r}.$$

showing that we do have

$$\frac{d\mathbf{p}(t)}{dt} = \int_{\delta V} \mathbf{j}(\mathbf{r}, t)\, \mathrm{d}\mathbf{r}. \tag{2.29}$$

In the dipolar approximation where we retain only the first term of Eq. (2.25), we obtain:

$$\mathbf{A}(\mathbf{r}) = \frac{\mu_0}{4\pi} \frac{e^{i\frac{\omega}{c}r}}{r}(-i\omega \mathbf{p}_0). \tag{2.30}$$

It is important to stress here that the dipole approximation we have just derived shows that any charge distribution with arbitrary shape and arbitrary number of charges can be replaced by a dipole moment inasmuch as its interaction with light is concerned, provided that the typical size is much smaller than the wavelength. The fundamental importance of the dipole concept becomes clear as soon as one realizes that any large distribution of currents can be described, by using a fine mesh, as a collection of small-volume elements which are all equivalent to dipoles.

2.3.3 Dipole approximation in the near field

A situation of interest in near-field optics is the case of a distribution of charges confined in a subwavelength volume generating a field at a distance which is also subwavelength. This is typically the situation for two molecules or two nanoparticles separated by a distance r on the order of 10 to 100 nm. In that case, the retardation can be neglected so that the vector potential is given by:

$$\mathbf{A}(\mathbf{r}) = \frac{\mu_0}{4\pi} \int \frac{\mathbf{j}(\mathbf{r}')}{|\mathbf{r} - \mathbf{r}'|}\, \mathrm{d}^3\mathbf{r}'. \tag{2.31}$$

Assuming that $r \gg L$, we can replace $|\mathbf{r} - \mathbf{r}'|$ by r. The integral is simplified and we obtain:

$$\mathbf{A}(\mathbf{r}) = \frac{\mu_0}{4\pi} \frac{-i\omega \mathbf{p}_0}{r}. \tag{2.32}$$

2.3.4 Magnetic dipole moment

Let us now consider cases where the concept of magnetic dipole moment and quadrupole moment are useful. If we consider a ring with a current circulating in the ring, it can be split in a collection of small (electric) dipoles. Yet, the overall dipole is zero. In that case, it is essential to evaluate the next term in the expansion given by:

$$\mathbf{A}(\mathbf{r}) = \frac{\mu_0}{4\pi} \frac{e^{i\frac{\omega}{c}r}}{r} \int \mathbf{j}(\mathbf{r}')[-i\frac{\omega}{c}\mathbf{u}_r.\mathbf{r}']\, \mathrm{d}^3\mathbf{r}'. \tag{2.33}$$

We can rearrange the integrand using

$$\mathbf{j}(\mathbf{u}_r \cdot \mathbf{r}') = \frac{1}{2}[\mathbf{u}_r \times (\mathbf{j} \times \mathbf{r}')] + \frac{1}{2}[\mathbf{j}(\mathbf{u}_r \cdot \mathbf{r}') + \mathbf{r}'(\mathbf{u}_r \cdot \mathbf{j})]. \tag{2.34}$$

When inserting this form in Eq. (2.33), the first term in brackets yields the magnetic dipole contribution. Indeed, we recognize the magnetic dipole moment of the current distribution:

$$\mathbf{m} = \frac{1}{2}\int \mathbf{r}' \times \mathbf{j}(\mathbf{r}')\,\mathrm{d}^3\mathbf{r}'.$$

The corresponding contribution to the vector potential is:

$$\mathbf{A}(\mathbf{r}) = \frac{\mu_0}{4\pi}\frac{e^{i\frac{\omega}{c}r}}{r}\left(i\frac{\omega}{c}\mathbf{u}_r\right) \times \mathbf{m}. \tag{2.35}$$

The second term of Eq. (2.34) yields the quadrupolar electric contribution.

2.3.5 Discussion

Both the far-field approximation and the dipole approximation correspond to an approximation of the path difference. To illustrate their meaning, we have displayed circles centred on the observation point $\mathbf{r}$ and separated by a wavelength. All the points located on a circle correspond to the same path. Assuming that the current density is uniform in the source so that the local phase of these points is the same, the fields produced by these points interfere constructively at the observation point. We now consider three cases (see Fig. 2.2).

- The source sketched in black: it has an extension much smaller than the wavelength. It is seen that all the points interfere constructively. The dipole approximation is thus valid for this source.
- The source sketched in medium grey: It is seen that the path difference between different points cannot be neglected, especially along the line of sight to the detector. Yet, all circles are well approximated by their tangent lines (difference $\ll \lambda$). This corresponds to the approximation $R \approx r - \mathbf{u}_r \cdot \mathbf{r}$. Indeed, the latter term $(\mathbf{u}_r \cdot \mathbf{r})$ is the distance between a point and a line. The far-field approximation is valid for that source. It is important to note that when the far-field approximation is valid, the spherical wave can be replaced by a plane wave with a wavevector parallel to $\mathbf{u}_r$.
- Finally, the source sketched in light grey is too large for the far-field approximation to be valid. It is seen that for points located on the same line perpendicular to $\mathbf{u}_r$, $r - \mathbf{u}_r \cdot \mathbf{r}$ takes the same value, whereas their distance to the observation point is clearly different–in this case by about a half-wavelength.

2.3.6 Electric and magnetic fields in the far field

In this section, we will derive the magnetic field and the electric field for a monochromatic source in the far field. The simplest method consists in first computing the magnetic field from the potential vector and then using Maxwell equations to derive

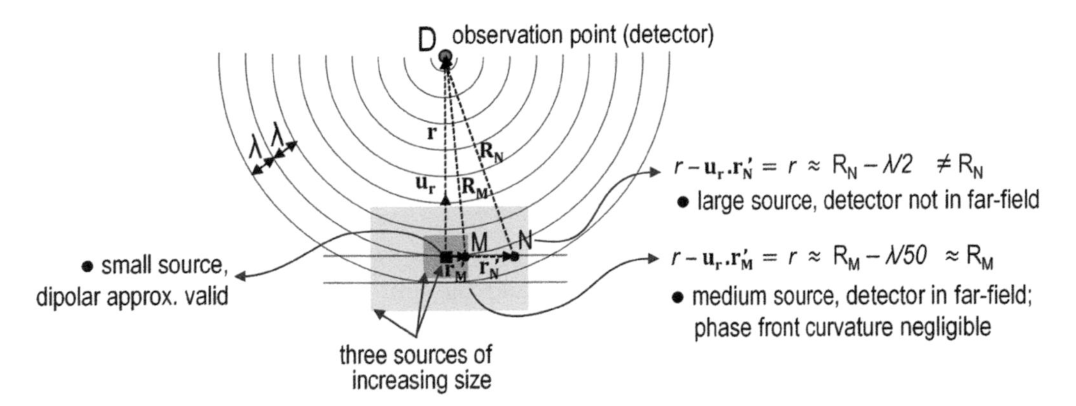

Fig. 2.2 Dipolar vs far-field approximation. For a source in the black zone, $R \approx r$, the dipolar approximation is valid. If in the grey zone, $R \approx r - \mathbf{u}_r.\mathbf{r}'$, so that the far-field approximation is valid. If in the light grey, the far-field approximation is invalid. At point N, for instance, $r - \mathbf{u}_r.\mathbf{r}'_\mathbf{N} = r$, but obviously $R_N = r + \lambda/2$, hence a phase difference.

the electric field from the magnetic field. Since we derive the quantities in the far field, we only keep terms decaying as $1/r$ and neglect faster-decaying terms:

$$\mathbf{B}(\mathbf{r}) = \nabla \times \frac{\mu_0}{4\pi} \int \frac{\exp i\dfrac{\omega}{c}R}{R} \mathbf{j}(\mathbf{r}')\, \mathrm{d}^3\mathbf{r}' = \frac{\mu_0}{4\pi} \int \nabla \left(\frac{\exp i\dfrac{\omega}{c}R}{R} \right) \wedge \mathbf{j}(r')\, \mathrm{d}^3\mathbf{r}'. \tag{2.36}$$

The nabla operates on the observer variable $\mathbf{r}$ only so that integration and derivation can be exchanged. We have used the formula $\nabla \times (f\mathbf{X}) = f\nabla \times \mathbf{X} + \nabla f \times \mathbf{X}$ and $\nabla_\mathbf{r} \times \mathbf{j}(\mathbf{r}') = 0$. The gradient is easily computed by noting that for a function f of x, y, and z that only depends on the distance r, we have $\nabla f(r) = \partial f/\partial r\ \nabla r$ and $\nabla r = \mathbf{r}/r = \mathbf{u}_r$.

$$\mathbf{B} = \frac{\mu_0}{4\pi} \int \left[\frac{-\mathbf{u}_R}{R} + i\frac{\omega}{c}\mathbf{u}_R \right] \wedge \mathbf{j}(\mathbf{r}') \frac{e^{i\frac{\omega}{c}R}}{R}\, \mathrm{d}^3\mathbf{r}' \simeq \frac{\mu_0}{4\pi} \int i\frac{\omega}{c}\mathbf{u}_R \wedge \mathbf{j}(\mathbf{r}') \frac{e^{i\frac{\omega}{c}R}}{R}\, \mathrm{d}^3\mathbf{r}' \tag{2.37}$$

No approximation was made until Eq. (2.36). If we now use the far-field approximation, we can replace the unit vector $\mathbf{u}_R$ by the unit vector $\mathbf{u}_r$, which is no longer dependent on the integration variable. We have also neglected the term $1/R$ as compared to $\omega/c = 2\pi/\lambda$, so that the validity condition is $R \gg \lambda$. By introducing a 'local' wavevector $k = \omega/c\mathbf{u}_r$, the magnetic field can be cast in the form:

$$\begin{aligned} \mathbf{B}(\mathbf{r}) &= i\frac{\omega}{c}\mathbf{u}_r \wedge \mathbf{A} \\ &= i\mathbf{k} \wedge \mathbf{A}. \end{aligned} \tag{2.38}$$

This result is the same as what would be obtained for a plane wave. It shows that in the framework of the far-field approximation, the field has a plane-wave structure locally. It follows that the calculation rules valid for a plane wave can be applied by using the local wavevector. Hence, we can easily derive the electric field:

$$\begin{aligned}\mathbf{E}(\mathbf{r}) &= -c\mathbf{u}_r \wedge \mathbf{B}(\mathbf{r}) \\ &= i\omega \left[\mathbf{A}(\mathbf{r}) - (\mathbf{u}_r.\mathbf{A}(\mathbf{r}))\,\mathbf{u}_r\right] \\ &= i\omega \mathbf{A}_\perp(\mathbf{r}).\end{aligned} \tag{2.39}$$

Note that in the far field, taking the curl of the vector potential twice amounts to performing a double vector product of $\mathbf{A}$ and the unit vector $\mathbf{u}$. It is important to notice that this amounts to projecting $\mathbf{A}$ on the plane perpendicular to $\mathbf{u}$. The result is denoted by $\mathbf{A}_\perp$. Hence, we recover *in the far field* the transverse structure of the plane waves. Note also that $|\mathbf{A}_\perp| = |\mathbf{A}| \sin\theta$, where θ is the angle between $\mathbf{A}$ and $\mathbf{u}$ varying between 0 and π. We also note that when the observation direction $\mathbf{u}$ coincides with the direction of $\mathbf{A}$, the angle θ is null, so that the radiated field is null. The result of Eq. (2.39) can also be retrieved using:

$$\mathbf{E} = -\nabla V - \frac{\partial \mathbf{A}}{\partial t} \tag{2.40}$$

and performing similar approximations to derive the scalar potential V. It turns out that the scalar potential contribution cancels exactly the longitudinal contribution of the vector potential $-\partial \mathbf{A}/\partial t$. Finally, we note that the electric field can be written in time domain in the form:

$$\mathbf{E}(\mathbf{r}, t) = -\frac{\partial \mathbf{A}_\perp(\mathbf{r}, t)}{\partial t}. \tag{2.41}$$

2.3.7 Electric and magnetic fields in the non-retarded region: Dipole approximation

In this section, we briefly outline the derivation of the electric field in the non-retarded region. We shall see in Chapter 10 that the non-retarded region is the near-field region. Again, we use the potential vector as a starting point. It is given by Eq. (2.32). The electric and magnetic fields are given by:

$$\mathbf{B}(\mathbf{r}) = \frac{\mu_0}{4\pi} \nabla \frac{1}{r} \times (-i\omega \mathbf{p}_0) \;\; ; \;\; \mathbf{E}(\mathbf{r}) = \frac{c^2}{-i\omega} \nabla \times \mathbf{B}(\mathbf{r}) = \frac{1}{4\pi\varepsilon_0} \nabla\nabla \cdot \frac{\mathbf{p}_0}{r}. \tag{2.42}$$

We have used the identity $\nabla \times \nabla \times \mathbf{X} = \nabla\nabla \cdot \mathbf{X} - \nabla^2 \mathbf{X}$, $\nabla^2(1/r) = 0$ for $r > 0$. It is seen that we recover the form of the electrostatic field, although we have derived the electric field from the magnetic field using Maxwell equations. The connection between the near field and the quasi-static field will be discussed in Chapter 10.

2.3.8 Power radiated in far field

The power radiated by the source is obtained by computing the flux of the Poynting vector through a sphere with radius r chosen so that the far-field approximation is valid.[3] Hence, we can use the simple form of the electric and magnetic field given in

[3] It is also possible to use the energy conservation relation and compute the work done by the source current on the field. Here, the field is the field radiated by the dipole on itself, which is called

terms of the vector potential, Eqs (2.38) and (2.39). If the dipole approximation is valid, we can use Eq. (2.30).

In order to derive the power radiated in a given solid angle $d\Omega$ subtended by the area $d\mathbf{S}$, we compute the flux of the Poynting vector through $d\mathbf{S}$:

$$\begin{aligned}\boldsymbol{\Pi}.d\mathbf{S} &= \Pi\mathbf{u}.d\mathbf{S} \\ &= \Pi r^2 d\Omega \\ &= \Pi r^2 \sin\theta \, d\theta \, d\varphi.\end{aligned} \tag{2.43}$$

Using Eqs (2.38) and (2.39), we find:

$$\Pi = \frac{\varepsilon_0 c}{2}\,\omega^2\,|\mathbf{A}_\perp|^2. \tag{2.44}$$

In the dipolar approximation, using Eq. (2.30), we find:

$$\Pi = \frac{\mu_0}{32\pi^2 c}\frac{\omega^4|\mathbf{p}_\perp|^2}{r^2}. \tag{2.45}$$

Note that the radiated power varies like ω^4. This is a key feature of radiation. It has very important and practical consequences. Indeed, while a metallic wire can be used as a waveguide to transport electrical energy over hundreds of kilometres at 50 Hz, a wire with a length of only one metre dissipates energy through radiation when operated at 100 kHz. In other words, the same metallic wire behaves as a waveguide or as an antenna depending on the operating frequency. This is not a surprise, as the radiated power varies by twelve orders of magnitude between 100 Hz and 10^5 Hz.

For a dipole with non-monochromatic time dependence, the power radiated is given by:

$$\Pi = \frac{\mu_0}{16\pi^2 c}\frac{|\ddot{\mathbf{p}}_\perp|^2}{r^2}. \tag{2.46}$$

After integration over all space, we obtain:

$$\begin{aligned}P = \int \Pi.d\mathbf{S} &= \frac{\mu_0}{8\pi c}|\ddot{\mathbf{p}}|^2 \int_0^\pi \sin^3\theta \, d\theta \\ &= \frac{\mu_0}{6\pi c}|\ddot{\mathbf{p}}|^2 = \frac{1}{4\pi\varepsilon_0}\frac{2|\ddot{\mathbf{p}}|^2}{3c^2}.\end{aligned} \tag{2.47}$$

Note that when using CGS units, the term $1/4\pi\varepsilon_0$ (right-most expression) disappears.

Considering a monochromatic dipole oscillating at the frequency ω, the mean radiated power is given by the Larmor formula, which follows from Eq. (2.47):

$$\langle P\rangle = \frac{1}{4\pi\varepsilon_0}\frac{\omega^4\, p_0^2}{3\, c^3}. \tag{2.48}$$

'radiation reaction'. This procedure was first introduced by Brillouin (Brillouin, 1922). The calculation will be performed in Chapter 19.

2.4 Antennas: Basic concepts

In the previous sections, we introduced a general framework that allows the electromagnetic fields to be computed from the knowledge of the current density **j** or the dipole moment. This is usually the formalism used in the context of optics. However, it is possible to deal with radiation and detection without performing explicit calculations of integrals over the whole volume of the emitting system. Such a formalism has been developed to deal with emission and detection of radiowaves and microwaves. In that case, sources and detectors are characterized by intensities and voltages, not by watts per unit area. The key tools for emitting or detecting electromagnetic waves are antennas which are fed by lines with a characteristic impedance. In what follows, we will give a brief introduction to the key concepts needed to characterize an antenna. A more extensive treatment can be found in (Staelin et al., 1994; Balanis, 2016).

2.4.1 Radiation resistance, efficiency, superradiance

Let us first consider an antenna used to emit radiowaves. From the electrical circuit point of view, the antenna is a two-pole device connected to a circuit, Fig. 2.3. It is characterized by an impedance whose real part is the antenna resistance R. The power dissipated by the antenna is either transformed into heat in the antenna wires or radiated. Two resistances are introduced to account for these two processes:

$$R = R_R + R_{NR}, \tag{2.49}$$

where R_R is the radiation resistance and R_{NR} accounts for the non-radiative losses. The fraction of dissipated power which is converted into radiation is proportional to the antenna efficiency:

$$\eta_{ant} = \frac{R_R}{R_R + R_{NR}}. \tag{2.50}$$

From the feed line point of view, the wave propagating along the line to the two-port device is characterized by an impedance so that the transmission is unity if the impedances are matched. The question is, therefore, *how does one increase the radiation resistance?*

Here, we do not attempt to derive a rigorous and general answer. Instead, we want to pinpoint the very basic physical phenomenon responsible for the increase in radiated power produced by an antenna as compared to a simple wire or formally a single electron. To start this qualitative discussion, we consider a small antenna wire with length $L \ll \lambda$, where a current with intensity I is flowing at frequency $\omega = ck$. The emitted power is given by (see Exercise 2.2).

$$P_R = \frac{\eta_0}{12\pi}(kL)^2 I^2 = \frac{1}{2} R_R I^2, \tag{2.51}$$

where $\eta_0 = \mu_0 c = 377\,\Omega$ is the vacuum impedance and $k = 2\pi/\lambda$ is the wavevector modulus. It is clearly seen that the field increases linearly with L. Let us estimate the maximum antenna resistance. It is tempting to increase the length, L. Yet, when L becomes larger than half a wavelength, destructive interferences will appear so that the power emitted in some directions will be cancelled. As a matter of fact, when increasing

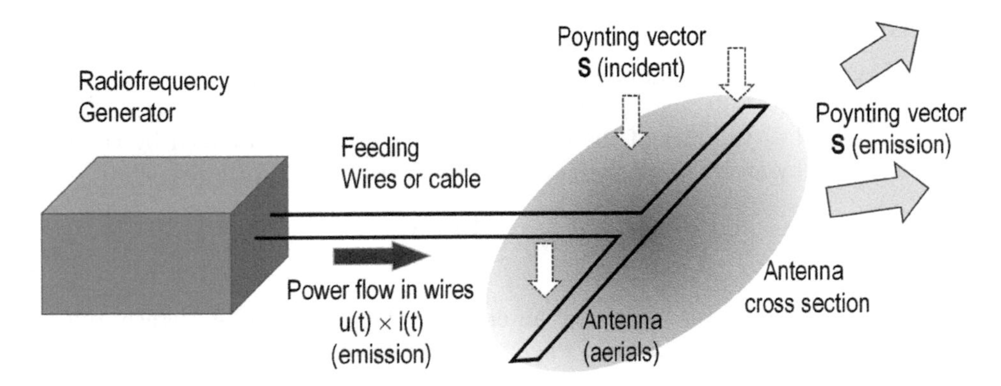

Fig. 2.3 Generator feeding an antenna. The antenna cross section is the grey ellipse.

the antenna length, the emitted power will be concentrated in a few directions but the overall radiated power will not be significantly affected, as it can be checked by a direct calculation of the total power emitted by a linear wire. *The consequence of this discussion is that a half-wavelength antenna with a current $I(z') = I\cos(kz')$ is close to the optimum resistance in terms of emitted power. It has a radiation resistance, $R_R = 2P_R/I^2$, which is given by $R_R = 73\,\Omega$.* A consequence of this statement is that the impedance of lines used to feed antennas should be on the order of 70 Ω. We now discuss the power emitted by a collection of N antennas which are within a subwavelength volume and are driven by the same voltage so that they are emitting fields which are in phase. The radiated field is N times larger, so that the radiated power is N^2 times larger. This simple effect is a direct consequence of *constructive interferences.* This effect is called collective radiation, or *superradiance* in the context of light emission by an ensemble of N atoms in the low excitation regime.[4] The superradiance can also be observed in acoustics. Two guitar strings or two organ pipes perfectly tuned will radiate more power and will therefore decay faster. In summary, the radiation resistance is not an intrinsic property of each antenna. The radiation resistance depends on the environment.

We now move to a nanophotonics object: a metallic nanosphere. The power radiated by the dipole moment of a metallic nanoparticle is proportional to the square of its dipole moment and therefore to the square of its volume, or, in other words, to the square of the number of electrons. By realizing that a metallic particle with a 50 nm radius contains $7.5\,10^6$ free electrons, it is clear that such a particle is a much better emitter than a single atom or molecule. The power radiated by such a particle is expected to be twelve orders of magnitude larger than for a single atom. This size effect is the key property of antennas in terms of increasing the radiated power.

[4] This classical superradiance effect corresponds to the spontaneous emission of one excited atom among $N-1$ non-excited identical atoms. This effect should not be confused with the superfluorescence effect observed when most of the N atoms are all excited. In that case, a burst of light is observed on a time scale given by τ/N.

2.4.2 Gain

Another key feature of an antenna is the control of the angular radiation pattern. To characterize the ability of an antenna to radiate light in direction (θ, ϕ), the gain $G(\theta, \phi)$ is defined as the flux of the Poynting vector $\mathbf{\Pi}(\mathbf{r})$ at $\mathbf{r}$ divided by the power density radiated by an isotropic antenna that would emit a uniform flux $P/4\pi r^2$:

$$G(\theta, \phi) = \frac{\mathbf{\Pi}(\mathbf{r}) \cdot \hat{\mathbf{r}}}{P/4\pi r^2}, \tag{2.52}$$

where P is the total power *fed* to the antenna by the generator. A slightly different quantity called directivity is defined when normalizing to the actual *radiated* power—that is, to $\eta_{ant}P$. The directivity accounts for the antenna efficiency

$$D(\theta, \phi) = \frac{G(\theta, \phi)}{\eta_{ant}}, \tag{2.53}$$

so that it verifies

$$\int_0^{2\pi} \mathrm{d}\phi \int_0^{\pi} \sin\theta \mathrm{d}\theta D(\theta, \phi) = 4\pi. \tag{2.54}$$

It is worth discussing briefly the conditions required to obtain a directional antenna. Let us consider the radiation by an antenna described by a monochromatic current density $\mathbf{j}(\mathbf{r}, \omega)$ in a finite volume V. We use the following definition of Fourier transform:

$$\mathbf{j}(\mathbf{r}, t) = \int_{-\infty}^{\infty} \mathbf{j}(\mathbf{r}, \omega) \exp(-i\omega t) \frac{\mathrm{d}\omega}{2\pi}. \tag{2.55}$$

The retarded vector potential is given by:

$$\mathbf{A}(\mathbf{r}, \omega) = \frac{\mu_0}{4\pi} \int_V \mathbf{j}(\mathbf{r}', \omega) \frac{\exp(i\frac{\omega|\mathbf{r}-\mathbf{r}'|}{c})}{|\mathbf{r}-\mathbf{r}'|} d^3\mathbf{r}'. \tag{2.56}$$

This form clearly displays the fact that different elements of the source produce spherical waves. The resulting total potential is due to the interferences between these waves. To discuss directivity, we need to move to the far field. In the far-field (Fraunhofer) approximation, valid provided that $r \gg L^2/\lambda$ where L is a characteristic length of the current distribution, the retarded potential can be simplified:

$$\mathbf{A}(\mathbf{r}, \omega) = \frac{\mu_0}{4\pi} \frac{\exp(i\frac{\omega}{c}r)}{r} \int_V \mathbf{j}(\mathbf{r}', \omega) \exp\left(-i\frac{\omega}{c}\frac{\mathbf{r}}{r} \cdot \mathbf{r}'\right) d^3\mathbf{r}'. \tag{2.57}$$

In this approximation, the interference appears clearly: the vector potential results from adding the contributions of all the volume elements but accounting for a path difference given by $\frac{\mathbf{r}}{r} \cdot \mathbf{r}'$. This equation shows explicitly how the vector potential depends on the emission direction $\mathbf{r}/r$. It is seen that this is given by the spatial Fourier transform of the current density. Hence, it follows immediately from the general property of the Fourier transform $\Delta x \Delta k_x \geq 2\pi$ that the radiation in the far field can be restricted to small solid angles if and only if the antenna is much larger than a wavelength.

Assuming an extension L along the x-axis, we find that the radiated field has a spatial spectrum with width $\Delta k_x \approx 2\pi/L$ corresponding to an angular width $\delta\theta \approx \frac{\lambda}{L}$. We recover the well-known result of diffraction theory. This is no surprise, as diffraction theory is based on the Huygens–Fresnel principle, which amounts to transforming the diffraction problem into a radiation problem by introducing secondary sources. As this discussion is based on the retarded potential solution of Maxwell equations and the far-field approximation, the conclusion is very general and can be applied to any radiating system. It follows that any subwavelength antenna cannot be highly directional. Note, however, that some degree of directivity from subwavelength radiators can be obtained due to either polarization effects (e.g. a $\sin^2\theta$ dependence for dipolar radiation) or interferences between dipolar and quadrupolar modes.

2.4.3 Receiving antenna: Effective antenna area

When an antenna is used to collect incident power and funnel it into a detector, an interesting quantity is the *effective antenna area* $A_{eff}(\theta,\phi)$ also called *capture cross section.* This effective area is defined by the equation

$$P_{abs}(\theta,\phi) = \Pi_{inc} A_{eff}(\theta,\phi), \tag{2.58}$$

where $P_{abs}(\theta,\phi)$ is the absorbed power in the load and Π_{inc} is the Poynting vector amplitude of the incident wave, assuming that the antenna is impedance-matched with the load. This effective area is close to the naive geometrical cross section $\sim L^2$ of the antenna for very large antennas with aerials of size L. By contrast, for small resonant antennas, both areas can differ a lot. In particular, for a dipolar electromagnetic antenna on resonance, $A_{eff}(\theta,\phi)$ is given by:

$$A_{eff}(\theta,\phi) = \frac{3\lambda^2}{8\pi}\sin^2\theta. \tag{2.59}$$

Note that this value is universal and does not depend on the size and exact shape of the antenna. In this particular example, $A_{eff}(\theta,\phi)$, the gain $G = 3\sin^2(\theta)/2$ and λ^2 are simply related by

$$A_{eff}(\theta,\phi) = \frac{\lambda^2}{4\pi} G(\theta,\phi), \tag{2.60}$$

which turns out to be a general result valid for any antenna, as can be shown using the reciprocity theorem (Staelin et al., 1994).[5] This is a fundamental property of antennas, which shows that emitting and receiving properties are connected. It also clearly shows that the directivity of an antenna is fundamentally related to its direction-dependent effective area. This relation can be understood qualitatively by recognizing that $\lambda^2/A_{eff} = (\lambda/L)^2$ is the solid angle of the emission pattern by a planar antenna with area $A_{eff} = L^2$. Hence, the gain can be cast in the form

$$G(\theta,\phi) = \frac{4\pi}{\Delta\Omega}, \tag{2.61}$$

where we recognize the ratio of two solid angles: the solid angle 4π of an isotropic antenna and the solid angle $\Delta\Omega$, where the power is concentrated by the antenna.

[5] See also the derivation of the maximum absorption and scattering cross section of a resonant dipolar scatterer in Complement 14.A.

2.4.4 Absorption by small antennas: The Chu–Wheeler limit

In the previous section, we saw that the effective area of a dipolar antenna *on resonance* is independent of its size. This apparently contradicts the intuition that a large antenna absorbs more efficiently than a smaller antenna. To reconcile these two ideas, we need to consider the absorption spectrum. A very small resonant antenna has a small radiation resistance and thus a larger quality factor. Hence, although the peak absorption on resonance is independent of the size, *the absorption spectrum is narrower* for small antennas (see also complement 14.B).

The figure of merit is the quality factor Q. It has been shown that for an antenna inside a sphere with radius a, Q has a minimum value known as the Chu–Wheeler limit (see exercise 2.7):

$$Q = \frac{1}{(ka)^3}. \tag{2.62}$$

It is seen that the quality factor decays as the antenna size increases.

2.A Complement: Derivation of the Green function of the propagation equation

In this complement, we derive the Green function of the scalar Helmholtz equation and the Green function of the scalar propagation equation in three dimensions following the procedure outlined in section 2.2.2. In order to transform the partial differential equation Eq. (2.15) into an algebraic equation, we introduce the Fourier representation:

$$G(\mathbf{r}, t) = \int \frac{\mathrm{d}^3\mathbf{k}}{8\pi^3} \int \frac{\mathrm{d}\omega}{2\pi} G(\mathbf{k}, \omega) \exp(i\mathbf{k} \cdot \mathbf{r} - i\omega t). \tag{2.63}$$

Inserting Eq. (2.63) in Eq. (2.15), we get:

$$\left(-\mathbf{k}^2 + \frac{\omega^2}{c^2}\right) G(\mathbf{k}, \omega) = -1. \tag{2.64}$$

This equation can be solved for real values of $\mathbf{k}$ and ω by adding a small imaginary part $i\eta$ to $k_0 = \omega/c$ in order to avoid the divergence. We perform the calculation and take the limit $\eta \to 0$ at the end of the derivation. We then get:

$$G(\mathbf{k}, \omega) = \frac{1}{\mathbf{k}^2 - k_0^2}. \tag{2.65}$$

We now compute $G(\mathbf{r}, \omega)$ using spherical coordinates with the polar axis along $\mathbf{r}$:

$$G(\mathbf{r}, \omega) = \int \frac{\mathrm{d}^3\mathbf{k}}{8\pi^3} \frac{\exp(i\mathbf{k} \cdot \mathbf{r})}{\mathbf{k}^2 - k_0^2} = \frac{1}{8\pi^3} \int_0^{\pi} \sin\theta \mathrm{d}\theta \int_0^{2\pi} \mathrm{d}\phi \int_0^{\infty} k^2 \frac{\exp(ikr\cos\theta)}{k^2 - k_0^2} \mathrm{d}k. \tag{2.66}$$

Carrying out the integrations over the angles ϕ and θ using the change of variable $u = \cos\theta$, we get:

$$G(\mathbf{r}, \omega) = \frac{1}{4\pi^2} \int_{-\infty}^{\infty} \frac{k}{(k - k_0)(k + k_0)} \frac{\exp(ikr)}{ir} \mathrm{d}k. \tag{2.67}$$

Using the residue theorem and a contour in the complex plane k closing the contour in the upper half-plane ($\mathrm{Im}(k) > 0$), we finally get:

$$G(\mathbf{r},\omega) = \frac{\exp(ik_0 r)}{4\pi r}, \tag{2.68}$$

which is the Green function of the Helmholtz equation defined by:

$$\nabla^2 G(\mathbf{r},\omega) + \frac{\omega^2}{c^2} G(\mathbf{r},\omega) = -\delta(\mathbf{r}). \tag{2.69}$$

The Green function of the propagation equation is obtained by performing the integral over frequencies:

$$G(\mathbf{r},t) = \int_{-\infty}^{\infty} \frac{\mathrm{d}\omega}{2\pi} \frac{\exp\left(i\frac{\omega}{c} r\right)}{4\pi r} \exp(-i\omega t) = \frac{\delta(t - r/c)}{4\pi r}. \tag{2.70}$$

Exercises

Exercise 2.1 Power radiated by an electric dipole. We consider an electron oscillating at frequency $6\,10^{14}$ Hz corresponding to a wavelength in vacuum of 500 nm. The amplitude of the electron oscillation is typically 10^{-11} m. What is the power radiated by the classical dipole? We now make a qualitative estimate of the lifetime of an excited state of an atom. Assuming that the energy of the excited state is given by $h\nu$, derive the order of magnitude of the lifetime of the atom. What is the frequency dependence of the lifetime? Compare the lifetime in the visible $6\,10^{14}$ Hz, and in the mid-infrared $6\,10^{13}$ Hz.

Exercise 2.2 Dipole antenna. We consider a metallic wire with length L and diameter $d \ll \lambda$, where λ is the wavelength in vacuum. The current density in the wire is denoted as $\mathbf{j}(x,y)\mathbf{e}_z \exp(-i\omega t)$. Using Eq. (2.29), show that the dipole moment can be cast in the form:

$$\mathbf{p} = \frac{IL}{-i\omega}\mathbf{e}_z,$$

where I is the electric current intensity. Using Eq. (2.45), derive Eq. (2.51). Prove that the gain is given by $G(\theta) = 3\sin^2(\theta)/2$.

Exercise 2.3 Radiative losses of a metallic wire. We consider the metallic wire described in Exercise 2.2. Its length L is 1 cm, its cross section is 10^{-6} m^2, and the resistivity ρ is 10^{-6} Ωm.

(1) Compare the ohmic resistance with the radiative resistance, assuming that it is dipolar. At which frequency ω_c are these two resistances equal? In other words, when does the ohmic resistor become an antenna?

(2) The antenna operates at this frequency. Discuss what the dominant loss mechanism is as a function of the antenna length L. A similar trend is observed for metallic nanoparticles.

Exercise 2.4 Dipolar magnetic source. We study the radiation by a magnetic dipole. We consider an intensity I flowing in a circular ring with radius a at frequency ω. We also assume $\lambda \gg a$. The observation point $\mathbf{r}$ is in the far field.

(1) Give the general form of the vector potential in the far-field approximation as a function of the monochromatic current density $\mathbf{j}(\mathbf{r})$. Assuming the diameter of the wire to be negligible compared to the ring radius, show that the vector potential is given by an integral along the wire and is proportional to the intensity I.

(2) Show that the electric dipole moment is zero.

(3) We assume $r \gg \lambda$. Expand the phase to first order and show that the magnetic moment $\mathbf{m} = I\mathbf{S} = I\pi a^2 \mathbf{e}_z$ can be factorized in the expression of the vector potential by using Kelvin's formula:

$$\oint_C f \mathrm{d}\mathbf{l} = -\int_S \mathbf{grad} f \times \mathrm{d}\mathbf{S}.$$

(4) Derive the electric and magnetic fields.

(5) Derive the power radiated in the solid angle $\mathrm{d}\Omega$, i.e. the radiated power density.

(6) Derive the time-averaged radiated power integrated over 4π sr (full space).

(7) What is the radiated power if N loops with identical current are implemented (namely a coil)? Explain the dependence on N physically.

Exercise 2.5 Radiation by a surface current. Reflection by a perfect conductor. We study the field radiated by a homogeneous surface current density,

$$\mathbf{j}_0\, \mathbf{e}_y \, \exp(-i\omega t).$$

The complex amplitude of the monochromatic vector potential is given by:

$$\mathbf{A}(\mathbf{r}) = \frac{\mu_0}{4\pi} \int \mathbf{j}(\mathbf{r}') \, \frac{\exp\left[i\frac{\omega}{c}|\mathbf{r} - \mathbf{r}'|\right]}{|\mathbf{r} - \mathbf{r}'|} \, d^3\mathbf{r}'.$$

(1) Explain why the far-field approximation cannot be used to compute this integral.

(2) Derive the vector potential amplitude using

$$\int_{-\infty}^{\infty} dx' \int_{-\infty}^{\infty} dy' \frac{\exp\left[i\frac{\omega}{c}|\mathbf{r} - \mathbf{r}'|\right]}{|\mathbf{r} - \mathbf{r}'|} = \frac{2i\pi c}{\omega} \exp\left[i\frac{\omega}{c}|z - z'|\right]. \tag{2.71}$$

(3) Derive the electric field radiated for $z < 0$ and for $z > 0$.

(4) We now consider that the homogeneous current density is induced by a plane wave normally incident on a perfect conductor with amplitude $\mathbf{E}_{inc}$. The field inside the perfect conductor ($z > 0$) is the linear superposition of the incident field and the field radiated by the surface current. Using the fact that the field in the metal is zero, derive the amplitude of the induced surface current (this is a form of the extinction theorem). Derive the amplitude of the field radiated towards $z < 0$ (reflected field).

Exercise 2.6 Radiation by a finite size planar antenna: Radiation, reflection, diffraction, or scattering? The purpose of this exercise is to show that reflection, diffraction, and scattering are particular cases of interaction of light with matter. They can all be considered to be a particular case of a radiation problem. We first study the field radiated by a surface current on a finite size rectangular planar antenna: $\mathbf{j} = j_0\ \mathbf{e}_y\ \exp[i(\alpha_i x + \beta_i y - \omega t)]$, where the planar antenna is a rectangular $a \times b$ patch (see Fig. 2.4).

(a) Derive the electric field radiated in the far field. Derive the power radiated in the solid angle $d\Omega$.

(b) The angular pattern has some lobes. We limit the analysis to the plane (x, z) and to the case of $\alpha_i = \beta_i = 0$.

What is the angular width of the central lobe defined by the first zero of the amplitude?

(c) Explain why the maximum of the radiated field is along the normal when $\alpha_i = \beta_i = 0$.

(d) Give a simple argument based on interferences to explain the position of the first zero.

Radiation, reflection, diffraction, and scattering

We now consider that the surface current density used in the previous calculation is induced by a plane wave, $\mathbf{E} = E_0\ \mathbf{e}_y\ \exp[i(\alpha_i x + \beta_i y - \gamma_i z - \omega t)]$, impinging on a perfectly conducting surface (see Fig. 2.4). The surface current amplitude j_0 is given by $j_0 = 2E_0/\mu_0 c$. This result is exact for an infinitely large surface. We neglect edge effects and assume that the surface current is uniform in the rectangular film. This is not an accurate approximation for a quantitative study but is enough for qualitative purposes. In what follows, we show that the radiation calculation which has been performed in the first section contains all the physics of the problem. We now explore the behaviour of the light radiated by the metallic patch depending on its size. By doing so, we will recover different regimes which are called

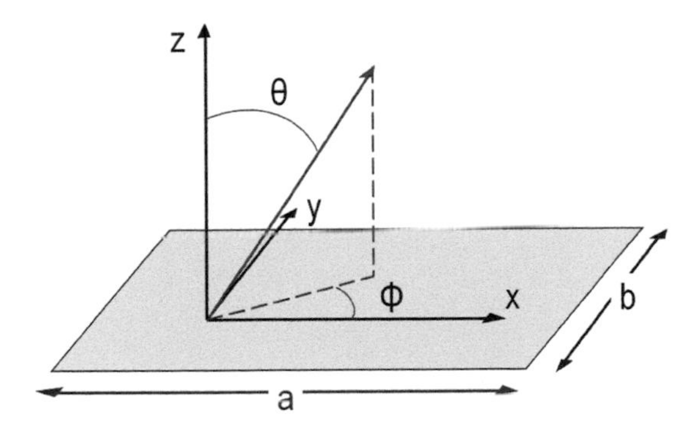

Fig. 2.4 Geometry of a rectangular, perfectly conducting antenna.

reflection, diffraction, and scattering. We want to make the point that these different words do not account for different physical phenomena. Instead, they describe different regimes of the same physical phenomenon, which is scattering, or in other words, radiation by the current density induced in an object by an incident field.

(a) If $a \gg \lambda$ and $b \gg \lambda$ and if λ is in the visible part of the spectrum, what would you call such a perfect conducting surface (e.g. in a bathroom)? Do you recover Snell's reflection laws? Explain qualitatively why the field emitted is highly directional.

(b) The reflected field has an angular acceptance. Compute the angular position of the first zero and compare with the usual diffraction angular width.

(c) We now use the same calculation and assume that $a \ll \lambda$. Discuss qualitatively the angular width of the emitted radiation. Show that the metallic film is equivalent to a dipole moment. Which word seems appropriate to describe this particular situation: reflection, diffraction, or scattering?

Exercise 2.7 Simple estimate of the Chu–Wheeler limit. The goal of this exercise is to derive a crude estimate of the Chu–Wheeler limit discussed in section 2.4.4 for a small antenna embedded in a sphere with diameter $2a$. We model a resonant dipolar antenna by an RLC circuit with a purely radiative resistance given by Eq. (2.51) where we take the antenna length $L = 2a$. The resonant frequency of the circuit is given by $LC\omega^2 = 1$. Its quality factor is $Q = L\omega/R = 1/RC\omega$. We estimate the capacitance using $C \approx \varepsilon_0 2a$. Show that $Q_{C-W} \approx \frac{3\pi}{4}\frac{1}{(ka)^3}$.

3

Electrodynamics in material media: Constitutive relations

3.1 Introduction

The interaction of electromagnetic waves and material media can be characterized by many different phenomena such as absorption, scattering, reflection, etc. Of course, these effects are strongly dependent on frequency and the nature of the materials. Visible light will be reflected by a metal and transmitted by glass. Glass is transparent in the visible but opaque in the infrared. A flat surface reflects light but a rough glass surface scatters light, etc. To account for these effects, it is necessary to include in Maxwell equations the relevant information concerning the materials. In Maxwell equations, the terms which couple matter to fields are the source terms: magnetization $\mathbf{M}$, polarization $\mathbf{P}$, charge density ρ, and current density $\mathbf{j}$. In some cases, the source terms are known. For example, a static charge q generates an electrostatic potential $q/(4\pi\varepsilon_0 r)$, a dipole moment generates an electric field, etc. In other cases, the sources are induced by the fields. A static external electric field induces a current density in a conductor medium or a polarization in a dielectric medium, an external magnetic field induces a magnetization in a paramagnetic material, etc. This raises the question of the connection between the induced sources and the fields. This connection is necessarily specific to a given material. For instance, some materials are conducting, and some are insulating. The equations quantifying these relations are called constitutive equations. Examples of such equations are given below:

$\mathbf{j} = \sigma\mathbf{E}$, Ohm's law,
$\mathbf{j} = -K\mathbf{A}$, London's law for superconductors,
$\mathbf{D} = \varepsilon_0\varepsilon_r\mathbf{E}$ for a linear isotropic dielectric medium,
$\mathbf{B} = \mu_0\mu_r\mathbf{H}$ for a linear isotropic magnetic medium.

Historically, the constitutive equations were introduced phenomenologically. In this chapter, we limit the discussion to linear response theory. Even though a pure phenomenological model is used (such as the examples quoted above), there are a number of requirements that need to be satisfied by the models. We start by defining a few general properties of the phenomenological laws. We then give an introduction to the models widely used to describe optical properties of gases, metals, and dielectrics. Although an accurate model requires in principle a full quantum treatment, the full quantum description is often untractable when dealing with condensed matter. Indeed, the very large number of particles leads to a many-body problem which cannot be

Introduction to Nanophotonics. Henri Benisty, Jean-Jacques Greffet, and Philippe Lalanne, Oxford University Press.
 DOI: 10.1093/oso/9780198786139.003.0003

solved in most cases. However, there are many simple models that can be used to fit the experimental data.

3.2 General properties of constitutive relations

In what follows, we summarize the main properties of the constitutive equations. This section is essentially a short vocabulary section.

3.2.1 Stationarity and homogeneity

A medium is stationary if its properties do not depend on time. For instance, a metallic mirror is stationary. By contrast, a gas sustaining an acoustic wave has a time-dependent refractive index so that it is not stationary. A medium is homogeneous if its properties do not depend on space. In most cases, media are piece-wise homogeneous. For instance, the equation

$$\mathbf{P}(\mathbf{r}) = \varepsilon_0 \chi \mathbf{E}(\mathbf{r}),$$

with χ independent on $\mathbf{r}$, corresponds to a homogeneous medium. We have introduced the susceptibility χ to characterize the polarization of the medium. Note that the susceptibility is related to the dielectric permittivity. Indeed, by definition of $\mathbf{D}$, $\mathbf{D} = \varepsilon_0 \mathbf{E} + \mathbf{P}$ and by definition of ε_r, $\mathbf{D} = \varepsilon_0 \varepsilon_r \mathbf{E}$, so that $\chi = \varepsilon_r - 1$.

3.2.2 Isotropy

A medium is isotropic if its properties do not depend on the orientation of the field. For instance:

$$\mathbf{P}(\mathbf{r}) = \varepsilon_0 \chi \mathbf{E}(\mathbf{r}),$$

where χ is a scalar for an isotropic medium. It means that the polarization $\mathbf{P}(\mathbf{r})$ induced by the field $\mathbf{E}(\mathbf{r})$ is the same in all directions. This is typically expected for a liquid or a cubic crystal. By contrast, a medium with a microscopic structure which is not isotropic, such as graphite or mica, both of which are made of a superposition of monoatomic layers has a polarization which depends on the orientation of the external field. It follows that χ needs to be a tensor:

$$\mathbf{P}(\mathbf{r}) = \varepsilon_0 [\chi] \mathbf{E}(\mathbf{r}).$$

Using the Einstein notation,[1]

$$P_i(\mathbf{r}) = \varepsilon_0 \chi_{ij} E_j(\mathbf{r}).$$

3.2.3 Linearity

A medium is linear if the equation between $\mathbf{P}$ and $\mathbf{E}$ is linear. An example of a linear relation is given by:

$$P_i = \varepsilon_0 \chi E_i.$$

An example of a non-linear and anisotropic medium is given by:

[1] In Einstein notation, there is a sum over the subscript which is repeated. For example, $\varepsilon_0 \chi_{ij} E_j(\mathbf{r})$ stands for $\varepsilon_0 \sum_{j=1}^{3} \chi_{ij} E_j(\mathbf{r})$.

$$P_i = \varepsilon_0[\chi_{ij}E_j + \chi_{ijk}E_jE_k].$$

In this example, it is seen that if the field tends to zero, the linear term gives the leading contribution. Non-linear terms are observed for large electric fields. The standard approach to generating a large electric field is to use a powerful laser and focus the energy on a tiny spot. However, with the advent of nanophotonics, it is possible to confine an electromagnetic mode in a small volume, so that the electric field can become very large for a small energy. This remark follows directly from the fact that the energy density associated with the electric field is $\varepsilon_0 E^2/2$. If we have a single photon in a confined volume V, the electric field is then given by $E \approx \sqrt{\frac{\hbar\omega}{\varepsilon_0 V}}$.

An example of the consequence of non-linearity is second harmonic generation. By driving an electron at ω_1, the dipole oscillation will have a component at frequency $2\omega_1$ due to the non-linear term. Hence, the induced dipole will also radiate energy at $2\omega_1$. By illuminating a non-linear medium with two fields at two different frequencies, the non-linear term will introduce terms with the frequency sum and the frequency difference. These coherent effects are widely used to generate laser light in spectral regions where there are no efficient gain materials.

3.2.4 Dispersion and spatial dispersion

The most general linear relation between the electric field and the polarization can be cast in the form:

$$\mathbf{P}(\mathbf{r}, t) = \varepsilon_0 \int \mathrm{d}t' \int \mathrm{d}^3\mathbf{r}' \chi(\mathbf{r}, \mathbf{r}', t, t')\mathbf{E}(\mathbf{r}', t'). \tag{3.1}$$

For a homogeneous and stationary medium, the susceptibility depends only on $\mathbf{r}-\mathbf{r}'$ and on $t - t'$. It follows that the previous relation becomes a convolution product:

$$\mathbf{P}(\mathbf{r}, t) = \varepsilon_0 \int \mathrm{d}t' \int \mathrm{d}^3\mathbf{r}' \chi(\mathbf{r} - \mathbf{r}', t - t')\mathbf{E}(\mathbf{r}', t'). \tag{3.2}$$

The convolution product accounts for the non-instantaneous and non-local relation between the excitation (the field) and the response (the polarization). In order to analyse the spectral response of the medium, we introduce the Fourier transform of the polarization with respect to t. We get:

$$\mathbf{P}(\mathbf{r}, \omega) = \varepsilon_0 \int \mathrm{d}^3\mathbf{r}' \chi(\mathbf{r} - \mathbf{r}', \omega)\mathbf{E}(\mathbf{r}', \omega). \tag{3.3}$$

Note that in time domain, $\chi(\mathbf{r}, t)$ is a real function, so that $\chi(\mathbf{r}, -\omega) = \chi^*(\mathbf{r}, \omega)$. Similarly, if we introduce the spatial Fourier transform, we get:

$$\mathbf{P}(\mathbf{k}, \omega) = \varepsilon_0 \chi(\mathbf{k}, \omega)\mathbf{E}(\mathbf{k}, \omega).$$

Hence, it is seen that the non-instantaneous and non-local relation entails a dependence of the susceptibility on frequency and wavevector. If the susceptibility depends on the frequency ω, the medium is dispersive. Similarly, if the susceptibility depends on the wavevector $\mathbf{k}$, the medium is spatially dispersive. For a non-dispersive medium, the susceptibility is a constant in frequency domain, so that in time domain, it is given by $\chi\delta(t-t')$. Similarly, a medium which is spatially non-dispersive is given in space domain by $\chi(\mathbf{r}-\mathbf{r}',\omega)=\chi(\omega)\delta(\mathbf{r}-\mathbf{r}')$. By inserting these forms in Eq. (3.3), we get:

$$\mathbf{P}(\mathbf{r},t)=\varepsilon_0\chi E(\mathbf{r},t). \tag{3.4}$$

We see that the link between the polarization and the electric field is instantaneous and local. This is why a non-spatially dispersive medium is often called a local medium. Conversely, a spatially dispersive medium is often called a non-local medium.

We now discuss briefly the physical origin of dispersion. The key aspect is that the microscopic processes responsible for the polarization of the media (be it electronic deformation of electronic wave functions, rotations of molecules in a liquid, vibrations in a crystal, collective motion of free electrons in a plasma) have characteristic times and characteristic lengths. If the characteristic length and time of the medium are much smaller than those of the exciting fields, the behaviour of the system is instantaneous and local. In other words, if the excitation is slow, the system can follow instantly. The key point of the previous comment is to specify what "slow" means. The period of the excitation has to be larger than any characteristic time of the response of the system. It turns out that in optics, the characteristic times of the materials (vibrational degrees of freedom, relaxation times, some electronic transitions) are often longer than the optical period, which is on the order of a few femtoseconds. Hence, the media are usually dispersive.

By contrast, the typical non-local length scale of the materials is on the order of 1 nm, so that the wavelengths are much larger and non-local effects can be ignored in most cases. However, when dealing with nanophotonics devices where the length scales of the electromagnetic field becomes on the order of a few nanometres, the local approximation may fail. The derivation of a non-local model of the permittivity of an electron gas using a hydrodynamic model is given in Complement 3.C.

3.3 Current density and polarization: Is it the same?

At first glance, the above question is strange, as a conductor is characterized by a current density, whereas an electric insulator, often called a dielectric, is characterized by its polarization. While this is a major difference for static electric fields, the situation is totally different for time-dependent fields. Indeed, a dielectric is an insulator when a DC voltage is applied. However, when placing a dielectric in a capacitor, the AC current across the capacitor is non-zero. The purpose of this section is to clarify the difference between the static case and the dynamic case when dealing with currents and polarization.

3.3.1 DC regime

We have seen in Eq. (1.35) that the current density in a material medium is the sum of three contributions: $\partial\mathbf{P}/\partial t$, $\mathbf{j}_{\rm net}$, and $\nabla\times\mathbf{M}$. The later is null for a non-magnetic material. When working in the static regime, conduction is only due to $\mathbf{j}_{\rm net}$, as $\partial\mathbf{P}/\partial t$ is obviously null. The electrical conductivity σ is defined as $\mathbf{j}_{\rm net}=\sigma\mathbf{E}$, where σ is a positive real number. The static polarization of a medium is described by $\varepsilon_r^{pol}(\omega=0)$.

These two quantities, σ and ε_r^{pol}, can be measured separately. The conductivity can be derived from a resistance measurement and the permittivity can be derived from a capacitance measurement.

3.3.2 AC regime

The situation changes in the AC regime. It is no longer possible to measure the permittivity and the conductivity separately. Indeed, both net charges and polarization charges oscillate when driven by a field. Hence, their effects are added and cannot be separated experimentally. As a consequence, only the sum of their contributions is a relevant quantity. This can be seen clearly when writing the current density in harmonic regime:

$$\sigma(\omega)\mathbf{E}(\mathbf{r},\omega) - i\omega\mathbf{P}(\mathbf{r},\omega) = \{\sigma(\omega) - i\omega\varepsilon_0[\varepsilon_r^{pol}(\omega) - 1]\}\mathbf{E}(\mathbf{r},\omega). \tag{3.5}$$

It is seen that the total current density is proportional to the electric field. It can be cast in the form of an equivalent current density or in the form of an equivalent time-dependent polarization density:

$$\begin{aligned}\{\sigma(\omega) - i\omega\varepsilon_0[\varepsilon_r^{pol}(\omega) - 1]\}\mathbf{E}(\mathbf{r},\omega) &= \sigma_{eq}\mathbf{E}(\mathbf{r},\omega)\\ &= -i\omega\varepsilon_0[\varepsilon_{r,eq}^{pol}(\omega) - 1]\mathbf{E}(\mathbf{r},\omega).\end{aligned}$$

The equivalent conductivity is thus given by $\sigma_{eq}(\omega) = \sigma(\omega) - i\omega\varepsilon_0[\varepsilon_r^{pol}(\omega) - 1]$. This is the conventional choice used when working at radio frequencies and microwaves. In the infrared and visible domain, the choice is different. The equivalent permittivity is used. It is given by:

$$-i\omega\varepsilon_0[\varepsilon_{r,eq}(\omega) - 1] = \sigma(\omega) - i\omega\varepsilon_0[\varepsilon_r^{pol}(\omega) - 1],$$

so that we have

$$\varepsilon_{r,eq}(\omega) = \varepsilon_r^{pol}(\omega) + i\frac{\sigma(\omega)}{\omega\varepsilon_0}.$$

For frequencies lower than 1 GHz, dispersion can usually be neglected, so that both the conductivity and the permittivity are real numbers equal to their static value. In that case, the above equation can be viewed as providing the real part and the imaginary part of the effective permittivity.

This is not the case for higher frequencies. Then, both the conductivity and the permittivity (i.e. the contributions of net charges and polarization charges) are complex. In summary, we can always write the following for time-dependent fields:

$$\mathbf{j}_{eq}(\mathbf{r},t) = \frac{\partial\mathbf{P}_{eq}(\mathbf{r},t)}{\partial t}. \tag{3.6}$$

The subscript *eq* used in this section will be omitted hereafter. It is always implicit that a permittivity or a conductivity accounts for all types of charges. The frequency-dependent material data found in the literature are either permittivities or conductivities.

3.3.3 Summary

In summary, the current density can be represented using either an effective conductivity or an effective permittivity:

$$\mathbf{j}_{\text{net}} + \frac{\partial \mathbf{P}}{\partial t} = \sigma_{eq}\mathbf{E} = -i\varepsilon_0 \left[\varepsilon_{r,eq}(\omega) - 1\right]\mathbf{E}. \tag{3.7}$$

This convention is always used so that the subscript *eq* will no longer be used. The choice between conductivity or permittivity depends on the frequency range for historical reasons. The data for materials are given in terms of permittivities (or refractive index) in infrared and visible light, in terms of conductivities in radiofrequency and microwaves.

3.4 Energy in material media

When dealing with a material medium, care has to be taken when writing the energy of the system. While the form of energy for non-dispersive media is similar to the energy in vacuum, it turns out to be quite different for dispersive media. We start from Eq. (1.59) derived in Chapter 1.

3.4.1 Energy in dispersive and non-dispersive media

Let us consider the case of a linear dielectric medium with real non-dispersive permittivity ε_r and permeability $\mu_r = 1$. Replacing $\mathbf{D}$ by $\varepsilon_0\varepsilon_r\mathbf{E}$ in Eq. (1.59) yields:

$$\frac{\partial}{\partial t}\left(\frac{\varepsilon_0\varepsilon_r\mathbf{E}^2}{2} + \frac{\mathbf{B}^2}{2\mu_0}\right) = -\mathbf{j}_{ext}\cdot\mathbf{E} - \nabla\cdot(\mathbf{E}\times\mathbf{H}). \tag{3.8}$$

The left term of the equation is the time derivative of the energy density. Hence, the energy density is similar to the energy density in vacuum. The permittivity ε_0 is simply multiplied by the relative permittivity ε_r. This type of formulation can be used for fields oscillating at frequencies below the lowest material resonances, which are typically on the order of a few GHz. The situation is much more involved when accounting for losses and dispersion. To derive the form of the energy accounting for dispersion, it is necessary to introduce the complex permittivity:

$$\frac{\partial}{\partial t}\left(\frac{\mathbf{D}\cdot\mathbf{E}}{2}\right) = \frac{1}{2}\frac{\partial}{\partial t}\left(\int\frac{\mathrm{d}\omega}{2\pi}\int\frac{\mathrm{d}\omega'}{2\pi}\varepsilon_0\varepsilon_r(\omega)\mathbf{E}(\mathbf{r},\omega)\cdot\mathbf{E}(\mathbf{r},\omega')\exp[-i(\omega+\omega')t)]\right). \tag{3.9}$$

Obviously, due to dispersion, the effect of the material depends on the frequency and the energy is more conveniently discussed in the frequency domain. When considering a quasi-monochromatic beam, it can be shown (Agranovich and Ginzburg, 1984; Brillouin, 1960) that the electromagnetic energy density can be cast in the form:

$$\varepsilon_0\frac{\partial\,\omega\varepsilon'_{ij}(\omega)}{\partial\omega}E_i(\mathbf{r},\omega)E_j^*(\mathbf{r},\omega) + \frac{B_i(\mathbf{r},\omega)B_i^*(\mathbf{r},\omega)}{2\mu_0}, \tag{3.10}$$

where $\varepsilon'_{ij}(\omega)$ is the real part of the permittivity. We note that for the particular case of a metal, the real part of the permittivity may be negative. However, Eq. (3.10) provides a positive contribution to the energy, as can be checked using the Drude form of the metal permittivity.

3.5 Absorption in material media

3.5.1 Absorbed power

In this section, we derive the form of the absorbed power for a non-magnetic material in the frequency domain. We start from Eq. (1.55). Let us assume that we choose to describe the medium using an effective electrical conductivity, σ. The time-averaged power for a monochromatic field is then given by:

$$< \frac{\partial P}{\partial V} >= \frac{1}{2}\,\mathrm{Re}(\mathbf{j}\cdot\mathbf{E}^*) = \frac{\mathrm{Re}(\sigma)}{2}|\mathbf{E}|^2, \tag{3.11}$$

where the brackets indicate time average. The result is the usual Joule effect rate in its local form. An alternative choice consists in using an effective permittivity. The current density is then replaced by a time-dependent polarization, $\mathbf{j} = \partial\mathbf{P}/\partial t$. We use the convention $\exp(-i\omega t)$ for monochromatic fields, so that $\mathbf{j} = -i\omega\varepsilon_0(\varepsilon_r - 1)\mathbf{E}$. We finally obtain:

$$< \frac{\partial P}{\partial V} >= \frac{\omega\varepsilon_0}{2}\,\mathrm{Im}(\varepsilon_r)|\mathbf{E}|^2. \tag{3.12}$$

This result clearly indicates that losses are proportional to the imaginary part of the permittivity. Since at thermodynamic equilibrium there cannot be gain, the loss term is positive, so that *the imaginary part of the permittivity is positive.* It is important to note that this conclusion depends on the conventional choice of the sign of the frequency in the Fourier transform. Indeed, the result is proportional to the frequency ω. It follows that if the sign convention in the Fourier transform is changed in $\exp(+i\omega t)$, the sign of the absorbed power in the formula will also be changed and the imaginary part of the permittivity will have to be negative. We introduce in the next sections some classical microscopic models of the permittivities of metals and dielectrics. It will be seen that the imaginary part of the permittivity always vanishes if we do not introduce damping processes in the microscopic models.

3.5.2 Energy balance in the monochromatic regime

Here, we revisit the energy conservation Eq. (1.59) in the specific case of a monochromatic regime. The electromagnetic energy stored in the materials during the transient growth of the fields is given by Eq. (3.10). It does not appear in the energy balance in the monochromatic regime, and we only have a balance between power transferred from the field to the matter $< -\mathbf{j}_{ext}\cdot\mathbf{E} >$ and power dissipated by the matter into heat given by $< \frac{\partial P_{abs}}{\partial V} >$ or radiated given by the divergence of the Poynting vector $\nabla\cdot(<\mathbf{E}\times\mathbf{H}>$. We get:

$$< -\mathbf{j}_{ext}\cdot\mathbf{E} >=< \frac{\partial P_{abs}}{\partial V} > +\nabla\cdot(<\mathbf{E}\times\mathbf{H}>). \tag{3.13}$$

Without external sources, the energy conservation equation can be cast in the form:

$$0 =< \frac{\partial P_{abs}}{\partial V} > +\nabla\cdot(<\mathbf{E}\times\mathbf{H}>), \tag{3.14}$$

where the absorption term is given by:

$$< \frac{\partial P_{abs}}{\partial V} >=< \mathbf{E} \cdot \frac{\partial \mathbf{P}}{\partial t} - \mathbf{M} \cdot \frac{\partial \mathbf{B}}{\partial t} >= \frac{\omega \varepsilon_0 \operatorname{Im}(\varepsilon_r)}{2} |\mathbf{E}|^2 + \frac{\omega}{2} \frac{\operatorname{Im}(\mu_r)}{|\mu_r|^2 \mu_0} |\mathbf{B}|^2. \tag{3.15}$$

The first term accounts for dielectric losses and Joule losses, as the permittivity accounts for both polarization charges and net charges. The second term accounts for magnetic losses. This form will be used in Chapter 12 when studying scattering and absorption by a body illuminated by an incident beam.

3.6 Permittivity of a gas: Lorentz–Lorenz model

3.6.1 Lorentz–Lorenz permittivity

The goal of this section is to derive a simple model for the permittivity of a gas. This model is based on classical mechanics. Its simplicity allows us to discuss very basic properties such as the origin of dispersion, the role of losses, and the dependence of permittivity on the density and frequency. A more detailed treatment can be found in ref. (Siegman, 1986). The connection with a quantum model can be found in (Cohen-Tannoudji et al., 1991; Allen and Eberly, 1987). To derive the permittivity, we simply write that the polarization density is given by:

$$\mathbf{P}(\mathbf{r}, t) = n_0 \mathbf{p} = n_0 \alpha \varepsilon_0 \mathbf{E}(\mathbf{r}, t), \tag{3.16}$$

where n_0 is the number of atoms per unit volume and α is their polarizability. Upon comparison with $\mathbf{P} = \varepsilon_0 [\varepsilon_r - 1]\mathbf{E}$, we get

$$\varepsilon_r = 1 + n_0 \alpha. \tag{3.17}$$

In order to derive the polarizability, we use the model of an electron elastically bounded with a single transition at frequency ω_0. The restoring force is $-m\omega_0^2 \mathbf{r}(t)$, where $\mathbf{r}(t)$ denotes the electron position and m its mass. We thus get

$$m \frac{d^2 \mathbf{r}}{dt^2} = -m\omega_0^2 \mathbf{r}(t) - m\nu \mathbf{v} - \mathbf{E}_{\text{micro}} e^{i[\mathbf{k}.\mathbf{r}(t) - \omega t]} - e\mathbf{v} \wedge \mathbf{B}_{\text{micro}} e^{i[\mathbf{k}.\mathbf{r}(t) - \omega t]}, \tag{3.18}$$

where $(\mathbf{E}_{\text{micro}}, \mathbf{B}_{\text{micro}})$ is the electromagnetic field of a plane wave propagating in the atomic vapour and where $-m\nu\mathbf{v}$ is a phenomenological force accounting for energy losses due to either radiation or collisions. In the above model, we neglect the phase term $\mathbf{k}.\mathbf{r}(t)$, because the charge displacement is much smaller than the wavelength. We also neglect the magnetic force as

$$\left| \frac{\mathbf{v} \wedge \mathbf{B}}{E} \right| \approx \frac{v}{c} << 1,$$

for non-relativistic electrons. Finally, we neglect the difference between the microscopic field illuminating the atom and the average macroscopic field appearing in Maxwell equations. Accounting for the difference amounts to introducing the local field correction discussed below. The amplitude of the monochromatic movement induced by the electric field is given by:

$$\mathbf{r}(\omega) = \frac{e\mathbf{E}}{m} \frac{1}{\omega^2 - \omega_0^2 + i\nu\omega}. \tag{3.19}$$

The monochromatic polarization density is thus given by:

$$\mathbf{P} = n_0 \mathbf{p} = -\frac{n_0 e^2}{m} \frac{\mathbf{E}}{\omega^2 - \omega_0^2 + i\nu\omega}, \tag{3.20}$$

so that the permittivity can be cast in the form:

$$\varepsilon_r = 1 - \frac{n_0 e^2}{m} \frac{1}{\omega^2 - \omega_0^2 + i\nu\omega}. \tag{3.21}$$

Introducing the plasma frequency $\omega_p^2 = \frac{n_0 e^2}{m\varepsilon_0}$, we obtain:

$$\varepsilon_r(\omega) = 1 - \frac{\omega_p^2}{\omega^2 - \omega_0^2 + i\nu\omega} = 1 - \frac{(\omega^2 - \omega_0^2)\omega_p^2}{(\omega^2 - \omega_0^2)^2 + \nu^2\omega^2} + i\frac{\nu\omega\omega_p^2}{(\omega^2 - \omega_0^2)^2 + \nu^2\omega^2}. \tag{3.22}$$

We note that if we suppress the damping term by choosing $\nu = 0$, the imaginary part of the permittivity cancels $\mathrm{Im}[\varepsilon_r(\omega)] = \varepsilon''(\omega) = 0$. This is consistent with the fact that the dissipated power $< \mathbf{j}.\mathbf{E} >$ is proportional to $\varepsilon''(\omega)$.

It is important to note that close to the resonance frequency ω_0, we can use the approximation $\omega^2 - \omega_0^2 \approx 2\omega_0(\omega - \omega_0)$, so that the permittivity is given by a Lorentzian profile:

$$\varepsilon_r = 1 - \frac{n_0 e^2}{2m\varepsilon_0\omega_0} \frac{1}{\omega - \omega_0 + i\frac{\nu}{2}}. \tag{3.23}$$

3.6.2 Physical origin of the dispersion

Let us emphasize that this model accounts for a single transition of the atoms. We learn many features from Eq. (3.22). First of all, it is seen that the permittivity tends to the vacuum permittivity when $n_0 \to 0$, as expected. A similar trend is obtained for frequencies ω much larger than the resonance frequency ω_0. This is a very basic feature of any oscillator: the amplitude decays as $1/\omega^2$ for $\omega \gg \omega_0$. Conversely, it is important to notice that for frequencies much smaller than the resonance frequency, the system is almost frequency-independent—namely, dispersionless. The physical content is very simple: a system can follow an excitation (stay in phase with the excitation), provided that the excitation frequency is smaller than the resonance frequency. Beyond that frequency, the amplitude decays, and there is a π phase shift when crossing the resonance. This feature, which appears to be a feature of a specific oscillator model, is a general property that can be derived from Kramers–Kronig relations, as shown in section 3.11.

3.6.3 Spectral broadening

In the Lorentz–Lorenz model, the broadening is described through the introduction of a damping term, which accounts for several relaxation phenomena. These include the radiative relaxation by spontaneous emission, which gives rise to the so-called natural width. Another contribution can be due to energy transfer during a collision between atoms. These two factors are the same for all the atoms and contribute to the so-called homogeneous broadening. A different mechanism is due to the Doppler effect. Here, for a given molecule, there is no broadening but a shift in frequency. However, when performing a frequency measurement of the interaction of an ensemble of atoms with light, either in absorption, emission, or transmission, a broadening is observed due to the addition of the contributions of different atoms with different Doppler shifts. This broadening mechanism is called inhomogeneous broadening. It is an ensemble property. Inhomogeneous broadening can also be observed for molecules embedded in solids. In this case, the spectral shift of an individual molecule is not due to the Doppler effect but to the local interaction with the environment, the so-called crystal field.

We now address briefly and qualitatively a simple question: how can we reconcile the line structure of optical properties of atoms with the broad spectral structure of solids and liquids, given that solids and liquids are made of atoms? From the Lorentz–Lorenz formula, we would simply expect that a denser medium has the same spectrum and a larger permittivity proportional to the number of atoms per unit volume. This is not the case, and the permittivity saturates and broadens. It turns out that when increasing the density, the mean distance between atoms decreases, so that the interaction between atoms can no longer be neglected. As a consequence, atoms are coupled and the resulting energy levels are shifted. So when interactions between atoms are no longer negligible, the spectrum broadens. This will be very clearly seen in the model of permittivity of an ionic crystal based on a linear chain of interacting atoms (see section 3.10). This is also clearly seen for metals when deriving the band structure of electronic energy levels, starting with a linear combination of atomic orbitals.

3.7 Local-field correction

3.7.1 Local field

In the Lorentz–Lorenz model, we have derived the dipole moment of each atom by writing that it is induced by the electromagnetic field illuminating the atom. This is called the local field $\mathbf{E}_{loc}$. In a microscopic situation, this illuminating field is the sum of the incident field and the field scattered by all the other atoms. As discussed in Chapter 1, the field appearing in Maxwell equations results from a spatial average of the field and differs from the field actually illuminating a particular atom. Indeed, the spatially averaged field includes the averaged contribution of the atom itself to the field in its surrounding. This distinction between Maxwell field and local field was introduced by H.-A. Lorentz (Lorentz, 1916). In order to derive the expression of the local field, we need to subtract the contribution of the atom itself to the Maxwell field $\mathbf{E}$ (i.e. the spatially averaged field). Following an argument due to Aspnes (Aspnes, 1982), we evaluate the contribution of a dipole $< \mathbf{E}_{dip} >$ to the spatially averaged field $\mathbf{E}$ around this dipole by spatially integrating the field produced by the dipole in a sphere surrounding the dipole with a volume given by the volume per atom $v = 1/n_0$:

$$< \mathbf{E}_{dip} >= \frac{1}{v} \int_{sphere} \frac{-1}{4\pi\varepsilon_0} \nabla \left(\frac{\mathbf{p} \cdot \mathbf{r}}{r^3} \right) \mathrm{d}^3\mathbf{r} = n_0 \int_{surface} \frac{-1}{4\pi\varepsilon_0} \frac{\mathbf{p} \cdot \hat{\mathbf{r}}}{r^2} \mathrm{d}S\hat{\mathbf{r}}, \tag{3.24}$$

where[2] $\mathrm{d}S = r^2 \sin\theta \mathrm{d}\theta \mathrm{d}\phi$. The only component that survives the integral over ϕ is along the dipole moment $\mathbf{p}$. Choosing the polar axis of the spherical coordinates along the dipole, we get the field amplitude

$$< E_{dip} >= n_0 \frac{-p}{4\pi\varepsilon_0} \int \cos^2\theta \sin\theta \mathrm{d}\theta \mathrm{d}\phi = -n_0 \frac{p}{3\varepsilon_0} = \frac{-P}{3\varepsilon_0}. \tag{3.25}$$

The local field is found by subtracting this contribution:

$$\mathbf{E}_{loc} = \mathbf{E} - < \mathbf{E}_{dip} >= \mathbf{E} + \frac{\mathbf{P}}{3\varepsilon_0}. \tag{3.26}$$

It is clearly seen that the correction is negligible when the medium is very dilute ($n_0 \to 0$).

3.7.2 The Clausius–Mossotti formula

We now revisit the form of the permittivity in Eq. (3.16) accounting for the local field correction. The polarization field is now given by:

$$\mathbf{P}(\mathbf{r}, t) = n_0\alpha\mathbf{E}_{loc} = n_0\alpha\varepsilon_0 \left(\mathbf{E} + \frac{\mathbf{P}}{3\varepsilon_0} \right), \tag{3.27}$$

and can be cast in the form:

$$\mathbf{P}(\mathbf{r}, t) = \frac{n_0\alpha}{1 - \frac{n_0\alpha}{3}} \varepsilon_0 \mathbf{E}(\mathbf{r}, t). \tag{3.28}$$

Here again, we observe that the correction term $1/(1 - n_0\alpha/3)$ tends to one for dilute media. Identifying $\mathbf{P}$ with the form $\varepsilon_0(\varepsilon_r - 1)\mathbf{E}$, we get the Clausius–Mossotti formula:

$$\frac{\varepsilon_r - 1}{\varepsilon_r + 2} = \frac{n_0\alpha}{3}. \tag{3.29}$$

This formula connects the macroscopic property ε_r with the microscopic polarizability of a molecule α.

3.8 Permittivity of an electron gas: The Drude model

3.8.1 Derivation of the Drude model using a classical model

The goal of this section is to derive a simple model of permittivity for a free electron gas. Here again we use classical mechanics. An alternative derivation that includes

[2] We have used the identity $\int_{volume} \nabla f(\mathbf{r})\mathrm{d}^3\mathbf{r} = \int_{surface} f(\mathbf{r})\mathrm{d}^2S\, \hat{r}$.

non-local effects is given in Complement 3.C. A more detailed discussion of the physics of electrons is given in Complement 3.B. To derive the permittivity, we simply write that the polarization density is given by:

$$\mathbf{P}(\mathbf{r}, t) = n_0 \mathbf{p} = -n_0 e \mathbf{r}, \tag{3.30}$$

where n_0 is the number of electrons per unit volume and $-e\mathbf{r}$ is their dipole moment. We use the model of a free electron:

$$m\frac{d^2\mathbf{r}}{dt^2} = -m\nu\mathbf{v} - e\mathbf{E}^{-i\omega t}, \tag{3.31}$$

where $\mathbf{E}$ is the electric field of a plane wave propagating in the free electron gas, $-m\nu\mathbf{v}$ is a phenomenological force accounting for energy losses, and ν is the inverse of a typical damping time. In the above model, we have neglected the phase term $\mathbf{k}.\mathbf{r}(t)$ and the magnetic force. Note that the magnetic force can be responsible for non-linear effects, as both the velocity and the magnetic field are proportional to the electric field. Hence, such a term is the source of radiation at 2ω. While these effects are essential for very large fields, we neglect them for most optical situations. The amplitude of the movement induced by the electric field is then given by:

$$\mathbf{r}(\omega) = \frac{e\mathbf{E}}{m}\frac{1}{\omega^2 + i\nu\omega}. \tag{3.32}$$

The polarization density is thus given by:

$$\mathbf{P} = \varepsilon_0(\varepsilon_r - 1)\mathbf{E} = n_0\mathbf{p} = -\frac{n_0 e^2}{m}\frac{\mathbf{E}}{\omega^2 + i\nu\omega}, \tag{3.33}$$

and the permittivity can be cast in the form:

$$\varepsilon_r = 1 - \frac{\omega_p^2}{\omega^2 + i\nu\omega}. \tag{3.34}$$

3.8.2 High-frequency limit

At very large frequencies compared to the damping rate ν, we get

$$\varepsilon_r \approx 1 - \frac{\omega_p^2}{\omega^2}. \tag{3.35}$$

In this regime, it is seen that the permittivity is negative for frequencies smaller than ω_p. Losses are negligible, but waves cannot propagate, as will be seen in the next chapter.

We also note that if $\omega \ll \omega_p$, the permittivity approaches one, For these frequencies, a metal behaves as a dielectric with a refractive index smaller than one. This is why ultraviolet (UV) and X radiation can penetrate metals.

3.8.3 Low-frequency limit

At low frequencies where $\omega < \nu$, we obtain

$$\varepsilon_r \approx 1 + \frac{i\omega_p^2}{\omega\nu} \approx \frac{i\omega_p^2}{\omega\nu}. \tag{3.36}$$

Comparing this form with the expression $i\sigma/(\omega\varepsilon_0)$ valid for non-dispersive metals (where the conductivity does not depend on frequency), we can identify the DC conductivity $\sigma = \varepsilon_0\omega_p^2/\nu$. It is very important to stress that ν is a parameter that can be used to fit the experimental data for the permittivity. As physical mechanisms responsible for losses are very different in the DC regime (collisions with phonons and with crystal defects) and in the optical regime (optical transitions in the band structure), by no means should one consider that the effective parameter ν takes the same value for all frequencies. A more detailed discussion can be found in Complement 3.B. Note also that we have not included a local field correction. For dilute systems like plasma, the correction is negligible. For dense systems such as metals or semiconductors, an appropriate treatment involves dealing with Bloch wave functions, which are delocalized in space. The Drude model can then be justified in the framework of a semiclassical approximation where an effective mass of the electrons is introduced. The local correction is not relevant in this context.

3.9 Permittivity of a polar liquid: Orientation polarization

A polar liquid is a liquid containing molecules with a permanent electric dipole moment such as water molecules. In the absence of an applied electric field, the orientation of the molecules is random, so that the average dipole moment of a volume element is null. When applying a static electric field, the molecules tend to be aligned in order to reduce the interaction energy $-\mathbf{p}\cdot\mathbf{E}$. The resulting dipole moment results from a competition between thermal random motion, which tends to destroy the alignment, and electric interactions. The derivation of the permittivity can be found in Exercise 3.5. It is shown that the permittivity can be cast in the form

$$\varepsilon = \varepsilon_\infty + \frac{\varepsilon_s - \varepsilon_\infty}{1 - ix}, \tag{3.37}$$

where ε_s is the static value of the permittivity at $\omega = 0$, ε_∞ is the value of the permittivity at frequency $\omega \gg 1/\tau$ where τ is the relaxation time of the polarization, and $x = \dfrac{\varepsilon_s + 2}{\varepsilon_\infty + 2}\omega\tau$.

3.10 Permittivity of an ionic crystal: The Lyddane–Sachs–Teller relation

In this section, we consider ionic crystals, namely crystals containing atoms which carry a net electric charge like NaCl. This charge can be one electron, as in the case of NaCl, or a fraction of electron like in SiO_2. In the latter case, the molecular binding is due to a combination of electrostatic binding and covalent binding, a situation described as iono-covalent. All oxides belong to that class of materials. The key point is that optical phonons, namely phonons where atoms A and B in the same cell are out of phase, generate electric dipoles oscillating at the vibration frequency. This results in resonances in the infrared spectrum. The typical form of the permittivity for this type of solid can be shown as follows (see Exercise 3.6):

$$\varepsilon(\omega) = \varepsilon_\infty \frac{\omega_L^2 - \omega^2}{\omega_T^2 - \omega^2}. \tag{3.38}$$

The Lyddane–Sachs–Teller relation connects the transverse ω_T and the longitudinal ω_L frequencies:

$$\omega_L^2 = \frac{\varepsilon_s}{\varepsilon_\infty} \omega_T^2 \tag{3.39}$$

where ε_S is the static permittivity $\varepsilon(0)$ and ε_∞ is the permittivity for high frequencies $\omega \gg \omega_L$. We note that this model predicts a negative permittivity when $\omega_T < \omega < \omega_L$. In this frequency range corresponding to the range of existence of optical phonons, the medium behaves as a metal. It screens the field and waves cannot propagate.

The above model predicts a pure real permittivity, so that there are no losses. There are two mechanisms producing losses: scattering of phonons on defects of the lattice such as vacancies, interstitial atoms, impurities; and non-linearity of the interaction between atoms. The above model is built assuming linear interactions between atoms. Accounting for non-linearities mixes the different modes, thereby introducing an effective decay. This effect is larger for high temperatures. When including losses in the model using a phenomenological damping force, the following form of the permittivity can be derived:

$$\varepsilon(\omega) = \varepsilon_\infty \frac{\omega_L^2 - \omega^2 + i\nu\omega}{\omega_T^2 - \omega^2 + i\nu\omega}. \tag{3.40}$$

This analytical form of the permittivity can be used to fit the experimental data with very good accuracy.

3.11 Kramers–Kronig relations

3.11.1 Derivation

The goal of this section is to establish relations which follow directly from causality. We will show that knowledge of the real part of the susceptibility *for all frequencies* allows the imaginary part of the susceptibility to be derived and vice versa. Interestingly, Kramers–Kronig relations apply to any system with a linear and causal response, such as a mechanical oscillator, an electric circuit, a dielectric, or a magnetic material. We consider a linear relation between a response X and a force F given by:

$$X(t) = \int_{-\infty}^{\infty} \chi(t - t')F(t')\,\mathrm{d}t'. \tag{3.41}$$

Causality implies that the response $X(t)$ cannot depend on forces applied at times $t' > t$. Hence, we have:

$$\chi(t) = 0 \quad \text{if} \quad t < 0. \tag{3.42}$$

We now examine the consequences of causality on the properties of its Fourier transform $\chi(\omega)$. Causality implies the following identity:

$$\chi(t) = \chi(t)H(-t), \tag{3.43}$$

where $H(t)$ is the Heaviside step function:

$$H(t) = 1 \text{ for } t \geq 0, \quad H(t) = 0 \text{ if } t < 0. \tag{3.44}$$

Since the Fourier transform of a product is the convolution product of the Fourier transforms, we have:

$$\chi(\omega) = \int_{-\infty}^{\infty} \frac{\mathrm{d}\omega'}{2\pi} \chi(\omega')H(\omega - \omega'). \tag{3.45}$$

The Fourier transform of the Heaviside function is a generalized function given by

$$H(\omega) = \left[\pi\delta(\omega) + PV \frac{i}{\omega}\right], \tag{3.46}$$

where PV stands for the principal value of the integral. Inserting this form in Eq. (3.45) we get:

$$\chi(\omega) = \int_{-\infty}^{\infty} \frac{\mathrm{d}\omega'}{2\pi} \chi(\omega') \left[\pi\delta(\omega - \omega') + PV \frac{i}{\omega - \omega'}\right] = \frac{\chi(\omega)}{2} + \frac{1}{i2\pi} PV \int_{-\infty}^{\infty} \frac{\chi(\omega')}{\omega' - \omega} \mathrm{d}\omega'. \tag{3.47}$$

Finally, we obtain:

$$\chi(\omega) = \frac{1}{i\pi} PV \int_{-\infty}^{\infty} \frac{\chi(\omega')}{\omega' - \omega} \mathrm{d}\omega'. \tag{3.48}$$

By taking the real part and the imaginary part of this relation, we find the Kramers–Kronig relations:

$$\chi'(\omega) = \frac{1}{\pi} PV \int_{-\infty}^{\infty} \frac{\chi''(\omega')}{\omega' - \omega} d\omega' \quad ; \quad \chi''(\omega) = \frac{-1}{\pi} PV \int_{-\infty}^{\infty} \frac{\chi'(\omega')}{\omega' - \omega} d\omega'. \tag{3.49}$$

Using the relation

$$\chi(-\omega) = \chi^*(\omega), \tag{3.50}$$

this can be cast in the form:

$$\chi'(\omega) = \frac{2}{\pi} PV \int_{0}^{\infty} \frac{\omega\chi''(\omega')}{\omega'^2 - \omega^2} \mathrm{d}\omega'. \tag{3.51}$$

3.11.2 Discussion: Dispersion and causality

Eq. (3.51) is very interesting because it yields $\chi'(\omega)$, satisfying the causality principle when starting from an approximate form of χ''. This is a very powerful technique to obtain the real part of the susceptibility (dispersion information) from an absorption spectrum. For the sake of illustration, let us consider the real part of the susceptibility $\chi'(\omega)$ produced by an absorption line at ω_r due to resonance of the system. We can write the imaginary part of the susceptibility as

$$\chi''(\omega') = K\delta(\omega' - \omega_r), \tag{3.52}$$

where K is a constant. Using Eq. (3.51), we derive χ:

$$\chi'(\omega) = \frac{2\omega_0}{\pi}\frac{K}{\omega_r^2 - \omega^2}. \tag{3.53}$$

This form shows that for $\omega \ll \omega_r$, the real part of the susceptibility $2\omega_0 K/\pi(\omega_r^2 - \omega^2) \approx 2\omega_0 K/\pi\omega_r^2$ is almost non-dispersive. By contrast, for $\omega \gg \omega_r^2$, the susceptibility decays as $1/\omega^2$, a very general behaviour. In between these two asymptotic regimes, the susceptibility *increases locally* when ω approaches the resonance frequency ω_r from below, a behaviour called anomalous dispersion. All these behaviours are a consequence of causality, so that they are observed for all materials and also for many other different linear physical systems. Fig. 3.1 represents the real and imaginary part of the permittivity of water. All these trends are clearly observed. The permittivity takes a value 81 between 0 and 1 GHz. It starts declining in the GHz regime due to the

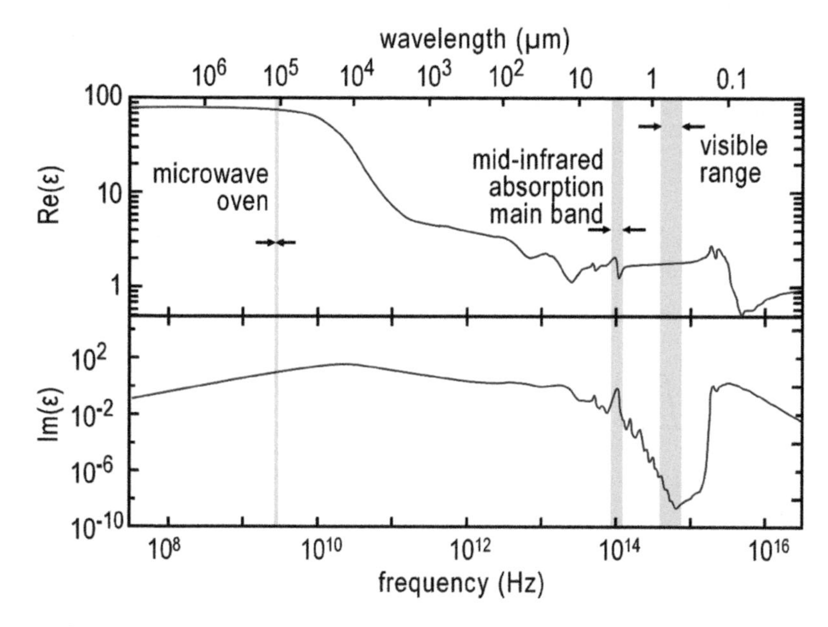

Fig. 3.1 Real part and imaginary part of the permittivity of water as a function of frequency, with some common frequency ranges indicated. Data taken from ref. (Palik, 1997).

presence of the absorption associated with the rotational movement (see Exercise 3.5). Anomalous dispersion bands corresponding to absorption peaks can be observed. Two of them have been highlighted in light grey. Finally, it is seen that the remarkable transparency window in the visible region (400–700 nm or 428–750 THz) corresponds to a low dispersion region.

3.A Complement: Microscopic interpretation of absorption

The purpose of this section is to provide a brief overview of the microscopic vision of losses and compare it with the macroscopic vision. In this chapter, losses were introduced in the framework of macroscopic Maxwell equations. The coupling between field and matter is given by the term $\mathbf{j} \cdot \mathbf{E}$, which accounts for the power transferred from the field to matter. This term accounts for both the electromagnetic energy stored in the system and the energy transformed into heat, which corresponds to losses. To identify the heat source, we have considered a system in stationary regime, so that the time-averaged electromagnetic energy does not change. To ensure stationarity, we need to assume that the system is in contact with a thermostat at fixed temperature. Hence, all the power supplied by the field to the system (work) is equal to the heat supplied by the system to the thermostat according to the first principle of thermodynamics. This provides a consistent picture of losses in the framework of electrodynamics and thermodynamics: the work done by the field on the system is transformed into heat and supplied to a thermostat.

If we now look for a microscopic vision of heat, we need to introduce a statistical description of the system. The degrees of freedom of the system (electrons, phonons in a solid, vibrations in a molecule, etc.) are randomly excited with a probability that can be derived in the framework of statistical physics. For example, the electromagnetic energy of an incident plane wave absorbed at a vacuum–metal interface is transformed into thermal excitations of the degrees of freedom in the metal. The forms of energy of the metal are the electronic energies given by the band structure and the spectrum of the phonons. These metal excitations are populated according to Fermi–Dirac and Bose–Einstein statistics, respectively, which depend on temperature.

We now revisit the question of energy exchange between the field and matter with a microscopic and deterministic point of view. It is possible to formally write a Hamiltonian accounting for all degrees of freedom of the system, namely all the electrons and nuclei, treated as point-like particles (Cohen-Tannoudji et al., 1991). Although this is intractable in practice for very large systems, it is formally possible to derive a Hamiltonian description of the system accounting explicitly for the kinetic and potential energy of all particles and fields in a system. In such a model, the description is deterministic and the energy is conserved so that there are no losses. Hence, we see that the concept of losses is inherently connected to the introduction of two subsystems, one of which plays the role of a thermostat. The usual macroscopic form of Maxwell equations indeed separates the quantities showing up in Maxwell equations (fields, charges, currents) and the rest of the material system, which plays the role of a thermostat. By contrast a purely microscopic formulation of Maxwell equations, where all the electrons and nuclei are treated with classical mechanics as point-like particles, leads to a Hamiltonian description where energy, momentum, and angular momentum are conserved.

The Hamiltonian description introduced above seems to be at odds with the irreversibility expected in a thermodynamic point of view. In this paragraph, we discuss the origin of irreversibility associated with the energy transfer from a microscopic subsystem (say an excited molecule) to its environment. We start the discussion with two molecules. We consider a molecule $M1$ (the system) initially excited by a linear coupling to another molecule $M2$ located nearby, which can be viewed as a very simple example of an environment. The distance separating the two molecules is such that they can exchange energy through dipole–dipole coupling. Due to this coupling, the new states of the coupled system are linear superpositions of the initial molecule states, with slightly different energies because of the level splitting produced by the interaction. In such a regime, the energy is exchanged periodically between the two molecules at a frequency called Rabi frequency. In a classical picture, the molecules can be viewed as two coupled pendulums and the Rabi oscillation is the beating between the two pendulums. We now increase the size of the environment. We introduce a linear coupling between molecule $M1$ and two neighbouring molecules with different coupling strengths, so that the two Rabi frequencies are different. The dynamics of the system is still deterministic, but more complicated. The energy is now shared between three molecules, so that the time-averaged energy of the initially excited molecule $M1$ is reduced. When increasing the coupling to a very large number of molecules, many Rabi oscillations with many different frequencies are generated. These Rabi oscillations are not synchronized, so that their linear superposition washes out gradually. It can be shown that this process leads to an exponential decay of the energy stored in molecule $M1$. This is a model of the irreversible relaxation process. Let us come back to the pendulum and consider a linear chain of pendulum as the example of a large number of coupled oscillators. The modes are collective modes with a phonon structure, and the excitation of all these modes becomes equivalent to the problem of radiation. Here, we can start realizing that if the system is finite, the energy may come back, but if the chain length increases so that the number of pendulum increases, the energy will not come back. The key in the transition between the periodic case and the exponential decay is the introduction of a continuum of states. The formal proof of this statement is the Wigner–Weisskopf treatment of the coupling between a discrete quantum state and a continuum of final states. When dealing with a macroscopic system, the order of magnitude of microscopic excitations is given by the Avogadro number, so that the system plays the role of a continuum. In the context of quantum physics, the Rabi oscillation between two molecules is a coherent energy-transfer process. By contrast, when the molecule is coupled to a large number of other molecules with many different Rabi oscillations, the superposition of these oscillations washes out the oscillation. This is called decoherence.

3.B Complement: Light–matter interaction in metals: Physical mechanisms

The goal of this complement is to go beyond the simplistic electron gas model, which may be misleading when dealing with a real metal. In particular, the Drude model can be successfully used to fit the metal permittivity from a few Hz to the optical regime. However, the light–matter interaction processes are different when using radiowaves or visible light. It follows that the decay term ν accounts for different microscopic processes in different frequency ranges and therefore should be considered to be a frequency-dependent parameter. We provide some background on the physical microscopic mechanisms hidden by the simple model. We stress that ν takes similar values in the DC regime and in the optical regime but has a very different temperature dependence, due to the fact that the microscopic processes are different.

3.B.1 Electron transport: Collision time and mean free path

We start by giving the order of magnitude of the collision time for an electron in a metal. Let us first define what is meant by electron collision time. The historical vision of Drude was that electrons would collide with atoms. Of course, the eigenstate of an electron in a periodic crystal is a Bloch wave, which propagates without scattering. The effect of the periodic potential is to modify the dispersion relation. Taking as a starting point the band structure of the energy of an electron in a perfect crystal, any departure from the model of an independent electron in a perfect crystal results in defects scattering the electrons. Hence, collisions may be due to: (i) electron–electron interactions which were not taken into account in the band structure calculation; (ii) electron–phonon interaction, as a phonon is the consequence of the time-dependent displacement of the atoms from their position in the periodic crystal; or (iii) scattering by defects in the crystal such as interstitials atoms, dislocations, or grain boundaries. These collision processes are two-body collisions, in marked contrast with the intraband photon absorption processes discussed later which are three-body collision processes. The electron collision time is typically the mean duration between two collisions. It is on the order of 10–40 fs for crystalline noble metals (see Table 3.1). When multiplying this time by the Fermi velocity, one finds the mean free path, which is typically on the order of 20–50 nm. It follows that the response of a system is expected to be different if the typical object size is below an electron mean free path.

3.B.2 Typical decay time for plasmons

We now discuss the decay time of a bulk plasmon due to losses. We will show that it is related to the ν effective parameter of the Drude model. From a macroscopic point of view, we can search the complex eigenfrequency of the bulk plasmon mode. This is a quasinormal mode as will be discussed in chapter 4 and 7. Its imaginary part contains the information on the decay time. For instance, for a bulk plasmon, the complex frequency is the solution of $\varepsilon(\omega) = 0$. Let us assume that we can use a Drude

Table 3.1 Fermi velocity, collision time, and mean free path for noble metals at ambient temperature. Data taken from ref. (Gall, 2016).

Material	Gold	Silver	Aluminium	Copper
Fermi velocity (10^5m/s)	13.8	14.5	16	11.1
Collision time (fs)	27.3	36.8	11.8	36
Mean free path (nm)	38	53	19	40

model, so that $\omega^2 + i\omega\nu - \omega_p^2 = 0$. The root of this equation is $\sqrt{\omega_p^2 - \nu^2/4} - i\nu/2$. Hence, the relaxation time of the plasmon appears to be given by ν. The corresponding microscopic process is the generation of an electron-hole pair. So we see that in this case, the parameter ν of the Drude model accounts for a different process, so that we should not expect the value of ν derived from resistivity measurement at low frequencies to match with the parameter ν derived from absorption at optical frequencies.

3.B.3 Absorption processes in bulk metals

The absorption of electromagnetic energy in the DC regime or at radio frequencies is due to electron collisions. In the low-frequency regime, the classical picture used to describe transport is valid: (i) the electromagnetic field transfers energy to the electrons; (ii) the electrons lose their energy through collisions. Among the microscopic mechanisms responsible for collisions cited above, the electron–phonon interaction is the dominant mechanism. As a consequence, the collision rate depends on temperature and decays as the temperature is reduced.

We now turn to the absorption of energy of an electromagnetic wave at visible or infrared frequencies. This process corresponds to an electron–hole pair generation. There are different mechanisms: intraband absorption assisted by phonons, interband absorption, and Landau damping.

1. Optical intraband absorption. For low values of the wavevector $k < \omega/v_F$ and low values of the frequency, the leading absorption mechanism is intraband absorption. Indeed, if the photon energy is much smaller than the interband energy, the electron can only be promoted to a state within the same band. It is seen that the momentum conservation cannot be satisfied because an electronic state with larger energy has a different momentum. As the photon momentum is negligible as compared to the electron momentum change, this process needs to be a three-body interaction, with a third partner providing momentum. Hence, this process is possible if a phonon is either emitted or absorbed. The temperature dependence of this process differs from the temperature dependence of the collision time. One of the practical conclusions of this paragraph is that metal losses at optical frequencies cannot be suppressed when reducing the temperature. The reader will find more information on electron losses in refs (Del Fatti, 1999; Kaveh and Wiser, 1984; Allen, 1971; Smith and Ehrenreich, 1982). The three-body absorption process can also be assisted by another electron or by the presence of an interface, so that momentum conservation is no longer imposed. Nevertheless, the phonon-assisted absorption is the leading process (Del Fatti, 1999; Kaveh and Wiser, 1984; Allen, 1971; Smith and Ehrenreich, 1982).
2. Interband absorption. The second absorption mechanism for an electromagnetic wave in a metal is related to the interband absorption. For noble metals, electrons in d-shells can be promoted to states above the Fermi level and with the same wavevector (so-called vertical transition) if the frequency is above a threshold. This mechanism is well known (Ashcroft and Mermin, 1976) and accounts for the colour of copper or gold. It is not accounted for by the standard Drude form. It can be included in the Drude model by adding a specific term of the dielectric

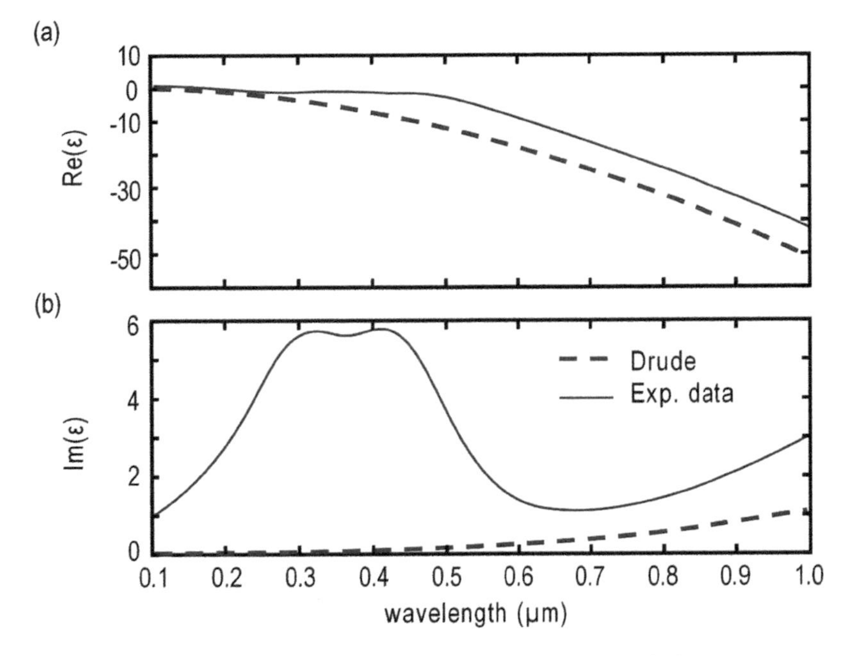

Fig. 3.2 Comparison of experimental data and a Drude fit of the dielectric constant for gold for (a) the real part of the dielectric constant; (b) the imaginary part of the complex dielectric constant.

constant. Fig. 3.2 illustrates the large difference in the optical part of the spectrum between a Drude model obtained by fitting experimental data and a more detailed fit of the data reported by Johnson and Christy for gold (Johnson and Christy, 1972).

The absorption band observed in the imaginary part of the dielectric constant between $0.3\,\mu$m and $0.4\,\mu$m is due to absorption by d-band electrons. This can be accounted for by developing fits of the measured dielectric constant, as reported in several references (Etchegoin et al., 2006; Vial et al., 2005; Laroche and Girard, 2006; Hao and Nordlander, 2007). It is, of course, essential to use these realistic models when using time-domain calculations.

3. Landau damping. The Landau damping mechanism is observed only for large wavevectors such that $k > \omega/v_F$. It can be understood using either a wave or particle picture. The wave picture is due to Landau and has been introduced to explain the absorption in plasmas. When the electron velocity v_F is equal to the phase velocity ω/k of the plasmon field, the electron velocity and the field are always in phase so that the energy transfer is very efficient. The particle picture considers that the absorption of the electromagnetic excitation allows an electron–hole pair to be generated. This process is possible if energy and momentum are conserved, and we have seen before that a third body was necessary to ensure momentum conservation. However, if the electromagnetic excitation (the plasmon) has a momentum large enough, the two-body absorption process becomes possible. Let us consider the absorption of an electromagnetic excitation with energy $\hbar\omega$

and momentum $\hbar\mathbf{k}$ by an electron with initial momentum $\hbar\mathbf{q}$ and initial energy $\hbar^2q^2/2m$. The electron is close to the Fermi surface, so that its energy is the Fermi energy $E_F = \hbar^2k_F^2/2m = mv_F^2/2$ and its momentum is $\hbar k_F = mv_F$. The energy and the momentum are conserved so that we have:

$$\begin{aligned}\hbar\omega + \frac{\hbar^2}{2m}\mathbf{q}^2 &= \frac{\hbar^2}{2m}(\mathbf{q}+\mathbf{dq})^2, \\ \hbar\mathbf{q} + \hbar\mathbf{k} &= \hbar(\mathbf{q}+\mathbf{dq}).\end{aligned} \tag{3.54}$$

It follows that this process is possible if the energy of the electromagnetic excitation belongs to the interval given by:

$$\begin{aligned}\frac{\hbar\omega_{max}}{E_F} &= \left(\frac{k}{k_F}\right)^2 + 2\frac{k}{k_F}, \\ \frac{\hbar\omega_{min}}{E_F} &= \left(\frac{k}{k_F}\right)^2 - 2\frac{k}{k_F}.\end{aligned} \tag{3.55}$$

For usual electromagnetic excitations, this process is forbidden, as the electromagnetic wavevector $k \approx \omega/c$ is much smaller than the electron wavevector. Yet, bulk plasmons or surface plasmons may have large wavevectors, and therefore this process can take place for large values of k. When trying to satisfy this relation with the smallest possible wavevector, k, we get $\hbar\omega \approx 2kE_F/k_F$. By inserting $E_F = \hbar^2k_F^2/2m$ and $\hbar k_F = mv_F$, we find that the condition is equivalent to $\omega/k \approx v_F$ and we recover the phase-matching condition between a wave and a particle. We note here that this process can also take place when a dipole is close to a metal interface at a distance d smaller than v_F/ω, as its near field contains large wavevectors. In summary, for values of k larger than ω/v_F, the field is damped, as it can relax by generating an electron–hole pair. When it is allowed, this process is so efficient that the fields are overdamped and virtually do not exist any longer. Clearly, this process introduces a spatial frequency cutoff for the fields. More information on the non-local description of the optical properties of solids can be found in refs (Ashcroft and Mermin, 1976; Ziman, 1972; Ford and Weber, 1984; Pines, 1964). For noble metals, the typical Landau damping length scale v_F/ω is on the order of 0.5 nm.

3.C Complement: Hydrodynamic model of the permittivity of an electron gas: Non-local effects

The purpose of this section is to derive the form of the non-local dielectric constant of a metal in the framework of the hydrodynamic model. We start by writing the Euler equation with an additional friction force $-n_em\nu\mathbf{v}$:

$$n_em\left[\frac{\partial\mathbf{v}}{\partial t} + \mathbf{v}\cdot\nabla\mathbf{v}\right] = -\nabla P_e - n_ee\mathbf{E} - n_em\nu\mathbf{v}, \tag{3.56}$$

where $\mathbf{v}$ is the electron flow velocity, n_e is the number of electrons per unit volume, and P_e is the electronic pressure. The particle conservation can be written as

$$\nabla \cdot (n_e \mathbf{v}) = -\frac{\partial n_e}{\partial t}, \tag{3.57}$$

and the pressure variation $P_1 = P_e - P_0$ is connected to the density variation $n_1 = n_e - n_0$ by the compressibility:

$$P_1 = \beta^2 m n_1, \tag{3.58}$$

where $\beta = \sqrt{3/5}v_F$ (Halevi, 1995). We introduce the Fourier transform $n_e(\mathbf{k}, \omega)$ and $\mathbf{v}(\mathbf{k}, \omega)$ of $n(\mathbf{r}, t)$ and $\mathbf{v}(\mathbf{r}, t)$. We now proceed to linearize the equations by keeping first-order terms in n_1, P_1, and $\mathbf{v}$. Projecting on the directions parallel and perpendicular to $\mathbf{k}$, we find:

$$\begin{aligned} -i\omega n_0 \mathbf{v}_{\parallel} &= -i\mathbf{k}\frac{P_1}{m} - n_0 \frac{e}{m}\mathbf{E}_{\parallel} - n_0 \nu \mathbf{v}_{\parallel}, \\ -i\omega n_0 \mathbf{v}_{\perp} &= -n_0 \frac{e}{m}\mathbf{E}_{\perp} - n_0 \nu \mathbf{v}_{\perp}. \end{aligned} \tag{3.59}$$

The linearization of the particle conservation equation yields $\frac{n_1}{n_0} = \frac{\mathbf{v}\cdot\mathbf{k}}{\omega}$, showing that n_1 and $\mathbf{v}$ are perturbations of the same order. Inserting $\frac{P_1}{m} = \beta^2 n_1 = \beta^2 n_0 \frac{\mathbf{v}\cdot\mathbf{k}}{\omega}$ in Eq. (3.59), we find:

$$\begin{aligned} v_{\parallel} &= \frac{-i\omega e}{m}\frac{E_{\parallel}}{\omega^2 + i\omega\nu - \beta^2 k^2}, \\ v_{\perp} &= \frac{-e}{m}\frac{E_{\perp}}{\nu - i\omega}. \end{aligned} \tag{3.60}$$

Using the definitions $\mathbf{j} = -n_1 e \mathbf{v} = -i\omega \mathbf{P} = -i\omega\varepsilon_0(\varepsilon_r - 1)\mathbf{E}$ yields:

$$\begin{aligned} \varepsilon_{r,\parallel} &= 1 - \frac{\omega_p^2}{\omega^2 + i\omega\nu - \beta^2 k^2}, \\ \varepsilon_{r,\perp} &= 1 - \frac{\omega_p^2}{\omega^2 + i\omega\nu}. \end{aligned} \tag{3.61}$$

The last equation shows that the permittivity decays as ω^{-2} for frequencies larger than ω_p. This is why metals become transparent in the deep UV and in the X-ray regime. When the frequency of the field is larger than the natural frequency of the system, the system does not respond. Similarly, it is seen that if the wavevector is larger than ω_p/β, the permittivity tends to zero. This means that the fields cannot be shielded anymore by the induced polarization if k is larger than $\omega_p/\beta \approx \omega_p/v_F$. In other words, electromagnetic fields varying on a length scale smaller than v_F/ω_p cannot be screened by the metal. This is the screening length discussed hereafter.

3.D Complement: Non-local effects: Beyond the hydrodynamic model

The purpose of this complement is to give an overview of non-local effects beyond metals and beyond the hydrodynamic model. More elaborated models than the hydrodynamic model accounting for the k-dependence of the dielectric constant (i.e. non-local models) are discussed in refs. (Ashcroft and Mermin, 1976; Ziman, 1972; Pines, 1964; Ford and Weber, 1984). The key idea is that spatial dispersion becomes significant when the size of the system becomes comparable to some relevant length scales. We thus start by introducing the microscopic length scales that may play a role. We then remark that the magnetic field is obtained by taking the curl of the electric field, so that there is a fundamental connection between magnetism and non-local description. This is particularly important in the context of magnetic metamaterials. Finally, we will discuss what happens when dealing with interfaces which break the translational invariance, so that the description given in Eq. (3.2) is no longer valid.

3.D.1 Typical length scales for non-locality

The mean free path l of a charge carrier in a system is an important quantity. Consider, for instance, a slab with thickness t. Depending on the ratio t/l, the behaviour of the carrier will be either ballistic if $l \gg t$ or diffusive if $l \ll t$. Accordingly, the optical response of the system will depend on t/l. A typical mean free path for electrons in a metal at ambient temperature varies between 30 nm and 50 nm. This value may be reduced if the metal has a polycrystalline structure, so that electrons are scattered at grain boundaries.

Another typical length scale is the screening length. In standard macroscopic electrodynamics, the normal component of an electric field is not continuous at an interface. Instead, εE_z is continuous, so that there is a jump of the electric field. In practice, the discontinuity takes place over a given length, which is the screening length. In other words, imagine that an external charge is introduced in a metal. The electronic density is modified around this charge in such a way that the charge is screened. The resulting potential around this charge is no longer a Coulomb potential varying as $1/r$. It has the structure of a Yukawa potential $\exp(-r/\lambda_{TF})/r$ where the decay length scale is the Thomas–Fermi length given by $\sqrt{\frac{v_F^2}{3\omega_p^2}}$, as shown in (Ziman, 1972). The order of magnitude for a metal is on the order of 0.1 nm.

The skin depth is the decay length of a wave in the metal. It is typically 30 nm in the visible and in the infrared. It changes in the GHz range, as predicted by the Drude model. Indeed, it is given by $\lambda/(2\pi n'')$, where it can be seen from the Drude model that $n'' \approx \omega_p/\omega$ for optical and IR frequencies and $n'' \approx \omega_p/(\sqrt{2\gamma\omega})$ for lower frequencies, so that the skin depth becomes on the order of 2 μm at 1 GHz.

An important length scale is v_F/ω. This length has been introduced in the previous section. It plays a role in the condition of absorbing a photon by generating an electron–hole pair (Landau damping). This is possible only for electromagnetic fields with wavevectors larger than ω/v_F.

For a metamaterial consisting in a periodic array of structures, a relevant length scale is the period a. The metamaterial will not diffract waves with wavelengths larger than $2a$. However, it may be required to use a non-local model to account for the optical response of this medium.

The metal length scales (screening length v_F/ω, skin depth) are much smaller than a typical wavelength in the visible range. This is why non-local effects are usually negligible for metals. Note, however, that they may play a role when dealing with objects whose *size* is comparable to theirs.

3.D.2 Non-locality and magnetism

In this section, we establish an equivalence between a local description of a magnetic material using a local permeability and a local permittivity and a non-local description in terms of a permittivity only. This is possible because the magnetic field can be expressed in terms of the curl of the electric field. Hence, the contribution to the current density $\nabla \times \mathbf{M}$ is proportional to $\nabla \times \nabla \times \mathbf{E}$. This is a non-local relation. As for a time variable, we will introduce a Fourier description which is more convenient for dealing with non-local relations. We start with the definition of the current density:

$$\mathbf{j}(\mathbf{r}, t) = \frac{\partial \mathbf{P}}{\partial t} + \nabla \times \mathbf{M}(\mathbf{r}, t). \tag{3.62}$$

Moving to Fourier domain in space and time, this relation becomes:

$$\mathbf{j}(\mathbf{k}, \omega) = -i\omega \mathbf{P}(\mathbf{k}, \omega) + i\mathbf{k} \times \mathbf{M}(\mathbf{k}, \omega). \tag{3.63}$$

For the sake of simplicity, we assume an isotropic linear and local constitutive relation for the magnetization:

$$\mathbf{M}(\mathbf{k}, \omega) = (\mu_r - 1)\mathbf{H} = \frac{\mu_r - 1}{\mu_0 \mu_r}\mathbf{B}. \tag{3.64}$$

We now use Maxwell equation to express $\mathbf{B}$ in terms of $\mathbf{E}$:

$$\mathbf{M}(\mathbf{k}, \omega) = \frac{\mu_r - 1}{\mu_0 \mu_r} \frac{\mathbf{k} \times \mathbf{E}}{\omega}. \tag{3.65}$$

Inserting this result in Eq. (3.63), we get:

$$\mathbf{j}(\mathbf{k}, \omega) = -i\omega \mathbf{P}(\mathbf{k}, \omega) + i\frac{\mu_r - 1}{\mu_0 \mu_r \omega}\mathbf{k} \times \mathbf{k} \times \mathbf{E}, \tag{3.66}$$

where we see that the contribution of the magnetic field to the current density is purely transverse. Introducing a local and linear permittivity to connect $\mathbf{P}$ and $\mathbf{E}$, we get:

$$\mathbf{j}(\mathbf{k}, \omega) = -i\omega \varepsilon_0 (\varepsilon_r - 1)\mathbf{E} + i\frac{\mu_r - 1}{\mu_0 \mu_r \omega}\mathbf{k} \times \mathbf{k} \times \mathbf{E}. \tag{3.67}$$

Hence, we see that the *total* current density can be related to the electric field only, provided that we introduce a permittivity that depends on the wavevector, namely a non-local permittivity. Using $\mathbf{k} \times \mathbf{k} \times \mathbf{E} = k^2(\mathbf{E} - \hat{\mathbf{k}}\hat{\mathbf{k}}\mathbf{E}) = k^2 E_l \hat{\mathbf{k}}$, we see that this permittivity has a tensor structure, so that we need to introduce a longitudinal and a transverse permittivity. This is actually not surprising, as even though we have assumed an isotropic permittivity ε_r and permeability μ_r, rotational symmetry is broken by the fact that we deal with a specific $\mathbf{k}$. Finally, we have:

$$\mathbf{j}(\mathbf{k},\omega) = -i\omega\varepsilon_0(\varepsilon_l - 1)E_l\hat{\mathbf{k}} - i\omega\varepsilon_0(\varepsilon_t - 1)(\mathbf{E} - E_l\hat{\mathbf{k}}), \tag{3.68}$$

where

$$\begin{aligned} \varepsilon_l &= \varepsilon_r, \\ \varepsilon_t &= \varepsilon_r + \frac{\mu_r - 1}{\mu_r}\frac{c^2k^2}{\omega^2}. \end{aligned} \tag{3.69}$$

A further discussion can be found in ref. (Agranovich and Ginzburg, 1984).

3.D.3 Interfaces, confinement, and non-locality

So far, we have introduced the relation connecting the field to the polarization and we have discussed the relevant length scales that can be found in the linear response. We have assumed a homogeneous medium that is translationally invariant. We now discuss another issue: the role of interfaces. Indeed, close to an interface, the connection between the current density and the field can be affected. For instance, an electron can be reflected by an interface and contribute to the current density.

More generally, the presence of an interface breaks the translational symmetry, so that the general linear relation between the current density $\mathbf{j}(\mathbf{r}')$ and the field $\mathbf{E}(\mathbf{r})$ depends on $\mathbf{r}$ and $\mathbf{r}'$ and not only on $\mathbf{r} - \mathbf{r}'$. The knowledge of the bulk non-local response is therefore not sufficient to describe the current density close to an interface. The modification due to the presence of the surface is also needed.

A typical example of a physical effect due to the presence of the interface is the *anomalous skin effect* observed in microwaves. This effect is observed when the skin depth is smaller than the mean free path of the electrons. This may happen in the GHz range where the skin depth is on the order of 10^{-6} m. The mean free path can be larger than this value at low temperature for pure metals. In that situation, our derivation of the Drude formula based on the assumption that the amplitude of the field is uniform is no longer valid. The electrons with a velocity parallel to the interface can stay within the skin depth and interact efficiently with the electric field (Ziman, 1972), whereas the electrons with a velocity perpendicular to the surface will escape very rapidly from the skin depth. Hence, the absorption of the incident wave is mostly due to the electrons with a velocity parallel to the surface lying within the skin depth. This effect can be used to gain information on the Fermi surface of the metal (Ziman, 1972; Ashcroft and Mermin, 1976).

Another interesting system illustrating the role of surfaces in modifying the relation between the field and the current density comprises metallic nanospheres with *radii smaller than the mean free path.* We remind the reader that the mean free path is typically on the order of 30–50 nm. For a particle with a diameter smaller than the mean free path, so typically below 10 nm, the electrons do not experience collisions, so that the *local Drude model* is no longer valid. The Schrodinger equation has to be solved to find the electronic states which are *delocalized over the sphere.* The energy spectrum depends on the geometry—that is, the physics of the interaction between the electron and the field is modified by the geometry of the particle, so no local relation exists between a current and a field. Within this framework, it is possible to derive the polarizability of the particle (Hache et al., 1986). Surprisingly, the final form of the polarizability does not change much as compared to a macroscopic model based on a permittivity, except that an extra term Av_F/a has to be added to the scattering rate $\nu = 1/\tau$. In this formula, a is the particle radius, v_F is the Fermi velocity, and A is a constant on the order of unity used to fit the data. In a classical picture, the dephasing term proportional to v_F/a is attributed to a surface-scattering process. These modifications were first reported by Kreibig (Kreibig and Genzel, 1985).

We finally mention another surface effect: the decay time of molecules in close proximity to a surface. The decay is due to the non-radiative energy transfer between a molecule and an electron. The possibility of directly generating an electron–hole through the Landau damping mechanism affects the decay. Apart from the broadening of the metallic nanoparticles, this is one of the few optical experiments where non-local effects are unambiguously observed (Eagen et al., 1980) in the optical regime.

Finally, we consider the case of the enhancement of the field close to a surface, such as the apex of a tip. If we consider a radius of curvature on the order of the screening length, we may expect to observe a deviation from local electrodynamics. However, many effects can be accounted for by merely shifting the position of the interface by a quantity called the Feibelman parameter (Feibelman, 1982). Hence, a local model with a modified interface position can be used to fit the results. This has been confirmed by a recent quantum calculation of the fields between two nanowires with a diameter of 9.8 nm (Teperik et al., 2013). The key effect for gaps below 0.7 nm is the onset of a tunnel effect which suppresses the surface charges. For distances larger than 0.7 nm, the fields can be modelled using local electrodynamics with distances corrected by the Feibelman parameter (Teperik et al., 2013).

Exercises

Exercise 3.1 Spatial and temporal dispersion. We consider three different media characterized by the following constitutive relations. Are they temporally dispersive and spatially dispersive?

(1) $\varepsilon = 1 - \frac{\omega_p^2}{\omega^2 - Dk^2}$.

(2) $\mathbf{j}(\mathbf{r}, t) = \sigma \mathbf{E}(\mathbf{r}, t)$, where σ is a constant.

(3) $\mathbf{j}(\mathbf{r}, t) = \int \mathrm{d}t' \sigma(t - t') \mathbf{E}(\mathbf{r}, t')$.

Exercise 3.2 Modelling dispersion in time domain. The susceptibility of a dielectric medium is given by $\varepsilon_0 \chi(\omega) = \varepsilon_0[\varepsilon_r(\omega) - 1] = \frac{-\varepsilon_0 \omega_p^2}{\omega^2 - \omega_0^2 + i\omega\nu}$. We denote by $\chi(t)$ the susceptibility in the time domain. Write the relation between $\mathbf{P}(t)$ and $\mathbf{E}(t)$. This relation is not practical in the time domain, as it involves computing an integral at each step. Show that in the time domain, the polarization $\mathbf{P}(t)$ satisfies

$$\frac{\partial^2 \mathbf{P}}{\partial t^2} = -\omega_0^2 \mathbf{P} - \nu \frac{\partial \mathbf{P}}{\partial t} - \varepsilon_0 \omega_p^2 \mathbf{E}.$$

This equation can be used in finite-difference time-domain numerical schemes to avoid evaluating an integral at each time step.

Exercise 3.3 Screening charge dynamics and plasma oscillation. The purpose of this exercise is to study the decay dynamics of a charge in a metal. We first assume, incorrectly, that dispersion can be neglected. We then include dispersion in the model and find a very different behaviour.

(1) We assume that $\mathbf{j} = \sigma_0 \mathbf{E}$, where σ_0 is taken to be a constant. Using the Maxwell–Gauss equation and the charge conservation equation, prove that

$$\frac{\partial \rho}{\partial t} + \frac{\sigma_0}{\varepsilon_0}\rho = 0.$$

A typical decay time is thus given by ε_0/σ_0, which is roughly $8.8510^{-12}/10^7 \approx 10^{-18}$s, where $\sigma_0 \approx 10^7 SI$ units. This non-dispersive model incorrectly predicts a vanishing of the charge on an attosecond scale.

(2) We now include dispersion in the model and use $\sigma(\omega) = \frac{\sigma_0}{1-i\omega\tau}$, where τ is a relaxation constant on the order of 10^{-14}s. Show that

$$\tau\frac{\partial^2 \rho(t)}{\partial t^2} + \frac{\partial \rho(t)}{\partial t} + \frac{\sigma_0}{\varepsilon_0}\rho = 0.$$

(3) We seek a solution in the form $\exp(-i\omega t)$. Prove that

$$\omega \approx \sqrt{\frac{\sigma_0}{\varepsilon_0 \tau}} + i\frac{1}{2\tau}.$$

For a metal, the conductivity can be cast in the form $\sigma_0 = ne^2\tau/m$, where e is the electron charge, m the electron mass, and n the number of electrons per unit volume. It follows that $\frac{\sigma_0}{\varepsilon_0\tau} = \frac{ne^2}{m\varepsilon_0} = \omega_p^2$ is the square of the plasma frequency. Hence, the model predicts an oscillation of the charge at the plasma frequency and a vanishing of the charge on a time scale given by τ.

Exercise 3.4 Drude model. Using the model given in section 3.8, derive the current density $\mathbf{j} = -n_0 e\mathbf{v}$ and the electrical conductivity. Check directly the equality in Eq. (3.6).

Exercise 3.5 Polarization orientation. Microwave absorption by water. The purpose of this exercise is to model the permittivity of a polar fluid such as water. With this model, we will be able to derive the rate of electromagnetic heating of water in a microwave (2.4 GHz) oven. We consider water containing n_0 molecules per unit volume.

(1) Each molecule has a permanent electric dipole moment $\mathbf{p}_0$. This dipole moment is randomly oriented at equilibrium but tends to be aligned to an external electric field. This phenomenon is called orientation polarization. We aim at computing the polarization induced by a static electric field $\mathbf{E}$ at temperature T to first order in the parameter E_l/kT, where $\mathbf{E}_l$ is the local field. In order to derive the average polarization, we use

the probability that a dipole $\mathbf{p}_0$ makes an angle θ with the field $\mathbf{E}_l$ is proportional to $\exp{(E_l p_0 \cos\theta)/k_B T}$, where k_B is the Boltzmann constant.

In the microwave range, the electronic polarizability of each molecule is a real number α, so that $p_{el} = \alpha\varepsilon_0 \mathbf{E}_l$. Show that:

$$\mathbf{P} = \mathbf{P}_{el} + \mathbf{P}_{or} = n_0 \mathbf{p} = n_0 \varepsilon_0 \left[\alpha + \frac{p_0^2}{3k_B T \varepsilon_0}\right] \mathbf{E}_l. \tag{3.70}$$

(2) The second term is the orientation polarization. We now seek the frequency-dependent permittivity. We first derive an equation for the dynamic behaviour of the polarization. To this aim, we note that when turning off an electric field, it is observed that the orientation polarization decays exponentially in time with a time constant τ. Show that the orientation polarization satisfies the equation

$$\frac{\mathrm{d}\mathbf{P}_{or}}{\mathrm{d}t} + \frac{1}{\tau}\mathbf{P}_{or}(t) = \mathbf{0}. \tag{3.71}$$

(3) We now consider a monochromatic incident electric field with circular frequency ω: the local field can be written as $\mathbf{E}_l = \mathbf{E}_0 e^{-i\omega t}$. Show that the polarization satisfies:

$$\frac{\mathrm{d}\mathbf{P}_{or}}{\mathrm{d}t} + \frac{1}{\tau}\mathbf{P}_{or} = C\mathbf{E}_l, \tag{3.72}$$

where C is a constant to be determined by considering the static case. Show that

$$C\tau = \frac{n_0 p_0^2}{3k_B T}. \tag{3.73}$$

(4) The local field is given by $E_l = \mathbf{E} + \frac{\mathbf{P}}{3\varepsilon_0}$. Show that:

$$\frac{\varepsilon_r - 1}{\varepsilon_r + 2} = \frac{n_0}{3\varepsilon_0}\left[\alpha + \frac{p_0^2}{3k_B T\varepsilon_0}\frac{1}{1 - i\omega\tau}\right]. \tag{3.74}$$

This form of the permittivity is given in terms of unknown microscopic parameters $p(0)$ and α. We now introduce macroscopic parameters $\varepsilon_s(\omega = 0)$ and $\varepsilon_\infty = \varepsilon(\omega \to \infty)$. Show that the permittivity can be cast in the form

$$\varepsilon = \varepsilon_\infty + \frac{\varepsilon_s - \varepsilon_\infty}{1 - ix}, \tag{3.75}$$

where $x = \dfrac{\varepsilon_s + 2}{\varepsilon_\infty + 2}\omega\tau$.

Exercise 3.6 Permittivity of an ionic crystal. The goal of this exercise is to derive the permittivity of an ionic crystal. We take as a model a linear chain of positive ions with charge q and mass m_+ and negative ions with charge $-q$ and mass m_-. Positive ions are located at $x_{2n} = 2na$ and negative ions are at $x_{2n+1} = (2n+1)a$, where n is an integer. The interactions between ions are limited to interactions with the first neighbour and given by the potential

$$V(\mathbf{r}_n) = \frac{C}{2}\left[(\mathbf{r}_n - \mathbf{r}_{n-1})^2 + (\mathbf{r}_n - \mathbf{r}_{n+1})^2\right], \tag{3.76}$$

where $\mathbf{r}_n$ is the displacement of the ion n with respect to its equilibrium position along the x-axis.

We start by studying the movement of the ions in the absence of external electric fields in order to identify the modes of the system.

(1) Write the equations satisfied by the ions located at x_{2n} and x_{2n+1}.

(2) We seek solutions of the form:

$$\begin{aligned} \mathbf{r}_{2n} &= \mathbf{r}_+ \exp[ikx_{2n} - i\omega t], \\ \mathbf{r}_{2n+1} &= \mathbf{r}_- \exp\left[ik(x_{2n+1} - a) - i\omega t\right]. \end{aligned} \tag{3.77}$$

Solving these equations yields the elastic waves propagating in the linear chain. In particular, it is possible to find the sound velocity for low values of the wavevector. If the energy of these elastic waves is quantized, one obtains phonons. In what follows, we only consider the modes that can be coupled to light, namely the modes with a non-zero electric dipole moment. Since we are interested in waves in the optical or infrared domain, the wavevectors $k \approx \omega/c$ satisfy $k \ll \pi/2a$, so that the phase term ka is negligible. Hence, we solve the problem in the limit $k = 0$. Show that $\mathbf{r}_+ - \mathbf{r}_-$ is an eigenfunction of the problem with an eigenfrequency given by $\omega_p^2 = 2C/\mu$, where $\mu = m_+ m_- / m_+ + m_-$.

(3) We now study the polarizability of the system. A plane wave propagates in the system. It is given by:

$$\mathbf{E}(\mathbf{r}, t) = Re\left\{\mathbf{E}_i \exp[i(\mathbf{k}.\mathbf{r} - \omega t)]\right\}. \tag{3.78}$$

The local field can thus be written as

$$\mathbf{E}_{loc}(\mathbf{r}, t) = Re\left\{\mathbf{E}_{loc} \exp[i(\mathbf{k}.\mathbf{r} - \omega t)]\right\}. \tag{3.79}$$

Show that the dipole moment of a cell is given by

$$\mathbf{p} = -\frac{q^2}{\mu} \frac{\mathbf{E}_{loc}}{\omega^2 - \omega_p^2}. \tag{3.80}$$

(4) We now introduce the electronic polarizability of the ions α^+ and α^-. We assume that the electronic and the vibrational contributions can simply be added. Derive the relation

$$\frac{\varepsilon(\omega) - 1}{\varepsilon(\omega) + 2} = \frac{n}{3\varepsilon_0}\left[\alpha_+ + \alpha_- - \frac{q^2}{\mu}\frac{1}{\omega^2 - \omega_p^2}\right]. \tag{3.81}$$

(5) We now seek a form of the permittivity which is independent on the microscopic parameters by introducing the static permittivity ε_S and the high-frequency permittivity ε_∞. Note that the high-frequency permittivity is not, strictly speaking, the permittivity for infinite frequencies, which is ε_0. It is the permittivity when the vibration contribution has been subtracted, namely the electronic permittivity:

$$\frac{\varepsilon(\omega) - 1}{\varepsilon(\omega) + 2} = \frac{\varepsilon_\infty - 1}{\varepsilon_\infty + 2} + \frac{\omega_p^2}{\omega_p^2 - \omega^2}\left[\frac{\varepsilon_0 - 1}{\varepsilon_0 + 2} - \frac{\varepsilon_\infty - 1}{\varepsilon_\infty + 2}\right]. \tag{3.82}$$

Show that this form can be further simplified to get:

$$\varepsilon(\omega) = \varepsilon_\infty \frac{\omega_L^2 - \omega^2}{\omega_T^2 - \omega^2}, \tag{3.83}$$

where $\omega_L^2 = \omega_T^2 \frac{\varepsilon_s}{\varepsilon_\infty}$ and $\omega_T^2 = \omega_p^2 \frac{\varepsilon_\infty + 2}{\varepsilon_s + 2}$.

4 Propagation

In its simplest 1D form, a solution of a propagation equation can be cast in the form $f(x - vt)$ where v is the wave velocity. In this chapter, we generalize this elementary definition to include propagation in a *3D vacuum* and discuss the connection with diffraction. We also introduce the concept of evanescent waves. We then consider propagation in a *dispersive and lossy medium.* This leads to the introduction of both a complex refractive index in the frequency domain and a dispersion relation. We discuss several definitions of the velocity. We finally introduce the quasinormal modes in lossy media.

4.1 Propagation in vacuum: Angular spectrum

4.1.1 Statement of the problem

In this section, we deal with the propagation in vacuum of monochromatic radiation with a real circular frequency ω (we will simply say frequency) coming from a source which is not specified. We assume that we know the electric field on the plane $z = 0$. It could be produced by a laser beam that illuminates the plane $z = 0$ or by an antenna (or any other current distribution) placed at a point $z < 0$. We assume that the propagation takes place in vacuum. We seek an explicit expression of the electric field at any point $z > 0$ (see Fig. 4.1).

This formulation of a propagation problem can be applied to the field behind an aperture in a planar opaque screen illuminated by a plane wave inasmuch as the field

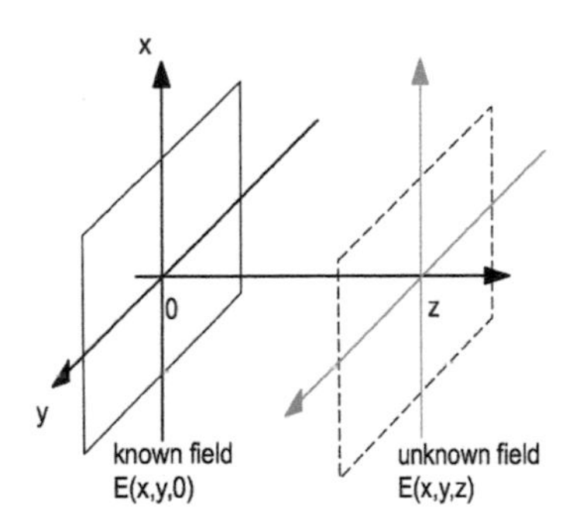

Fig. 4.1 We assume that a scalar field is known in the plane $z = 0$ and want to find the field for any point with $z > 0$. This is a propagation problem. Diffraction by a screen on plane $z = 0$ is a particular case of the general problem.

Introduction to Nanophotonics. Henri Benisty, Jean-Jacques Greffet, and Philippe Lalanne, Oxford University Press.
 DOI: 10.1093/oso/9780198786139.003.0004

behind the aperture is assumed to be the incident field. It also includes the study of any beam with finite section propagating in vacuum. Hence, it is seen that the Huygens–Fresnel treatment of diffraction appears to be a particular case of the propagation problem.

Historically, the situation was different. Geometrical optics was the theoretical framework, and diffraction by a small aperture was an unexplained phenomenon. This unexplained phenomenon deserved to have a name (diffraction) and a principle had to be added to the theory in order to be able to describe it. We now know that the Maxwell equations provide a framework which embodies both geometric optics and wave propagation. Diffraction is thus merely a convenient word to refer to a particular class of propagation problems. In what follows, we will use the framework of Maxwell equations. We will find a general solution of the problem in the form of a linear superposition of plane waves with the same real frequency and different wavevectors corresponding to different propagation directions. This is the angular spectrum representation of the field. It will be seen that the transverse size of the beam increases upon propagation, a phenomenon called natural diffraction which is associated with a damping of the amplitude of the field. In Chapter 6, we will introduce the concept of the 'waveguide', which allows us to prevent natural diffraction and therefore control propagation between two points. This has major applications in energy and information transport.

To solve the propagation problem, we simply solve a partial differential equation with the relevant boundary conditions. For simplicity, we consider a scalar amplitude. This can be considered to be a cartesian component of a vector field, so that the extension to the vector case is straightforward. The field obeys the propagation equation in vacuum:

$$\nabla^2 E(\mathbf{r},t) - \frac{1}{c^2}\frac{\partial^2 E(\mathbf{r},t)}{\partial t^2} = 0 . \tag{4.1}$$

Such a partial derivative equation in time domain can be solved, provided that initial conditions are given. Here, instead, we consider that there is a monochromatic incident field that can be cast in the form $E(\mathbf{r})\exp(-i\omega t)$. The propagation equation for the field $E(\mathbf{r},t)$ now becomes a Helmholtz equation for the complex amplitude $E(\mathbf{r})$. They are related by $E(\mathbf{r},t) = \mathrm{Re}[E(\mathbf{r})\exp(-i\omega t)]$. The Helmholtz equation is written as follows for the scalar electric field in vacuum:

$$\nabla^2 E(\mathbf{r},\omega) + \frac{\omega^2}{c^2} E(\mathbf{r},\omega) = 0 . \tag{4.2}$$

In the monochromatic regime, there is no need for initial conditions. The boundary conditions are given by the knowledge of the field in the plane $z = 0$ and by the Sommerfeld radiation condition (Jones, 1989; Schot, 1992):

$$\lim_{r\to\infty} r\left(\frac{\partial E}{\partial r} - i\frac{\omega}{c}E\right) = 0, \tag{4.3}$$

r being the distance from the origin. Note that the Sommerfeld radiation condition entails that the wave behaves asymptotically as an outgoing spherical wave. A vectorial form of this condition known as the Silver-Müller condition can be found in ref. (Schot, 1992).

4.1.2 Field propagation: The angular spectrum

To proceed, we note that the field $E(x, y, z)$ may be considered as a function of x and y in a fixed plane, z. We introduce the Fourier decomposition of E according to x and y only.

$$E(x, y, z) = \iint E(k_x, k_y, z) \exp[i(k_x x + k_y y)] \frac{\mathrm{d}k_x}{2\pi} \frac{\mathrm{d}k_y}{2\pi} , \tag{4.4}$$

where k_x and k_y are real because it is a Fourier transform (see Fig. 4.2). It is always possible to introduce the Fourier transform of the field because the field is square-integrable over the plane (x, y). This follows from the fact that the energy flux across the plane $z = 0$ is finite. Thus the problem of determining $E(x, y, z)$ amounts to finding the Fourier transform $E(k_x, k_y, z)$. Noting that a partial derivative $\partial/\partial x$ becomes an algebraic multiplication by ik_x, we get:

$$\nabla^2 E(k_x, k_y, z) \exp[i(k_x x + k_y y)] = (-k_x^2 - k_y^2 + \frac{\partial^2}{\partial z^2}) E(k_x, k_y, z) \exp[i(k_x x + k_y y)], \tag{4.5}$$

and we obtain the equation satisfied by $E(k_x, k_y, z)$ by inserting Eq. (4.4) into Eq. (4.2):

$$\iint \left\{ \frac{\partial^2 E(k_x, k_y, z)}{\partial z^2} + \left(\frac{\omega^2}{c^2} - k_x^2 - k_y^2 \right) E(k_x, k_y, z) \right\} \exp[i(k_x x + k_y y)] \frac{\mathrm{d}k_x}{2\pi} \frac{\mathrm{d}k_y}{2\pi} = 0 . \tag{4.6}$$

Since the plane waves form an orthogonal basis, the integrand is null for any (k_x, k_y), so that:

$$\frac{\partial^2 E(k_x, k_y, z)}{\partial z^2} + \left(\frac{\omega^2}{c^2} - k_x^2 - k_y^2 \right) E(k_x, k_y, z) = 0. \tag{4.7}$$

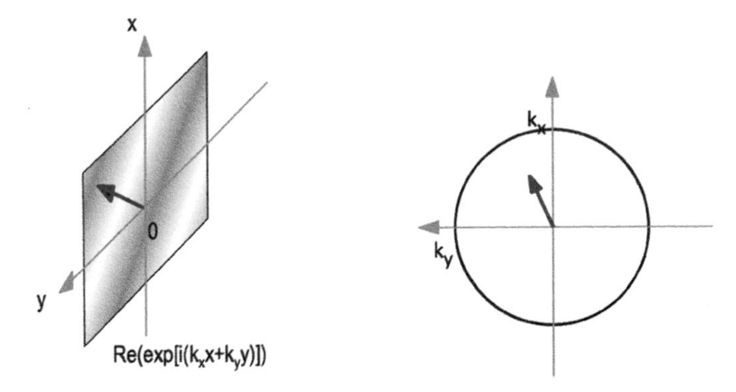

Fig. 4.2 Illustrating the decomposition of the field in plane waves. The grey shaded area in the left figure illustrates the periodic amplitude in the (x, y) plane of a plane wave $\exp(ik_x x + ik_y y)$. The wavevector (k_x, k_y) plays the role of a vectorial spatial frequency. The period of the oscillations is inversely proportional to $k_\parallel = \sqrt{k_x^2 + k_y^2}$: rapidly varying structures of the field correspond to a small period and large $k_\parallel$. The amplitude of the plane wave does not vary perpendicular to (k_x, k_y).

The general solution of this second-order differential equation may be written in the form of two exponentials. Let us introduce the notation:

$$\gamma^2 = \frac{\omega^2}{c^2} - k_x^2 - k_y^2. \tag{4.8}$$

We need to specify a choice for the determination of the square root:

$$\begin{aligned} \gamma &= \sqrt{\frac{\omega^2}{c^2} - k_x^2 - k_y^2} \quad \text{for} \quad \frac{\omega^2}{c^2} \geq k_x^2 + k_y^2, \\ \gamma &= i\sqrt{k_x^2 + k_y^2 - \frac{\omega^2}{c^2}} \quad \text{for} \quad \frac{\omega^2}{c^2} < k_x^2 + k_y^2 \,. \end{aligned} \tag{4.9}$$

With this choice, the general solution can be cast in the form:

$$E(k_x, k_y, z) = A(k_x, k_y)\exp[i\gamma z] + B(k_x, k_y)\exp[-i\gamma z]\,. \tag{4.10}$$

We now use the boundary conditions to determine A and B. We first observe that the term $B(k_x, k_y)\exp[-i\gamma z]$ corresponds to a wave propagating along negative z if γ is real and to a wave exponentially diverging for positive z if γ is imaginary. This term does not satisfy Sommerfeld's radiation condition, so that $B = 0$. A is then found from the knowledge of the field $E(x, y, 0)$ in the plane $z = 0$. We simply write the expression of the field in the plane $z = 0$ using Eqs (4.4) and (4.10):

$$E(x, y, 0) = \iint A(k_x, k_y)\exp[i(k_x x + k_y y)]\,\frac{\mathrm{d}k_x}{2\pi}\frac{\mathrm{d}k_y}{2\pi}\,. \tag{4.11}$$

By inverting this equation, we find:

$$A(k_x, k_y) = \iint E(x, y, 0)\exp[-i(k_x x + k_y y)]\,dx\,dy. \tag{4.12}$$

This expression simply shows that the complex amplitude $A(k_x, k_y)$ is nothing but the Fourier transform of the field in the plane $z = 0$:

$$A(k_x, k_y) = E(k_x, k_y, 0). \tag{4.13}$$

Finally, the field at any point (x, y, z) can be cast in the form:

$$E(x, y, z) = \iint E(k_x, k_y, 0)\exp[i(k_x x + k_y y + \gamma z)]\,\frac{\mathrm{d}k_x}{2\pi}\frac{\mathrm{d}k_y}{2\pi}\,. \tag{4.14}$$

From Eq. (4.14), it is clear that the field is written in the form of a superposition of plane waves whose wavevectors have components (k_x, k_y, γ). γ is given by Eq. (4.9), so that the dispersion relationship in vacuum is satisfied:

$$k_x^2 + k_y^2 + \gamma^2 = \frac{\omega^2}{c^2}\,. \tag{4.15}$$

It should be noted that Eq. (4.14) *is equally valid in the far field and in the near field* and that no approximation has been made so far. This exact result should not be confused with the far-field or Fraunhofer approximation introduced in the standard diffraction theory. The far-field limit can be recovered from the angular spectrum Eq. (4.14) by performing an asymptotic analysis for $z \to \infty$, as shown in section 4.2.2. Summarizing the last steps, it is seen that the field at any plane z is obtained by:

(i) decomposing the field in the plane $z = 0$ into plane waves propagating towards positive z;
(ii) propagating each plane wave by simply multiplying by the factor $\exp[i\gamma z]$;
(iii) adding all the plane waves.

The result summarized by Eq. (4.14) is also called the angular spectrum of the field. We are now going to discuss the physical content of this result.

4.1.3 Propagation in the far field is a (spatial) spectrum analyser

In this section, we introduce the concept of spatial frequency. As an example, let us consider a periodic grid with period d printed on paper with lines perpendicular to $[k_x, k_y, 0]$ (see Fig. 4.3). The associated spatial frequency is $\sqrt{k_x^2 + k_y^2}/2\pi = 1/d$. We can also introduce a spatial frequency along x. For example, the component k_x may also be written in the form $2\pi f_x$, where f_x is a spatial frequency. More generally,

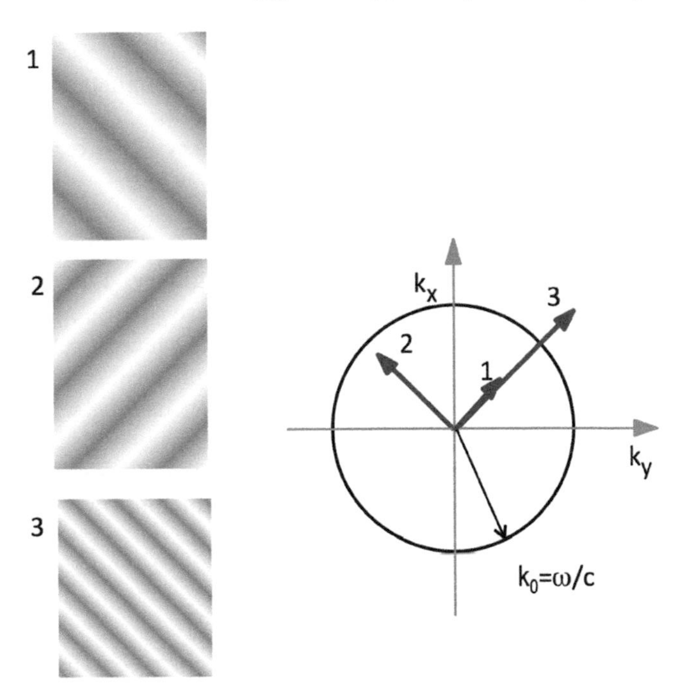

Fig. 4.3 Illustrating the decomposition of the field in plane waves. A wavevector describes both the orientation and the spatial period of the periodic field.

a square-integrable function $f(x,y)$ can be represented using a Fourier transform as a superposition of harmonic functions characterized by their wavevector (k_x, k_y). The knowledge of the Fourier transform $\tilde{f}(k_x, k_y)$ provides a spectral analysis of the function $f(x,y)$.

We have seen that the field $E(x, y, z = 0)$ can be represented as a sum of plane waves. Taking its Fourier transform with respect to x and y amounts to decomposing the field $E(x, y, 0)$ into its spectral components with amplitude $E(k_x, k_y, z = 0)$. To perform a spectral analysis of the field in terms of spatial frequencies amounts to measuring $E(k_x, k_y, z = 0)$. According to Eq. (4.14), this quantity is the amplitude of a plane wave propagating with a wavevector given by (k_x, k_y, γ). It is thus possible to measure the spectrum of the field by separating the different plane waves just like a prism allows the separation of different temporal frequencies.

Separating spatial frequencies can be done by letting the field propagate over large distances. Indeed, if we let propagation take place between the origin towards a point $\mathbf{r}$ in the far field, it turns out that the field is proportional to the amplitude of the plane wave with wavevector parallel to $\mathbf{r}$.[1] Alternatively, this can be achieved by observing the intensity in the back focal plane of a converging lens. While at $z = 0$ all the plane waves interfere to produce the field distribution, in the far field (or in the back focal plane), they are spatially separated. Hence, *propagation plays the role of a spectrum analyser.*

Let us now return to the case of an infinite periodic structure such as the grid discussed previously. Let us assume that this periodic structure is illuminated by a beam. This is the familiar situation of a grating diffracting light and producing several orders (see Chapter 15). From the point of view of the angular spectrum, we have a field in the plane of the object that is periodic. Therefore, its Fourier transform is discrete. The amplitude in the far field is then zero everywhere except for a discrete set of values: the grating orders.

4.1.4 Propagating waves and evanescent waves

We have seen that γ can be either real or purely imaginary. The former case corresponds to a propagating plane wave, while the second case corresponds to the so-called evanescent wave. Evanescent waves do exist close to the plane $z = 0$ but decay exponentially away from this plane. Let us stress that their physical origin is very different from the exponentially decaying wave encountered when dealing with propagation in a metal or an absorbing system, for instance. Here, the presence of evanescent waves results from the fact that (i) a field with subwavelength variations contains large spatial frequencies and (ii) these large spatial frequencies cannot be provided by propagating waves with wavevector satisfying $\sqrt{k_x^2 + k_y^2} \leq \omega/c$.

4.2 The Huygens–Fresnel principle revisited

The purpose of this section is to show the equivalence between the rigorous solution outlined at the beginning of the chapter—the Kirchhoff–Sommerfeld formulation of

[1] This statement is rigorously supported by the asymptotic far-field expansion of the field, as shown in section 4.2.2.

diffraction—and the Huygens–Fresnel principle. To establish a connection, we first note that the field was shown to be a superposition of plane waves in the previous section, whereas the Huygens–Fresnel principle describes the field as a superposition of spherical waves. Hence, we are trying to establish a link between two different representations of the field. The problem is therefore likely to be solved if we know how to decompose a spherical wave on a basis of plane waves. We start by introducing the plane wave expansion of a spherical wave known as the Weyl expansion (see Complement 4.A):

$$\frac{\exp(ik_0 r)}{r} = 2i\pi \int \frac{1}{\gamma} \exp[i(k_x x + k_y y + \gamma|z|)] \frac{dk_x}{2\pi}\frac{dk_y}{2\pi}, \tag{4.16}$$

where γ is a function of k_x and k_y defined by Eq. (4.9) and $r = (x^2 + y^2 + z^2)^{1/2}$. We will need the Fourier transform of $\exp[i\gamma(k_x, k_y)|z|]$. Taking the z-derivative of Eq. (4.16) for $z > 0$, we obtain:

$$\frac{\partial}{\partial z}\left[\frac{\exp(ik_0 r)}{r}\right] = -2\pi \int \exp[i\gamma(k_x, k_y)z] \exp[i(k_x x + k_y y)] \frac{dk_x}{2\pi}\frac{dk_y}{2\pi}. \tag{4.17}$$

4.2.1 Connection between the angular spectrum, the Kirchhoff–Sommerfeld formulation, and the Huygens–Fresnel principle

Let us return to the starting point: the expression of the field in Eq. (4.14). Note that the field at the point (x, y, z) is given by the Fourier transform of the product of two functions: $E(k_x, k_y, 0)$ and $\exp[i\gamma(k_x, k_y)z]$. We know that we may write it in the form of a convolution product of the two Fourier transforms. We have just derived the Fourier transform of $\exp[i\gamma(k_x, k_y)z]$. Hence, by using Eq. (4.17), we obtain:

$$E(x, y, z) = \frac{-1}{2\pi} \iint E(x', y', 0) \frac{\partial}{\partial z}\left[\frac{\exp(ik_0 R)}{R}\right] dx'dy', \tag{4.18}$$

where $R^2 = (x - x')^2 + (y - y')^2 + (z - 0)^2$. This expression, which encompasses a near-field term varying as $1/r^2$, is identical to the Kirchhoff–Sommerfeld formulation obtained using the Green theorem (Goodman, 2005) but is different from the original Huygens–Fresnel principle. The latter is recovered by doing two approximations: (i) neglecting the near-field term; (ii) using a paraxial approximation, replacing the z-derivative by $ik_0 \cos\theta$, where θ is the angle between the observation direction and the Oz-axis. Note that Eqs (4.14) and (4.18) are exact for any z, including the Fresnel region and the near field.

4.2.2 Far-field or Fraunhofer approximation, stationary phase approximation, and natural diffraction

Despite the presence of a Fourier transform in Eq. (4.14), the angular spectrum should not be confused with the far-field approximation (or Fraunhofer approximation). Starting from Eq. (4.18), we may derive an asymptotic form of the field diffracted by an aperture with a characteristic length L. This expression is valid at large distances $k_0 r \gg 1$.[2] The same asymptotic result may be derived directly from Eq. (4.14) using an asymptotic form of the integral, valid for large distances, given by the so-called stationary phase approximation. For integrals with the structure

$$U(x, y, z) = \iint a(m, p) \exp[ik_0(mx + py + qz)]dmdp, \tag{4.19}$$

it can be shown (Born, 1999; Jones, 1989) that:

$$U(x, y, z) \approx -\frac{2i\pi}{k_0}\frac{z}{r} a\left(\frac{x}{r}, \frac{y}{r}\right) \frac{\exp(ik_0 r)}{r}\ . \tag{4.20}$$

This form is often useful for calculating fields scattered at large distances. Here, we apply it to the asymptotic evaluation of the field $E(x, y, z)$ given by Eq. (4.14). We see that, in the far field, the amplitude $E(x, y, z)$ is proportional to the Fourier transform $E(k_x = k_0\frac{x}{r}, k_y = k_0\frac{y}{r}, 0)$.

We can now discuss the concept of natural diffraction. It is well known that a plane wave illuminating an aperture in a screen with a typical size L has an angular aperture on the order of λ/L. From our analysis, this turns out to be a very general property of propagating fields with a limited aperture L. Indeed, repeating the discussion made in section 2.4.2 for the directivity of antennas which was based on the Fourier transform structure of the radiated field by a current distribution with finite size L, we arrive at the same conclusion. It follows that any field propagating in vacuum with finite size has an angular divergence. This is often called natural diffraction. We emphasize that the key feature is not the presence of a screen or an edge but the fact that the field is spatially limited. As the energy of any physical beam is finite, it has a limited spatial transverse extension and therefore an angular divergence.

4.3 Propagation equation in homogeneous non-dispersive media

In order to study the propagation of electromagnetic waves in material media, we need to specify the media. We have introduced the constitutive relations in the previous chapter and we have discussed a few widely used models for different types of materials (metals, gases, plasmas, ionic crystals). As we have seen, the relation depends on the frequency. Indeed, it is well known that glass is transparent in the visible but opaque in the infrared, metals transmit X-rays but reflect visible light, etc. However, dispersion can be neglected if the frequency of the electromagnetic field is lower than the lowest absorption frequency, or, in other words, lower than any characteristic frequency of the material. This is usually a good approximation for radiowaves. In optics, there is always an absorption region, so that, according to Kramers–Kronig relations, there is always some dispersion.

[2] Note that the concept of far field is defined here by the condition $r \gg L^2/\lambda$, where L is a typical transverse extension of the field in the plane $z = 0$. Moreover, it is identical to the far-field concept defined in radiation theory. Note that the concept of near field is not defined as the opposite of far field. Near field is defined as the region where evanescent waves have a non-negligible amplitude. Thus, near field is confined to distances typically less than $\lambda/10$. A detailed introduction to near field is given in Chapter 10.

4.3.1 Model

We study propagation of electromagnetic waves in a medium which is assumed to be linear, homogeneous, and isotropic. We assume that the constitutive relations can be cast in the form:

$$\mathbf{D}(\mathbf{r},t) = \varepsilon_0\varepsilon_r\mathbf{E}(\mathbf{r},t), \tag{4.21}$$
$$\mathbf{j}(\mathbf{r},t) = \sigma\mathbf{E}(\mathbf{r},t), \tag{4.22}$$
$$\mathbf{B}(\mathbf{r},t) = \mu_0\mu_r\mathbf{H}(\mathbf{r},t), \tag{4.23}$$

where the dielectric permittivity or dielectric constant ε_r, the magnetic permeability μ_r, and the conductivity σ are constant real numbers.

4.3.2 Differential equation for E

Assuming a neutral material, we have $\nabla\cdot\mathbf{D} = 0$. Since the medium is homogeneous and isotropic, we have $\nabla\cdot(\varepsilon\mathbf{E}) = \varepsilon\nabla\cdot\mathbf{E} = 0$. Here, we exclude the possibility of a zero dielectric constant, so that we seek a transverse electric field. From Maxwell equations coupling **E** and **B**,

$$\nabla\times\mathbf{E} = -\frac{\partial\mathbf{B}}{\partial t}, \tag{4.24}$$

we derive:

$$\begin{aligned}\nabla\times(\nabla\times\mathbf{E}) &= -\frac{\partial}{\partial t}(\nabla\times\mathbf{B}),\\ \nabla(\nabla\cdot\mathbf{E}) - \nabla^2\mathbf{E} &= -\mu_0\mu_r\frac{\partial}{\partial t}\left(\sigma\mathbf{E} + \frac{\partial\mathbf{D}}{\partial t}\right),\\ \nabla^2\mathbf{E} - \frac{\varepsilon_r\mu_r}{c^2}\frac{\partial^2\mathbf{E}}{\partial t^2} &= \mu_0\mu_r\sigma\frac{\partial\mathbf{E}}{\partial t}.\end{aligned} \tag{4.25}$$

This equation is not a propagation equation because of the presence of the term proportional to the conductivity. It is called the 'telegrapher's equation. We will see later that this term accounts for losses. We recognize on the left hand side a propagation equation where the phase velocity is given by $v = \frac{c}{n} = \frac{c}{\sqrt{\varepsilon_r\mu_r}}$.

4.4 Propagation in homogeneous dispersive media and the Helmholtz equation

4.4.1 Model

We now turn to the more general case of local and dispersive media. The propagation equation in a non-magnetic medium is given by:

$$\nabla\times\nabla\times\mathbf{E}(\mathbf{r},t) = -\mu_0\left[\frac{\partial\mathbf{j}_{net}}{\partial t} + \frac{\partial^2\mathbf{D}(\mathbf{r},t)}{\partial t^2}\right]. \tag{4.26}$$

To avoid manipulating convolution products in the time domain (as in Eq. (3.2)) to relate $\mathbf{E}(\mathbf{r},t)$ and $\mathbf{D}(\mathbf{r},t)$, we introduce the Fourier transforms

$$\mathbf{E}(\mathbf{r},t)=\int_{-\infty}^{\infty}\frac{d\omega}{2\pi}\mathbf{E}(\mathbf{r},\omega)\exp(-i\omega t)\quad;\quad\varepsilon_r(t)=\int_{-\infty}^{\infty}\frac{d\omega}{2\pi}\varepsilon(\omega)\exp(-i\omega t)\quad;\quad\text{etc.}\tag{4.27}$$

Moving now to Fourier space, the convolution product Eq. (3.2) is replaced by a simple product:

$$\mathbf{j}(\mathbf{r},\omega)=\sigma(\omega)\mathbf{E}(\mathbf{r},\omega)\quad\text{and}\quad\mathbf{D}(\mathbf{r},\omega)=\varepsilon_0\varepsilon_r(\omega)\mathbf{E}(\mathbf{r},\omega).\tag{4.28}$$

4.4.2 Helmholtz equation: An equation in the frequency domain

Maxwell equations are modified when moving from $\mathbf{E}(\mathbf{r},t)$ to $\mathbf{E}(\mathbf{r},\omega)$. This transformation corresponds to two situations. Either one introduces a Fourier transform or one considers a system excited by a monochromatic source. In practice, it amounts to replacing a time derivative by a multiplication by $-i\omega$. For instance, we have

$$\nabla\times\mathbf{B}(\mathbf{r},\omega)=\mu_0\mathbf{j}(\mathbf{r},\omega)-i\omega\varepsilon_0\mu_0\varepsilon_r(\omega)\mathbf{E}(\mathbf{r},\omega).\tag{4.29}$$

It is important to realize that time derivatives cannot be used anymore, as $\mathbf{E}(\mathbf{r},\omega)$ is a function of ω and not of t. The telegrapher equation Eq. (4.25) becomes:

$$\nabla^2\mathbf{E}(\mathbf{r},\omega)+\varepsilon_r(\omega)\mu_r(\omega)\frac{\omega^2}{c^2}\mathbf{E}(\mathbf{r},\omega)=-i\omega\mu_0\mu_r(\omega)\sigma(\omega)\mathbf{E}(\mathbf{r},\omega).\tag{4.30}$$

It is remarkable to observe that the two last terms can be grouped to produce a Helmholtz equation:

$$\nabla^2\mathbf{E}(\mathbf{r},\omega)+\underline{\varepsilon}_r(\omega)\mu_r(\omega)\frac{\omega^2}{c^2}\mathbf{E}(\mathbf{r},\omega)=0,\tag{4.31}$$

where we have used the generalized permittivity $\underline{\varepsilon}_r(\omega)=\varepsilon_r(\omega)+\frac{i\sigma(\omega)}{\varepsilon_0\omega}$ introduced in Chapter 3. It accounts for both the net charges and the polarization charges. This is a key feature of the analysis in frequency domain. While current density and polarization are two different phenomena in the DC regime, the difference is no longer significant in the AC regime: under the action of the time-dependent field, both net charges and polarization charges are driven by the field and oscillate. Only their total effect can be observed. From the structure of the Helmholtz equation, it is seen that the different properties of matter appear through the group $\underline{\varepsilon}_r(\omega)\mu_r(\omega)$. This leads us to introduce the refractive index defined by $n^2(\omega)=\underline{\varepsilon}_r(\omega)\mu_r(\omega)$ and obtain:

$$\nabla^2\mathbf{E}(\mathbf{r},\omega)+n^2(\omega)\frac{\omega^2}{c^2}\mathbf{E}(\mathbf{r},\omega)=0.\tag{4.32}$$

We know that the permittivity is generally a complex number, so that the polarization may be out of phase with the electric field. Hence, the refractive index is complex in general, and can be cast in the form $n=n'+in''$. In this equation, the frequency ω has been introduced using a Fourier transform, so that it is real.

4.4.3 Propagation equation in reciprocal space

Here we repeat the derivation of the propagation equations in Fourier space. This section has two goals: (i) to emphasize the connection between the propagation equation and the dispersion relation; (ii) to emphasize the existence of longitudinal electromagnetic waves in material media as opposed to vacuum. The nabla operator is replaced by $i\mathbf{k}\times$, and the time derivative is replaced by $-i\omega$.

$$\begin{cases} \nabla\times\mathbf{E} = -\dfrac{\partial\mathbf{B}}{\partial t} \Rightarrow \mathbf{k}\times\mathbf{E} = \mu_0\mu_r\omega\mathbf{H}, \\ \nabla\times\mathbf{H} = \dfrac{\partial\mathbf{D}}{\partial t} \quad \Rightarrow \mathbf{k}\times\mathbf{H} = -\varepsilon_0\varepsilon_r\omega\mathbf{E}. \end{cases} \tag{4.33}$$

Eliminating the magnetic field gives:

$$\begin{aligned} \mathbf{k}\times(\mathbf{k}\times\mathbf{E}) &= -\mu_0\mu_r\varepsilon_0\varepsilon_r\omega^2\mathbf{E}, \\ \mathbf{k}(\mathbf{k}.\mathbf{E}) - k^2\mathbf{E} &= -\frac{\omega^2}{c^2}\varepsilon_r\mu_r\mathbf{E}. \end{aligned} \tag{4.34}$$

This equation can be split into two equations, one for the transverse part and one for the longitudinal part of the electric field. We take the dot product of Eq. (4.34) and $\mathbf{k}$ to derive the propagation equation for the longitudinal field defined in Eq. (1.11),

$$0 = -\frac{\omega^2}{c^2}\varepsilon_r(\omega)\mu_r(\omega)(\mathbf{k}.\mathbf{E}).$$

If the electric field has a component parallel to the $\mathbf{k}$ vector, $\mathbf{E}_\parallel \neq 0$, the condition $\varepsilon_r(\omega) = 0$ or $\mu_r(\omega) = 0$ needs to be fulfilled for non-zero frequencies. The dispersion relation for the longitudinal waves is thus given by

$$\varepsilon_r(\omega)\mu_r(\omega) = 0. \tag{4.35}$$

Another solution is possible if $\omega = 0$. This is nothing but the electrostatic field.
The *transverse* solutions $\mathbf{E}_\perp$ are characterized by $\mathbf{k}.\mathbf{E} = 0$. Eq. (4.34) becomes $\mathbf{E}_\perp[k^2 - \varepsilon_r(\omega)\mu_r(\omega)\frac{\omega^2}{c^2}] = 0$, and the dispersion relation for transverse fields is then given by:

$$k^2 = \varepsilon_r(\omega)\mu_r(\omega)\frac{\omega^2}{c^2}. \tag{4.36}$$

We stress that imposing $\nabla\cdot\mathbf{E} = 0$ *in a medium* is not a consequence of a null charge density. It is a choice of eliminating longitudinal solutions given by $\varepsilon_r = 0$ in order to keep transverse solutions.

4.5 Mode of the Helmholtz equation in a homogeneous medium

In this section, we discuss the main properties of the fields propagating in a medium. We define the homogeneous problem as the Helmhotz equation without sources. The solution of this problem is called a mode.

4.5.1 1D propagation

We seek a solution of Eq. (4.32) in the form of a plane wave $\mathbf{E} = E(x, \omega)\mathbf{e}_y$. We have:

$$\frac{\partial^2 E(x,\omega)}{\partial x^2} + n^2 \frac{\omega^2}{c^2} E(x,\omega) = 0. \tag{4.37}$$

The solution has the form:

$$E(x,\omega) = A\, e^{ikx} + B\, e^{-ikx}, \tag{4.38}$$

where the wavevector must satisfy the dispersion relation:

$$k^2 = n^2(\omega)\frac{\omega^2}{c^2}. \tag{4.39}$$

This is the general form of the dispersion relation for a linear, isotropic, and local medium. Let us consider a wave propagating only towards positive x. Then, the solution can be written as:

$$\begin{aligned} E(x,\omega) &= A\, e^{i(n'+in'')\frac{\omega}{c}x}, \\ &= A\, e^{in'(\omega)\frac{\omega}{c}x} e^{-n''(\omega)\frac{2\pi}{\lambda_0}x}. \end{aligned}$$

As $n^2 = \varepsilon_r$ for a non-magnetic medium, we need to select a root determination, $\pm[\varepsilon_r]^{1/2}$. Let us write the permittivity in the form $|\varepsilon_r|e^{i\phi_\varepsilon}$. We will show that the imaginary part of ε_r is always positive with the convention $\exp(-i\omega t)$. Hence, ϕ_ε is in the interval $[0,\pi]$. It follows that the refractive index is given by $n = \pm\sqrt{|\varepsilon_r|}\exp(i\phi_\varepsilon/2)$. We choose $n = \sqrt{|\varepsilon_r|}\exp(i\phi_\varepsilon/2)$, so that a wavevector $k_x = n\omega/c$ propagates towards positive x. Note that both the real part and the imaginary part of the refractive index are positive. We note that this may not be the case if we deal with a magnetic material such that $|\mu_r|e^{i\phi_\mu}$ differs from one. Here again, the imaginary part of μ_r is always positive in passive media with the convention $\exp(-i\omega t)$, so that ϕ_μ is in the interval $[0,\pi]$. It follows that the phase of the refractive index $[\phi_\varepsilon + \phi_\mu]/2$ belongs to the interval $[0,\pi]$. This makes it possible to have a refractive index with a negative real part![3]

4.5.2 The physical meaning of the real and imaginary parts of n and ε_r

The physical meaning of n' and n''. Let us emphasize that the complex refractive index does not appear in Maxwell equations. It has been introduced when deriving the Helmholtz equation. Hence, its physical meaning must be related to properties of the monochromatic waves which are solutions of the Helmholtz equation.

[3] Such a refractive index would lead to negative ray refraction according to Snell's law (see Chapter 18). This effect is never observed in optics with natural media because μ_r is generally close to one at optical frequencies. Diamagnetism exists at optical frequencies, but ferromagnetism has a typical cutoff frequency in the GHz range. However, it is now possible to build metamaterials with large artificial magnetic-like permeabilities at optical frequencies.

Table 4.1 Summary of physical meanings of real and imaginary parts of the refractive index and permittivity.

Symbol	Physical meaning
n'	phase velocity $v = c/n'$
n''	amplitude decay length $\lambda/(2\pi n'')$
ε_r	charge screening
ε_r''	absorbed power per unit volume $\omega\varepsilon_0\varepsilon_r''\|E\|^2/2$

- The real part n' is related to the phase velocity $v_\varphi = \frac{c}{n'(\omega)}$.
- The imaginary part n'' is related to the decay of the wave: $E \propto e^{-x/\delta}$. The characteristic decay length is

$$\delta = \frac{\lambda_0}{2\pi n''(\omega)},$$

where λ_0 is the wavelength in vacuum. This decay length can be due to different physical mechanisms: losses or reflection. We stress that in the case of a wave illuminating a metal, the exponential decay in the metal is mostly due to the fact that the wave cannot penetrate in the metal, so that the energy of the incident wave is reflected and not absorbed. The decay length is called skin depth for metals. In summary, the decay of a wave should not be confused with absorption, which is proportional to ε''.

We now discuss the order of magnitude of n''. Let us consider a refractive index given by $n = 1.5 + i10^{-3}$ at a wavelength of 500 nm. The imaginary part of the refractive index seems to be small as compared to the real part, and that might suggest that the decay is negligible and, therefore, that the medium is transparent. A closer inspection shows that the decay length is given by

$$\delta = \frac{\lambda_0}{2\pi n''(\omega)} \approx \frac{5\,10^{-7}}{10\,10^{-3}} \approx 5\,10^{-5}\,\mathrm{m} = 50\,\mu\mathrm{m}. \tag{4.40}$$

In other words, a thickness of 200 μm of such a material is almost opaque!

Physical meaning of ε' and ε''. While the permittivity and the complex refractive index are closely related for a non-magnetic material ($\varepsilon_r = n^2$), their physical meanings are very different. We first remind the reader that the permittivity accounts for the screening of net charges, as can be seen from Maxwell equation $\nabla \cdot \mathbf{E} = \rho/(\varepsilon_0\varepsilon_r)$. The same charge ρ produces a field divided by ε_r when going from a vacuum to a medium. The power absorbed per unit volume is proportional to the imaginary part of the permittivity, as shown in Eq. (3.12).

4.6 3D propagation in a lossy medium

We now consider the angular spectrum representation of a beam with finite section propagating in a lossy homogeneous medium. The procedure is the same as in section 4.1. The only difference is that the dispersion relation is now given by

$$\mathbf{k}^2 = n^2(\omega)\frac{\omega^2}{c^2}, \tag{4.41}$$

where ω is the real frequency introduced in the Helmholtz equation. Clearly, for a complex refractive index, a plane wave with real wavevector cannot be a solution. Repeating the derivation of section 4.1 using the Fourier transform along x and y with real k_x and k_y and remembering that the frequency is real for the Helmholtz equation, we find a z-component of the wavevector $\gamma^2 = n^2k_0^2 - k_x^2 - k_y^2$ and get a solution as a superposition of modes of the form:

$$\mathbf{E}\exp(ik_x x + ik_y y + i\gamma z - i\omega t), \tag{4.42}$$

where the choice of solution is given by the condition $Re(\gamma) + Im(\gamma) > 0$ (see Complement 4.B). Note that the propagation direction $(k_x, k_y, Re(\gamma))$ does not coincide with the decay direction (z-axis), since we stick to real k_x and k_y.

4.7 Quasinormal modes of uniform media

In the previous section, we studied the modes $\mathbf{E}(\mathbf{r}, \omega)$ of the Helmholtz equation with real frequency, since we considered a spectral analysis with temporal harmonics. Alternatively, we may seek a solution of time-dependent Maxwell equations without sources (homogeneous problem) with a prescribed real wavevector $\mathbf{k} = (k_x, k_y, k_z)$. To proceed, we make the ansatz $\mathbf{E}(\mathbf{r}, t) = \mathbf{E}\exp(i\mathbf{k}\cdot\mathbf{r})\exp(-i\tilde{\omega}t)$. For a lossy medium, this leads to the introduction of solutions with complex frequencies $\tilde{\omega}$. These modes will be called quasinormal modes in the book and are documented in Chapter 7, which discusses the modes of optical cavities and resonators. The physical meaning of the imaginary part of the frequency is simple to understand by observing that the field varies as $\mathbf{E}(\mathbf{r})\exp(-i\tilde{\omega}'t)\exp(-\tilde{\omega}''t)$, so that $1/(2\tilde{\omega}'')$ is the mode lifetime. Hence, the mode pattern decays exponentially in time everywhere in the same way. In Complement 4.C, we study the fields emitted by a monochromatic 1D source and show that the mode of the Helmholtz equation with complex wavevectors can be seen as a superposition of quasinormal modes. We then study the 1D real field emitted by a current with the form $I\delta(x)\delta(t)$ and show that it can be represented using either the quasinormal modes or the modes with complex wavevector.

4.8 Group, phase, and energy velocity

In this section, we will discuss different possible definitions for the propagation velocity of electromagnetic waves. In particular, we will address the issue of supraluminic propagation, its physical origin, and its connection with special relativity.

4.8.1 Special relativity and light velocity

Special relativity is based on two postulates:

- Principle of relativity: The laws of physics are the same in all inertial frames of reference.
- Invariance of c: Light propagation velocity is c in all inertial frames of reference.

Let us consider two events, (x, t) and $(x+\Delta x, t+\Delta t)$, in the inertial frame called the laboratory frame. In this frame, the event (x, t) happens before the event $(x+\Delta x, t+\Delta t)$ and can be considered to be its cause. According to Lorentz transformation for an observer in a frame with relative velocity v, the time interval between the two events in the observer frame is given by

$$\Delta t' = \gamma_{Lo}\left[\Delta t - \frac{v}{c^2}\Delta x\right], \tag{4.43}$$

where $\gamma_{Lo} = 1/\sqrt{1-v^2/c^2}$. Let us assume that the second event is triggered by a signal generated by the first event. Then the signal propagation velocity is defined by: $v_s = \frac{\Delta x}{\Delta t}$. Here, we do not make any assumptions about the type of interaction generating the information transfer. The time interval between the two events measured by the observer can be cast in the form:

$$\Delta t' = \gamma_{Lo}[\Delta t][1 - \frac{v\, v_s}{c^2}]. \tag{4.44}$$

If $v_s > c$, $\Delta t'$ can have the sign opposite to Δt. In other words, causality is not the same in both frames, in contradiction with the *relativity principle.*

Hence, a signal velocity larger than c is not compatible with the relativity principle. In particular, a light signal propagating in a medium cannot propagate faster than light in vacuum. In what follows, we will discuss different definitions of light velocity. Some of them may be larger than c. This does not mean that they are incorrect. It simply means that they do not correspond to a signal that can be used to exchange information.

4.8.2 Phase velocity

Let us consider a plane wave with an electric field:

$$\mathbf{E}(x, t) = \mathbf{E}_0 \exp[i(kx - \omega t)] = \mathbf{E}_0 \exp[ik(x - \frac{\omega}{k}t)].$$

The phase velocity is defined by considering the velocity of a plane of constant phase

$$v_\varphi = \frac{\omega}{c}.$$

The wavevector can be cast in the form

$$k = n(\omega)\frac{\omega}{c} = [n'(\omega) + in''(\omega)]\frac{\omega}{c}. \tag{4.45}$$

We observe that $v_\varphi = \frac{c}{n'(\omega)}$ is larger than c if the real part of the refractive index n' is smaller than one. This is, for example, the case for X-rays in any material (although it is very close to one) or for ultraviolet in metals.

4.8.3 Group velocity

The group velocity is defined as the velocity of the centre of a wave packet. Let us establish its connection to the refractive index. We consider a field given by $E_0(t)$

at $x = 0$ and propagating along positive x. We first introduce its Fourier transform $E(x,t) = \int E(x,\omega)e^{-i\omega t}\frac{\mathrm{d}\omega}{2\pi}$, so that $E(x,\omega)$ satisfies a Helmholtz equation:

$$\frac{\partial^2 E(x,\omega)}{\partial x^2} + n^2(\omega)\frac{\omega^2}{c^2}E(x,\omega) = 0. \tag{4.46}$$

The solution can be cast in the form $E(x,\omega) = Ae^{ik(\omega)x}$, where $k(\omega) = n(\omega)\omega/c$. For $x = 0$, $E(0,\omega) = E_0(\omega) = A(\omega)$, so that

$$E(x,t) = \int E_0(\omega)e^{i[k(\omega)x-\omega t]}\frac{\mathrm{d}\omega}{2\pi}. \tag{4.47}$$

We now assume that $E_0(\omega)$ has a narrow spectrum centred on ω_0. It follows that we can expand $\mathrm{Re}[k(\omega)] = k'(\omega)$ in the vicinity of ω_0. This is particularly possible if the field spectral width is smaller than any resonance width. Thus, we get:

$$k'(\omega) = k'(\omega_0) + (\omega-\omega_0)\left(\frac{\mathrm{d}k'}{\mathrm{d}\omega}\right)_{\omega_0} + (\omega-\omega_0)^2\left(\frac{\mathrm{d}^2k'}{\mathrm{d}\omega^2}\right)_{\omega_0} + \dots \tag{4.48}$$

By inserting the first two terms of the expansion in the solution, we get:

$$E(x,t) = \underbrace{e^{i[k'(\omega_0)x-\omega_0 t]}}_{\substack{\text{phase velocity}\\ v_\varphi = \frac{\omega_0}{k(\omega_0)}}} \underbrace{e^{-k''(\omega_0)x}}_{\text{decay}} \int \frac{\mathrm{d}\omega}{2\pi}\tilde{E}_0(\omega)\underbrace{e^{i[\left(\frac{\mathrm{d}k'}{\mathrm{d}\omega}\right)_{\omega_0}x-t](\omega-\omega_0)}}_{\substack{\text{group velocity}\\ v_g = \left(\frac{\mathrm{d}\omega}{\mathrm{d}k'}\right)}}.$$

The wave is given by the product of three terms. The first one accounts for the wave phase. The second term accounts for the decay of the amplitude. The third accounts for the envelope whose velocity is by definition the group velocity. The group velocity can be cast in a slightly different form:

$$v_g = \frac{1}{\frac{\mathrm{d}k'}{\mathrm{d}\omega}} = \frac{c}{\frac{\mathrm{d}(n'\omega)}{\mathrm{d}\omega}}, \tag{4.49}$$

$$= \frac{c}{n' + \omega\frac{\mathrm{d}n'}{\mathrm{d}\omega}} = v_\varphi\frac{1}{1+\frac{\omega}{n'}\frac{\mathrm{d}n'}{\mathrm{d}\omega}}. \tag{4.50}$$

As $\frac{\mathrm{d}n'}{\mathrm{d}\omega}$ can be negative, we observe that the group velocity can be either negative or tend to infinity or be zero. At a first glance, this seems to be in contradiction with the relativity principle. A wave packet is a light pulse. This is exactly how information is transmitted along optical fibres. However, a detailed analysis shows that there are no contradictions. This issue had been first discussed by Brillouin and Sommerfeld (Brillouin, 1960) shortly after the publication of the special relativity theory.

Before briefly addressing this issue, we note that the higher-order term $\frac{k''(\omega_0)}{2}(\omega-\omega_0)^2$ can be included in the expansion by introducing an effective form of the wavevector:

$$k'(\omega_0)_{eff} = k'(\omega_0) + \frac{k''(\omega_0)}{2}(\omega-\omega_0). \tag{4.51}$$

When accounting for this dispersion effect, we obtain a group velocity that depends on ω. Hence, different frequency bands of the wave packet have different group velocities and travel different distances over a given time interval. As a consequence, there is a broadening of the pulse over time, with spatial separation of different frequencies. This effect can be used notably to perform a spectral analysis of the pulse.

4.8.4 Signal velocity and relativity (Brillouin 1920)

The key idea of Brillouin and Sommerfeld amounts to carefully defining the time zero of a signal. The apparent contradiction between relativity and group velocity larger than c is based on the fact that the beginning of a wavepacket is not clearly defined. Let us consider, for instance, a Gaussian wavepacket. In the time domain, a Gaussian wave packet exists between $t = -\infty$ and $t = +\infty$ without being strictly zero; it is thus impossible to define exactly when the detector detects the signal. It is arbitrary to define the maximum of the wavepacket. To remove any ambiguity associated with the definition of a detection threshold, Brillouin studied the propagation of a signal that is strictly null for $t < 0$. An example of this is a sinusoidal signal starting at $t = 0$ given by $S(t) = H(t)\sin(\omega t)$, where $H(t)$ is the Heaviside step function.

It is possible to compute exactly the form of the signal transmitted by a medium using an integral superposition of the monochromatic incident waves. A numerical evaluation shows that the signal is, indeed, zero for $x > ct$. Brillouin (Brillouin, 1960) performed an asymptotic analysis showing that the field consists of (i) a signal called the precursor or forerunner that propagates at the light velocity in vacuum, whatever the medium; (ii) the rest of the signal which follows. There is a very simple physical explanation as to why the precursor propagates at velocity c in any medium. It is based on the fact that the transmitted wave is the sum of the incident field propagating at velocity c and the field radiated by the currents induced in the medium by the incident wave. Before the precursor reaches a given point, the charges are at rest and therefore do not radiate any field. In other words, the precursor propagates in a medium where charges do not yet play any role and which is therefore equivalent to a vacuum. This means that no material is truly opaque. Indeed, it takes some time to polarize the material. On that time scale, the medium is partially transparent. This is illustrated in Fig. 12.2, where a short pulse with a duration of 300 ns propagating through an ensemble of cold atoms is shown as a function of time. It is seen that the transmission factor is almost one during the first 30 ns, which is the time needed to polarize the atoms.

4.8.5 Transmission experiments, supraluminic propagation, and Wigner delay

It is possible to study the light velocity experimentally by measuring the time of flight of a transmitted light pulse. Depending on the system, it is possible to observe group velocities much smaller than c (slow light) or larger than c (supraluminic propagation). These experiments can be explained using the concept of group velocity or in terms of propagation duration by introducing a delay time. Let us introduce the transmission factor $t(\omega) = |t(\omega)|\mathrm{e}^{i\Phi(\omega)}$ by $E_t(x,\omega) = t(\omega)E_{inc}(x,\omega)$ for a monochromatic field. The transmitted field can be cast in the form:

$$E_t(t) = \int t(\omega)\tilde{E}_0(\omega)e^{-i\omega t}\frac{\mathrm{d}\omega}{2\pi}. \tag{4.52}$$

We now expand the phase of the transmission factor around the central frequency of the beam $\Phi(\omega) = \Phi(\omega_0) + \left[\frac{d\Phi}{d\omega}\right]_{\omega_0}(\omega - \omega_0)$ and assume that the amplitude does not vary, so that $t(\omega) \approx t(\omega_0)$:

$$E_t(t) = |t(\omega_0)|e^{i[\Phi(\omega_0) - \frac{d\Phi}{d\omega}\omega_0]} \int \tilde{E}_0(\omega) e^{-i\omega[t - \frac{d\Phi}{d\omega}]} \frac{d\omega}{2\pi}$$
$$= |t(\omega_0)|e^{i[\Phi(\omega_0) - \frac{d\Phi}{d\omega}\omega_0]} E\left[t - \left[\frac{d\Phi}{d\omega}\right]_{\omega_0}\right].$$

It is seen that the transmitted field has the same structure as the incident field but is translated in time by the so-called Wigner delay given by $\frac{d\Phi}{d\omega}$. Depending on the sign of this shift, the propagation is either slower or faster than c.

Let us give an example of a supraluminic propagation of a light pulse to understand why the group velocity may be larger than c. We consider a Fabry–Perot cavity made of two metallic layers with large reflectivity. Such a cavity can be tuned to have a very large transmission factor and a very small reflectivity factor. The high transmission is due to constructive interferences in transmission or, equivalently, destructive interferences in reflection. Let us consider a cavity tuned on resonance, such that the transmission is large. When studying the group velocity of a reflected pulse much longer than the distance d between mirrors, it is found that it is larger than c. In other words, the centre of the wavepacket travels faster than c.

To understand this effect, we further analyse the reflection as the pulse hits the first mirror. The beginning of the pulse for times smaller than $2d/c$ is reflected with a very high efficiency because there are no interferences yet. Light experiencing two internal reflections in the cavity with thickness d arrives with a time delay given by $2d/c$. After this time, destructive interferences can start or, in other words, energy starts being stored in the cavity. It follows that the beginning of the pulse is reflected, whereas the end of the pulse is transmitted. Hence, when measuring the reflected pulse, one measures a small part of the incident pulse, extracted from its beginning (see Fig. 4.4). This time selection results in an apparent supraluminic propagation.

4.8.6 Energy velocity

Energy velocity is defined from the energy flux. In a non-lossy medium, the energy velocity coincides with the group velocity $\mathbf{v}_g$; see, for instance, the derivation of the equivalence of the energy and group velocities for waveguides in Chapter 6. However, this is not always the case. In particular, for lossy dispersive media, they are different. The energy velocity can be defined by analogy with the drift velocity of conduction

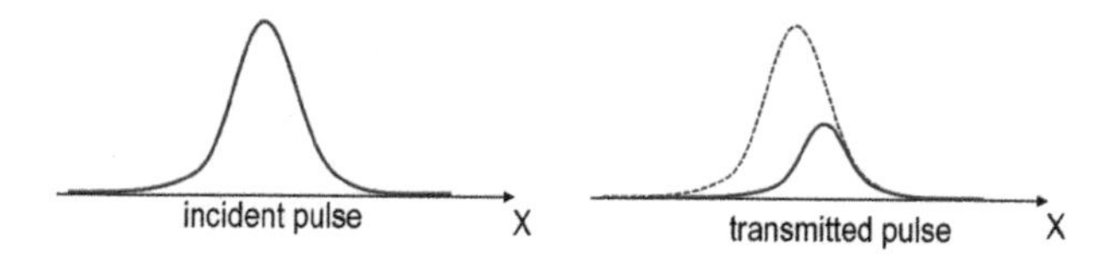

Fig. 4.4 Comparison of the incident pulse and the transmitted pulse. It is seen that the centre of the transmitted wavepacket appears to have a speed faster than c.

charges connected to the current density by $\mathbf{j}_q = \rho_q \mathbf{v}$. The time-average Poynting vector $\langle \mathbf{\Pi} \rangle$ plays the role of the current density and the time-average energy density $\langle u_E \rangle$ plays the role of the charge density. Thus, we define an energy velocity $\mathbf{v}_E$ by $\langle \mathbf{\Pi} \rangle = \langle u_E \rangle \mathbf{v}_E$. In other words, the energy velocity is given by:

$$\mathbf{v}_E = \frac{\langle \mathbf{\Pi} \rangle}{\langle u_E \rangle}.$$

Note that this velocity decays abruptly close to material resonances as the energy density increases (see also Chapter 18).

4.A Complement: Weyl expansion

The purpose of this section is to derive the Weyl formula, which gives a representation of a spherical wave as an angular spectrum. The starting point is the Fourier representation of the Green function (see Complement 2.A):

$$G(\mathbf{r}, \omega) = \frac{\exp(ik_0 r)}{4\pi r} = \int \frac{\mathrm{d}^3\mathbf{k}}{8\pi^3} \frac{\exp(i\mathbf{k} \cdot \mathbf{r})}{\mathbf{k}^2 - k_0^2}. \tag{4.53}$$

To proceed, we integrate over k_z, noting that $\mathbf{k}^2 - k_0^2 = (k_z - \gamma)(k_z + \gamma)$, where γ is defined in Eq. (4.9):

$$G(\mathbf{r}, \omega) = \int \frac{\mathrm{d}k_x \mathrm{d}k_y}{4\pi^2} \exp(ik_x x + ik_y y) \int_{-\infty}^{\infty} \frac{\mathrm{d}k_z}{2\pi} \frac{\exp(ik_z z)}{(k_z - \gamma)(k_z + \gamma)}. \tag{4.54}$$

For positive values of z, we integrate over k_z using the residue theorem and choosing a contour that runs along the real axis and is closed by a semicircle in the upper half-space. Introducing losses ($k_0(1 + i\eta)$), the pole γ moves to the upper half-space so that the contour encloses one pole. Finally, we find, after taking the limit $\eta \to 0$,

$$\frac{\exp(ik_0 r)}{r} = 2i\pi \int \frac{\mathrm{d}k_x \mathrm{d}k_y}{4\pi^2} \exp(ik_x x + ik_y y) \frac{\exp(i\gamma z)}{\gamma}. \tag{4.55}$$

A similar derivation yields for negative values of z:

$$\frac{\exp(ik_0 r)}{r} = 2i\pi \int \frac{\mathrm{d}k_x \mathrm{d}k_y}{4\pi^2} \exp(ik_x x + ik_y y) \frac{\exp(-i\gamma z)}{\gamma}, \tag{4.56}$$

so that the result can be cast in the form

$$\frac{\exp(ik_0 r)}{r} = 2i\pi \int \frac{\mathrm{d}k_x \mathrm{d}k_y}{4\pi^2} \exp(ik_x x + ik_y y) \frac{\exp(i\gamma |z|)}{\gamma}. \tag{4.57}$$

4.B Complement: Choice of the square-root solution

When studying the propagation in a lossy medium, the z-component of the wavevector is complex: $\gamma = \gamma' + i\gamma''$. It is given by $\gamma^2 = z = \varepsilon k_0^2 - k_x^2$, where z is a complex number. In this section, we discuss how to define the square root of a complex number, z. This can be written as $z = |z| \exp(i\theta)$. It follows that its square root can be

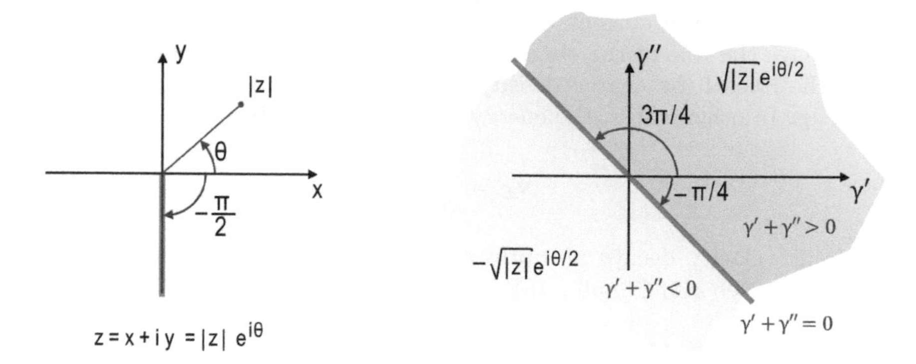

Fig. 4.5 Choice of a determination for the definition of γ, solution of $z = \gamma^2$. Left figure: Introduction of a cut in the complex plane $z = x + iy$. Right figure: Complex plane γ. It is seen that the choice $\gamma = +\sqrt{|z|}\exp(i\theta/2)$ corresponds to the zone $\gamma' + \gamma'' > 0$.

defined as $\pm\sqrt{|z|}\exp(i\theta/2)$. It is seen that if θ is increased by 2π, z is unchanged, but $\pm\sqrt{|z|}\exp(i\theta/2)$ is multiplied by $\exp(i\pi) = -1$, so that it changes sign and we end up with two possible values of the function for a given variable z. In order to have a well-defined function, this rotation is forbidden by introducing a cut in the complex plane $z = x + iy$, where the function is defined. We use as a cut the half-axis $y < 0$ or in other terms, the line corresponding to $\theta = -\pi/2$ so that the angle θ varies between $-\pi/2$ and $3\pi/2$.[4] The origin of the cut is called branch point. As a consequence, the value of γ has an argument (see Fig. 4.5) that varies between $-\pi/4$ and $3\pi/4$ if we choose $\gamma = \sqrt{|z|}\exp(i\theta/2)$ and between $3\pi/4$ and $7\pi/4$ if we choose $\gamma = -\sqrt{|z|}\exp(i\theta/2) = \sqrt{|z|}\exp(i(\theta/2 + \pi))$. The choice is made in such a way that if we consider a real permittivity, a real frequency, and a real wavevector, the conditions $\gamma' > 0$ and $\gamma'' > 0$ that had been introduced for the non-lossy case are recovered. Hence, according to Fig. 4.5, the choice is $\gamma = \sqrt{|z|}\exp(i\theta/2)$. With the cut that we have chosen, this determination corresponds to the half-plane $\gamma' + \gamma'' > 0$. We stress that the condition used for real quantities ($\gamma' > 0$ and $\gamma'' > 0$) can still be used in the particular case of a real frequency, a real wavevector, and a complex permittivity, given that for a lossy medium, $\varepsilon'' > 0$. Yet, when searching modes with complex frequencies or with complex wavevectors, this simple condition cannot be used any longer, and we have to use $\gamma' + \gamma'' > 0$. If the form of the cut in the (k_x', k_x'') plane needs to be found out, one has to solve the equation $z = \varepsilon k_0^2 - k_x^2 = iy$, where y takes any real negative value.

4.C Complement: Modes of the propagation equation in a lossy homogeneous medium

The purpose of this section is to introduce the modes of the *propagation* equation. As opposed to the Helmholtz equation, we do not assume that the system is driven

[4] Other choices are possible. This choice turns out to be convenient.

by a monochromatic source imposing a $\exp(-i\omega t)$ dependence with real ω. Since the fields are real, we are searching real functions of $\mathbf{r}$ and t in a source-free setting. For a lossy medium, it will be seen that the modes decay in time. This is similar to the modes of a lossy cavity, a RLC circuit, or a damped mechanical harmonic oscillator. The corresponding modes are described by a complex frequency whose imaginary part characterizes the time decay. These modes will be called quasinormal modes in the book and are documented in Chapter 7. The ensemble of solutions is the frequency spectrum of the system. We first seek the quasinormal modes of the telegrapher equation as a simple example. We then generalize the procedure to arbitrary dispersive media.

4.C.1 Quasinormal modes of the telegrapher equation

The telegrapher equation deals with the propagation of an electric field in a 1D non-magnetic material with real positive permittivity ε_r and conductivity σ. The source-free equation for a scalar component of the electric field can be written in the form:

$$\frac{\partial^2 E(x,t)}{\partial x^2} - \frac{\varepsilon_r}{c^2}\frac{\partial^2 E(x,t)}{\partial t^2} - \mu_0\sigma\frac{\partial E(x,t)}{\partial t} = 0. \tag{4.58}$$

We seek real solutions with separate variable dependence of the form $f(x)g(t)$. Inserting this ansatz in the partial differential equation yields:

$$\frac{1}{f(x)}\frac{\partial^2 f(x)}{\partial x^2} = \frac{\varepsilon_r}{c^2}\frac{1}{g(t)}\frac{\partial^2 g(t)}{\partial t^2} + \mu_0\sigma\frac{1}{g(t)}\frac{\partial g(t)}{\partial t}. \tag{4.59}$$

The function on the left-hand side depends on x, while the function on the right-hand side depends on t, so that both functions must be equal to some constant that we denote by $-k^2$. The function $f(x)$ obeys:

$$\frac{\partial^2 f(x)}{\partial x^2} + k^2 f(x) = 0. \tag{4.60}$$

Without imposing specific boundary conditions, there is no unique solution. We consider a translationally invariant medium and seek a solution of the form $E_0\exp(ikx)$, where k is real and chosen to be positive here. We now consider the equation obeyed by $g(t)$, keeping in mind that the product $f(x)g(t)$ is real:

$$\frac{\varepsilon_r}{c^2}\frac{\partial^2 g(t)}{\partial t^2} + \mu_0\sigma\frac{\partial g(t)}{\partial t} + k^2 g(t) = 0. \tag{4.61}$$

This is a linear differential equation of second order with constant coefficients. It is convenient to search a solution of the form $\exp(-i\tilde{\omega}t)$, where $\tilde{\omega}$ can be complex. Note that $\tilde{\omega}$ is not introduced through a Fourier transform or the modelling of a stationary monochromatic excitation but as a mathematical technique to find the form of the solution. We find the characteristic equation:

$$-\varepsilon_r\frac{\tilde{\omega}^2}{c^2} - i\mu_0\tilde{\omega}\sigma + k^2 = 0. \tag{4.62}$$

The corresponding frequency is thus a complex function of k denoted by $\tilde{\omega}(k)$. We choose the solution corresponding to a decay in time. In summary, the solution of the propagation equation without sources is the real part of the quasinormal modes given by

$$E_0 \exp\left[ikx - i\tilde{\omega}(k)t\right] \quad ; \quad \tilde{\omega}(k) = \frac{ck}{\sqrt{\varepsilon_r}}\sqrt{1 - \frac{\sigma^2}{4\varepsilon_r \varepsilon_0^2 c^2 k^2}} - i\frac{\sigma}{\varepsilon_0 \varepsilon_r}.$$

4.C.2 Quasinormal mode in a dispersive lossy medium

We now generalize the procedure outlined above to the case of an arbitrary dispersive medium characterized by a complex permittivity $\varepsilon(\omega)$, and we consider a translationally invariant medium. The propagation equation given by Eq. (4.26) involves convolution products and is not practical. To benefit from the simple structure of the relation between $\mathbf{D}$ and $\mathbf{E}$ in frequency domain, we introduce the ansatz $E(x,t) = E_0 \exp(ikx - i\tilde{\omega}t)$, where k is real to account for the translational invariance and $\tilde{\omega}$ is a complex frequency. As in the previous section, we seek a complex frequency $\tilde{\omega}$, the solution of the dispersion relation:

$$-k^2 + \varepsilon_r(\tilde{\omega})\frac{\tilde{\omega}^2}{c^2} = 0, \tag{4.63}$$

where $\varepsilon(\tilde{\omega})$ is the analytic continuation of the function $\varepsilon(\omega)$ in the complex-frequency plane. The continuation is unique, implying that $\varepsilon(\tilde{\omega})$ is unambiguously defined, although there is no simple physical meaning of $\varepsilon(\tilde{\omega})$ for complex frequencies.

4.C.3 Quasinormal mode (pole) expansion of the field radiated by a localized monochromatic source

Always keeping a 1D analysis for simplicity, we consider a monochromatic current source at frequency ω in the plane $x = 0$ radiating in an otherwise homogeneous lossy system with a complex permittivity ε. The source intensity per unit length along the y-axis is denoted by I in Am^{-1}. The system can be modelled by the partial differential equation

$$\frac{\partial^2 E(x,\omega)}{\partial x^2} + \varepsilon\frac{\omega^2}{c^2}E(x,\omega) = -i\mu_0\omega I\delta(x), \tag{4.64}$$

and we use the radiation condition at infinity.

Solving this equation amounts to calculating the monochromatic Green function in 1D. This is a standard problem. What we intend to do here is less standard. We will expand the Green function in the basis of the quasinormal modes with complex frequencies, therein connecting the actual solution of inhomogeneous problems to those of the related homogeneous ones.[5]

[5] Mathematically, this refers to pole expansions of meromorphic functions: see Chapter 11 in ref. (Arfken et al., 2013). The latter are analytic functions that have only isolated poles as singularities. Mittag-Leffler showed that, instead of making an expansion about a single regular point (a Taylor expansion) or about an isolated singular point (a Laurent expansion), it is also possible to make a completely different expansion, for which each term arises from a different pole. In 1D and uniform media (ε is independent of x), there are no branch cuts and the Green function is meromorphic.

In order to find the solution of this problem, it is convenient to introduce the spatial Fourier transform of the field

$$E(x,\omega) = \int \frac{\mathrm{d}k}{2\pi} E(k,\omega)\exp(ikx). \tag{4.65}$$

Inserting this representation in the differential equation yields

$$E(k,\omega)[-k^2 + \varepsilon_r \frac{\omega^2}{c^2}] = -i\mu_0\omega I. \tag{4.66}$$

The partial differential equation has been converted into an algebraic equation. For a permittivity with a non-zero imaginary part, there is no singularity and the solution can be written in the form

$$E(k,\omega) = \frac{-i\mu_0\omega I}{-k^2 + \varepsilon_r \frac{\omega^2}{c^2}}. \tag{4.67}$$

We now take the Fourier transform to obtain the solution in space

$$E(x,\omega) = \int \frac{\mathrm{d}k}{2\pi} \exp(ikx)E(k,\omega) = \int \frac{\mathrm{d}k}{2\pi} \exp(ikx)\frac{i\mu_0\omega I}{k^2 - \varepsilon_r \frac{\omega^2}{c^2}}. \tag{4.68}$$

It is important to note that this solution is a pole expansion over the modes $\exp(ikx)$ with real k's (the modes expand over the whole space), which are associated with a complex eigenfrequency. For each k, between the two possibilities $\tilde{\omega}(k) = \pm ck/(\varepsilon_r)^{1/2}$, we select the pole with a negative imaginary part, so that $\exp(-i\tilde{\omega}t)$ is exponentially damped in time. Eq. (4.68) is rewritten as

$$E(x,\omega) = \int \frac{\mathrm{d}k}{2\pi} \exp(ikx)\frac{C(k)}{\omega - \tilde{\omega}(k)}, \qquad \text{where} \qquad C(k) = \frac{-i\mu_0\omega I c^2}{\varepsilon_r(\omega + \tilde{\omega}(k))}, \tag{4.69}$$

to emphasize the pole contribution. $\frac{C(k)}{\omega - \tilde{\omega}(k)}$ is called the (resonant) excitation coefficient.

The integral over k can be performed using the residue theorem. For positive x, the contour chosen is a half-semicircle with positive imaginary part and there is a single surrounded complex pole $\tilde{k} = \sqrt{\varepsilon_r}\omega/c$ with a positive imaginary part. Finally, we find for positive x:

$$E(x,\omega) = \frac{-\mu_0\omega I}{2\tilde{k}} \exp(i\tilde{k}x), \qquad \text{with} \qquad \tilde{k} = \sqrt{\varepsilon_r}\omega/c. \tag{4.70}$$

For negative x, the pole is given by $\tilde{k}(\omega) = -\sqrt{\varepsilon_r}\omega/c$ and a similar expression is obtained; the Green function is an even function. Hence, we see that the field radiated by a monochromatic source in a lossy medium appears to be given by the modes of the Helmholtz equation with complex wavevector $\exp\left(i\frac{n\omega}{c}x\right)$ for positive x and $\exp\left(-i\frac{n\omega}{c}x\right)$ for negative x, so that the field decays away from the source. The same field can also be expanded over quasinormal modes with real wavevectors, as seen in Eq. (4.68).

4.C.4 Field radiated by a time-dependent surfacic current

We now consider a localized surface current source with arbitrary time dependence. We will see that the solution can be expanded using either quasinormal modes $\exp(ikx - i\omega(\tilde{k})t)$ with a complex frequency or modes of the Helmholtz equation with real frequency $\exp(i\tilde{k}x - i\omega t)$. To illustrate this idea, we compute the field radiated by a pulse current given by $I(t) = I\delta(t)$. We first expand the current source using a Fourier transform:

$$I(t) = \int \frac{\mathrm{d}\omega}{2\pi} I(\omega) \exp(-i\omega t). \tag{4.71}$$

For the particular case of a short pulse of current $I(t) = I_0\delta(t)$, we have $I(\omega) = 2\pi I_0$. Using Eq. (4.68), we then write the radiated field as a linear superposition:

$$E(x,t) = \int \frac{\mathrm{d}\omega}{2\pi} \int \frac{\mathrm{d}k}{2\pi} \exp(ikx - i\omega t) E(k,\omega) = \int \frac{\mathrm{d}k}{2\pi} \exp(ikx - i\omega t) \frac{i\mu_0 \omega I(\omega)}{k^2 - \tilde{k}^2(\omega)}. \tag{4.72}$$

This equation can be cast in a slightly different form displaying the poles $\tilde{\omega}(k)$:

$$E(x,t) = \int \frac{\mathrm{d}\omega}{2\pi} \int \frac{\mathrm{d}k}{2\pi} \frac{-i\mu_0 c^2 \omega I(\omega)}{n^2} \frac{\exp(ikx - i\omega t)}{\omega^2 - \tilde{\omega}^2(k)}. \tag{4.73}$$

In these integrals, the two variables k and ω were introduced using a Fourier transform, so that they were real. Upon integration over frequencies using the residue theorem, we get for positive t:

$$E(x,t) = \int \frac{\mathrm{d}k}{2\pi} \frac{\mu_0 c^2 \pi I_0}{n^2} \exp[ikx - i\tilde{\omega}(k)t]. \tag{4.74}$$

Similarly, we can choose to integrate using the residue theorem over wavevector k for positive x. The result comes from the contribution of the pole $\tilde{k}(\omega) = n\omega/c$:

$$E(x,t) = \int \mathrm{d}\omega \frac{-\mu_0 c \pi I_0}{n} \exp[i\tilde{k}(\omega)x - i\omega t]. \tag{4.75}$$

Eq. (4.74) shows that the field $E(x,t)$ produced by a Dirac source is well described by a sum of quasinormal modes $\exp[ikx - i\tilde{\omega}(k)t]$, each of them having its own decay time $1/\omega''(k)$. The long time behaviour is thus dominated by the modes with long decay times. Alternatively, Eq. (4.75) shows that the field can be described as a superposition of the modes of the Helmholtz equation $\exp[i\tilde{k}(\omega)x - i\omega t]$.

Exercises

Exercise 4.1 Decay length. We consider two different slabs. Compare the damping of a wave propagating through these slabs with $\lambda = 600$ nm.

(1) A lossy glass with $n = 1.5 + i0.001$ and a thickness $t = 10^{-3}$ m.

(2) A silver film with $n = 0.05 + i4$ and a thickness $t = 10\,\mathrm{nm}$.

Exercise 4.2 Skin depth of a metal: Radiowave and optical regimes. The skin depth characterizes the exponential decay of a wave in a metal. We use the Drude model (Eq. (3.34)) and consider two limit cases: (i) the optical regime $\omega \gg \nu$ and (ii) the radiowave regime $\omega \ll \nu$. Show that the skin depth in the optical and IR regimes does not depend on the frequency and is given by $\lambda_p/2\pi = c/\omega_p$. Show that the skin depth in the low-frequency regime is given by $\sqrt{\mu_0 \sigma \omega}$, where we have used the relation $\nu = \varepsilon_0 \omega_p^2/\sigma$.

Exercise 4.3 Wave attenuation: Absorption or reflection? In this exercise, we examine the physical mechanism responsible for the decay of a wave propagating in a medium. We consider a monochromatic plane wave propagating along the x-axis. The field can be written as $E_0 \exp(ikx - i\omega t)$ and is linearly polarized along the y-axis. The medium is non-magnetic (i.e. $\mu_r = 1$) and has a refractive index, n.

(1) What is the form of k imposed by the dispersion relation for a transverse wave?

(2) We consider two media. Medium 1 has a refractive index given by $n = 1.5 + i0.01$, corresponding to a lossy dielectric. Medium 2 has a purely imaginary refractive index given by $2i$, corresponding to a non-lossy metallic case. What is the decay length for the field amplitude for both cases?

(3) What is the power absorbed per unit volume for both cases?

(4) What is the flux of the Poynting vector for both cases?

(5) Check that the local form of energy conservation is satisfied in both cases. It appears that for a lossy glass, losses are responsible for the decay. For a non-lossy metal, the origin of the decay is due to reflection. Direct calculations are given in Exercise 4.4 for a thin film and in Exercise 12.1 for a half-space.

Exercise 4.4 Physical origin of the refractive index. We study the light transmitted and reflected by a thin film illuminated at normal incidence by a linearly polarized plane wave. The thin film contains a linear, homogeneous, isotropic material between the planes $z = 0$ and $z = -d$. The thickness d is assumed to be sufficiently small, so that the field can be considered to be homogeneous. The material is non-magnetic, so that its permeability is μ_0. Its susceptibility is denoted by χ. The incident electric field has the form

$$E_0 \, \mathbf{e}_y \, \exp(-i\frac{\omega}{c}z - i\omega t).$$

The radiated vector potential by a monochromatic current density $\mathbf{j}(\mathbf{r})$ induced by the incident field is given by:

$$\mathbf{A}(\mathbf{r}) = \frac{\mu_0}{4\pi} \int \mathbf{j}(\mathbf{r}') \frac{\exp\left[i\frac{\omega}{c}|\mathbf{r} - \mathbf{r}'|\right]}{|\mathbf{r} - \mathbf{r}'|} \, d^3\mathbf{r}'.$$

(1) Explain why the far-field approximation cannot be used when dealing with an infinite planar source along x and y.

(2) The current density in the film is given by $\mathbf{j} = -i\omega\chi\varepsilon_0\mathbf{E}$, where $\mathbf{E}$ is the field in the film. For a very thin film, the electric field inside the film is given by the incident field. Derive the vector potential $\mathbf{A}$ using the formula

$$\int \frac{\exp\left[i\frac{\omega}{c}|\mathbf{r}-\mathbf{r}'|\right]}{|\mathbf{r}-\mathbf{r}'|}\, dx'\, dy' = \frac{2i\pi c}{\omega}\exp\left[i\frac{\omega}{c}|z-z'|\right]. \tag{4.76}$$

(3) Noting that $\mathbf{A}$ is a plane wave, derive the electric field E_1 radiated by the film for $z < -d$ and for $z > 0$ if $d \ll \lambda$, where $\lambda = 2\pi c/\omega$. Give the result to the lowest order in d/λ.

(4) From the linearity of Maxwell equations, the field for $z > 0$ is the linear superposition of the field radiated by the sources (the incident field) and the field radiated by the induced currents in the thin film. Show that depending on the phase of the complex susceptibility χ, the field radiated by the film can be in quadrature or out of phase with the incident field. Compare the case of a non-lossy dielectric (real and positive χ) and a non-lossy metal (real and negative χ).

(5) Both the incident field and the field radiated by the film propagate with a phase velocity, c. We consider a dielectric with positive real permittivity. Show that the phase of the field $E_0 + E_1$ can be modelled by assuming that the phase velocity in the medium is $c/\sqrt{\chi+1}$.

Exercise 4.5 Talbot effect. We consider the field transmitted by a periodic array of slits in the plane $z = 0$, translationally invariant along the y-axis, illuminated at normal incidence by a monochromatic plane wave. The field in the plane $z = 0$, behind the screen is assumed to be given by $E(x)\exp(-i\omega t) = E_0 t(x)\exp(-i\omega t)$, where $t(x)$ is a periodic function with period d, $t(x+d) = t(x)$. In the interval $[0, d]$, we have $t(x) = 1$ if $x > d/2$ and $t(x) = 0$ if $x \leq d/2$.

(1) Show that the angular spectrum contains a discrete set of plane waves of the form $E_0\exp(in\frac{2\pi}{d}x + i\gamma_n z)$, where $\gamma_n = \sqrt{\frac{\omega^2}{c^2} - n^2\frac{4\pi^2}{d^2}}$.

(2) We consider propagation close to the z-axis, so that $\frac{\omega}{c} \gg \frac{2\pi}{d}$. Give an approximate form of γ_n.

(3) Show that the field in the planes given by $z = p\frac{d^2}{\lambda}$, where p is an integer reproduces the field in the plane $z = 0$ apart from the contribution of the evanescent waves. In other words, the system has the property of self-imaging in some particular planes.

Exercise 4.6 Light speed of a guided field. In this exercise, we study the light speed of a guided wave.

(1) We consider a waveguide consisting of two perfectly reflecting mirrors at $y = \pm b/2$. Check that the electric field given by $\mathbf{E}(y, z, t) = E_0\mathbf{e}_x\cos(ky)\exp(i\gamma z - \omega t)$ is a solution of the propagation equation and satisfies the boundary conditions, provided that $k = (2p+1)\pi/b$ where p is an integer and $\gamma = \sqrt{\omega^2/c^2 - k^2}$.

(2) Show that the phase velocity for $p = 0$ is $c/\sqrt{1 - \pi^2c^2/(\omega b)^2}$, the group velocity v_g is given by $c\sqrt{1 - \pi^2c^2/(\omega b)^2}$, and the energy velocity $< \Pi_z > / < u >= v_g$.

(3) Give a physical explanation of the fact that the guided mode has a velocity smaller than c in vacuum.

Exercise 4.7 Light speed of a non-diffracting beam. In this exercise, we introduce the concept of the non-diffracting beam. We study its domain of validity and its speed.

(1) We consider an axicon consisting of a conically shaped lens with axial revolution. All incident rays parallel to the optical axis are deflected and form an angle α with the optical axis. Upon transmission at normal incidence by an axicon, an incident plane wave is modelled by $E(k_x, k_y, z = 0) = E_0\delta(k_\parallel - k_0 \sin\alpha)$, where $k_\parallel = \sqrt{k_x^2 + k_y^2}$. Show that the field in a plane $z > 0$ is given by

$$E(x, y, z) = \frac{kE_0}{2\pi} J_0(k_0\rho \sin\alpha) \exp(ik_0 z \cos\alpha),$$

where $\rho = \sqrt{x^2 + y^2}$ and J_0 is the Bessel function or order zero. Use the formula $J_0(x) = \frac{1}{2\pi}\int_0^{2\pi} \exp(ix\cos\phi)d\phi$, and note that $k_x x + k_y y = k_\parallel \rho \cos\phi$.

(2) Check that the transverse structure of the field does not depend on z, so that the field is not diffracting.

(3) The non-diffracting property is in contrast with the discussion of section 4.2.2. This can be explained by realizing that a Bessel beam cannot be produced, as it carries an infinite power. To check this, use the asymptotic form of the Bessel function $J_0(x) \approx \sqrt{\frac{2}{\pi x}} \cos (x - \frac{\pi}{4})$ and assume that the energy flux is proportional to the integral of the square of the electric field.

(4) Show that the phase and group velocity of the beam are given by $c/\cos\alpha$. Compare with the plane-wave light velocity in vacuum. To check that the energy velocity is smaller than c, we consider, for the sake of simplicity, a 2D axicon and a linearly polarized electric field such that $E_x = E_0 \cos(k_0 y \sin\alpha) \exp(ik_0 z \cos\alpha)$. Check that the energy velocity (averaged over y) is given by $c\cos\alpha$. Compare with the waveguide case studied in Exercise 4.6.

5
Reflection and refraction at an interface

5.1 Introduction

In this chapter, we deal with reflection and refraction of electromagnetic waves at a planar interface between two homogeneous media described by a linear and isotropic permittivity and permeability. We will derive the general form of the fields in both media and derive the forms of their complex amplitudes, namely the Fresnel reflection and transmission factors.

Before proceeding to derive these laws and discuss their physical content, we start with a qualitative discussion of the physical phenomenon. The basic idea is that reflection and refraction can be described as a scattering process, as already discussed in Exercises 2.5 and 2.6. The reflection and transmission factor at an interface can also be computed from this point of view (see Exercise 12.1). Let us consider the simple case of an incident monochromatic wave propagating in vacuum and impinging on the interface with medium 2. The electric and magnetic fields of the incident wave induce forces on the charges in the medium. It follows that current densities (or equivalently time-dependent polarization and possibly magnetization for magnetic materials) are induced in the medium. In turn, these induced time-dependent current densities radiate waves at the frequency of the incident fields. These radiated fields produce the so-called reflected field in vacuum and two fields in medium 2: a first field that cancels the incident field in the medium and a second field that is called the refracted field. This mechanism corresponds to a powerful integral formalism that will be introduced in Chapter 12 and can be used for arbitrary geometries. Here, we limit the discussion to a simple plane interface. In that case, a different approach based on partial differential equations and boundary conditions is sufficient to solve the problem. We consider only incident plane waves. The case of any finite-sized beam can be recovered by linear superposition using the angular spectrum. An example of finite beam size effect is discussed in Exercise 5.6.

In what follows, we use complex refractive indices in order to include the losses in the discussion. It follows that the usual Snell law, $n_1 \sin\theta_1 = n_2 \sin\theta_2$, would require complex angles. As the physical meaning of such a complex angle is not well defined, we will avoid using angles and instead use the Cartesian components of the wavevectors. This has the great advantage of providing a single formalism that can handle dielectric and metals and lossy and non-lossy materials.

Let us consider the interface shown in Fig. 5.1. The electric field $\mathbf{E}$ satisfies a partial differential equation given by:

Introduction to Nanophotonics. Henri Benisty, Jean-Jacques Greffet, and Philippe Lalanne, Oxford University Press.
 DOI: 10.1093/oso/9780198786139.003.0005

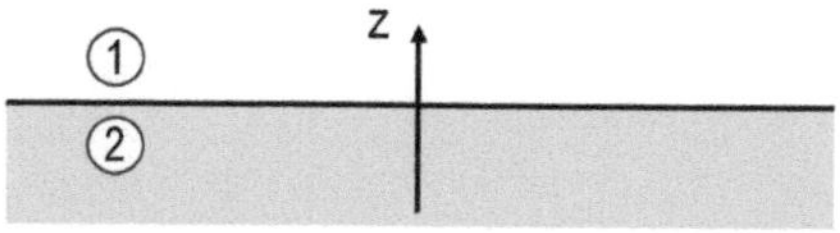

Fig. 5.1 Interface separating medium 1 ($z > 0$) from medium 2 ($z < 0$).

$$\nabla^2 \mathbf{E} + \varepsilon_r(\omega, z)\frac{\omega^2}{c^2}\mathbf{E} = 0, \tag{5.1}$$

which can be cast as two different equations:

$$\nabla^2 \mathbf{E}_1 + \varepsilon_{r1}(\omega)\frac{\omega^2}{c^2}\mathbf{E}_1 = 0 \quad z > 0, \tag{5.2}$$

$$\nabla^2 \mathbf{E}_2 + \varepsilon_{r2}(\omega)\frac{\omega^2}{c^2}\mathbf{E}_2 = 0 \quad z < 0. \tag{5.3}$$

This system of equations can be solved accounting for

- boundary conditions for $z \to \pm\infty$,
- continuity conditions at $z = 0$.

5.2 Continuity conditions

The continuity conditions can be derived from Maxwell equations. Let us first introduce some notations: the unit vector $\mathbf{n}$ normal to the interface coincides with the unit vector $\mathbf{e}_z$. The interface is the plane $z = 0$ which separates media 1 and 2 with dielectric constants ε_1 and ε_2. The dielectric constant is thus discontinuous at the interface. In what follows, we consider that discontinuous functions are limits of continuous functions varying very rapidly close to the interface. Hence, their derivatives can be defined. In particular, it is possible to define the z-derivative of $\varepsilon(z)$. It is a function which is non-zero in a tightly localized region of space $[-z_0, z_0]$. This is schematically shown in Fig. 5.2.

Given that the jump of the function is fixed, the derivative must satisfy:

$$\lim_{z_0 \to 0} \int_{-z_0}^{+z_0} \frac{\partial \varepsilon}{\partial z} \mathrm{d}z = \varepsilon_2 - \varepsilon_1.$$

By contrast, a similar integral performed with a function that has no singularity at the interface yields a zero contribution. Hence, a practical procedure to derive the continuity conditions is as follows: (i) write the Maxwell equations explicitly; (ii) identify terms that can potentially provide singularities at the interface:

- $\partial X/\partial z$ if X has a discontinuity,
- ρ_S if there is a surface charge, or
- $\mathbf{j}_S$ if there is a surface current density.

Finally, integrate from $-z_0$ to $+z_0$ and take the limit to move from $\partial X/\partial z$ to X.

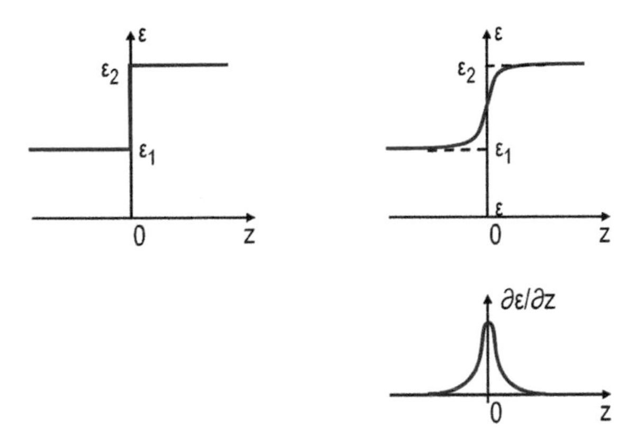

Fig. 5.2 Schematic representation of a discontinuous function at $z = 0$.

Continuity condition for D_z. We start from $\nabla \cdot \mathbf{D} = \rho_{\text{net}}$. We obtain $\frac{\partial D_x}{\partial x} + \frac{\partial D_y}{\partial y} + \frac{\partial D_z}{\partial z} = \rho_{\text{net}}$. Only $\partial D_z/\partial z$ and ρ may have a singularity $\rho_S \delta(z)$. After integration, we get:

$$\lim_{z_0 \to 0} \int_{-z_0}^{+z_0} \frac{\partial D_z}{\partial z} \mathrm{d}z = \lim_{z_0 \to 0} \int_{-z_0}^{+z_0} \rho_{\text{net}} \mathrm{d}z,$$

$$D_{n,1} - D_{n,2} = \rho_{S,\text{net}}, \tag{5.4}$$

where the subscript n indicates here a normal component (see Fig. 5.3).

Continuity condition for B_z. We start from $\nabla \cdot \mathbf{B} = 0$. We get $\frac{\partial B_z}{\partial z} = 0$, so that:

$$B_{n,1} - B_{n,2} = 0. \tag{5.5}$$

Continuity condition for E_{tg}. We start from $\nabla \times \mathbf{E} = -\frac{\partial \mathbf{B}}{\partial t}$. Keeping only terms that are potentially singular, we get $-\partial E_y/\partial z = 0$ and $\partial E_x/\partial z = 0$, so that $E_{y,1} = E_{y,2}$ and $E_{x,1} = E_{x,2}$ or

$$\mathbf{E}_{tg,1} = \mathbf{E}_{tg,2}. \tag{5.6}$$

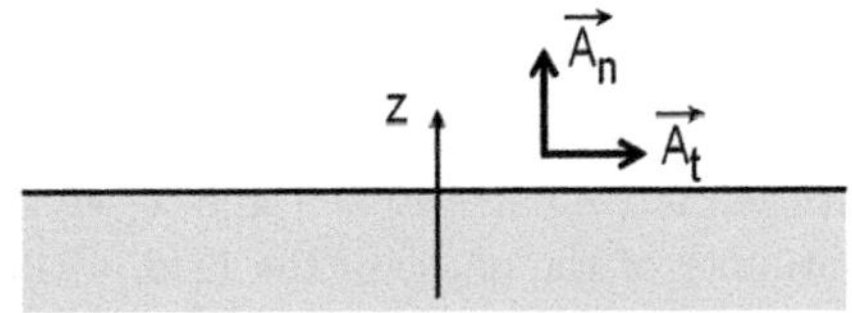

Fig. 5.3 Normal and tangential components of a vector **A**.

Continuity condition for H_{tg}. We start from $\nabla \times \mathbf{H} = \mathbf{j}_{\text{net}} + \frac{\partial \mathbf{D}}{\partial t}$. Keeping only potentially singular terms, we get $-\partial H_y/\partial z = j_{sx}$ and $\partial \mathbf{H}_x/\partial z = j_{sy}$.
Here, it is critical to realize that a surface current is usually a simplified model (perfect-conductor model) that requires the skin depth to vanish. It may also exist for a monoatomic layer of a 2D material such as graphene. In any other case, we have $j_{S,\text{net}} = 0$. In other words, the surface current is a limit of the current density on the skin depth, as the skin depth tends to zero when the conductivity tends to infinity for the perfect-conductor model.

Apart from these particular cases, the general continuity condition valid for any metal is $H_{y,1} - H_{y,2} = 0$ and $H_{x,1} - H_{x,2} = 0$, so that

$$\mathbf{H}_{tg,1} - \mathbf{H}_{tg,2} = 0. \tag{5.7}$$

In the particular case of a perfect-conductor model, we have $\mathbf{H}_{tg,2} = 0$ and the continuity condition becomes $\mathbf{H}_{tg,1} = \mathbf{j}_{S,\text{net}} \wedge \mathbf{n}$. For a monoatomic layer of a 2D material such as graphene, we have $\mathbf{H}_{tg,1} - \mathbf{H}_{tg,2} = \mathbf{j}_{S,\text{net}} \wedge \mathbf{n}$.

5.3 Snell's laws

Snell's laws are a consequence of the continuity of the phase of the monochromatic fields along the interface. This idea is schematically illustrated in Fig. 5.4.

We now derive Snell's laws from the continuity of the fields at an interface separating two homogeneous media. We consider an incident plane wave propagating along the x-axis. The electric field is linearly polarized along the y-axis. Electric fields at $z = 0$ are given by:

$$\begin{cases} E_{y,i}(z=0) = E_0 e^{ik_{x,i}x} \\ E_{y,r}(z=0) = rE_0 e^{ik_{x,r}x + ik_{y,r}y} \\ E_{y,t}(z=0) = tE_0 e^{ik_{x,t}x + ik_{y,t}y}. \end{cases}$$

The y-component of the electric field is tangential to the interface. It is continuous across the interface, so that:

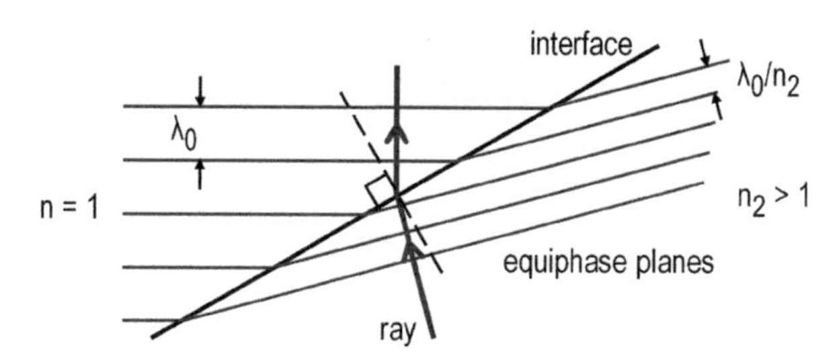

Fig. 5.4 Snell's laws and phase continuity. The distance between two planes with the same phase is λ_0 in a medium with refractive index $n_1 = 1$ and λ_0/n_2 in a medium with refractive index n_2. To ensure the continuity of the phase of the field, a tilt of medium 2 is needed in order to increase the spacing between two planes with the same phase. This tilt corresponds to the difference between the refraction angle and the incident angle. The same argument leads to the conclusion that the reflected angle is equal to the incident angle, as both beams are in the same medium.

$$\mathbf{E}_{y,1}(z=0) = \mathbf{E}_{y,2}(z=0),$$

$$E_{y,i}(z=0) + E_{y,r}(z=0) = E_{y,t}(z=0),$$
$$e^{ik_{x,i}x} + re^{ik_{x,r}x+ik_{y,r}} = te^{ik_{x,t}x+ik_{y,t}y}.$$

This equation can be cast in the form

$$t = e^{i(k_{x,i}-k_{x,t})x-ik_{y,t}y} + re^{i(k_{x,r}-k_{x,t})x-ik_{y,t}y}.$$

This equation must be satisfied for any value (x, y). It follows that the constant function on the left-hand side is equal to an exponential of the form $\exp(iK_x x + iK_y y)$ on the right-hand side. This is only possible if $K_x = K_y = 0$. Finally, we find that the phase continuity implies $k_{x,i} = k_{x,r}$ and $k_{y,i} = k_{y,r}$, which is the Snell reflection law, and $k_{x,i} = k_{x,t}$ and $k_{y,i} = k_{y,t}$, which is the Snell refraction law.

The above relation is general and can be used for both lossy and non-lossy media. It is a direct consequence of the continuity conditions that follow from Maxwell equations. We now specialize to the particular case of non-lossy media with a real positive refractive index (see Fig. 5.5):

$$\mathbf{k}_i \begin{cases} k_{x,i} = n_1\frac{\omega}{c}\sin\theta_i \\ k_{y,i} = 0 \\ k_{z,i} = -n_1\frac{\omega}{c}\cos\theta_i \end{cases} \quad \mathbf{k}_r \begin{cases} k_{x,r} = n_1\frac{\omega}{c}\sin\theta_i \\ k_{y,r} = 0 \\ k_{z,i} = n_1\frac{\omega}{c}\cos\theta_i \end{cases} \quad \mathbf{k}_t \begin{cases} k_{x,t} = n_2\frac{\omega}{c}\sin\theta_t = n_1\frac{\omega}{c}\sin\theta_i \\ k_{y,t} = 0 \\ k_{z,t} = -n_1\frac{\omega}{c}\cos\theta_t. \end{cases}$$

5.4 Reflection factor and TE polarization

5.4.1 Notations: transverse-electric and transverse-magnetic polarizations

The interface $z = 0$ separates two linear and isotropic media characterized by a complex refractive index n_1 and n_2. We use the notations

$$\mathbf{k}_i = \begin{pmatrix} k_{x,i} \\ k_{y,i} = 0 \\ k_{z,i} \end{pmatrix}.$$

By definition, the plane of incidence is the plane containing the two vectors $\mathbf{k}_i$ and $\mathbf{n} = \mathbf{e}_z$. A field with an electric vector *perpendicular* to the plane of incidence is

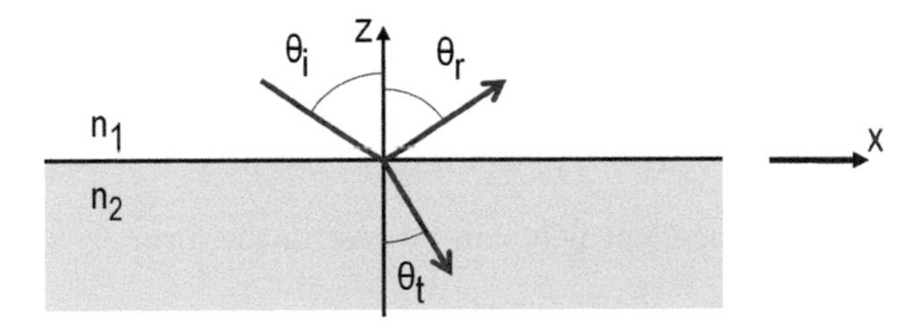

Fig. 5.5 Incidence, reflection, and refraction angles. The signs of the angles are chosen according to the equation $k_x = k_0 \sin(\theta)$.

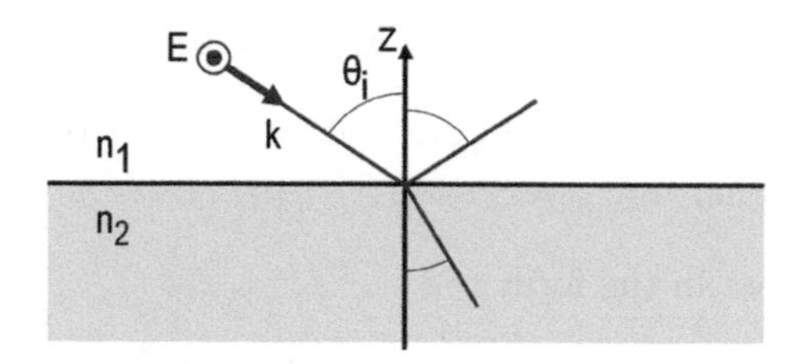

Fig. 5.6 Definition of transverse-electric polarization.

called transverse electric (TE) (see Fig. 5.6). A field with the electric field in the plane of incidence and the *magnetic* field perpendicular to the plane of incidence is called transverse magnetic and denoted by TM.

Depending on the communities (electrical engineering, optics, remote sensing), different notations are commonly used. A table of equivalence summarizes them:

TE	s	H	⊥
TM	p	V	∥

The 's' (*senkrencht* means perpendicular in German) and 'p' (parallel) notation is often used in optics, while the H (horizontal) and V (vertical) notation is often used in radar remote sensing. We also mention a possible source of confusion regarding the definition of TE and TM when dealing with gratings. A field can be perpendicular to a plane of incidence or to the lines of the grating, so that one should keep in mind that different conventions can be used.

5.4.2 Reflection and transmission factors

The procedure to derive the amplitude of the reflected and transmitted fields is a standard procedure for solving a system of partial differential equations. The successive steps are:

(i) Write the partial differential equations satisfied by the fields in the media.
(ii) Establish the general solution.
(iii) Use boundary conditions at infinity and continuity conditions at the interface.
(iv) Solve the resulting linear system.

(i) Partial differential equation:

$$\nabla^2 \mathbf{E}_1 + \varepsilon_{r,1}\frac{\omega^2}{c^2}\mathbf{E}_1 = 0 \quad z > 0, \tag{5.8}$$

$$\nabla^2 \mathbf{E}_2 + \varepsilon_{r,2}\frac{\omega^2}{c^2}\mathbf{E}_2 = 0 \quad z < 0. \tag{5.9}$$

(ii) General solution. The incident field can be cast in the form

$$\mathbf{E}_i = E_0 \mathbf{e}_y e^{i(k_{x,i}x - \gamma_i z)}, \tag{5.10}$$

where the z-component of the wavevector is derived from Eq. (5.8), $\gamma_i^2 = \varepsilon_{r,1}\frac{\omega^2}{c^2} - k_{x,i}^2$. The choice of determination is discussed in Complement 4.B. It is given by $\mathrm{Re}[\gamma_i] + \mathrm{Im}[\gamma_i] > 0$. This choice, associated with the choice of the time dependence $e^{-i\omega t}$, entails that a wave of the form $\exp[i(\gamma z - \omega t)]$ propagates along the z-axis towards positive values of z. We can cast the amplitudes of the reflected and transmitted field in the form

$$\mathbf{E}_r = rE_0\mathbf{e}_y e^{i(k_{x,i}x+\gamma_i z)},$$
$$\mathbf{E}_t = tE_0\mathbf{e}_y e^{i(k_{x,i}x-\gamma_t z)},$$

where $\gamma_r^2 = \varepsilon_{r,1}\frac{\omega^2}{c^2} - k_{x,i}^2$, $\mathrm{Re}[\gamma_r]+\mathrm{Im}[\gamma_r] > 0$ and $\gamma_t^2 = \varepsilon_{r,2}\frac{\omega^2}{c^2} - k_{x,i}^2$, $\mathrm{Re}[\gamma_t]+\mathrm{Im}[\gamma_t] > 0$. We do not introduce propagating waves of the form $e^{-i\gamma_t z}$ because there is no incident wave coming from medium 2. The incident wave coming from $+\infty$ has a known amplitude, E_0. It follows that there is a general solution satisfying boundary conditions at infinity and Snell's laws. Two unknowns remain: the amplitudes of the reflected and transmitted fields r, t. The continuity conditions will provide two equations.

(iii) Boundary conditions. The continuity of $\mathbf{E}_{tg}$ at $z = 0$ yields

$$1 + r = t.$$

Note that this equation has no energy conservation meaning; it is an amplitude continuity condition. Energy conservation is discussed in section 5.5. The continuity of the tangential magnetic field $\mathbf{H}_{tg}$ can be converted into a condition for the derivative of the field using the Maxwell–Faraday equation

$$B_x = \frac{1}{i\omega}(\nabla \times \mathbf{E})_x = \frac{1}{i\omega}\left(\frac{\partial E_z}{\partial y} - \frac{\partial E_y}{\partial z}\right) = -\frac{1}{i\omega}\frac{\partial E_y}{\partial z}.$$

The continuity of H_x implies the continuity of $-\frac{1}{\mu}\frac{\partial E_y}{\partial z}$, so that:

$$\frac{1}{\mu_1}(-\gamma_i + \gamma_i r) = -t\frac{\gamma_t}{\mu_2}.$$

We finally obtain a linear system:

$$\begin{cases} 1 + r = t \\ 1 - r = t\dfrac{\gamma_t}{\gamma_i}\dfrac{\mu_1}{\mu_2}. \end{cases}$$

(iv) Solution. Solving for the reflection and transmission factors r and t, we find:

$$\begin{cases} r = \dfrac{\mu_2\gamma_i - \mu_1\gamma_t}{\mu_2\gamma_i + \mu_1\gamma_t} \\ t = \dfrac{2\mu_2\gamma_i}{\mu_2\gamma_i + \mu_1\gamma_t} \end{cases} \quad \text{or, if } \mu_1 = \mu_2 = \mu_0 \quad \begin{cases} r = \dfrac{\gamma_i - \gamma_t}{\gamma_i + \gamma_t} \\ t = \dfrac{2\gamma_i}{\gamma_i + \gamma_t}. \end{cases}$$

In general, r and t are complex. The phase of the complex number provides the phase shift upon reflection or transmission.

5.4.3 Total internal reflection

Let us suppose that medium 1 has a real refractive index larger than the real refractive index of medium 2. This may be the case for light propagating in water at the air–water interface. If the incidence angle θ_i is such that $n_1 \frac{\omega}{c} \sin\theta_i > n_2 \frac{\omega}{c}$, the z-component of the wavevector in medium 2 is purely imaginary ($\gamma_t = i|\gamma_t|$), so that the wave decays exponentially away from the interface (see Fig. 5.7a) and does not carry energy along z. Accordingly, the reflection factor has a modulus equal to one. However, the field does exist in medium 2 in the close vicinity of the interface, and its amplitude can be computed. Let us consider $n_2 = 1$, as shown in Fig. 5.7a. If $\sin\theta_i > \frac{1}{n_1}$, $\gamma_t = i|\gamma_t|$ is imaginary. We have:

$$E_t = \frac{2\gamma_i}{\gamma_i + i|\gamma_t|} E_0 e^{ik_{x,i}x} e^{-|\gamma_t|z}, \tag{5.11}$$

where

$$\begin{aligned}\gamma_t &= \left[\frac{\omega^2}{c^2} - n_1^2 \frac{\omega^2}{c^2} \sin^2\theta_i\right]^{1/2} \\ &= \frac{\omega}{c}\left[1 - n_1^2 \sin^2\theta_i\right]^{1/2}.\end{aligned}$$

While the reflectivity has a unit modulus, the field is non-zero in medium 2. It propagates along the x-axis and decays exponentially away from the interface, with a decay length given by $1/|\gamma_t|$. We will see that the energy flux perpendicular to the interface is zero, despite the presence of a non-zero field in medium 2.

5.4.4 Reflection and transmission factors in TM polarization

A similar procedure leads to the definition of the reflection and transmission factors in TM polarization. They can be defined either as a ratio of some electric-field components or as the ratio of magnetic-field amplitudes. If we use the general form of the magnetic fields,

$$\begin{aligned}\mathbf{H}_r &= r_{TM} H_0 \mathbf{e}_y e^{i(k_{x,i}x + \gamma_i z)}, \\ \mathbf{H}_t &= t_{TM} H_0 \mathbf{e}_y e^{i(k_{x,i}x - \gamma_t z)},\end{aligned}$$

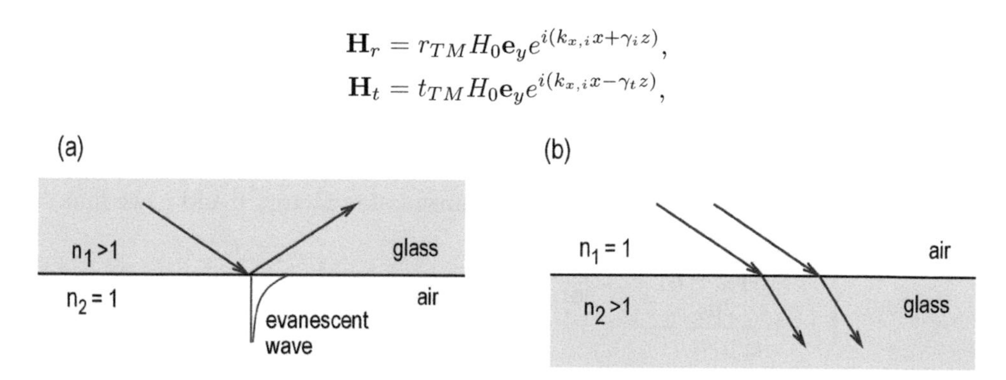

Fig. 5.7 (a) Total internal reflection with evanescent wave in the lower index medium; (b) refraction, with a propagative wave in the higher-index medium (reflection has been omitted).

we find for linear isotropic and non-magnetic media:

$$t_{TM} = \frac{2\varepsilon_2\gamma_i}{\varepsilon_2\gamma_i + \varepsilon_1\gamma_t}, \tag{5.12}$$

$$r_{TM} = \frac{\varepsilon_2\gamma_i - \varepsilon_1\gamma_t}{\varepsilon_2\gamma_i + \varepsilon_1\gamma_t}. \tag{5.13}$$

5.5 Intensity reflection factor

5.5.1 Definition

So far, we have derived the amplitudes of the reflected and transmitted fields. They are characterized by the Fresnel reflection and transmission factors r, t. We now address the issue of the fraction of energy reflected and transmitted at the interface. To define them, we start by computing the flux of the Poynting vector across the interface, namely, the z-component of the Poynting vector. We compute the fluxes corresponding to the reflected and transmitted waves and we normalize by the flux of the Poynting vector of the incident field:

$$R = \frac{|\langle S_z^r \rangle|}{|\langle S_z^i \rangle|} \qquad T = \frac{|\langle S_z^t \rangle|}{|\langle S_z^i \rangle|}. \tag{5.14}$$

We now seek a connection between the intensity (R, T) factors and the amplitude (r, t) factors.

5.5.2 Poynting vector flux

For TE polarization, the complex amplitudes of the fields are given by:

$$\mathbf{E}_i = E_0 \mathbf{e}_y e^{i(k_x x - \gamma_1 z)},$$

$$\mathbf{E}_r = r E_0 \mathbf{e}_y e^{i(k_x x + \gamma_1 z)}.$$

The time average of the z-component of the Poynting vector in medium 1 is:

$$\langle S_z^{(1)} \rangle = \frac{1}{2}\mathrm{Re}[\mathbf{E}^* \wedge \mathbf{H}].\mathbf{e}_z$$

$$= -\frac{1}{2}\mathrm{Re}[E_y^* H_x].$$

Assuming that medium 1 is non-magnetic, we have $\nabla \times \mathbf{E} = i\omega\mathbf{B} = i\omega\mu_0\mathbf{H}$, so that we obtain

$$H_x = -\frac{1}{i\omega\mu_0}\frac{\partial E_y}{\partial z}.$$

On the interface $z = 0$,

$$E_y = E_0(1 + r)$$

$$\frac{\partial E_y}{\partial z} = -i\gamma_1 E_0(1 - r).$$

The z-components of the Poynting vectors are thus given by:

$$\langle S_z^{(i)}(z=0)\rangle = \frac{1}{2\mu_0\omega}|E_0|^2\,\mathrm{Re}[\gamma_1], \tag{5.15}$$

$$\langle S_z^{(r)}(z=0)\rangle = -\frac{1}{2\mu_0\omega}|E_0|^2\,|r|^2\mathrm{Re}[\gamma_1], \tag{5.16}$$

$$\langle S_z^{(t)}(z=0)\rangle = \frac{1}{2\mu_0\omega}|E_0|^2\,|t|^2\mathrm{Re}[\gamma_2]. \tag{5.17}$$

Finally, the intensity reflection and transmission factors can be cast in the form:

$$R = |r|^2, \tag{5.18}$$

$$T = |t|^2\frac{\mathrm{Re}[\gamma_2]}{\mathrm{Re}[\gamma_1]}. \tag{5.19}$$

Note that in the case of total internal reflection, the transmitted field amplitude is non-zero and the field decays exponentially in the medium. However, the transmission intensity factor is zero because $\mathrm{Re}(\gamma_2) = 0$. It can be directly checked that $R = 1$ when γ_2 is imaginary.

When comparing the form of R and T, it is seen that R is merely the square of the amplitude reflection factor, whereas an additional factor $\mathrm{Re}[\gamma_2]/\mathrm{Re}[\gamma_1]$ shows up in the case of the transmission factor. Let us discuss the origin of this factor in the case of a non-dispersive medium where energy velocity and phase velocity are equal. Fig. 5.7b illustrates a refracted beam. On the one hand, the width of the beam is modified by a geometrical factor which is the ratio of the cosines of the angle between the wavevector and the surface normal. On the other hand, the index modification entails a modification of the energy velocity (divided by n) and an increase of the energy density (increased by n^2). The energy flux which is the product of energy density by light velocity is thus proportional to n. The factor $\mathrm{Re}[\gamma_2]/\mathrm{Re}[\gamma_1]$ accounts for these three effects connected to the refractive index change.

The connection between R and T is derived by studying the energy conservation between two planes on both sides of the interface. This energy balance is discussed in detail in Complement 5.A. The discussion is simple when the interface separates two non-lossy media illuminated by propagating waves and leads to $R+T = 1$. If the media are lossy or if evanescent waves generated by sources located at finite distance from the interface are involved, the energy balance is no longer given by $R + T = 1$.

5.5.3 Emissivity and absorptivity

A body at temperature T emits thermal radiation. The emitted radiance is given by the product of the black-body radiance and the emissivity of the body. Kirchhoff's law establishes that the emissivity is equal to the absorptivity A for a given direction, angle, and polarization.[1] The energy conservation of a plane wave propagating in a transparent medium impinging on an interface with an opaque medium yields $A = 1 - R$. For an interface separating a transparent medium from an opaque medium, we

[1] It can be shown that this property is a direct consequence of the reciprocity theorem.

have $R + T = 1$ when light is incident from the transparent medium. It follows that $1 - R = T$. Consequently, the absorptivity (and therefore the emissivity) is given by $1 - R = T$. It follows that both the absorptivity and the emissivity of a body are actually given by the intensity transmission factor of the interface. This point will be further discussed in Chapter 21, where we deal with thermal radiation.

5.6 Normal incidence

The purpose of this section is to discuss a few basic properties of reflection and refraction. For the sake of simplicity, we limit the discussion to normal incidence.

5.6.1 Reflection factor

At normal incidence, $k_x = 0$ and $\gamma = n\frac{\omega}{c}$. The reflection factor can be cast in the form

$$r = \frac{\gamma_1 - \gamma_2}{\gamma_1 + \gamma_2} = \frac{n_1 - n_2}{n_1 + n_2}.$$

For a glass–air interface, $n_1 \approx 1.5$, $n_2 = 1$, so that $r = 0.2$ (and $r = -0.2$ for an air–glass interface) and $R = |r|^2 = (0.2)^2 = 4\%$. In other words, the intensity transmission factor is $T = 96\%$. This may seem a large number at first glance. However, an optical system with five lenses has ten interfaces, so that the intensity transmission is only $0.96^{10} = 0.66$. This clearly indicates the need for anti-reflection coatings.

5.6.2 Anti-reflection coatings

There are several concepts that can be used to reduce the reflectivity.

Magnetic media. For magnetic media, the reflection factor can be cast in the form

$$r = \frac{\gamma_1/\mu_1 - \gamma_2/\mu_2}{\gamma_1/\mu_1 + \gamma_2/\mu_2}.$$

At normal incidence, we find $\gamma_1 = n_1\frac{\omega}{c} = \sqrt{\varepsilon_1\mu_1}\frac{\omega}{c}$. It is seen that the reflection factor is zero if

$$\sqrt{\frac{\varepsilon_1}{\mu_1}} = \sqrt{\frac{\varepsilon_2}{\mu_2}}.$$

If a coating satisfies this condition, there will be no reflection at the interface. If, in addition, the medium is absorbing and its thickness is larger than the decay length, all the incident radiation will be absorbed.

Antireflection layer. By depositing a uniform layer of refractive index n' on the interface, a second interface is introduced. If the interference between the rays reflected by the two interfaces takes place destructively, the total reflection can be zero (see Fig. 5.8).

Two conditions must be fulfilled:

- Waves 1 and 2 should be dephased by an amount π. Hence, the thickness of the layer should be $d = \lambda/4n'$.

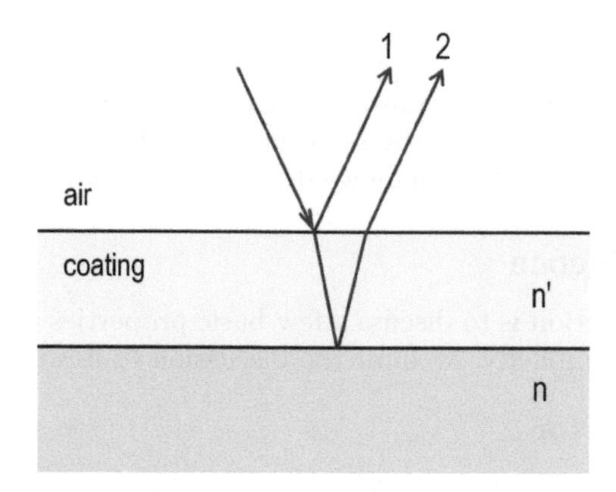

Fig. 5.8 Interference coating made of a single layer.

- Waves 1 and 2 should have the same amplitude, so that $\frac{1-n'}{1+n'} = \frac{n'-n}{n'+n}$. It follows that $n' = \sqrt{n}$. This simplified calculation was based on the assumption that only two rays provide a significant contribution, which is a reasonable assumption for low-index contrast. A full calculation (see Exercise 5.4) leads to the same result.

This type of anti-reflection coating is used for lenses and glasses. As the path difference depends on the angle, however, some colours can be seen at grazing incidence.

Index gradient. A key remark is that the reflection factor $\frac{n_1-n_2}{n_1+n_2}$ is proportional to the index contrast $n_1 - n_2$. Hence, the reflection factor decays if the refractive index contrast is small. It is possible to deposit a large number of layers with a small index difference. This is an approximation of a gradient of refractive index (see Fig. 5.9). Such a system is less sensitive to the frequency.

5.6.3 The paradox of the absorber

The paradox can be stated as follows: a good absorber does not absorb. Of course, such a statement can only be true if two different definitions of what is a good absorber are used. Let us consider an interface-separating vacuum from an absorbing medium with a complex refractive index denoted by $n = n' + in''$. A first definition of a good absorber is to consider that n'' is a large number, so that the decay length in the absorbing

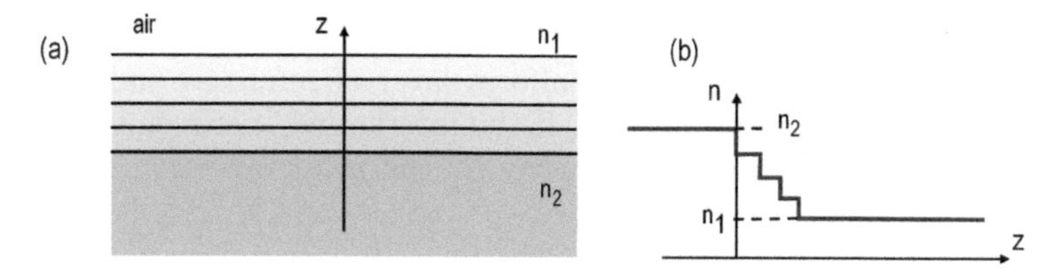

Fig. 5.9 A multilayer coating simulating an index gradient: $n(z)$ has small steps.

medium $\lambda/(2\pi n'')$ is very small. In other words, the absorption coefficient $\alpha = 4\pi n''/\lambda$ is very large.

We now consider the reflection factor of the interface. It is given by:

$$R = \left|\frac{n-1}{n+1}\right|^2 = \frac{(n'-1)^2 + n''^2}{(n'+1)^2 + n''^2}.$$

Assuming that $n'' >> 1$, it is seen that the reflection factor $R \to 1$. Hence, the medium reflects most of the light, so that there is little absorption.

In practice, to absorb light efficiently, it is necessary to ensure that light penetrates the absorbing material; therefore, one should not have a large dielectric contrast. Then, a second requirement is to make sure that the thickness of the absorbing medium is larger than the decay length in order to ensure absorption. A very important message arising from this discussion is that absorption is not only a material property. It depends on the system property. We have seen that for a planar system, we need to design an anti-reflection coating for the interface and optimize the thickness of the medium. We will see in later chapters that absorption can be tuned by inserting the absorbing material into a resonator.

5.6.4 Colours

The previous concepts can be used to discuss the origin of colours for some materials. We start by discussing coloured inks. These are solutions in transparent solvents of pigment which absorb some particular frequencies. When ink is deposited on white paper, light propagating through the ink is partially absorbed. The frequencies which are not absorbed can emerge from the absorbing medium and produce the complementary colour. Chlorophyll is absorbing in the visible spectrum except in the 500–600 nm interval so that green is not absorbed. The colour of the sea is also due to absorption of the red part of the spectrum. The absorption coefficient of sea water being much smaller than that of chlorophyll, it requires a larger thickness. Let us now consider gold. The values of the refractive index are given below at 0.5 μm (green) and 0.8 μm (near-infrared).

Au optical properties	$n = n' + in''$	$\varepsilon_r = \varepsilon_r' + i\varepsilon_r''$
$\lambda = 0.5\,\mu$m	0.84 + i 1.84	-2.68 + i 3.09
$\lambda = 0.8\,\mu$m	0.16 + i 4.84	-23.4 + i 1.55

It is seen that the modulus of the refractive index is larger in the near-infrared than in the green part of the spectrum. Hence, gold is a better reflector at 0.8 μm. This explains the origin of the yellowish colour of gold in reflection. The colour is different if one deposits a thin film of gold (on the order of the decay length in gold) on a transparent substrate and observes in transmission: then, gold is green. This may seem surprising, as the intrinsic loss term given by ε_r'' is larger in the green part of the spectrum, so that one might think that this part of the spectrum should vanish due to absorption.

Actually, two effects play a role: firstly, reflection at the interface is lower for green, so that transmission is larger; secondly, the decay length of the wave through propagation in the metal is proportional to the imaginary part of the refractive index. The wave is less attenuated in the green! Hence, both factors yield the same effect—a larger transmission in the green—although the intrinsic loss term given by ε_r'' is larger in the green. This again points to the fact that attenuation and absorption are two different phenomena, as discussed in Chapter 4.

5.7 Angle dependence of the Fresnel factors

5.7.1 Reflection and transmission factors between dielectric media

The graphs of Fig. 5.10 show the angular dependence of the Fresnel intensity reflection factor at a vacuum–glass interface as a function of the angle of incidence i for the two polarizations. Note that for the glass–air interface, the reflection factor is 1 for any angle larger than the critical angle i_c; this is the total internal reflection discussed in section 5.5.2. It is also of interest to note that the reflection factor takes the same value for both polarizations at normal incidence. This is a direct consequence of the fact that the medium is assumed to be isotropic. It follows that a lens working in the so-called Gauss conditions (low-incidence angles) will not depolarize the light. A remarkable property is that the reflection factor is zero in TM polarization for a particular angle called the Brewster angle, i_B.

5.7.2 Brewster angle

This property can be used to produce polarized light. Light reflected by an interface at the Brewster angle i_B is totally polarized in the TE polarization. The physical origin of this effect is the anisotropy of the dipole radiation pattern. Let us consider an incident plane wave impinging on an interface with an angle of incidence θ_i. The reflected wave will propagate along the direction $\theta_r = \theta_i$ according to Snell's law. The physical origin of this wave is the radiation of the induced dipoles in the substrate. As the field radiated by the induced dipoles is zero along their axis, it turns out that if the orientation of the induced dipoles coincide with the emission direction determined by Snell's law, the reflected field will be zero. Let us consider the case of a vacuum–glass

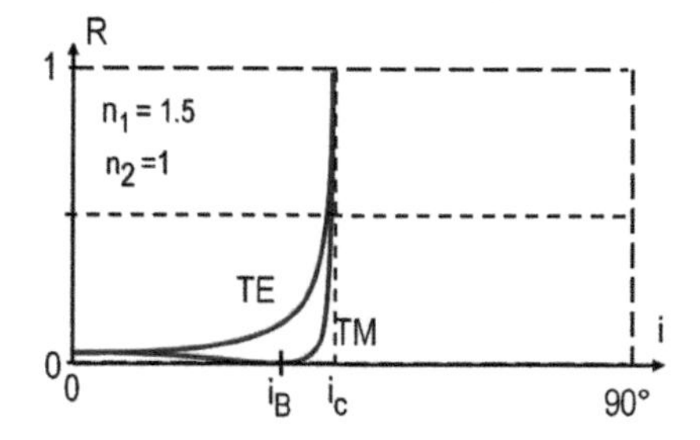

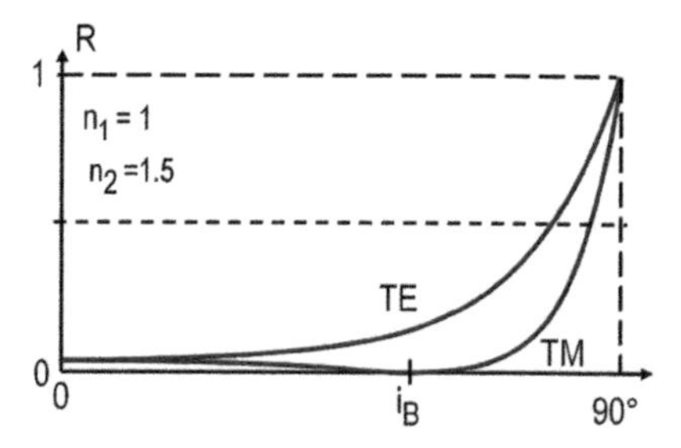

Fig. 5.10 Angle dependence of the intensity reflection factor at a vacuum–glass interface when impinging either from the glass side (left) or the vacuum side (right).

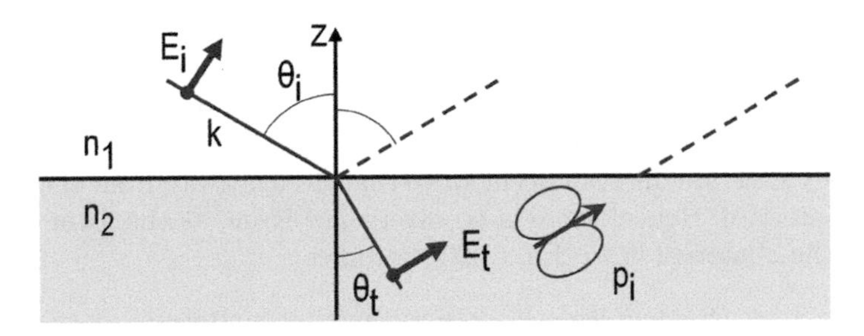

Fig. 5.11 The Brewster angle seen as a zero of dipole radiation induced in the second medium.

interface. We denote by n the refractive index of glass. If the induced dipoles $\mathbf{p}_i$ are parallel to the propagation direction of the reflected wave, we see from Fig. 5.11 that:

$$\pi = i_B + \frac{\pi}{2} + \theta_t.$$

It follows that

$$\theta_t = \frac{\pi}{2} - i_B.$$

Using Snell's law, $\sin i_B = n \sin\theta_t$, we find $\sin i_B = n\cos i_B$, so that the Brewster angle is given by the simple relation

$$\tan i_B = n.$$

5.A Complement: Energy conservation at an interface

When dealing with non-lossy materials characterized by real dielectric constants, it is a simple matter to check that $R + T = 1$ is satisfied for a monochromatic wave after averaging over time. This relation can be seen as the energy conservation in a thin volume containing the interface at $z = 0$. However, this is not a general form of the energy conservation and it may be wrong in several cases. To rigorously analyse this issue, we now proceed to write the energy conservation Eq. (1.60) in stationary regime in a volume comprised between two planes at $z = \delta$ and $z = -\delta$, where δ can be arbitrarily small. As the volume tends to zero, assuming that there is no singular surface current we obtain:

$$\langle S_z^{(1)}(z=0)\rangle = \langle S_z^{(2)}(z=0)\rangle. \tag{5.20}$$

For a plane wave, we obtain:

$$\langle S_z^{(1)}(z=0)\rangle = \frac{1}{2\mu_0\omega}|E_0|^2\,\mathrm{Re}[(1+r^*)(1-r)\gamma_1] \tag{5.21}$$

$$= \frac{1}{2\mu_0\omega}|E_0|^2\,\{(1-|r|^2)\,\mathrm{Re}[\gamma_1] + \mathrm{Re}[(r^*-r)\gamma_1]\} \tag{5.22}$$

$$= \frac{1}{2\mu_0\omega}|E_0|^2\,\{(1-|r|^2)\mathrm{Re}[\gamma_1] + 2\mathrm{Im}[\gamma_1]\mathrm{Im}[r]\}. \tag{5.23}$$

Note that we have not made any approximation in the derivation. If γ is imaginary, the wave decays or increases away from the interface. A wave that increases exponentially as the distance to the interface increases seems to diverge at infinity. However, this is exactly what happens if a source is located at a finite distance d above the interface. The source may generate an evanescent wave that decays away from the source before reaching the interface. Hence, there is no divergence issue, as the form of the wave is only valid in the z-interval $[0, d]$. Eq. (5.20) becomes

$$(1 - |r|^2)\mathrm{Re}[\gamma_1] + 2\,\mathrm{Im}[\gamma_1]\mathrm{Im}[r] = |t|^2\mathrm{Re}[\gamma_2],$$

so that $R + T = 1$ is valid only if $\mathrm{Im}[\gamma_1] = 0$. The general energy conservation law is:

$$R + T = 1 + 2\,\frac{\mathrm{Im}(\gamma_1)\mathrm{Im}(r)}{\mathrm{Re}(\gamma_1)}. \tag{5.24}$$

The additional term is known as the mixed Poynting vector. It results from an interference term between the incident and the reflected waves. We now discuss two cases where $\mathrm{Im}(\gamma_1)$ is non-zero.

First case: Medium 1 is lossy.

If medium 1 is lossy as shown in Fig. 5.12a, then $\gamma_1 = [n_1^2\frac{\omega^2}{c^2} - k_x^2]^{1/2}$ has a non-zero imaginary part. In that case, the amount of energy absorbed in medium 1 depends on the interference between the incident field and the reflected field. Modifying the phase of the reflection factor modifies the spatial distribution of the intensity in medium 1. This intensity distribution depends both on the phase and the decay, due to losses.

Second case: The incident wave is evanescent.

It is possible to illuminate with an evanescent wave by using a prism, as depicted in Fig. 5.12b. Another possibility for generating an evanescent wave illuminating an interface is to use a subwavelength source such as a tip, a small aperture, or a dipolar source, as shown in Fig. 5.12c. According to Eq. (4.14), a highly confined field contains evanescent waves. If the source is close enough to the interface, the evanescent waves can contribute to the illumination of the interface.

In both cases, the illuminating wave decays exponentially in medium 1 with a z-component of the wavevector given by:

$$\gamma_1 = [n_1^2\frac{\omega^2}{c^2} - k_x^2]^{1/2} = i|\gamma_1|.$$

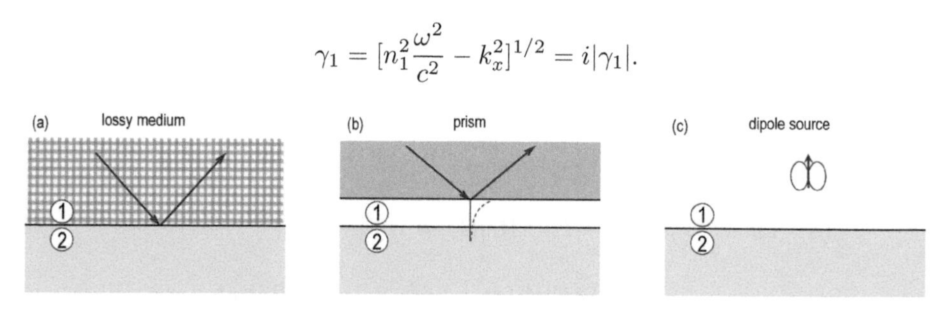

Fig. 5.12 Energy conservation differs from $R + T = 1$. We illustrate this case with three examples; (a) the presence of losses in the incident medium; (b) illumination by an evanescent wave produced by a prism; (c) illumination by a dipolar source at finite distance.

As already discussed, an evanescent wave in an infinite medium corresponds to a zero flux, $\langle S_z^i(z=0)\rangle = \frac{1}{2\mu_0\omega}|E_0|^2 \,\mathrm{Re}[\gamma_1] = 0$. However, the contribution of evanescent waves between two interfaces is non-zero. This is formally identical to the tunnel effect in quantum mechanics.

In summary, we have identified two cases where the simple relation $R + T = 1$ is not valid. In the first case, the interference between the incident and the reflected waves takes place in a lossy medium. In that case, the losses depend on the phase of the reflection factor. The second case deals with the presence of sources or scatterers in the near field of the interface, so that evanescent waves are involved in the energy flux. We will discuss in Chapter 21 the impact of evanescent waves on the radiative heat transfer at the nanoscale.

5.B Complement: Scattering by a slightly rough surface and perturbative treatment

In this complement, we study the effect of the roughness of an interface. This is an important issue, as most interfaces are not flat. The reader can check that the paper surface or the screen surface on which this book is displayed is not a perfectly flat surface and therefore scatters light instead of reflecting it specularly. The reader is invited to check around him/her how many surfaces are planar and can be described using Fresnel factors (mirror) and how many are diffusing (clothes, walls, ground, ceiling, furniture). It is thus of interest to get some insight into surface scattering, which is ubiquitous. For the sake of simplicity, we deal with a perfectly conducting metal lying in the half-space $z < S(x)$. We assume that the problem is translationally invariant along the y-axis, so that the fields depend on x, z only. We consider a linearly polarized electric field along the y-axis, so that the problem becomes a scalar problem. We aim at deriving the scattered field by a rough surface of arbitrary profile. In other words, we aim to find the connection between the surface profile and the scattered field. The problem cannot be solved analytically for any surface profile. Here, we discuss only the case of a surface profile $z = S(x) = \delta s(x)$, where $s(x) \leq 1$ and δ, the height scale of the roughness verifies $\delta \ll \lambda/2\pi$. This assumption is useful to look for a solution as a perturbation of the plane interface case.

The y-component of the incident electric field propagating in vacuum is given by $E_{\mathrm{inc}} \exp(ik_{\mathrm{inc}}x - i\gamma_{\mathrm{inc}}z)$. The field is null in the metal and satisfies the Helmholtz equation in vacuum, so that:

$$\nabla^2 E(x,z) + \frac{\omega^2}{c^2} E(x,z) = 0. \tag{5.25}$$

The general solution of the problem in vacuum above the surface profile can be written as $E(x,z) = E_{\mathrm{inc}}(x,z) + E_s(x,z)$, where the scattered field $E_s(x,z)$ can be written as a sum of plane waves,

$$E_s(x,z) = \int \frac{dk_x}{2\pi} E(k_x) \exp[i(k_x x + i\gamma z)],$$

where $\gamma(k_x) = [\omega^2/c^2 - k_x^2]^{1/2}$ with $Re(\gamma) > 0$ and $Im(\gamma) > 0$. The problem is thus to find the unknown amplitudes of the scattered plane waves $E(k_x)$. To proceed, we write the continuity condition at the surface. The electric field is tangential, so that it is continuous. Hence,

$$E_{\text{inc}}[x, z = S(x)] + E_s[x, z = S(x)] = 0. \tag{5.26}$$

This equation assumes that the plane-wave expansion is valid everywhere along the surface. It is a general solution of a Helmholtz equation in vacuum so that it is valid above the highest point of the surface. To use the expansion inside the groove is *a priori* an approximation known as Rayleigh hypothesis. This approximation is valid in the context of slightly rough surfaces with $\delta \ll \lambda/2\pi$. Inserting the form of the fields in the continuity conditions yields

$$E_{\text{inc}} \exp[ik_{x,\text{inc}}x - i\gamma_{\text{inc}}S(x)] + \int \frac{dk_x}{2\pi} E_s(k_x) \exp[i(k_x x + i\gamma S(x))]. \tag{5.27}$$

This equation is non-linear in $S(x)$. It can be linearized by expanding $\exp[-i\gamma_{\text{inc}}\delta s(x)] \approx 1 - i\gamma_{\text{inc}}\delta s(x)$ and $\exp[i\gamma S(x)] \approx 1 + i[\gamma_{\text{inc}}\delta]\frac{\gamma}{\gamma_{\text{inc}}}s(x)$. The perturbation parameter is clearly $X = \gamma_{\text{inc}}\delta$. We now formally expand the field in powers of X:

$$E_s(x, z) = E_s^{(0)}(x, z) + XE_s^{(1)}(x, z) + ... \tag{5.28}$$

We can now expand the continuity condition to zero and first order in X:

$$\begin{aligned} E_{\text{inc}}e^{ik_{x,\text{inc}}x} + \int \frac{dk_x}{2\pi} E_s^{(0)}(k_x)e^{ik_x x} &= 0, \\ E_{\text{inc}}e^{ik_{x,\text{inc}}x}[-is(x)] + \int \frac{dk_x}{2\pi} e^{ik_x x} \left[E_s^{(1)}(k_x) + E_s^{(0)}(k_x)[i\frac{\gamma}{\gamma_{\text{inc}}}s(x)]\right]&. \end{aligned} \tag{5.29}$$

The first equation yields:

$$E_s^{(0)} = -2\pi E_{\text{inc}}\delta(k_x - k_{x,\text{inc}}). \tag{5.30}$$

Inserting this solution into the second equation, we get

$$E_s^{(1)}(k_x) = 2iE_{\text{inc}}s(k_x - k_{x,\text{inc}}), \tag{5.31}$$

where $s(k_x)$ is the Fourier transform of the surface profile:

$$s(x) = \int \frac{dk_x}{2\pi} s(k_x) \exp[i(k_x x)]. \tag{5.32}$$

We finally get the following solution to first order in the roughness of the surface:

$$E_s(x, z) \approx -E_{\text{inc}} \exp(ik_{x,\text{inc}}x + i\gamma_{\text{inc}}z) + \gamma_{\text{inc}}\delta \int \frac{dk_x}{2\pi} s(k_x - k_{x,\text{inc}}) \exp[i(k_x x + i\gamma z)]. \tag{5.33}$$

The first term is the specular reflection term corresponding to a flat surface. The second term provides the scattered field. It shows that the angular distribution of the scattered field contains the information on the spectrum of the surface. Note, in particular, that if the surface profile contains low spatial frequencies as compared to ω/c or, in other words, the transverse typical lengths are larger than the wavelength, the scattered field will be concentrated around the specular component. By contrast, if the surface profile contains small features on the scale of the wavelength, the field is scattered almost isotropically. Note also that the scattered field contains evanescent waves corresponding to $|k_x| > \omega/c$.

We can also note that to zero order, energy is conserved as it should be. The reflection factor is unity, and there is no transmission or absorption. Adding the first-order correction amounts to increasing the flux of the scattered field, so that the first-order solution does not conserve energy. This is a usual problem of first-order solutions. As the power is a quadratic form of the field, adding a first-order correction to the fields amounts to adding a second-order correction to the Poynting vector. It is thus required to also expand the field solution up to second order in order to recover the crossed terms involving a second-order correction of the electric field and a zero-order contribution to the magnetic field, for instance. From a physical point of view, this comment means that the amplitude of the specular term has a second-order correction in order to ensure energy conservation.

If we consider a sinusoidal grating, the surface profile is given by $S(x) = \delta \cos\left(\frac{2\pi x}{d}\right)$. Its Fourier transform is thus given by

$$s(k_x) = \frac{1}{2}\left[\delta\left(k_x - \frac{2\pi}{d}\right) + \delta\left(k_x + \frac{2\pi}{d}\right)\right].$$

We see that the scattered field is actually what is called the diffracted field in the context of a periodic surface. To first order in X, we find the first order of the grating characterized by the wavevector $k_x = k_{x,\text{inc}} \pm \frac{2\pi}{d}$ in agreement with the grating law. Here, we have found not only the angular position of the orders but also their amplitudes. We see that a non-sinusoidal grating would produce higher orders even in a first-order expansion in the surface-profile height. Interestingly, it is seen that the scattered field can be evanescent waves.

Exercises

Exercise 5.1 Electrostatic limit of reflection factor. We consider an interface between two non-magnetic media with dielectric permittivity ε_1 and ε_2. Derive the limit of the Fresnel reflection factor for s and p polarization when the wavevector $k_x \gg \frac{\omega}{c}$. Compare with the electrostatic reflection factor used when introducing an electrostatic image $r = \frac{\varepsilon_1 - \varepsilon_2}{\varepsilon_1 + \varepsilon_2}$.

Exercise 5.2 Transverse-magnetic (TM) reflection factors. We consider an interface between two non-magnetic media with dielectric permittivity ε_1 and ε_2.

(1) Derive the Fresnel reflection factor for p polarization given by Eq. (5.12).

(2) We now use a different definition of the reflection factor for the same polarization. We use the ratio of the x-component of the electric field. Derive the corresponding Fresnel reflection factor.

Exercise 5.3 Metallic reflection and surface impedance. The goal of this exercise is to compare different models often used to deal with reflection by a metal, Fig. 5.13. The rigorous approach leads to the modelling of the fields above the metal and in the metal. The simplest approach is the perfect-conductor model. Within this model, the number of unknowns is divided by two as the fields are zero in the metal. An intermediate approach retains the simplicity of the perfect-conductor model, as the fields in the metal are no longer studied, but allows us to account for losses. This model uses the concept of surface impedance and surface current density. In what follows, we compare the three models in TM polarization.

Perfect conductor model

We consider an interface separating vacuum $z > 0$ from a perfect conductor $z < 0$ characterized by an infinite conductivity, σ. Hence, the electric field is zero for $z < 0$.

(1) The incident magnetic field can be cast in the form

$$\mathbf{H}_i = H_0\mathbf{e}_y \exp[i\mathbf{k}_i.\mathbf{r} - i\omega t] \quad \text{where} \quad \mathbf{k}_i = (\alpha_i, 0, -\gamma_i).$$

Derive the electromagnetic field. The reflection factor is defined by $r = H_r/H_i$, where H_r is the amplitude of the reflected field.

(2) Derive the surface charge density σ_q and the surface-current density $\mathbf{j}_s$ as a function of H_0. Check that the charge conservation law is satisfied.

Finite conductivity metal. We now solve the reflection problem without any approximation. We seek an exact expression of the field and current density in the metal.

(3) Write the Helmholtz equation using the generalized dielectric constant $\varepsilon_r = 1 + i\frac{\sigma}{\omega\varepsilon_0}$.

(4) Write the general form of the fields $r = H_r/H_i$ et $t = H_t/H_i$. Derive the form of the skin depth δ as a function of $\text{Im}(\gamma_t)$.

(5) Derive r and t using the continuity condition of the fields at the interface. Note that there is no surface current, as the metal is not perfectly conducting.

Surface impedance. In this section, we introduce the concept of surface impedance. The basic idea is to consider that the skin depth is much smaller than any other relevant length scale.

(6) Using the results of questions 4 and 5, derive the current density as a function of x and z.

(7) We assume that the electrical conductivity is large, so that $\dfrac{\sigma}{\omega\varepsilon_0} \gg 1$. Derive an asymptotic form of ε_r, γ_t, and δ.

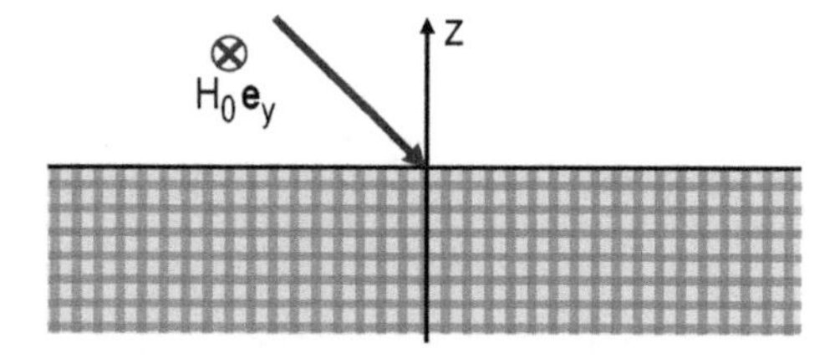

Fig. 5.13 Wave impinging on a metal surface.

(8) We introduce a surface current density $\mathbf{j}_s$ defined as follows:

$$\mathbf{j}_s = \int_{-\infty}^{0} \mathbf{j}\, dz.$$

Show that the current is tangential in the limit of a good conductor. The surface impedance Z_s is defined by:

$$\mathbf{j}_s = \frac{\mathbf{E}_{tg}}{Z_s},$$

where $\mathbf{E}_{tg}$ is the tangential component of the electric field parallel to the plane $z = 0$. Derive the form of Z_s.

(9) Show that in the plane $z = 0$:

$$\mathbf{E}_{tg} = -Z_s \mathbf{H}_{tg} \times \mathbf{e}_z. \tag{5.34}$$

This equation provides a link between the incident field and the reflected field. It can be used as a boundary condition, so that it is no longer necessary to compute the transmitted field.

(10) We now compute the losses in the metal. It will be seen that the impedance provides the result. Using the full solution, derive the absorption in the metal and compare to the flux of the Poynting vector and the formula

$$\langle P \rangle = \frac{1}{4} \frac{\mathrm{Re}(Z_s)}{|Z_s|^2} |\mathbf{E}_{tg}(z = 0)|^2.$$

(11) Check that the losses tend to zero in the limit of a perfect conductor. This is another form of the paradox of the absorber: as the imaginary part of the dielectric constant tends to infinity, the absorption tends to zero. To understand the physical origin of this behaviour, analyse the amplitude of the transmitted field and the skin depth as the conductivity tends to infinity.

Exercise 5.4 Reflection by two parallel interfaces. Anti-reflection coating. We consider a thin film of thickness d and refractive index n' deposited on a substrate with refractive index n (see Fig. 5.14). Both media are linear, homogeneous, isotropic, and non-magnetic ($\mu = \mu_0$). An incident plane wave with frequency ω and linearly polarized along the y-axis propagates in vacuum and illuminates the system at normal incidence. Its electric field is given by $\mathbf{E} = E_0\, \mathbf{e}_y\, \exp[-i(kz + \omega t)]$, with $k = \omega/c$.

(1) Following the general procedure used to derive the Fresnel factor in TE polarization, derive the reflection ρ and transmission τ factors of the two-layer system. To proceed, write the general solution of the fields in the three media. Establish the form of the Fresnel factors:

$$\rho = \frac{r_{12} + r_{23}\, \exp(2in'kd)}{1 - r_{21}r_{23}\, \exp(2in'kd)},$$

with $r_{ij} = (n_i - n_j)/(n_i + n_j)$ (reflection factor at the interface $i - j$ at normal incidence),

$$\tau = \frac{t_{12}t_{23}\, \exp(in'kd)}{1 - r_{21}r_{23}\, \exp(2in'kd)}\, \exp(-inkd),$$

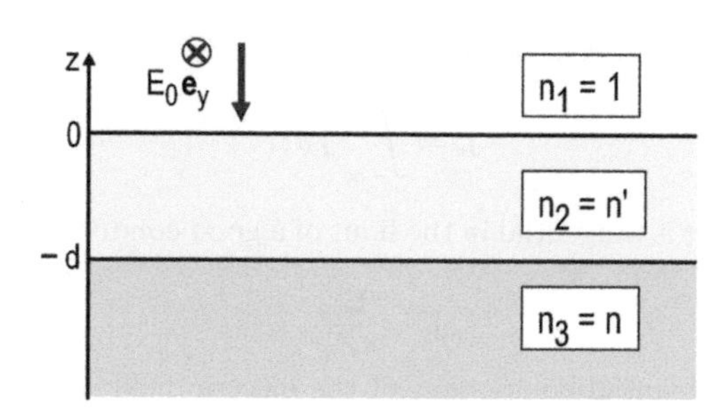

Fig. 5.14 Two parallel interfaces illuminated at normal incidence (anti-reflection coating).

where $t_{ij} = 2n_i/(n_i + n_j)$ is the transmission factor at the interface $i - j$ at normal incidence.

(2) Show that the reflection factor can be zero by properly choosing n' and d.

Exercise 5.5 Absorption by a thin film. In this exercise, we consider a thin metallic film with thickness d located between the planes $z = d/2$ and $z = -d/2$. Here, thin means that the thickness of the film is smaller than the skin depth of the metal. The goal of the exercise is to show that such a thin film can absorb half of the power of an incident plane wave at normal incidence. We denote by σ the conductivity of the metal. The optical properties of the metal are given by $\varepsilon = 1 - \frac{i\sigma}{\omega\varepsilon_0}$.

Derivation of r and t, first method.

(1) Use the result of the previous exercise to derive:

$$r = \frac{E_r}{E_{\text{inc}}} = \frac{y}{2+y} \; ; t = \frac{E_t}{E_{\text{inc}}} = \frac{2}{2+y} \; ; y = \mu_0 c\sigma d. \tag{5.35}$$

Derivation of r and t, second method.

(2) What is the form of the power dissipated per unit volume as a function of conductivity and electric field?

(3) Write the conservation of the electromagnetic energy in the volume contained between two planes, $z = h$ and $z = -h$. Prove that

$$1 = |r|^2 + |t|^2 + y|t|^2. \tag{5.36}$$

To establish the above equation, we have assumed that the field in the metal is uniform and equal to the amplitude of the transmitted field. Explain why this is a valid assumption.

(4) Prove that $1 + r = t$.

(5) Derive the value of r and t.

Derivation of r and t, third method. In this section, we take the point of view that the reflected field is the field radiated by the induced currents in the film. Similarly, the transmitted field is the superposition of the incident field and the field radiated by the induced currents in the film. We will thus derive the amplitude of the reflected and transmitted waves by computing the field radiated by the currents induced in the thin film.

(6) Prove that the electric field radiated by a monochromatic and uniform current density $\mathbf{j}_0$ in the thin film is given by $-\frac{\mu_0 c d}{2}\mathbf{j}_0 \exp(-i\frac{\omega}{c}|z|)$. We remind the reader that for a plane wave, $\mathbf{E} = i\omega\mathbf{A}$, and that the general form of the potential vector is given by

$$\mathbf{A}(\mathbf{r},t) = \frac{\mu_0}{4\pi}\int \mathbf{j}\frac{\exp(i\frac{\omega}{c}R)}{R}\mathbf{dr}', \tag{5.37}$$

where $R = |\mathbf{r} - \mathbf{r}'|$. Use the following integral:

$$\int dx' \int dy' \frac{\exp(i\frac{\omega}{c}R)}{R} = \frac{-2\pi\omega}{c}\exp(i\frac{\omega}{c}|z - z'|). \tag{5.38}$$

(7) The film is illuminated by a plane wave at normal incidence with an electric field denoted by E_{inc}. From the linearity of Maxwell equations, the field is everywhere the sum of the incident field and the field radiated by the induced currents in the film. Write the fields for $z > d/2$ and $z < -d/2$. We denote by E_0 the amplitude of the field in the film. Derive the relation $E_{\text{inc}} = E_0[1 + \frac{\mu_0 \sigma c d}{2}]$. Derive the reflection and transmission factors.

Maximizing the absorption. Prove that the power absorbed in the film per unit area is given by

$$P_{abs} = \frac{\varepsilon_0 c}{2} E_{\text{inc}}^2 \frac{4y}{(2 + y)^2}. \tag{5.39}$$

(8) The absorptivity is the ratio of the power absorbed P_{abs} per unit area divided by the incident Poynting vector $\frac{\varepsilon_0 c}{2}E_{\text{inc}}^2$. Prove that there is a maximum absorptivity $A = 0.5$ for a thickness d given by $2\varepsilon_0 c/\sigma$. If such an optimized film is placed on top of a perfectly conducting mirror at a distance of $\lambda/4$, the system becomes totally absorbing. This is known as the Salisbury screen. This configuration is used to increase the absorptivity of bolometers.[2]

Exercise 5.6 Goos–Hänchen effect. The purpose of this exercise is to study the reflection of a beam with a finite size. We will show that the reflected beam is shifted along the interface. A finite beam can be described as a linear superposition of plane waves or, in other words, a wavepacket. As the reflection factor amplitude depends on the angle of incidence, the different plane waves composing the beam are reflected differently. This introduces a distortion of the reflected beam. A phase distortion can be modelled as a lateral shift called the Goos–Hänchen shift (Fig. 5.15), whereas an amplitude distortion can be modelled as an angular shift.

(a) For the sake of simplicity, we consider a 2D monochromatic beam which depends on x and z. We consider a TM-polarized beam and denote by $H_y^{\text{inc}}(x, z)$ the y-component of the magnetic field. The beam propagates towards negative z and can be represented as follows:

$$H_y(x, z) = \int \frac{dk_x}{2\pi} H_y(k_x)\exp(ik_x x - i\gamma z). \tag{5.40}$$

[2] See a discussion in S. Bauer, 'Optical properties of a metal film and its application as an infrared absorber and as a beam splitter', S. Bauer, *Am. J. Phys.* **60**, 257 (1992).

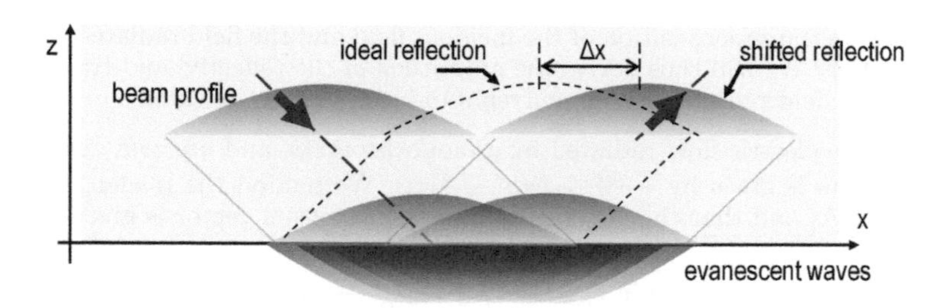

Fig. 5.15 Apparent lateral shift of the reflected beam (Goos–Hänchen shift).

(b) The beam impinges on an interface with a reflection factor denoted by $r(k_x)$. Give the form of the reflected field as an integral over k_x.

(c) In this question, we assume that the beam is well collimated and has a central propagation direction characterized by k_{x0}. We further assume that there is total internal reflection, so that the reflection factor has a unit amplitude and can be described as a phase factor $\exp[i\Phi(k_x)]$.

(*i*) We first assume that the beam is so well collimated that all the plane waves have the same reflection factor $\exp[i\Phi(k_{x0})]$. Show that the reflected field is given by $\exp[i\Phi(k_{x0})]H_y^{\mathrm{inc}}(x,-z)$.

(*ii*) We now explore the deviations due to the dependence of the reflection factor on k_x. Expand to first order the phase around the central direction k_{x0} and prove that the reflected field can be written as $\exp[i\Phi(k_{x0})]\exp(i\delta)\,H_y^{\mathrm{inc}}(x-\Delta x,-z)$. Give the form of δ and Δx. The shift along the x-axis is called the Goos–Hänchen shift.

6 Guided modes

An optical waveguide can be defined with some generality as any structure with a translational invariance along a given axis used for guiding the flow of electromagnetic waves along this axis. It allows attenuation due to diffraction of a beam propagating in a non-lossy homogeneous space to be prevented. It is a key component used for transmission over 100 km with optical fibres, or, for shorter ranges, on chips in the field known as integrated optics. Waveguides transport light by guiding it within a finite set of optical modes which correspond to discrete solutions of Maxwell equations with a transverse spatial intensity pattern that remains constant along the waveguide axis.

In section 6.1, we first introduce different types of waveguide structures used for radio- or optical waves. Then the concept of optical waveguide modes is discussed qualitatively, key results being presented with minimal proofs. In section 6.2, modes of the dielectric planar waveguides are studied as a useful example to substantiate important concepts such as dispersion, effective index, group velocity, cutoff, and confinement. Section 6.3 presents important properties of modes, their orthogonality, which differs for lossy and non-lossy waveguides, and, alongside several complements, it presents some key results on light transport of guided waves, e.g. the equivalence between the group and energy velocities in lossless waveguides.

Comprehensive textbooks exist on optical waveguides: see (Snyder and Love, 1983), for instance. We thus mostly try to catch the key concepts that are important for nanophotonics.

6.1 Introduction to guided modes

In nanophotonics, three general types of waveguide structures (second row in Fig. 6.1) have an impact on applications. The first type involves dielectric materials only, and especially high-index dielectric (H) cores surrounded by low-index dielectric (L) claddings, such as semiconductor photonic wires or photonic-crystal waveguides. On chip, the typical attenuation is $\simeq 0.1$–$10\,\mathrm{dB/cm}$ and the transverse mode is about one wavelength. The second type involves metal–dielectric–metal (MHM or MLM) structures. The mode is essentially confined to the thin dielectric layer, and its size is much smaller than the wavelength; the attenuation is much larger, at $\simeq 1\,\mathrm{dB}/\mu\mathrm{m}$. The last type is a hybrid of the two previous types, involving MLH structures with intermediate mode sizes and attenuations.

Although they are all translation-invariant in one or two directions and thus all sustain various modes that convey electromagnetic waves between two endpoints, these structures are completely different. To understand why nanophotonics offers such a wide variety of guiding structures, it is relevant to start with waveguides for radiowaves.

Introduction to Nanophotonics. Henri Benisty, Jean-Jacques Greffet, and Philippe Lalanne, Oxford University Press.
 DOI: 10.1093/oso/9780198786139.003.0006

6.1.1 From radiowaves to photonic waveguides

At radiowave frequencies, the wavelength λ is about 10 m. Even if dielectrics with ultra-high indices n and low absorption exist, waveguides composed of only dielectric materials with transverse dimensions of $\sim\lambda/n$ with $n \sim$ 10–100 cannot find broad applications, owing to their lack of compactness. Thus, waveguides for radiowave should confine light into modes with transverse sizes that are much smaller than λ. This turns out to be possible only with waveguides that confine the field into a dielectric sandwiched between two metallic materials.

Fortunately, metals are very good conductors at low frequencies and metallic waveguides have low attenuations. The top panel in Fig. 6.1 sketches the most famous radiofrequency waveguides.

The simplest guides are the two-wire line and its miniaturized equivalent on microwave chips, the microstrip line. The two conductors are separated by an insulating dielectric. In addition to absorption, attenuation upon propagation is caused by radiation in the cladding.

In coaxial cables, modes are essentially confined to a dielectric, generally a flexible polymer, surrounding a central metal wire. The shielding outer conductor is often a (flexible) dense mesh of thin wires. There is no frequency cutoff, in the sense defined hereafter, and coaxial cables work well from dc to a few GHz. The outer conductors prevent the system from radiating outside. Thus, loss is only due to the conductor's ohmic loss at the periphery of the metal parts on a skin-depth thickness, and from the polymer residual dielectric losses. At, say, 2 GHz, even the best coaxial cables have a $\sim$50 dB loss and can hardly be used above 100 m.

At around 3 GHz ($\lambda \simeq$ 10 cm), guided modes can propagate in reasonably small hollow conductors of comparable size. Conversely, in the low-frequency limit, there is a cutoff owing to the impossibility of generating a voltage (a nearly dc signal is essentially a voltage) across two points of such a *conducting* connex waveguide. Hence, in retrospect, there were good reasons to have two independent conductors in the previous coaxial example. The way a mode in the hollow waveguide bounces on the four conducting walls as it propagates is closely related to the example of Complement 6.D on a slab waveguide with perfectly conducting walls.

All these waveguides, even 'coaxial cables', which remain a challenge for nanofabrication and offer deep subwavelength confinements, have been studied in nanophotonics. The microstrip mode, called a metal–dielectric–metal waveguide in nanophotonics, plays a key role in advanced studies, as it supports gap-surface-plasmon modes with slow group velocities (with strong attenuation, too) when the dielectric gap thickness becomes smaller than 10 nm (see Chapter 13). Embodiments of metal–dielectric–metal waveguides are the V-grooves in metal substrates or nanowires deposited on metallic substrates covered by an ultra-thin dielectric layer (see Fig. 6.1).

Low attenuation is obtained with purely dielectric waveguides with transverse sizes $\sim \lambda/n$. Guidance is guaranteed by total internal reflection of light in a high-index *core* surrounded by *claddings* with lower indices. Much of the early days of integrated optics have focused on cores fabricated by local crystal or glass modifications, e.g. by ion migration under adequate chemistry on lithium niobate or on silica, implementing tiny index variations. In this way fully functional low-loss devices, with reasonable small

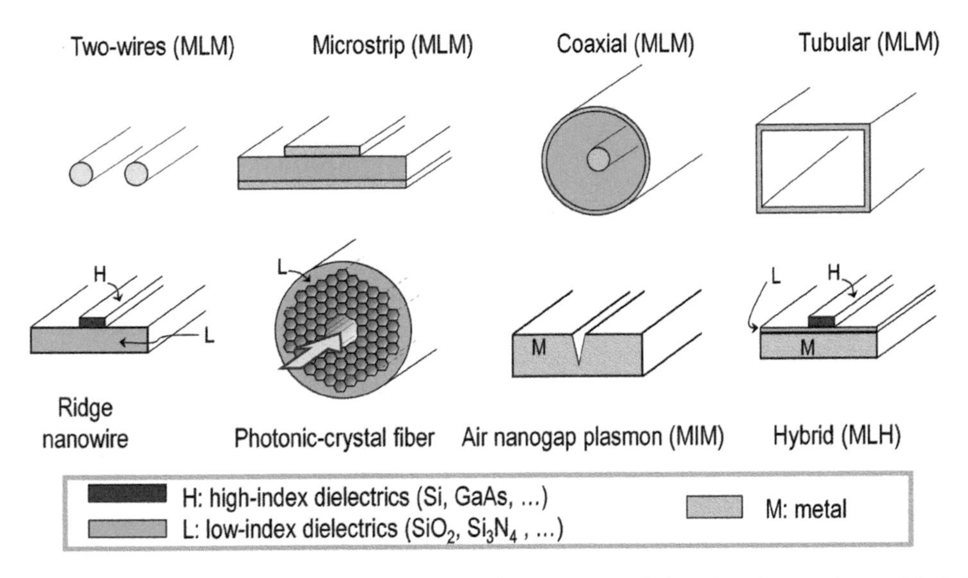

Fig. 6.1 Top row: radiofrequency waveguides; bottom row: dielectric, plasmonic, and hybrid nanowaveguides. Code for material types (H, L, M) is at the bottom. H and L types are often called I or insulator types.

mode transverse sizes ($\sim$2–5λ for the near infrared) were developed. Modern guidance techniques rely on high-index contrasts between the core and cladding materials (notably obtained with semiconductor cores), casting different compromises between ease and flexibility of fabrication, and other parameters such as index contrast, mode size, residual losses, and sensitivity to imperfections and interface roughness. Famous examples are semiconductor wires or ridge waveguides, as sketched in Fig. 6.1, or photonic-crystal waveguides that are not translational-invariant but periodic in the direction of propagation (Chapter 17).

Total internal reflection is not the sole guidance mechanism: Bragg reflection (see Chapter 16) may also be used. Fig. 6.1 depicts a case of guidance in a low-index material (air) with a photonic-crystal cladding composed of a mesh of holes in silica to ensure a mechanical stability and forbid lateral radiation.

6.1.2 Waveguide modes: General definition

We begin by considering a waveguide with an arbitrary cross section (e.g. any of those in Fig. 6.1). We denote the invariant axis by z. We assume that the permeability $\mu = \mu_0$ everywhere, so the guide is fully described by its dielectric permittivity tensor $\varepsilon(x, y)$.

Formally, there are infinitely many ways in which the electric and magnetic fields can arrange themselves to form a field distribution whose intensity remains stable (if absorption is neglected) upon propagation along z at a given frequency ω, but only a finite subset has a field pattern that decays exponentially in the outer spaces (the claddings) surrounding the waveguide core. All those distributions deserve to be called 'modes'; however, hereafter, we will mainly consider the finite subset of *guided* modes that propagate over long distances (many wavelengths), once launched in the waveguide.

By definition, a mode, labelled by a running subscript n, is a solution of the source-free Maxwell equations

$$\nabla \times \mathbf{E}_n(\mathbf{r}) = i\omega\mu_0\mathbf{H}_n(\mathbf{r}), \tag{6.1}$$
$$\nabla \times \mathbf{H}_n(\mathbf{r}) = -i\omega\varepsilon(x,y)\mathbf{E}_n(\mathbf{r}). \tag{6.2}$$

In the next chapter, on cavities, we will look for the solutions of Eqs. (6.1) and (6.2) for complex-valued frequencies. Cavities are 3D systems with all dimensions finite. The present geometry is not of finite dimension, as one direction, z, is invariant. Thus it features an additional 'quantum number', the z-component of the wavevector (further denoted by β) that is conserved in the whole space, implying that we may solve Eqs. (6.1) and (6.2) for a real frequency ω and compute $\beta(\omega)$. Waveguide modes are thus, by definition, modes with an infinite lifetime and an infinite spatial extension in the z-direction if material absorption is neglected (modes have an exponentially decaying behaviour otherwise). Owing to the spatial and temporal invariances, the actual fields $\mathbf{E}_n(\mathbf{r})$ and $\mathbf{H}_n(\mathbf{r})$ are

$$\mathbf{E}_n(\mathbf{r}) = \mathsf{E}_n(x,y)\ \exp(i(\beta_n z - \omega t)), \tag{6.3}$$
$$\mathbf{H}_n(\mathbf{r}) = \mathsf{H}_n(x,y)\ \exp(i(\beta_n z - \omega t)), \tag{6.4}$$

where $\mathsf{E}_n(x,y)$ and $\mathsf{H}_n(x,y)$ are three-component vectors with a distinct typeface to distinguish them from the actual fields $\mathbf{E}, \mathbf{H}$. They describe the mode structure or its *profile*, i.e. the spatial distribution of the electromagnetic field across the core and the claddings.

The factor β_n is a propagation constant. It relates the phase at different points just like the wavevector modulus $k = n_0\omega/c$ does in the similar expression $E = E_0\ \exp(i(kz - \omega t))$ for a plane wave propagating along the z-axis in a medium of refractive index n_0. For this very reason, one can introduce a convenient dimensionless quantity that captures the phase evolution along z in the usual sense of a refractive index

$$n_{\text{eff,n}} = \frac{\beta_n}{\omega/c}, \tag{6.5}$$

which is known as the *effective index* of the nth mode. The denomination 'effective' broadly means averaged; the electromagnetic field of the mode covers both the core and the cladding, and sees an 'average' medium as it propagates, with an effective index between the refractive index of the core and that of the claddings.

To determine the propagation constants β_n and the mode profiles $\mathsf{E}_n(x,y)$ and $\mathsf{H}_n(x,y)$, we conveniently inject Eqs. (6.3) and (6.4) into Eqs. (6.1) and (6.2) to obtain

$$\nabla \times \mathsf{E}_n(x,y) - i\omega\mu_0\mathsf{H}_n(x,y) = -i\beta_n\,\mathbf{u}_z \times \mathsf{E}_n(x,y), \tag{6.6}$$
$$\nabla \times \mathsf{H}_n(x,y) + i\omega\mu_0\mathsf{E}_n(x,y) = +i\beta_n\,\mathbf{u}_z \times \mathsf{H}_n(x,y). \tag{6.7}$$

Just like for plane waves in bulk media in Chapter 4, when a pulse propagates in a waveguide with a central frequency ω, it is convenient to define the group velocity of a guided-wave packet in a given mode $v_{g,m} = \partial\omega/\partial\beta_m$. Using $\beta_n = n_{\text{eff,n}}\omega/c$ and omitting the mode index n, it reads

$$v_g = \frac{c}{n_{\text{eff}} + \omega\left(\partial n_{\text{eff}}/\partial\omega\right)} = \frac{c}{n_g}, \tag{6.8}$$

where the group index $n_g = n_g(\omega)$ is also conveniently defined.

Symmetry plays a key role in waveguide theory and integrated optics design. In addition to the usual planar symmetry with respect to a plane parallel to the z-axis (e.g. planes $x = 0$ or $y = 0$) which implies that the modal field components are either symmetric or antisymmetric with respect to this plane, it is noteworthy that z-invariant waveguides possess a mirror symmetry with respect to any transverse plane parallel to the plane $z = 0$. We refer the reader to Complement 6.A, where we demonstrate the existence of pairs of counter-propagative modes. This 'intuitive' result stipulates that, for any given forward-propagating mode having a $\exp(i\beta z)$ dependence with $\beta > 0$, a backward-propagating mode with an $\exp(-i\beta z)$ dependence exists. This is, however, guaranteed by the unintuitive condition that the waveguide materials are reciprocal (the dielectric tensor obeys $\varepsilon^{\mathbf{T}} = \varepsilon$).

Only a few geometries can be solved analytically, e.g. the step-index fibre, the dielectric slab of section 6.2, the dielectric slab bounded by two perfect-conductor 'hard-wall' planes of Complement 6.D. Thus, in practice, one should resort to numerical computations. Many mode solvers exist, based on different numerical schemes (finite element, Fourier methods). Most solvers implement Eqs. (6.6) and (6.7) as a generalized eigenproblem whose eigenvalues are the β_n's and whose eigenvectors are the field profiles. Note that the eigenoperators depend on the frequency ω and the permittivity tensor. The numerical implementation does not present any particular difficulty, but attention should be paid to the fact that a finite transverse domain is in general implemented, whereas the physical problem is an open one. Thus, one should satisfy additional outgoing wave conditions at the boundaries of the numerical domain: no power should flow towards the core. There are an infinity of eigenvectors (as many as the dimension of the numerical basis used), and most of them correspond to the so-called radiation modes, whose fields are not confined near the core; only a finite set corresponds to the guided modes with mode profiles that exponentially decay in the cladding.

Before considering the general properties of guided modes again in section 6.3, it is valuable to consider an analytical case to get familiar with what a mode is, why a guided mode decays exponentially in the nature of the cladding and which values of the effective index are allowed. We will use the classical example of a dielectric slab surrounded by dielectric claddings. Exercise 6.3 prompts the reader to implement a program to compute the modes of the slab. This program is used in the following exercises to catch all the main properties of guided modes on their own, including their symmetry, orthogonality, and the equivalence between their group and energy velocities that is documented in section 6.3.

6.2 The slab waveguide

In this section, we consider the examples of a dielectric slab waveguide, an example that is universally considered in all textbooks, to unveil the most important characteristics

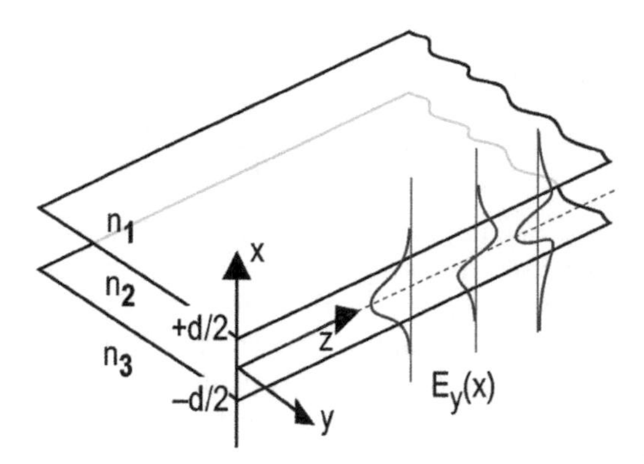

Fig. 6.2 Slab waveguide with core index n_2, with different cladding layers of lower indices $n_1 < n_3 < n_2$. On the side, sketch of the field profiles $\mathsf{E}_y(x)$ for the three lower-order transverse-electric modes, assuming propagation along z.

linked to confinement in optical waveguides. The geometry, shown in Fig. 6.2, consists in a dielectric slab waveguide with core index n_2, with two different claddings of lower indices $n_1 < n_3 < n_2$. The system possesses two translation invariances along the y- and z-axes, so that we need to introduce two spatial quantum numbers and look for modes with a factored expression $\exp(i(k_y y + k_z z - \omega t))$ for a mode propagating in the in-plane direction $[k_y, k_z]$. For simplicity reasons only, we will assume that $k_y = 0$ and again denote k_z by β.

We encourage the reader to show that the group of the six-curl Maxwell equations of Eqs. (6.1) and (6.2) splits into two independent groups. One group, designated as TM (transverse-magnetic or H-parallel), has **H** along y and **E** with two components, one along x and the other along the propagation direction z. Conversely, the other group, designated as TE (transverse-electric or E-parallel), has **E** along y while **H** has two components, one along x and the other along z. Hereafter, we study TE guided modes only.[1]

6.2.1 Field expressions

For TE polarization, the electric field **E** of the modes has a single non-vanishing component $E_y(x, z)$, and its knowledge amounts to knowing all field components since the two non-zero magnetic field components, H_x and H_z, can be derived as partial spatial derivatives of E_y according to Eq. (6.2).

Solving the source-free Maxwell equations (6.1) and (6.2) amounts to solving the Maxwell equations in three uniform media with refractive indices n_1, n_2, n_3 and satisfying the field-continuity conditions at the interfaces. For TE polarization, the electric field $E_y(x, z)$ and its normal derivative $\partial E_y/\partial x$ have to be continuous (due to the H_z continuity).

[1]Confusion arises regarding the names TE and TM, since the names are different for waveguides and photonic crystals, not to mention the simple reflection at a planar interface. The name E- or H-parallel is preferable, as it refers to the field that is parallel to the interfaces.

We do know the solutions of source-free Maxwell equations in uniform media. They are plane waves, implying that, in every medium labelled $p = 1, 2, 3$, the electric field takes the form $\propto \exp(ik_{x,p}x + ik_{z,p}z)$ with $k_{x,p}^2 + k_{z,p}^2 = (\omega/c)^2 n_p^2$. Importantly, we note that $E_y(x, z)$ should be continuous at the two interfaces, i.e. for $|x| = d/2$ and for *any* z. The only way that $\exp(ik_{z,1}z) \propto \exp(ik_{z,2}z) \propto \exp(ik_{z,3}z)$ for any z is for $k_{z,p}$ to be the same in every layer. It will be denoted β hereafter for consistency. Thus, in a simple example, the general factored form adopted in Eqs. (6.3) and (6.4) based on translation invariance is justified by the field-continuity condition over infinite planes. One often says that the *parallel wavevector components* are unchanged across the three media, much as was the case at a single interface illuminated by a plane wave in Chapter 5: this is nothing but Snell's law.

Therefore, the electric field can be written

$$\begin{aligned}
&\text{for } x < -d/2,\ E_y(x,z) = D\exp(i\beta z)\exp(ik_{x,3}(x+d/2))\,,\\
&\text{for } |x| < d/2,\ E_y(x,z) = \exp(i\beta z)[B\cos(k_{x,2}x) + C\sin(k_{x,2}x)]\,,\\
&\text{for } x > +d/2,\ E_y(x,z) = A\exp(i\beta z)\exp(ik_{x,1}(x-d/2))\,,
\end{aligned} \tag{6.9}$$

where, for convenience, we introduce a sum of sine and cosine functions instead of a sum of two exponential functions with opposite k_x for $|x| < d/2$, allowing partly stationary waves with cosinusoidal or sinusoidal fields to be formed. Owing to the fact that the source-free Maxwells equations are satisfied in every medium, we have

$$\text{for } p = 1, 2, 3, \quad \beta^2 + k_{x,p}^2 = (\omega/c)^2 n_p^2. \tag{6.10}$$

For $p = 1, 3$, we have two possibilities for $k_{x,p}$, according to Eq. (6.10). If $\beta^2 < (\omega/c)^2 n_p^2$, $k_{x,p}^2 > 0$, implying that $k_{x,p}$ is a real number: the field in the cladding is a progressive plane wave whose amplitude does not decay as the distance to the core increases. Conversely, if $\beta^2 > (\omega/c)^2 n_p^2$, $k_{x,p}^2 < 0$: $k_{x,p}$ is a purely imaginary number. This is exactly the condition we require for a truly guided mode with a field confined near the core (of course, with the right sign of $k_{x,p}$). This conclusively demonstrates that a necessary condition for guidance is $n_{\text{eff}} > \max(n_1, n_3)$.

As the mode results from a field arrangement that remains stable upon propagation by covering both the core and cladding media, it is difficult to imagine an effective index exceeding the largest refractive index n_2 of the core, a value that can be expected to be approached when the field marginally sees the claddings and is nearly completely confined in the core. We thus expect that the effective indices of guided modes satisfies the condition

$$n_2 > n_{\text{eff}} > max(n_1, n_3), \tag{6.11}$$

a condition that is expected to be generally verified when the guiding mechanism is total internal reflection.

The electric field distribution of Eq. (6.9) can be written

$$\begin{aligned}
&\text{for } x < -d/2,\ E_y(x,z) = D\exp(i\beta z)\exp(s(x+d/2))\,,\\
&\text{for } |x| < d/2,\ E_y(x,z) = \exp(i\beta z)[B\cos(hx) + C\sin(hx)]\,,\\
&\text{for } x > +d/2,\ E_y(x,z) = A\exp(i\beta z)\exp(-q(x-d/2))\,,
\end{aligned} \tag{6.12}$$

with $q = (\beta^2 - (n_1\omega/c)^2)^{1/2}$, $s = (\beta^2 - (n_3\omega/c)^2)^{1/2}$, $h = ((n_2\omega/c)^2 - \beta^2)^{1/2}$, and $+q$ and $-s$ solutions discarded to grant confinement. The field is exponentially decaying (evanescent) in the cladding and is a superposition of two progressive plane waves with wavevectors $[\beta, h]$ and $[\beta, -h]$ in the core. The distinction between progressive and evanescent waves will also be crucial in explaining the regimes of near-field and far-field in Chapter 10.

The remainder of the calculation is less interesting. Three constants among A, B, C, D can be determined by writing the continuity equations for E_y at boundaries $x = d/2$ and $x = -d/2$ and for H_z or equivalently $\partial E_y/\partial x)$ at boundary $x = d/2$. The procedure provides three equations for four unknowns and results in C, A, D being expressed as functions of B,

$$\begin{aligned}
&\text{for } x < -d/2,\ E_y(x,z) = B\frac{h\cos(hd) + q\sin(hd)}{h\cos(hd/2) + q\sin(hd/2)}\exp(i\beta z)\exp(s(x + d/2))\,,\\
&\text{for } |x| <\ d/2,\ E_y(x,z) = B\exp(i\beta z)[\cos(hx) + \frac{h\sin(hd/2) - q\cos(hd/2)}{h\cos(hd/2) + q\sin(hd/2)}\sin(hx)]\,,\\
&\text{for } x > +d/2,\ E_y(x,z) = B\frac{h}{h\cos(hd/2) + q\sin(hd/2)}\exp(i\beta z)\exp(-q(x - d/2))\,.
\end{aligned} \tag{6.13}$$

The free parameter B can further be used for mode normalization (it may be fixed by setting the mode power flow to unity, for instance). What remains to be matched is a single field continuity for H_z at boundary $x = -d/2$. Since q, h, and s in Eq. (6.13) are all dependent on the sole unknown β, this additional field continuity results in a final equation in which only the propagation constant β shows up. We let the reader derive this transcendental equation that can be solved only with the help of a computer (the literature proposes a graphical solution, too). The equation being non-linear, there are several solutions for β in general, β_1, β_2 ... β_N corresponding to N guided modes that may propagate at frequency ω.

In Exercise 6.3, we investigate a numerical method for solving the TE electromagnetic mode of thin-film stacks. The method is similar to that used in chapters 15 and 16 on Bloch modes; it does not rely on solving a transcendental equation (i.e. solving for the zeros or poles of a non-linear equation), but relies on solving an eigenproblem, much like most numerical-solver softwares do. The numerical method will also be further explored to check the main theoretical results of the chapter on the orthogonality issue.

Figure 6.3 shows the squared electric field profile of three truly guided modes: the first (fundamental) mode and the second and fourth modes (often called higher-order modes). The profiles are computed for a symmetric waveguide with $n_1 = n_3$. The profile of the fundamental mode is symmetric, the profile of mode two is antisymmetric with one node in the slab centre, the profile of mode four is also antisymmetric with three nodes. Importantly, the fundamental mode has the smallest k_x in the core, and this means that it has the largest β and the largest n_{eff} of all guided modes. All other modes have smaller effective indices, in sequence of course.

Let us now briefly discuss the higher-order modes. When many modes are allowed, in broader guides, the mth mode must nearly fit its m lobes within the core, thus

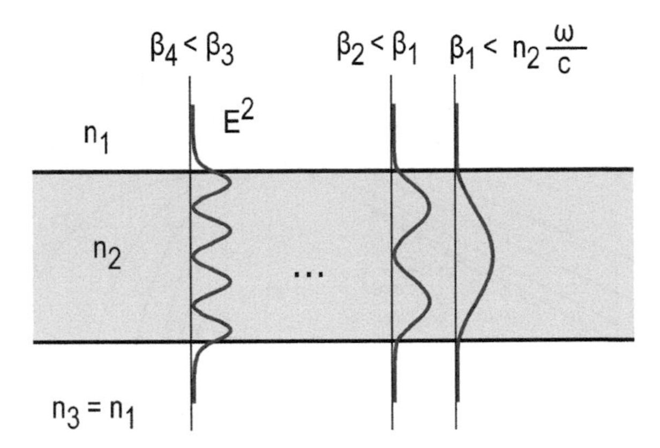

Fig. 6.3 Mode profiles $E_y(x)$ of the fundamental modes of a symmetric slab waveguide $n_1 = n_3$ at a given frequency, ω. The fundamental mode ($m = 1$), with the lowest β value (β_1) possesses no node of the electric field. The second mode $m = 2$ (with the propagation constant β_2) has one node, ... Clearly, $k_{x,2}$ increases in the core as m increases, and owing to the fact that $\beta^2 + k_{x,2}^2 = (\omega/c)^2 n_2^2$, we find that β decreases with m. We also observe that as m increases, the field decreases faster in the claddings. Again, this is realized from $q = s = (\beta^2 - (n_1\omega/c)^2)^{1/2}$, therein implying that q or s increases with β.

$k_{x,2}d \simeq m\pi$: the guide becomes closer to a hard-wall cavity with zero-field boundary conditions (see Complement 6.D) than a single-mode case of a narrower guide. The series of effective indices can then be approximated by

$$n_{\text{eff,m}}^2 \simeq n_2^2 - m^2\pi^2/(k_0 d)^2 = n_2^2 - m^2(\lambda/2d)^2. \tag{6.14}$$

This means that the effective indices series $n_{\text{eff,m}} \simeq n_2 - m^2(\lambda/2d)^2/(2n_2)$ samples an arithmetic series (of pitch $(\lambda/2d)^2/(2n_2)$); hence, some remarkable relations arise between their phases after a given length z along the waveguide, as we will suggest later. Note that the same conclusion holds for TM modes.

6.2.2 Dispersion diagram

The procedure of the previous section should be repeated to find the dispersion of every individual guided mode. A powerful representation is the $\omega(\beta)$ dispersion diagram shown in Fig. 6.4. The diagram visualizes the pulsation ω (or sometimes $k_0 = \omega/c$, or the dimensionless quantity $k_0 d/2\pi \equiv d/\lambda$) as a function of momentum β. This representation is directly inspired from the band diagram of electrons in crystals (note that by multiplying ω and β by $\hbar$, we then plot the energy as a function of the in-plane momentum). Fig. 6.4 contains two dashed lines, $\omega/c = \beta/n_2$ and $\omega/c = \beta/n_1$, which are straight lines because we neglect material dispersion (n_2 and n_1 are assumed to be frequency-independent). These two lines, often called the light lines of the core and cladding materials, delineate a sector in which all guided modes must reside.

The lowest curve represents the fundamental guided mode. This bound solution can exist down to $\omega = 0$ for the symmetric case studied here. We then say that there is no cutoff for this $m = 1$ mode, implying that the waveguide is monomode for frequencies

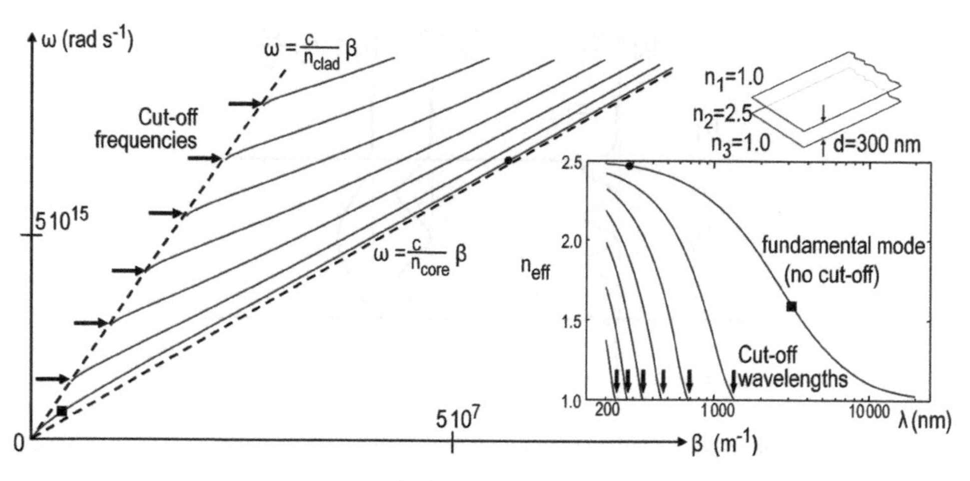

Fig. 6.4 Dispersion relation of the transverse-electric guided modes of a symmetric slab ($d = 300\,\mathrm{nm}$, $n_1 = n_3 = 1$, $n_2 = 2.5$). A few modes are represented. Dashed lines are the light line of the core and cladding. Note the cutoff for all but the fundamental mode, highlighted with short arrows. In the bottom right inset, the mode effective indices are represented as a function of wavelength on log scale. The dot and square symbols added to the fundamental mode lines of main plot and inset help grasping the $n_{\text{eff}} \leftrightarrow \beta$ and $\lambda \leftrightarrow \omega$ correspondences.

below the cutoff of the second mode (for an asymmetric waveguide with $n_1 \neq n_3$, however, the fundamental mode has a cutoff). It is striking that the fundamental mode dispersion is asymptotically approaching the cladding light line at low frequencies and the core light line at high frequencies. This behaviour will be analysed in the next section. We also note that a similar trend is valid for the higher-order modes, the other blue curves. However, the dispersion curve of these modes reaches the cladding light line, where we have a cutoff at some frequency. This idea of a cutoff is very important for practical applications: it is often desirable to use a single guided mode for signal transmission. Only below the cutoff frequency of the second mode can this single mode operation be safely ensured. Above the cutoff frequency, at least two guided modes exist; in practice, even if we only excite a single mode, fabrication imperfections or waveguide bends commonly couple different modes together. The transmitted intensity then results from an interference with all the risks brought by uncontrolled phase differences.

In a waveguide, each mode m has its own effective index dispersion $n_{\text{eff,m}}(\omega)$. Therefore, the modes have different group velocities. Part of these variations comes from the guidance mechanism itself, as was shown in Fig. 6.4. Another part of the dispersion, often the minority one, is the material dispersion itself; in other words, $\partial\varepsilon/\partial\omega$. An important consequence is that an optical pulse of short duration, made from the superposition of wavepackets from different modes $\{m\}$ and different frequencies $\{\omega\}$ centred around some value ω_0, splits into subpackets associated with the specific group velocities of individual modes, instead of remaining a single packet as it would in a uniform and gently dispersive medium. For this reason, it is often preferable to use a single-mode waveguide, for which all but one mode, $m = 1$, are cut off.

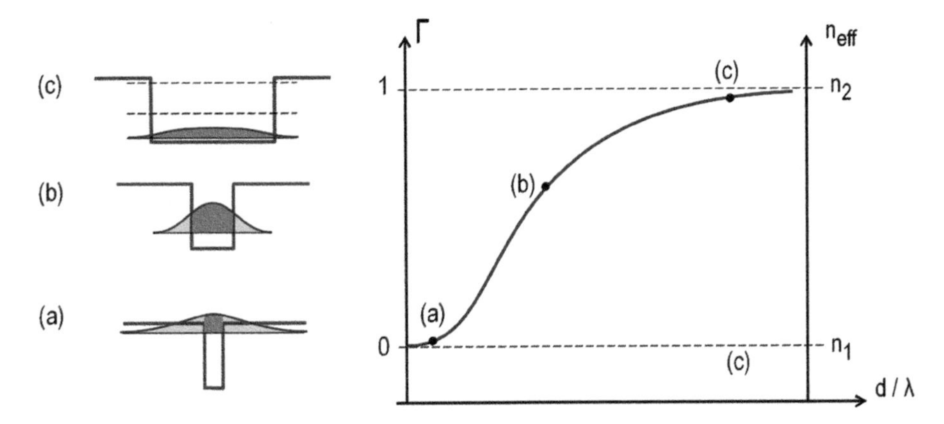

Fig. 6.5 Qualitative evolution of the confinement factor Γ and of the effective index n_{eff} for the fundamental guided mode as a function of the ratio d/λ (a single curve is used for clarity); the three sketches on the left show the mode profile (the shade highlights the waveguide core from $-d/2$ to $d/2$) for three emblematic values of the ratio: (a) $\lambda/d \gg 1$: the mode spreads far away in the claddings (it is said to be deconfined); (b) $\lambda \sim d$ (more precisely $\lambda \sim d\sqrt{n_2^2 - n_1^2}$): there is an appropriate balance of the field intensity between the core and the cladding, and Γ and n_{eff} vary rapidly as λ is varied; (c) $\lambda/d \ll 1$, the mode is essentially confined in the core, and higher-order modes (not shown, suggested by dashed lines above the mode profile sketch) must exist.

6.2.3 Mode-profile evolution

The dispersion of a mode effective index vs wavelength is accompanied by an evolution of its mode profile. Given the scaling of the problem with λ/d for non-dispersive materials, it is equivalent to describe this profile evolution vs d at a fixed wavelength λ. The qualitative evolution of n_{eff} is shown in Fig. 6.5 for the fundamental mode, together with the confinement factor Γ, defined as the fraction of the squared field in the core, $\Gamma = \frac{\int_{-d/2}^{d/2} |E_y|^2\, dx}{\int |E_y|^2\, dx}$, analogous to the probability of the presence of a particle in the well or the barrier region of a quantum well.

When $\lambda \gg d$ (low-frequency limit in Fig. 6.4), $n_{\text{eff}} \gtrsim n_1$. The mode spreads away in the claddings with an exponentially decaying tail much larger than the core width. In this deconfinement regime, the field profile resembles a plane wave with an asymptotically vanishing spatial amplitude modulation. One cannot squeeze the light of a given wavelength into an ordinary dielectric of vanishing thickness.

When the wavelength progressively decreases, n_{eff} grows, the tail size is reduced, and the confinement factor increases. At a given wavelength (that can be shown to scale as $d\sqrt{n_2^2 - n_1^2}$), n_{eff} then lies roughly midway between the core and cladding indices, and changes rapidly with the wavelength: the so-called modal dispersion is large.

If the wavelength is further decreased, the guide will certainly become multimodal. The fundamental mode becomes mostly localized in the core. Now we have the trend $n_{\text{eff}} \lesssim n_2$ and a confinement factor of $\Gamma \to 1$. The profile tails play a negligible role.

6.3 Mode orthogonality

Mode orthogonality is perhaps the most important property of waveguide theory. It allows modal reflection and transmission coefficients (scattering matrices) at the interface between two guiding structures to be derived, a key result for designing components in integrated optics such as efficient mode converters, splitters, or reflectors. It takes different forms. If one neglects absorption, the orthogonality is defined in the sense of energy, as will be shown below, and stipulates that the total power carried through any cross section of the waveguide is the sum of the powers carried by every individual mode. This orthogonality based on energy consideration is fully derived in this section. For lossy materials, the orthogonality does not allow simple interpretation and is treated in Complement 6.C. The latter also addresses the case of the orthogonality of radiation modes.

Before considering modes and waveguides, we start with the reciprocity theorem (see Chapter 1) for source-free Maxwell equations and consider a closed surface S defining a volume Ω without sources. We consider two field solutions labelled 1 and 2 with different frequencies ω_1 and ω_1, permeabilities μ_1 and μ_2, and permittivities ε_1 and ε_2, all described with tensors

$$\nabla \times \mathbf{E_1}(\mathbf{r}) = i\omega_1\mu_1\mathbf{H_1} \quad \text{and} \quad \nabla \times \mathbf{H_1}(\mathbf{r}) = -i\omega_1\varepsilon_1\mathbf{E_1}, \tag{6.15}$$

$$\nabla \times \mathbf{E_2}(\mathbf{r}) = i\omega_2\mu_2\mathbf{H_2} \quad \text{and} \quad \nabla \times \mathbf{H_2}(\mathbf{r}) = -i\omega_2\varepsilon_2\mathbf{E_2}. \tag{6.16}$$

Using the Green–Ostrogradsky theorem (see Chapter 1) and the curl equations (6.15) and (6.16) we get

$$\iint_S (\mathbf{E_2} \times \mathbf{H_1}).\mathbf{dS} = \iiint_\Omega i(\omega_1\mathbf{E_2}.\varepsilon_1\mathbf{E_1} + \omega_2\mathbf{H_1}.\mu_2\mathbf{H_2})\, d\Omega. \tag{6.17}$$

By subtracting the equivalent relation with 1 and 2 swapped, we get

$$\iint_S (\mathbf{E_2} \times \mathbf{H_1} - \mathbf{E_1} \times \mathbf{H_2}).\mathbf{dS} = i \iiint_\Omega [\omega_1(\mathbf{E_2}.\varepsilon_1\mathbf{E_1} - \mathbf{H_2}.\mu_1\mathbf{H_1}) \\ -\omega_2(\mathbf{E_1}.\varepsilon_2\mathbf{E_2} - \mathbf{H_1}.\mu_2\mathbf{H_2})]\, d\Omega. \tag{6.18}$$

6.3.1 Mode orthogonality in the sense of energy for lossless waveguides

Equipped with the general relation (6.18), we now consider two guided modes of the same waveguide ($\varepsilon_1 = \varepsilon_2 = \varepsilon$ and $\mu_1 = \mu_2 = \mu$) at the same frequency $\omega_1 = \omega_2 = \omega$. The waveguide geometry is arbitrary (see Fig. 6.6). The guided modes are labelled 1 and 2. We assume that the materials are reciprocal (the tensors ε and μ are symmetric) and lossless ($\varepsilon^\star = \varepsilon$). We thus have $\mathbf{E_1}.\varepsilon\mathbf{E_2} = \mathbf{E_2}.\varepsilon\mathbf{E_1}$ and $\mathbf{H_1}.\mu\mathbf{H_2} = \mathbf{H_2}.\mu\mathbf{H_1}$, therein implying that the volume integral on the right-hand side of Eq. (6.18) is identically zero. We obtain

$$\iint_S (\mathbf{E_1} \times \mathbf{H_2} - \mathbf{E_2} \times \mathbf{H_1}).\mathbf{dS} = 0. \tag{6.19}$$

Now, we specifically consider a parallelepiped box containing the guide as the volume Ω, with two surfaces normal to its axis $z = z_1$ and $z = z_2$, whose surfaces extend sufficiently far away from the guide core so that the exponentially damped fields of

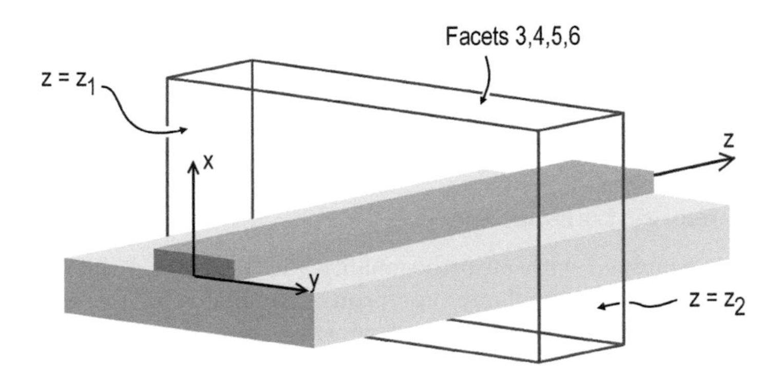

Fig. 6.6 Parallelepiped box used for deriving the orthogonality for lossless waveguides. The box has six facets, labelled $1, 2 \ldots 6$; facets 1 and 2 correspond to the transverse cross-section planes $z = z_1$ and $z = z_2$, respectively. The surface integrals of Eq. (6.19) on facets 3 to 6 vanish as the facets are asymptotically pushed away from the core.

guided modes vanish on the four parallelepiped facets parallel to the z-axis (labelled 3, 4, 5, 6 in Fig. 6.6). Thus, the four surface integrals of Eq. (6.19) on these facets are zero and Eq. (6.19) becomes

$$\iint_{z=z_1 \cup z=z_2} (\mathbf{E_1} \times \mathbf{H_2} - \mathbf{E_2} \times \mathbf{H_1}).\mathbf{dS} = 0. \tag{6.20}$$

We now explicitly consider two modes. The first one, corresponding to field 1, is labelled n: $\mathbf{E}_n(\mathbf{r}) = \mathsf{E}_n(x, y) \exp(i(\beta_n z - \omega t))$ and $\mathbf{H}_n(\mathbf{r}) = \mathsf{H}_n(x, y) \exp(i(\beta_n z - \omega t))$. The second one, corresponding to field 2, is the mode defined by: $\mathbf{E}^*_m(\mathbf{r}) = \mathsf{E}^*_m(x, y) \exp(-i(\beta_m z - \omega t))$ and $-\mathbf{H}^*_m(\mathbf{r}) = -\mathsf{H}^*_m(x, y) \exp(-i(\beta_m z - \omega t))$.[2]

Noting that for $z_1 < z_2$, the unit vector normal to plane $z = z_1$ is $-\mathbf{u}_z$, whereas the unit vector normal to plane $z = z_2$ is $+\mathbf{u}_z$, and injecting these solutions into Eq. (6.20), we obtain

$$\mathbf{E}^*_m(\mathbf{r}) \ = \mathsf{E}^*_m(x, y) \exp(-i(\beta_m z - \omega t)), \tag{6.21}$$

$$-\mathbf{H}^*_m(\mathbf{r}) = -\mathsf{H}^*_m(x, y) \exp(-i(\beta_m z - \omega t)). \tag{6.22}$$

Eq. (6.19) on the parallelepiped surface now reduces to integrals on two planes at $z = z_1$ and $z = z_2$ as it vanishes on the four other faces. The two surface integrals factor and the result reads

$$[\exp(i(\beta_n - \beta_m)z_2) - \exp(i(\beta_n \quad \beta_m)z_1)] \iint_{S'} (\mathsf{E}^*_m \times \mathsf{H}_n + \mathsf{E}_n \times \mathsf{H}^*_m).\mathbf{u}_z \, dx \, dy = 0, \tag{6.23}$$

where S' is any cross section plane $z = Cte$. The left-hand bracket terms are either zero if $m = n$ or not zero if $m \neq n$. In this case, the right-hand integral vanishes. This gives

[2] As shown in Complement 6.A.2, for transparent materials: if $[\mathsf{E}_m, \mathsf{H}_m] \exp(i\beta z - \omega t)$ is a mode, then $[\mathsf{E}_m, -\mathsf{H}_m] \exp(-i\beta z - \omega t)$ is also a mode. This holds because of our hypothesis of transparent media, $\varepsilon = \varepsilon^*$.

us an expression for the orthogonality product (in the sense of the energy or Poynting vector):

$$\iint_{S'} \mathrm{Re}(\mathbf{E}_n \times \mathbf{H}_m^*).\mathbf{u}_z \, dx \, dy = \delta_{m,n} \iint_{S'} \mathrm{Re}(\mathbf{E}_n \times \mathbf{H}_n^*).\mathbf{u}_z \, dx \, dy. \tag{6.24}$$

6.3.2 Modal energy transport

Now that we are equipped with an orthogonality (in the sense of bilinear forms), we can deduce that the energy flux of any superposition of modes boils down to the sum of individual modal flux contributions (like plane waves in uniform media). A full proof would include a demonstration that any field can be represented as a superposition of all the modes, in both directions, and with confined modes as well as radiation modes. This 'principle of completeness' is, however, delicate to prove, and we will restrict ourselves to a superposition of guided modes only. Thus, let us assume that the total field $[\mathbf{E}, \mathbf{H}]$ propagating in a waveguide is a sum of guided modes $[\mathbf{E}_m, \mathbf{H}_m]$ (a_m will denote the modal coefficients of the generic expansion), and check that the total Poynting vector flux F across a cross section S' is then a sum of the modes:

$$\begin{aligned} F &= \frac{1}{2} \iint_{S'} \mathrm{Re}\left[(\mathbf{E} \times \mathbf{H}^*).\mathbf{u}_z\right] dx \, dy \,, \\ F &= \frac{1}{2} \mathrm{Re} \iint_{S'} \left(\sum_m a_m \mathbf{E}_m \right) \times \left(\sum_n a_n \mathbf{H}_n \right)^* \exp(i\beta_m z - i\beta_n z).\mathbf{u}_z \, dx \, dy \,, \\ F &= \frac{1}{2} \sum_{m,n} \mathrm{Re} \iint_{S'} a_m a_n^* \exp(i(\beta_m - \beta_n)z)(\mathbf{E}_m \times \mathbf{H}_n^*).\mathbf{u}_z \, dx \, dy \,. \end{aligned} \tag{6.25}$$

Let us now take out of the integrand quantities that are independent of x and y. In the case of lossless media, the transverse components of the electric and magnetic field modes are real (see the end of Complement 6.A); thus, $(\mathbf{E}_m \times \mathbf{H}_n^*).\mathbf{u}_z$ is real and we have

$$F = \frac{1}{2} \sum_{m,n} \mathrm{Re}\left[a_m a_n^* \exp(i(\beta_m - \beta_n)z)\right] \iint_{S'} \mathrm{Re}(\mathbf{E}_m \times \mathbf{H}_n^*).\mathbf{u}_z dx \, dy. \tag{6.26}$$

We may now use the orthogonality of Eq. (6.24). Since the integral amounts to $\delta_{m,n}$, all indices n can be renamed m and, notably, $\beta_m - \beta_n$ is zero. We obtain

$$F = \frac{1}{2} \sum_m |a_m|^2 \iint_{S'} \mathrm{Re}(\mathbf{E}_m \times \mathbf{H}_m^*).\mathbf{u}_z \, dx \, dy. \tag{6.27}$$

At this stage, the substitution of field profiles $\mathbf{E}, \mathbf{H}^*$ with fields that incorporate the z-dependence (e.g. Eq. (6.4)) can be made, since the associated exponential terms $\exp(\pm i\beta_m z)$ have unity product. This form implies that the power carried through any cross section of the waveguide can be decomposed as a sum of partial powers carried by individual waveguide modes. This important result promotes mode concepts at the heart of energy transport by guided waves.

Note that the proof was obtained using an expansion over only the guided modes, since the mode orthogonality was established for guided modes. The mathematics for the derivation of the orthogonality of radiation normal modes are not obvious. However, the orthogonality condition has been established for a number of limited cases, including in particular slab waveguides that are invariant in two directions.

6.3.3 Mode orthogonality for lossy waveguides

Complement 6.C is devoted to the more complex case of lossy systems. Instead of allowing a simple $\delta_{m,n}$ orthogonality rule between modes of the same direction, it is shown that a similar property can be obtained for two modes of opposite directions only. Numbering forward- and backward-propagating modes through the index sign $\pm m$, we essentially get an orthogonality relation with a $\delta_{m,-n}$ Kronecker factor, instead of $\delta_{m,n}$ in Eq. (6.24). There are various conditions to be verified to reach this more general but far less intuitive result.

6.A Complement: Pairs of counter-propagating modes for reciprocal materials

6.A.1 Reciprocal tensor and counter-propagating modes

The existence of symmetries imposes certain conditions, not only on the interface geometry and the spatial permittivity distribution, but also on the tensor constitutive relations. For instance, any cross-section plane $z = z_0$ is a symmetry plane in a waveguide. It imposes that out of the six non-diagonal tensor elements of ε, the only non-zero elements left are ε_{xy} and ε_{yx}, while all those tied to z vanish: $\varepsilon_{xz} = \varepsilon_{yz} = \varepsilon_{zx} = \varepsilon_{zy} = 0$. This can be seen, for instance, by considering a symmetry-invariant E_x component: a non-zero ε_{zx} would then impose a non-zero P_z polarization, which would inconsistently transform into $-P_z$ under the symmetry plane.

Let us assume a general tensor at the moment (this will be refined later on) and a mode of the form $(\mathsf{E}_x, \mathsf{E}_y, \mathsf{E}_z, \mathsf{H}_x, \mathsf{H}_y, \mathsf{H}_z)\exp(i\beta z - i\omega t)$. The Maxwell equations then imply

$$\begin{aligned}
&\partial \mathsf{E}_z/\partial y - i\beta \mathsf{E}_y = i\omega\mu_0 \mathsf{H}_x\,,\\
&i\beta \mathsf{E}_x - \partial \mathsf{E}_z/\partial x = i\omega\mu_0 \mathsf{H}_y\,,\\
&\partial \mathsf{E}_y/\partial x - \partial \mathsf{E}_x/\partial y = i\omega\mu_0 \mathsf{H}_z\,,\\
&\partial \mathsf{H}_z/\partial y - i\beta \mathsf{H}_y = -i\omega(\varepsilon_{xx}\mathsf{E}_x + \varepsilon_{xy}\mathsf{E}_y + \varepsilon_{xz}\mathsf{E}_z)\,,\\
&i\beta \mathsf{H}_x - \partial \mathsf{H}_z/\partial x = -i\omega(\varepsilon_{yx}\mathsf{E}_x + \varepsilon_{yy}\mathsf{E}_y + \varepsilon_{yz}\mathsf{E}_z)\,,\\
&\partial \mathsf{H}_y/\partial x - \partial \mathsf{H}_x/\partial y = -i\omega(\varepsilon_{zx}\mathsf{E}_x + \varepsilon_{zy}\mathsf{E}_y + \varepsilon_{zz}\mathsf{E}_z)\,,
\end{aligned} \tag{6.28}$$

and with some rearrangements to introduce $-\beta$ and deliberate sign changes for $(\mathsf{E}_z, \mathsf{H}_x, \mathsf{H}_y)$, we obtain

$$
\begin{aligned}
&\partial(-\mathsf{E}_z)/\partial y - i(-\beta)\mathsf{E}_y = i\omega\mu_0(-\mathsf{H}_x)\,,\\
&i(-\beta)\mathsf{E}_x - \partial(-\mathsf{E}_z)/\partial x = i\omega\mu_0(-\mathsf{H}_y)\,,\\
&\partial\mathsf{E}_y/\partial x - \partial\mathsf{E}_x/\partial y = i\omega\mu_0\mathsf{H}_z\,,\\
&\partial\mathsf{H}_z/\partial y + i(-\beta)(-\mathsf{H}_y) = -i\omega(\varepsilon_{xx}\mathsf{E}_x + \varepsilon_{xy}\mathsf{E}_y + (-\varepsilon_{xz})(-\mathsf{E}_z))\,,\\
&i(-\beta)(-\mathsf{H}_x) - \partial\mathsf{H}_z/\partial x = -i\omega(\varepsilon_{yx}\mathsf{E}_x + \varepsilon_{yy}\mathsf{E}_y + (-\varepsilon_{yz})(-\mathsf{E}_z))\,,\\
&\partial(-\mathsf{H}_y)/\partial x - \partial(-\mathsf{H}_x)/\partial y = -i\omega((-\varepsilon_{zx})\mathsf{E}_x + (-\varepsilon_{zy})\mathsf{E}_y + \varepsilon_{zz}(-\mathsf{E}_z))\,.
\end{aligned}
\tag{6.29}
$$

We directly see that for $(\mathsf{E}_x, \mathsf{E}_y, -\mathsf{E}_z, -\mathsf{H}_x, -\mathsf{H}_y, \mathsf{H}_z)\exp(-i\beta z - i\omega t)$ to be a solution of the Maxwell equation for the same permittivity tensor as in Eqs. (6.28), it is sufficient that the tensor elements associated with sign changes in the last three equations of Eqs. (6.29) are zero. In other words, the mirror-symmetry condition $\varepsilon_{xz} = \varepsilon_{yz} = \varepsilon_{zx} = \varepsilon_{zy} = 0$ tells us about the existence of pairs of counter-propagating modes with entangled profiles.

Thus the set of normal modes of optical waveguides is composed of pairs with opposite propagation constants, β and $-\beta$, and it is sufficient to know the mode profile of one element of the pair to know the other.

6.A.2 Transparent materials

Additionally, if the waveguide materials are transparent ($\varepsilon = \varepsilon^*$), we can directly complex conjugate Eq. (6.28). Introducing $-\beta$, $-\mathsf{E}_z$, and -H_z where needed, we may repeat essentially the same reasoning as in the previous subsection and show that if $(\mathsf{E}_x, \mathsf{E}_y, \mathsf{E}_z, \mathsf{H}_x, \mathsf{H}_y, \mathsf{H}_z)\exp(i\beta z - i\omega t)$ is a mode, then $(\mathsf{E}_x, \mathsf{E}_y, \mathsf{E}_z, -\mathsf{H}_x, -\mathsf{H}_y, -\mathsf{H}_z)^*\exp(-i\beta z - i\omega t)$ is also a mode with opposite β.

Comparison with the rule of the previous subsection shows that for a lossless waveguide, we have

$$
\begin{aligned}
&\mathsf{E}_x = \mathsf{E}_x^*,\ \ \mathsf{E}_y = \mathsf{E}_y^*,\ \ \mathsf{E}_z = -\mathsf{E}_z^*\,,\\
&\mathsf{H}_x = \mathsf{H}_x^*,\ \ \mathsf{H}_y = \mathsf{H}_y^*,\ \ \mathsf{H}_z = -\mathsf{H}_z^*\,.
\end{aligned}
\tag{6.30}
$$

This shows that the field components of modes are all either purely real or purely imaginary for lossless waveguides.

6.B Complement: Equivalence between the group and energy velocities in lossless waveguides

This complement demonstrates the intuitive property that the energy and group velocities are identical for lossless waveguides. For this purpose, we consider two solutions of the source-free Maxwell equations of an arbitrary waveguide. For solution 1, we take the normal mode, $[\mathbf{E}_m, \mathbf{H}_m]\exp(i\beta_m(\omega_1)z)$, which is a solution of Maxwell equations at frequency ω_1 for a tensor map $\varepsilon_1 = \varepsilon(\mathbf{r}, \omega_1)$, and for solution 2, the corresponding counter-propagating normal mode, $[\mathbf{E}_{-m}, \mathbf{H}_{-m}]\exp(-i\beta_m(\omega_2)z)$, which is also a solution of Maxwell equations for the same waveguide, but for another frequency ω_2, hence with a dispersion-modified dielectric map $\varepsilon_2 = \varepsilon(\mathbf{r}, \omega_2)$ because of material dispersion.

Carrying out a derivation similar to that followed in section 6.3 where we derive the mode orthogonality with a surface integral that involves only two cross sections S' at $z = z_1$ and $z = z_2$, we obtain

$$\left[e^{i(\beta_m-\beta_{-m})z_2} - e^{i(\beta_m-\beta_{-m})z_1}\right] \iint_{S'} (\mathbf{E}_{-m} \times \mathbf{H}_m + \mathbf{E}_m \times \mathbf{H}_{-m}).\mathbf{u}_z \, dx\, dy = \frac{e^{i(\beta_m-\beta_{-m})z_2} - e^{i(\beta_m-\beta_{-m})z_1}}{\beta_m - \beta_{-m}}$$

$$\iint_{S'} [\mathbf{E}_{-m}.(\omega_1\varepsilon_1 - \omega_2\varepsilon_2)\mathbf{E}_m - \mathbf{H}_{-m}.(\omega_1\mu_1 - \omega_2\mu_2)\mathbf{H}_m] \, dx\, dy. \tag{6.31}$$

The first z-dependent factors have been kept to trace the origin of the calculation, but they obviously simplify. We now consider the case in which $d\omega = \omega_2 - \omega_1$ vanish and both frequencies tend towards ω. The denominator becomes a small quantity $d\beta$. The fields can then be taken, in that limit, such that the x, y-components satisfy $\mathbf{E}_{-m} = \mathbf{E}_m = \mathbf{E}_T$, where T stands for transverse but the z-component satisfies $\mathbf{E}_{z,-m} = -\mathbf{E}_{z,m} = -\mathbf{E}_z$. The converse rules hold for the magnetic field: $\mathbf{H}_{-m} = -\mathbf{H}_m = -\mathbf{H}_T$ and $\mathbf{H}_{z,-m} = +\mathbf{H}_{z,m} = \mathbf{H}_z$. Then the two terms of the left-hand side of Eq. (6.31) add up for the transverse part and cancel for the z-components. As for the right-hand side, the vanishing ω-dependent quantities can be expressed as $d\omega \frac{d}{d\omega}$, so that we may factor, for instance, $\frac{d\omega}{d\beta}$. Doing so, we get (leaving it more general than will be needed below)

$$\frac{1}{2}\iint_{S'} (\mathbf{E}_T \times \mathbf{H}_T).\mathbf{u}_z \, dx\, dy =$$

$$\frac{1}{4}\frac{d\omega}{d\beta}\iint_{S'} \left[\mathbf{E}_T.\frac{\partial(\omega\varepsilon)}{\partial\omega}\mathbf{E}_T - \mathbf{E}_z.\frac{\partial(\omega\varepsilon)}{\partial\omega}\mathbf{E}_z + \mathbf{H}_T.\frac{\partial(\omega\mu)}{\partial\omega}\mathbf{H}_T - \mathbf{H}_z.\frac{\partial(\omega\mu)}{\partial\omega}\mathbf{H}_z\right] dx\, dy. \tag{6.32}$$

In our case (see Complement 6.C for the general case), with a vanishing absorption, we have $\mathbf{E}_T = \mathbf{E}_T^*$ and $\mathbf{E}_z = -\mathbf{E}_z^*$, as well as $\mathbf{H}_T = \mathbf{H}_T^*$ and $\mathbf{H}_z = -\mathbf{H}_z^*$, so that we obtain an expression with all four signs turning positive on the right-hand side, allowing vector terms to be recast simply in two terms,

$$\frac{1}{2}\iint_{S'} (\mathbf{E}_T \times \mathbf{H}_T^*).\mathbf{u}_z \, dx\, dy = \frac{1}{4}\frac{d\omega}{d\beta}\iint_{S'} \left[\mathbf{E}.\frac{\partial(\omega\varepsilon)}{\partial\omega}\mathbf{E}^* + \mathbf{H}.\frac{\partial(\omega\mu)}{\partial\omega}\mathbf{H}^*\right] dx\, dy. \tag{6.33}$$

The right-hand term is reminiscent of the electromagnetic energy density in dispersive and weakly lossy materials. Strictly speaking, if one neglects the absorption losses, materials dispersion should be neglected as well. In this limit case, we have $\frac{\partial(\omega\varepsilon)}{\partial\omega} = \varepsilon$ and $\frac{\partial(\omega\mu)}{\partial\omega} = \mu$. This gives a form that we easily recognize:

$$\frac{1}{2}\iint_{S'} (\mathbf{E}_T \times \mathbf{H}_T^*).\mathbf{u}_z \, dx\, dy = \frac{1}{4}\frac{d\omega}{d\beta}\iint_{S'} [\mathbf{E}.\varepsilon\mathbf{E}^* + \mathbf{H}.\mu\mathbf{H}^*] \, dx\, dy. \tag{6.34}$$

The left-hand side is the Poynting vector flux, and the right-hand side is the electromagnetic energy density. This equation thus reflects an energy conservation law, and further states that the group velocity $\frac{d\omega}{d\beta}$ and energy velocity coincide with our assumptions.

6.C Complement: Mode orthogonality in lossy/amplifying waveguides and radiation mode treatment

The mode orthogonality relation of Eq. (6.24) is derived for lossless waveguides in the sense of energy flux. What happens for lossy or amplifying waveguides? Another difficulty is that of radiation modes. Since they are not bounded and form a continuum, their theoretical treatment is awkward, requires an analytical continuation for complex propagation constants, and is mathematically solved with rigour only for specific simple geometries, notably the planar waveguide. In this supplementary section, we aim to show a mathematical and numerical formalism that solves both issues.

To remove the theoretical difficulty introduced by radiation modes, mathematicians consider analytical continuations by prolonging, in the complex plane, key variables that are usually real, such as the propagation constant or the frequency (the next chapter considers complex frequencies). Instead of considering an actual open waveguide with semi-infinite claddings, it is elegant to consider a companion closed waveguide geometry, obtained by surrounding the actual geometry with perfectly matched layers (PMLs).

The latter have become very popular, not least because they are convenient reflection-less boundary conditions in finite-difference time-domain simulation, but also, more generally, because they represent a beautiful mathematical tool to map finite numerical domains onto open infinite spaces. The application of PMLs defines a companion waveguide geometry with finite claddings. At their edges, they have new complex-valued permittivity and permeability distributions ($\mathrm{Im}(\varepsilon)$ and $\mathrm{Im}(\mu)$ are both non-zero in the PML), in which the field, purposely, is exponentially damped. The guided modes remain basically unchanged upon application of the PML, but the radiation modes of the initial waveguide with infinite claddings become square-integrable owing to the exponential damping in the PML, hence the direct relationship between lossy or amplifying waveguides on the one hand and radiation mode treatments on the other hand.

The derivation of the general form of the mode orthogonality for lossy or amplifying waveguides with non-zero $\mathrm{Im}(\varepsilon)$ and $\mathrm{Im}(\mu)$ is very similar to the derivation of section 6.3. Again, starting from Eq. (6.19), which is valid for reciprocal materials, we inject two normal modes, $[\mathbf{E}_m, \mathbf{H}_m]$ and $[\mathbf{E}_n, \mathbf{H}_n]$, which may be guided or radiative or leaky (leaky modes are often seen as the analytical continuation of a guided mode of the discrete spectrum that is below its cutoff frequency). We use a parallelepiped surface bounded by the cross-section planes $z = z_1$ and $z = z_2 > z_1$ and by the four outer boundaries now consisting of PMLs away from the waveguide core, but only at some finite distance. Owing to the property that the modes vanish at the PML *outer* boundaries, again only the integral over the cross-section planes is non-zero, and we have:

$$\left[e^{i(\beta_m+\beta_n)z_2} - e^{i(\beta_m+\beta_n)z_1}\right] \iint_{S'} (\mathbf{E}_n \times \mathbf{H}_m - \mathbf{E}_m \times \mathbf{H}_n).\mathbf{u}_z \, dx \, dy = 0, \qquad (6.35)$$

where S' represents any finite transverse cross section of the waveguide which attains the PMLs.

The first term on the left side of Eq. (6.35) is identically zero if and only if $\beta_m + \beta_n = 0$ (i.e. if $n = -m$); so whenever $n \neq -m$, the integral is zero. Note that $\beta_m = -\beta_n$ in our complex-valued case means that we consider two modes that have opposite phase velocities and that both decay (or are amplified) along the z propagation axis.

We are now going to show that the integral of Eq. (6.35) for $n = -m$ is $-2\iint_{S'}(\mathsf{E}_n \times \mathsf{H}_n).\mathbf{dS}$, allowing us to assemble arbitrary n, m cases under the single final orthogonality equation

$$\iint_{S'} (\mathsf{E}_n \times \mathsf{H}_m - \mathsf{E}_m \times \mathsf{H}_n).\mathbf{dS} = -2\delta_{n,-m} \iint_{S'} (\mathsf{E}_n \times \mathsf{H}_n).\mathbf{u}_z \, dx \, dy, \tag{6.36}$$

which is a specific form of orthogonality: it is a bilinear form allowing the decomposition of any field distribution in the complete basis formed by the modes of the 'PMLized' waveguide.

To get the above result, we may consider the following equalities: $\mathsf{E}_{x,-n} = \mathsf{E}_{x,n}$, $\mathsf{E}_{y,-n} = \mathsf{E}_{y,n}$ and $\mathsf{H}_{x,-n} = -\mathsf{H}_{x,n}, \mathsf{H}_{y,n} = -\mathsf{H}_{y,n}$. Since the surface element vector $\mathbf{u}_z$ is pointing along z, only these transverse $\mathsf{E}_{x,y}$ and $\mathsf{H}_{x,y}$ field components give non-zero contributions in the flux integral onto the cross section S' (for instance, x- and z-components and the curl operator gives y-component, which is normal to $\mathbf{u}_z$. It follows that $(\mathsf{E}_n \times \mathsf{H}_{-n}).\mathbf{u}_z$ and $(\mathsf{E}_n \times \mathsf{H}_n).\mathbf{u}_z$ have opposite signs. Hence the desired result, $(\mathsf{E}_n \times \mathsf{H}_{-n} - \mathsf{E}_{-n} \times \mathsf{H}_n).\mathbf{u}_z = (-\mathsf{E}_n \times \mathsf{H}_n - \mathsf{E}_n \times \mathsf{H}_n).\mathbf{u}_z$, which is indeed valid locally when the z-projection is included.

6.D Complement: Metal–insulator–metal waveguides with perfect metals: TE modes

Metal–insulator–metal waveguides are composed of a dielectric slab (the core with a refractive index n) bounded by two metal films at positions $x = -d/2$ and $x = d/2$, with d the slab thickness. They are used over the entire electromagnetic spectrum for TM polarization (magnetic field parallel to the metal planes) for values of d that are much smaller than the wavelength ($d \sim \lambda/100$). At microwave frequencies, the mode field weakly penetrates in the metal and low-loss guidance is achieved, with many applications in the microwave for printed circuit boards, e.g. microstrips. At optical frequencies, the field significantly penetrates in the metal and modes are highly damped by absorption. Nevertheless, the fundamental TM mode, called gap-surface-plasmon in section 13.5, is involved in many devices of current interest, e.g. nanocavities or metallic slit gratings.

Hereafter, for consistency with section 6.2 on slab waveguides, we consider TE polarization (electric field $E_y(x)$ parallel to the metal films) and assume the perfect conductor approximation, which is approximately valid for microwaves. We let the reader treat TM waves with very similar considerations.

6.D.1 Field solutions

Translation invariance in the propagation direction z leads us to look for modes labelled by integer m in the form

$$E_y^{(m)}(x, z) = E_0 \sin(k_{x,m}(x + d/2)) \exp(i\beta_m z) = \mathsf{E}^{(m)}(x) \exp(i\beta_m z). \tag{6.37}$$

This is the superposition of the two local plane waves $\exp(\pm k_{x,m}x)\exp(i\beta_m z)$, as discussed in section 6.2.1 and illustrated in Fig. 6.7b. $E_y^{(m)}$ already satisfies the perfect-conductor condition ($E_y = 0$) for $x = -d/2$. To satisfy the same condition for $x = d/2$, we must have $\sin(k_{x,m}d) = 0$, therein implying that $k_{x,m} = m\pi/d$ with $m = 1, 2...$ The boundary conditions quantify the transverse propagation constant, just like for the infinite potential barriers of wells in quantum mechanics.

6.D.2 Dispersion

To calculate the β_m for a fixed value of the frequency ω, we simply write that the local plane waves satisfy the source-free Maxwell equations in a uniform medium of refractive index n, $\beta_m^2 + k_{x,m}^2 = n^2(\omega/c)^2$, providing us with the dispersion relation

$$\beta_m = \left((n\omega/c)^2 + (m\pi/d)^2\right)^{1/2}. \tag{6.38}$$

The dispersion curves are shown in Fig. 6.7a with a family of hyperbola sharing the same asymptote $\omega/c = \beta/n$ (or equivalently, $\beta = nk_0$). There is no 'light line' related to the outer medium as discussed in the case of the slab waveguide with dielectric boundaries, and therefore there is no cutoff. The values at the origin for $\beta = 0$ are $\omega_m/c = m\pi/(nd)$. They are evenly spaced, with a spacing $\Delta\omega/c = \pi/(nd)$. This quantity is also known as the free-spectral range of an ideal cavity of length d, usually given in terms of the frequency, $\Delta\nu = \frac{c}{2nd}$.

6.D.3 Group and phase velocities

The next quantities of interest in this model system are the phase and group velocities, respectively defined for each mode m as $v_\varphi^{(m)} = \omega/\beta$ and $v_g^{(m)} = \partial\omega/\partial\beta$. A simple calculation shows that

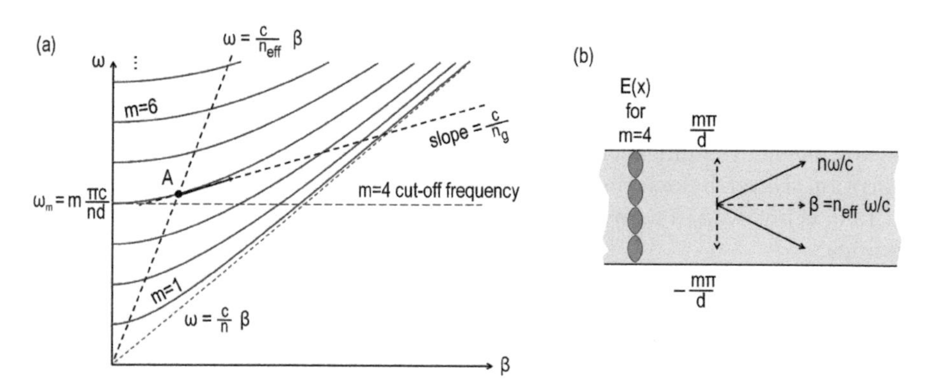

Fig. 6.7 (a) Dispersion relation branches $m = 1, 2, \ldots$ of the transverse-electric modes of metal–insulator–metal waveguide with perfectly conducting walls. At the outlined point A of the $m = 4$ branch, the dashed lines showing the effective index and the group velocity ($\equiv c/n_g$) are drawn. (b) Sketch of the $E_y(x)$ field profile for $m = 4$, and decomposition into two local plane waves having wavevectors $[n_{\text{eff}}\frac{\omega}{c}, \pm\frac{m\pi}{d}]$.

$$v_\varphi^{(m)} = \frac{c}{n}\left(1 + (m\pi/(\beta_m\, d))^2\right)^{+1/2}, \tag{6.39}$$

$$v_g^{(m)} = \frac{c}{n}\left(1 + (m\pi/(\beta_m\, d))^2\right)^{-1/2}. \tag{6.40}$$

The product of the two velocities for a fixed β_m is the same for all branches: $v_\varphi^{(m)} v_g^{(m)} = c^2/n^2$. This is intuitively understood by considering that larger m values correspond to more pronounced 'zig-zag' modes, with more transverse momentum $k_{x,m} = m\pi/d$. This slows down the group velocity, as the energy effectively has to travel up and down ($\pm x$) a longer distance as the mode goes along the guide axis z. But at the same time, the phase velocity increases. The wavefronts of the two underlying plane waves $\mathbf{k} = \beta\mathbf{u}_z + \pm m(\pi/d)\mathbf{u}_x$ evolve towards those of two counter-propagating waves as m grows. In such a case, the crossing point of two wavefronts, which defines the local field value, travels much faster than the medium's light velocity, c/n. This is no mystery: consider the phase velocity of sea waves arriving on a shore or on a beach at near normal incidence; the crest collapses almost simultaneously at distant points, hence the 'superluminal' phase velocity.

6.E Complement: Application example: The multimode interferometer

Most devices in integrated optics are monomode, owing to the fact that any waveguide imperfection, even a tiny surface roughness, couples modes together. An exception is the multimode interferometer (MMI), a device that splits or combines different waves. It is a waveguide designed to support a large number of modes (typically > 3) which relies on a self-imaging principle well known for diffraction grating at finite distance (hence *not* a Fraunhofer regime) referred to as the Talbot effect. In order to launch light into and recover light from the MMI, a number of access waveguides (usually single-mode) are placed at its entrance and its exit. Such devices are generally referred to as $N \times M$ MMI couplers, where N and M are, respectively, the number of input and output waveguides, also called 'ports'.

A 1×2 MMI is sketched in Fig. 6.8. In this example, the MMI of length L is fed at its input $z = 0$ by a single single-mode waveguide (often referred to as an input 'port') centred at a position x and two symmetrical output ports at $z = L$.

The analysis of MMI devices can be performed qualitatively in a simple way. Owing to the large transverse dimension ($t \gg \lambda/ncore$), we may use the slab notations to treat the MMI as a 2D geometry, even though it is actually also confined in the direction normal to the figure. And we further assume that the mode profiles are those of the perfect-conductor waveguide just studied in Complement 6.D, $\mathsf{E}_m(x) \simeq d^{-1/2}\sin(\frac{m\pi x}{d})$, since the evanescent tails in the cladding negligibly contribute to the MMI modes. The excitation coefficients of the forward-propagating MMI modes by the input port can be obtained analytically using the orthogonality condition, Eq. (6.24), for lossless waveguides and assuming that the total field in the input single-mode channel and the MMI can be expanded on the sole basis of guided modes, i.e. neglecting radiation modes. The same reasoning applies for the output ports.

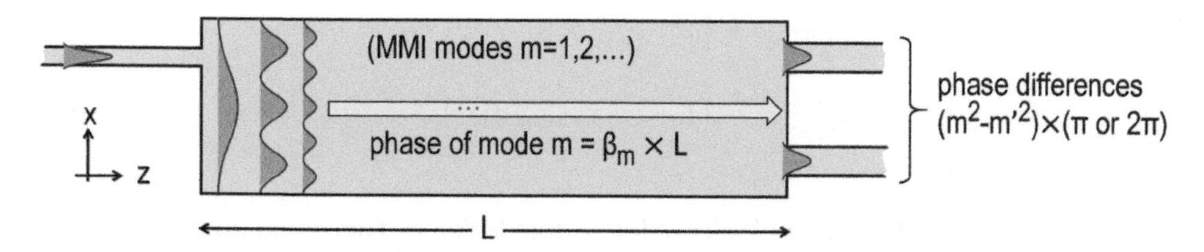

Fig. 6.8 Sketch of a multimode interferometer: a mode from a single-mode waveguide on the left excites many MMI modes of a broad waveguide section. Owing to the fact that the phase differences are all multiples of a simple phase value, a specific interference pattern is formed on the right: here two symmetric images. Hence we have a 1×2 beam splitter.

The operation principle of an MMI device relies on the fact that the propagation constants of all the lower-order modes, those with the larger β, sample an arithmetic series of constant $\delta\beta = (\beta_2 - \beta_1)/3$, as shown when discussing Eq. (6.14). The relevant adaptation of this equation is here:

$$\beta_m \simeq k_0 n_{core} - m^2(\pi/t)^2/(2k_0 n_{core}). \tag{6.41}$$

Let us analyse the phases $\varphi_{m,L} = \beta_m L$ taken by the different modes m upon a single pass L. In addition to a constant phase term $k_0 n_{core} L$, they feature integer multiples of $\phi_L = -L(\pi/t)^2/(2k_0 n_{core})$ (the factoring integers being $1^2, 2^2, 3^2, \ldots$). It is now clear that if L is such that ϕ_L is a multiple of 2π, all $\varphi_{m,L}$ also differ by multiples of 2π (they sample an arithmetic series). Thus, their interferences at $z = L$ are exactly reproducing the same amplitudes and phases as those at $z = 0$. The device then essentially images the input field at the plane $z = L$ (with an x resolution limited by the largest transverse mode momentum).

A similar analysis can be carried out, targeting a different distance L, such that the phase difference $L\delta\beta \equiv L(m^2 - m'^2)(\pi/t)^2/(2k_0 n_{core})$ is a multiple of $\pi, 2\pi/3$, or $\pi/2, \ldots$ instead of a multiple of 2π. It can then be shown (see Exercise 6.9) that one obtains multiple images that are evenly distributed across the waveguide section $0 < x < d$ because partial sets of modes have adequate values of $(m^2 - m'^2)$ to cause a field build-up at the corresponding x values. As long as the MMI interference is dominantly carried out by lower-order modes (smaller m's) that sample the arithmetic series precisely enough, the input port is eventually split into $N = 2, 3, 4,\ldots$ ports at $z = L$. We thus implement an integrated $1 \times N$ splitter. Higher m's would deviate in phase and blur images, causing wavefront distortion.

The MMI is an instructive example because it involves ray and wave aspects. In the ray picture, rays from the input port are multiply reflected by sidewalls, therein creating multiple evenly spaced image ports that replicate the initial port at $z = L$. This is why the situation is then similar to the Talbot imaging effect of gratings: the MMI sidewalls at $x = \pm t/2$ act as two face-to-face mirrors, so that rays between them are reflected over and over, losses allowing.

Exercises

Exercise 6.1 Dielectric planar waveguide. We consider a symmetric slab waveguide (core: n_2, claddings: n_1). The dispersion relation of the fundamental mode is $\beta(\omega)$. Consider the plane $\beta - \omega/c$ (the x-axis being β). n_1 and n_2 are real and independent of the frequency.

(1) Plot the lines corresponding to the dispersion relation of plane waves propagating in uniform media of refractive indices n_1 and n_2. Justify that guided modes necessarily lie between the two lines.

(2) Owing to the symmetry, there is no cutoff for the fundamental mode. Sketch its dispersion relation, paying special attention to the asymptotic limits for small and large frequencies and the dispersion curvature, downward or upward, in these regions.

(3) Assuming that the first excited mode has a cutoff, sketch its dispersion relation.

Exercise 6.2 Perfect-conductor waveguide modes. Calculate the cutoff frequency and wavelengths of the two first transverse-electric (TE) modes $m = 1$ and $m = 2$ for (a) a guide of thickness $d = 1\,\mu$m and index $n = 1.5$ (glass); (b) a guide of thickness $d = 0.1\,\mu m$ and index $n = 3.5$ (silicon).

Exercise 6.3 2D layer-stack waveguide: Program. Our goal is to implement a simple, effective program to study the guidance of 2D waveguides. We also study the limitations of the numerical approach based on Fourier expansion. In the following exercises, the program will help us to study the physics of guided modes and check numerically the main theorem of the chapter. The program may also further help with studying Bloch modes of periodic stacks in Chapter 15.

As shown in Fig. 6.9a, we consider an optical stack composed of a series of N layers deposited on a substrate. The relative permittivities of the layers are $\varepsilon_1, \varepsilon_2 \ldots \varepsilon_N$ and those of the substrate and superstrate are ε_0 and ε_{N+1}. The abscissae of the layer interfaces are $x_1, x_2 \ldots\ x_N$. All layers are non-magnetic. We consider TE modes $E_y(x, z) = \mathsf{E}_y(x)\ \exp(i(\beta z - \omega t))$.

(1) From Maxwell curl equations, show that

$$\left[\frac{\partial^2}{\partial x^2} - \beta^2\right] \mathsf{E}_y(x) + (\omega/c)^2 \varepsilon(x) \mathsf{E}_y(x) = 0. \tag{6.42}$$

(2) **Supercell approach.** By definition, $\mathsf{E}_y(x)$ is exponentially damped in the substrate and superstrate, implying that it is virtually null at large enough distances D_1 and D_2 in the claddings. The ansatz implies that $\mathsf{E}_y(x)$ can be looked at as a periodic function (it is null at $x = x_1 - D_1$ and $x = x_N + D_2$ and is thus periodic of period $a = x_N - x_1 + D_1 + D_2$) that can be calculated with a Fourier expansion. a is called the supercell period (or the numerical period) of the periodized waveguide illustrated in Fig. 6.9b. If D_1 and D_2 are large enough, we will nearly recover the actual guided modes of the real waveguide of Fig. 6.9a among the set of numerical modes of the periodic structure, the accuracy being dictated by the residual amount of field at the supercell boundaries.

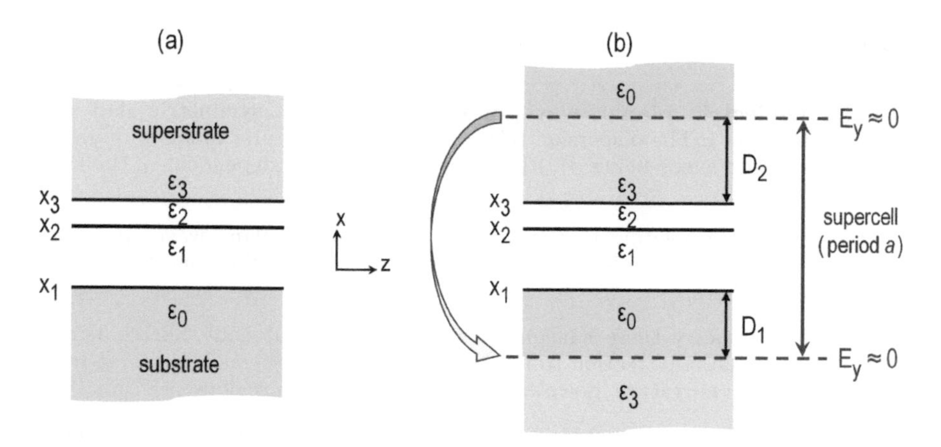

Fig. 6.9 Supercell approach. (a) Actual waveguide with semi-infinite claddings; (b) supercell made periodic at the dashed lines. For sufficiently large distances D_1 and D_2, the guided-mode main field, E_y for transverse-electric or H_y for transverse-magnetic, is almost zero in the remainder of the claddings. If the field is null, it also means that it is periodic with a period $a = x_{N+1} - x_1 + D_1 + D_2$, and so we can compute it by expanding it in a Fourier series. In the program, the chosen supercell period a should be large enough.

(2a) The relative permittivity becomes a periodic function of x and we expand it into the Fourier series $\varepsilon(x) = \sum_p \hat{\varepsilon}_p \exp(ipK\,x)$, with $K = 2\pi/a$. Similarly, $\mathsf{E}_y(x) = \sum_p S_p \exp(ipK\,x)$. From Eq. (6.42), show that

$$(\beta^2 + p^2K^2)\,S_p = (\omega/c)^2 \sum_p \hat{\varepsilon}_{p-q}\,S_q. \tag{6.43}$$

Our goal is now to implement a program that computes, for a fixed frequency, the β (as an eigenvalue) and the S_p's (as the elements of an eigenvector) of all guided modes. However, we cannot deal with an infinite number of S_p's; we thus truncate the Fourier series, retaining a total number of $2M + 1$ Fourier harmonics in the expansion of E_y from $p = -M$ to $p = M$. M is called the truncation rank.

(2b) In the program, we introduce the permittivity vector perm defined by $\mathsf{perm}(p) = \varepsilon_{p-1}$, $p = 1, 2, \ldots\ N + 2$ and the interface coordinate vector xx defined by $\mathsf{xx}(1) = x_1 - D_1, \mathsf{xx}(2) = x_1, \mathsf{xx}(3) = x_2, \ldots\ , \mathsf{xx}(N) = x_N, \mathsf{xx}(N+1) = x_N + D_2$. The supercell period is $\mathsf{xx}(end) - \mathsf{xx}(1)$. Using the definition of the Fourier coefficient $\hat{\varepsilon}_m = a^{-1}\int_0^a \varepsilon(x)\exp(imKx)\,dx$, verify that

$$\hat{\varepsilon}_0 = a^{-1}\sum_{q=1}^{N+2} [\mathsf{xx}(q+1) - \mathsf{xx}(q)]\,\mathsf{perm}(q), \tag{6.44}$$

$$\hat{\varepsilon}_p = \sum_{q=1} [\exp(-ipK(xx)(q)) - exp(-ipK(xx)(q+1))]\,(perm)(q)/(2i\pi p), \quad p \neq 0. \tag{6.45}$$

Considering Eq. (6.43), show that a total number of $4P+1$ Fourier coefficients of $\varepsilon(x)$, from $-2P$ to $2P$, have to be calculated to compute $P = 2M + 1$ Fourier coefficients of E_y, from S_{-M} to S_M. Build a function that computes the Fourier coefficients $\hat{\varepsilon}_p$

and organize them in a vector denoted permF, with permF$(2P+1) = \hat{\varepsilon}_0$ (the cyclic boundary condition).

(2c) To test your program, plot the permittivity distribution $\sum_{-M}^{+M} \hat{\varepsilon}_p \exp(ipKx)$ for the example perm $= [1, 3, 1]$ xx $= [-1, 1]$. Compare with the actual discontinuous function $\varepsilon(x)$. What happens at every interface? Try to increase the parameter M to study the convergence, paying attention to sampling the space coordinate finely enough. Please refer to the 'Gibbs phenomenon' in the literature to explain what you are observing.

(3) **Program implementation.** The implementation is very similar to that of the well-known rigorous coupled wave analysis (RCWA). We recommend you consider reading the TE case of the popular article *J. Opt. Soc. Am. A* 12, 1068 (1995) if necessary. We first transform Eq. (6.43) into an eigenproblem, with eigenvalues β^2 and eigenvectors the Fourier coefficients S_p, $p = -M, \ldots, M$:

$$(\omega/c)^2 \sum_q \hat{\varepsilon}_{p-q} S_q - p^2 K^2 S_p = \beta^2 S_p. \tag{6.46}$$

(3a) We introduce the $P \times 1$ vector $\mathbf{S} = [S_{-M}, S_{-M+1} \ldots \; S_0 \ldots \; S_M]$. Show that the β^2's can be computed as the eigenvalues of the square matrix $\mathbf{A}$,

$$\mathbf{A} = (\omega/c)^2 \mathbf{E} - \mathbf{K}^2, \tag{6.47}$$

where $\mathbf{K}$ is a $P \times P$ diagonal matrix defined by $\mathbf{K}(p,p) = (-M - 1 + p)K$ and $\mathbf{E}$ is also a $P \times P$ matrix for which you will determine the coefficients $\mathbf{E}(p,q)$ as a function of the Fourier coefficient of the relative permittivity. Note that we mostly follow the notations of *J. Opt. Soc. Am. A* 12, 1068 (1995).

(3b) For a symmetric 400 nm-thick planar waveguide and for $\lambda = 1\ \mu m$ (perm $= [1, 3.5^2, 1]$, xx $= [-1.2, -0.2, 0.2, 1.2]$), compute the eigenmodes. What are the values of D_1 and D_2 in this example?

Inspired by (2c), write a function that visualizes the mode profiles. Among the $2M+1$ modes that are computed by diagonalization, only a few correspond to the true guided modes of the actual waveguide. What is the double-inequality condition that should be satisfied by the effective index of guided modes?

Implement the condition to select the true modes, and additionally verify that their intensities $|E_y|^2$ at the supercell boundaries xx(*end*) and xx(1) are much smaller than their maximum intensities. Is the waveguide monomode for the considered wavelength?

Exercise 6.4 Confinement factor.

(1) We again consider the same 400-nm-thick planar waveguide. To further test the program, try to compute the dispersion relation of Fig. 6.4. Plot the field profiles of the fundamental mode for $n_{\text{eff}} \approx n_1$ and $n_{\text{eff}} \approx n_2$. Also plot the profile for an intermediate value.

(2) Why is it difficult (or even impossible?) to compute the dispersion curve for large wavelengths with the program? Try to plot the field profile of the second mode close to the cutoff.

(3) The confinement factor in the core is defined as $\Gamma = \int_{-d/2}^{d/2} |\mathsf{E}(x)|^2\, dx$, the integral $\iint |\mathsf{E}(x)|^2\, dxdy$ being unity. Start by plotting Γ and $1 - \Gamma$ (the 'deconfinement' factor) as a function

of d/λ for the fundamental mode. You should reproduce the trends of Fig. 6.5: asymptotes and monotonous rise of Γ. In particular, show that the lower Γ bound is zero.

What is modified for a higher-order mode in this case?

(4) For an asymmetric waveguide (take an index sequence perm $= [1.5^2, 2^2, 1]$, typical of a silica/silicon nitride/air sequence), the lower limit of Γ should not be zero. Track the roles of the substrate and superstrate by estimating their respective contributions as $1 - \Gamma$.

Exercise 6.5 Mode orthogonality. We still consider the same 400-nm-thick symmetric planar waveguide.

(1) For $\lambda = 1\ \mu m$, plot all the field components, E_y, H_x, H_z, and verify that they are either purely real or purely imaginary; this represents a verification of Eq. (6.30).

(2) Consider the fundamental and first-order modes. Check that the orthogonality condition of Eq. (6.24) for lossless waveguides is satisfied.

(3) Let us consider a lossy case: perm $= [1, 3.5^2 + 0.1i, 1]$. Verify that the orthogonality condition of Eq. (6.24) is no longer valid. Check that the orthogonality relation of Eq. (6.36) remains valid.

Exercise 6.6 Intuitive picture of the mode orthogonality. Our goal is to intuitively illustrate the meaning of the orthogonality energy relation of Eq. (6.24). Uniform media being also z-invariant, the modal theory developed in the chapter should apply to plane waves as a special case. Consider the orthogonality relation in the sense of the Poynting vector of Eq. (6.24), which holds for two modes at the same frequency.

(1) Consider two plane waves that propagate along the z-direction with the same propagation constant β. Show that they are orthogonal in the sense of Eq. (6.24) if and only if they are orthogonally polarized.

(2) Consider two plane waves with different wavevectors $\mathbf{k}_1$ and $\mathbf{k}_2$, which are are not propagating in the same direction. Show that they are orthogonal in the sense of Eq. (6.24), whatever their polarization.

Exercise 6.7 Group velocity. For the dispersion curves of the first question of Exercise 6.4, compute the group velocity of the fundamental mode as $v_g = d\omega/d\beta$. Plot the group index $n_g = v_g/c$, and the effective index n_{eff} on the same graph, as a function of the wavelength.

The computation of the group velocity with the derivative $d\omega/d\beta$ requires the computation of two modes with slightly different β for two slightly different frequencies. By using Eq. (6.34), compute v_g for $\lambda = 1\,\mu\text{m}$ with a single computation.

Exercise 6.8 Comparison of TE-guided modes and perfect-conductor modes, Fabry-Perot modes, and TM modes. This exercise aims to compare the dispersion branches of different systems and their relative shifts.

(1) Compare the perfect-conductor waveguide and the slab waveguide based on the field profiles. For a given branch number and a given frequency, indicate which case gives the largest.

(2) In the region above the light line of the cladding, $\beta < \omega c/n_{clad}$, the perfect-conductor waveguide modes cannot be compared to the slab modes. But we can compare them to Fabry–Perot modes. Using the trend of Fresnel reflection, find out whether this is more justified close to or far from the cladding light line. This is another introduction to resonances with variable quality factor which, for real frequencies, lead to a 'blurred' dispersion relation associated with the finite photon lifetime.

(3) Compare TE modes and TM modes: check that they are staggered using the total internal reflection phase, a phase which is also associated with the Goos–Hänchen shift of total internal reflection.

Exercise 6.9 Multimode interferometer (MMI) length and Talbot diffraction of a grating. We use the depiction of the MMI device shown in Complement 6.E, with width t and index n and operate at wavelength λ.

(1) Calculate the MMI length L vs t, n, and λ: what is the condition on the MMI length L for implementing a 2π phase difference between the first two modes?

(2) Relate the results to Talbot diffraction: at distance L from a transmission grating of period a illuminated under normal incidence, each period being assimilated to a point scatterer at position $x_p = pt$, calculate the distances L at which the phase differences from the scatterers as seen from a point P of coordinates $x = x_p, z = L$ are multiples of 2π. Repeat the calculation for $x = x_p + t/2$. This is actually a study of the Fresnel diffraction regime.

7 Basics of resonators and cavities

7.1 Introduction

In the previous chapter, the concept of modes was introduced in relation to waveguides. Modes were defined as source-free solutions of the Maxwell equations, obtained by calculating the propagation constant β in one of the dispersion branches for a fixed real frequency. They hold for systems which are invariant, or periodic, as will be discussed in Chapter 17 on periodic waveguides, in at least one direction of space.

In this chapter, we consider modes of systems with finite dimensions. These are also source-free solutions of the Maxwell equations but are completely different in nature from the guided modes. Because the notion of dispersion branch disappears, our new modes are no longer characterized by a propagation constant (i.e. a momentum $\hbar\beta$), but only by a frequency ω (i.e. an energy $\hbar\omega$), similar to most usual quantum systems modelled with an Hamiltonian. Consider the examples of an acoustic resonator, a tuning fork, and a guitar string (RLC circuits (see Exercise 7.1), or simple gravity pendulums are equally interesting examples), and suppose that the resonator is suddenly struck. The initial transient response is rather complicated, but if we wait enough, we can hear a very pure tone, in general at 440 Hz, which lasts longer than the other overtone ones. The purity is a signature of the excitation of a natural resonance, the fundamental resonance built up from a self-sustained exchange between kinetic and elastic energies. However, since perpetual motion is forbidden, the energy exchange cannot last forever, at least for a system with a finite dimension, and the fork dissipates energy by either warming its own steel or leaking soundwaves into the surrounding air. This implies that the resonance oscillation at 440 Hz is not a pure sine wave and that the exchange of energy is not perfectly self-sustained. The oscillation is factored by an exponentially damped envelope and can be accounted for by a *complex frequency*

$$\tilde{\omega} = \omega - i\Gamma/2, \tag{7.1}$$

Γ being the decay rate, i.e. the inverse of the mode lifetime $\tau = 1/\Gamma$. The minus sign is due to our convention $\exp(-i\tilde{\omega}t)$: $-i \times -i = -1$ and the factor 2 accounts for the difference between amplitude and energy decays. The frequency (i.e. the energy $\hbar\omega$ in our analogy with quantum mechanics) is prevented from being real-valued, sometimes by absorption and sometimes—or, in the case of finite systems, always—by leakage. The physics of dissipative systems is non-Hermitian, and to emphasize the difference with the well-known *normal modes* of Hermitian systems (classical or quantum), we call the natural modes of resonant systems with complex frequencies *quasinormal modes* throughout the book. Other names in the literature are resonance states, morphology-dependent resonances, Mie mode, and localized plasmon resonance.

Introduction to Nanophotonics. Henri Benisty, Jean-Jacques Greffet, and Philippe Lalanne, Oxford University Press.
 DOI: 10.1093/oso/9780198786139.003.0007

In theoretical physics, quasinormal modes are formal solutions of linearized differential equations. They play a role in the general relativity theory of black holes or in dramatic events such as the collapse of macroscopic structures, a famous example being the collapse of the Broughton Suspension Bridge near Manchester in 1831. They also play a key role in striking optical phenomena such as the Purcell effect, an emblematic effect named after Edward Mills Purcell, who showed in 1946 that the spontaneous emission rate of excited emitters can be strongly enhanced by placing atoms inside a cavity in interaction with a resonance mode that modifies the spontaneous emission rate of quantum emitters.

In nanophotonics, the quasinormal modes of nanoresonators and microcavities are much the same as those of our acoustic example. They are either Mie resonances for dielectric structures or localized surface plasmons for metallic ones. They play a pivotal role because even tiny absolute amounts of stored energy can yield very large field intensities when the field is confined during a long time or at a deep subwavelength scale. Writing, for instance, the electric-field part of the energy of a cavity mode as $\mathcal{U} = \frac{\varepsilon_0}{2}E^2 \times V$, where V is the geometric volume of the cavity, we have $E^2 \propto \mathcal{U}/V$. If $\mathcal{U}$ is expressed as a function of a photon number $\mathcal{U} = N_{\text{ph}}\hbar\omega$, we then get $E^2 \propto N_{\text{ph}}\hbar\omega/V$. Provided V is small enough, large fields can be obtained with few photons. A typical macroscopic cavity features $V \sim 1\,\text{cm}^3$, whereas a dielectric microcavity of index n may offer V's limited by the diffraction, $V \sim (\lambda/2n)^3 \sim 10^{-14}\,\text{cm}^3$, and a plasmonic nanocavity may even confine the field at a deep subwavelength scale. So, nanophotonics offers a real opportunity for studies on light–matter interaction relying on a few photons.

In the following, we start with an overview of micro- and nanoresonators and their respective modes. We identify the difference between normal and quasinormal modes, highlighting that the former are simple stationary patterns which correspond to the eigenstate of an Hermitian operator, while the latter are the eigenstate of a non-Hermitian operator. We study the emblematic case of Fabry–Perot cavities, for which the quasinormal modes are known analytically. We then define the main characteristic parameters of quasinormal modes, their quality factor Q, and mode volume V in a general non-Hermitian framework. From there on, we discuss the relationship between the modal picture and the linear response, e.g. the classical Lorentzian frequency response of resonators. We then use the simplicity of the modal approach to get insights into many important phenomena: critical coupling, strong coupling, and a large section on *temporal* two-mode coupled-mode theory. The complements tackle both 'ends' of the issues, and notably: (A) the Fabry–Perot with lossy mirrors; (B) the relationship to the other flavour of coupled-mode theory, the *spatial one*—a classic in waveguide and non-linear optics textbooks.

7.2 Overview of micro- and nanoresonators

There are two emblematic examples of optical quasinormal modes: the Fabry–Perot modes for which light bounces back and forth between two parallel mirrors and whispering-gallery modes (WGMs) for which light travels along the periphery of concave surfaces. Both of these are found in Mie resonances of dielectric spheres or more complex geometries. Fabry–Perot quasinormal modes will be analysed in section 7.4

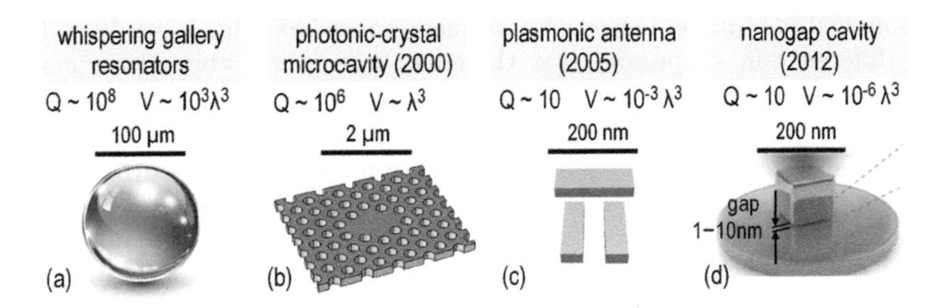

Fig. 7.1 Milestone nanostructures that have led to extreme light confinement, with either high Q's or small V's. (a) Optical whispering-gallery dielectric spheres or discs; (b) photonic-crystal microcavities; (c) plasmonic nanoantennas with metallic nanoparticles; (d) nanocavities with tiny dielectric gaps formed by a nanoparticle on a metal substrate.

and complement 7.B. Originally discovered for soundwaves in the Whispering Gallery of St Paul's Cathedral, WGMs are encountered in optics for various geometries, the simplest case being Mie resonances of a sphere or disc of radius R and refractive index n in air (see Fig. 7.1a).

Like Fabry–Perot modes, WGMs are qualitatively understood with ray optics (see Fig. 7.2a). Consider a ray of light propagating inside a body, hitting the surface with angle of incidence i. If $i > \arcsin(1/n)$, then total internal reflection (TIR) occurs. Because of rotational symmetry, all subsequent angles of incidence are the same, and the ray is trapped. If this ray strikes the surface at near-glancing incidence ($i \approx \pi/2$) and R is much larger than the resonance wavelength λ_r, the trapped ray propagates close to the surface and travels a distance $\approx 2\pi R$ in a round trip. If one round trip exactly equals m wavelengths in the medium (m is an integer), then one expects a constructive interference to occur for $2\pi Rn/\lambda_r \simeq m$. The number of wavelengths in the circumference can be identified as the angular momentum in the usual sense. During the round trip, there is a decay caused by the fact that, unlike a flat surface, the internal reflection from a curved surface cannot be total and leads to a radiation leakage into air. That is where the complex frequency comes in to compensate for these losses with a physics very similar to that encountered for the Fabry–Perot slab in Eq. (7.8) of section 7.4.

Optical whispering-gallery resonators are typically dielectric silica microspheres with an excellent surface finish and only a few nanometres or less of surface roughness. With the bulk optical loss from silica also being exceptionally low, it is not uncommon to see microspheres with diameters of $\sim 100\,\mu$m ($V \approx 10^3\lambda^3$) and Q-factors of 10^9, implying that the mode lifetime $1/\Gamma$ is in the microsecond range, i.e. that light makes 10^6 rotations each $10^3\,\lambda$ long before fading away.

Fabry–Perot and whispering-gallery resonators are a hundred years old, but optical cavities have experienced a rapid evolution since the year 2000, owing to the importance of confining light in subwavelength volumes. The topic initially started with micropillar GaAlAs cavities that played a major role in initial applications of the Purcell effect, soon followed by semiconductor photonic-crystal microcavities that confine light in volumes $V \sim \lambda^3$ with Q-factors of 10^5–10^6, thanks to a precise control of the position and size of tiny holes etched in semiconductor wires or membranes (see Fig. 7.1b). In spite of their apparent difference in shape and size, the confinements in Fabry–Perot macroscopic cavities formed by two metallic mirrors separated by thousands of wavelengths and in photonic-crystal microcavities are essentially the same, because the fields of interest are lying in dielectric media. In either configuration, the electromagnetic quasinormal mode is a standing wave pattern composed of progressive waves that are bouncing back and forth. Similarly, inside the resonator, the energy is then transferred every half-period between the electric-field energy $u_E = (1/2)\varepsilon E^2$ and the magnetic-field energy $u_H = (1/2)\mu_0 H^2$ (like for progressive waves). Because the system size is larger than the diffraction limit, the energy is almost conserved while it is transferred from one form of the energy to the other, and nearly self-sustaining oscillations with ultra-high Q's are achieved. More details are found in chapter 16. However, further shrinking of the confinement is not possible with dielectric materials, owing to the fact that tiny pieces of dielectric materials essentially behave as scatterers that radiate energy rather than store it.

A paradigm shift occurred around 2005 with the advent of metallic nanoantennas. These devices are similar to radiowave and microwave antennas and can take various unusual forms (spheres, rods, polyhedrons, or small assembly of them), owing to the strong shape and material dependence of their resonance properties. Nanoantennas exploit the unique properties of metal nanostructures to confine light in deep-subwavelength volumes $V \sim \lambda^3/10^3$ (see Fig. 7.1c). Even smaller confinements down to volumes $V \sim \lambda^3/10^6$ are achieved with structures that trap light in nanometre-sized gaps between two metal surfaces (see Fig. 7.1d), enabling optical spectroscopy on subfemtolitre volumes and giving routine access to the single-molecule regime. In such small volumes, self-sustaining oscillations are no longer possible between the electric-field and magnetic-field energies. With metals, however, the energy can also be stored by the free electrons in the form of kinetic energy, u_K. Even for tiny pieces of metal, oscillations may again take place between u_K and u_E and the diffraction limit can be beaten. This success, however, comes at a price: the energy stored in the form of kinetic energy is lost at a rate commensurate with the scattering rate of electrons in metal, which occurs on the very quick scale of 10 fs. Thus, as the optical mode becomes deeply sub-wavelength in all three dimensions, independent of its shape, the Q-factor is limited to $\sim$10.

7.3 Modes in conservative and non-conservative contexts

In this section, a mathematical definition of quasinormal modes in electromagnetism is provided and the difference between undamped normal modes with real frequencies and quasinormal modes with complex frequencies is clarified.

7.3.1 Definition of electromagnetic quasinormal modes

As said before, quasinormal modes with temporally decaying settings are defined as a formal solution of linearized differential equations. In electromagnetism, the Maxwell equations beautifully offer a unique framework for modelling all kinds of electromagnetic modes, be it related to photonic microcavities made of dielectrics or nanoantennas made of metals. In all cases, the quasinormal modes can be found by solving the time-harmonic source-free Maxwell equations

$$\begin{bmatrix} 0 & i\varepsilon^{-1}(\mathbf{r},\tilde{\omega})\nabla\times \\ -i\mu^{-1}(\mathbf{r},\tilde{\omega})\nabla\times & 0 \end{bmatrix} \begin{bmatrix} \widetilde{\mathbf{E}}(\mathbf{r}) \\ \widetilde{\mathbf{H}}(\mathbf{r}) \end{bmatrix} = \tilde{\omega} \begin{bmatrix} \widetilde{\mathbf{E}}(\mathbf{r}) \\ \widetilde{\mathbf{H}}(\mathbf{r}) \end{bmatrix}, \tag{7.2}$$

with frequency-dependent permittivity ε and permeability μ. One should additionally consider that Eq. (7.2) should be solved by considering *outgoing-wave boundary conditions* that stipulate that, away from the resonator, there are no incoming waves bringing energy to the system.

Eq. (7.2) is general, and takes the form of an eigenproblem, $\tilde{\omega}$ and $[\widetilde{\mathbf{E}}(\mathbf{r}), \widetilde{\mathbf{H}}(\mathbf{r})]$ being the eigenvalues and eigenvectors. The eigenproblem is linear for non-dispersive materials, but for dispersive ones, such as nanoresonators, the permittivity and permeability depend on frequency and the eigenproblem becomes non-linear. This also implies that the permittivity and permeability are defined at complex frequencies, although the material constants are measured at real frequencies. There exist various effective methods for inferring the permittivity and permeability values at complex frequencies from data measured at real frequencies over a finite spectral range, the most obvious one being to consider an analytic model deduced from a microscopic description of charges motion with Drude or Drude–Lorentz models. In general, any dispersive material can be modelled with several Lorentz poles, since the optical properties are arising from electronic resonances, even when a quantum description is adopted (Ashcroft and Mermin, 1976).

Quasinormal modes can be calculated analytically only for simple geometries, e.g. planar Fabry–Perot cavities, as will be done in section 7.4. For other geometries, they have to be computed numerically. Efficient algorithms and many commercial Maxwell's mode solver packages exist to solve Eq. (7.2) for virtually all geometries and materials (Ashcroft and Mermin, 1976). Figure 7.2 shows three examples of quasinormal

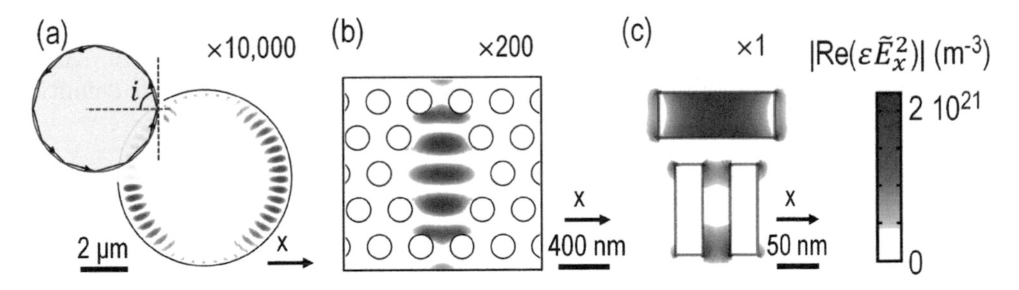

Fig. 7.2 Three examples of quasinormal modes. (a) WGM of a silica disc in air ($\tilde{\omega} = 1.26\,10^{15} - i\,4.60\,10^{9}$ rad/s). The inset shows a ray-optics picture. (b) Photonic-crystal cavity fabricated by etching circular holes in a GaAs membrane in air ($\tilde{\omega} = 1.38\,10^{15} - i\,1.35\,10^{11}$ rad/s). (c) Nanoantenna composed of three silver nanorods ($\tilde{\omega} = 2.68\,10^{15} - i\,6.82\,10^{11}$ rad/s). In all plots, we show the real part of $\varepsilon\widetilde{E}_x^2$, and since the modes are normalized such that the numerator integral in Eq. (7.15) is unity, the plots allow a direct visual comparison of the interaction strength of the mode with point-like dipoles (molecules) linearly polarized along the x-direction (see Section 7.6). Note the multiplicative scaling factors $\times 10,000$, $\times 200$, and $\times 1$ in each panel.

mode distributions for a whispering-gallery silica disc, a photonic-crystal cavity, and a nanoantenna composed of three silver nanorods.

7.3.2 Normal vs quasinormal modes

Let us first consider the textbook case of perfect, closed Hermitian cavities without dissipation. Then $\tilde{\omega}$ and ε are real numbers, and a solution for Eq. (7.2) can be found with $\widetilde{\mathbf{E}}$ and $\widetilde{\mathbf{H}}$ fields that are valued as pure real and pure imaginary, respectively. Remembering that we are in complex notations and that we should take the real parts of the complex fields, the true electric and magnetic fields are

$$\begin{aligned} \mathrm{Re}(\widetilde{\mathbf{E}} \exp(-i\tilde{\omega}t)) &= \mathrm{Re}(\widetilde{\mathbf{E}}(\mathbf{r}))\cos(\omega t)\,, \\ \mathrm{Re}(\widetilde{\mathbf{H}} \exp(-i\tilde{\omega}t)) &= \mathrm{Im}(\widetilde{\mathbf{H}}(\mathbf{r}))\sin(\omega t)\,. \end{aligned} \tag{7.3}$$

As expected, we end up with a well-known stationary pattern with a separable dependence on t and $\mathbf{r}$, classical examples being the Fabry–Perot resonator with two perfect mirrors and a dissipation-free string held at its two extremities.

In contrast, for dissipative non-Hermitian systems, $\tilde{\omega}$, ε, $\widetilde{\mathbf{E}}$, and $\widetilde{\mathbf{H}}$, all become complex-valued, and all the components (for instance the x-components of the electric and magnetic fields) of the true fields

$$\begin{aligned} \mathrm{Re}(\widetilde{E}_x \exp(-i\tilde{\omega}t)) &= \|\widetilde{E}_x(\mathbf{r})\| \cos(\omega t + \Phi_{E_x}(\mathbf{r})) \exp(-\Gamma t/2)\,, \\ \mathrm{Re}(\widetilde{H}_x \exp(-i\tilde{\omega}t)) &= \|\widetilde{H}_x(\mathbf{r})\| \cos(\omega t + \Phi_{H_x}(\mathbf{r})) \exp(-\Gamma t/2) \end{aligned} \tag{7.4}$$

are no longer described by a stationary pattern. The quasinormal of non-Hermitian systems cannot be expressed as a product of two independent functions that depend on $\mathbf{r}$ and t. They have space-dependent phase factors $\Phi_{E_x}(\mathbf{r})$ and $\Phi_{H_x}(\mathbf{r})$ that reveal the embodiment of some progressive waves in the mode pattern. This progressive character is a signature of non-hermiticity, which results in a finite lifetime and provides a marked difference from the stationary normal modes of Hermitian systems (Wu et al., 2021). Later, we will see its embodiment in the mode volume $\tilde{V}$ and in the energetic definition of the Q-factor.

7.3.3 The exponential growth in space of a quasinormal-mode field

In this section, we consider the spatial form of a quasinormal mode that is characterized by a complex frequency, $\tilde{\omega}$, instead of a real one, ω. For simplicity, we consider that our resonator is embedded in a media assumed to be non-dispersive (refractive index n independent of the frequency) and uniform, and we focus on the field far away from the resonator in the open space. Since such a field satisfies the outgoing-wave boundary condition in a uniform medium and results from a body located far away (as seen in chapter 2), it takes the asymptotic simple form:

$$\widetilde{\mathbf{E}}(r \to \infty) \,\propto\, \frac{1}{r}\exp(-i\tilde{\omega}(t - nr/c)). \tag{7.5}$$

It is a spherical progressive wave. The most stringent characteristic of Eq. (7.5) is the exponential growth of the mode field away from the resonator: the sign of $Im(\tilde{\omega})$

being such that the exponential function is decaying in time, the exponential function diverges in space as $r \to \infty$, owing to the minus sign in $(t - nr/c)$. The exponential growth is illustrated in Fig. 7.3a for the simplest case of a Fabry–Perot cavity and for two real temporal signals at time t and $t' > t$. Note that the individual right-going and left-going fields also exhibit amplified patterns inside the slab itself.

Is there any inconsistency caused by the space divergence? The answer is no. Remember our initial tuning fork example that is suddenly struck at, say, $t = 0$. Based on our daily experience, we admit that if we are at a fixed position $\mathbf{r}$, we may hear an exponentially damped tone, reminiscent of the fundamental quasinormal mode that decays by leaking its energy away into the air. However, we will not hear the quasinormal mode in its totality, and in particular its exponential growth $\exp(-i\omega t)\exp(-t/\tau)$ as $t \to -\infty$. The observation of the exponential decay is constrained to a time interval, here $t > 0$, which prevents the observation of any time divergence. Exactly the same limitation holds for the space divergence.

Now consider a fixed time $t > 0$ after the fork is struck, and look at the field distribution in space. Admittedly, it is not our daily experience, but indeed since the quasinormal mode is excited, the space divergence for fields at position r further and further away from the fork is observed. But if we look at infinity, $r \to \infty$, the divergence is not observed because it would imply that the sound would have been emitted from the fork at a time $t' = t - r/c$ that tends towards minus infinity for $r \to \infty$. Causality precludes such an observation and implies that the exponentially growing field with r cannot be observed for all r and is constrained to a space interval, here $r < ct$. It is interesting to estimate the characteristic distance from which the quasinormal-mode exponential growth is significantly perceived. Away from the resonator, since the field in vacuum diverges as $\exp[-\text{Im}(\tilde{\omega})r/c]$, a characteristic distance can be formulated as $-c/\text{Im}(\tilde{\omega})$ (remember that $\text{Im}(\tilde{\omega}) < 0$ with our convention) or, equivalently, if we introduce the quality factor $Q = -\frac{\text{Re}(\tilde{\omega})}{2\text{Im}(\tilde{\omega})}$ (see Section 7.5.1), it is $\approx Q\lambda$, with λ the resonance wavelength equal to $\text{Re}(\frac{2\pi c}{\tilde{\omega}})$. For metallic nanoresonators, $Q \approx 10$ at visible wavelengths and the divergence is noticeable even in the near field of the nanoresonator. In contrast, for photonic-crystal microcavities, Q can readily exceed 10^4, implying that the divergence is perceived only in the far field.

No matter whether it occurs at small or large characteristic distances, the exponential growth cannot be ignored and is rooted in the essence of quasinormal modes. In general physics, the explicit exponential divergence is characteristic of quasinormal-mode fields in the open space, and sharply contrasts with the form of 'plain', truly harmonic normal modes that are the modes of conservative and bounded systems of Hermitian physics (atomic energy levels are among the best known of the physical examples).

7.4 Quasinormal modes of Fabry–Perot resonators

The exponential growth in space for $r \to \infty$ and in time for $t \to -\infty$ has hindered the development of quasinormal-mode theory in the past, as compared to the development of guided-mode theory. In particular, it has been difficult to understand how to normalize quasinormal-mode fields (Sauvan et al., 2022) or to quantify how much they can be excited by, for instance, a pulse coming from the far field. Indeed, these difficulties are now overcome for arbitrary geometries thanks to initial works on the 1D Fabry–Perot resonator (or the tunnel junction in quantum mechanics) and on spherical potentials encountered in Mie resonators (Lalanne et al., 2018).

Given the simplicity of our example of the Fabry–Perot cavity, our objective is less ambitious, but we nonetheless hope to gain much insight into the profound meaning of complex frequencies and to clarify the link between quasinormal modes, laser modes, and the normal modes of waveguides.

7.4.1 Fields, frequencies, and damping rates

We consider a slab of index n_2 and thickness L sandwiched between two half-spaces of refractive index $n_1 = 1$. The truly guided modes of this geometry have been studied in the previous chapter and, consistent with this, we denote by x the axis perpendicular to the slab.

The system has two translational invariances along the y- and z-axes. Since these two axes can be identically treated, we do not consider any dependence on y. In addition, the translation invariance along z implies that the z-dependence of the field can be factored out, and takes the form $\exp(i\beta z)$, β being a real number: see chapter 4.

With this exponential prefactor, we look for the source-free solutions of the Maxwell equation, considering the TE-polarization case for simplicity. Thus the quasinormal modes are entirely determined by their electric-field y-component $\tilde{E}_y$, the magnetic field being obtained with the Maxwell–Faraday equation. The unusual point is that we solve the equations for a complex frequency $\tilde{\omega}$ with the time dependence $\exp(-i\tilde{\omega}t)$. In each medium, $|x| < L/2$ or $|x| > L/2$, the solutions of the Maxwell equations are plane waves, since the permittivity is uniform, and therefore the wavevector x-component $\tilde{k}_p$ ($p = 1$ for the claddings and $p = 2$ for the slab) is complex:

$$(\tilde{k}_p^2 + \beta^2) = \tfrac{n_p^2}{c^2}\tilde{\omega}^2, \quad p = 1, 2. \tag{7.6}$$

The outgoing-wave conditions impose that we have only *outgoing* waves in the outer parts, implying that we use only $+\tilde{k}_1$ (with positive real part) for $x \to +\infty$ and only $-\tilde{k}_1$ for $x \to \infty$. Inside the slab, we have a superposition of two counterpropagative plane waves with opposite $\tilde{k}_2$. In summary, the electric field component $\tilde{E}_y$ of the quasinormal mode can be written as follows:

$$\tilde{E}_y(x, z, t) = \begin{cases} B\exp(i\tilde{k}_1 x) & x > L/2, \\ \exp(i\beta z - i\tilde{\omega}t)A[\exp(i\tilde{k}_2 x) \pm \exp(-i\tilde{k}_2 x)] & |x| < L/2, \\ \pm B\exp(-i\tilde{k}_1 x) & x < -L/2. \end{cases} \tag{7.7}$$

The field amplitude B outside the cavity is related to the amplitude A inside it by $B\exp(i\tilde{k}_1 L/2) = tA[\exp(i\tilde{k}_2 L/2) \pm \exp(i\tilde{k}_2 L/2)]$, with r and t being the real reflection and transmission coefficients at each interface; r and t depend on β and are independent of frequency for non-dispersive media: see chapter 5. The pluses (minuses) correspond to symmetric (antisymmetric) quasinormal modes.

The *required condition to build up a resonance* in the cavity is the Fabry–Perot condition (see Complement 7.B): the field amplitude A should be recovered after one round trip in the resonator, i.e. after one reflection on each interface $A = Ar^2 \exp(2i\tilde{k}_2 L)$. Since the cavity loses energy via leakage, $|r|^2 < 1$, the resonance condition can be satisfied only for complex $\tilde{k}_2$, which therefore verifies

$$1 - r^2 \exp(2i\tilde{k}_2 L) = 0. \tag{7.8}$$

The imaginary part of the complex frequency has a clear physical role here, illustrated in Fig. 7.3a for $\beta = 0$: it restores a consistent state in the lossy cavity by imposing a spatially amplified profile for each wave propagating between the interfaces (amplification refers here to the spatial envelope of what we represent, which is a snapshot of the field; the overall time dependence at any point is a temporal decay, as the two field patterns and the clocks remind us). Similarly, the plane waves are amplified in the claddings and exponentially diverge away from the slab towards $x = \pm\infty$. Eq. (7.8) has a simple discrete set of solutions $\mathrm{Re}(\tilde{k}) = m\frac{\pi}{L}$ and $\mathrm{Im}(\tilde{k}) = -\frac{1}{2L}\ln\left(\frac{1}{r^2}\right)$ with $m = 0, 1, 2, \ldots$ We can deduce the complex frequencies of the quasinormal modes

$$\mathrm{Re}(\tilde{\omega}_m) = m\,\frac{\pi c}{n_2 L} \tag{7.9}$$

and

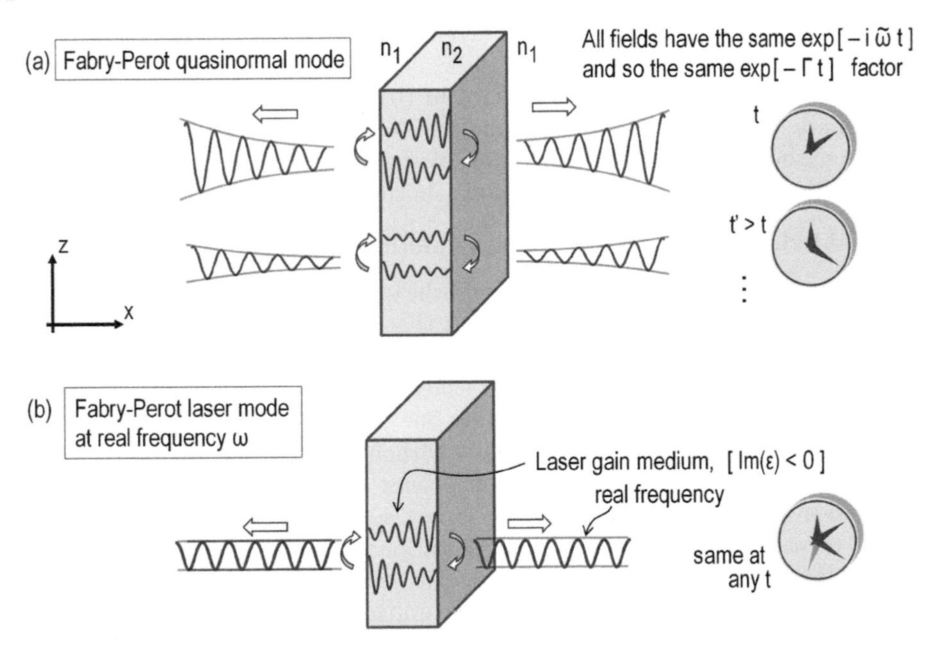

Fig. 7.3 (a) Fabry–Perot quasinormal mode of a slab for $\beta = 0$: the mode field grows everywhere as it 'propagates' along x because the frequency is complex. (b) Fabry–Perot laser mode emitting light at a real frequency. The field grows only in the slab because the slab has a complex-valued relative permittivity with $\mathrm{Im}(\tilde{\varepsilon}) < 0$ and amplifies light. There is no need to have a complex frequency to compensate for the loss incurred at the reflection on each mirror. Thus a laser mode is a special case of quasinormal mode with infinite Q.

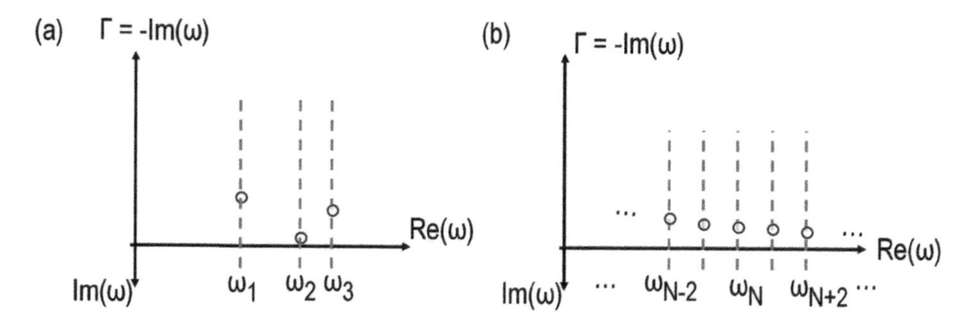

Fig. 7.4 Complex plane and quasinormal-mode complex frequency (poles): (a) generic example of a system having three quasinormal modes. Only mode 2 has a high quality factor ($Q_2 \gg 1$) due to the small $\mathrm{Im}(\tilde{\omega}_2)/\mathrm{Re}(\tilde{\omega}_2)$ ratio. (b) A situation more representative of the FP case, with many modes of regular spacing and slowly evolving quality factors.

$$\frac{\Gamma_m}{2} = -\mathrm{Im}(\tilde{\omega_m}) = \frac{c}{n_2 L}\ln\left(\frac{1}{r^2}\right). \tag{7.10}$$

Hence, in the complex plane, all the complex frequencies (the poles of the response of the system) are evenly distributed along the real axis, spaced by the so-called free spectral range (FSR), and all lie at the same imaginary distance from the real axis $-\frac{\Gamma}{2}$. In Fig. 7.4, we present the complex plane with a generic case of three quasinormal modes in (a) and a case closer to the Fabry–Perot in (b). In later subsections, we interpret these results in terms of quality factors and finesse of the cavity. Exercise 7.3 proposes an extension that considers the dispersion of the slab and deepens the analysis by additionally deriving the Q-factor of the quasinormal mode.

7.4.2 Guided, leaky, and quasinormal modes of invariant systems

The guided modes of the previous chapter and the quasinormal modes of the present one have much in common, starting from the fact that they are both source-free solutions of the Maxwell equations that satisfy the outgoing-wave conditions, and it is legitimate to wonder: how far might the analogy be pushed?

For 3D systems like spheres, there is only one 'quantum number' that can be used to find a solution of the Maxwell equations without source: the frequency $\tilde{\omega}$ (or the energy $\hbar\tilde{\omega}$), in the sense of a solution consistency along the time coordinate. Their quasinormal modes form an infinite discrete set. For 2D or 1D systems that possess one or two translation invariances, we have additional quantum numbers. Owing to the invariance, the solution should conserve the wavevector components that are parallel to the interfaces (the parallel photon momentum in the quantum analogy $\hbar k_{//}$): the β's for 2D waveguides such as fibres or the k_y and k_z for the Fabry–Perot slab of section 7.4.1. For every value of the parallel momentum, there is an infinite discrete set of quasinormal modes, and in total we have a continuum, since the parallel momentum varies continuously.

In particular, this implies that for the source-free Maxwell equations of waveguides, we have additional flexibilities. For fibres, we may fix ω and compute β to obtain guided modes, or we may fix β and compute ω to obtain quasinormal modes. For Fabry–Perot slabs, we may fix k_y and k_z and compute ω to obtain quasinormal modes, or we may fix one k-component and ω and compute the other k-component to obtain guided modes. All these modes are different.

There is, however, a special case in which they coincide. Going back to Eq. (7.6), let us consider β values such that total internal reflection occurs for the plane waves that bounce back and forth inside the slab. Then $|r| = 1$ and the logarithm is equal to zero in Eq. (7.10). There is no leakage and frequencies become real, and coming back to Eq. (7.7) with real $\tilde{k}_1$ and $\tilde{k}_2$, we find that the quasinormal modes are nothing but the truly guided modes of Chapter 6. For this special case of infinite systems with translation invariances, quasinormal modes with an infinite lifetime exist and they can be interpreted as truly guided modes. A similar discussion could be provided for periodic waveguides (see chapter 17).

For β values such that total internal reflection does not occur, the guided modes of the slab are found by assuming a real frequency and looking for a solution with a complex β. From Eq. (7.6), we directly infer that $\tilde{k}_p$ is also complex with an imaginary part whose sign is opposite to that of $Im(\beta)$, implying that if the mode is exponentially damped as it propagates along the slab in the z-direction, it exponentially diverges away from the waveguide in the transverse x-direction. We enter the realm of leaky modes that were briefly addressed in Complement 6.C. Like quasinormal modes, these modes are observed only in a finite region of space in experiments.

We are thus facing a variety of modes: quasinormal ones, normal ones that are truly guided, and leaky guided ones. These modes are cross-functional topics in nanophotonics and in the book and have different meanings. Consistently through the book, the modes computed by fixing ω are called guided modes (they are either truly guided or leaky) and those computed by fixing k are called quasinormal modes; the latter have a special significance, as they represent the states in the sense of solid-state physics: counting them amounts to quantifying the density of electromagnetic states.

The existence of a variety of modes will be revisited in chapters 15–17 for Bloch modes of periodic systems, for which, even in the absence of absorption, resonances with complex frequencies and real Bloch propagation constants along the waveguide axis offer a wide variety of interesting effects. Quasinormal modes also exist for metal surfaces, in the form of surface plasmon modes. They are very similar to the slab modes and have a real propagation constant $-\infty < \beta < \infty$ along the interface axis and a complex frequency due to absorption. Like the quasinormal modes of the slab with arbitrary β, they possess a sinusoidal spatial dependence parallel to the surface (no spatial damping along the slab) and a finite lifetime proportional to the inverse of $\text{Im}(\tilde{\omega})$ (see chapter 13). Finally, let us note that surface plasmon modes also exist for a real frequency and a complex propagation constant. They are the analogue of the guided modes and the imaginary part of the propagation constant just accounts for an exponential damping due to metal absorption.

7.4.3 Fabry–Perot laser modes

A reader familiar with the basics of lasers has certainly spotted the near-identity of Eq. (7.8) with the textbook equation providing the *laser threshold gain.* The analogy is partly misleading, as we illustrate by the hopefully heuristic comparison between Figs 7.3a and b. For lasers, the amplitude-matching condition of Eq. (7.8) is fulfilled because of *gain* in the slab (the relative permittivity $\tilde{\varepsilon}$ is complex with $\mathrm{Im}(\tilde{\varepsilon}) < 0$ and not positive, as for metals) and not because of the complex frequency. Thus the waves grow up as they propagate in the slab only and the outgoing waves in the open space are not exponentially diverging and are internally fuelled by energy conversion from an electrical or optical pump. The snapshot of partial waves inside the slab is deceptively similar, but the time dependence is stationary in a continuous-wave (cw) laser: a laser mode may be considered as a quasinormal mode with a real frequency, owing to the gain in the slab that exactly compensates for the mirror losses, therein allowing self-consistent oscillations.

Note that quasinormal modes may also be defined for slabs with a complex relative permittivity causing amplification, at least when amplification is not strong enough to reach the laser threshold. The complex frequency provides the lacking 'amplification' to reach threshold and again the field exponentially grows in the open claddings.

7.5 Cavity Q's and $\tilde{V}$'s

Two characteristic parameters, which figure prominently in the physics and device applications of cavities, quantify the capability of cavity modes to boost light–matter interactions, the quality factor Q, and the mode volume $\tilde{V}$. The latter is denoted by a tilde to emphasize that it is a complex quantity, as will be explained later on. It is fundamental to be familiar with the meaning of the two characteristic parameters. The issue is important, as large Q's and small $|\tilde{V}|$'s promote strong coupling regimes with small transition dipole moments and single-molecule levels. Q governs the temporal interaction. $\tilde{V}$ governs the spatial interaction and deserves much attention, especially when dealing with non-Hermitian systems such as metallic nanocavities.

7.5.1 Energy balance and cavity Q

In the broadest sense, the Q-factor measures the sharpness of the response of a resonator to an external excitation. In textbooks (Jackson, 1975), it is often defined as the frequency times the ratio of the time-averaged energy stored in the cavity to the power loss per cycle, $Q \equiv \omega \times$ (Energy stored in the cavity)/(Power loss). For low-loss systems, when many energy oscillations take place in the cavity before damping (slowly varying amplitudes limit with $\mathrm{Re}(\tilde{\omega}) \ll \mathrm{Im}(\tilde{\omega})$), there is general agreement on this definition. But for larger loss rates (when the energy is dissipated within a few oscillation periods), such a definition of Q is no longer evident, and this gives rise to an ambiguous $\tilde{V}$.

Actually, this definition, albeit classical (Jackson, 1975), requires much clarification, as the energy stored in the cavity is not straightforwardly defined. Consider, for instance, the whispering-gallery disc of Fig. 7.1a; the energy inside the disc can be perfectly defined, but why should we constrain the stored energy to a geometrical volume defined by the disc? There is indeed some energy around the disc in air, and why should we not include it as well? We note that it is not an easy task. Actually, the fields $\tilde{\mathbf{E}}$ or $\tilde{\mathbf{H}}$ exponentially diverge in the cladding and any integral of the form

$\iiint \varepsilon|\widetilde{\mathbf{E}}|^2\, dV$ or $\iiint \mu|\widetilde{\mathbf{H}}|^2\, dV$ also exponentially diverges. The same difficulty is encountered with the Fabry–Perot cavity of section 7.4 or the photonic-crystal cavity of Fig. 7.1b, and the difficulty is compounded for metallic nanoantennas for which energy should additionally be defined in the metal.

Fundamentally, the difficulty is not linked to a specific geometry but is inherent in non-Hermitian systems with energy dissipation. To clarify the purpose, we rather define the Q-factor as

$$Q = -\frac{\mathrm{Re}(\tilde{\omega})}{2\,\mathrm{Im}(\tilde{\omega})}, \tag{7.11}$$

which is unambiguously defined for any cavity, be it photonic or plasmonic, and we further give an energetic interpretation to the definition through the Poynting theorem written at complex frequency (Lalanne et al., 2018).

Let us consider an arbitrary system and a first quasinormal mode $(\widetilde{\mathbf{E}}, \widetilde{\mathbf{H}}, \tilde{\omega})$. For the sake of simplicity, we assume here that the resonator is made of isotropic non-magnetic materials, but the following derivation is easily extended. $(\widetilde{\mathbf{E}}, \widetilde{\mathbf{H}}, \tilde{\omega})$ satisfies the source-free Maxwell equations of Eq. (7.2). Conjugating this equation, one finds another solution $(\widetilde{\mathbf{E}}^*, \widetilde{\mathbf{H}}^*, -\tilde{\omega}^*)$ with the same permittivity and permeability distribution, since $\varepsilon^*(\mathbf{r},\omega) = \varepsilon(\mathbf{r}, -\omega^*)$ due to the Hermitian symmetry of real Fourier transforms (see chapter 1). Note that this second quasinormal mode also satisfies the outgoing-wave conditions, since its Poynting vector is identical to the first one, $\mathbf{P} = \frac{1}{2}\mathrm{Re}\left(\widetilde{\mathbf{E}} \times \widetilde{\mathbf{H}}^*\right) = \frac{1}{2}\mathrm{Re}\left(\widetilde{\mathbf{E}}^* \times \widetilde{\mathbf{H}}^{**}\right)$.

Applying the divergence theorem to the vector $\widetilde{\mathbf{E}}^* \times \widetilde{\mathbf{H}} + \widetilde{\mathbf{E}} \times \widetilde{\mathbf{H}}^*$ on an *arbitrary* closed surface, Σ, that defines a finite volume V (see Fig. 7.5) leads to

$$\int_\Sigma \mathrm{Re}\left(\widetilde{\mathbf{E}} \times \widetilde{\mathbf{H}}^*\right) \cdot d\mathbf{S} = -\int_V \left[\mathrm{Im}\left[\tilde{\omega}\,\varepsilon(\tilde{\omega})\right] \widetilde{\mathbf{E}} \cdot \widetilde{\mathbf{E}}^* + \mu_0\,\mathrm{Im}(\tilde{\omega})\,\widetilde{\mathbf{H}} \cdot \widetilde{\mathbf{H}}^*\right] dV, \tag{7.12}$$

which is conveniently rewritten as

$$\begin{aligned} &\frac{1}{2}\int_\Sigma \mathrm{Re}\left(\widetilde{\mathbf{E}} \times \widetilde{\mathbf{H}}^*\right) \cdot d\mathbf{S} + \frac{\mathrm{Re}(\tilde{\omega})}{2}\int_V \mathrm{Im}\left[\varepsilon(\tilde{\omega})\right] |\widetilde{\mathbf{E}}|^2\, dV = \\ &-\frac{\mathrm{Im}(\tilde{\omega})}{2}\int_V \left[\mathrm{Re}\left[\varepsilon(\tilde{\omega})\right] |\widetilde{\mathbf{E}}|^2 + \mu_0 |\widetilde{\mathbf{H}}|^2\right] dV\,. \end{aligned} \tag{7.13}$$

Insertion of Eq. (7.11) and a few elementary algebraic steps lead to

$$Q = -\frac{\mathrm{Re}(\tilde{\omega})}{2\,\mathrm{Im}(\tilde{\omega})} - \frac{\mathrm{Re}(\tilde{\omega})\,\frac{1}{4}\int_V \left[\mathrm{Re}\left[\varepsilon(\tilde{\omega})\right] |\widetilde{\mathbf{E}}|^2 + \mu_0 |\widetilde{\mathbf{H}}|^2\right] dV}{\frac{1}{2}\int_\Sigma \mathrm{Re}\left(\widetilde{\mathbf{E}} \times \widetilde{\mathbf{H}}^*\right) \cdot d\mathbf{S} + \frac{\mathrm{Re}(\tilde{\omega})}{2}\int_V \mathrm{Im}\left[\varepsilon(\tilde{\omega})\right] |\widetilde{\mathbf{E}}|^2\, dV}. \tag{7.14}$$

Equation (7.14) is valid for any surface enclosing a finite volume. It may surround the whole cavity (Σ_1 in Fig. 7.5) or only part of it (surface Σ_3). It may even be located outside the physical resonator in the cladding (Σ_2). The surface may potentially enclose absorbing dielectric materials. The denominator of Eq. (7.14) corresponds to the sum of the power flowing out of the surface plus the power dissipated by absorption. If the

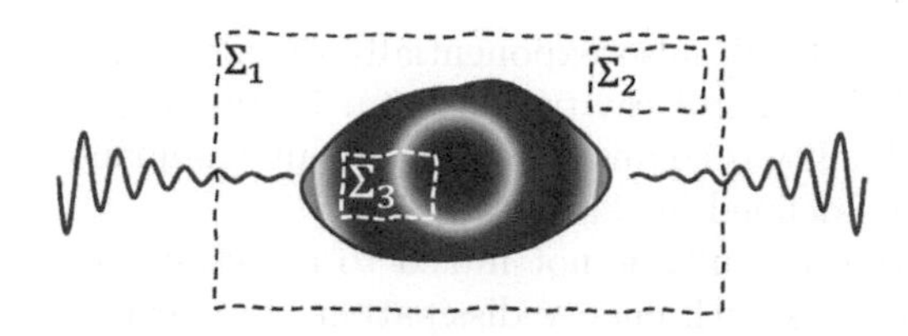

Fig. 7.5 Sketch of an electromagnetic resonator (a photonic-crystal cavity, a plasmonic antenna . . .) with its quasinormal-mode distribution. The Q-factor of the mode can be derived from the field distribution inside any closed surface that either surrounds the cavity (surface Σ_1), or is located entirely outside the resonator in the cladding (surface Σ_2), or encloses part of the mode (surface Σ_3).

surface encloses only non-dispersive materials (ε and μ are frequency-independent), the volume integral in the numerator can be interpreted as the time-averaged electromagnetic energy stored in the volume V multiplied by $\mathrm{Re}(\tilde{\omega})$. It is important to keep in mind that the numerator does not define the energy stored in the cavity but just establishes an energy balance of the electromagnetic field distribution for the energy stored in an arbitrary box. Additionally, it is important to realize that the energy balance is achieved for field distributions at complex frequencies, so that the energy balance is only approximately satisfied with true field distributions obtained by illuminating resonators at real frequencies $\mathrm{Re}(\tilde{\omega})$.

If the surface Σ encloses dispersive materials, a difficulty arises with the definition of the energy (Jackson, 1975). When the variations in the permittivity and permeability of the materials are small over the bandwidth of the resonance, it is possible to generalize Eq. (7.14) to dispersive materials (Lalanne et al., 2018). For highly damped and dispersive materials, the demonstration requires us to consider an analytic form of the temporal dispersion, e.g. Drude or Drude–Lorentz materials, and additionally brings to light the kinetic energy of the electrons in the energy balance. Exercise 7.3 shows an example of the application of Eq. (7.11) by retrieving the classical formula of the Q-factor of Fabry–Perot cavities as the function of the mirror reflectance, cavity length, and group index of the bouncing waves.

7.5.2 Mode volumes

Q governs the temporal interaction and is unambiguously defined as the number of cycles before damping. $\tilde{V}$ governs the spatial interaction and deserves more attention. To figure out the significance of $\tilde{V}$, imagine a cavity perturbed by a tiny foreign object, a molecule, for instance, that can be represented by a polarized point-like dipole. The interaction is mutual: on one side, the object perturbs the cavity modes, therein modifying their complex frequencies, and on another side, the modes polarize the object. The strength of this two-way interaction, which depends on the position $\mathbf{r}_0$ of the object and the polarization direction $\mathbf{u}$ of the point-like dipole, measures the mode's ability to strengthen the interaction between light and matter.

Such interactions were initially studied with microwave cavities in the context of either cavity perturbation in classical physics or spontaneous emission control in quantum physics, with the celebrated work by Purcell (Purcell, 1946) on cavity quantum electrodynamics. Because of their high Q's, microwave cavities weakly leak and their quasinormal modes may be approximately considered as normal modes of Hermitian systems, and a definition of the mode volume based on energy is often adopted (Yokoyama and Ujihara, 1995; Taflove et al., 2013): $V_{\mathcal{E}}(\mathbf{r}_0) \simeq \frac{\iiint \left[\varepsilon|\tilde{\mathbf{E}}|^2 + \mu_0|\tilde{\mathbf{H}}|^2\right] dV}{2\varepsilon(\mathbf{r}_0)|\tilde{\mathbf{E}}(\mathbf{r}_0)\cdot\mathbf{u}|^2}$, where $\mathbf{u}$ is the polarization direction and the subscript $\mathcal{E}$ holds for 'energy'. $V_{\mathcal{E}}$ is a real quantity that suggests that the strongest interaction is implemented at the field-intensity maximum (Yokoyama and Ujihara, 1995).

In this energy-based expression, the phase of the mode is absent, implying that all the essence of non-hermiticity is absent (see Section 7.3.2). Actually, $V_{\mathcal{E}}$ only incompletely models high-Q photonic crystal cavities and is completely unfounded for plasmonic nanocavities. Non-hermiticity can be taken into consideration, provided that the mode volume be defined as a complex-valued quantity,

$$\tilde{V}(\mathbf{r}_0) = \frac{\iiint \left[\frac{\partial \varepsilon\omega}{\partial \omega}\tilde{\mathbf{E}}\cdot\tilde{\mathbf{E}} - \mu_0\tilde{\mathbf{H}}\cdot\tilde{\mathbf{H}}\right] d^3\mathbf{r}}{2\varepsilon(\mathbf{r}_0)\left[\tilde{\mathbf{E}}(\mathbf{r}_0)\cdot\mathbf{u}\right]^2}, \tag{7.15}$$

for non magnetic materials. We note that the phase of the mode explicitly appears in this definition due to the replacement of $|\tilde{\mathbf{E}}|^2$ by $\tilde{\mathbf{E}}^2$. The normalization integral in the numerator requires some care, owing to the fact that the field is exponentially divergent away from the resonator in the claddings. The mathematical demonstration of the convergence of the integral for arbitrary geometries is nowadays well established and implemented in most quasinormal-mode solvers (Lalanne et al., 2018).

Indeed, in the limit of conservative systems ($Q \to +\infty$), Eq. (7.15) reduces to $V_{\mathcal{E}}$. We leave it to the reader to assess it by injecting the expressions of the electric and magnetic fields of Eq. (7.3) obtained in the high-Q limit into Eq. (7.15).

The real part of $\tilde{V}(\mathbf{r}_0)$ retains an intuitive meaning and quantifies the coupling strength of a dipole with the resonator. Returning to Fig. 7.2, all the quasinormal modes are normalized such that the numerator integral in Eq. (7.15) is unity, and thus the plots of $\mathrm{Re}(\varepsilon\tilde{E}_x^2)$ provide a direct visual comparison of the interaction strength of the mode with point-like dipoles (molecules) linearly polarized along the x-direction. It is the nanocavity that provides the strongest interaction.

Initially seen as a mathematical abstraction, complex mode volumes have gradually shown their usefulness for analysing microcavities and nanoantennas (Wu et al., 2021). Indeed, $\mathrm{Im}(\tilde{V})$ has its roots in many important phenomena of light-matter interaction in photonic and plasmonic systems, leading to new effects that are not anticipated with a Hermitian approach. Complement 7.A lays out the premises of a quasinormal treatment of the concept of local density of electromagnetic states (LDOS), with an emphasis on the link between the complex mode volume and the Purcell effect. Similarly, section 13.4.1 provides a quasinormal treatment of the LDOS of a flat metallic interface. For a Hermitian treatment of the LDOS, the reader is directed to chapter 12.

7.6 Resonant external excitation

7.6.1 Quasinormal-mode expansion

So far, we have only looked at the natural time evolution in $\exp(-i\tilde{\omega}t)$ of resonator quasinormal modes. Now, let us attempt to excite the resonator from outside with a driving field (or current). It is well known that the most general motion of a system is a superposition of its modes, or eigenstates. For conservative systems with a Hermitian operator, the temporal evolution describing the motion can be described in the language of linear algebra with normal modes, i.e. eigenvectors of the Hermitian operator.

The most classical textbook example for a physicist is offered by quantum mechanics. The time-dependent Schrödinger equation predicts that wave functions can form standing waves, called stationary states, which can be found by a simpler form of the Schrödinger equation, the time-independent Schrödinger equation. The stationary states are particularly important, as their individual study later simplifies the task of solving the time-dependent equation by describing the evolution of the system in the basis of its stationary states.

For non-conservative systems, it is no longer enough to invoke linear algebra to justify such a description. However, our intuition leads us to believe that the motion could still be described with quasinormal modes. Consider, for instance, the guitar. When the string is plucked normally, one tends to hear the fundamental frequency most prominently, but the overall sound is also coloured by the presence of various overtones, or frequencies greater than the fundamental frequency. The fundamental frequency and its overtones are perceived by the listener as a single note; however, different combinations of overtones can be obtained by slightly different initial excitations and give rise to noticeably different notes.

It is beyond the scope of this chapter to describe under which conditions the free evolution of an electromagnetic resonator, once its excitation has occurred, can be expanded into the set of its quasinormal modes and to discuss whether the set is complete or needs to be augmented by other modes for completeness (Lalanne et al., 2018), and we will simply admit that the field scattered by the resonator can be expanded as

$$\mathbf{E}_S(\mathbf{r},t) = \mathrm{Re}\left(a_1(t)\,\widetilde{\mathbf{E}}_1(\mathbf{r}) + a_2(t)\,\widetilde{\mathbf{E}}_2(\mathbf{r}) + a_3(t)\,\widetilde{\mathbf{E}}_3(\mathbf{r}) + \ldots\right), \tag{7.16}$$

where a_m's, $m = 1, 2\ldots$, are the time-dependent modal excitation coefficients, which describe how the quasinormal modes are loaded by the driving field and release their energy (for a musical insight, see Exercise 7.2). Note that, unlike the total field, $\mathbf{E}_S(\mathbf{r},t)$ satisfies the outgoing-wave condition, which is what makes the projection on quasinormal modes possible.

Fig. 7.6 illustrates the predictive force of Eq. (7.16) for a dolmen-like nanoantenna composed of three gold nanorods. The antenna is assumed to be illuminated by a short pulse, and its response is dominated by three modes in the visible. The contributions of every individual mode are shown in the left bottom panel and their sum according to Eq. (7.16) is plotted in the right panel, therein recovering the reference data obtained from finite-difference time-domain (FDTD) computations.

The modal approach clarifies the origin of the complex temporal response predicted (but not interpreted) by the FDTD method as the beating of the contribution of three

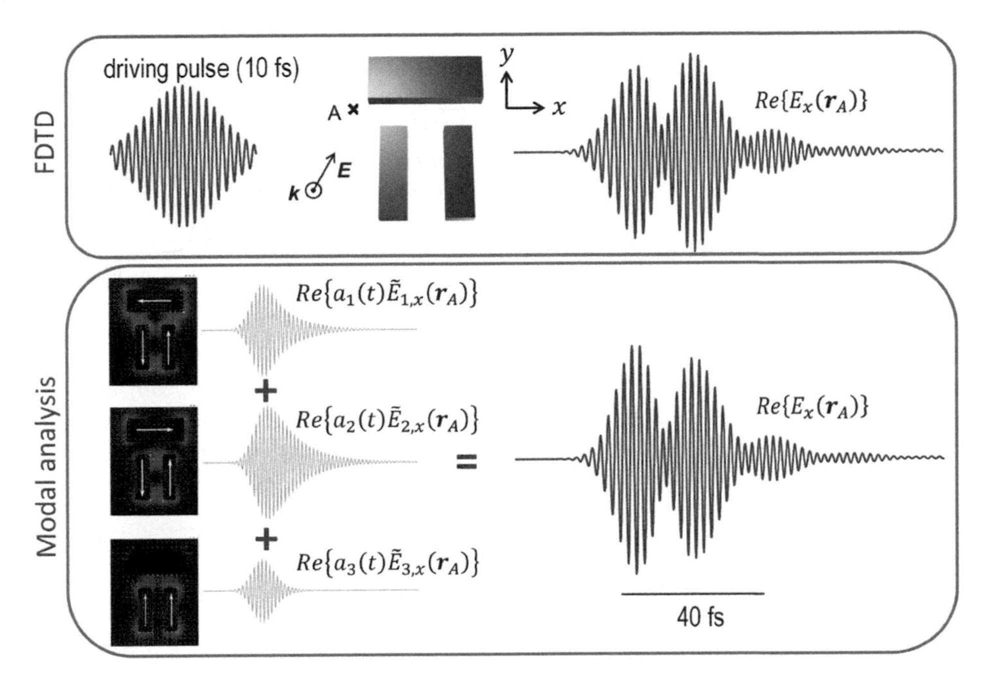

Fig. 7.6 Temporal response of a dolmen-like nanoantenna composed of three gold nanorods under illumination by a short linearly polarized 10-fs pulse incident from the z-direction with a central frequency $\omega = 2.9 \times 10^{15}$ rad/s. (a) Reference data for $E_x(A)$ as computed with the finite-difference time-domain method at point A situated 12 nm away from the bottom-left corner of the 128-nm-long longitudinal rod. (b) Reconstruction of $E_x(A)$ with three dominant quasinormal modes, $m = 1, 2, 3$. Although many high-energy modes are neglected in the expansion of Eq. (7.16), the deviation from the reference data is very weak. Adapted from ref. (Lalanne et al., 2018).

dominant modes. Another strength of the approach is the capability to rapidly predict the response for other pulses with different duration, polarization or incidence, owing to the fact that once the three modes are computed, the new time-dependent modal excitation coefficients are known analytically from spatial-overlap integrals between the driving beam and the quasinormal-mode fields.

7.6.2 Phenomenological approach

Later in this chapter, we adopt a *phenomenological* approach: we assume that the main dynamical equation of the interaction between a quasinormal mode and a generic excitation can be written as follows:

$$\frac{da_m}{dt} = -\, i\tilde{\omega}_m\, a_m - i\rho_m u(t) = -\, i\omega_m\, a_m - \frac{\Gamma_m a_m}{2} - i\rho_m u(t), \tag{7.17}$$

where ρ_m defines the coupling to the generic excitation u. Such an excitation could be either due to a far-field source like in the previous example (then u is a coefficient related to the driving plane-wave amplitude) or to a near-field source (then u is a

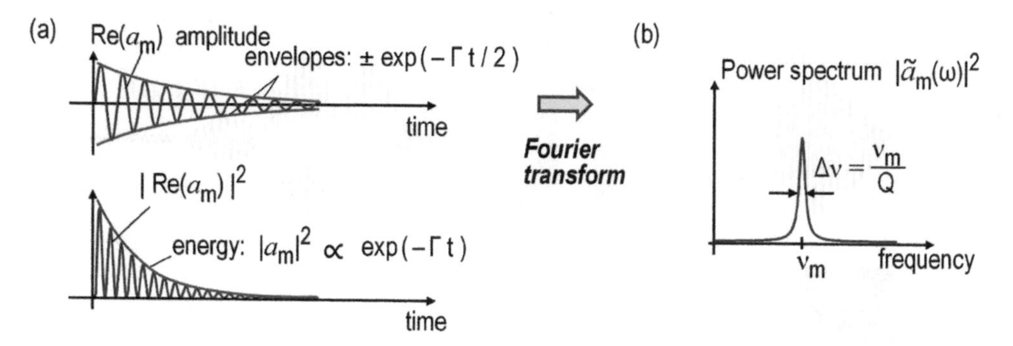

Fig. 7.7 (a) Amplitude decay after an excitation pulse (impulse response, top) and energy decay at twice the rate of amplitude. (b) The power spectrum is obtained from the squared modulus of the Fourier transform of the amplitude. With some care, the same power spectrum is obtained by a slow frequency scan of the energy (quasi-cw excitation).

coefficient related to the field generated by the source). It is convenient to assume that u is of the same electromagnetic nature as a (thus ρ is a frequency), and that for this reason, adding the i factor allows them to be treated more similarly. We abandon any ambition to derive a rigorous expression for the coupling rate ρ_m,[1] a complex-valued coefficient that requires much attention and will be simply assumed to be fitted, just like in the coupled-mode theory of the next section.

The archetypal excitation is the impulse excitation, a very short pulse that can be mathematically taken as $u = u_0\delta(t)$. The damped oscillator response, the so-called Green function, can be shown to be equal to the product of the step function by $i\,u_0 \exp(-i\tilde{\omega}_m t)$ for $\Gamma_m \ll \omega_m$ (see Fig. 7.7a). Exercise 7.6 proposes to recover the Lorentzian power-density spectrum. The latter can also be rigorously found by assuming harmonic dependence of the generic excitation $u = u_0 \exp(-i\omega t)$ and looking for a harmonic solution at the same frequency $a_m(t) = \tilde{a}_m \exp(-i\omega t)$. From Eq. (7.17), we directly get $-i\omega\,\tilde{a}_m = -i\tilde{\omega}_m\,\tilde{a}_m - i\rho_m\,u_0$. The Lorentzian power density spectrum, sketched in Fig. 7.7b, reads

$$|\tilde{a}_m(\omega)|^2 = \left|\frac{\rho_m\,u_0}{\omega - \omega_m + i\Gamma_m/2}\right|^2 = \frac{|\rho_m|^2\,u_0^2}{(\omega-\omega_m)^2 + \Gamma_m^2/4}\,. \tag{7.18}$$

The resonance full-width at half-maximum (FWHM) is Γ_m. The relative width is thus Γ_m/ω_m. This leads us to another convenient expression for Q,

$$Q = \frac{\omega_m}{\Gamma_m} \approx \frac{\omega_m}{\Delta\omega_{FWHM}} = \frac{\nu_m}{\Delta\nu_{FWHM}}, \tag{7.19}$$

where we also introduced the genuine frequency $\nu_m = \omega_m/(2\pi)$. Equation (7.19) is particularly useful when a spectrum contains a well isolated and marked resonance with a Lorentzian shape. But it may become confusing when the resonance interferes with a strong continuum or another nearby resonance, thereby leading to Fano lineshapes.

[1]Analytical expressions for ρ exists. Furthermore, Eq. (7.17) may incorporate an additional term proportional to $du(t)/dt$ for resonators made of dispersive materials, see *Phys. Rev. B*, **97**, 205422 (2018).

7.7 Critical coupling of a single resonator

We now exploit the phenomenological picture of section 7.6.2 to address the condition known as 'critical coupling'. It concerns the non-intuitive optimal way to excite a resonator mode. The critical-coupling condition can be generalized to an arbitrary number of modes, as well as the interaction of the modes with other extraneous degrees of freedom (purely electronic or mechanical, for instance), but we restrict ourselves to monomode resonators. Throughout this section, for simplicity, we denote $\gamma = \frac{\Gamma}{2}$.

To figure out what critical coupling is, let us consider photodetectors made from thin quantum wells whose absorption band is not achievable with bulk semiconductor structures; for further examples, see chapter 8. These wells being weak absorbers, they absorb a small fraction of the incident power in a single pass, and a multiple-pass scheme must be implemented to boost the detection efficiency. Absorption is best enhanced by locating these wells in an antinode of the cavity mode, in the centre for instance, see Fig. 7.8. The important question arises of how strong the mirror transmission should be to efficiently detect light. Is there an optimal mirror transmission which maximizes absorption? It might be of interest to consider a non-symmetric detection for which the incident light is impinging from one side only, so the mirror on the other side could be chosen with $R = 1, T = 0$, but for the sake of simplicity we stick to the symmetric case and let the reader explore the asymmetric cavity. So we must find the common transmission T of the two mirrors.

Intuitively, on the one hand, a high transmission is desirable for transferring energy from the outside driving field to the wells in the resonator, but on the other hand, large transmission prevents light from field enhancement through multiple passes. Thus there must be a 'critical coupling' for a prescribed value of $(1 - R) = T$ that maximizes the desired rate of incoming energy.

To find the critical-coupling condition, we split the cavity loss into two physical channels with subscripts M for mirror and A for well absorption,

$$\frac{da_m}{dt} = -i\omega_m a_m - \gamma_A a_m - \gamma_M a - i\rho_m u \; . \tag{7.20}$$

For a monochromatic incident wave at frequency ω, we insert $\frac{da_m}{dt} = -i\omega a$ and get

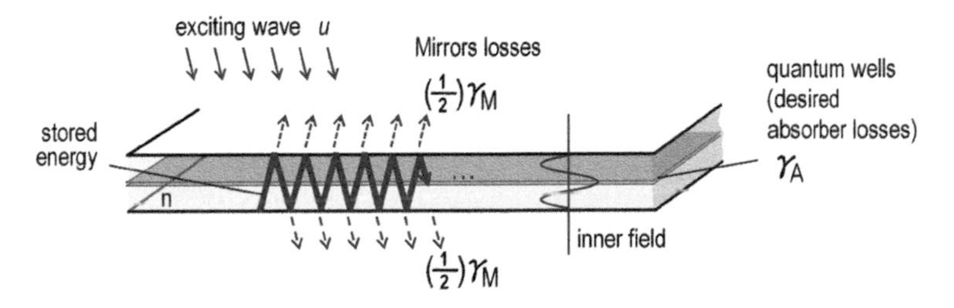

Fig. 7.8 Critical coupling: under what conditions does an external wave impinging on the resonator deliver maximum absorption inside the resonator? The absorber (consisting of quantum wells aimed at photodetection, rate γ_A) competes with mirror losses (rate γ_M). Critical coupling, i.e. selecting $\gamma_M = \gamma_A$, maximizes the power dissipated in the absorber.

$$a_m(\omega) = \frac{\rho_m\, u_0}{\omega - \omega_m + i(\gamma_A + \gamma_M)}\,. \tag{7.21}$$

The coefficient ρ_m describes the coupling rate between the incoming field amplitude u_0 and the field amplitude inside the cavity. The coupling to the outside world being monitored by the *amplitude* transmission of the mirrors, we have $\rho_m \propto t$. For good mirrors, since $|t| \approx \sqrt{1-R}$, we have

$$|\rho_m| \;\propto\; |t| \;\approx\; \sqrt{1-R} \;\propto \sqrt{1/Q_M} \;\propto\; \sqrt{\gamma_M}, \tag{7.22}$$

where Q_m is the quality factor of the empty cavity without wells. The proportionality between $\sqrt{1-R}$ and $\sqrt{1/Q_M}$ is a classical result derived in Complement 7.B. It holds for Fabry–Perot cavities with large mirror reflectances, i.e. large Q's, a reasonable hypothesis for our problem with small absorption.

Using Eq. (7.22), we obtain the key expression:

$$a_m(\omega) \propto \frac{\sqrt{\gamma_M}}{\omega - \omega_m + i(\gamma_A + \gamma_M)}\, u_0\,. \tag{7.23}$$

Equation (7.23) contains all mirror-related γ_M dependences. We look for the maximum $|a_m|^2$ at resonance ($\omega_m = \omega$) by maximizing $\xi(\gamma_M) = \gamma_M/(\gamma_A + \gamma_M)^2$. The resulting critical-coupling condition is

$$\gamma_M = \gamma_A\,. \tag{7.24}$$

In other words, the mirror losses should let as much energy go away as goes into the desired absorption process. Then, including the mirror-screened excitation, the energy delivered to that process is maximized. We see that we cannot get all the energy in the desired process (here, our quantum well absorption) but only half of it. This is no big surprise, as it is rooted in the way energy is shared once in the resonator.

It is not the end of the story, however: we noted here that the outgoing wave represents an energy loss. But if, say, some of the exciting wave amplitude undergoes a more complex fate and is allowed to interfere destructively with these outgoing waves, the conclusions could be very different. This shows the importance of spending more time examining a variety of outside situations. Among them, the case of resonant waveguide gratings (see chapter 17) can expand the insight proposed here.

7.8 Coupled-mode theory

7.8.1 Coupled resonators

We have prioritized the simplicity of the motion equation of oscillators as set above. The phenomenological coefficients or parameters such as ω_m, γ_m, ρ were kept to a minimum, with only limited developments to link with actual parameters: in the simplest case of the Fabry–Perot cavity, such parameters were mirror properties such as r or the finesse F. Here, we try to exploit with similar simplicity the analysis of a set of two coupled oscillators. It is a logical next step compared to a single oscillator, and we will even refrain from using any explicit driving excitation, instead focusing on the new quasinormal modes of the coupled system arising from the mutual interaction of the individual resonators: we will only consider leakage from the resonators to outgoing waves, but we need not specify these radiation channels explicitly.

The equation of motions for two coupled resonators described with modal amplitudes a_1 and a_2 is

$$\begin{aligned}\frac{da_1}{dt} &= -i\omega_1\, a_1 - \gamma_1 a_1 + iga_2,\\ \frac{da_2}{dt} &= -i\omega_2\, a_2 - \gamma_2 a_2 + iga_1,\end{aligned} \tag{7.25}$$

where ω_1 and ω_2 are the (generally distinct) resonance frequencies of the two uncoupled resonators, γ_1 and γ_2 their associated amplitude loss rates, and g the coupling constant. With $g = 0$, we would just have the two individual poles $\omega_1 - i\gamma_1$ and $\omega_2 - i\gamma_2$. Anticipating the remainder of this section, we can already guess from the complex plane representation that the interesting cases arise when poles are close to each other. We will focus on poles with two close real frequencies, $\omega_1 \approx \omega_2$, and will look at various cases for the proximity of the two imaginary parts. The above evolution equation, Eq. (7.25), can be written in matrix form,

$$\frac{d}{dt}\begin{bmatrix} a_1 \\ a_2 \end{bmatrix} = \begin{bmatrix} -i\omega_1 - \gamma_1 & -ig \\ -ig & -i\omega_2 - \gamma_2 \end{bmatrix}\begin{bmatrix} a_1 \\ a_2 \end{bmatrix} = M\begin{bmatrix} a_1 \\ a_2 \end{bmatrix}, \tag{7.26}$$

where M is a matrix describing the time evolution. If we shift the i so as to start the left-hand side with $i\frac{d}{dt}$, we have a full analogy with Schrödinger's equation in quantum mechanics $i\hbar\partial\psi/\partial t = H\psi$, with the column vector $[a_1\ a_2]^\dagger$ (the superscript $\dagger$ represents the transpose) playing the role of the wavefunction ψ, implying that the new quasinormal modes of the coupled system are described as a linear combination of each individual quasinormal mode. For energy conservation, hermiticity of iM is required. Hence there are opposite signs of non-diagonal terms $\pm g$ that become $\pm ig$ in iM. However, if necessary, we may still allow the non-Hermitian terms related to $\gamma_{1,2}$ to dissipate in energy in exactly the way they did in individual oscillators: the non-trivial constraints are on non-diagonal terms only. To solve this differential equation, we look for harmonic behaviour, whereby the whole column vector undergoes a harmonic motion,

$$[a_1(t), a_2(t)] = [a_1(t_0), a_2(t_0)]\exp(-i\omega[t - t_0])\,. \tag{7.27}$$

The solution is then the result of an 'eigenproblem',

$$-i\omega\,[a_1\ \ a_2]^\dagger = M[a_1\ \ a_2]^\dagger\,. \tag{7.28}$$

That is, we simply need to diagonalize a 2×2 matrix with complex coefficients. There are five parameters (two frequencies, two loss rates, and coupling g), so we shall split the study. First, we will consider a lossless case and focus on the frequency difference, the so-called detuning $\delta = \omega_2 - \omega_1$ between the two resonators, thereby reducing the parameters to only two, δ and g. These parameters are at the core of any coupling problem.

7.8.2 Anticrossing of two dispersion curves

Let us consider the lossless case $\gamma_1 = \gamma_2 \equiv 0$ and denote by $\delta = \omega_2 - \omega_1$ the *detuning* frequency. We take ω_1 as the reference here, but one could select another variable (e.g. $(\omega_1 + \omega_2)/2$) yielding some differences in algebra only. The matrix to be diagonalized becomes

$$M = \begin{bmatrix} -i\omega_1 & -ig \\ -ig & -i(\omega_1 + \delta) \end{bmatrix} . \tag{7.29}$$

It is straightforward to find the determinant $\det[M - (-i\omega)I]$, I being the identity matrix, to obtain the two roots $\omega_\pm$ of the resulting quadratic equation

$$\omega_\pm = \omega_1 + \frac{\delta \pm \sqrt{\delta^2 + 4g^2}}{2} . \tag{7.30}$$

We thus have two new eigenfrequencies for the coupled system, differing by $\omega_+ - \omega_- = \sqrt{\delta^2 + 4g^2}$. We exploit this generic result to introduce the notion of anticrossing. Let us just plot the eigenfrequencies of Eq. (7.30) as a function of δ, for a typical range of a few g. We get two branches, as shown in Fig. 7.9a. One branch starts asymptotically at ω_1 and turns upwards at around $\delta = 0$ within a typical interval $\delta \sim g$. The other branch starts with the asymptotic slope of the diagonal term $\omega_1 + \delta$, but turns and flattens out at around $\delta = 0$. Actually, for $|\delta| \gg g$, the two asymptotic branches of the plot correspond to the initial frequencies ω_1 and $\omega_1 + \delta$ of the uncoupled oscillators.

The dashed lines simply correspond to the two frequencies on the diagonal of the matrix or, in other words, the frequencies of the uncoupled oscillators. In our choice of variables, a non-dispersive line ω_1 is traversed by the ascending ω_2 one at $\delta = 0$. The actual branches (solid lines) do not cross each other. Following one of them from a given left-most asymptote $\delta \ll +g$, the branch attains the other asymptote at $\delta \gg +g$, and in the central region $\delta \lesssim g$, the two branches do not cross each other. This feature is called an *anticrossing*. We will meet this term in two other chapters in relation to the electronic and photonic bands (see chapters 8 and 16, respectively). At the anticrossing, the branches are maximally distant from their asymptotes.

But, conversely, the frequency distance between the two branches, $\omega_+ - \omega_-$, is smallest at $\delta = 0$, where it reaches the simple value $\Omega = 2g$. The avoided crossing, or anticrossing, is certainly a ubiquitous feature. This difference is also called the *splitting* of the initial mode frequencies; such splitting studies are a tenet of quantum mechanics textbooks. Figure 7.9b represents another graphically familiar configuration. We leave the reader to first formulate the asymptotes (diagonal) vs δ and next determine the anticrossing branches.

In many practical cases, the convenient branches that anticross are those that describe some known dispersion, such as $\omega = \omega(k)$, and are not even straight lines. It suffices to say that ω_1 and ω_2 are parametrized by some other quantity in a way that forms two branches that cross to get an anticrossing at the point of equal frequencies. In exercise 7.7, we suggest an example with branches that have an appearance often seen in polariton physics, with one parabolic branch and one straight branch.

Another interesting aspect in anticrossing is the eigenvectors, i.e. the expansion of the new eigenmodes in terms of the original uncoupled resonator quasinormal modes.

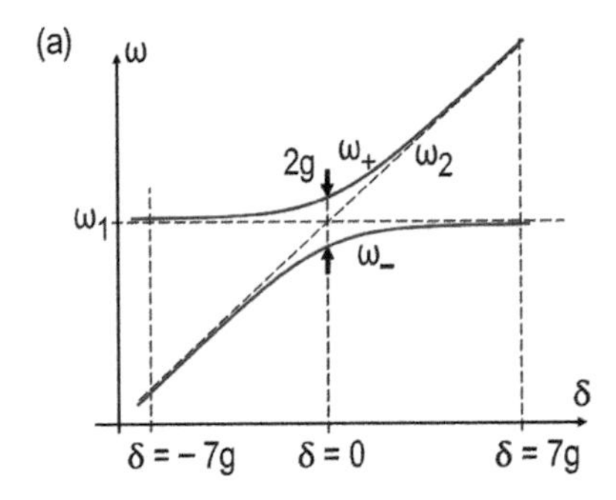

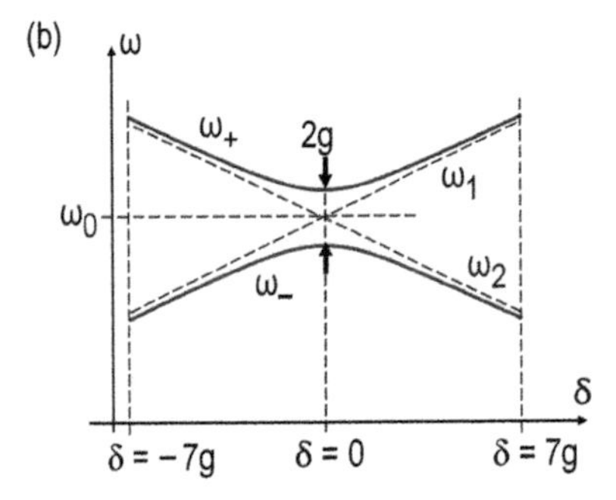

Fig. 7.9 Anticrossing. (a) Solid lines: plot of the two eigenfrequencies as a function of the detuning δ, the difference between the resonance frequencies ω_1 and ω_2 (shown with the dashed asymptotes) of the uncoupled resonators. (b) Same plot when the two frequencies evolve symmetrically vs a parameter, here δ itself. This latter version is more familiar in the framework of gap opening in band dispersion diagrams (see chapter 15).

Writing, for instance, the eigenmode 1 as $c_1\tilde{E}_1+c_2\tilde{E}_2$, the two coefficients $(c_1,\ c_2)$ tell us in what proportions the two original modes (the asymptotes) contribute to the coupled modes (the branches). In exercise 7.8, it is shown that the mixing coefficient (defined as the smallest one, $\min(c_1, c_2)$, assuming a normalized combination $c_1^2 + c_2^2 = 1$), is obviously 0.5 at $\delta = 0$ and vanishes when $|\delta| \ll g$, but also falls off with a behaviour one order of magnitude less steep than the eigenfrequency coupling-induced shifts $\omega_+ - \max(\omega_{1,2})$ or $\omega_- - \min(\omega_{1,2})$. In brief, in the asymptotic tails, the wavefunction of the problem varies more smoothly than the eigenvalues do. It is a special case of the general first-order perturbation theory for Hermitian operators.

7.8.3 Light or Rabi oscillation

We can now look at what happens to the system for a particularly simple excitation. A classical case is to consider, for two identical resonators ($\delta = 0$), an initial excitation, given $a_1(t=0) = A$, $a_2(t=0) = 0$. Only resonator 1 is excited at $t = 0$. The key point is that such an excitation is no longer a mode (an eigenvector) of the coupled system. Under our assumptions of identical resonators, the two eigenvectors are a symmetric and an antisymmetric combination of the two initial resonators (akin to the binding and antibinding states in the quantum-mechanical description of chemical bonds within identical atoms):

$$\begin{aligned} [a_1(t)\ \ a_2(t)]^\dagger &= [+a\ \ \ +a]^\dagger \exp(-i\omega_+ t), \\ [a_1(t)\ \ a_2(t)]^\dagger &= [+a\ \ \ -a]^\dagger \exp(-i\omega_- t). \end{aligned} \tag{7.31}$$

Thus we write the initial condition at $t = 0$ (hence $\exp(-i\omega_\pm t) = 1$) as a sum of the two eigenvectors

$$[a_1(0)\ \ a_2(0)]^\dagger = [A\ \ 0]^\dagger = \frac{1}{2}[A\ \ \ A]^\dagger + \frac{1}{2}[A\ \ \ -A]^\dagger\,, \tag{7.32}$$

so that

$$[a_1(t)\ \ a_2(t)]^\dagger = \frac{1}{2}A\left[\left(e^{-i\omega_+ t} + e^{-i\omega_- t}\right)\ \ \left(e^{-i\omega_+ t} - e^{-i\omega_- t}\right)\right]^\dagger . \tag{7.33}$$

In this expression, both parentheses have a so-called beating period: after some time, the two exponential terms are out of phase, the first component vanishes, and the second is maximal, and after an identical interval, they are in phase in the first component, while being destructive again in the second. By factoring $\left(e^{-i\omega_+ t} \pm e^{-i\omega_- t}\right) = e^{-i\omega_+ t}\left(1 \pm e^{-i\Omega t}\right)$, a modulation of each resonator by $\cos(\Omega t)$ or $\sin(\Omega t)$ appears (we saw that $\Omega = 2g$), with a beating period T_{beat} given by

$$T_{\mathrm{beat}} = \frac{2\pi}{\omega_+ - \omega_-} = \frac{2\pi}{\Omega}. \tag{7.34}$$

In other words, over a period of duration T_{beat}, the motion amplitude slowly sloshes from its initial position (resonator 1) to the other one (resonator 2) and back, as sketched in Fig. 7.10.

This kind of oscillation of coupled resonators is well known. In the domain of quantum mechanics or atom physics, when a quantum system possesses only two states, which are no longer eigenstates due to a perturbation coupling (e.g. a cw laser coupling two atomic levels of electrons in a given atom), the time to flip from an initial state (a ground state, for instance) to the final state (an excited state) is known as the Rabi period, and the similar phenomenon is known as a Rabi oscillation. It is also invoked in the case of a polariton, a hybrid electronic–photonic excitation discussed in Chapter 8, for which an initial excitation in one of the forms gives a slow temporal

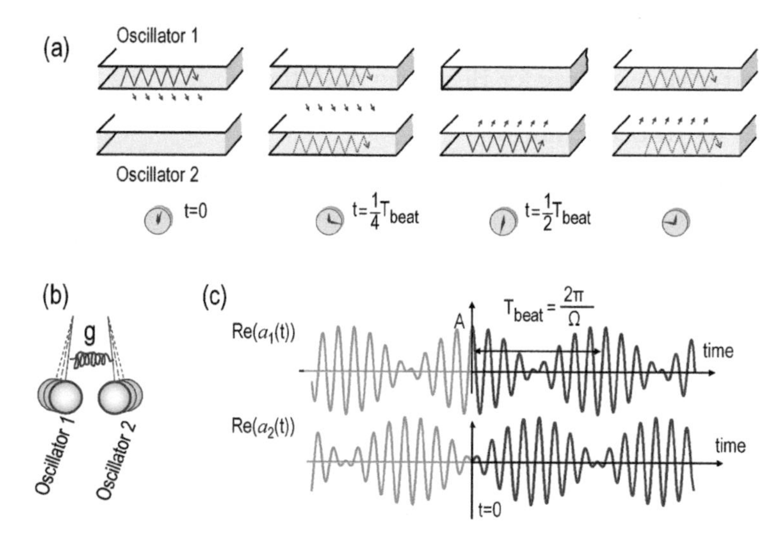

Fig. 7.10 (a) Oscillation of two identical coupled resonators in the time domain; at $t = 0$, only oscillator 1 is loaded and the second has zero amplitude; at $T_{\mathrm{beat}}/4$, transfer from 1 to 2 is at half-point, and at $T_{\mathrm{beat}}/2$, energy is entirely in the second one, before going back to the initial situation; (b) scheme of two coupled pendulums in mechanics; (c) time-domain modal amplitudes showing the beating. The $t < 0$ area has been shaded to correspond to our formal boundary condition $[a_1 = 1,\, a_2 = 0]$ at $t = 0$.

oscillation between the two forms. It is of particular importance for 2D excitons coupled to planar cavity quasinormal modes, the so-called *cavity polariton* for which the coupled-mode splitting was evidenced in 1992 (Weisbuch et al., 1992).

7.8.4 Strong and weak coupling

We now remain focused on the special case of zero-detuning (the resonators have the same frequencies denoted by ω_0) and consider loss or leakage to see how the Rabi oscillation survives in such circumstances. In doing so, we will meet the important distinction between *strong coupling* and *weak coupling.*

We thus start from the matrix

$$M = \begin{bmatrix} -i\omega_0 - \gamma_1 & -ig \\ -ig & -i\omega_0 - \gamma_2 \end{bmatrix}, \tag{7.35}$$

with g a real coefficient. Note that the introduction of a damping rate is phenomenological and that matrix M does not rigorously describe the non-Hermitian Hamiltonian of the problem. In particular non-hermiticity arises by considering complex-valued g's that arise from the embodiment of radiative waves in the quasinormal modes of the individual resonators (see section 7.3.2).

The eigenvalues $\tilde{\omega}$ are obtained from the roots of $\det(M + i\omega I) = 0$. A simpler expression is obtained by using $\Delta\tilde{\omega} = (\tilde{\omega} - \omega_0)$ as a convenient *complex* variable

$$\Delta\tilde{\omega}^2 + i\Delta\tilde{\omega}(\gamma_1 + \gamma_2) - (g^2 + \gamma_1\gamma_2) = 0, \tag{7.36}$$

whose discriminant, denoted by Δ_c, reads as

$$\Delta_c = 4g^2 - (\gamma_1 - \gamma_2)^2 . \tag{7.37}$$

Noting that the sign of Δ_c flips when $2g = |\gamma_1 - \gamma_2|$, we get the following conclusions:

$$2g > |\gamma_1 - \gamma_2| \;\rightarrow\; \Delta\tilde{\omega} = -i\,\frac{\gamma_1 + \gamma_2}{2} \pm \frac{\sqrt{\Delta_c}}{2} \;\rightarrow\; \text{decay+oscillation} \quad = \text{strong coupling,} \tag{7.38}$$

$$2g < |\gamma_1 - \gamma_2| \;\rightarrow\; \Delta\tilde{\omega} = -i\,\frac{\gamma_1 + \gamma_2}{2} \pm i\frac{\sqrt{|\Delta_c|}}{2} \;\rightarrow\; \text{decay only} \quad = \text{weak coupling.} \tag{7.39}$$

The overall behaviour of $\tilde{\omega}$ vs g is plotted in Fig. 7.11, showing for real and imaginary parts, respectively, a parabola for $g > g_{\text{crit}} = \frac{1}{2}|\gamma_1 - \gamma_2|$ and an ellipse for $g < g_{\text{crit}}$. It is a constant common to both new modes on the complementary g regions. This diagram passes through a so-called exceptional point at $g = g_{\text{crit}}$ and $\tilde{\omega} = \omega_0 + i\frac{1}{2}|\gamma_1 + \gamma_2|$, where eigenvalues are mathematically degenerate and the matrix is not diagonalizable.

The corresponding results for the same one-sided excitation as above are shown in Fig. 7.12 for various cases. In Fig. 7.12d, the location of the uncoupled and coupled eigenvalues in the complex plane are given for all examined cases as well as the lossless case, i.e. a canonical anticrossing. In Fig. 7.12a, we show a *strong coupling* case where $|\gamma_1 - \gamma_2| < 2g$ and $\gamma_2 = 0$. The beating is maintained, and the coupled-mode decay

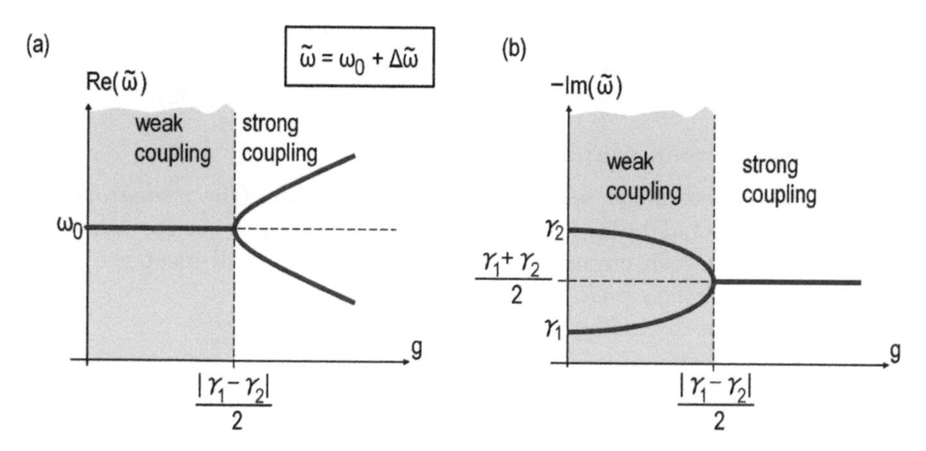

Fig. 7.11 Behaviour of complex frequencies through the weak and strong coupling regimes. (a) The real part is not split until the critical value $g_{\mathrm{crit}} = \frac{1}{2}|\gamma_1 - \gamma_2|$. (b) Conversely, the imaginary part is split below this critical value, but not above.

is logically provided by $\gamma_1/2$, since, upon beating, it feels γ_1 only half of the time. In Fig. 7.12b we show a *weak coupling* case where $|\gamma_1 - \gamma_2|$ exceeds $2g$ but still with $\gamma_2 = 0$. Both new modes have imaginary frequencies (see Fig. 7.12d) which differ by $\sqrt{\Delta_c}$ (see Eq. (7.39)). Hence if the two new modes coexist, we have a combination of quick and slow decays. What we observe for our initial excitation is indeed the initial strong decay, related to the stronger damped mode, acting here before oscillation can even take place. At longer times, the strongly damped mode has entirely vanished and both oscillator envelopes (magnified part) follow the decay associated with the smaller damped new mode, the logical 'surviving' mode.

In Fig. 7.12c, we show a less obvious *strong coupling* case where $|\gamma_1 - \gamma_2| < 2g$, but both γ_1 and the average $(\gamma_1 + \gamma_2)/2$ exceed $2g$, showing explicitly that even strong average losses do not prevent the system from expressing the Rabi oscillation, as mathematically predicted. Physically, both modes tend to decay at similar paces, so that the amplitude and phase conditions for energy exchange are satisfied even though the overall decay takes place: there is a hardly visible beating (arrows point at its nodes). In log scale, we would just see a beating pattern similar to the lossless one but slanted downwards. In Fig. 7.12d, we show where the uncoupled and coupled eigenvalues lie in the complex-frequency plane for a lossless case (canonical anticrossing). We also qualitatively show their location for the three cases [a], [b], and [c] for the general example of quasinormal modes in the complex plane (see Fig. 7.4).

Overall, the beating (the strong coupling signature feature) is a matter of eigenvalues being close enough in the complex plane, especially along the imaginary axis. This remark justifies the insight that we can obtain through the tools of quasinormal modes, notably the complex plane.

7.A Complement: LDOS and quasinormal modes

The concept of density of *electromagnetic* states is the photonic analogue of the density of states (DOS) introduced in solid-state physics to study the population of electrons or

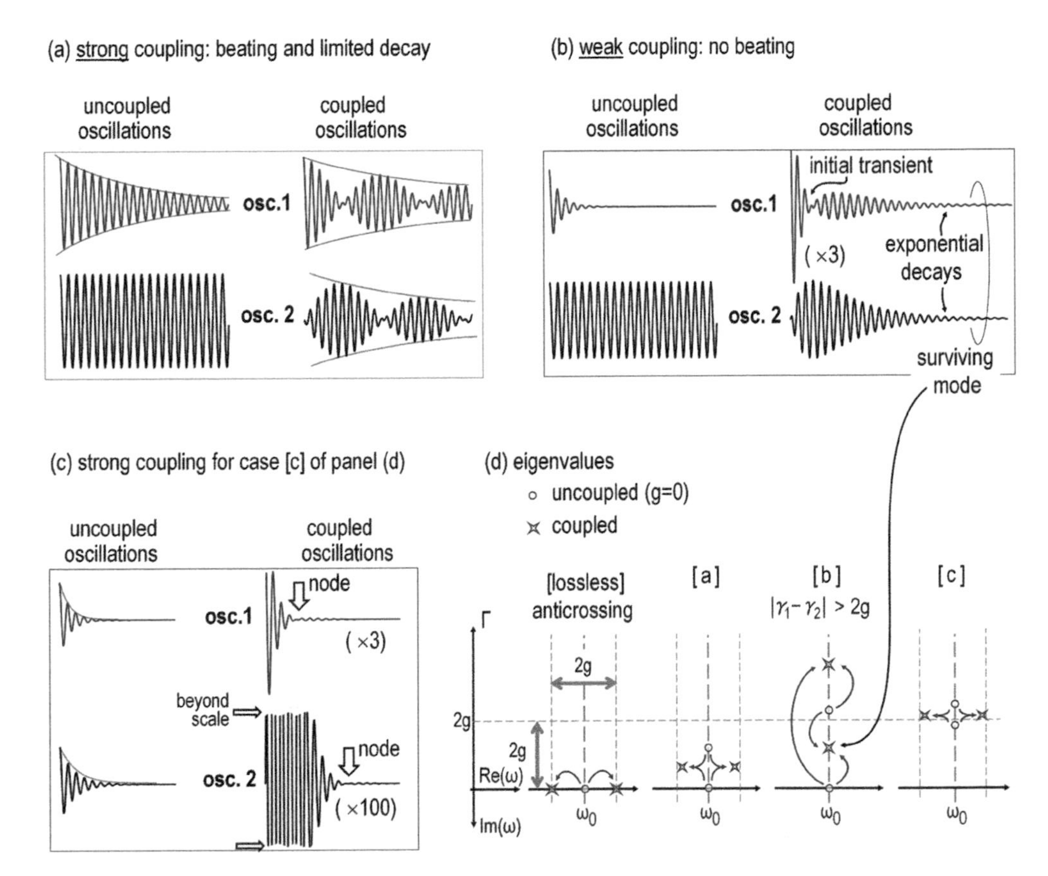

Fig. 7.12 (a) Strong coupling: uncoupled and coupled evolution of two oscillators; one is lossy (loss $\gamma_1 < 2g$) and the other is lossless. Note that the beating results in a slower decay at rate $\gamma_1/2$; (b) weak coupling (loss $\gamma_1 > 2g, \gamma_2 = 0$) with $\times 3$ magnification as indicated: no beating; in the modal combination, the smallest $|\text{Im}(\tilde{\omega})|$ is the long-time survivor; (c) revival of strong coupling when γ_2 approaches γ_1, even though their average exceeds $2g$. The arrows indicate the tiny beating (the overall decay is very strong, note the $\times 100$ magnifications); (d) complex plane with uncoupled and coupled eigenvalues for the lossless case (anticrossing) and for (a),(b), and (c). For (b), weak coupling, all eigenvalues have the same real part, and the smallest $|\text{Im}(\tilde{\omega})|$ dominates at long times.

phonons (see sections 8.2.1 and 12.5.3). For non-uniform systems, it is more appropriate to deal with the LDOS, which gives the density of states at a given position $\mathbf{r}_0$ in the system. The LDOS plays a key role for understanding many optical phenomena related to the interaction of light with matter. In many optics textbooks, the DOS is introduced in a conservative Hermitian context by counting the number of normal modes available in a closed box, like the 'particle in a box' approach used in solid-state physics. It is this simplified approach relying on Hermitian mode volumes that will also be adopted in chapter 12 and in chapter 19 when studying light emission assisted by cavities and nanoantennas.

This short Complement lays out the premises of the study of the electromagnetic LDOS directly from the eigenspectrum, i.e. the quasinormal modes. We will take the example of the modification of the spontaneous emission rate of a 'monochromatic' dipolar emitter coupled to a resonator, considering that the emission linewidth is much smaller than the linewidths of the quasinormal modes of the resonator. Thus the escape time of emitted photons from the cavity is much shorter than the radiative lifetime, and reabsorption is negligible. In this so-called weak coupling regime, spontaneous emission is usually derived by applying Fermi's Golden Rule at first order in time-dependent perturbation theory (see Chapter 19).

Spontaneous emission is a quantum effect, but the normalized emission rate $\frac{\gamma}{\gamma_0}$ (γ_0 is the emission rate of the same emitter placed in a bulk medium of refractive index n), like the LDOS, is a classical quantity that can be derived from the power radiated by classical electric dipoles. The QNM formalism of nanoresonators offers a natural platform for 'counting' the states. It can be shown that γ/γ_0 takes the form (Sauvan et al., 2013)

$$\frac{\gamma(\mathbf{r}_0,\omega)}{\gamma_0(\omega)} = \Sigma_m F_m \frac{\Omega_m^2}{\omega^2} \frac{(\Gamma_m/2)^2}{(\omega-\Omega_m)^2+(\Gamma_m/2)^2}\left[1+2Q_m\frac{\omega-\Omega_m}{\Omega_m}\frac{\mathrm{Im}(\tilde{V_m})}{\mathrm{Re}(\tilde{V_m}))}\right], \tag{7.40}$$

where $m = 1, 2, \ldots$ labels the quasinormal modes, Q_m is the quality factor, $\tilde{V_m}$ is the mode volume defined in Eq. (7.15) for an emitter polarized along the unit vector $\mathbf{u}$. F_m is the Purcell factor given by

$$F_m = \frac{3}{4\pi^2}\left(\frac{\lambda_m}{n}\right)^3 Re\left(\frac{Q_m}{\tilde{V}_m}\right), \tag{7.41}$$

which is exactly the expression provided by Purcell in 1946 (Purcell, 1946), except for the real part.

Equation (7.40) quantifies the contribution of every individual mode to the LDOS. Note that, owing to the dipole polarization, only the modes that do couple to the polarization direction $\mathbf{u}$ of the emitter contribute to the expansion. The total LDOS is the sum over three orthogonal polarizations. It may even include the contribution of magnetic fields excited by magnetic dipoles (see Chapter 12 ; see ref. (Wu et al., 2021) for the definition of the mode volume for magnetic dipoles).

The modal expansion of Eq.(7.40) is similar to the classical expression derived in a Hermitian context for closed systems. All the difference lies in the definition of the mode volume: its real part is different from $V_\mathcal{E}$ and its imaginary part gives rise to the square-bracket term. For significant values of $\mathrm{Im}(\tilde{V})$, as compared to $\mathrm{Re}(\tilde{V})$, certain modes feature a marked Fano-like spectral lineshape, which contrasts with the classical Lorentzian lineshape of the Hermitian approach adopted in chapter 12 for simplification. These modes may additionally contribute negatively to the local density of optical states if the second term in the square bracket dominates (Sauvan et al., 2013). Therefore, a QNM may increase the lifetime of quantum emitters instead of shortening it or, equivalently, adding more modes does not necessarily result in an increase in the LDOS. This is why LDOS inverse design is often counterintuitive and difficult. More insight, including an experimental evidence of Eq. (7.40), has been recently summarized in (Wu et al., 2021).

7.B Complement: The Fabry–Perot cavity and the finesse

7.B.1 Resonance, basic mirror properties, and finesse

Here, we treat with some generality the Fabry–Perot cavity. Its ingredients, a spacer and two plane mirrors, are familiar, being those of laser models. But allowing them some freedom (asymmetry, losses, dispersion) provides heuristic views to several micro- and nanophotonics issues, notably the so-called finesse. We purposely present it without connection to the Fabry–Perot dielectric slab. The latter is a constrained two-parameter system, namely thickness and index.

The system under consideration has two mirrors $j = 1, 2$, taken formally as very thin but dispersive of the form $r_j = R_j \exp(i\phi_j(\omega))$, and comprises a dispersive medium of length L and real positive but dispersive index $n(\omega)$. Furthermore, the transmission of the two mirrors, t_j, does not complement the reflection: $|r_j|^2 + |t_j|^2 = 1 - W_j \leq 1$, where W_j typically represents an absorptive loss in a thin metallic layer or an imperfect reflection due to light scattering in the cladding of an integrated mirror. We will take dispersion into account only up to first order around a reference frequency $\omega_0 = 2\pi c/\lambda_0$, with adequate definitions given later.

All the fields are indicated in Fig. 7.13a, their phase reference being to the closest mirror on the left. Incidence is from the left with field E_s. All fields can be deduced by solving elementary scattering matrices (outflows vs inflows):

$$\begin{pmatrix} E_f \\ E_r \end{pmatrix} = \begin{pmatrix} t_1 & r_1 \\ r'_1 & t_1 \end{pmatrix} \begin{pmatrix} E_s \\ E_b \end{pmatrix} \quad \text{and} \quad \begin{pmatrix} E_t \\ E_b\, e^{-ikL} \end{pmatrix} = \begin{pmatrix} t_2 & r'_2 \\ r_2 & t_2 \end{pmatrix} \begin{pmatrix} E_f\, e^{+ikL} \\ 0 \end{pmatrix}. \tag{7.42}$$

All scattered and inner fields are seen by elementary algebra (line 1 of matrix 1 and line 2 of matrix 2) to share the same denominator, $1 - r_1 r_2 \exp(2ikL)$. For instance, the inner and transmitted fields are:

$$E_f = \frac{t_1}{1 - r_1 r_2 \exp(2ikL)} E_s \qquad E_t = \frac{t_1 t_2 \exp(ikL)}{1 - r_1 r_2 \exp(2ikL)} E_s. \tag{7.43}$$

The second term provides the Fabry–Perot transmission $t_{\text{FP}} = E_t/E_s$. Its squared modulus is

$$T_{\text{FP}} = \frac{T_1 T_2}{1 + R_1 R_2 - 2(R_1 R_2)^{1/2} \cos\phi}, \tag{7.44}$$

where we introduce also $|r_j|^2 = R_j$, $r_j = |r_j| \exp(i\phi_j)$ and $\phi = 2kL + \phi_1 + \phi_2$ as the total round-trip phase. The usual method here is to rewrite the denominator as $[1 - (R_1 R_2)^{1/2}]^2 + 2(R_1 R_2)^{1/2}(1 - \cos\phi)$, and from $1 - \cos\phi = 2\sin^2\frac{\phi}{2}$ and basic algebra, we obtain:

$$\begin{aligned} T_{\text{FP}} &= \frac{T_1 T_2}{[1-(R_1 R_2)^{1/2}]^2} \times \frac{1}{1 + \frac{4(R_1 R_2)^{1/2}}{[1-(R_1 R_2)^{1/2}]^2} \sin^2\frac{\phi}{2}} \\ &= T_{max} \times \text{Airy}(\sqrt{R_1 R_2}, \phi). \end{aligned} \tag{7.45}$$

The first term, $T_{max} \leq 1$, is an upper bound to the transmission. The second term, known as the Airy function, obviously obeys $\text{Airy}(\sqrt{R_1 R_2}, \phi) \leq 1$. It contains all the

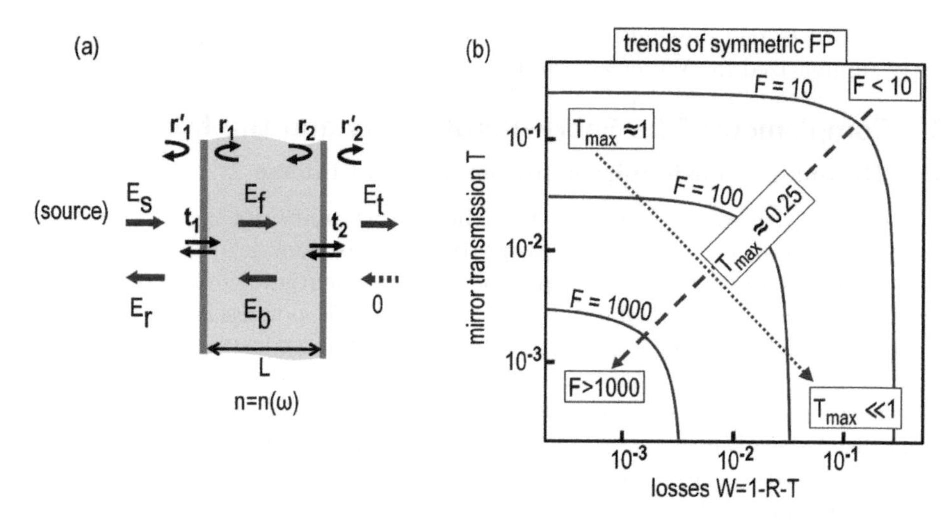

Fig. 7.13 (a) Ingredients of a general asymmetric Fabry–Perot cavity. (b) Symmetric case: diagram in the $\ln(W)/\ln(T)$ plane, where $W = 1 - R - T$ are the mirror losses. Three equi-finesse contours are drawn, showing the trend from top right to bottom left (dashed arrow) for increasing F, as well as the trend for the maximum transmission T_{max}, which decreases along the top-left to bottom-right diagonal (dotted arrow) according to $T^2/(T + W)^2$.

resonant and dispersive effects, displaying resonance maxima when $\sin(\phi/2) = 0$, which we now examine. Starting from a given resonance stated by [$\sin(\phi/2) = 0$, Airy= 1], thus at some integer value $m = \phi/2\pi$ (the resonance *order*), we can see that the Airy function *considered in terms of phase* has a well-defined FWHM $\delta\phi$: when its denominator is 2 instead of 1, assuming a small phase and a large prefactor of the $\sin(\ldots)$, we have, using $\sin\left((2m\pi \pm \delta\phi/2)/2\right) \simeq \pm\delta\phi/4$,

$$\frac{4(R_1R_2)^{1/2}}{[1-(R_1R_2)^{1/2}]^2}\left(\frac{(\delta\phi)}{4}\right)^2 \simeq 1. \tag{7.46}$$

Note that in this limit where a small phase suffices to reach the FWHM interval limit, the Airy function locally resembles a Lorentzian function of $\delta\phi$, which is of course reminiscent of Eq. (7.18): a Lorentzian spectrum follows if a linear phase/frequency relation holds. Here, in the generic phase vision of Eq. (7.46), a definition of the *finesse* F is obtained by comparing $\delta\phi$ to 2π (the similar and customary comparison of spectral linewidth to FSR will be done later):

$$F = \frac{2\pi}{\delta\phi} = \pi\frac{(R_1R_2)^{1/4}}{1-(R_1R_2)^{1/2}} \qquad \text{general case,} \tag{7.47}$$

$$F = \frac{2\pi}{\delta\phi} = \pi\frac{\sqrt{R}}{1-R} \qquad \text{symmetric case.} \tag{7.48}$$

So it is obvious that the finesse is a property of only the mirror *inner reflectivities in the cavity* and nothing else. This is very precious indeed. Among other things, it is one of the few ways to reliably measure very high reflectivities approaching unity for outstanding mirrors such as those of the complex gravitational-wave interferometers (VIRGO and LIGO).

Let us spend some time on the symmetric 'imperfect' Fabry–Perot whereby $T_1 = T_2 = T$, $R_1 = R_2 = R$. Then $T_{max} = [T/(1-R)]^2$. Ideally, both T and $1-R$ are small quantities, but we defined $R+T = 1-W$ above. Hence $T_{max} = [T/(T+W)]^2$ is another way to present this upper bound. So, while the mirror is, experimentally speaking, a single embodiment, we see a clear distinction between its ingredients: on the one hand, R defines the resonance *finesse* through Airy(R, ϕ) and we will confirm below that it is the sole mirror ingredient that plays a role in the quasinormal-mode lifetime through the quality factor Q. On the other hand, the pair (T, W), only constrained by $T + W = 1 - R$, dictates what will be observed in terms of power. Generalizing, the pair (T, W) sets the magnitude of the structure scattering coefficients. This is part of a general idea about quasinormal modes: the response magnitude depends on how the inner fields inherent to the structure (i.e. those that see only R) are 'addressed' by the specific exciting field (through T). A highly resonant Fabry–Perot with very lossy mirrors ($W \gg T$, but still $W + T \ll 1$) obviously features $T_{max} \ll 1$, and we leave it as an exercise for the reader to find that it also features $R_{min} \simeq 1$, stressing only the physical intuition that whatever constructive interference the inner rays build up, the penalty of $W \gg T$ prevents them from causing a sizeable destruction of the outside reflection. Practically speaking, the Fabry–Perot mostly behaves as a 'white' high reflector, with only faint features in its reflection betraying the inner resonances. Its transmission is also faint in absolute terms, but remains much more spectrally contrasted.

The diagram of Fig. 7.13b shows the trends in a log-log diagram of T and W, notably the T_{max} trend just tackled, but also the shape of three equi-finesse contours. Several phenomena sought in nanophotonics may have a non-trivial behaviour at various points of this diagram: absolute field concentration for non-linear effects, Purcell effect, other mode–volume effects, enhancement of light extraction, and others addressed elsewhere in this book.

7.B.2 Roles of cavity ingredients in the spectral response

If we now turn to actual measures of resonance width, we are doing spectral measurements. So let us articulate the phase vs frequency or wavelength relation and say more on the roles of cavity ingredients in the spectral response. We shall take advantage of the circumstances to add a few specifically signalled 'nanophotonic grains of salt' where appropriate.

We assume that a first-order ω dependence is sufficient to accurately describe the variations of mirror phases and index, which enter into ϕ around a given frequency $\omega_0 = 2\pi c/\lambda_0$:

$$n(\omega) = n(\omega_0) + \frac{\partial n}{\partial \omega}(\omega - \omega_0) \quad \text{and} \quad \phi_j(\omega) = \phi_j(\omega_0) + \frac{\partial \phi_j}{\partial \omega}(\omega - \omega_0) \ (j = 1, 2). \quad (7.49)$$

Since we lie around the m-th resonance, that is

$$2m\pi = 2n(\omega_0)L\omega_0/c + \phi_1(\omega_0) + \phi_2(\omega_0), \quad (7.50)$$

from Eq. (7.48) with the use of standard expansion, we find that the FWHM peak verifies:

$$\delta\phi = \frac{2\pi}{F} = 2\frac{L}{c}\left(n + \omega\frac{\partial n}{\partial \omega}\right)_{\omega_0} \delta\omega + \left[\frac{\partial \phi_1}{\partial \omega} + \frac{\partial \phi_2}{\partial \omega}\right] \delta\omega \,. \tag{7.51}$$

The first right-hand-side contribution involves the *group index* $n_g = n + \omega\frac{\partial n}{\partial \omega}$. If m is large enough and the first-order expansion valid enough, an entirely similar relation holds by substituting $\delta\phi \to 2\pi$ $\delta\omega \to \Delta\omega$, where the FSR $\Delta\omega$ is the frequency separation between the m-th order resonance and its $m \pm 1$ neighbours. The peak FHWM frequency width $\delta\omega$ can be expressed in terms of the *quality factor* using the convenient frequency definition:

$$Q = \frac{\omega_0}{\delta\omega} = \frac{\omega_0}{2\pi}\left(2n_g(\omega_0)\frac{L}{c} + \left[\frac{\partial \phi_1}{\partial \omega} + \frac{\partial \phi_2}{\partial \omega}\right]\right) \times F. \tag{7.52}$$

In this equation, the non-dispersive case would result in the factor of F on the right-hand side being simply $2nL\omega_0/(2\pi c) = m$ (see Eq. (7.50)). This would give the textbook result:

$$Q = m\, F. \tag{7.53}$$

Conversely, we note that the mirror-phase derivative can be seen as a *penetration depth*: forcing the corresponding terms on the right-hand side of Eq. (7.52) to resemble the optical-path-related term $n_g L/c$ naturally gives two lengths according to

$$L_j^{\text{pen}} = \frac{c}{2n_g}\frac{\partial \phi_j}{\partial \omega} \tag{7.54}$$

whereby, in the basic case $\phi_1(\omega_0) = \phi_2(\omega_0) = 0$, we can see the cavity as if it had an expanded length $L + L_1^{\text{pen}} + L_2^{\text{pen}}$, and would just be bounded by non-dispersive mirrors instead of the actual ones (see more details in chapter 16). This being said, we can now take a broader view. We must here insist that Eq. (7.53) helps when distinguishing the Q in electronics and optics. In electronics, it is customary to associate Q with overvoltage. But the resonator in electronics is generally a subwavelength object (except in some microwave cases), so that, qualitatively, $m \to 1$. In other words, there are only two physical times: the oscillation period $2\pi/\omega_0$ and the decay time. In an optical resonator, even one made of complex materials, these two time scales of course exist, so that we could define the Q-factor from the ratio of $\text{Re}(\tilde{\omega})$ and $\text{Im}(\tilde{\omega})$.

But the 'round-trip time' $\tau_{\text{round-trip}} = 2n_g L/c$ is a third time that typically falls between the optical oscillation period and characteristic decay time. To study the overvoltage—essentially how much the field is enhanced over that of a travelling wave in the same material but deprived from mirrors around—the number of round trips, essentially given by the finesse F, is the good quantity. Alternatively, one can compute the E_f/E_s ratio at resonance, yielding $t_1/(1 - |r_1 r_2|)$. Thus, to reach strong fields in atom physics with ordinary lasers, for instance, high-finesse cavities F are used. Any macroscopic cavity (say, centimetres long) has an order $m \equiv L/(n\lambda/2) \sim 10^5$. So even with very low finesses (say, $R \sim 0.7, F < 3$), a large Q-factor of $Q > 200{,}000$ can be obtained. But the field is not much enhanced at the resonances. For truly large finesses, say $F > 1,000$, we assume $Q \sim 10^8$, and hence a resonance width in the few MHz range.

A last point regards the growing quasinormal-mode fields outside the cavity. For $F \gg 1$, these fields remain low when quite far away and are easier to ignore than in a low-F case. It is a logical complement to the enhancement effect of F on the E_f/E_s ratio: anyhow, a good cavity enhances the inside/outside field ratio at resonance. In practice, it means that integration taken as for a closed cavity shall yield safer results than in a general case, where such an approximation may cause large and unwanted effects. Nevertheless, it can be shown that modifications of Q due to perturbations must take those fields into account.

When we go to nanophotonic systems through micro-optical ones (microcavities), we tend to diminish m to a few units, again putting Q and F on a similar footing. Also, the two dispersive contributions that we have evidenced (embodied in the group index and in the penetration lengths) may display a very different balance at small L and at large L. Somehow, these two issues, if explored in real cases, highlight two typical challenges of nanophotonics.

The first one is really about design: how to design a mirror that does not add too much penetration length (hence a more metallic character seems good, note, by the way, that a negative value stemming from $\frac{\partial \phi_j}{\partial \omega} < 0$ is also possible), but that does not impact the losses W (hence a more dielectric character is good). Furthermore, when light circulates in guides and is confined in the transverse dimensions (see Chapter 6), the issue of finding a good 'modal reflector' for a waveguide mode without radiating too much outside the guides is a tough one.

The second challenge is about physical understanding. In a complex nanophotonic system (say, the metamaterials seen in chapter 18, notably using metallo-dielectric parts, where the field structure becomes complex), identifying a *bona fide* mirror may become an issue. For instance, there might be a couple of extra peaks at hundreds of nm of a $\lambda_0 = 2\pi c/\omega_0$ target resonance. But without proper knowledge of the spatial nature of associated modes, it is too presumptive to use the interval between λ_0 and these peaks as a FSR and to inject it into mode finesse F to deduce a modal reflectivity. Conversely, if the scheme laid out above works for some identified sub-part, for instance if it does behave as a mirror with a predictable and well-understood phase behaviour, then, insight from the simple Fabry–Perot model may still usefully guide intuition.

7.C Complement: Coupled-mode theories: Spatial vs temporal

We stress in this Complement that the temporal 'coupled-mode theory' presented above differs from what the optoelectronic community usually denotes by the same words, but which refers to a *spatial* 'coupled-mode theory'. In such theories, pioneered by E. Kogelnik and recurrently used in A. Yariv's textbooks (Kogelnik, 1969; Yariv, 1990; Yariv, 1989) and in the quantum electronics community, the coupled modes of the field are general and are rarely those of resonators. In most cases, a mode of the given coupled set of modes has a field described by a slowly varying spatial envelope multiplying an otherwise canonical field pattern (oscillatory for periodicity or confined for waveguides, or both). A perturbation such as the presence of an extra photonic system (waveguide) in the surroundings, a non-linear effect, or a periodic modulation of dielectric constant then couples the various spatial modes and dictates the *spatial* evolution of their envelope ($\frac{d}{dz}$ and not $\frac{d}{dt}$). The matrix form is used, hence mathematical tools around eigenmodes may evoke the approaches used above in various respects. In the area of X-ray diffraction, the interested reader may also find similar concepts under the entry 'dynamical diffraction' in the literature; this theory explains what happens when the simpler 'kinematic diffraction' fails (Authier, 2001).

While the above paragraph sounded like a 'please do not confuse' warning, there are, nevertheless, touchpoints of both flavours of 'coupled modes', especially as regards waveguides: if we extend the modes considered in slabs to those that are not x- and y-invariant, in other words, those that feature non-zero in-plane wavevector components $\mathbf{k}_{//}$ in their Fourier spatial decomposition, we can go continuously from a resonator to a waveguide when increasing $k_{//} = \beta$ and crossing the relevant light lines of the guide dispersion diagram. This will be exploited in Chapter 17, once we have more knowledge on periodicity.

Exercises

Exercise 7.1 The quasinormal modes of the RLC oscillator. The most famous resonator in 'electromagnetism' broadly speaking is the series RLC circuit, Fig. 7.14, having inductance L, resistor R, and capacitance C with charge q and current $I(t) = q'(t)$. It is fed by voltage V_{in} and outputs $V_{\text{out}} = q/C$. The voltage drops are in series. Establish the governing differential equation shown below by substituting the constitutive equation for each of the three elements into Kirchhoff's series voltage law:

$$I''(t) + \frac{R}{L}I'(t) + \frac{1}{LC}I(t) = \frac{1}{L}V'_{\text{in}}(t). \tag{7.55}$$

We can define a scattering operator, which is nothing but the circuit admittance, linking the Laplace-transformed current $I(s)$ and voltage $V_{\text{in}}(s)$. Show that $I(s) = V_{\text{in}}(s)\ (Ls^2 + Rs + 1/C)^{-1}$. The modes correspond to the poles of the scattering operator. They are also equivalently defined as the solutions of the homogeneous differential equation derived from Eq. (7.55) for $V_{\text{in}} = 0$, which you can establish as

$$I''(t) + 2\alpha I'(t) + \omega_0^2 I(t) = 0, \tag{7.56}$$

with $\alpha = \frac{R}{2L}$ the neper 'frequency' (or attenuation) and $\omega_0 = (LC)^{-1/2}$ the angular resonance frequency. Look at solutions of the form $\exp(st)$ with the characteristic second-order equation on s.

In the $R = 0$ limit ($\alpha = 0$), show that one has $s = i\omega_0$ or $s = -i\omega_0$. Solutions $I(t)$ for an initial condition $[I(t = 0)$ and $I'(t = 0)]$ are linear combinations of $\exp(i\omega_0 t)$ and $\exp(-i\omega_0 t)$. Show that in this ideal case, energy is periodically exchanged at a frequency of

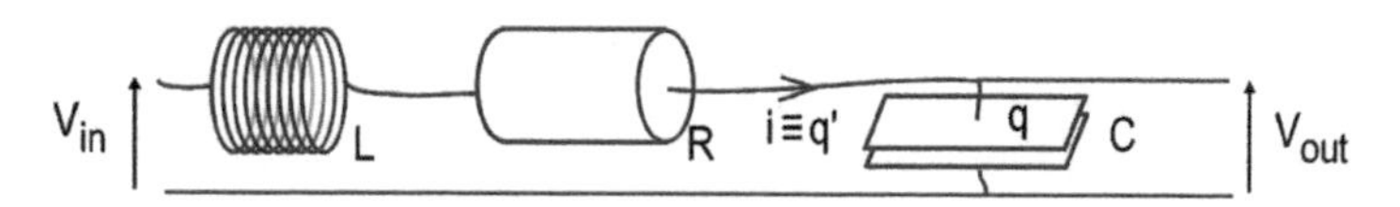

Fig. 7.14 Scheme of an RLC circuit. The charge of capacitance C is $q(t)$ and the current is $q'(t)$.

$2\omega_0/2\pi$ between capacitive $\frac{q^2}{2C}$ and inductive $\frac{LI^2}{2}$ forms. This is the equivalent of the energy oscillation between electric and magnetic-field energies for photonic resonators.

For $R \neq 0$, classify the different regimes: the overdamped response without oscillations, the critically damped response with the fastest possible decay, and the underdamped response (the oscillator retains energy during many periods before releasing its energy). Argue why it is this latter that is of main interest in electromagnetism of resonators. In this regime, show that the two *complex frequencies* can be expressed as

$$\tilde{\omega}_1 = \omega_0\left(1 - \frac{\alpha^2}{\omega_0^2}\right)^{1/2} - i\alpha \qquad \tilde{\omega}_2 = -\omega_0\left(1 - \frac{\alpha^2}{\omega_0^2}\right)^{1/2} - i\alpha. \tag{7.57}$$

(1) The frequencies $\mathrm{Re}(\omega_{1,2})$ are slightly smaller than the lossless value $(\omega_0/2\pi)$. Is it intuitively correct?

(2) Do the two solutions have different or similar exponential decays?

(3) A damped system and a nearly undamped system are left to decay in parallel after a common very short pulse excitation at $t = 0$. Assuming, for instance, that their outputs are fed to two oscilloscope channels, at what time t_π would a phase difference of π appear?

(4) Expand the resonance FWHM up to order three in $\frac{\alpha}{\omega_0}$.

Exercise 7.2 String harmonics. A short musical excursion: for a piano or a guitar, the string resonant frequencies ω_m are all multiples of the fundamental ω_1, giving the note's harmonics and the instrument-specific 'timbre'. Resonators of different shapes will break this property. Explore percussive instruments (bells, etc.: you can look at the Chinese 'Zenghouyi Bells' from 433 BC for instance) and relate their perceived sound to their harmonic content.

Exercise 7.3 Link between Fabry–Perot Q and complex frequency. As shown by Eq. (7.2), the quasinormal modes and their complex frequencies involve a prolongation at complex frequencies of the real physical quantities. This exercise deepens the analysis of section 7.4 of Fabry–Perot cavities without dispersion. We thus again consider a Fabry–Perot dielectric slab with a refractive index $n_2(\omega)$ and a reflection coefficient $r(\omega) = |r|\exp(i\phi(\omega))$, with the classical approximation that $|r|$ is independent of ω. The round-trip resonance condition $1 - r^2\exp(2i\frac{\omega}{c}n_2L) = 0$ can be satisfied only for a complex frequency $\tilde{\omega} = \omega' + i\omega''$, implying that all the physical quantities, r and n_2, involved in the resonance condition should be considered at complex frequency (prolongation).

(1) Show that the resonance condition becomes: $1 - r(\tilde{\omega})^2\exp(2in_2(\tilde{\omega})\tilde{\omega}L/c) = 0$.

(2) We may assume that $r(\omega) = |r|\exp(i\phi(\omega))$, with $|r|$ independent of ω. Under the high-Q assumption that $\omega'' \ll \omega'$, and using a first-order Taylor series expansion around the real frequency ω', show that $n_2(\tilde{\omega}) = n_2(\omega') + i\omega''\frac{dn_2}{d\omega}$, where the derivative is taken at ω'. Note that this implies that n_2 at complex frequencies is determined solely from the knowledge of n_2 at the real frequencies of the physicist. Derive a similar expression for $\phi(\tilde{\omega})$.

(3) By injecting the Taylor expansions of ϕ and n_2 into the resonance condition, show that the condition becomes $1 - |r|^2\exp(2in_2(\omega')\omega'L/c + 2i\phi(\omega'))\exp(-2\omega''(n_2(\omega') + \omega'\frac{dn_2}{d\omega})L/c) = 0$.

(4) By noting that the previous equation is satisfied if and only if $n_2(\omega')\omega' L/c + 2i\phi(\omega') = m\pi$ (round-trip phase matching) and $\exp(-2\omega''(n_2(\omega') + \omega' \frac{dn_2}{d\omega})L/c) = 1/|r|^2$, use the expression $Q = -\frac{\mathrm{Re}(\tilde{\omega})}{2\mathrm{Im}(\tilde{\omega})}$ to retrieve Eq. (7.52).

Exercise 7.4 Symmetric Fabry–Perot cavity.

(1) Find the transmission at normal incidence of a cavity formed from a slab of index n, thickness L, bounded by two symmetric lossless mirrors ($r^2 = R = 1 - T = 1 - t^2$), using the round-trip phase $\varphi = 2knL = 2nL\frac{\omega}{c}$ to add the amplitudes of the outgoing rays. You should find the Airy function of the form $\left(1 + \mathcal{M}\sin^2\left(\frac{\varphi}{2}\right)\right)^{-1}$ with $\mathcal{M} \propto F^2 = \pi^2 R/(1-R)^2$.

(2) The spacing between two resonances ($\Delta\varphi = 2\pi$) corresponds to a frequency interval $\Delta\nu_{\mathrm{FSR}}$ called the *free spectral range* (FSR). Express the FSR as a function of n, L, and c.

(3) Express the approximate width (FWHM) of a resonance when $F \gg 1$ as either $\delta\varphi$ or $\delta\nu_{\mathrm{res}}$. Show that $\delta\varphi = 2\pi/F$ (the original finesse definition) and compare $\delta\nu_{\mathrm{res}}$ to FSR. Should a microcavity (or a nanocavity) have a vanishing FSR or a diverging FSR?

(4) At oblique incidence θ in air, thus θ' in the slab, show that the resonances are shifted to higher frequencies (blue-shifted) by a $(\cos\theta')^{-1}$ factor. Further check that a good approximation of $(\cos\theta')^{-1}$ is $1 + \frac{\theta^2}{2n^2}$. Find the angle limit for a 10%-error limit of the validity of this correction, first for $n = 1.5$ (silica) and next for $n = 3.5$ (silicon).

(5) Evaluate the intensity enhancement inside the cavity vs outside; it should scale with F.

Exercise 7.5 Building an explicit first-order amplitude evolution equation for a hard-wall resonator. We mimic here Haus's introduction (Haus, 1984) of the first-order time equation. It circumvents the global approach of Eq. (7.2) by focusing on a particular case. Instead of Haus's choice of RLC circuit and conjugate variables (intensity/current), we propose (electric/magnetic) fields of the modes of an elementary planar PEC-bounded resonator, Fig. 7.15 (PEC meaning perfect electric conductor with zero-electric-field Dirichlet-like condition).

(1) Justify the spatial/temporal field separation proposed here for the m-th mode (m being a strictly positive integer, $m = 1$ in Fig. 7.15):

$$E_y = \sin(k_m x)\, Q_E(t), \text{ and } B_z = \frac{n}{c}\cos(k_m x)\, Q_B(t). \tag{7.58}$$

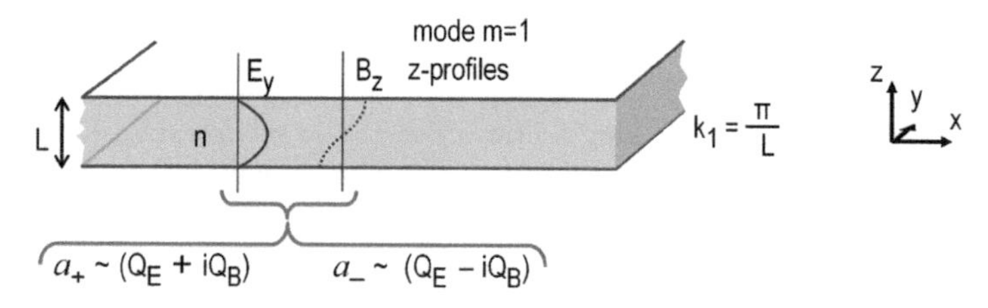

Fig. 7.15 Resonator with index n between perfect metallic slabs: Electric- and magnetic-field profiles of the non-zero components of mode $m = 1$.

(2) Inject these forms in Maxwell curl equations to relate Q_E and Q_B to each other's time derivative.

(3) By studying the two linear combinations $a_\pm = Q_E \pm iQ_B$, show that the new variables $a_\pm$ obey the desired form, $da_\pm/dt = \pm\omega_m a_\pm$.

(4) Study the energy in more depth: compare its temporal fate between electric/magnetic forms to the fate of kinetic/potential energy exchanges of the spring-and-mass case in mechanics.

A remark in relation to quantum mechanics: in *Second Quantization*, the introduction of creation/annihilation operators starts with a similar operator combination of a variable $\pm i\times$ its conjugate; this formulation yields a kind of square root of the constitutive second-order relations.

Exercise 7.6 Impulse response and Fourier spectrum. Assume that the pulse response of a system for $t > 0$ is an exponentially decaying oscillation that we measure through the real field $\mathcal{E}(t) = E_0 \exp(-\frac{\Gamma}{2}t)\cos(\omega_0 t)$ with $\Gamma \ll \omega_0$. Calculate explicitly the Fourier transform $E(\omega) = \int \mathcal{E} \exp(+i\omega t)\, dt$ and show that $E^2(\omega)$ for $\omega > 0$ is a Lorentzian. Relate the quality factor Q to Γ and check your physical intuitions.

Note also that there is the same information in the $\omega > 0$ and $\omega < 0$ parts of the Fourier spectrum, as we start from a real-valued signal. The field obtained by the truncated inverse Fourier transform of $E(\omega)$ on only the $\omega > 0$ part of the spectrum is thus a complex field $E(t) \neq \mathcal{E}(t)$ but still *with all the information*. It is sometimes called the 'analytical field'. It has, within a factor of 2, the same real part as $\mathcal{E}$. Essentially, adding the imaginary part just builds what is classically taught as the 'complex representation of the real field', mostly in connection with solutions of second-order differential equations.

Exercise 7.7 Anticrossing for polariton-type dispersion. The usual parameter of cavity polariton physics is an in-plane wavevector $k_{//}$ (1D or 2D; see ref. (Weisbuch et al., 1992)). Consider two oscillators, one with fixed frequency $\omega_1 = \omega_0$ independent of $k_{//}$ (the 'heavier' exciton dispersion with some effective mass; see Chapter 8) and a parabolic one: $\omega_2(k_{//}) = \omega(k_{//}) + \delta + (k_{//}/k0)^2$ with δ the detuning and k_0, a normalization wavevector. Plot through a computer the dispersion of the new polariton modes $\omega_\pm$ induced by their coupling (non-diagonal $\pm g$ matrix terms), exploring δ over $\pm$ a few g's. Find a physical value for k_0 by tracking the frequency of an oblique cavity mode (item 4 of exercise 7.4), using $k = n\frac{\omega}{c} = (m^2\frac{\pi^2}{L^2} + k_{//}^2)^{1/2}$, which holds notably for a PEC boundary condition and can still be deemed valid enough for idealized mirrors with $T < 1$.

Exercise 7.8 How much do the coupled modes mix the uncoupled modes?

(1) Study in detail the eigenvalues and eigenvectors for lossless coupling. The diagonal elements of M only differ by the detuning term $i\delta$ (M_{11} can be 0 and $M_{22} = i\delta$), whereas $M_{12} = g$ and $M_{21} = -g$. Show that the asymptotic *relative* trends of eigenvalues denoted by $i\omega_\pm$ for large $|\delta|$ vs their asymptote goes like δ^{-2} (study the ratio $\to \omega_+/|\delta|$ for $\delta < 0$, for instance). The mixing coefficient can be defined in general as $C_{\text{mix}}{=}\min(C_{11}, C_{21})$, the smallest of the non-diagonal components of the eigenvector (the two-column eigenvectors forming a matrix C). Show that the asymptotic trend of C_{mix} has a δ^{-1} trend (on computer and by adequate analytic expansions). The vectors are more perturbed by the coupling than the eigenvalues.

(2) Based on this study, explore Rabi oscillation in more depth: firstly what happens at $\delta = 0$ for an arbitrary boundary condition (random complex amplitudes in both resonators)?

(3) Secondly, what happens, for the ideal starting condition of zero amplitude in the second resonator and maximal amplitude in the first one, if $\delta \neq 0$? Based on the result, propose ways to manipulate signals and pulses by controlling the detuning of two resonators.

Part II

Optical properties of confined electrons

8
Semiconductors and quantum wells

Semiconductors are not only at the origin of twentieth-century technology revolutions; they have also inspired physicists, paving the way to nanostructures that were explored to exploit their electronic confinement. Here we will be mostly concerned with the photonic properties of semiconductors. We shall need the notion of bands, the basic description of the extended states offered to electrons.

We will, for this first step, start from atoms. Next, we will address the crucial issues of band occupancy and carrier densities. This is the key point in electronics: the ability to change the nature and density of charge carriers at different points of the same semiconductor chip and, furthermore, to control these densities and the current by applied voltages. Being oriented towards photonic aspects, we will, however, stick to that simpler case where a sufficient degree of local electronic equilibrium is ensured, which notably discards situations with permanent electric fields or with space-charge regions.

The next step will be to address a basic photonic experiment: photoluminescence. This will involve an account of the light–matter interaction: absorption and stimulated and spontaneous emission. They will be treated in a simplified fashion at this stage, with some additional material in Exercise 8.13. Mastering the basics of photoluminescence is also useful to grasp the operation of light-emitting diodes (LEDs) and laser diodes, the two most popular photonic semiconductor devices nowadays. But these devices also rely on spatial confinement of carriers. Therefore, we will be concerned with the issue of semiconductor interfaces. At this point, it will be useful to provide a reminder of some semiconductor growth considerations, notably epitaxy methods.

We will then be equipped to give a first sensible account of the light-emitting diode (LED), thanks to the study of the *double heterojunction* (DH), a structure proposed by Z. I. Alferov and H. Kroemer independently in 1963, earning them the Nobel Prize in 2000 (jointly with J. S. Kilby who won it for his work on integrated circuits). The quantum well (QW) structure is in the logical continuity of these structures and is also a perfect bridge for the study of nanostructures, allowing us to outline key attractive aspects of electronic confinement.

The particular issue of the threshold of edge-emitting lasers based on slab waveguides comprising a gain layer provides us with the first 'smart' combination of nanophotonics: the separate confinement heterostructure (SCH), whereby the guiding properties of the DH and the specific density of states (DOS) of the QW are associated with lowering the threshold current density.

Introduction to Nanophotonics. Henri Benisty, Jean-Jacques Greffet, and Philippe Lalanne, Oxford University Press.
 DOI: 10.1093/oso/9780198786139.003.0008

8.1 From wavefunctions to band formation

8.1.1 Band of a chain of atoms

Let us start with the formation of a solid from individual isolated atoms (Fig. 8.1). Each atom has several well-defined atomic levels. For simplicity, we consider a single kind of atom (hence an elemental semiconductor such as Si or Ge) and a single non-degenerate level (although there are many degenerate levels, s, p, in a real atom).

We then regularly align the atoms in a 1D chain, and we change the distance between the atoms gradually from an infinite distance to some smaller distance a. The localized wavefunctions $\psi_b(x)$ of the individual atoms can then interact through their decaying tails (see Fig. 8.1b). Formally, we have a Hamiltonian $H = H_o + H_{\text{coupling}}$, with H_o accounting for the binding potential of each electron to its ionic core ($\sim$ nucleus) and H_{coupling} accounting for the influence of this ionic core on the electron of a neighbouring atom, which, of course, depends on a. As the distance a gets smaller, electrons can eventually hop from one atom to the next. Due to this extra motion, their energy can take different values, depending on the momentum of electrons. The simplest way to account for this momentum in one dimension is the wavevector $k \equiv k_x$ here.

In Exercise 8.1 the steps to build the total Hamiltonian for a large chain of M coupled atoms are proposed, with Born–von Kármán periodic boundary conditions—i.e. atoms $\#(M-1)$ and $\#0$ are coupled. The general solution of such a coupling Hamiltonian between first neighbours has the form of an electronic Bloch mode:

$$\psi_k(x) = \frac{1}{\sqrt{M}} \times \sum_{n=0}^{M-1} \exp(ikna)\, \psi_b(x - na), \tag{8.1}$$

where $\psi_b(x - na)$ is the wavefunction of an electron in some energy level b of one atom centred at position $x_n = na$. Note that this solution requires an adiabatic approximation: electron and nucleus motions, e.g. due to thermal agitation, are assumed to be uncorrelated. It is easily checked that $\psi_k(x)$ is pseudo-periodic, i.e. that we have $\psi_k(x + a) = \exp(ika)\, \psi_k(x)$, with a periodic part ψ_b in every period, and a space-dependent phase term which is discretized and evolves stepwise in the above presentation, $\exp(ikna)$. But this term could equally be written with a continuous function, $\exp(ikx)$, with the term carrying the phase difference $\exp(i(kna - x))$ artificially attributed to the periodic term ψ_b.

The simplest explanation of H_{coupling} is to say that it describes the interaction between neighbours, with its matrix element $\langle \psi_b(x - na) | H_{\text{coupling}} | \psi_b(x - na + a) \rangle$ being a constant that we define as $\frac{1}{2}C(a)$. The eigenvalues of H are then deduced by calculating $E_k = \langle \psi_k | H | \psi_k \rangle$, where two terms $\exp(\pm ika)$ appear, their combination leading to the simple form

$$E_k = E_b + C(a)\cos(ka). \tag{8.2}$$

The resulting band is plotted in Fig. 8.1c as a function of k. It is interesting for physical understanding to examine explicit forms of the wavefunctions at the two band extrema $k = 0$ and $ka = \pi$. For $k = 0$, all coefficients $\exp(ikna)$ are unity. We can make a mechanical analogy of coupling with a spring connection between adjacent masses

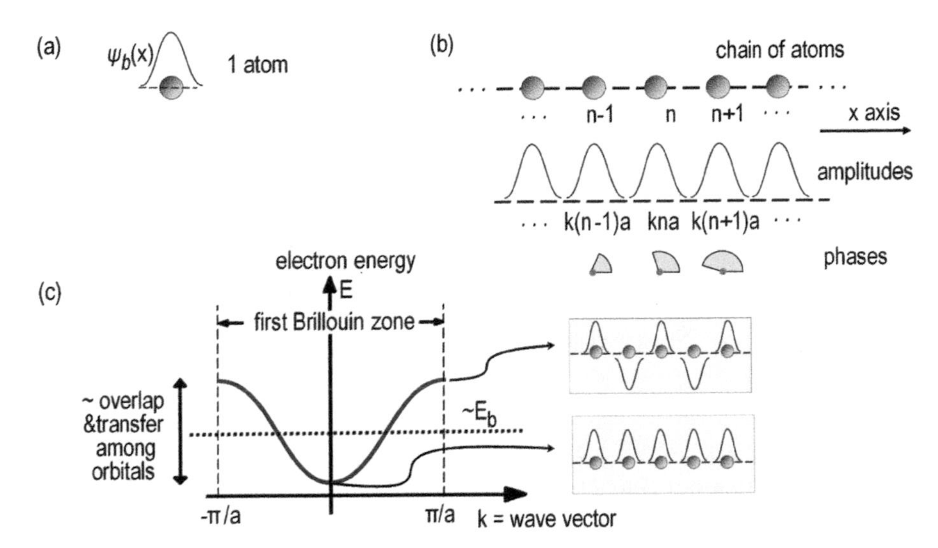

Fig. 8.1 From atoms to solids: (a) single atom and the wave function $\psi_b(x)$ of a given energy level E_b in it; (b) 1D chain, with similar amplitudes for all atoms but a phase factor that varies linearly with position; (c) dispersion $E(k)$ for this tight-binding system. The first Brillouin zone is defined by the dashed lines at $k = \pm\pi/a$.

constrained to moving longitudinally along the sole x-axis of the system: if all the masses have the same phase in their motion, none of the springs are solicited; thus, the energy is minimal, $E_{k=0} = E_b - C(a)$. Conversely, for $ka = \pi$, the coefficients $\exp(ikna)$ form an alternate series $(-1)^n$. Hence neighbours are in phase opposition. The spring between adjacent masses is then maximally solicited. Its average potential energy thus adds to the system, and the total energy is therefore maximal, $E_{k=\pi/a} = E_b + C(a)$.

If we increase k beyond π/a, the energy goes along the $\cos(ka)$ curve again and again, periodically. One usually restricts the range of k to the first Brillouin zone of the periodic system, in order to ensure a non-redundant description of its states. Here, the first Brillouin zone is simply the interval $]-\pi/a, \pi/a]$. The same reasoning is used in Chapter 15 about the electromagnetic fields in photonic periodic lattices.

8.1.2 Multiple bands

We may now generalize the work done above for a single level b to a set of several energetically distant levels $b1$, $b2$, $\cdots$ (Fig. 8.2a). We assume that the corresponding electron states have specific symmetries that forbid coupling from bi to bj states when $i \neq j$. Each of these sets then gives rise to its own energy band (in the naïve view that overlapping bands are not an issue):

$$E_{k,j} = E_{b,j} + C_j(a)\cos(ka). \tag{8.3}$$

Note that the coupling constants $C_j(a)$ for each state bj have no obvious relation. We must thus expect changes of strength as well as changes of *sign* between, for instance, the coupling constant $C_1(a)$ and $C_2(a)$. We can therefore meet the case where the

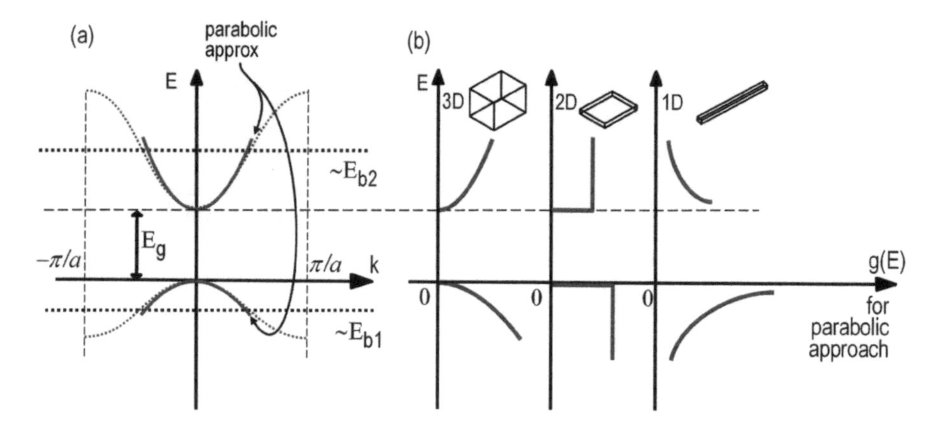

Fig. 8.2 Bands and density of states: (a) two bands and the bandgap separation E_g, with parabolic approximation in the bands for the dispersion around $k = 0$ (band edges); (b) DOS as a function of dimensionality, from a 3D large cube (left) to a 1D 'wire' (right).

lower band $b1$ has a maximum at $k = 0$ (hence $C_1(a) < 0$), while the second band has a minimum at k=0 ($C_2(a) > 0$), and there is a separation between them (see Fig. 8.2a). This configuration corresponds very much to semiconductors of utmost importance in optoelectronics, namely the three III-V compounds GaAs, InP, and GaN. In Fig. 8.18a, b of Complement 8.A, a more precise account is given on bands and their relation to the underlying atomic orbitals, accounting for the overlap of bands.

Due to the way the bands are filled (that will be explained later), the important elements here are the separation between the two bands and the two opposite local curvatures of the bands dispersion relation $E_k \equiv E(k)$. The separation denoted E_g is a forbidden energy region for electron states: E_g is the famous *energy bandgap* of semiconductors. The curvatures of each band at $k = 0$ can be described by coefficients that are commonly called effective masses $m^*_{c,v}$, the suffixes c or v referring to conduction or *v*alence bands. From now on, we denote the band '1' by 'v' and the band '2' by 'c' (Fig. 8.18 of complement 8.A explains that the 4 valence electrons are mainly in the 'v' band for a column IV element). The form $C_{1,2}(a)\cos(ka) \equiv C_{v,c}\cos(ka)$ in Eq. (8.3) involves, with the usual algebra ($\cos\eta \simeq 1 - \eta^2/2$), that in a sizeable energy range around the bandgap, the energies are well approximated by

$$E_{c,v}(k) = E_{c,v}(k=0) + \frac{1}{2}\,C_{c,v}(a)k^2a^2. \tag{8.4}$$

We can identify the last term with the canonical form of kinetic energy of a massive particle $\frac{\hbar^2k^2}{2m^*}$ if we introduce the two adequate effective masses, namely $m^*_v = a^2/\hbar^2C_v(a)$ for the top edge of the valence band (VB) and $m^*_c = a^2/\hbar^2C_c(a)$ for the bottom edge of the upper conduction band (CB). We see that $m^*_c > 0$, whereas $m^*_v < 0$, for our archetypal case (known as the direct gap; see below).

The reader can refer to Exercise 8.3 to check that this definition of effective masses retrieves a semiclassical picture of electron motion whereby this mass does play the role of the particle rest mass.

It is customary to refer to these effective masses by scaling them to the free electron mass m_0. For instance, for GaAs, the effective mass values are $m_v^* \sim 0.7\, m_0$ and $m_c^* \sim 0.07\, m_0$, a value that is ten times 'lighter'. These masses have important consequences for the DOS of electrons.

The situation that we have favoured with common extrema at $k = 0$ is called a *direct bandgap* zone-centre situation and is central to the optical properties of most useful III-V and II-VI compounds. Logically, the situation of separated extrema, for instance, the maximum of the VB at $k = 0$, and the minimum of the CB at $k = \pi/a$, is called an *indirect bandgap*. In Complement 8.A, Fig. 8.18c illustrates the fact that GaAs has a direct bandgap, while AlAs has an indirect bandgap. We can now proceed to study how these bands are filled.

8.2 Bands and carriers

We now give some rules on the carrier population in the VB and CB introduced above.

8.2.1 Semiconductor DOS

A cornerstone for dealing with electron population is to count the number of electronic states available in a given energy range in the considered system. The concept is the well-known DOS $g(E) = dN_{\mathrm{av}}/dE$, whereby dN_{av} is the number of states available in the energy interval dE (the reader may find many entries in this book on *photonic* DOS, the cousin concept for photons; see index). The system has to be finite at least formally: at a later stage, densities per unit volume can be introduced. The DOS will be of importance because we can write the total number of carriers N in a given band as the integral:

$$N = \int dN = \int_{\text{energy range}} g(E)\, f(E)\, dE, \tag{8.5}$$

where $f(E)$ is the occupancy number of those states, so $dN = dN_{\mathrm{av}} f(E)$. The relevant choice is presently a Fermi–Dirac statistics of occupancy ≤ 1, which can then be considered as a probability of occupancy for fermions. This is obviously at variance with bosons such as photons.

We outline here firstly how the density $g(E)$ is obtained (more can be found in many textbooks (Kittel, 2004; Ashcroft and Mermin, 1976)) in a generic D-dimensional system ($D = 1, 2, 3$); we shall turn to the occupancy $f(E)$ later only. For convenience, the system considered is a parallelepipedic box (e.g. a rectangle in dimension 2). The actual shape of the system has no effect on the final result in the large volume limit so that surface effects can be discarded. The box that we study in dimension D has volume V_{box} and has D periodic boundary conditions (sometimes called Born-von-Kármán conditions). We study the Bloch wavefunction $\psi(x, y, z) = u_{\mathbf{k}} \exp(i\mathbf{k} \cdot \mathbf{r})$, where $\mathbf{k}$ is a D-dimension vector. These boundary conditions cause the wavevector to be quantized along each of the D directions, with a given, energy-independent pitch along each k-axis. For instance, along direction y, we have

$$k_y = m_y \times \frac{2\pi}{L_y}, \tag{8.6}$$

with m_y an integer. Furthermore, m_y is bounded to a value such that k_y remains in the first Brillouin zone, but we shall not need an explicit formulation in this abbreviated introduction to semiconductor bands.

Due to this quantization of $\mathbf{k}$, and omitting the suffix 'av', the number dN of states available between two infinitesimally close values k and $k + dk$ of the modulus $k = |\mathbf{k}|$, takes the form of the differential of the total number of states with wavevector $\mathbf{k}'$ of modulus less than k, $k' < k$ —i.e. the total number of states inside the generalized sphere of radius k, whose volume we denote by $V_{\text{sph}}(k)$. With the help of Eq. (8.6), we can count the discretized points and find that there are $\frac{V_{\text{sph}}(k)\,V_{box}}{(2\pi)^D}$ total states. $V_{\text{sph}}(k)$ is the main dimensionality-dependent ingredient here. $V_{\text{sph}}(k)$ is thus $2k$ for $D = 1$ (segment), becoming simply πk^2 for $D = 2$ (disc), and reading as follows for $D = 3$ (sphere):

$$V_{\text{sph}} = \frac{4\pi k^3}{3}. \tag{8.7}$$

Here we develop only the $D = 3$ case further and let the reader derive the formulations for other dimensions in Exercise 8.4. In this case of 3D space, dN, obtained by differentiating $V_{\text{sph}}(k)$, takes the form

$$dN = 4\pi \times k^2 \times \frac{V_{box}}{(2\pi)^3}\, dk\,, \tag{8.8}$$

for a particle with a mass and a parabolic dispersion, $E = \hbar^2 k^2/2m^*$. We can retain some more generality and assume only that the dispersion of our particles remains *isotropic* and that it follows a generic power-law form with exponent s:

$$E(k) = A\, k^s. \tag{8.9}$$

Think of other excitations, phonons, or excitons, or of the special dispersion of electrons in graphene, and photons in some simple cases. For the parabolic electron dispersion, $s = 2$ and $A \propto 1/m^*$.

Combining Eq. (8.9), or more precisely its differential form $dE(k) = s\, A\, k^{(s-1)}\, dk$, with Eq. (8.8), we get the following result for the DOS $g(E)$ denoted here by $g_{3,s}(E)$ in dimensionality $D = 3$ and with an exponent s in the dispersion:

$$g_{3,s}(E) = \frac{dN}{dE} = \frac{dN}{dk} \times \frac{dk}{dE} = 4\pi k^2 \frac{V_{box}}{(2\pi)^3} \times s^{-1} A^{-1} k^{-(s-1)}, \tag{8.10}$$

and substituting E to k:

$$g_{3,s}(E) = E^{\frac{3-s}{s}}\, \frac{V_{box}}{2\pi^2}\, s^{-1} A^{-\frac{3}{s}}. \tag{8.11}$$

Hence the DOS is generally an s-dependent power function of the energy. For the linear dispersion of free photons in a closed box, $s = 1$, and A^{-1} is now proportional to the velocity. We now give specific results for $s = 2$, where A^{-1} is proportional to the mass.

The exponent of E is $\frac{3}{2}-1=\frac{1}{2}$ and, since $A \propto 1/m^*$, the exponent of m^* is $+\frac{3}{s}=\frac{3}{2}$. Therefore, for a bulk semiconductor case $D=3$, we have:

$$g_{D,s}(E) = g_{3,2}(E) = 4\pi \; E^{\frac{1}{2}} \frac{V_{box}}{(2\pi)^3} \; \frac{1}{2} \; A^{-\frac{3}{2}}. \tag{8.12}$$

We can thus list the results also illustrated in Fig. 8.2b as follows (see also Exercise 8.2 for D = 1 and D = 2):

- for $D=1$, $g_{1,2}(E) \propto E^{-\frac{1}{2}} \frac{V_{box}}{(2\pi)} \; m^{*\frac{1}{2}}$, and $V_{box} \equiv$ length of the system;
- for $D=2$, $g_{2,2}(E) \propto \frac{V_{box}}{(2\pi)^2} \; m^*$, and $V_{box} \equiv$ area of the system;
- for $D=3$, $g_{3,2}(E) \propto E^{+\frac{1}{2}} \frac{V_{box}}{(2\pi)^3} \; m^{*\frac{3}{2}}$.

So in total, $g_{D,2} \propto E^{\frac{D}{2}-1} m^{*\frac{D}{2}}$. We leave it as an exercise to the reader to get the full coefficients by factoring the various numerical constants. Also, the factor of two for spin degeneracy of electrons has to be included. We note that in all these three cases, the band edge is singular. However, the singularity is a pure divergence in 1D, a jump in 2D, and only a vertical slope in 3D. Also, the 2D case can greatly simplify further calculations, since the DOS is energy-independent.

8.2.2 Electron population at equilibrium: the Fermi–Dirac function

Let us now discuss the population of the band, and hence the occupancy of the states we have counted. Let us, in particular, assume that the carriers in this band are at equilibrium with other electrons elsewhere. It is usually said that these other electrons form a reservoir that 'clamps' their chemical potential or, in the usual words of electronics, that imposes their Fermi level E_F, at a temperature T. The Fermi–Dirac statistics obeyed by $f(E)$ thus reads:

$$f(E) = \frac{1}{1+\exp\left(\frac{E-E_F}{k_B T}\right)}. \tag{8.13}$$

Let us take this opportunity to remind the reader that the classical limit of this statistics is the Maxwell distribution function $f \propto \exp(-E/k_B T)$, valid when the argument of the exponential in Eq. (8.13) is large, i.e. for $(E-E_F) \gg k_B T$. When $f(E)$ is closer to unity, the system is said to be degenerate, as this situation implies notably near-unity occupancy number for states with energies close and under the Fermi level that verify $\exp[(E-E_F)/k_B T] \ll 1$, i.e. a very negative argument in the exponential. It follows that the energies of these states, as well as the Fermi level itself, are much larger than the thermal energy. It is logical that in such a situation, the exact quantum nature of the electron gas regains its importance; the gas cannot be considered as a classical one. It can be shown (see Exercise 8.5) that the ratio of interparticle distance to the *de Broglie wavelength* of electron sets the limits between the two regimes.

We first consider the case of a CB, with a positive mass, starting from some band edge value E_c. Suppress line break after E_c. We can now be much more specific than in Eq. (8.13) above and count the carriers in, say, the CB:

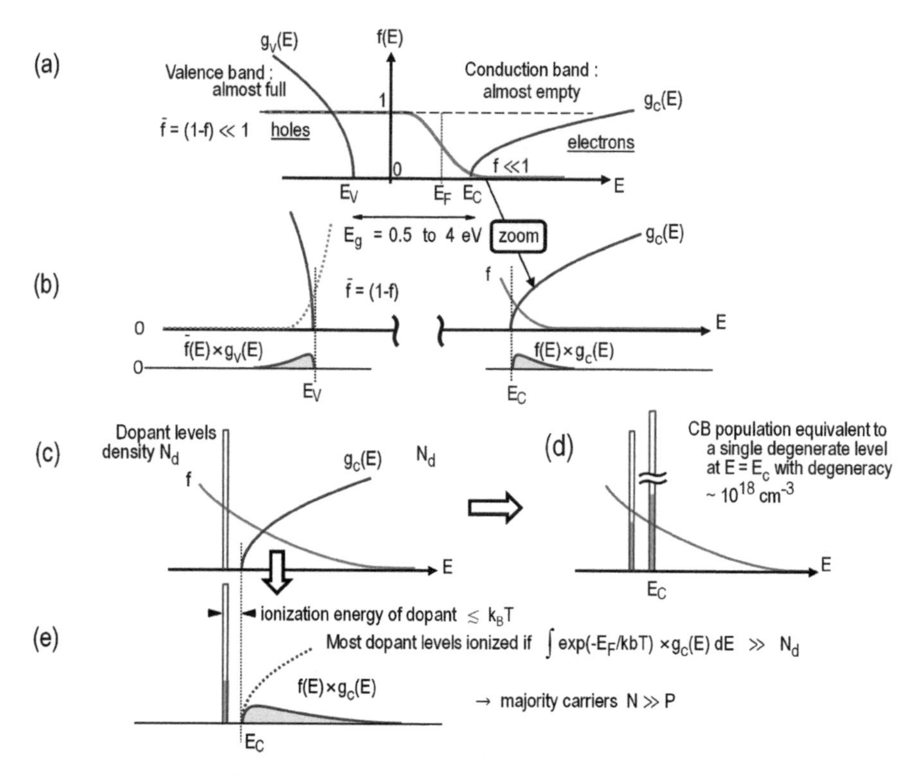

Fig. 8.3 Fermi–Dirac statistics and semiconductor carriers: (a) Fermi–Dirac function and Fermi level, plotted along with the electron density of states (DOS) in valence band $g_v(E)$ and in conduction band $g_c(E)$; (b) carrier population: electron and holes; (c) with the adjunction of dopants whose bound level lies in the bandgap near a band edge (here donors), majority carriers are formed by the almost complete ionization of these donor levels, which occurs provided that the equivalent DOS suggested in (d), given essentially by the integral shown in the population profile in (e), is much larger than the dopant density.

$$N \propto \int_{\text{band bottom}}^{\text{band top}} \frac{1}{1+\exp\left(\frac{E-E_F}{k_BT}\right)} E^{\frac{D}{2}-1}\, m^{*\frac{D}{2}}\, dE. \tag{8.14}$$

We see that the integrand is a combination of a power law with a function having an exponential decay (Fig. 8.3a) asymptotically. Provided E_F is low enough and the band has a width $\gg k_BT$, the latter dictates the behaviour. The decay of the Fermi–Dirac form Eq. (8.13) of $f(E)$, if it takes place in the parabolic region of the band dispersion, discards any role of the energy upper range. So, even though the $E \propto k^2$ dispersion cannot hold forever, as was learned in our tight-binding model, this can be of no consequence on carrier population in practice.

Even without performing the integral in Eq. (8.14), we can derive the temperature dependence of the band population when $f \propto \exp(-E/k_BT)$ [thus $(E - E_F) \gg k_BT$]. Introducing the dimensionless variable $u = E/k_BT$, we find that $N \propto (k_BT)^{\frac{D}{2}}$. The

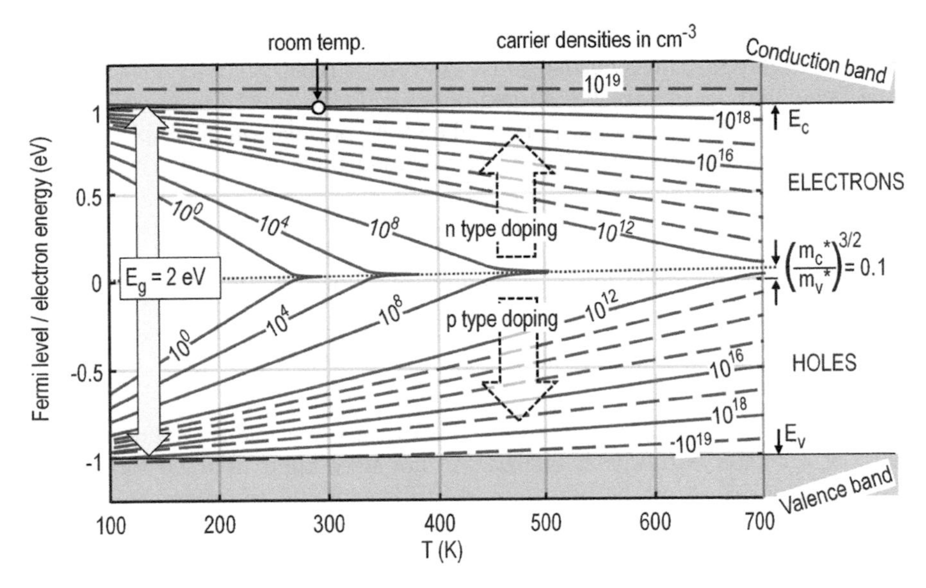

Fig. 8.4 Fermi level E_F as a function of temperature for various amounts of net carrier concentrations $|N|$ or $|P|$. This is an example case for an idealized semiconductor with a temperature-independent gap $E_g = 2\,\text{eV}$ and a mass ratio in the bands $(m_c^*/m_v^*)^{\frac{3}{2}} = 0.1$, reminiscent of $Ga_{0.75}Al_{0.25}As$. The conduction mass m_c^* is such that the equivalent density is $N_{eq}(293K) = 2\,10^{18}\,\text{cm}^{-3}$ at 293 K; hence, the electron density is $N = 10^{18}\,\text{cm}^{-3}$ when $E_F = E_c$ (white dot), since then $f(E_c) = \frac{1}{2}$. The mobile charge density of carrier is compensated in the bulk by the opposite charge of ionized dopant species, n or p as indicated. The neutrality (dotted line) is offset from the band middle by 1.15 k_BT due to the imbalance of masses. At $10^{19}\,\text{cm}^{-3}$, E_F enters the bands, starting the degenerate regime of electron gas.

concept of equivalent density $N_{eq}(T)$ is often used. It amounts to replacing the band by a single energy level situated exactly at E_c with a very large degeneracy, $N_{eq}(T)$. For instance, $N_{eq}(T) \propto T^{3/2}$ in three dimensions (see Fig. 8.3d). The resulting formulation, $N = N_{eq}(T)\exp[-(E_c - E_F)/k_BT]$, separates the role of E_F and also makes the power law and exponential trends in the temperature dependence of $N(T, E_F)$ in this regime more explicit.

We now have a clear connection between electron population N in the CB and Fermi level E_F. Concerning electrons, at least, the trends given above appear in the example case of Fig. 8.4, but we need to explain more about holes, and to explain next how dopants dictate those equilibrium densities and thus the Fermi level.

Finally, the wording 'density' rather than 'population' is not as ambiguous as it might look in our developments, since the extensive quantity V_{box} can be factored at will at any time in these formulas.

8.2.3 The population of electrons and holes

For the VB, we have to reason in terms of *holes* rather than electrons, as we anticipate that $f(E)$ will be essentially close to unity if the Fermi level E_F lies in the forbidden gap range above the VB (hence, as the VB name says, it hosts most of the valence electrons). Holes are just missing electrons. They are virtual particles that have exactly the same behaviour as positively charged ones, and, like electrons, they can carry a current and ensure all corresponding roles. The hole population is simply obtained by considering the complement to unity of $f(E)$ defined by:

$$\bar{f}(E) = 1 - f(E), \tag{8.15}$$

and therefore the hole population P is counted as follows in the VB:

$$P = \int_{\text{band energy range}} g_v(E)\bar{f}(E)dE, \tag{8.16}$$

where $g_v(E)$ counts the density of states of the VB (see Fig. 8.3b). Expressions $P_{eq}(T)$ similar to $N_{eq}(T)$ can be derived, but we will not need their detailed form here. The reader can also now look at the bottom part of Fig. 8.4 and compare the electron and hole densities, and the role of effective masses in the difference.

What we need now is a way to establish where the Fermi level falls. This can be simply done by requiring electronic neutrality; thus $N = P$ in the absence of any other charges or, more precisely, $N(E_F) = P(E_F)$ with the same E_F, at least if a system at equilibrium is being considered. For instance, with the large ratio of effective masses $m_v^*/m_c^* \sim 10$, neutrality is *not* obtained by setting $E_F = \frac{E_c+E_v}{2}$ at the exact middle of the two bands. It has to lie above that middle energy by $(\mathrm{Ln}(10)/2)k_BT$ in order to compensate the lower DOS $g_c(E)$ that intervenes in N compared to $g_v(E)$ intervening in P, by a factor $(m_c^*/m_v^*)^{\frac{3}{2}}$, ensuring a higher electron occupancy to hole occupancy ratio $f(E_c)/\bar{f}(E_v)$ by a factor of a few $\exp(k_BT)$, to eventually get $N = P$ (see Fig. 8.3b and Fig. 8.4; it is easily checked that the occupancy at band edges controls the whole population; that was the rationale in defining the equivalent densities above).

In the domain where the Fermi level is far enough from the band edge to retain exponential behaviours for $N \propto \exp(-(E_c - E_F)/k_BT)$ and $P \propto \exp(+(E_v - E_F)/k_BT)$, the product of the two equilibrium densities NP is a constant independent of the Fermi level. If we also invoke neutrality $P = N$, this constant can be written as $NP = N^2 = N^2_{\text{intrinsic}}$. The value $N_{\text{intrinsic}}$ is called the intrinsic carrier concentration. It is still temperature-dependent; it is the density of the 'landing point' of the curves of N and P in Fig. 8.4 near the middle of the bandgap. In this particular example, it is very low because $(\frac{1}{2}E_g)/k_BT \sim 40$ at 300 K; for instance, ($e^{40} = 2.3\,10^{17}$). For gaps of order 0.5 eV, however, (one-fourth of 2 eV, $(\frac{1}{2}E_g)/k_BT \sim 10$ and $e^{10} = 2.2\,10^4$), there is room for only a few decades between equivalent density and intrinsic density.

However, to put it simply, life with only these intrinsic values of N and P would be terribly boring.

8.2.4 Doping, fixed charges, space charges

To evolve from this situation, which seems too perfect, one can offset neutrality of *mobile* charge carriers by adding *fixed* charges. Such fixed charges are called *dopants* and are often incorporated during growth. These dopants can be atoms that are substituted at, e.g. Ga site in GaAs, and can then typically bring (or capture) one extra electron each, leaving a positive (resp. negative) net charge behind them. For instance, for a Si atom substituted at a Ga site of GaAs, and prone to lose its fourth valence electron to acquire a Ga-like configuration with its four neighbour As atoms, we can write $(\text{Si})_{\text{Ga}} \longrightarrow (\text{Si})^{+}_{\text{Ga}} + e^{-}$. In the absence of any electric field, a cloud of carriers would then surround a cloud of fixed charges. But there can be such an electric field, either imposed on outer electrodes, or built-in due to the presence of interfaces liable to develop some charges. When the carriers are attracted strongly enough elsewhere, the cloud of immobile ionized dopant atoms causes a space charge to be created, and the corresponding static electric field follows from the Poisson equation in the medium ($\nabla \cdot \mathbf{E} = \rho(x, y, z)/\epsilon_r\epsilon_0$), taking into account the static dielectric response.

This topic of space charges is of great practical importance, but, as mentioned earlier, we will not detail it here because we will manage to focus on situations where there are no such charges.

8.2.5 Ionization of dopants

Coming back to the bulk, we can therefore see that electron-rich dopants cause N to increase, provided they can be ionized. The energy level of the electron bound to the dopant can be calculated easily in a hydrogenoid model, i.e. by analogy with the hydrogen atom. Ionizing a dopant is similar to ionizing a hydrogen atom. Fortunately, whereas the energy to ionize a hydrogen atom, the Rydberg, is as much as 13.6 eV, an excess electron inside a semiconductor has a $\sim 10^3$ times lower binding energy, on the order of $E_{\text{dopant}} \sim 13\,\text{meV}$ (Fig. 8.3c), because of two factors: the ten times lower effective masses, and the static relative dielectric constant, $\epsilon_r \sim 10$–15 typically, which further intervenes through its square, $\epsilon_r^2 \sim 10^2$, since the hydrogenoid energy levels scale like m^*/ϵ_r^2. Therefore, at room temperature, with $k_B T \simeq 25\,\text{meV}$, *these levels tend to be mostly ionized.*

Given the low value of the intrinsic carrier density, even a small amount of dopants such as 10^{16} cm^{-3} (circa 1 ppm impurity fraction in chemical terms) suffices to largely offset the situation and cause the Fermi level to approach the CB edge (holes are then minority carriers). Typically, it would reach or even exceed this edge around a few 10^{18} cm^{-3} dopants (see Fig. 8.3d, e and Fig. 8.4). Exercise 8.7 is a simple calculation of Fermi level for a given amount of dopants N_{dopants} and a given equivalent density of the CB states $N_{eq}(T)$. It continues with the calculation of Fig. 8.4 curves. We can therefore remember that in many cases, the electron population and the density of dopants need not be distinguished, because these dopant atoms are often completely ionized. To get a physical view of the limit case of Fermi level at band edge, say $N = 10^{18}\,\text{cm}^{-3}$, we note that the corresponding average distance in the solid between two neighbouring dopant atoms is $10^{-6}\,\text{cm} = 10\,\text{nm}$. Compared to the unit cell of a tiny fraction of a nanometre, this still means a relative concentration of dopants below or around 10^{-4}. Indeed, this is the justification of the semiconductor concept, whereby the mobile carrier density is intermediate between the insulator ones (say, $< 10^6\,\text{cm}^{-3}$) and the metal ones ($\gtrsim 10^{22}\,\text{cm}^{-3}$) that we will see when dealing with plasmons. In the ample room left between, the intrinsic densities ($N_{\text{intrinsic}} = \sqrt{N\,P}$) corresponding to 'semi-insulating' material lie around the bottom (densities of $\simeq 10^{10}\,\text{cm}^{-3}$), whereas the highly doped ones (so-called n^+ doping, 10^{19}–$10^{20}\,\mathrm{cm}^{-3}$) lie around the top and entail incomplete ionization, albeit of little practical consequence.

8.2.6 Conductivity, donors, acceptors, relevant scales

Among the consequences, the conductivity σ should be discussed (note that the CB name gives it no privilege in providing conduction, the VB also does! only in true metals does conduction rely on a single partially filled CB). Using the classical Drude conductivity, $\sigma = \frac{Ne^2\tau}{m\epsilon_0}$, we see that conductivity and concentration, N, are directly related, but the so-called collision time, τ, could present large variations. Fortunately it does not do so, hence the extreme variations in conductivity in semiconductor electronics mostly reflect the huge variations in mobile carrier concentrations.

Next, concerning the very notion of doping, the 10 nm typical distance between dopants also means that if our semiconductor is structured below this length scale, the very notion of doping is inappropriate, or even invalid. For instance, the smallest confined electron systems that will be called a quantum box typically have dimensions of the order of magnitude of 10 nm (see Chapter 9). It is therefore not advisable to consider doping of such systems in the same plain way as doping of usual bulk or thick layer material.

In the above discussion, we outlined what an *n-doped* semiconductor looks like: based on the introduction of *donor dopants*, a large free electron density results at equilibrium at room temperature, and the cloud of fixed ionized dopants can either be fully screened by these electrons or it can create built-in electric fields prevailing in so-called space-charge regions (see section 8.2.4).

We can similarly consider the situation of *acceptor dopants* in *p-doped* semiconductors: they will capture electrons, get ionized, and cause a cloud of holes to be formed. Because of the generally larger DOS of holes (their heavier masses), it is more difficult to even bring the Fermi level to the VB edge. But apart from this shift of about an order of magnitude, the physics is essentially the same.

At this stage, if we had to deal with *electronics*, we would introduce the p–n junction of an ordinary diode and discuss the fate of the space-charge layers between the p- and n-doped regions upon the application of a bias at the electrodes of the diode. Excellent books already do that job—see, for instance, ref. (Sze, 2008)—and fortunately we can avoid it. We thus continue our path towards a more photonic playground by discussing a basic photoluminescence experiment. At variance with the previous settings, this will lead us to introduce a minimal amount of *out-of-equilibrium* physics.

8.3 Photoluminescence basics for direct bandgap configuration

8.3.1 Photon–semiconductor basics

There are essentially three kinds of photon–semiconductor interactions, similar to those known in atom physics, on account of the analogy between the CB and VB with two atomic levels (see Fig. 8.5a, b, c):

- Photon Absorption:
  1 photon + e^- in VB $\longrightarrow$ e^- in CB + hole in VB;

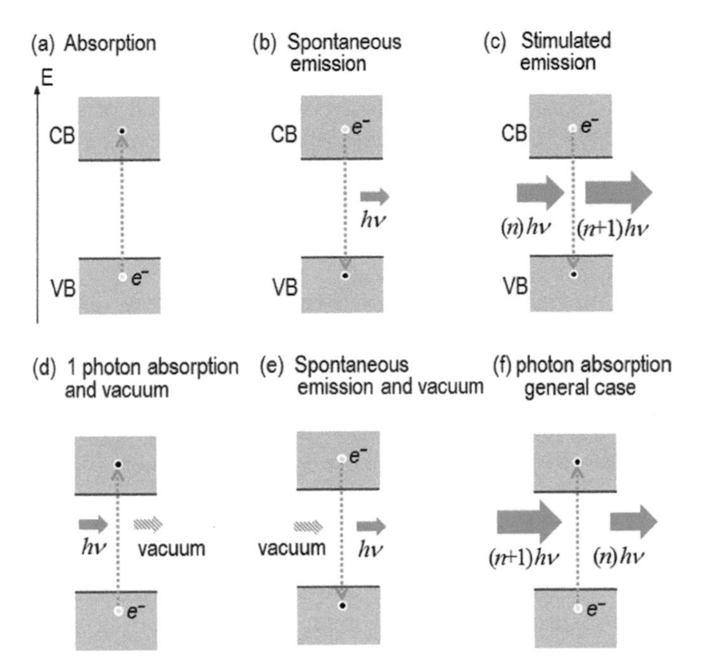

Fig. 8.5 Photon–semiconductor interactions: (a) absorption of a photon; (b) spontaneous emission; (c) stimulated emission; (d) absorption of one photon redrawn with the resulting vacuum photon state; (e) spontaneous emission as the 'stimulation' of a vacuum photon; (f) absorption of one photon out of $n+1$, reciprocal of stimulated emission.

- Photon Spontaneous Emission:
  e^- in CB + hole in VB $\longrightarrow$ 1 photon;
  (the electron lands in the VB, having filled the hole)
- Photon Stimulated Emission:
  n photons + e^- in CB + hole in VB $\longrightarrow$ $(n+1)$ photons.

Here, we are essentially interested in the first two 'reactions'. The reader can further refer to the treatment of Einstein coefficients in textbooks (Yariv, 1989). The three illustrations of Fig. 8.3d-f refer to the *quantum* picture whereby vacuum states are explicitly used for absence of photons and the reciprocity of absorption and stimulated emission is seen more explicitly (see Grynberg et al., 2010). The main point of these two latter processes is the conservation of energy and momentum. Denoting by $\hbar\omega$ the energy of a monochromatic photon, energy conservation implies, for all three processes,

$$E_e = \hbar\omega + E_h', \tag{8.17}$$

where E_h' is the energy of the electron in the VB, $E_h' = E_v(k=0) + \frac{\hbar^2 k^2}{2m_v^*}$, if a negative mass is used. For momentum conservation, it reads:

$$\hbar k_e = \hbar k_{\text{photon}} + \hbar k_h' = (\bar{n})\hbar\frac{\omega}{c} + \hbar k_h', \tag{8.18}$$

where $\hbar k'_h$ is the momentum of the electron in the VB (as above, this avoids having to justify the definition of a hole's momentum and energy and the correct signs for them). In this chapter alone, $\bar{n} \equiv \sqrt{\varepsilon}$ is the refractive index instead of n elsewhere. Let us look at the consequences, assuming the case of a *direct gap* semiconductor, where the interaction between carriers and photons is simpler and generally more efficient than in indirect bandgap ones (see Complement 8.C).

8.3.2 Vertical transitions

Firstly, the order of magnitude of the wavevectors are widely different: the electron wavevector can in principle reach a very large value $k \sim \pi/a \sim 10^{10}\ \mathrm{m}^{-1}$, but this corresponds to large energies of several eV. Nonetheless, even if we consider smaller electron energies on the order of the thermal energy, say 100 meV, due to the parabolic nature of the band edges, the wavevector is still very large: $k \sim$ a few $10^8\ \mathrm{m}^{-1}$.

Comparatively, the wavevector of the photons of concern, here photons with energies $\sim 1\,\mathrm{eV}$, and thus vacuum wavelength $\lambda \sim 1\,\mu\mathrm{m}$ and wavevectors $k \sim 2\pi 10^6 \bar{n} \sim 10^7\ \mathrm{m}^{-1}$, are much smaller. It is therefore simpler, and in most cases largely sufficient, to consider that $k_e \approx k'_h \equiv k$. This is known as the *rule of vertical transitions*, given its simple representation (see Fig. 8.6a). Only a few special cases have to take into account the photon momentum; for instance, those involving the 'photon drag' that the reader can retrieve in the literature. Another reason that will be considered in more detail later is the strong scattering experienced by electrons in solids, which 'blurs' their wavevectors. With this vertical transition rule, we now see that in Eq. (8.17) the energies can be written with the effective masses and the energy gap, so that, referring to Fig. 8.6, it also reads:

$$\frac{\hbar^2 k^2}{2m_c^*} + \frac{\hbar^2 k^2}{2\,|m_v^*|} + E_g = \hbar\omega. \tag{8.19}$$

This form naturally prompts the introduction of the reduced mass of the electron–hole pair, denoted $\mu = (m_c^{*-1} + |m_v^*|^{-1})^{-1}$ (the equivalent of resistances in parallel). The above equation becomes:

$$\frac{\hbar^2 k^2}{2\mu} + E_g = \hbar\omega. \tag{8.20}$$

8.3.3 Absorption and emission strength

Thanks to the above relations, we have established a simplified model for light–matter interactions, whereby an electron–hole pair with a given wavevector ($\mathbf{k}$ in a general case, a single component, say k_x, in our 1D model) interacts with a photon. Establishing the probability per unit time that one transforms into the other (in other words, absorption or emission) is, to first order, a time-dependent perturbation problem. Such problems are addressed with the Fermi Golden Rule, which predicts the transition rate from an initial state $|\psi_i\rangle$ of energy E_i to a final state $|\psi_f\rangle$ of energy $E_f = E_i + \hbar\omega$ by a perturbation at angular frequency ω:

$$W_{i\to f} = \frac{2\pi}{\hbar}\left|\left\langle\psi_f\left|\frac{e}{2m_0}\mathbf{A}\cdot\mathbf{p}\right|\psi_i\right\rangle\right|^2 \rho_f\,\delta(E_f - E_i - \hbar\omega), \tag{8.21}$$

where the light–matter interaction potential acting as a perturbation on the electronic system is the product of the vector potential $\mathbf{A}$, the ratio $\frac{e}{m_0}$, and the momentum operator $\mathbf{p} \equiv i\hbar\nabla$. It is seen that its matrix element intervenes by its square modulus. The next factor of Eq. (8.21), $\rho_f = \rho_f(E_f/\hbar)$, is the density of final states (counted in frequency rather than energy for consistency in this book). Since there is a continuum of final states and there are many final states for both the electron and the photon, this quantity is needed. The last factor is a Dirac δ-function distribution that enforces energy conservation.

However, in finding out what a *photon* absorption probability is, two features arise:

- We must first consider all possible initial electron states of different energies that are liable to undergo a transition induced by that photon of given energy. Fortunately, this does not require a true extra integration, thanks to the energy conservation factor, as developed below.
- We must account for an electron–hole pair with probabilistic occupancy of initial and final levels, instead of {full valence state/empty electron state} in absorption processes or, vice-versa, in emission ones. It can be admitted that the probabilistic character of occupancy translates into an additional factor $f(E_i)\,(1 - f(E_f))$, accounting for the fact that the initial state must be statistically full and the final one empty to achieve the transition. This factor is expressed as $f(E_i)\bar{f}(E_f)$ for an emission process (E_f in VB) and $\left(1 - \bar{f}(E_i)\right)(1 - f(E_f))$ for an absorption process (E_f in CB). The balance (i.e. the algebraic sum of transitions in both ways) leads to an interesting expression when we look at two reciprocal processes of identical 'strength' such as stimulated emission and absorption of Fig. 8.5c, f:

$$\text{Balance Term} = f(E_i)[1 - f(E_f)] - f(E_f)[1 - f(E_i)] = f(E_i) - f(E_f). \quad (8.22)$$

We will see how it relates to optical gain, for instance. We leave this point aside for a long development where we assume this term to be unity ($f(E_i) = 1, f(E_f) = 0$).

8.3.4 Joint DOS

We now come back to the issue of aggregating electron–hole pairs that are liable to absorb a photon to get a meaningful process rate. Instead of counting the density of final *electron* states, we have to count the density of *state pairs* connected by a given frequency or photon energy $\hbar\omega$. Since we can apply the vertical transition assumption and we can also assume that the effective mass description holds for both bands, the density of state pairs can be deduced in exactly the same way as we deduced earlier, in a single band, the electron DOS, except that we replace *electron* energies at a given wavevector modulus $k = |\mathbf{k}|$ by the density of *electron–hole state pairs* connected by a given *photon* energy. This follows from the fact that a given wavevector sets altogether the hole, electron, and photon energies. This reasoning is proposed in more detail in Exercise 8.8. The result involves just the same DOS as in Eq. (8.12), but with the reduced mass μ of Eq. (8.20) instead of the electron or hole effective mass and the photon energy above the bandgap $\hbar\omega - E_g$ instead of the electron energy above the band bottom. Hence in 3D (and with the factor 2 for spin states):

$$g_{\text{joint}}(\hbar\omega) = \frac{V_{box}}{2\pi^2}\left(\frac{2\mu}{\hbar^2}\right)^{\frac{3}{2}}\sqrt{\hbar\omega - E_g}. \quad (8.23)$$

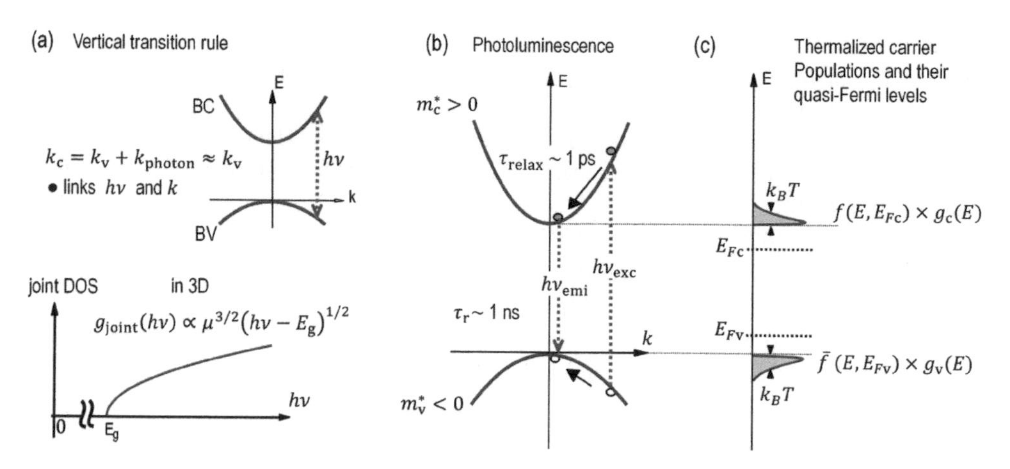

Fig. 8.6 Joint density of states (JDOS) and photoluminescence: (a) vertical transition rule and JDOS, related to section 8.3; (b) photoluminescence: carriers and lifetimes; (c) transient situation with local equilibrium in each band, and thus two quasi-Fermi levels. Panels (b, c) relate to section 8.4.

This DOS is called the *joint density of states.* It can be generalized to any dispersion relation $E(\mathbf{k})$ and is a basis for an important segment of light–matter interaction in solids. In the Fermi Golden Rule, we have to find $|\mathbf{A}.\mathbf{u}|^2$ with $\mathbf{u}$ a unit vector parallel to momentum matrix element (see section 8.3.5). We assume a local plane wave with field $\mathbf{E}$ and $\mathbf{A}$ parallel to $\mathbf{u}$ and the momentum. The moduli of those fields, $\mathcal{A}$ and $\mathcal{E}$, are related by $\mathcal{A} = \omega\mathcal{E}$ and are dictated by the photon density in the box, N_{ph}, itself related directly to the local intensity (e.g. Poynting vector flux for a plane wave in a basic absorption setting, but photon density is more convenient for a stationary wave than that from a laser). It suffices to write the total energy of the plane or stationary wave (i) as $\frac{1}{2}\, V_{box}\, \bar{n}^2\varepsilon_0\mathcal{E}^2$ (in the non-dispersive limit, otherwise $\bar{n}^2$ should be replaced by the product of phase and group indices) and (ii) naturally as $N_{ph}\hbar\omega$, to find that:

$$\mathcal{A}^2 = \frac{2\hbar}{\bar{n}^2\varepsilon_0\omega} N_{ph}. \tag{8.24}$$

This factor in the Fermi Golden Rule shows that the rate of transition is proportional to the intensity or photon density, as is natural. The factor $(e/2m_0)^2$ of the Fermi Golden Rule poses no problem. We are now left with the evaluation of a canonical momentum matrix element.

In atomic physics, a dipole matrix element is often used for the light–matter interaction. It contains an extra $m_0^2\omega^2$ term, so that the main trends in ω may have a different mathematical appearance, while the difference is compensated in the trend of the relevant matrix element (which is either the dipole, i.e. the position matrix element, or the momentum one). This is a subtle topic, as it has been argued in chapters 2 and 10 that the oscillator strength for a given dipole transition scales like ω^3. With the dipole matrix element, our presentation shows no more than a ω scaling. The distinction comes from the photon density of states, with a ω^2 scaling in 3D, which is 'absorbed' in the vertical transition rule in our case to cope with the huge number of degrees of freedom. The photon density of states is much more a concern in spontaneous emission. So it should not be such a surprise if the absorption coefficient does not show a scaling that is of interest for spontaneous emission. Conversely, for spontaneous emitters, the rate for small bandgaps is more in line with the expectations of ω trends from the oscillator strength of given isolated dipoles in vacuum or uniform dielectric media.

8.3.5 The matrix element

The rate of absorption transitions is thus proportional to the joint density of states, the field intensity, and the squared modulus of a matrix element. The final and initial wavefunctions in this formulation are the Bloch wavefunctions of electrons in CB and VB. Given the near-identity of the wavevectors of the Bloch wavefunction (vertical transitions), only the periodic part of Bloch functions plays a role and the normalized value can be calculated for any single unit cell. The core matrix element to be evaluated then reads $P_{cv,\mathbf{k}} = |\langle u_{c,\mathbf{k}} |\mathbf{p}| u_{v,\mathbf{k}} \rangle|^2$.

This matrix element turns out to be pretty independent of k in the vicinity of the band edge. For instance, in most III-V semiconductors of interest (As-based and P-based compounds), the $k = 0$ value of $P_{cv,\mathbf{k}}$ is customarily given through a formal electronic energy as

$$\frac{P_{cv,0}}{2m_0} \sim 20 - 30 \text{ eV}. \tag{8.25}$$

The limited range 20–30 eV shows that the exact nature of chemical bonds does not differ that much in the III-V family. Assembling the above elements, it means that the absorption rate essentially scales like the joint DOS (JDOS), and thus scales like $\sqrt{\hbar\omega - E_g}$ above the bandgap (being 0 for $\hbar\omega < E_g$!) with a similar prefactor for all compounds. An extra polynomial dependence in ω arises from Eq. (8.24), but the range of the matrix element is given here for E_g spanning a modest range (1.2–1.6 eV), not precluding larger ω^n scaling laws for different gaps. Indeed, the atomic dipole version involves an extra ω^2 scaling.

There are two practical consequences of this rather abstract value that we give here by anticipation: (i) the order of magnitude for the absorption coefficient at energies reasonably (a few k_BT) above the bandgap energy E_g is $10^4\ \text{cm}^{-1}$; (ii) the spontaneous emission lifetime for a fully inverted electron–hole pair (i.e. from full CB state to empty VB state) is ~ 1 ns, as shown in Fig. 8.6.

8.3.6 The absorption coefficient for full to empty bands

We can now write the absorption coefficient in the limit case of a full to empty band transition, whereby the balance term is $f(E_i) - f(E_f) = 1$. Transforming the temporal absorption rate into a spatial coefficient is achieved by comparing the number of photons absorbed from spatial and temporal points of view in a thin slice Δz along the light path: in the spatial view, it is $\Delta N_{ph} = \alpha \Delta z N_{ph}$ (differentiate $\exp(-\alpha z)$ to check), while in the temporal view it is $\Delta N_{ph} = W_{i\to f}\, V_{box}^{-1}\, \Delta t$ (V_{box}^{-1} is needed because, unlike N_{ph}, $W_{i\to f}$ is an extensive transition rate that scales like V_{box} from Eq. (8.23)). The ratio $\Delta z/\Delta t$ should be identified with the wave's group velocity $v_g = c/(\text{group index})$

$\simeq c/\bar{n}$ (this group index actually cancels out with that of Eq. (8.24) in a dispersive case, so we leave it as the phase index $\bar{n}$). The resulting expression is:

$$\alpha(\omega) = \frac{e^2 \hbar}{\varepsilon_0\, c\, m_0^2\, (\hbar\omega)\, (2\pi^2)}\, P_{cv,0} \left(\frac{2\mu}{\hbar^2}\right)^{\frac{3}{2}} \sqrt{\hbar\omega - E_g}. \tag{8.26}$$

We see that all the analysis above eventually causes the absorption coefficient $\alpha(\omega)$ in a bulk material to scale like the JDOS: it is simply zero for $\hbar\omega < E_g$, and scales like $\mu^{3/2}(\hbar\omega - E_g)^{1/2}$ for $\hbar\omega > E_g$.

If we neglect, for the time being, the difficulty of defining photons in 2D and 1D, an idealized 2D and 1D electronic system would just follow a similar rule of an absorption dictated by their respective JDOS (seen in Eq. (8.23)), thus scaling above the bandgap like μ in 2D and $\mu^{1/2}(\hbar\omega - E_g)^{-1/2}$ in 1D.

There are several deviations from the idealized absorption picture drawn above. We comment on common ones in Complement 8.B: so-called *excitons* formed from interacting holes and electrons; phonons with a large temperature dependence; and doping, which is more extrinsic. In Complement 8.C, we discuss how silicon specifically behaves. It is an indirect bandgap semiconductor whose absorption brings us everything from solar cells to everyday digital images and other light sensors.

8.3.7 An idealized semiconductor slab transmission spectrum

In Fig. 8.7, we show a practical result that summarizes several aspects of what we learned about semiconductor optical properties. We represent in log scale the transmission of various thicknesses of an idealized semiconductor slab, standing in air, hence

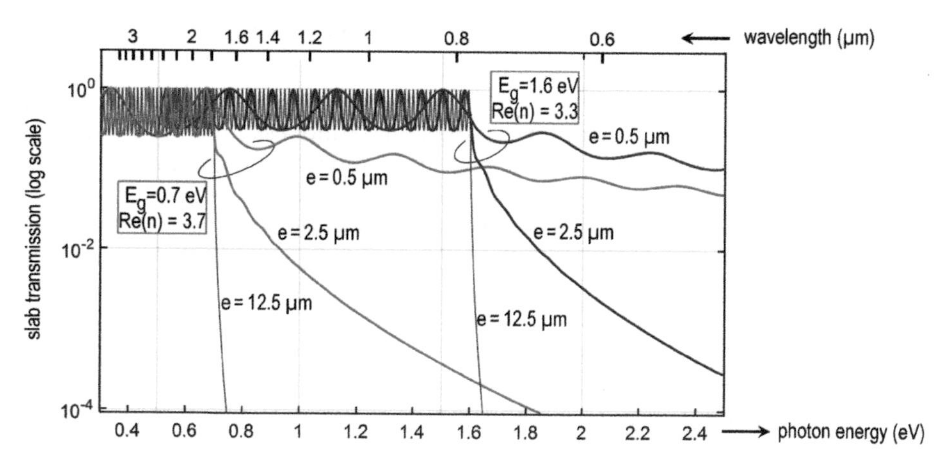

Fig. 8.7 Transmission spectra $T(\hbar\omega)$ of semiconductor slabs for idealized optical parameters, and three thicknesses $e = 0.5\,\mu$m, $e = 2.5\,\mu$m, and $e = 12.5\,\mu$m. The index is constant, and the absorption coefficient amounts to $\alpha = 10^4\,\text{cm}^{-1}$ at $\hbar\omega = E_g + 100\,$meV. The two bandgaps and indices chosen are: $E_g = 0.7\,$eV and $\bar{n} = 3.7$, similar to $Ga_{0.5}In_{0.5}As$; and $E_g = 1.6\,$eV and $\bar{n} = 3.3$, similar to $Ga_{0.88}Al_{0.12}As$. Note the below-bandgap Fabry–Perot oscillations, and the $-\sqrt{\hbar\omega - E_g}$ shape of $\ln T$ above bandgap for the most illustrative $e = 2.5\,\mu$m thickness ($\ln T \sim -\alpha e$), with damped oscillations in the bandgap vicinity.

with noticeable Fresnel reflection. We assume that it has a constant index of refraction around $\bar{n} = 3$ and a typical absorption coefficient following the $(\hbar\omega - E_g)^{1/2}$ scaling, with a prefactor such that absorption attains $10^4\,\mathrm{cm}^{-1}$ at 100 meV above E_g. Two examples close to archetypal III-V InGaAs and GaAlAs semiconductors are shown to emphasize the obvious role of E_g and to show that the index itself, being quite large, can play a large role through Fresnel reflections to shape the overall spectrum: Fabry–Perot oscillations have a noticeable contrast for such indices.

In a real case, the refractive index would be dispersive (fringe spacing and contrast would gently evolve) and the absorption edge would be smoothed or reshaped by various temperature-dependent higher-order effects such as phonon-mediated absorption and the excitons just mentioned.

8.3.8 From absorption to gain

It was seen in Fig. 8.5 that absorption and stimulated emission are reciprocal processes, both with transition rates proportional to the incident field in a semiclassical picture, with the same rate $W_{i\to f}^{\mathrm{abs}}$ for absorption transitions from full VB to empty CB and $W_{i\to f}^{\mathrm{emi}}$ from full CB to empty VB for stimulated emission. We thus call this single *radiative* rate W_{rad}.

We now look at the general scenario with any degree of occupancy in CB and VB. We are looking at reciprocal processes; thus, the rate incorporates the occupation factors according to Eq. (8.22). The resulting expression for the net number of *absorption* transitions (thus field decay), starting at the energy of a hole in the VB, $E_{\mathrm{hole}} \equiv E_i$, is $W^{\mathrm{net}}(\omega) = W_{\mathrm{rad}}[f(E_{\mathrm{hole}}) - f(E_{\mathrm{hole}} + \hbar\omega)]$. A negative net rate $W^{\mathrm{net}} < 0$ thus conversely means a temporal increase in wave amplitude, characteristic of amplified stimulated emission. Correspondingly, we now have a net positive or negative coefficient of spatial evolution:

$$\alpha^{\mathrm{net}}(\omega) = \alpha(\omega)[f(E_{\mathrm{hole}}) - f(E_{\mathrm{hole}} + \hbar\omega)] \; . \tag{8.27}$$

Depending on the sign of $[f(E_{\mathrm{hole}}) - f(E_{\mathrm{hole}} + \hbar\omega)] \in [-1, 1]$, the spatial evolution is a decay or an amplification; the latter is usually called the *medium gain* $g(\omega) = -\alpha^{\mathrm{net}}(\omega)$. The boundary case $\alpha^{\mathrm{net}}(\omega) = 0$ due to $f(E_{\mathrm{hole}}) = f(E_{\mathrm{hole}} + \hbar\omega)$ is called the *transparency* case. It can be seen as the onset of an *inversion regime*, in the terms of two-level systems. We will deal with the spectrum of gain/absorption (i.e. the curve of $\alpha^{\mathrm{net}}(\omega)$) in the case of QWs for laser diodes in section 8.7. We just note here that since, obviously, the maximum gain occurring when $[f(E_{\mathrm{hole}}) - f(E_{\mathrm{hole}} + \hbar\omega)] = -1$ is just the same as the maximum absorption of Eq. (8.26), a medium of diluted character, with few radiative state pairs per unit energy, has a limited gain.

8.3.9 Spontaneous emission

We now study the recombination from an occupied CB state to a hole (empty VB state), allowing a photon to be emitted. This process is especially interesting in non-equilibrium cases, e.g. in a diode, when excess injected electrons give rise to light. But it also exists at equilibrium and is at the origin of the black-body emission. Its discussion is conveniently treated by the use of so-called Einstein coefficients between an upper level 2 and a lower level 1. The radiative transitions due to the incident flux are generally denoted B_{12} (absorption), B_{21} (stimulated emission), against A_{21} (spontaneous emission), this latter featuring a transition rate independent of the flux. The reasoning is to write equilibrium with a black body (a photon gas) at temperature T. The generic result, in line with the symmetry between α and g seen above, is $B_{12} = B_{21} \equiv B$, while A_{21} is found to be proportional to B within a function of frequency that comes essentially from the surrounding photon DOS (see Chapter 12).

To better understand it when being out of equilibrium, we can also relate it to the Fermi Golden Rule above as it is a microscopic view. Apparently, in this case, there is no pre-existing photon flux. However, the Hamiltonian of interaction $\mathbf{A} \cdot \mathbf{p}$ is not strictly zero if one also adopts a quantum mechanics point of view for photons: even for zero photons as the initial state, the vacuum fluctuations (pictured in Fig. 8.5d, e) ensure a non-zero value for $\mathbf{A}$. In the simplest terms, it is found that the role of these fluctuations is the same as if there was a photon in the 'vacuum' initial state. Technically, one must turn to a quantized representation of the electromagnetic field. Then, the whole emission process is driven by a 'photon number' term that reads $(n_{ph} + 1)$, which does not vanish when $n_{ph} = 0$.

Gathering the various elements above, the two parameters that dictate the amount of spontaneous emission are:

- The value of the spontaneous emission lifetime or, equivalently, the A_{21} coefficient. For a fully inverted pair of states, the typical value to remember is $\tau_{\text{spont}} = 1\,\text{ns}$.
- The role of occupation factors, which comes here through the product $f(E_{\text{hole}} + \hbar\omega)[1 - f(E_{\text{hole}})] = f(E_{\text{hole}} + \hbar\omega)\bar{f}(E_{\text{hole}}) \in [0, 1]$. For weak electron and hole occupations, this term becomes very small and the lifetime very long. It is also compatible with common experience that at room temperature semiconductors left at equilibrium around us do not glow with any visible radiation at all. But, conversely, for a device bringing the product close to unity, an intense spontaneous emission can result. This is just what happens in common LEDs.

8.4 Photopumping and relaxation processes

8.4.1 Dynamics at the shortest times: Electron–electron and electron–phonon interactions

In this section, we take the situation of photopumping as a mean to describe relevant pictures of *out-of-equilibrium processes*. We include relaxation processes of either radiative (microscopically described above) or non-radiative nature.

We specifically study the situation of a laser beam at a photon energy $\hbar\omega_{\text{exc}}$ impinging on a piece of direct bandgap semiconductor such as GaAs, with a photon energy somewhat above the direct bandgap energy E_g (see Fig. 8.6b). The beam propagates along the z-axis, say, with absorption causing an $\exp(-\alpha z)$ decay. The impinging excitation promotes electrons from VB to CB, creating an initial electron + hole (e^-, h) population at a wavevector dictated by Eq. (8.20).

The electrons and holes are not at all in equilibrium at creation. The steps to equilibrium are the following, in order of increasing time scale:

- Thermalization among carriers: because electrons are charged particles, they tend to see each other strongly, even at modest concentrations. It typically takes tens of femtoseconds for electrons within CB to thermalize. They lose the memory of the hole momentum they originated from, and large momentum exchange among themselves broadens their energetic distribution.
- This excited electron gas starts an energy relaxation process through phonon interaction. Momentum conservation rules do not restrict this, given the many degrees of freedom of electron and holes in the bulk. The typical time scale of this relaxation process is the picosecond. Both optical and acoustic phonons are involved, although the literature rather privileged optical ones until recent detailed investigations. The important point is that this time scale is still much shorter than the 1 ns time scale of radiative rates.
- Thus, after a couple of picoseconds, electrons and holes have lost their excess energy to the lattice (note that this time is too short to give sizeable luminescence, and the so-called hot carrier luminescence is generally very weak in standard structures). They form *in each band* a relaxed electron gas, at the lattice temperature T. Their distribution can then be described by a pair of quasi-Fermi levels, $E_{F,c}$ and $E_{F,v}$; see Fig. 8.6c. This means, explicitly, that their occupancy is given by $f_c(E) = f(E, E_{F,c})$ for electrons and $f_v(E) \equiv f(E, E_{F,v})$ for holes, $f(E, E_F) = f_{\text{FD}} = (1 + (k_B T)^{-1} \exp(E - E_F))^{-1}$ being the Fermi–Dirac distribution. This is a general feature of optoelectronics of active or passive devices, and a tenet of their practical description. It appears here in the case of photopumping, but it is valid for most cases of electrical pumping of diodes, whereby $E_{F,c} - E_{F,v}$ will simply represent the voltage drop across a junction within a $-e$ factor.

8.4.2 The basics of decay dynamics: Spontaneous radiative decay

We have described the microscopic rate for a transition associated with a given photon, and we have described the cascade that leads within a few picoseconds to two thermalized populations of electrons and holes, N and P respectively. What is now the overall dynamics? We assume that a very short pulse excitation took place at $t = 0$ and that we look at times $t \gg 1\,\text{ps}$ in the thermalized regime.

We first examine in Fig. 8.8 the case of pure spontaneous radiative decay (see also Exercise 8.13). Since we deal with thermalized populations having quasi-Fermi levels, the dynamics of one kind of carriers can be captured through the decay of a single variable, its overall population (amounting to its density). Figs 8.8a, b are both cases in which populations do not fully fill the bands, so the shape of the distribution, e.g. for electrons above E_c, $f_c(E) \propto g_c(E) \exp(-E/k_B T)$, is universal, whatever the prefactor $\exp(E_F/k_B T)$. Also, having $f < 0.5$ prevents gain (the net absorption rate remains positive). If the semiconductor is undoped (see Fig. 8.8a), both populations are identical, $N = P$, and the decay rate is of the form $B\,NP = B\,N^2$, so it slows down during decay itself, as Fermi levels both approach mid-gap. Physically, the decreasing occupancy f means that carriers are rarefied, making it difficult for electrons to find a k-matching hole and vice-versa (vertical selection rule). In chemical terms, the reaction $e^- + h \rightarrow \hbar\omega$ has a kinetics of second order.

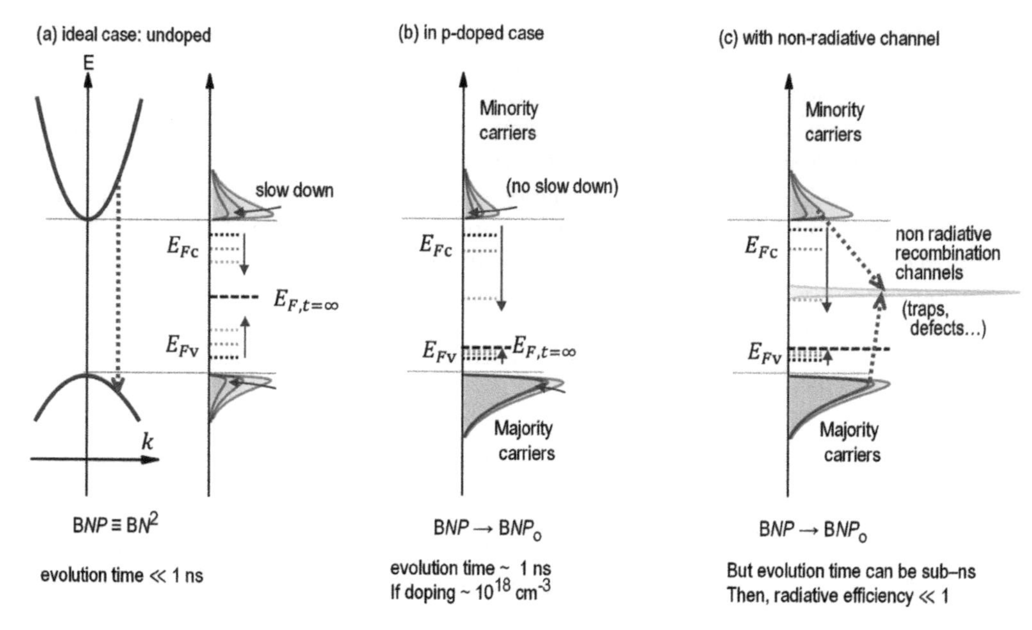

Fig. 8.8 (a) Decay dynamics in ideal non-doped case: $N = P$, the quasi-Fermi levels go back to the near-mid-gap equilibrium Fermi level, decay time slows down; (b) doped case: majority carriers, here holes at concentration $P_0 \gg N$, suffer little change, while minority carriers always find enough matching pairs to recombine at constant rate; (c) non-radiative recombination is present (exemplified here with states at mid-gap). It acts for both carriers like a set of final states with large density, dynamics can be much quicker than the typical nanosecond rate. Note that the relationship of quasi-Fermi level difference $(E_{F,c} - E_{F,v})$ to optical gain and the lasing threshold is discussed later in section 8.7.

Conversely, if we are in a doped semiconductor and at a modest enough excitation to create a minority carrier density, say, N, much smaller than the majority doping-induced carrier density P_0, then the approximation $P \simeq P_0$ can hold through the whole process. During this time, $E_{F,c}$ goes down to the low equilibrium value of E_F, which is close to E_v, whereas $E_{F,v}$ only has to marginally move up to reach the same point. The dynamics is now of first order, as there is an excess 'reagent species', the holes in our example. We can in this case define a radiative lifetime, τ_r. It will remain on the order of a few nanoseconds if P_0 is on the order of 10^{17}–10^{18} cm^{-3}, and the decay will follow the simple associated laws $dN(t)/dt = -N(t)/\tau_r$ and $N(t) \propto \exp(-t/\tau_r)$.

8.4.3 The basics of decay dynamics: Non-radiative decay

Fig. 8.8c shows a microscopic picture of non-radiative decay. Some defect levels in the bandgap of electrons are available, and each of them can capture an electron *and* a hole, simultaneously, and give the extra energy as heat (phonons). A lattice defect can indeed break the vertical transition rule both in terms of energy and momentum (wavevector).

The phenomenological description of the non-radiative channel is often made by using a single characteristic time; i.e., there is an extra decay contribution for the decay rate dN/dt of CB population N, for instance, which is $-N/\tau_{nr}$. The physics remains simple if the radiative rate is also first order in N, as was the case for minority carriers in Fig. 8.8b.

The pair of lifetimes τ_r and τ_{nr} then describe the overall dynamics and the quantum yield (efficiency), much in the same way as is seen for (nano)antennae (see Chapter 2):

$$\tau^{-1} = \tau_r^{-1} + \tau_{nr}^{-1}, \tag{8.28}$$

$$\eta_{QE} = \frac{\tau_r^{-1}}{\tau^{-1}} = \frac{\tau_r^{-1}}{\tau_r^{-1} + \tau_{nr}^{-1}}, \tag{8.29}$$

where τ is the observed decay rate: $N(t) \propto \exp(-t/\tau)$. Two limit trends can be drawn from this situation of competing radiative and non-radiative channels: (i) For a standard radiative time $\tau_r \sim 1$ ns, a faster apparent rate, say 10 GHz associated with a 0.1 ns time constant, can be obtained, thanks to a very high *non-radiative* rate, but at the expense of quantum yield—typically $\eta_{QE} = 0.1$ in this example; (ii) Some ways of managing emission features in advanced structures such as photonic crystals (Chapter 15), e.g. favouring emission directionality from an emitting dipole, eventually translate into a diminished radiative rate of only ~0.1 GHz, ten times less than the standard one. Then a non-radiative channel with, say, $\tau_{nr} \sim 3$ ns which was of limited harm at 1 GHz, becomes a large penalty for the non-standard rates.

8.5 The double heterojunction (DH)

8.5.1 Semiconductor interfaces, heteroepitaxy, and lattice-matching

Let us introduce here a major piece of semiconductor science, the heterojunction. In a nutshell, it was introduced to force electrons and holes to reside at the same place in high concentrations to favour radiative recombination, which is proportional to NP. If many electrons and holes are created in a small volume of a bulk medium, they will tend to diffuse out of this initial volume, causing lower decay rates and, even more importantly, lower quantum yields.

We thus need barriers for both carriers around a given region. The barrier efficiency depends on quantities known as the band discontinuities ΔE_c and ΔE_v at interfaces (see Fig. 8.9a). They describe how electronic levels are affected right at the metallurgic junction, the atomic plane at which semiconductor A ends and semiconductor B begins (in Fig. 8.9a, we pictured fictitious rectangle-shaped atomic potentials for both bands to remind the reader of where the bands came from). In the majority of cases, ΔE_c and ΔE_v follow the gap evolution ΔE_g: the CB goes down and the VB goes up if B has a smaller gap than A, $E_g^B < E_g^A$, and $\Delta E_g < 0$; this is known as *type I* discontinuity (or interface). However, the converse situation is possible, especially if A and B are of quite a different nature, having no common anions or cations, for instance. Such a case is known as a *type II* interface. The practical attainment of good-quality interfaces is by growth, usually not by bonding. Growth that respects crystal structure is known as *epitaxial* growth, and is a cornerstone of semiconductor optoelectronic industry.

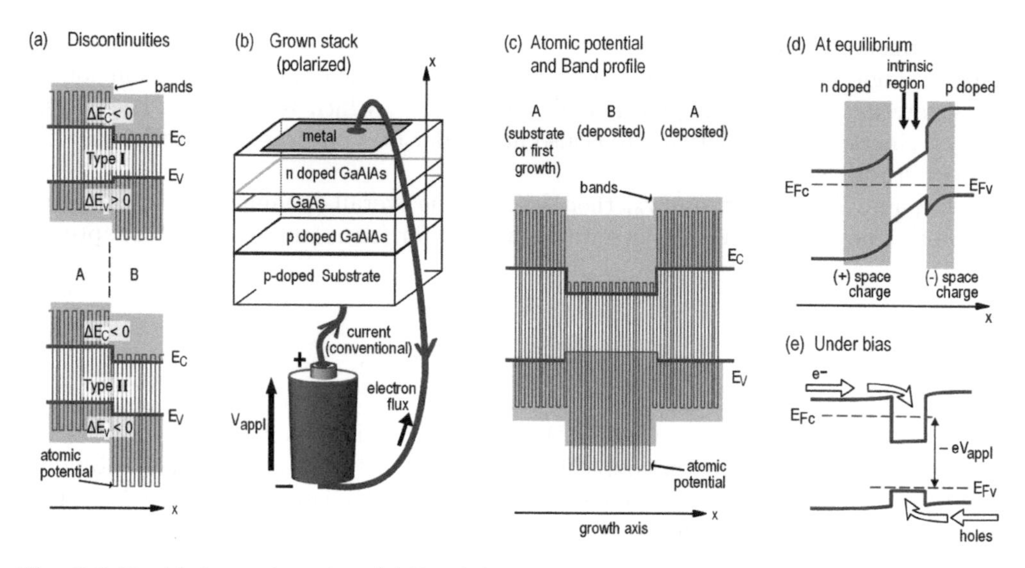

Fig. 8.9 Double heterojunction. (a) Band discontinuities for type I and type II junctions; (b) a schematic Ga(Al)As junction under bias V_{appl}; (c) bands and atomic potential as a function of x (type I), usually the growth axis, with B labelling the central region; (d) band diagram at equilibrium (zero bias), with space charges and a single Fermi level $E_F = E_{Fc} = E_{Fv}$; (e) band diagram under direct bias, with almost no space charge, but two distinct quasi-Fermi levels separated by $-eV_{\text{appl}}$, allowing the steady-state population to build up in the small-gap region.

Because the perfect crystalline nature of semiconductors is vital, notably to limit non-radiative recombination rates τ_{nr}^{-1} below the typical radiative ones ($1\,\text{ns}^{-1}$), two such semiconductors should match their crystalline lattice at the interface. If they have the same bulk lattice constant within 0.1% or 0.2%, the so-called lattice parameter a, as is the case for the GaAs and AlAs binary compounds that form the most successful III-V compound family, $\text{Ga}_{1-x}\text{Al}_x\text{As}$, this will be easily feasible.

However, this good match is rather rarely met in practice. Good substrates are binary compounds, notably GaAs and InP, whereas tuning the gap or another semiconductor property requires a ternary alloy: $\text{Ga}_{1-x}\text{In}_x\text{As}$, for instance. If we add the constraint of *lattice matching*, for instance of $\text{Ga}_{1-x}\text{In}_x\text{As}$ to the InP substrate, we are bound to a single value of x: here $x \simeq 0.5$. Thus, the only solution to get an epitaxial layer of given lattice parameter and adjustable bandgap energy is to recourse to *quaternary* compounds such as $\text{Ga}_{1-x}\text{In}_x\text{As}_{1-y}\text{P}_y$, abridged as GaInAsP, widely used on InP substrates to reach the much-desired telecom wavelengths $\lambda \simeq 1550\,\text{nm}$.

Many textbooks explain epitaxial growth techniques, with subtopics such as the ability to leave a known amount of strain in the layers. We will revive this topic, about self-organized quantum dot growth in Chapter 9.

In all cases, for a given AB junction, it can be remembered that the band discontinuities are *given fractions* of the bandgap discontinuity ΔE_g.

8.5.2 DHs and confinement

A typical DH, also known as 'double heterostructure', is shown in the form of a polarized Ga(Al)As stack in Fig. 8.9b. The substrate can be GaAs in practice, but let us call A the large-gap p-doped layer, even if here it is $Ga_{1-x}Al_xAs$, and B the small-gap layer, here GaAs.

A DH is elaborated by growing the sequence {B then A} onto semiconductor A. The resulting ABA sequence features a profile whose CB and VB are pictured in Fig. 8.9c. Both kinds of carrier are now confined by this potential configuration at the same location in space. They remain free in the two remaining directions of space, but in the x direction we have indeed created a well for each species, of depth ΔE_c for electrons and ΔE_v for holes. We need not invoke quantum aspects for this well if it is large enough (we will consider narrow wells in the next section). When the concept of the DH was proposed independently by Herbert Kroemer and Zhores Alferov in 1963, the idea of quantization was not present.

Adding doping to a spatially non-uniform situation might seem complex. However, we will be able to greatly simplify the picture. In order to inject electrons and holes in the well (semiconductor B), we manage to n-dope the left-side A material and to p-dope the right-side A material. In this way, without entering in the details of electronics, we have a path for electrons and another path for holes, and an ability to push electrons and holes to meet together in the DH, just as in ordinary *pn* diodes (the usual diode of electronics is a homojunction silicon diode).

The difficulty paradoxically comes if we insist on knowing the *equilibrium* situation of such a system. This amounts to requiring a single Fermi level for the whole structure: see Fig. 8.9d. This is a point without any flow, thus the point $I = 0$, $V = 0$ of its I–V characteristics (be it ohmic or not). Let us remind the reader that the quasi-Fermi level difference is nothing but the voltage V_{appl} applied between the device electrodes (within ohmic drops and a factor -e). Hence, with full equilibrium, there is a single Fermi level and no voltage drop. The Fermi level is in general the energy at which electrons are marginally exchanged, by definition: it is the chemical potential of electrons, and the chemical potential describes the energetic cost of particle exchanges with a reservoir. For quasi-Fermi levels, when applying a bias V_{appl}, this picture must be used separately for each wire of the battery of Fig. 8.9b.

8.5.3 DHs and polarization

So, for the equilibrium situation of Fig. 8.9d, the way to accommodate the fact that the Fermi level in each of the n-type and p-type sides must be close to the corresponding CB and VB is that a charged volume builds up around the B region, due to the ionized dopants not being compensated by their associated electron gas. The associated electrostatic potential lifts the energy bands with respect to the Fermi level. We have depletion regions (depletion of free carriers) around the B region, which are also known as space-charge regions or band-bending regions. In total, charges compensate (no outer field). The associated physics is well documented in electronics textbooks. But the reader can already guess that the situation is complex, with several parameters: two doping levels and thickness of the B material. It would be too sketchy to summarize here how to use the Poisson and Fermi–Dirac laws to describe the loop feedback in the chain [charge $\longrightarrow$ potential $\longrightarrow$ Fermi level $\longrightarrow$ population].

The main reason why we need not know the details is that if we polarize our stack, as illustrated in Fig. 8.9e, what we obtain are nearly 'flat bands'. Each side has its quasi-Fermi level aligned with the external potential of the corresponding electrode, and the power supply maintains a voltage drop V_{appl} between the electrodes, typically of the same order as the B bandgap energy, i.e. $E_g^B \simeq -eV_{\text{appl}}$. And internally, there are almost no space-charge regions, thus no band-bending. So the situation looks more like the basic, heuristic one of Fig. 8.9e than the equilibrium Fig. 8.9d situation did.

8.5.4 Injection in DHs and recombination

Let us now look at the B region in the case of Fig. 8.9e. What we could achieve, thanks to the DH under proper electrical bias, is a practical way to impose a quasi-Fermi level $E_{F,c}$ for electrons aligned close to B's CB, while in the same region there is another quasi-Fermi level for holes $E_{F,v}$ aligned close to B's VB. In the well of the DH, initially undoped, we thus accumulate a sizeable population of electrons and holes, so we master in which region of space the fluxes generated by the out-of-equilibrium situation (the applied bias) are produced. Spatially confined electrons and holes shall now recombine there to attempt restoring equilibrium, much in the same way as for photoluminescence. But because the concentrations remain large due to the lack of diffusion in the x-direction, a much more efficient photon generation occurs.

It is, for instance, the reason why, compared to the first semiconductor laser diode without DH (GaAs diodes realized by N. Holonyak in 1962), the laser diodes that incorporated a DH readily reduced the threshold current density from tens of kA/cm^2 to only a few kA/cm^2.

Let us, finally, expand our view to deal with the photonic aspect of the DH. Generally, a smaller gap goes along with a larger index of refraction (think of a metal as a zero-gap material, with infinitely polarizable electrons: when reducing E_g, polarizability and the dielectric constants, static or optical, shall increase). Here, $E_g^B < E_g^A$, thus $\bar{n}^B > \bar{n}^A$. Therefore, a DH also acts as a waveguide (see Chapter 5), comprising a high-index *core* layer sandwiched between two low-index *cladding* layers. Hence, not only are electrons and holes confined in a small volume, but photons are also confined in the same volume, thereby increasing their interaction with electron–hole pairs. This aspect is crucial to define the useful optical modes of laser diodes. More is said in chapters 6 and 21.

8.6 Quantum well

8.6.1 The envelope

The well formed by the DH leads in general to a quantization of the otherwise continuous band levels. We are going to detail this quantization and justify the fact that, at room temperature, it does not have sizeable effects until the thickness is reduced to a few tens of nanometres.

We remind the reader that x is the growth direction, and y, z are the invariant directions. In a potential that has lost translational invariance along x, the electron

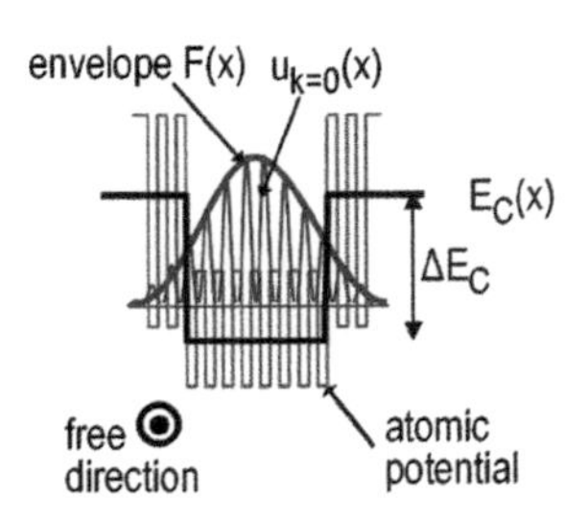

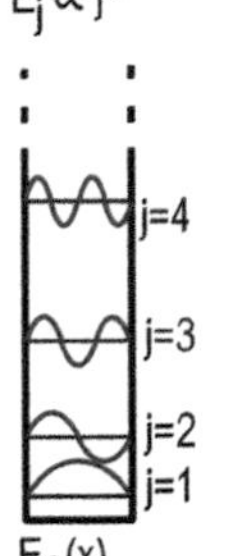

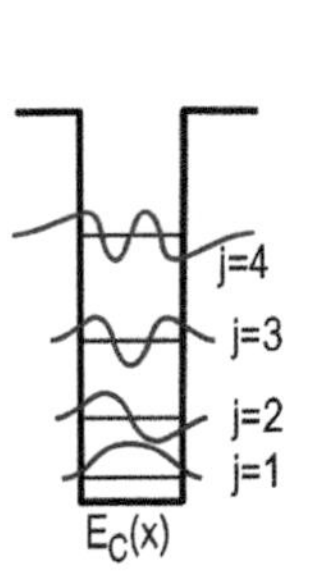

Fig. 8.10 Quantum well: (a) envelope and periodic wavefunctions; (b) levels of infinite well; (c) levels of finite well, wavefunctions penetrate and decay in the barrier.

wavefunction can be usefully approximated within the so-called *envelope* approximation. Compared to the formal Bloch wavefunction in an infinite system $\psi(\mathbf{r}) = u_{\mathbf{k}}(\mathbf{r})\exp(i\mathbf{k}\cdot\mathbf{r})$, we do not have a well-defined $\mathbf{k}$ anymore (no translational invariance along x). However, it turns out that the periodic part of the Bloch wavefunction $u_{\mathbf{k}}(\mathbf{r})$ does not depend much on $\mathbf{k}$. It is then accurate enough to write the general wavefunction for a well as the product of the $k = 0$ (Γ point) function, $u_{\mathbf{0}}(\mathbf{r})$, by a term $F(x)\exp(ik_y y + ik_z z)$ replacing the plane wave $\exp(i\mathbf{k}\cdot\mathbf{r})$: see Fig. 8.10a. In this term, $F(x)$ is an envelope wavefunction but plane wave terms $\exp(ik_y y + ik_z z)$ are retained for the two invariant directions:

$$\psi(\mathbf{r}) = u_{\mathbf{0}}(\mathbf{r}) \; \times F(x) \times \exp(ik_y y + ik_z z). \tag{8.30}$$

It is then shown from the (3D) Schrödinger equation that a 1D Schrödinger equation holds for $F(x)$, with the CB profile $V(x) \equiv E_c(x)$ serving as the potential (we focus on the CB). In other words, $u_{\mathbf{0}}(\mathbf{r})$ does the job of incorporating all the 'atomic' details, while the envelope accounts for the compositional changes along x. We admit here that the 1D wavefunction with eigenfunction $F_j(x)$ and eigenenergy E_j for the j-th quantized level obeys the following equation (see refs (Bastard, 1992; Zory, 1993)):

$$\frac{-\hbar^2}{2m_c^*}\frac{d^2F_j(x)}{dx^2} + E_c(x)F_j(x) = E_jF_j(x). \tag{8.31}$$

The main result is thus the quantization, leading to the appearance of subbands starting at the corresponding quantized levels, with energies given by:

$$E = E_j + \frac{\hbar^2(k_y^2 + k_z^2)}{2m_c^*}, \tag{8.32}$$

which are depicted as paraboloids in Fig. 8.12a.

A similar treatment holds for the VB. However, the situation is more complex due to the doubly degenerate VB depicted for the III-V and II-VI compounds (heavy and light holes). For the usual epitaxial growth direction on (100) planes of crystals with zinc-blende structure, heavy holes and light holes give two separate level sets, with corresponding heavy and light effective masses m^*_{hh}, m^*_{lh} in Eq. (8.30). We shall not give more details on the effective masses in the invariant y, z directions, which are well treated in specialized textbooks (Bastard, 1992; Zory, 1993).

8.6.2 The infinite well

The simplest case is the infinite well of width L ($-L/2 < x < L/2$), with energy at the well bottom conveniently taken as 0. Then, the Dirichlet condition applies to the wavefunction : $\psi(\pm L/2) \equiv 0$. The normalized quantized wavefunctions simply read:

$$F_j(x) = \sqrt{\frac{2}{L}} \cos\left(\frac{\pi j x}{L}\right) \quad \text{for } j = 1, 3, 5 \cdots \tag{8.33}$$

$$F_j(x) = \sqrt{\frac{2}{L}} \sin\left(\frac{\pi j x}{L}\right) \quad \text{for } j = 2, 4, 6 \cdots \tag{8.34}$$

and the energy levels are $E_j = \frac{\hbar^2 \pi^2}{2m^* L^2} j^2 = E_{\text{conf}} \times j^2$ with the *confinement energy* defined as $E_{\text{conf}} = E_1 = \frac{\hbar^2 \pi^2}{2m^* L^2}$. This is illustrated in Fig. 8.10b.

8.6.3 The finite well

In a QW with finite walls (Fig. 8.10c), the eigenfunctions extend into the barriers, the sine or cosine arches of $F(x)$ are less squeezed, and it naturally follows that energies E_j are lower than in the case of an infinite well. The level spacing also becomes smaller and j has a maximum value, j_{max}.

The sequence as a function of well width L, illustrated in Fig. 8.11, is similar to the fate of the confined TE-polarized guided mode in a slab waveguide of variable thickness seen in Chapter 5, since F_j and $\partial F_j/\partial x$ are continuous at the well boundaries. As the width L is reduced (from Fig. 8.11a to b), the energy $E_1 = E_{\text{conf}}$ of the $j = 1$ level increases. Correspondingly, the envelope wavefunctions extend more and more in the barriers ('claddings') and j_{max} diminishes. In the very thin well limit (Fig. 8.11c), there is always a confined mode (no 'cutoff'), but its energy is more and more approaching the barrier energy, and the exponentially decaying wings tend to mimic plane waves with a very smooth evolution of the spatial amplitude. This particular limit is called a wavefunction *deconfinement*, again analogous to a guide-mode profile deconfinement for very thin cores. The most confined envelope is thus found around case b.

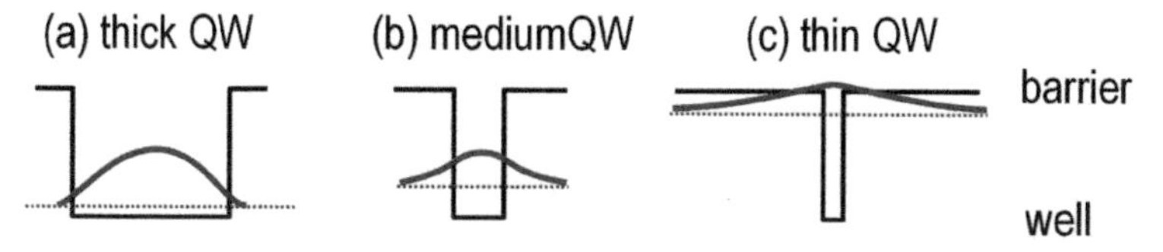

Fig. 8.11 Effect of quantum well width L on the degree of confinement of the E_1 level: (a) large well; (b) well with good confinement; (c) narrow well, wavefunction deconfinement.

8.6.4 Intersubband transitions

What is the possibility of a photon of energy $\hbar\omega = E_j - E_\ell$ being involved in transitions between subbands? In the context laid out for light–matter interaction between bands, what must be revisited, in brief, is not the role of occupation factors, but the role of vertical transitions and the role of envelopes in the matrix element. Vertical transitions still hold because there still exists a momentum in two transverse directions (Fig. 8.12a; see also below interband transitions in QWs). But since, at variance with the interband case, the masses are identical and of the same sign across subbands, this does not introduce a selection of a particular wavevector modulus (see Fig. 8.12b). All pairs of states with proper occupation (from full to empty for absorption, etc.) participate. This makes the transitions ideally monoenergetic, with narrow Dirac-like absorption lines at energies $\hbar\omega_{j\ell} = E_j - E_\ell$ (see Fig. 8.12c).

In the matrix element, the roles of the envelope (essentially $\exp[i\mathbf{k}.\mathbf{r}]$ in 3D) and the periodic part $u_{\mathbf{k}}(\mathbf{r})$ of the Bloch wavefunction are now reversed. It can be shown that in the wavefunction product contributions to the matrix element from $\langle\psi_f|$ and $|\psi_i\rangle$, the identical periodic parts vanish (unity product), while the x-product of the envelopes remain, bracketing the momentum-based interaction Hamiltonian. Thus, going from momentum to position operator, it is the electric dipole moment among envelope functions $p_{j,\ell} \equiv \langle F_j(x)| - ex|F_\ell(x)\rangle$ that dictates the interaction strength. For the infinite well (chosen to illustrate Fig. 8.12b, c), it cancels among levels of the same parity, and is maximal for the case $p_{1,2} \equiv \langle F_1(x)| - ex|F_2(x)\rangle$, yielding a transition dipole moment of about half the size of the well.

These intersubband transitions are mostly occurring in the mid-infrared to far-infrared range (indeed, down the THz range). They are used in the *quantum cascade laser* invented in 1994 (Faist et al., 1994), a device that represents arguably the most remarkable achievement of the branch of optoelectronics known as 'band engineering'. The main difficulty is in managing the quick relaxation to the ground state due to phonons, which tends to be on the order of few picoseconds only, as we mentioned for the intraband part of the relaxation processes.

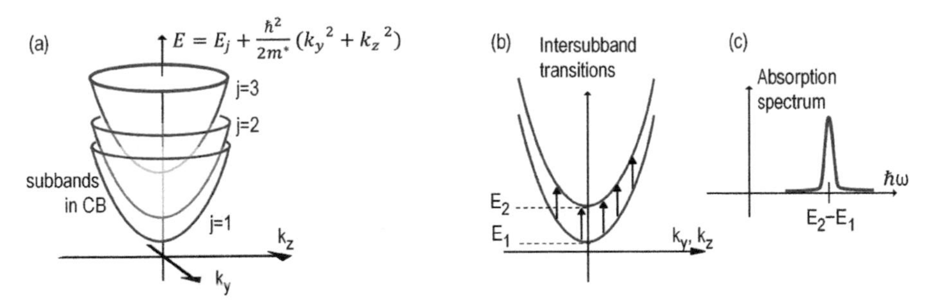

Fig. 8.12 Subbands in a quantum well: (a) free directions x and y are associated with kinetic energy, the subbands can be plotted as nested paraboloids $E = E(k_y, k_z)$; (b) intersubband transitions are vertical but all are at the same photon energy, here $E_2 - E_1$; (c) the spectrum is peaked at $\hbar\omega_{12} = E_2 - E_1$, depending, however, on how subbands are populated.

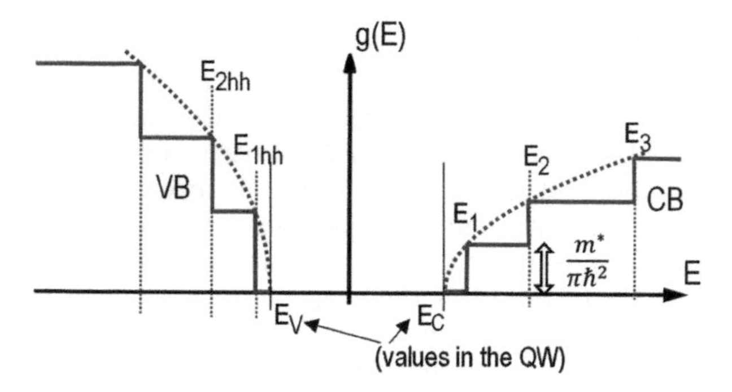

Fig. 8.13 Density of states $g(E)$ for quantum well: in the conduction band (CB), it has steps at E_j and in the valence band (VB) we assumed for simplicity a single heavy-hole (hh) band with a heavier mass.

8.6.5 DOS

According to the general development above, section 8.5, we have a set of *2D subbands* labelled by j. Each of them features a dispersion of the form $E(k_y, k_z) - E_c = E_j + \hbar^2(k_y^2 + k_z^2)/2m_c^*$ (see Fig. 8.12a). Therefore, each such subband has an associated DOS that is within a factor of the unit step function $g_j(E) = H(E - E_c - E_j) \times m^*/\pi\hbar^2$ (a factor of 2 comes from the spin, as mentioned earlier in section 8.2.1).

The overall DOS then reads $g = \sum g_j(E)$ and is a staircase function with steps of identical height, positioned in energy at the levels E_j, as illustrated in Fig. 8.13. An extension of Exercise 8.4, left to the reader, demonstrates that the *envelope* of this DOS $g(E)$ is an exact square-root law $g(E) \propto \sqrt{E - E_c}$ in the infinite well case.

Let us stress here that the step height is *only dictated by* m^*, and not by the well width L. This will have an important consequence on the density of carriers once we populate these bands with given quasi-Fermi levels, particularly for the laser threshold. Another point to note is that the way these levels are populated is specific to each subband. It is clear that the population of the upper subbands is lower than that of the lower subbands. Making this point quantitative is key in deciding to take into account this quantization or not, as discussed below.

8.6.6 Relevance of the confinement to electronic and optical properties

To discuss the need of the quantized picture, the population of the various subbands is the proper indication: if subband edges lie much closer apart than the thermal energy k_BT, there is little difference between the population of adjacent levels j and $j + 1$. Therefore, most of the properties deduced from a continuous 3D DOS would be similar to those deduced using the full details of the 2D DOS. Conversely, if the spacing between adjacent modes is larger than k_BT, there will be a predominance of the properties from the sole set of occupied subbands. This is all the more true if we consider the two lower subbands, $j = 1$ and $j = 2$. Indeed, such a QW can be considered as genuinely 2D when the $j = 1$ subband is the sole substantially populated. This picture is somewhat more complex to generalize in a non-equilibrium case. However, we will stick to the simple criterion below to decide that we are in the limit of a 2D gas:

$$k_B T \ll (E_2 - E_1) = 3\,\frac{\hbar^2 \pi^2}{2m^* L^2} = 3\,E_{\text{conf}}. \tag{8.35}$$

Quantitatively, considering the GaAs electron mass, $m_c^* \simeq 0.07\,m_0$, and room temperature, $k_B T \simeq 25\,\text{meV}$, we find that $(E_2 - E_1)$ reaches $4k_B T \simeq 100\,\text{meV}$ for a width of about 13 nm in the infinite well limit. So typical QWs are just a little narrower in practice. Note that it is much *larger* than the distance between two Ga planes, $\sim 0.3\,\text{nm}$. Conversely, it is much *smaller* than the half-wavelength in GaAs $\lambda/2\bar{n}$ at, say, $\lambda \simeq 1\,\mu\text{m}$, which amounts to 140 nm with a refractive index of $\bar{n}_{\text{GaAs}} \simeq 3.55$.

8.6.7 Multiple QWs

It is quite common to grow multiple QWs (MQWs) (see Fig. 8.14a). One interest is to affect the DOS $g(E)$, which, as noted above, does not depend on well thickness. The physics of MQWs depends very much on the width of the barriers left between them. For very large barriers, say 100 nm, many phenomena shall hinder any chances of coupling between confined electronic states. For usual degrees of confinement with levels well below the barrier energy, tunnelling will have a negligible rate. Then only the injection from the upper bands will be considered, not the coupling between QWs.

For intermediate values of the barrier thickness, say 20–30 nm, tunnelling becomes sizeable. The situation is then completely comparable to the coupling of optical waveguide or atomic states. For two wells, for instance, each level splits into a lower, symmetric level and an upper, antisymmetric one. For more than two or three coupled wells, a *miniband* appears in the growth direction (see Fig. 8.14a).

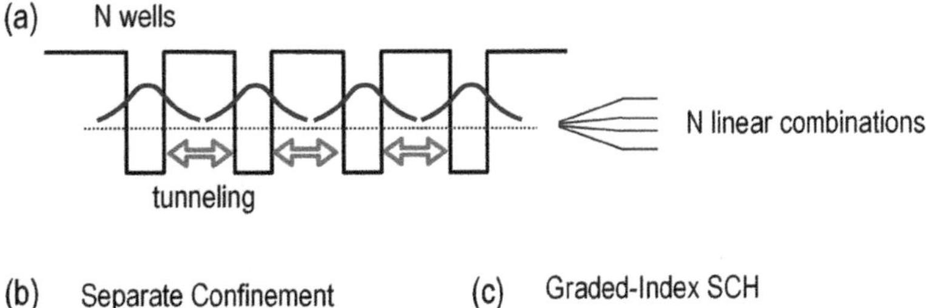

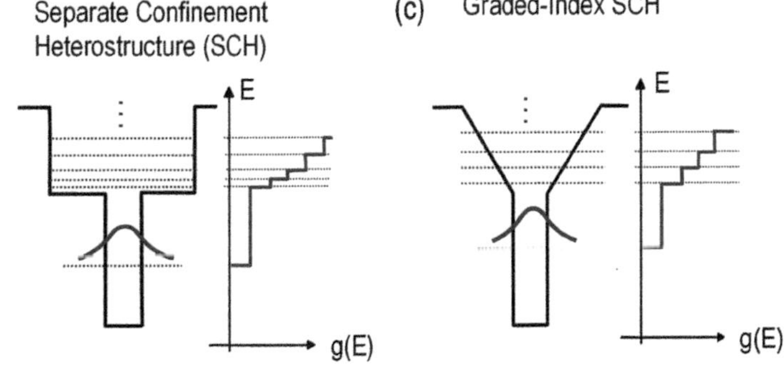

Fig. 8.14 Heterostructures for guided photonics: (a) Multi-quantum-well structure; (b) the separate confinement heterostructure (SCH) and its density of states; (c) the graded index version of the SCH or GRIN-SCH.

For even lower separation between wells, the coupling picture may not be relevant anymore. The structure belongs to the general category of superlattices, with an artificially shaped band dispersion along x. Much effort has been devoted to understanding the properties of these superlattices since the 1980s. One of the most remarkable results is the very subtle superlattices that operate at the heart of the above-mentioned intersubband quantum cascade laser (QCL) (Faist et al., 1994). Band engineering of superlattices for electronic properties can somehow be compared to efforts towards artificial metamaterials for optical properties. There are, for instance, some fascinating properties of superlattices whose band structure can be made to mimic a quasi-static electric field, $E(x)$, so as to bias the stack and play with the degree of carrier localization in it. The reader can look at the entry 'Wannier–Stark ladder' for this class of phenomenon.

8.6.8 The separate confinement heterostructure (SCH)

An important variant of QW and DH implementation is to combine both in the so-called separate confinement heterostructure (SCH) of Fig. 8.14b. For this, three different materials have to be stacked: for instance, GaAs for the narrow quantum well of width L_{QW}, $Ga_{0.8}Al_{0.2}As$ for the intermediate bandgap material with total width L_{DH}, and $Ga_{0.6}Al_{0.4}As$ for the higher barriers. The interesting feature of this system is that it properly confines photons as well as electrons, so as to make better laser diodes. This is discussed in more detail below (see section 8.7.7).

Note that in the electronic DOS of such a system, many levels are appearing at some energetic distance above the level of the QW of interest. This can be an issue if one wants to substantially populate this QW level without being obliged to add too many carriers at the upper levels. For this reason, the so-called GRaded-INdex SCH (GRIN-SCH) shown in Fig. 8.14 c), can be useful: the continuous variation of the CB makes a kind of triangular large well in which the levels are more spaced than in the SCH large well bottom (the large well is also called the 'optical cavity' in laser-diode practice because it will host the optical guided mode). Therefore, fewer high-energy carriers need to be introduced for the same amount of QW level population. However, it is a highly complex procedure to grow a graded material twice; furthermore, there is a price to pay, since the photon confinement shall not be as good if higher-bandgap material of a smaller index of refraction $\bar{n}(x)$ is used.

8.7 Laser threshold

8.7.1 Stimulated emission

Here, we discuss the importance of the DH and QW for the operation of a laser diode. The topic has been well known since the 1980s, but it is worth having it in mind when discussing further degrees of confinement of electrons or photons.

In the schematic account of light–matter interaction given above, we first want stimulated emission to dominate over absorption. The two mechanisms are strictly symmetric; see Fig. 8.5 and Eqs (8.22) and (8.27). Therefore, what we require is known as *inversion*. If we had a two-level system, we would simply want to have a greater population in the upper level (occupancy f_2) than in the lower level (occupancy f_1), so that the probability of absorption, which behaves like $f_1(1 - f_2)$, exceeds that of stimulated emission, which behaves like $f_2(1 - f_1)$: the net absorption/gain balance behaves like $f_2 - f_1$.

If we now have two continuums, one described by $g_v(E)$ instead of a single level 1, and the other by $g_c(E)$ instead of a single level 2, we can again make use of the JDOS and the vertical transition rule to identify whether at a given energy we have gain or loss. Assuming that two quasi-Fermi levels $E_{F,c}$ and $E_{F,v}$ describe the population of VB and CB states, we just have to transform the condition above. Therefore, the presence of absorption or gain (at this stage meaning stimulated emission, and not yet the usual spatial gain) at a given *photon* energy $E_{\rm phot} = \hbar\omega_{\rm phot}$ is given by

$$f_{\rm FD,c}(E_{\rm elec}) - f_{\rm FD,v}(E_{\rm hole}) > 0 \quad \text{for Stimulated Emission,} \tag{8.36}$$

where the energies of carriers are those that satisfy the vertical rule for the photon energy considered $E_{\rm phot}$. We now head towards a condition on the quasi-Fermi levels themselves. Denoting the lowest electron energy by E_1, the highest hole energy by E_{1hh}, and the effective gap by E_g (be it for the bulk or QW case), and reminding the reader that the reduced mass is denoted by μ, we get the relationships we need from Eq. (8.20):

$$E_{elec} = E_1 + \frac{\mu}{m_c^*}(E_{\rm phot} - E_g), \tag{8.37}$$

$$E_{hole} = E_{1hh} + \frac{\mu}{m_v^*}(E_{\rm phot} - E_g). \tag{8.38}$$

8.7.2 Transparency and the Bernard–Duraffourg condition

Let us now consider Fig. 8.15, exploiting the simplicity of the constant DOS in the case of a QW. From the electronic point of view (see Fig. 8.15a), we assume that a single subband is occupied and that there is no doping, so that $N = P$ due to neutrality. The difference in effective masses in CB and VB causes a substantial imbalance between the positions of quasi-Fermi levels: in the CB, for instance, $E_{F,c}$ can lie well above the subband edge E_1, whereas an identical hole density only requires the quasi-Fermi level $E_{F,v}$ to lie close to the band edge E_{1hh}, but still inside the bandgap (see section 8.4). The integral relating population to quasi-Fermi levels, e.g. $N = \int f_{\rm FD,c}(E) g_c(E) dE$, is tractable analytically and is proposed as Exercise 8.9.

Assuming a constant matrix element, the factor driving absorption is the product of the JDOS $g_{\rm joint}(E_{\rm phot})$ by the 'balance term' factor $[f_{\rm FD,c}(E_{\rm elec}) - f_{\rm FD,v}(E_{\rm hole})]$. The JDOS is also constant here, so that we essentially have the difference of two Fermi–Dirac functions, with Fermi energies correlated by the neutrality requirement, and with the 'deleveraging' effects of Eq. (8.37) and Eq. (8.38) between carrier and photon energies. As shown in Fig. 8.15b as a function of photon energy, these curves must lie between the two extreme values of this product, $\pm g_{\rm joint}(E_{\rm phot})$, and they must have an upward trend with increasing $E_{\rm phot}$, since the Fermi–Dirac factors evolve, in this case, towards 0 ($f_{\rm FD,c}$ in CB) and 1 ($f_{\rm FD,v}$ in VB).

For low carrier densities, we can guess that $[f_{\rm FD,c}(E_{\rm elec}) - f_{\rm FD,v}(E_{\rm hole})] \simeq 1$ over the relevant range, $E_{\rm phot} > E_g$. Absorption remains dominant: it is partly diminished but at this stage no stimulated emission is yet produced. The *transparency density* N_{tr}

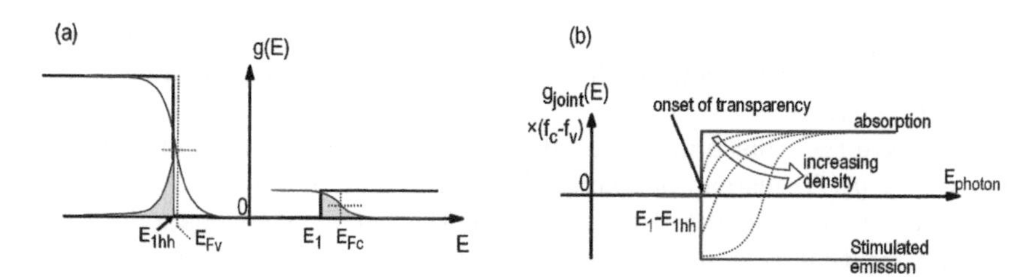

Fig. 8.15 Gain, absorption, and density of states (DOS): (a) quantum well DOS with typical near-transparency situation for carrier populations; (b) absorption and gain have the limits of the joint DOS, and evolve within these limits, as indicated by the dashed line.

is a very important quantity. It is the density at which stimulation emission begins. We can say from the fact that $[f_{\mathrm{FD,c}}(E_{\mathrm{elec}}) - f_{\mathrm{FD,v}}(E_{\mathrm{hole}})]$ is monotonous that this happens just at the lowest allowed transition, between band edges. We will come back in section 8.7.3 to this particular situation. At larger densities, stimulated emission appears in a given photon energy range above the bandgap E_g. It reverts to absorption beyond a density-dependent *transparency energy.* The typical size of the transition region is obviously related to the driving quantity of the Fermi–Dirac function, hence $k_B T$.

Thus, a gain spectrum of a gain medium shall generally end up relatively sharply on its high-energy side, reverting to absorption over a small range of typically 25 meV at room temperature. This does not preclude the ability to have a large gain spectrum (say 150 meV) with adequately strong pumping and engineered DOS, but the transition at the end of this range will be sharp. Let us come back to the transparency density N_{tr} and establish the condition for it: it requires $f_{\mathrm{FD,c}}(E_{\mathrm{elec}} = E_1) > f_{\mathrm{FD,v}}(E_{\mathrm{hole}} = E_{1hh})$, which can be inverted as:

$$1 + \exp\left(\frac{E_{1hh} - E_{F,v}}{k_B T}\right) > 1 + \exp\left(\frac{E_1 - E_{F,c}}{k_B T}\right). \tag{8.39}$$

Since $1 + \exp\left(\frac{E}{k_B T}\right)$ is a growing function of E, this implies:

$$E_{1hh} - E_{F,v} > E_1 - E_{F,c}. \tag{8.40}$$

This can eventually be recast as an equation known as the Bernard–Duraffourg condition:

$$E_{F,c} - E_{F,v} > E_1 - E_{1hh} \equiv E_g. \tag{8.41}$$

Therefore, the Bernard–Duraffourg requirement is that the *separation* of quasi-Fermi levels exceeds the bandgap energy. Fortunately, it does not require the VB quasi-Fermi level to reach the VB—that is, to exceed 0.5 occupancy of holes in the VB. Such a situation would demand a much higher carrier density. The choice made in Fig. 8.15a is such a case with less than 50% hole occupancy (i.e. over 50% electron occupancy in VB).

What we have learned from this analysis is that the transparency threshold shall correspond to a substantial occupancy of the CB for a GaAs QW. The analysis made in Exercise 8.9 indicates a population of typically triple the naïve value for 0.5 occupancy at the E_1 CB edge, itself on the order of $0.5 g_c k_B T$; thus, an electron population $\sim 1.5 g_c k_B T$, using the fact that the QW DOS $g_c = m^*/\pi\hbar^2$ is constant. So, here, unlike the JDOS, where the compound effect of different masses was simply cast in the reduced mass, a different combination occurs.

8.7.3 Transparency current density of a QW

We can now use the knowledge introduced for photoluminescence to understand how this density translates into the transparency current density, i.e. the current density J to be fed to 1 cm^2 of QW to make it transparent. Achieving sufficient gain to reach a lasing threshold generally does not cost much more in current density, but it depends on the laser photon losses and is the subject of the next subsection. The point is that at transparency, there is as yet no gain and no net stimulated emission. Thus, current only has to be provided in order to feed spontaneous emission (in a simplified and optimistic view). It is, then, straightforward to write a detailed balance of carriers in the well using for their decay the *spontaneous emission decay rate* τ_r^{-1} of 1 ns that we explained above. It is perfectly justified here, because occupancies are of order unity. The transparency current density reads:

$$J_{tr} = e\, N_{tr}\, \tau_r^{-1}. \tag{8.42}$$

The typical value that one can easily remember for our standard GaAs QWs is $\sim$ 100 A/cm^2, stemming from a typical transparency density N_{tr} of 0.6×10^{12} e$^-$/cm^2 and a rate $\tau_r^{-1} = 10^{-9}$ s^{-1}. It is remarkable that the only ingredients for such a practical quantity are the effective mass of electrons, the thermal energy, and the spontaneous emission lifetime, with a numerical factor for the asymmetrical CB and VB effective masses.

As for the current threshold itself (not the current density) of an actual edge-emitting laser diode, since it has a typical area of $3 \times 300\ \mu$m$^2 \sim 10^{-5}$ cm^2, it results in a $I_{tr} \sim 1$ mA threshold current range.

8.7.4 Double heterostructure transparency

We make use here of elements not only from the above discussion on QW transparency, but also from the slab waveguide discussed in Chapter 6 to see a limitation of the DH and introduce a clever solution. We have discussed the JDOS using the vertical transition rule, thus assuming implicitly that a plane wave is interacting with an electron–hole pair. However, the role of the double heterostructure was not only to concentrate electrons and holes at the same location in space, but also to confine photons within a guided mode travelling along it, and therefore having a privileged interaction with electron–hole pairs in the DH.

Following the same path as that followed for QWs, we can guess that now, achieving inversion in a DH still boils down to $(f_c(E_{\text{elec}}) - f_v(E_{\text{hole}}) > 0)$, and so it does not change the targeted occupancy factors: they must be, like for QW, about one-half at band edges, typically more at the CB edge and less at the VB edge if we have the usual trends in effective mass ratio. Grossly, the resulting density remains on the order of the equivalent density of states at the temperature in question, say $N_{c,eq}(T)$. But to get a DH current to be compared with the QW case, we must think of the transparency density of carriers *per unit area* in our DH rather than its bulk density. This density *per unit area* becomes obviously proportional to its width L_{DH}. Again with a constant radiative rate in the spontaneous regime, $\tau_r^{-1} \sim 1\ \text{ns}^{-1}$, we reach the conclusion that the transparency current density J_{tr} (flowing normal to the DH plane) scales like L_{DH}.

8.7.5 The weak confinement factor: The problem of the thin DH

We now want to describe the interaction of a guided wave with our active region, QW or DH (or MQW, etc.), particularly in order to study transparency and spatial gain. The overlap of the guided wave and the active part is the quantity that describes this. In the absorption we calculated, we dealt with plane waves. Replacing them with guided waves of profile $E_{guid}(x)$ leads to an overlap integral of the guided waves squared field with the active region, known as the confinement factor Γ (see Chap. 6):

$$\Gamma = \frac{\int_{\text{active region}} |E_{guid}(x)|^2\, dx}{\int_{\text{everywhere}} |E_{guid}(x)|^2\, dx}. \tag{8.43}$$

This depends on the guided mode considered, of course (including polarization). Let us study the trend of Γ for a DH of variable width. In a very broad DH, the field is almost entirely in the low gap active region, there are only small evanescent tails in the cladding ($\simeq$ barriers for a DH, low refractive index coincides with large bandgap), so $\Gamma \simeq 1$.

However, if we diminish L_{DH}, the guided mode profile $E_{guid}(x)$ follows a sequence analogous to that of the wavefunction profile in a QW of shrinking width (see Fig. 8.11). Below a width L_{DH} of order $\lambda/2\bar{n} \sim 150\,\text{nm}$ around $\lambda = 1\ \mu\text{m}$, the profile enters into the deconfinement regime, becoming broader again. This latter trend is very strong. If we look at typical widths of QWs, say 10–20 nm, see Fig. 8.16a, Γ can reach extremely low values, $\Gamma \ll 1$. This 'only' means that gain and absorption are diminished by a strong factor, not that gain disappears. So we must study what kind of actual spatial gain we want in an acceptable device to see if values that are very low, $\Gamma \ll 1$, are indeed unacceptable. More precisely, we must see if the modal gain $g_{\text{mod}} = \Gamma g$ (or the modal absorption $\alpha_{\text{mod}} = \Gamma\alpha$) is large enough, with g or α being the values that would hold for a wave fully interacting with the system as in the bulk case (this is only a photonic view, not an electronic one, so α or g can be those of QWs).

8.7.6 Threshold current density

The usual gain requirement in a Fabry–Pérot-type laser cavity of length L_{cav} is to balance 'mirror losses' (see Chapter 7) stemming from the cavity's mirror reflectivity product $r_1 r_2$ in amplitude terms, thus giving the well-known formula:

$$r_1\, r_2\, \exp(g_{\mathrm{mod}} L_{cav}) = 1. \tag{8.44}$$

Notice however, that we have replaced gain, g, by modal gain, g_{mod}, because we want to describe round trips of a guided wave: see Fig. 8.17 and Chapter 21. Usually, reflectivities are not that high ($|r|^2 < 0.9$, often $|r|^2 \sim 0.3$). A typical device length for low enough threshold current at the few 100 A/cm^2 value seen above is $L_{\mathrm{cav}} = 300\,\mu$m (the width of the laser must be introduced, typically, say, 2.5 μm, to connect L_{cav} and J to amperes, say to 10 mA). This means that the demand for modal gain is at least $-\ln\left(r_1 r_2\right)\left(L_{\mathrm{cav}}^{-1}\right) \sim$ 5–30 cm^{-1}. Since a bulk gain of 300–500 cm^{-1} is already not so easy to achieve, it is seen that $\Gamma < 10^{-2}$ becomes a big hurdle in the way of reaching that goal.

Therefore, we can guess that for a DH, as the width L_{DH} is reduced, even though the *transparency* current density J_{tr} diminishes, the *threshold* current density J_{th} does not diminish forever. It goes through a minimum when the trend in Γ offsets the reduction in volume. This typically happens around $L_{DH} = 100$ nm. For DH widths on the order of 20–50 nm, which are formally large QWs—but still too large to get level separation larger than $k_B T$ at 300 K—there is not enough gain, and this will remain true for thinner QWs, $L_{QW} \sim 10$ nm, which are more genuinely acting as 2D systems, since only their fundamental level is sizeably populated when gain appears at room temperature. The minimum *threshold* current density J_{th} for DH can be calculated to be on the order of 2 kA/cm^2 from a typical threshold density 10^{18} cm^{-3} and still the ~1 ns radiative time. So it is about 20 times the 100 A/cm^2 transparency current we calculated for a single QW.

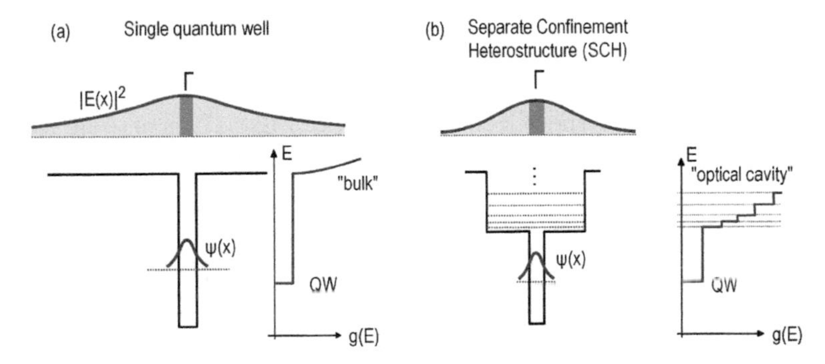

Fig. 8.16 Gain and confinement factor. (a) Single QW situation: the guided mode profile is very broad, Γ is small; on the right side, the electronic density of states (DOS) is plotted; (b) SCH: the guided mode profile is optimally narrow, Γ is as large as possible. The DOS becomes of 3D nature when the QW broadens into a DH.

8.7.7 Threshold in separate confinement heterostructure lasers

What does the SCH buy us in this context? It provides the attainment of *threshold* current densities on the order of *transparency* current densities, $J \sim 100\ \mathrm{A/cm^2}$. It sacrifices part of the span in materials (alloys) to get optical confinement while retaining enough span to get a decent quantized QW operation with good enough electronic barriers.

The optimum point is to get the confinement factor Γ to the highest possible value that can be reached between a guided mode and the QW (see Fig. 8.16b). There are some subtle factors behind the typical waveguide width value $\lambda/2\bar{n}$ that provides optimal optical confinement, related to index contrast, but the situation is rather 'tractable'. If we decrease the index step of a slab waveguide by a factor of two, for instance, the optical confinement is less, but not much less: Γ drops only by a factor of $\sqrt{2}$ or so, as can be guessed from the role of cladding indices in formulas such as Eq. (6.14) on guided modes in Chapter 6.

Hence, in the SCH, although the range of alloys available has to be shared roughly 50/50 between QW confinement and waveguide confinement, a good optical confinement is obtained, and the QW confinement factor Γ_{QW} retains nearly the ideal value $\Gamma_{QW} \sim 0.05$–0.1,which is sufficient to attain a large enough modal gain: Γ_{QW} is the value derived by approximating the profile to a half-sine period of average $\frac{1}{2}$, as made explicit below:

$$\Gamma_{QW} = \frac{\int_{\text{QW region}} |E_{guid}(x)|^2}{\int_{\text{everywhere}} |E_{guid}(x)|^2} \simeq \frac{E_{\max}^2 \times L_{QW}}{\frac{1}{2}E_{\max}^2 \times L_{DH}} \simeq \frac{2L_{QW}}{L_{DH}}. \tag{8.45}$$

This philosophy of confining the optical mode also explains the term 'optical cavity', used historically by SCH practitioners to denote the waveguide itself, and not the lasing cavity formed by facets or feedback gratings. So, a SCH laser shall reach its threshold at current densities in the range 120–200 $\mathrm{A/cm^2}$ per QW, which, in turn, grants device thresholds in the 2–10 mA range.

8.7.8 SCH and optimized light–matter interaction

At this point, we can see that the SCH laser diode (see Fig.8.17) is in some sense the first sophisticated nano-optical device dimensioned to optimize light–matter interaction on

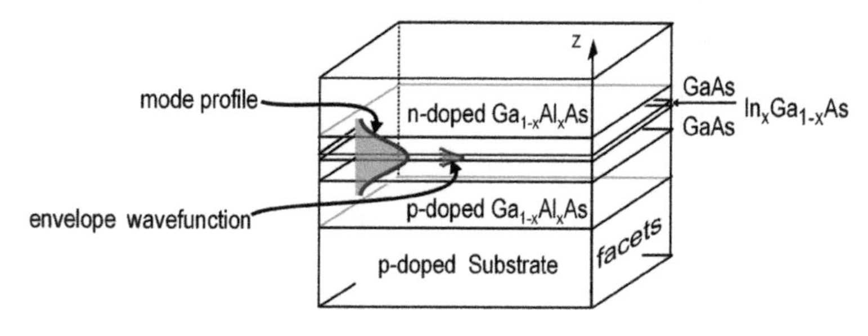

Fig. 8.17 Perspective view of a (In)Ga(Al)As-based device comprising a separate confinement heterostructure, with mode profile and QW envelope wavefunction sketched.

a very fundamental basis. Indeed, there is no lower JDOS than the QW one in 2D, given by the reduced effective mass μ. We do not detail here many efforts, leveraging strain notably, to diminish the mass of heavy holes, the 'bad' mass, which helps to reduce the threshold. But these factors are secondary with respect to the main message: the QW with its size-independent DOS is an optimum *per se*, and we will see in Chapter 9 that more quantized nanostructures only beat the QW-based ones in rather specific cases.

The design of a laser having a single QW for gain then has to take care that an optical mode interacts optimally with the QW. To retain sufficient confinement, the SCH is the best solution. Refinements, such as the GRIN-SCH of Fig. 8.14c again play at the margin: to minimize the possible population of the DH in the SCH, the graded version substantially diminishes the DOS in the intermediate energy region, since the levels are more spaced, whereas on the photonic side, the reduction of waveguide confinement when 'bevelling' the square-index landscape of the DH is not too severe, often ensuring an overall advantage for the GRIN-SCH.

When it comes to commercial devices, of course, there may be too much difficulty in producing the GRIN version, and a SCH with a few QWs (a MQW) is often a good and popular compromise.

To conclude, it is rightly an enticing thought that the workhorse of an optical technology, here the laser diode, can be broken into such fundamental quantities as the electronic DOS and guided wave confinement factor of its constituents. This example should thus be kept in mind when examining the merits of the various nanophotonics approaches invoked elsewhere in this book.

8.A Complement: Band structure of zinc-blende materials

Fig. 8.18 pictures key aspects of bands for the cubic zinc-blende crystal structures, to which GaAs and AlAs belong (but not GaN, which is usually of hexagonal wurtzite form): energy level diagram, element classification, and a sketch of their band structures that outlines direct (GaAs) or indirect (AlAs) bandgaps and their similar VB structure (see Fig. 8.18 caption).

8.B Complement: Absorption beyond the basic sharp-band-edge picture

8.B.1 Excitons

An important deviation met to the basic sharp-band-edge picture of absorption and JDOS (hence in 3D $\sqrt{\hbar\omega - E_g}$ if $\hbar\omega > E_g$ or 0 if $\hbar\omega < E_g$) is due to the so-called *excitons*: we have used uncorrelated Bloch states for the electron and holes, but these two particles interact and can form a bound state—the excitons. Much like the electron and the proton in the hydrogen atom, their energy comprises the attractive interaction. Here, it can thus be less than the minimum energy of a free electron–hole pair, thus $E < E_g$. This plays a role mostly at the singular point of the spectrum, namely the absorption edge at $\hbar\omega = E_g$, which acquires extra features. The binding energy of excitons is comparable with that of shallow donors, typically 10 meV (see next paragraph), so the spectrum of pure samples is often compatible with an hydrogenoid model at low temperatures (say, 77 K or 4 K). But details are smeared at room temperature or due to impurities (see Complement 8.B.2). Nevertheless, even at room temperature, the actual absorption curves of bulk semiconductors and, even more so, of QWs (whose binding energy for excitons is larger due to the extra confinement), has a clear bump just after the edge, explained by excitonic effects. It makes the square-root shape less obvious (see, for instance, the closely related excitation spectrum of a colloidal quantum dot: Fig. 9.8b), but the trends in absorption values are still correct.

Let us take this opportunity to bridge this chapter with the next, dealing with more confined systems. If, in a solid, a potential well captures a delocalized electron, the well becomes negatively charged and attracts opposite charges, such as holes in semiconductors. This is true in organic molecules of some size as well: there are many orbitals and they can be grouped in terms of energy, much as in a semiconductor. The role of the CB is played by the so-called LUMO (lowest unoccupied molecular orbital) and the role of the VB by the HOMO (highest occupied molecular orbital). Yakov Frenkel indeed developed the concept of the exciton in 1931, and the Frenkel exciton nowadays relates to a version with localized orbitals where modest screening prevents the orbitals to extend over many atoms, while Gregory Wannier and Sir Nevill Francis Mott extended the exciton concept to a delocalized version in covalent solids in 1937. In usual semiconductors (Si, III-V, II-VI), the strong dielectric constant

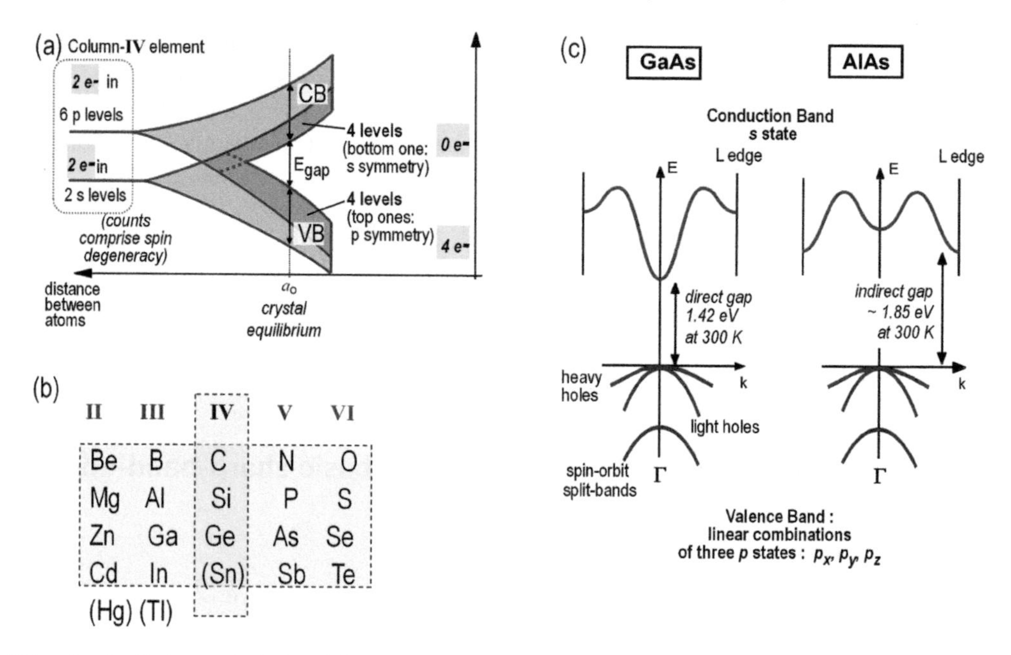

Fig. 8.18 GaAs and AlAs: (a) band formation from s and p states and the corresponding electrons in them; notice that the actual equilibrium lattice parameter is beyond the crossing point. Thus, upon thermal expansion, the parameter increases and the gap diminishes; (b) the part of the Mendeleiev classification relevant to III-V and II-VI compounds. These compounds generally have a local sp^3 arrangement and have either the zinc-blende cubic structure (GaAs, AlAs, InP) or the hexagonal wurtzite structure (GaN, AlN); (c) band structures of GaAs (direct gap at the Γ point) and AlAs (larger indirect gap at the L point of the actual 3D Brillouin zone edge).

is key to screening the Coulombic interaction and reduces the binding energy down to the ~10 meV range, a crucial feature to support Wannier excitons. Specifically, the two ingredients of the hydrogenoid model of the exciton that enter the modified Rydberg energy $R_X = \mu Ry/\varepsilon_r^2$ (Ry=13.6 eV) are: (i) the reduced effective mass $\mu = m_c^* m_v^*/(m_c^* + m_v^*)$, close to the lighter electron mass ratio m_c^*/m_0; and (ii) the screening term ε_r^{-2}, using the static constant, hence often less than 10^{-2}. Hence the total reduction of Ry from 13.6 eV to ~10 meV, which is accompanied by an increase of the Bohr radius ($a_0 = 0.0529$ nm) by a factor ε/μ bringing the exciton radius a_X to the 10 nm range.

Carrier localization by impurities in this context resembles doping, and rules for ionization can be deduced from those given in section 8.2.5 (see also just below in 8.B.2).

8.B.2 Doping and absorption

Doping acts by providing more free carriers, which mostly behave as a Drude gas. Empty donors can also accept electrons promoted from the VB and give rise to subbandgap absorption. All these phenomena are heavily temperature-dependent, since the degree of ionization of donors or the carrier mobility are strongly dependent on temperature, the former through Fermi–Dirac statistics, the latter mostly through phonon scattering. The overall result is that the band-edge singularities are smoothed in doped semiconductors. Excitons at low temperature have a more complex fate, as they can interact with impurities, a feature that greatly helped the study of the physics of impurities in semiconductors.

8.B.3 Phonons

When we put forward the role of the vertical transition rule, we actually considered it the main mechanism. However, it is also possible to envision absorption or emission processes that involve one or several phonons. But these are higher-order processes that are in general weaker by orders of magnitudes, unless clever observations and configurations are selected.

Phonon energies in semiconductors are typically $\hbar\omega_{phon} \sim$ 20–50 meV. The general effect of such phonons is to allow transitions below the bandgap and to smooth singularities just on this energy scale, a scale that is quite comparable to $k_B T$ at 300 K. Note that even at zero temperature, it is possible for a photon to create a phonon and an electron–hole pair.

8.C Complement: Silicon optical properties

Silicon is not a direct bandgap semiconductor; nor is germanium, the next column IV element. The silicon bandgap at 300 K is about 1.1 μm, or nearly 1.1 eV (the product is $hc/e \simeq 1.24$ eVμm), the germanium one being 0.66 eV.

The silicon VB top is at point $\Gamma(k = 0)$. But the conduction valleys lie at a wavevector k close to the edge of the first Brillouin zone (FBZ), along the 6 ΓX FBZ directions.

Another indirect gap arises at about 2 eV, with valleys lying at the eight L points. The direct band gap at Γ point is large, about 3.5 eV (near ultraviolet). So, for any photon with $\hbar\omega < 3.5$ eV, an extra *phonon* is needed to bridge the VB electron wavevector at Γ and the CB one at some point far towards the X or L edges. This causes the absorption curve to rise very smoothly from 1.1 eV on.

Subsequently, the characteristic value $\alpha \sim 10^4\,\text{cm}^{-1}$ of III-V semiconductors at $\hbar\omega = E_g + 100$ meV becomes much lower in the near-infrared and red part of the spectrum. The $10^4\,\text{cm}^{-1}$ value is reached for green light (520 nm), whereas at 1.22 eV (1020 nm) the absorption is only $\alpha \sim 80\,\text{cm}^{-1}$. An important consequence is that a silicon solar cell aimed at a good absorption of 1000 nm radiation in a single pass requires a thickness of $\gtrsim 100\,\mu$m, thus fitting a typical wafer thickness. But it brings many extra difficulties, e.g. ensuring collection of both electron and hole carriers across such large distances.

Conversely, in the near-ultraviolet, silicon absorbs in about 10 nm thickness only. Carrier collection is equally an issue in such conditions. Fortunately, for mainstream sensors in the visible, the absorption is in the range of a few microns, so ordinary 'p-i-n' photodiode

Exercises

Exercise 8.1 Chain of identical atoms with neighbour interaction. In a chain of M atoms $j = (0, ..., M-1)$ separated by distance a along axis x, consider on atom number j an orbital $|j\rangle$ of energy E_0 when the atom is isolated: $H_0 = E_0|j\rangle\langle j|$ is the single-atom Hamiltonian. The interaction between neighbours gives rise to an energy term $\langle j-1|H|j\rangle = \frac{C}{2} = \langle j|H|j+1\rangle$, and we have a ring, so $\frac{C}{2} = \langle (M-1)|H|0\rangle$.

(1) Verify that $|\psi_k\rangle = M^{-1/2}\sum_0^{M-1}\exp(ikja)|j\rangle$ is a Bloch function of the form $u_k(x)\exp(ikx)$ by adequately forming and dispatching $\exp(ikja - ikx)$.

(2) Show that the energy $\langle\psi_k|H|\psi_k\rangle$ involves a factor $\cos(ka)$. Thus you have the key ingredient of the band structure. Check the band energy width $2C$.

(3) What ingredients would be needed for dispersions other than a pure cosine shape?

Exercise 8.2 Density of states (DOS) in 1D and 2D. Use the adequate expression of V_{sph} to find out the DOS in 1D and 2D for dispersions of the form $E = Ak^s$ for $s = 1$ and $s = 2$. Check $s = 2$ using the schemes found in Fig. 8.2b.

Exercise 8.3 Effective mass and electron dynamics.

(1) We consider an electron of velocity v_x along x, subjected to a force F_x during time δt. Express the energy δE it has gained.

(2) Using the relation of group velocity $v_x = \hbar^{-1}dE/dk_x$, relate F_x and dk_x/dt.

(3) Now we study $\frac{dv_x}{dt}$. Differentiating v_x above, you can introduce $\frac{d^2E}{dk_x^2}$. Using the first result, you can now prove that $F_x = m^* \frac{dv_x}{dt}$ with $m^* = \hbar^2 \left(\frac{d^2E}{dk_x^2}\right)^{-1}$. This means that m^* does describe the response to external forces in the presence of forces due to the lattice potential.

Exercise 8.4 DOS 3D and 2D comparisons.

(1) Factor the constants in section 8.2.1 to give the exact coefficient for a band of electrons in 3D, starting at $E_c = 0$, of effective mass m_c^*, taking into account the factor of 2 for spin degeneracy.

(2) Deduce the electron density (in cm^{-3}) N_{100} in a range of $[E_c, E_c + E_0]$ with $E_0 = 100\,\text{meV}$ for $m_c^* = 0.07\,m_0$. Compare to the density of atoms of GaAs, 4.43 $10^{22}\,\text{cm}^{-3}$.

(3) Perform the same comparison for 1 cm^2 of the first subband of a QW of identical mass, for the same 100 meV range, yielding N_{100}^{2D}.

(4) Check that the ratio N_{100}^{2D}/N_{100} is a length $L \sim 10\,\text{nm}$. Calculate the quantized energy levels of the infinite well for this length, $\hbar^2(\pi/L)^2/2m_c^* \times (1, 2, \ldots)^2$ and interpret the result with the help of Fig. 8.13.

Exercise 8.5 Degenerate regime, density, and de Broglie wavelength λ_{dB}. Use Exercise 8.2 to get the constant DOS g_c of 2D electrons of effective mass m_c^* without spin. What is the population per unit area for $E_c = E_F$? Deduce the mean interdistance particle $\langle d \rangle$ at this limit case. For particles with two degrees of freedom, the de Broglie thermal wavelength $\lambda_{dB} = 2\pi\hbar/p$ is derived from the 'thermal p' defined by $E = 2\frac{k_BT}{2} = p^2/2m_c^*$. You should find $\langle d \rangle \sim \lambda_{dB}$.

Exercise 8.6 Equivalent densities.

(1) Consider a CB with DOS $g_c(R)$ and population density $N = \int_{E_c}^{\infty} g_c(E) f_{\text{FD}}(E)\, dE$. Using the Boltzmann-like expression for $f_{\text{FD}}(E)$ in the regime of low enough Fermi energy, $(E_c - E_F) \gg k_BT$, show that N can be put to a form equivalent to a single level at E_c with degeneracy N_{eq} given by $N_{eq} = \int_{E_c}^{\infty} g_c(E) \exp(-[E - E_c]/k_BT)\, dE$.

(2) Formulate the result in the ideal 3D case ($g_c(E) \propto E^{1/2}$) as a function of $(k_BT)^{3/2} \times$ a tabulated dimensionless integral, $\Gamma(3/2) \simeq 0.886$.

(3) For an ideal 2D case $g_c(E) =$ constant, the integral is easy; show that $N_{eq}^{2D} = g_c k_BT$.

(4) In the 2D case, the general form $N = \int_{E_c}^{\infty} g_c f_{\text{FD}}(E)\, dE$ is also calculable. Find its expression and its two asymptotic forms depending on whether $(E_c - E_F) \ll k_BT$ (which should be consistent with the above) or $(E_F - E_c) \gg k_BT$. Also check that for the case $(E_c - E_F) = 0$, $N = g_c \ln(2)$. Interpret the latter result as band-filling analogous to the Fermi energy in metals.

Exercise 8.7 Doping and Fermi level for different bandgaps. The net doping densities $|N - P|$ of Fig. 8.4 are calculated based on the two parameters E_F and T using integrals of $g_c \times f(E - E_F, T)$ and $g_v(E)f(E - E_F, T)$ (f = Fermi–Dirac function) on both bands up to 1 eV away from band edges, and asking MATLAB for contours.

(1) By analytical means, adapt all values to $E_g = 1\,\text{eV}$ instead of $E_g = 2\,\text{eV}$, keeping the same masses, m_c^* and m_v^*.

(2) What is the influence of a larger mass ratio, say $(m_c^*/m_v^*)^{3/2} = 0.05$ instead of 0.10?

(3) Make a script to compute $|N - P|$ on a fine mesh, draw its contours to check the above.

Exercise 8.8 Joint DOS. Note that Eq. (8.20) is similar to the dispersion of electrons if one reads it as $\hbar\omega = \hbar\omega(k)$.

Express the resulting JDOS $g_{\text{joint}}(E)$, in line with the approach for electrons. This is the quantity entering the Fermi Golden Rule, given the energy conservation term. Check which of m_c^* or m_v^* dominates μ in a III-V standard semiconductor.

Exercise 8.9 Transparency calculations. Consider a single QW and only its two $n = 1$ subbands, the conduction band (CB) and valence band (VB), with edges denoted by E_c and E_v and separation E_g. The effective masses are supposed to differ by a typical factor of $u = m_v^*/m_c^* = 10$. Populations are given by two distinct quasi-Fermi levels, $E_{F,c}$ and $E_{F,v}$.

(1) Make use of the results of Exercise 8.6 to find $E_{F,v}$ when $E_{F,c} = E_c$ (0.5 band edge occupancy), and similarly to find $E_{F,c}$ when $E_{F,v} = E_v$, using neutrality $N = P$. What should the ratio of the population of the first case to the population of the second case be?

(2) Tabulate or plot a few values of band-edge occupancies $f_c(E_c)$ and $f_v(E_v)$ for a couple of intermediate occupancy cases. Forming the balance term $f_c(E_c) - f_v(E_v)$, characterize the transparency conditions in terms of both bands: what is $N_{\text{transp}}/(g_c k_B T)$ (hints: it is unity when $u = 1$, and it scales sub-linearly with u), and what are the individual values $f_c(E_c)$ and $f_v(E_v)$? Also check that the Bernard–Duraffourg condition, Eq. (8.41), can be properly applied.

Exercise 8.10 Quantum well (QW) basics.

(1) Calculate the energy difference between the first few adjacent levels of an infinite QW of width L with mass $m^* = 0.07m_0$ and $m^* = 0.7m_0$. Which of these cases describes a typical III-V CB, and which a typical III-V heavy-hole VB?

(2) Let E_{12} be the energy difference of the two first levels of the CB. What should L be to ensure that in the Boltzmann regime (E_F well below E_C) at 293 K, the ratio of the subband edge occupation is $\exp(-4) \simeq 0.018$? What, then, is the ratio of the two subband populations (see Exercise 8.6)?

(3) What is the new L value for a similar request on hole population in the heavy-hole VB?

Exercise 8.11 QW levels (square well, even case). We consider a square well with bottom potential $V = 0$ and a barrier at V_B, of width L, and an electron of mass m^* inside. We look for even envelope wavefunctions $F(x)$. We write $F(x) = A\cos(k_w x)$ in the interval $[-\frac{L}{2}, \frac{L}{2}]$, and $F(x) = A'\cos(k_w \frac{L}{2})\exp(-K[x - \frac{L}{2}])$ for $x > \frac{L}{2}$ and $F(-x) = F(x)$.

(1) Relate k_w and K to the energy E of the electron.

(2) Justify the way we wrote $F(x)$ in the barrier by establishing A', thanks to the continuity of $F(x)$ at $x = \frac{L}{2}$.

(3) $\frac{dF}{dx}$ must also be continuous (flux conservation). Deduce a relation between K and k_w and show that $\tan(k_w \frac{L}{2})$ can be written as a function ζ_B of k_w, m^* and V_B.

(4) The solution to the above often requires graphical (k_w solves $\tan(k_w \frac{L}{2}) = \zeta_B(k_w)$) and related methods. Identify the limit cases so as to find the number $n \geq 1$ of confined even levels.

(5) Show that the adaptation to the odd-level case $F(-x) = -F(x)$ simply amounts to $\tan \to \cot$. Plot both families together with ζ_B.

Exercise 8.12 Confinement factor, transparency, and gain for single QW and multi QWs (MQWs).

(1) Consider a rather broad SCH optical cavity (akin to a DH) with $L_{DH} = 1\ \mu$m. We insert a single QW of width $L = 8$ nm at the antinode. What is the confinement factor Γ as defined in Eq. (8.43)?

(2) Going to a MQW by stacking 8 nm wells with 8 nm barriers, how many wells can be valuably added around the guided mode antinode, retaining as a criterion that the worst well should contribute at least 83.3% of the central one? What is roughly the resulting confinement factor of the MQW ensemble?

(3) Assuming a transparency current of 100 A/cm^2 for each well, how does the total transparency current compare with the optimal DH case, 2–3 kA cm^{-2}?

(4) Using inspiration from Fig. 8.15, compare how the *photon energy of maximum gain* evolves in the MQW case vs a DH case with a $\sqrt{E - E_g}$ JDOS.

Exercise 8.13 Decay after excitation; radiative and Auger recombinations. We start from a situation with $N(t)$ and $P(t)$ as the carrier densities and two quasi-Fermi levels. The decay dynamics of a population of electron–hole pairs is described in many devices by the following equation:

$$\frac{dN}{dt} = -(AN + BN^2 + CN^3). \tag{8.46}$$

The exercise is aimed at justifying the various terms.

(1) Express $\frac{dN}{dt}$ with an integral over energy and $\frac{df_c(E)}{dt}$, and do the same for holes, noting the hole occupancy, $\bar{f}_v(E) = 1 - f_v(E)$.

(2) Use the vertical selection rule to relate electron and hole energies to a single photon energy, $E_{\text{photon}} = \hbar\omega$.

(3) In the regime of weak occupancy, write the integral of each photon energy slice with the $\exp(-(E - E_{F,c/v}/k_B T)$ forms of f factors. In the spirit of Exercise 8.6, deduce that $\frac{dN}{dt} \propto -f_{0,c} f_{0,v} \times I_p$ where the occupations are those of band edges, and what remains can be seen as the time-independent integral I_p. Deduce that $\frac{dN}{dt} = -BNP$ with a constant B.

(4) In the case of negligible doping, why does this lead to the $-BN^2$ term?

(5) In the case of strong doping, there is a large equilibrium population of, say, holes $P = P_0$. We cannot neglect it, unlike what we did above (where the equilibrium population was the very weak intrinsic value; see Fig. 8.4). In principle, one should rewrite the equations with the departures from equilibrium ΔN, Δf, etc. But we can assume that we are in a regime $P \simeq P_0$. Show that this accounts for the AN term.

(6) In which order are the AN and the BN^2 term appearing after an intense short pulse and for, say, $P_0 = 10^{17}\,\text{cm}^{-3}$?

(7) At densities near or above transparency, what is the trend of *spontaneous* emission?

(8) Discuss the kinds of defects or impurities that can cause *non-radiative channels* with the decay laws $A_{nr}N$ and $B_{nr}N^2$.

(9) At high densities, the *Auger* recombination mechanism accounts for the CN^3 term. It is the main contender for the 'droop' of the efficiency of blue GaN-based LEDs at high current densities (hence high e–h pair densities). It also limits the performances of telecom lasers at 1550 nm, among others.

An example Auger process is $(e^-)_g + (e^-)_g + \text{hole}_g \longrightarrow (e^-)_u$. Suffix X_g means a carrier close to band edge (usual CB and VB populations), while suffix X_u means (some) upper level.

By applying energy conservation, show that with an adequate energy of $(e^-)_u$, there is no need for a photon in this process. What constraint does momentum conservation bring? By considering the case of $(e^-)_g + \text{hole}_g + \text{hole}_g$ and the complexity of the VB, propose which Auger channels are likely to be the strongest. Justify from the above 'chemical reaction' the cubic dependence on N (or P).

9

More confined electrons: Quantum dots and quantum wires

9.1 Electronic description of quantum dots (QDs) and quantum wires

9.1.1 Tailoring optical properties by confining electrons at nanoscale in 2D and 3D

We have seen in Chapter 8 how confinement of electrons along one dimension, the quantum well, could radically alter their density of states (DOS). One direct influence of confinement is the transition energy, whose dependence on size is extremely welcome, since spectral properties (such as fluorescence emission) can be adjusted by size rather than through the more delicate choice of alloy composition. We were also reminded that this DOS is an important handle for a series of key photonic and optoelectronic properties, related, broadly speaking, to device efficiency. The advent of low-threshold quantum-well laser diodes in the late 1970s and early 1980s comes mostly from a reduction of the threshold current *density*, an *intensive* quantity. This kind of improvement is distinctly different from the trend of miniaturization accompanying Moore's law, whereby the diminishing energy and switching delays of electronic elements scale *extensively*, like the length or surface of the device.

Therefore, it seems desirable to go further in terms of electronic quantization by considering quantum dots (QDs) and quantum wires. As synthesis methods of QDs, notably, can be simple, the size dependence of spectral properties can be leveraged even more easily than in delicate epitaxial growth of quantum wells (QWs). We start with a description of the effect of the extra confinement dimensions in generic terms in this section. We also give some hints on the reasons why quantum wires are not so easily implemented. We will consider in detail in the next two sections (sections 9.2 and 9.3) two kinds of QDs that have become the most popular: self-organized indium arsenide (InAs) QDs obtained by epitaxy techniques and colloidal II-VI nanocrystals (CdS, CdSe, ...), which are more easily obtained. We eventually emphasize differences between QD emitters and organic emitters for visible or near-infrared emission.

The first Complement, 9.A, depicts the specific growth mechanism of InAs QDs, and a second one, 9.B, gives two instances of silicon-related nanostructures: Ge-Si systems and porous silicon.

Introduction to Nanophotonics. Henri Benisty, Jean-Jacques Greffet, and Philippe Lalanne, Oxford University Press.

DOI: 10.1093/oso/9780198786139.003.0009

9.1.2 Quantum dots

The simplest way to describe confined electrons in a small particle of semiconductor is to start from a picture of non-interacting electrons in a given band and confine them. We start with electrons in a conduction band (CB), where they have a positive effective mass m^*. We first assume that they lie in a box with hard walls, $V = \infty$, of dimensions L_x, L_y, L_z along the three Cartesian axes, with a convention for the potential energy $V(x, y, z)$ such that $V = E_c$ inside the box. Here E_c is the CB edge level. This picture holds because of the envelope function approximation used for the quantum well.

Then, since the potential for electrons is separable (one can write $V(x, y, z) = W_x(x) + W_y(y) + W_z(z)$), the envelope wavefunction is a product of three functions $\psi(x, y, z) = \psi_x(x)\psi_y(y)\psi_z(z)$ along the three axes. These three functions are just the cosine or sine functions seen in the infinite well because the potential is infinite and separable. With the same convention of a centred box extending from $-L_x/2$ to $L_x/2$, we have, for instance, along x for odd-numbered levels:

$$\psi_x(x) = \sqrt{\frac{2}{L_x}} \cos\left(\frac{\pi l x}{L_x}\right) \quad \text{for } l = 1, 3, 5, \cdots \tag{9.1}$$

For even-numbered levels ($l = 2, 4, 6, \cdots$), just replace the cosine by the sine. Energy levels are then numbered by three quantum numbers l, m, n. They can be written as follows:

$$E_{l,m,n} = \frac{\hbar^2\pi^2}{2m^*}\left(\frac{l^2}{L_x^2} + \frac{m^2}{L_y^2} + \frac{n^2}{L_z^2}\right) . \tag{9.2}$$

We obtain a set of discrete levels, much as in atoms. Compared to the Born–von Kármán periodic boundary conditions used to compute the DOS, the present $V = \infty$ boundary condition is a Dirichlet condition for the wavefunction (ψ_x must cancel). Both give very similar expressions for energy, but the DOS calculation was done to deal with dense levels, hence high quantum numbers l, m, n, whereas here we are interested in low quantum numbers l, m, n of order unity.

We have also made an important choice in our description if we compare it to the familiar description of atoms around a nucleus with spherical symmetry: the potential felt by the electron was taken as independent of other electrons. In other words, we have no Coulombic repulsion between electrons in this simple model. It is therefore adapted to a small (or even a single) electron population in a box, and cannot be valid for a number of carriers that is too large. However, for a large number of carriers, a knowledge from atom physics would point to a symmetry-related organization of the levels, with the filling of successive orbitals $(s, p, \cdots)$: we would not be in unknown territory.

9.1.3 Role of the shape

From the scaling $1/L^2$ of level spacing, we can guess that the ratio of the three dimensions of the dot matters a lot. If one dimension is smaller than the two others, its quantization energy is much larger. In Fig. 9.1, we compare the quantized levels of (a) a quasi-cubic dot to (b) those of dots with platelet shape or (c) wire shape. These are the main remarkable categories within parallelepipeds other than the arbitrary one: that two sides or a single side are larger than the remaining ones. If we had started from

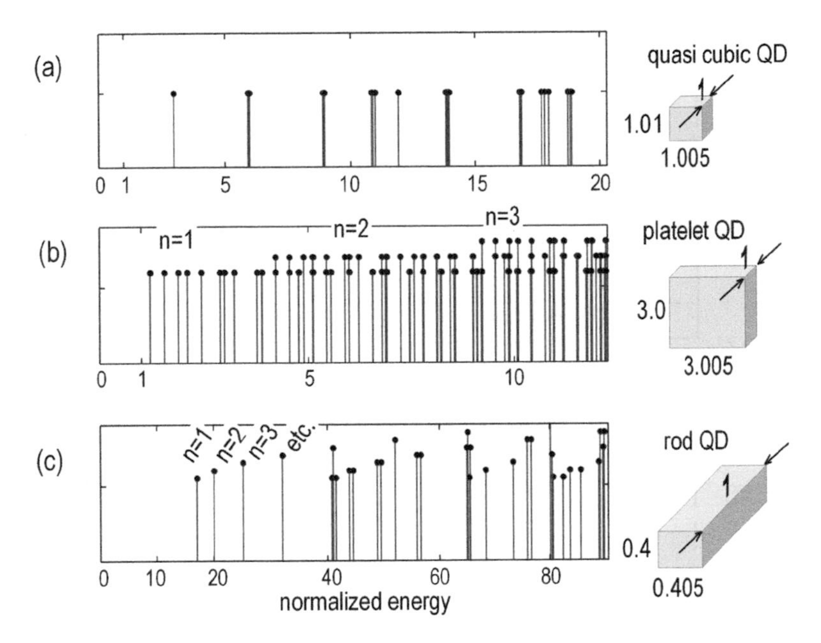

Fig. 9.1 Typical level arrangements for exemplary quantum-dot (QD) shapes. The parallelepiped dimensions are parametrized as $(\frac{L_x}{L} : \frac{L_y}{L} : \frac{L_z}{L})$, whereby energies $E_{l,m,n}$ of Eq. (9.2) are scaled to $\hbar^2\pi^2/(2m^*L^2)$. (a) Nearly cubic QD (1.01:1.005:1); (b) platelet-shaped (3:3.005:1); (c) rod-shaped (0.4:0.405:1). Note that the platelet and rod are two categories of affine distortions of the cube, much as canonical affine distortions of the sphere yield, respectively, oblate and prolate ellipsoids.

a sphere and had made a comparable affine transform, we would obtain the so-called oblate and prolate ellipsoids. The first excited levels are then of a 2D or 1D nature–i.e. one can forget about the small dimension(s) in describing the first few levels of the dots. Other typical shapes entering this category are flat, round discs. If two dimensions are small and the third is large, then we have a range of energy where the excited levels are only associated with the large dimension. Indeed, if we have a very elongated cylinder or rod, whatever its cross section, we are approaching the quantum-wire case: the remaining long dimension has much denser modes than the two others; these modes may be considered as a continuum.

In practice, the separation of levels also has to be compared to the thermal energy k_BT, which is about 25 meV at room temperature (see Fig. 9.2). The same was true for the passage from the double heterostructure (DH) to quantum well (QW) in Chapter 8. If there are many states within such a k_BT range associated with increments of one of the three quantum numbers l, m, n, the corresponding direction can be considered as a continuum. With this temperature criterion, therefore, when separations are too small, it is rather useless to consider a box as a QD. Conversely, for a very small box, the levels have large separations. If we replace the idealized potential barrier $V = \infty$ with an actual finite barrier height V_{barrier}, it might be that even the first excited level gets cut

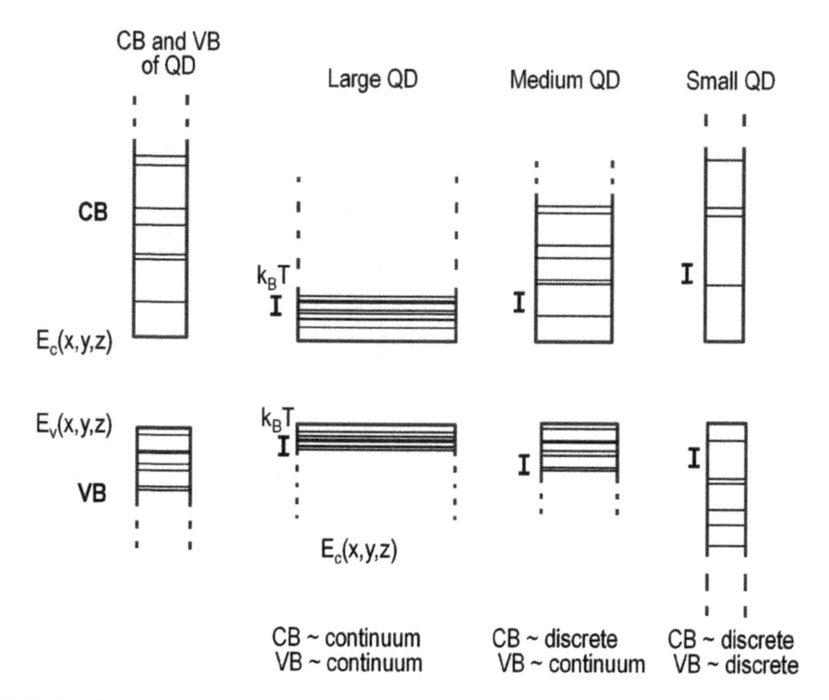

Fig. 9.2 Role of temperature: (a) generic levels in the valence band (VB) and conduction band (CB) of a quantum dot (QD), with heavier mass in the VB; (b) for a large QD, both bands look like a continuum. (c) For some medium size, the spacing of CB levels becomes larger than k_BT, thus justifying a discrete description, but not the VB spacing; (d) for a small enough QD, both bands behave as a set of discrete levels.

off, it would lie above the barrier V_{barrier}. Therefore, the DOS in a band, while formally consisting of a set of discrete Dirac states, $g(E) = \Sigma_j \delta(E - E_j)$, can exhibit a variety of patterns as regards the actual observation of quantization at a given temperature.

9.1.4 Confinement of holes and confinement of electrons

The description made above for electrons is extended to holes in the VBs in much the same way as was done for QWs in Chapter 8. A negative mass has to be used or, in other words, the electron confinement energies are negative. There is more complexity for the VB structure, so that a single negative mass is not enough for a description of a generic QD.

However, for many popular semiconductor QDs (notably those discussed in this chapter), the *heavy-hole* band is the most important one, as it determines the fundamental level in the sense of the VB, and it may also determine the next excited level. Then, the spacing of the modes is dictated by the specific heavy-hole mass m^*_{hh}, which is typically 10 times heavier than the CB effective mass m^*_c in most III-V semiconductors.

Therefore, a situation can arise for a particular QD whereby quantization and confinement have to be considered for electrons, but due to the smaller spacing of hole levels, it can be neglected in the VB: a continuum description can suffice for them (see also section 9.1.6, where this assertion is revisited along the different confinement axes).

9.1.5 Excitons, QDs, and molecules

Excitons are a delicate topic that we evoked in Complement 8.D.1. A QD is a potential well for both carriers, and thus can produce charging effects when capturing delocalized carriers, which in turn favours capture of an opposite charge. The common presence of an electron and a hole in a QD is therefore an all-too-natural effect. The neutral situations always have an energetic advantage. So, expanding the picture, there are biexcitons when two electrons and two holes are captured altogether, and so on. Unbalanced situations can also be termed 'charged excitons', e.g. those with two electrons and one hole, and so on (these are also favourable situations for the Auger effect, a non-radiative recombination mechanism; see Exercise 8.13).

The binding energy that comes into play between free particles and a bound electron–hole system can be devised in the same way as the binding energy of the hydrogen atom. In Fig. 9.3, we give an extremely simplified and pictorial description of a hydrogen-like electron–hole pair, and compare it to an electron bound to a fixed positive charge and, further, to the QD case (see legend).

It is useful here to still have the hydrogenoid model in mind and exploit its mass and dielectric constant dependencies, μ/ε_r^2 for the Rydberg and ε_r/μ for the Bohr radius;

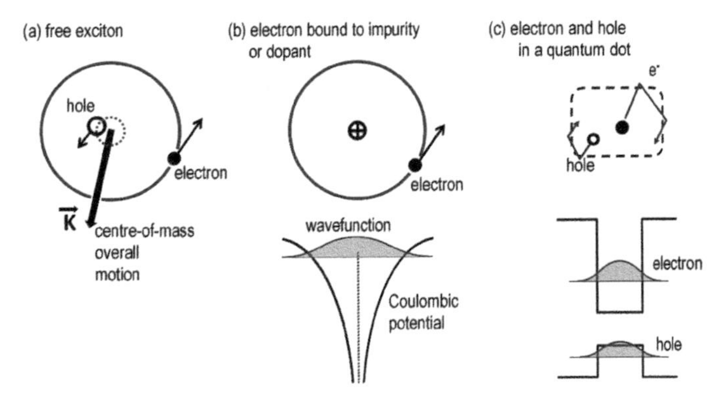

Fig. 9.3 Excitons and quantum dots (QDs): (a) pair formed by an electron and a hole analogous to a hydrogen atom, having its overall centre-of-mass motion. This can be referred to as a Wannier exciton; (b) electron bound to a positively charged impurity. The simple Coulombic picture applies, with a potential and a wavefunction sketched below. However, the radius of the trajectory (or, equivalently, the size of the wavefunction) depends on the dielectric constant of both the background and the electron mass, which differ greatly according to whether the context is the impurity of a solid or an electron in vacuum bound around, say, a heavy nucleus; (c) in the QDs considered here, the confining potential comes from the heterostructure, which is rendered by the different levels of the bands due to the different atomic nature of the lattice. The Coulombic attraction between an electron and a hole is secondary for confinement purposes.

they can both vary a lot between narrow-gap column-IV elements (Si, Ge) and wider gap II-VI compounds such as ZnO. If the confinements of the electron and hole are both much tighter than the Bohr radius, the potential well of the QD plays a larger role than the electrostatic effects. On the other hand, in a QD larger than the Bohr radius (a situation more likely for a small Bohr radius), an electron and hole interact more together than with the dot boundaries

As a conclusion to this short mention, it is hard to avoid the exciton concept when going any deeper into QD physics. It can be quite subtle, and the use of the word itself (vs just 'electron–hole pair') is not standardized: extrapolating from the discussion made above on the different forms of confinement of electron and holes, and adding the physics of Coulombic repulsion in combination with the confinement by barriers, no short description can hold.

9.1.6 Quantum wires

We have seen in Fig. 9.1 that the aspect ratio of a parallelepipedic box would greatly affect the resulting energy spectrum due to the $1/L^2$ scaling of the energy spacing. A wire can thus be obtained by elongating a dimension of a box to such a point that none of the relevant energy spacing between adjacent modes can be resolved within the range of interest (in Exercise 9.8, it is seen that the n^2 scaling (along z) of energy levels E_n results in a 'local' spacing around energy E_n of $(E_n E_{\text{conf}})^{1/2}$, with $E_{\text{conf}} = \hbar^2\pi^2/(2L^2 m^*)$ the confinement energy along the long axis). Adopting such a continuum description along the elongated dimension, the DOS takes the shape of successive 'spikes' arising at each of the confined levels in the two x, y confined directions. These levels are, for the infinite well approximation,

$$E_{l,m} = \frac{\hbar^2\pi^2}{2m^*}\left(\frac{l^2}{L_x^2} + \frac{m^2}{L_y^2}\right). \tag{9.3}$$

The DOS for a single 1D branch $E(k_z) = \frac{\hbar^2 k_z^2}{2m^*} + E_{l,m}$ was deduced in the generic calculation of Chapter 8 (see Fig. 8.2):

$$g(E) = \frac{2\pi}{L_x}\left(\frac{2m^*}{\hbar^2}\right)^{1/2} (E - E_{l,m})^{-1/2}\ \Theta(E - E_{l,m}), \tag{9.4}$$

where Θ is the Heaviside unit step function and a factor of two for the spin degeneracy is included. With this result, we can now picture in Fig. 9.4 the generic shape of the DOS of a semiconductor quantum wire, with the simplifying assumption of a single kind of hole in the VB, usually the heavy holes.

Our discussion of the continuum and the DOS shape holds for the VB. It is essentially a mirror of the CB situation, thus with the continuum in energy extending towards the low-energy side of the energy axis from each l, m confined level. But we must take into account the heavier effective mass. The various subbands then appear with a smaller separation compared to Eq. (9.4) but only a modestly higher prefactor ($m^{*1/2}$ scaling of g) on the VB side.

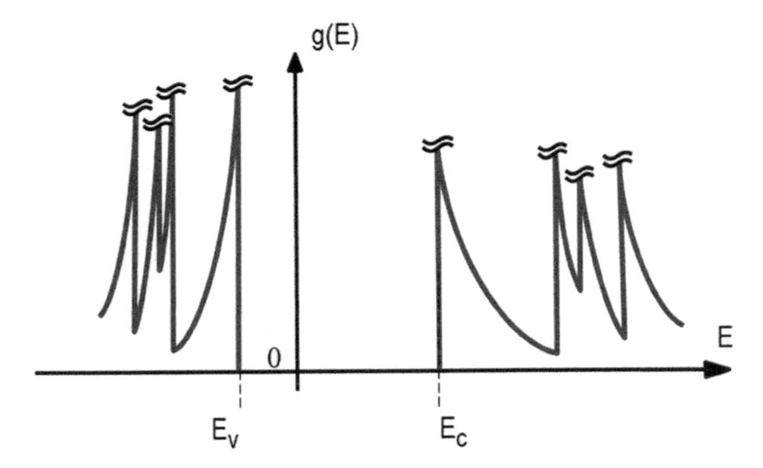

Fig. 9.4 Generic shape of the density of states for a semiconductor quantum wire. The peaks formally diverge. Note also that here we have denoted the quantum wire band edges (not the underlying bulk semiconductor band edges) by E_v and E_c

9.1.7 Carrier capture among wells, wires, and dots

When faced with the issue of actually feeding electrons or holes to dots or wires, we will see that very often, carriers are not injected directly from a metallic contact into these nanostructures. Much more commonly, one proceeds from known situations for contacts, whereby a metallic contact is made onto a layer of sufficient thickness to avoid quantum effects, typically thicker than 200 nm. Injected carriers are then first captured in adequate QWs, and only then, by a next step, in wires or dots. All this involves a downward energy cascade at the same time as a gradual freezing of degrees of freedom.

There are thus two issues used to describe the fate of carriers. Here we address the first issue: the capture from a higher-dimensional to a lower-dimensional structure. In section 9.1.8, transport along the wire itself is discussed. Granted, the transport issue formally arises for QWs, but it is not critical, resulting in standard conduction diffusion behaviour.

The crucial point is that the capture process requires an energy-loss mechanism to dive into the well. Much as was argued in the photoluminescence description in Chapter 8, phonon emission provides such a mechanism at any density. The Auger process, which involves two carriers, one of which is excited by taking advantage of the energy loss of the other, is another intraband mechanism liable to act at higher densities. However, some care must be exerted when comparing similar mechanisms in various dimensionalities. The scarcity of final electron states and the change from a continuum to a single state in a dot case also means that the mechanism has to comply with the corresponding restrictions. Broadly speaking, here we are tackling the two issues of (a) cross section—, i.e. how well does the mechanism work for a given set of impinging carriers and (b) selection rules.

Selection rules derive from three kinds of consideration: energy conservation, momentum conservation, and symmetry. The two latter considerations appear in the Fermi Golden Rule through the matrix element.

We have already incorporated energy conservation by saying, for instance, that phonon emission ensures energy loss of absolute value E_{if}: the energy lost is that of a phonon, typically $\hbar\omega_{\text{phonon}} = 15$–$50\,\text{meV}$ for optical phonons.

Momentum conservation is less intuitive. The phonons of interest for small energy losses are acoustic phonons. A first approximation is to consider these phonons as bulk ones characterized by an effective sound velocity, c_s. For nanostructures such as epitaxial InAs QDs, the c_s value of the surrounding GaAs is an acceptable start. Such embedded QDs do not localize these low-energy phonons because their wavelength λ_s is too large at the relevant energies E_{if}. Taking $\lambda_s = 2\pi/q$ and the wavevector q as $q = \omega/c_s = \hbar\omega/\hbar c_s = E_{if}/\hbar c_s$, we simply get $\lambda_s = hc_s/E_{if}$. Numerically, for $E_{if} = 5$ meV and a relatively slow $c_s = 3\,000\,\text{m/s}$, we get $\lambda_s \simeq 25$ nm, which is larger than the typical size of a QD of, say, $L_x \leq 15\,\text{nm}$.

These values also tell us an estimate of the trend of the electron–acoustic phonon interaction, and of its drastic *intrinsic* reduction among closely separated individual levels of QD (Benisty et al., 1991; Benisty, 1995). Skipping most prefactors, the relevant matrix element for electron–phonon interaction contains the following squared product $\Theta_s(q)$:

$$\Theta_s(q) = |\langle\psi_f(\mathbf{r})\,|\exp(i\mathbf{q}\cdot\mathbf{r})|\,\psi_i(\mathbf{r})\rangle|^2\,. \tag{9.5}$$

$\Theta_s(q)$ is an integral on the 3D position of a single electron. The exponential means that Θ_q is nothing but the squared modulus Fourier transform of the $\psi_f^*(\mathbf{r})\,\psi_i(\mathbf{r})$ product. For instance, in the case of free electrons, $\psi_f^*(\mathbf{r})\,\psi_i(\mathbf{r})$ amounts to $\exp(i\mathbf{k_f})\,\exp(i\mathbf{k_i})$ $= \exp(i[\mathbf{k_i} - \mathbf{k_f}])$. So the integral amounts to a Dirac function. It just tells the momentum selection rule $\mathbf{q} = \mathbf{k_f} - \mathbf{k_i}$, the addition rule in momentum space. However, for localized wavefunctions in QDs that have only a couple of oscillations in the largest direction, say x (more generally a direction where degrees of freedom are discretized but not frozen to the fundamental), we can infer that $\Theta_s(q)$ has peaks at a wavevector associated the typical oscillation pseudo-period of the product $\psi_f^*(\mathbf{r})\psi_i(\mathbf{r})$. This is typically a vector with a modulus q_{peak} of a few $2\pi/L_x$. Furthermore, beyond this peak, a power-law decay $\Theta_s(q) \propto (q/q_{\text{peak}})^{-n}$ with n up to 4 takes place (the more gently the wavefunction decays, the smaller the Gibbs phenomenon in Fourier space). So at the $q = E_{if}/\hbar c_s$ value dictated by *energy* conservation (and in the limit of delocalized bulk-like phonons), $\Theta_s(q)$ quickly collapses if L_x is reduced: it decays as L_x^n for a given level pair. Of course, the largest energy separations E_{if} among several discretized states are the first targeted by this decay. In other words, even though the electronic wavefunctions of interest are not plane waves anymore, their k-content is limited, being still concentrated at some modest value, and falls off quickly at large values. Unlike in the bulk, where there are always states in the continuum that fulfil the requirement of a large enough $\Theta_s(q)$, this is no longer true for QDs.

Overall, considering relaxation over several hundreds of meV, as is common in injected real-life devices, the trend explained above means that there is a 'phonon bottleneck', of intrinsic origin when an electron attempts to relax its high injected energy down the cascade of several levels of a typical QD: the rate of transition slows down, and slows most acutely at the largest separations E_{if} in the cascade, due to this collapse of $\Theta_s(q)$ at large q (Benisty et al., 1991; Benisty, 1995). While the phenomenon itself is hard to evidence in actual experiments due to many competing channels that eventually achieve energy relaxation, it is of heuristic interest: reduction of dimensionality means that vague beliefs inherited from continuous systems, such as 'after some rather short time, the electron will cascade down' (such a time is on the order of 1 ps in the bulk) must be carefully revisited: they are no longer granted, even loosely.

Finally, symmetries are also involved, which should come as no surprise. However, in the case of a transition from a higher dimensionality to a lower dimensionality, a larger variety of states is generally available on the higher-dimensionality side, so that full inhibition by symmetry is quite unlikely.

Before closing this discussion on capture, it must be said that the reciprocal process, upward promotion of a carrier into the higher-dimensional system, also exists and that some of the factors listed above also modulate it. One can speak of carrier 'evaporation', stressing the fact that if the balance is negative for the most confining system, it means that we are in an out-of-equilibrium situation.

9.1.8 Transport in quantum wires

The transport process in quantum wires is another delicate point to handle in low-dimensional active nanophotonic systems. If we are concerned with genuine wires with one or very few modes, i.e. few (l, m) states to transport electrons or holes, it can be considered that impurities will play a much more critical role than in higher dimensionality. In one dimension, any perturbation can be seen as a reflector of some strength acting on the impinging plane wave $A\exp(ik_z z)$, one of the components of a wavepacket describing the motion of the electron in the wire. The resulting effect of a series of such impurities can be quite dramatic, stopping transport unless the Fermi level is raised and states are populated that lie above most such barriers. Such issues are commonly tackled within the physics of localization, kickstarted by the pioneering work of P. W. Anderson in 1958 (about this, see ref. (Brandes and Kettemman, 2003), and the recent studies on Bose–Einstein condensates that have triggered a large renewal of the literature in this area. Note also that the popularity of Anderson localization should not hide the many phenomena that generically show transitions due to the change of transport of waves, either electrons or photons, through disordered media. One basic criterion for sorting among this rich phenomenology is whether the coherence length of the wave is longer than the decay length of localization, a necessary condition to be relevant to the Anderson localization framework).

In other words, quantum wires are prone to developing carrier localization, which can cause a behaviour dramatically different from the ideal one. Only nearly ideal wires, rarely available, can safely escape such effects. However, the specific case of whisker growth (Dubrovskii et al., 2009) does produce such nearly perfect wires. It works by catalysing a vapour-phase decomposition on the outer surface of metal droplets that themselves lie on a proper substrate.

9.1.9 Carbon nanotubes and graphene nanoribbons

The remark closing the previous paragraph is one of several things that spurred much interest in carbon nanotubes since the 1990s. The so-called single-wall carbon nanotubes (SWNT), in particular, are made of a sheet of graphene rolled in a precise and continuous fashion. Fig. 9.5a illustrates such a nanotube. Here we remind the reader that graphene is the name given to a single sheet of graphite; A. Geim and K. Novoselov were awarded a Nobel Prize in 2010 for their study of this material that revealed its extraordinarily high electronic mobility (Geim, 2009). Due to the atomically regular arrangement, its electronic states are very well defined. They depend only on the way in which the graphene sheet, with its three-fold symmetry, has been rolled up, generally diminishing its symmetry. A tutorial exercise in solid-state physics is to show that rolling arrangements of different kinds—technically of different *chirality*—are either metals, semiconductors with a bandgap, or semimetals.

Note that a flat and very elongated graphene sheet with well-defined boundaries can also be seen as a quantum wire. It is conventionally termed a 'GNR', or graphene nanoribbon (Fig. 9.5b). The two GNR of highest symmetries exhibit boundaries which are either of the 'armchair' or of the 'zig-zag' type, this latter case being illustrated here. Both nanotubes and nanoribbons can be synthesized in single-sheet and multiple-sheet versions. It is interesting to note that for the 'father' 2D system—i.e. for a perfect graphene sheet—due to the conical bands around the six equivalent Dirac K points of the graphene band structure, a spectrally flat absorption results, $A = \pi\alpha \simeq 2.3\%$,

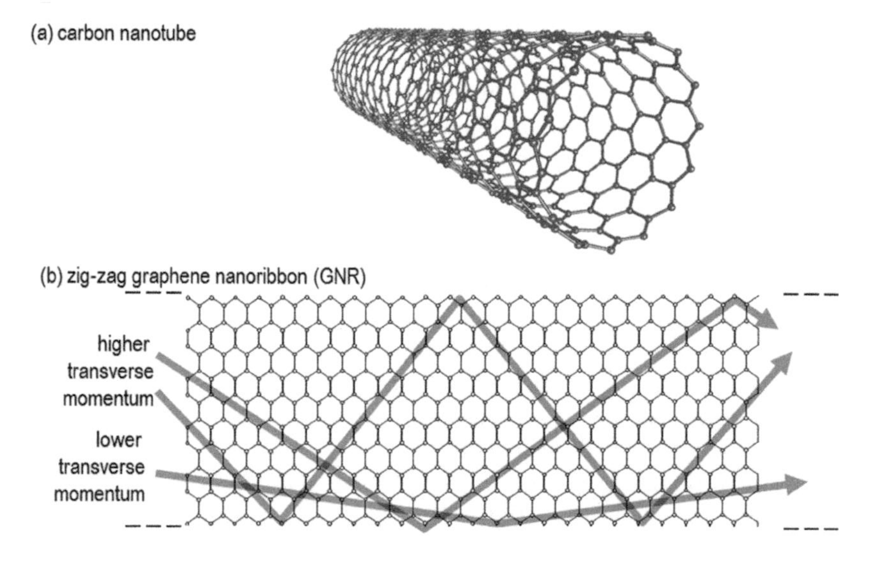

Fig. 9.5 Low-dimensional carbon: (a) single-walled carbon nanotube; (b) graphene nanoribbon (GNR) with electrons of variable transverse momentum shown in a ray picture.

which is dictated by the sole fine-structure constant $\alpha \simeq 1/137$ (itself defined by $4\pi\varepsilon_0\hbar c\alpha = e^2$). Exercise 9.7 gives a simple path to this surprisingly simple result. Other interesting properties of GNR and graphene dictated by the large conductivity, such as special types of plasmons, also prompt a lot of interest in these systems, but these plasmons are more delicate to grasp than those of the noble metals.

In the next sections, we study the specific kinds of QDs that have been popular in the last decades.

9.2 Self-organized QDs

9.2.1 Growth, mismatch, and elastic energy

In this section, we deal with the fabrication of QDs; it is important to understand a number of their individual and collective properties. An interesting case is when the synthesis of such dots spontaneously occurs due to the physical separation of the various constituent elements. Of course, the experimentalist has carefully prepared a favourable out-of-equilibrium situation, but none of the details of growth are determined.

Self-organization refers to an ability of such an out-of-equilibrium system to develop a spontaneous spatial organization, having a given characteristic size or more generally favouring given characteristic correlation lengths in its final spatial distribution (instead of completely random ones). Here we discuss such a phenomenon in the case of semiconductor epitaxial growth, and especially molecular-beam epitaxy of III-V arsenide compounds (Goldstein et al., 1985; Brandt et al., 1991; Marzin et al., 1994).

Epitaxial growth of semiconductors and the related issue of lattice-matching between substrate and deposited layers were presented in section 8.5.1. In summary, if a layer is deposited epitaxially onto a substrate (i.e. with one-to-one correspondence of atoms through, for example, the sp^3 valence bonds for many elements of columns III, IV, V), it adopts the substrate lattice in the growth plane and develops compressive or extensive strain if this lattice is respectively smaller or larger than its own (bulk form) lattice (see Fig. 9.9a). Strained layers grown in this manner are anisotropic even if the bulk crystal of the deposited material is isotropic, because the lattice is stressed only along the growth plane, leaving to the third direction the capability to compensate the in-plane strain as much as dictated by Poisson's modulus, which relates stress along one axis to strain along other axes.

For lattice mismatches on the order of $\Delta a/a = 1\%$ only, however, the elastic energy accumulated while a homogeneous layer grows quickly becomes very large. Using the analogy of an in-plane spring, for a layer thickness h, the effective spring becomes stiffer in proportion to h, while the amount of stress is constant, being given by the lattice mismatch. Thus, elastic energy grows linearly with h. Elastic energy is also proportional to the square of the strain and thus to $(\Delta a/a)^2$.

There are limits to the accumulation of strain, however, due to various forms of atom motions that relax this strain. We detail in Complement 9.A the way in which an organized array of InAs islands (the QDs) eventually emerges through such mechanisms and the associated peculiarities such as the wetting layer and the specific elongated InAs QD shape.

9.2.2 Specific optical properties: Emission range

Optical properties of InAs QDs have been much studied since individual ones could be measured (Marzin et al., 1994). We will start with a generic description and then focus on the properties of collection of QDs. These collections obtained by growth have some scatter in their properties due mainly to size dispersion, as hinted above. The scheme of reference is that quantization in the CB is large even at room temperature or somewhat above it. In the VB, quantization is less obvious due to the complexities of the three bands intervening and the heavier masses. In addition, there is a large strain and some anisotropy, and the shape of the QD is not so specific: it is not a disc, for instance.

However, for nanophotonic applications, the main property is that interband recombination favourably takes place between the pair of fundamental levels $(l, m, n) = (1, 1, 1)$ of CB and VB, corresponding to electrons and heavy holes. However, under a strong pumping condition, it is common to observe sizeable luminescence from the excited CB level and even gain and laser action. It is not important to provide much detail on which VB levels are involved, for the reasons given above. What the presence of excited transition means is that the relaxation to the ground CB state is not as quick or strong as in a quantum well. This was more or less expected due to the scarcity of final states in the intraband cascade and the difficulty in finding, for instance, phonons that efficiently assist such transitions. The consensus is that even if there is less relaxation in some regimes, there are nevertheless extra functional mechanisms that provide sufficiently large relaxation rates in InAs QDs without many special precautions, notably Auger mechanisms.

Let us examine how the structure described above indicates what range of photon energy is made available by InAs QDs on GaAs. An upper bound is that of the wetting layer, 20–40 meV below the gap of GaAs, which lies at 1.42 eV at room temperature, thus setting the limit at 1.37 eV or $\lambda \simeq 930$ nm. The lower bound is unfortunately not as low as suggested by the narrow gap of bulk InAs ($E_g \simeq 0.35$ eV), the difference being accounted by the strong vertical quantum confinement: InAs QDs cannot grow too long due to the associated strain accumulation. Therefore, their thickness (along the growth axis) remains minimal, at a couple of monolayers. Correspondingly, the picture of vertical confinement of electrons implies a large confinement energy E_{conf} above the InAs CB edge of typically several hundred milli-electron-volts.

This topic has been investigated in some depth because of the desire to reach the optical fibre telecom domain, $\lambda = 1300$ nm (lowest fibre dispersion), or even, preferably, 1550 nm (lowest fibre absorption), and thus photon energies $\hbar\omega$ of 0.8–0.9 eV. This domain usually requires growth on expensive and frailer InP substrates, so any large-scale growth on GaAs substrates would spur a welcome industrial avenue. After many efforts, recently InAs QDs grown on GaAs have been able to be turned into commercial 1300 nm laser devices. This was notably thanks to the efforts of Y. Arakawa in Tokyo, himself a coauthor of a pioneering 1982 paper (Arakawa and Sakaki, 1982) on the QD laser. No commercial InAs-on-GaAs-based device at 1550 nm is envisioned as yet, though.

9.2.3 Dynamics of QD ensembles for gain media

It is interesting to look at further properties of QD lasers. These can be classified into dynamical properties and static properties. Dynamical properties are related to the relaxation times between excited and fundamental states suggested above. Essentially, due to the long relaxation times, the excited carriers act as a reservoir for lower-energy carriers once they are populated. This can stabilize the low-energy carrier population. For instance, after a strong light transient associated with a large stimulated emission rate in a laser diode, there is a transient depletion of carriers, and thus a modulation of the optical gain. If this gain modulation is weaker because the reservoir opposed its inertia, the device is considered better. Another set of dynamical properties is related to the change in index of refraction when carriers are added in, or depleted from, the active area. When looking at an ordinary (QW) laser emission and its spectral content, these refractive index changes cause a change in the round-trip phase, and thus transform amplitude noise (the fluctuation of carrier density that translates into a variation of photon flux) into phase noise. It is also found that a laser using only a few layers of QDs can diminish such effects because carriers in the QDs behave less as an electron gas than carriers in a QW.

If we look at static properties, we may hope for the threshold current density to be lowered. In the idealized prediction of Arakawa (Arakawa and Sakaki, 1982), it was thought that the lowering of the dimensionality, with a comb-like DOS structure for QDs instead of a stepwise DOS for QWs (see Chapter 8) would result in a better use of the carriers and a lower threshold current density: more exactly, all carriers at energies that differ from the laser energy are 'useless' in a laser diode. They do not participate to gain in a spectrally useful region.

Thus, by concentrating the DOS in the useful photon energies, much more efficient use is made of electron-to-photon conversion, and more photons in the future lasing modes are emitted, thus considerably lowering threshold carrier density and, in turn, threshold current density. However, it is difficult to get a distribution of QD energies much narrower than $\simeq 15\,\mathrm{meV}$, which is of the same order as thermal energy $25\,\mathrm{meV}$. This makes a QD layer still not so close to an ensemble of narrow-line atoms used in gas lasers, for instance. Therefore, the expected reduction in threshold gain expected from the naïve argument of a more structured DOS does not have an impressive effect in practice. Granted, some QD devices have been elaborated in such a way as to actually observe low thresholds in the 10s of $\mathrm{A/cm^2}$ range in standard edge-emitting diodes, but this resulted in some drawbacks such as a lower maximum power.

9.2.4 Temperature independence of optical properties

The most obvious advantage of the QD laser to date is not so much the initial physicist's dream of shaping atom states into semiconductors to mimic a known atomic gain medium, as temperature insensitivity. Unlike QW lasers, where the larger spread of the Fermi–Dirac distribution at higher temperature $T > T_{\mathrm{ambient}}$ diminishes the carrier densities and involves higher injection, in a QD laser there is no longer a continuum of states and thus no more extra unwanted states that have to be populated at relevant energies, typically at 50–100 meV (i.e. 2–4 k_BT) above the Fermi level: with a near-

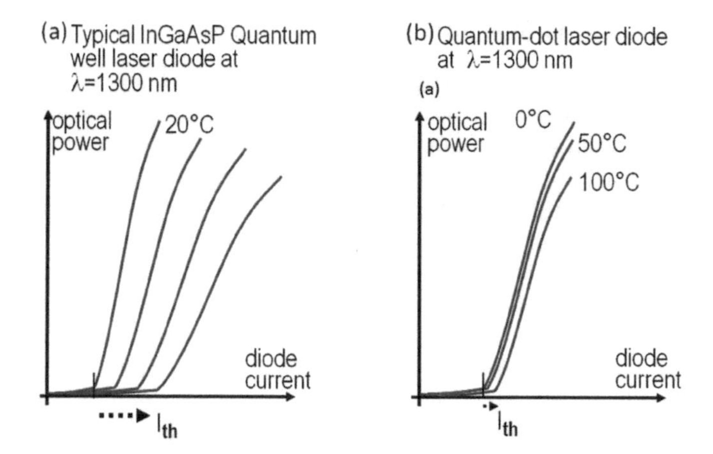

Fig. 9.6 (a) Optical output power as a function of diode current in a standard quantum-well-based InGaAsP laser diode emitting around 1300 nm for increasing temperatures: the threshold current increases; (b) same quantities for a properly designed quantum-dot-based laser: much less variation occurs.

constant current density between 0°C and 100°C, a QD laser provides nearly the same degree of population inversion among its CB and VB lasing levels, $(l, m, n) = (1, 1, 1)$.

The measure of that performance is an empirical quantity called the 'T_0' temperature of the laser. It is a practical measure of the drift of the threshold current vs T, determined by fitting the drift with an empirical law,

$$I_{\text{threshold}}(T) = I_0 \exp(T/T_0). \tag{9.6}$$

A good laser model can retrieve the trend and the value of T_0 from laser parameters, so, calling this quantity empirical is not a strict qualification, but there is no simple analytical formulation of T_0, as it depends on the DOS in complex regions of high energy. In a usual QW laser at $\lambda = 1300$ nm, based on growth on an InP substrate, T_0 lies between 60 K and 100 K, meaning that the threshold current certainly doubles for a change of $\simeq$65 K in T. This is illustrated in Fig. 9.6.

With a QD laser at $\lambda = 1300$ nm, based on growth on a GaAs substrate, there is very little change over such a range; thus T_0 is found to be around 300 K or so (see (QDLASER, 2021)).

It is thus interesting to witness that, after intensive physics work to understand the nature of the DOS of the QW laser, moving towards the Holy Grail of a 'zero-threshold' laser diode, the final success comes from a feature that was not initially deemed as so crucial: the temperature dependence. In a context where laser diodes are to be deployed in uncooled outdoor systems, there is an obvious advantage to a low-temperature dependence.

After this optimistic 'happy ending' to the QD laser story, we now leave behind III-V compounds and consider in the next paragraphs the specific features of colloidal QDs based on the very different II-VI compounds.

9.3 Nanocrystals and colloidal QDs

9.3.1 General properties

The name 'colloid' is most often applied to colloidal suspensions, which refer to suspensions of small particles in a liquid. It particularly applies when particles have been formed in the liquid itself, nucleating and growing through chemical reactions. For our purposes, colloidal QDs, the initial chemistry is secondary. More important is the possibility of forming nanocrystals of controlled size for either III-V or II-VI compounds, which are typically between 2 and 20 nm size (Guyot-Sionnest, 2008). III-V compounds grown in this way have not been deemed as interesting as II-VI ones for various reasons (spectra, role of defects at the nanocrystal outer interface, etc.) and are not further discussed. As for II-VI compounds (also known as *chalcogenides*), those of interest are based on cadmium and zinc for column II, and on oxygen, sulfur, selenium, or tellurium for column VI.

According to the rule that bandgap and atom mass vary inversely, the range of bandgaps of the *bulk* constituent ranges from near-ultraviolet for ZnO to near-infrared for CdTe (going to mid-infrared for the $Hg_{1-x}Cd_xTe$ alloy used for mainstream photovoltaic infrared detectors). The visible range is very well represented, comprising ZnS, ZnSe, CdS, and CdSe. The bulk crystal structure of the lattice inside these colloidal dots can be either zinc-blende (cubic family) or wurtzite (hexagonal family). But there is not much need to interrogate crystallography as regards the main photonic properties, since in both cases we have direct gap semiconductors at the zone centre. This is the 'canonical' case, with very good radiative recombination of electron–hole pairs in both bands once they are in states with adequate overlap, and here we consider the overlaps of the ground states of each band with each other.

The most important photonic property of these dots is their quantum confinement energy, which has been discussed above in the infinite well limit (Eq. (9.2)). Even if our calculation was done for a cubic dot and not for a more realistic shape (e.g. more spherical or a platelet type) the $\frac{1}{L^2}$ trend as a function of size holds. Hence, several tens or hundreds of meV are added to the bulk confinement energy (much as in the case of self-organized InAs QDs on GaAs). Colloidal QDs may thus act as strong sources of photoluminescence in the visible range, with an energy that can be tuned at will by changing their size.

This is, of course, easier said than done, but there are many techniques to stop growth in a controlled manner or to sort QDs by size based on a wealth of physicochemical mechanisms that allow a batch of QDs to be produced with a decently reduced dispersion in size and transition energy, typically around 20 meV (or about 5–10 nm in wavelength). The ability to produce such monodisperse solutions greatly helps with a variety of applications.

9.3.2 Core–shell QDs

The quantum yield of colloidal QDs is an important parameter. For a basic QD—i.e. a mere nanocrystal, as described above—whose surface is as good as it can be, the quantum yield can be quite poor (say, a few per cent), on account of the numerous defects at the surface (Fig. 9.7a). For these very confined systems that, unlike

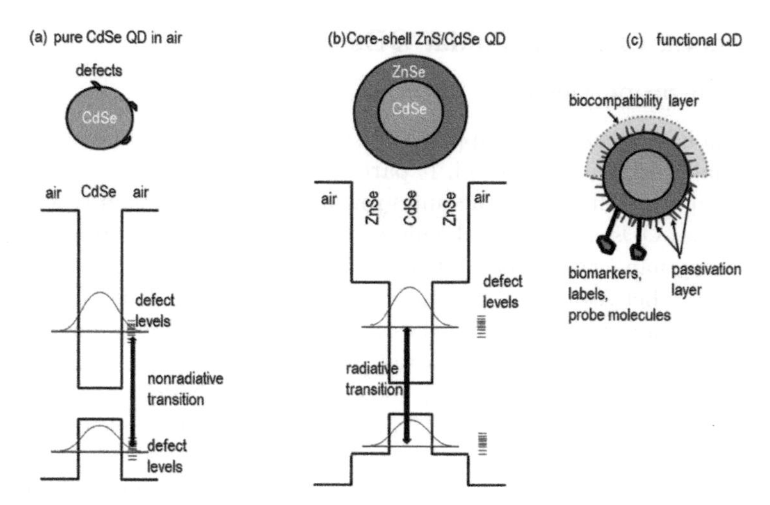

Fig. 9.7 Colloidal quantum dots (QDs): (a) Simple CdSe QD. The confined electron and holes can be captured by defects (broken bonds, interstitial atoms, etc.) that induce non-radiative relaxation (dissipation by heat); (b) core–shell QD. The extra ZnSe layer protects the electron and hole wavefunctions from visiting the defect-rich outermost layers and thus leaves enough time for the radiative transition to take place, typically nanoseconds; (c) a colloidal QD for use in biology. Various coatings are applied to passivate and functionalize the QD with respect to the media in which it will operate. Markers for specific biological reactions can also be added with proper chemistry.

epitaxial growth, do not benefit from the perfection of the underlying monocrystal, there are numerous such defects. Very large and specific efforts are needed to tame them.

These defects act on the ground transition because the tail of the electronic wavefunction easily extends to the surface of the QD, and especially strongly for the smaller ones.

A simple idea to control the situation is to then add to the core nanocrystal a shell of material that provides electronic confinement but nevertheless prolongs the crystal as neatly as possible (Fig. 9.7b). The result of this strategy is known as a core–shell QD and is considered as the only viable QD for most applications. The shell can be made of a material of larger bandgap, grown around the core in adequate amounts. A simple example is a CdSe core with a larger-bandgap shell made of a companion material: either a ZnSe shell if the cation is changed or a CdS shell if the anion is changed. Needless to say, the synthesis of such core–shell QDs requires at least one additional step compared to single-material QDs.

If we compare this kind of 'heteroepitaxial' core–shell growth with that of epitaxial QDs grown on a substrate, the lattice-mismatch issue is generally much less severe, given the small size of the system: it can accommodate distorted bonds in its periphery and is not constrained by a substrate, for instance. If growth methods grant sufficient freedom in terms of size, the choice of the transition energy can be made with just two pure binary compounds, as in the CdSe/CdS example provided. There is no need to attempt growing well-defined alloys for the core or the shell or both: obviously, this would be a difficult task.

The important advantage of the core–shell QD—the considerable increase in the quantum yield—does not solve, unfortunately, all the issues of QDs. With many colloidal core–shell QDs, experimentalists are faced with the phenomenon of 'blinking': when continuously pumping a QD, luminescence is observed only intermittently. There are dark periods during which nothing happens and whose random duration follows some less-common distribution. It has been observed that the presence of shells may significantly reduce the blinking (Mahler et al., 2008), turning dark states into grey states. The origin of this blinking phenomenon is possibly due to the presence of an external charge in the outer layers of the QD, which triggers a fast Auger relaxation (Javaux et al., 2013). The electron–hole recombination is then quenched by this faster relaxation path. In brief, a charged QD is dark and a neutral QD is bright. The shell acts positively by separating those outer layers from the useful electrons and holes confined in the core.

9.3.3 Nanoplatelets

A recent interesting alternative to spherical QDs of chalcogenides (II-VI materials) was proposed in the form of nanoplatelets with very flat interfaces (Tessier et al., 2012). Their thickness is controlled to such a point that the number of monolayers involved is constant across the nanoplatelets, with atomically flat interfaces to the outside. In such a situation, the wavefunctions and their specific properties, for instance their modified oscillator strength and their possible anisotropies, can be much better controlled. The obtainment of excellent optical properties (spectra with narrow lines and high intensity) and limited blinking has experimentally confirmed the benefits of this approach.

Such 'flat' systems are also useful from a heuristic point of view: they bridge the situation of InAs epitaxial QDs (also elongated in a plane and with neat interfaces) with the situation of self-assembled colloidal QDs. The nearly perfect shape control with two large flat surfaces and a thickness much smaller than the two other dimensions, makes it possible to push modelling (electronic wavefunctions) and also to better assess the relationship between individual QDs and ensembles. The study of the long tail of time-resolved luminescence, for instance, has provided a clear picture of the system dynamics (Rabouw et al., 2016), which is helpful for progressing on several concepts around excitons.

9.3.4 The difference between QDs and organic emitters

An important use of colloidal QDs is to attach them as fluorescent 'labels' to various biochemical species or biological systems (cells, DNA, organites). The mineral aspect of QDs is at variance with the organic aspect of most classical fluorescent systems, notably those used for marking biological samples. There are two main differences. First, mineral systems are robust against bleaching and can even be immune when protected by coatings. In the latter case, it is assumed that the coatings act as a barrier against oxygen diffusion. By contrast, virtually all organic molecules suffer in their excited states from the presence of oxygen or moisture, even in small amounts, and eventually undergo chemical modifications that are darkening their luminescence. Much of the science of dyes is indeed concerned, by and large, with the minimization

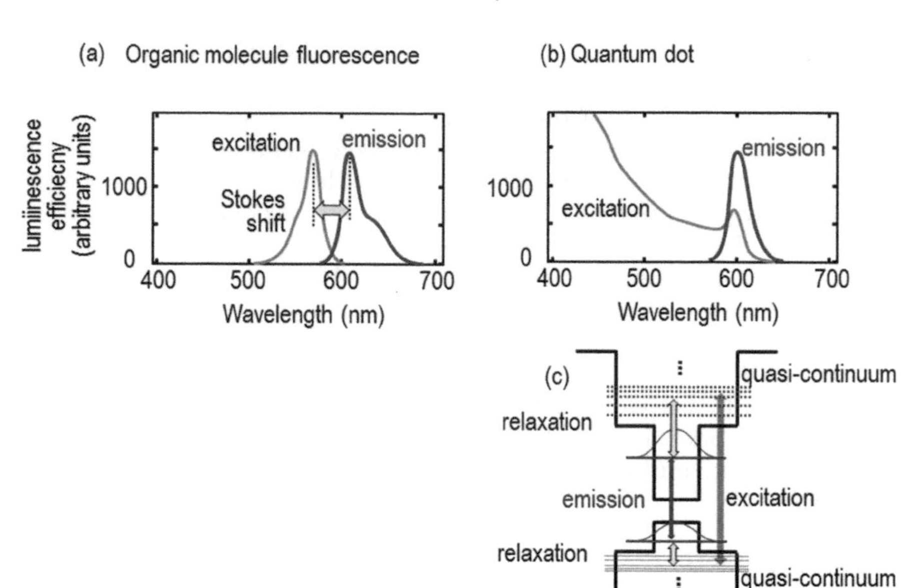

Fig. 9.8 Luminescence of organic molecules and QDs. (a) In an organic molecule, excitation of photoluminescence takes place only in a limited spectral range. The Stokes shift between the centre of this range and the luminescence centre is often small, a few tens of nm. (b) Case of a colloidal QD. A large absorption is present from the fundamental transition (where luminescence occurs) to the blue and even the ultraviolet. (c) Role of the upper excited states in absorbing excitation radiation. Relaxation to the ground states of conduction band and valence band is shown by the light grey arrows.

of this problem, thanks to some molecular control of the dye's surrounding chemistry. A typical organic molecule will emit 'only' a few million times before 'dying', which is not so much in practice. Minerals capitalize on different kinds of orbitals to produce luminescence, and accordingly they can be completely immune to this effect. A specific and common case is that of the rare-earth compounds used in the fluorescent lighting we see around us or in white light-emitting diodes (cerium, terbium, europium) to convert blue light down to yellow and red. Another specific aspect as compared to molecules is the ability to excite QDs at photon energies much higher than their fundamental transition energy (Fig. 9.8). This arises because it is not a specific set of molecular orbitals that undergoes absorption, but rather any of the many excited states of the QD, which form a quasi-continuum at energies 0.5–1 eV above the fundamental energy: remember that a 3D solid's electronic DOS scales like $(E - E_c)^{1/2}$ and thus increases with E as long as the underlying CB is parabolic.

As seen in Fig. 9.8, the absorption spectrum is blue-shifted compared to the emission spectrum. This is due to the fact that after excitation, the excited electrons relax on time scales smaller than 1 ps to the lower-lying electronic states in the CB, whereas the radiative electron–hole recombination takes place on a few nanoseconds. Hence, the occupation of the states is thermalized and the discrepancy between emission and absorption spectra is due to the population of the states discussed in section 8.3.3.

The factor between emission and absorption cross sections is given by the McCumber relation, $\sigma_{\rm em}(\omega, T) = \sigma_{\rm abs}(\omega, T)\,\exp(-\hbar(\omega - \mu^{\omega}_{ph}(T))/k_B T)$. The reader can recognize the similarity of $\mu^{\omega}_{ph}(T)$ with a chemical potential (Band and Heller, 1988), albeit accounted in frequency units rather than in energy units.

Coming back to the QD/molecule comparison, the scarcity and large separation of QD states at low energies is no longer found at high energies (say, for electrons 1 eV above band edges). See also Exercise 9.4 for the statistics of level spacing in a band with two loosely confined degrees of freedom. This is very interesting because in any fluorescence experiments, a large energy separation between excitation and luminescence photons (the so-called *Stokes shift*) makes the task of filtering out excitation photons much easier, and all the more so because what is generally desired is to observe weakly luminescent situations, thus requiring nearly total extinction of the strong excitation light. For a small Stokes shift of say, 100–200 meV (50 nm in wavelength), it is difficult to get filters that do this job without incurring sizeable rejection of the luminescence photons themselves, because the need for filtering is very high and optical densities >6 (10^{-6} rejection) are required. This is easier for shifts $\gtrsim 300$ meV.

A last aspect relates to the biochemistry needed to attach a QD to a biological system in practice, and/or make it soluble in the proper fluid (Fig. 9.7c). There are many accounts of the specific coatings, surfactants, functionalizing layers, or molecules that are bound around QDs to further perform the desired attachment/immobilization. These important items of the biotechnology toolbox are, however, well beyond our present scope.

9.3.5 Display applications

Even though Chapter 20 is devoted to devices, in this chapter we address the use of QDs in commercial displays, which became a reality in 2013. We do so because there are no geometric complexities other than a plane of QDs in these displays. This is at variance with Chapter 20, which explores modal geometries for several applications, whereas displays just make good use of QDs' spectral properties. We assume the reader is familiar with the three-colour (RGB) subpixel layout of displays.

There are two essential uses of QDs. They both target the colour-rendering issue, i.e. how a three-colour system manages to render the whole gamut of possibly perceived colour (represented as a semi-elliptical fraction of the $[0, 1] \times [0, 1]$ square of the 2D plane that you can best visualize by searching for 'colour gamut diagram' online). The diagram limits, the so-called 'pure colours', are colours from monochromatic light.

In the first and rather established use, the *photo-emissive* use, the aim of the display industry is to produce three colours of great purity from LED light. LED light is usually white light cleverly homogenized through a so-called back-light system in modern LCD (liquid crystal displays), whereby the liquid crystal serves as the variable 'light-valve'. The historic solution of placing filters on this white light is far from ideal: filters are either not narrow enough or become too absorbing and expensive if narrowed, dimming the screen. Their unavoidable spectral width degrades colour purity, especially at the top left of the diagram, for green to blue-green edges. In the QD solution, white light is first replaced by blue light at no cost (bare GaN LEDs are used instead of phosphor-coated ones). Blue can serve directly in its dedicated subpixels, but for green and red pixels, QDs photoluminescence down-converts the blue photons. Hence only the narrow-emission lines of QDs are re-emitted—say, 40–50 nm-wide green and red lines. In the first products, this narrow light was still passed through the filters existing at the subpixel level. They nevertheless exhibited a purer green, and much less penalty in terms of dimmer output and diminished energy efficiency than white light due to the narrow spectra. As a result, displays could boast a good gamut and high luminance (well above 1000 cd/m^2, known previously as *nits*, such values still being much smaller than their equivalent atop an LED face).

In the next version, the subpixel RGB filters are directly replaced by locally deposited QD layers, e.g. as inks. Energy-wise, the rules of photoluminescence seen in Chapter 8 do apply: only the quantum defect from blue excitation to red/green emission is lost. A sketchy reasoning indicates that $> 50\,\%$ of blue photons are used to produce a pure colour, instead of about 25% in the case of filtered white light.

Switching to the second envisioned use, called the *electro-emissive* use, this is less established. QD-based LED emitters directly produce all RGB photons. Ideally, this is done inside the subpixel itself, avoiding many LCD layers. Heat is also spread compared to QDs at the back-light stage and the detrimental effect of heating on QD efficiency is avoided, while luminance can be dramatically large.

For both uses, cadmium-based QDs are virtually banned due to the toxicity of cadmium, a heavy metal. Zinc remains the favourite column-II element in the QD core, and indium compounds are used as shells, for instance. But there is little doubt that displays are the first large-scale QD application used in everyday life. QDs currently competes with OLED (organic LED) based emitters in displays, for which nanophotonic issues also exist but are not tackled here.

9.A Complement: Growth of epitaxial InAs QDs

9.A.1 Strain relaxation with Stranski–Krastanov mechanism

As said at the end of section 9.2.1, when too much elastic energy is present, it becomes more energetically favourable to create defects by moving atoms; these can be atoms with an improper placement or an improper number of bonds, or, more generally, dislocation lines (which we have invoked as a major source of non-radiative recombination). A single defect in bulk has a higher energy cost than a perfectly located atom. But if it helps compensating the strain in its vicinity, there can be an offset. What happens in practice is that after a critical thickness, h_c, defects appear; for instance, adatoms that have freshly arrived on the surface could be incorporated at locations far from exact lattice alignment (atoms that just start to react after landing on the surface do not have all their bonds established within the substrate lattice but only part of them; such so-called adatoms are thus mobile because they can hop along the surface).

From the above energy arguments, we can guess that this transition is attained when the cost of strain energy, which scales like $h(\Delta a/a)^2$, exceeds the energy cost of defect formation (correctly including the strain relaxation caused by the defect in its vicinity). Since the mismatch is dictated by the substrate, while the defects form at the growing surface, we can infer that the critical thickness h_c scales like $(\Delta a/a)^{-2}$, a

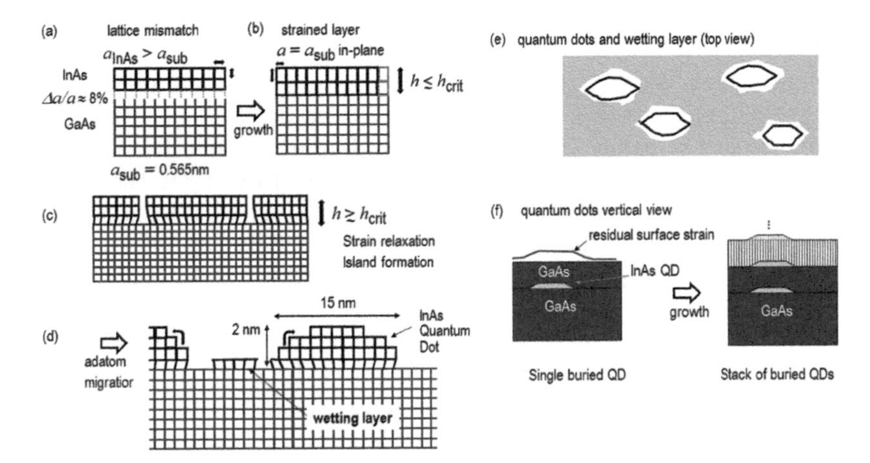

Fig. 9.9 Self-organized quantum dot (QD) formation (Stranski–Krastanov mechanism): (a) lattice mismatch of InAs vs GaAs; (b) growth below the critical thickness is epitaxial and strains the growing InAs layer with in-plane compression; (c) beyond the critical thickness, strain is relaxed and the layer breaks into islands; (d) migration of In atoms to the top of islands (in excess As adatoms) diminishes strain and surface energy, but a one-monolayer-thick wetting layer remains; (e) top view of QD with non-isotropic in-plane shape and the 'holey' wetting layer; (f) vertical growth of GaAs above InAs (left) leaves a surface strain above the QD, on which one QD and more subsequent QDs can grow in vertical alignment.

large dependence. For reference, the exact mechanism of defect formation that we will discuss for InAs QDs is known as the *Stranski–Krastanov* mechanism.

The critical thickness has an important consequence for the formation of electronic confinement by QDs: if we deal with modest mismatches and critical thicknesses of the order of 50–100 nm, we cannot hope for much more than some roughening of the growing layer near its surface but not its full reshaping.

In the 1980s (Goldstein et al., 1985; Brandt et al., 1991), it was discovered that, in the case of the low-energy-gap InAs growth on gallium arsenide (GaAs), since the mismatch, amounting to $\Delta a/a = 8\%$, is the largest one in the arsenide family, the critical thickness can be as small as two monolayers of InAs, and even slightly less. Thus, when attempting to grow two or three monolayers of InAs, typically by molecular-beam epitaxy techniques, an abrupt transition can be seen in the surface morphology, whereby the InAs layer transforms (by adatom migration) into a set of small islands of a few monolayers in height that cover only a small fraction of the surface (see Fig. 9.9b, c, d). At the top of these islands, of typical lateral size 10–20 nm, the strain of InAs is much more relaxed than at the interface. These islands are the future self-organized InAs QDs. To become useful QDs, they generally have to be buried again by a growth step consisting of GaAs. This restores the strain at the top of the islands, but instead of a fully highly strained layer that would arise with a uniform layer, we now have local strain regions localized around each InAs cluster (each QD) and not compromising the overall integrity of the grown layer (no peeling shall occur, for instance).

9.A.2 The wetting layer

A last complication is the fact that the very first InAs layer on GaAs plays a 'wetting' role (see Fig. 9.9d, e). For this layer, with one side towards GaAs but the other towards the vacuum, the energy balance between strain and interfaces is still largely favourable. Therefore, on the GaAs surface, and in between the InAs QDs that have formed following the growth instability, a so-called wetting layer remains, which consists of a partial coverage of a single InAs monolayer. Such a monolayer thickness, once buried in GaAs, acts as a thin QW confining carriers some 20–40 meV below the bands of GaAs.

Thus, the overall picture that emerges upon careful growth of InAs on GaAs, further embedded by a GaAs growth step, is the presence of a high density of InAs QDs, typically 400 QDs per square micron (an average centre-to-centre distance of $\sim$ 50 nm), having a height of 4–10 nm and a lateral extent of 10–25 nm, with a 'holey carpet' of InAs wetting layer around them. The dispersion of QD sizes has been intensively studied. It can be mastered through growth condition (temperature, fluxes, sequence timing, etc.): it can be selected between a rather narrow size distribution (giving a narrow spectrum: see earlier in Chapter 9 on optical properties of QD collections) and a deliberately broad distribution (giving a more polychromatic spectrum).

In order to use such InAs QDs individually and deterministically, for instance, to place one of them of given transition energy at the antinode of the mode of a resonant nanocavity (see chapters 7, 17 and 20), it is desirable to have a physical trace of their buried presence beneath the top GaAs layer. One generic way is to look at the photoluminescence under a microscope, since far-field optical techniques limit the accuracy of a single-object location not to the setup resolution but to a much lower value, often around 5–10 nm (see Chapter 11). By using clever lithography techniques based on the same microscope setup (Dousse et al., 2008) to locally expose a resist layer, a pattern centred on a QD is created, which serves as the basis for further technology steps, e.g. to create micropillars or photonic-crystal microcavities.

9.A.3 Stacks and self-organized growth

It is possible to grow several successive layers of vertically aligned QDs. When the GaAs spacing between these layers is thin enough, the strain field of the first layer of QDs is still significant when the second-layer growth is started (Fig. 9.9f), and it tends to increase the local lattice constant right above the buried InAs QDs. This favours the nucleation of the second layer of QDs exactly on top of the first layer ones. And, iteratively, a full set of QD stacks situated above the first QD array can be obtained. This is pursued until a layer that will be left to the air is grown, for instance up to the top layer of a membrane. The bumps at the top of this layer are useful topological traces of the otherwise invisible QD stacks. If, cleverly, all QDs but the first have been grown with a smaller size and thus a larger gap, they will even act to assist capture and injection of carriers into the first lower-energy QD (see section 9.1.7); even without extra help in capture, their photoluminescence range will be spectrally separated from that of the initial bottom QD of interest.

9.B Complement: Silicon-related nanostructures in photonics

9.B.1 Germanium QDs on silicon

The importance of silicon has brought focus to the possibility of growing germanium QDs on a silicon substrate. The lattice mismatch is about +4.2% in favour of germanium, which is very comparable to the GaAs/Si lattice mismatch but less than the 8% InAs/GaAs lattice mismatch. The same Stranski–Krastanov mechanism can thus operate to form Ge QDs on Si or, with adequate growth conditions, alloyed GeSi dots. Due to the smaller mismatch compared to InAs/GaAs, however, the critical thickness at which the transition from planar 'wetting' growth to island formation takes place is much larger. As a result, Ge dots on Si are not flat elongated flakes but rather tend to take the shape of huts with sizeable slopes of their 'roofs'. Confinement energies are not as large as in InAs with respect to the vertical dimension. As a result, there is room to create transitions in the range 0.65–1.0 eV where silicon is unable to absorb substantially. However, there is not such a big need for these dots for detection, as the growth of strained uniform SiGe QWs suffices for this scope. The critical issue is light emission, and thus the possible overcoming of the indirect gap of Ge due to electronic confinement effects that, in a picture similar to that which we meet in photonic crystals, fold the bands inside the Brillouin zone. Also, silicon often makes the best nanophotonic devices, e.g. photonic-crystal membranes with defect photonic modes of an extremely high quality factor (see chapters 16, 17). nanophotonic structures is an attractive avenue for continuing the development of such nanostructures.

9.B.2 Porous silicon

It is a delicate topic to conceive silicon nanostructures, and an apparent paradox, since silicon exists under so many forms (e.g. amorphous) and its chemistry and metallurgy have been mastered more than any other mineral element of columns III to V. In addition to expected conventional forms of nanostructures such as Si QDs and Si quantum wires, we have to add microporous silicon (often termed PSi) to the gallery.

One main photonic incentive to Si nanostructures is to circumvent the inhibition of the emission process associated with the indirect bandgap of bulk silicon. Here, by reminding the reader that the indirect bandgap lies along the ΓX-axis of the first Brillouin zone at $\simeq 1.1$ eV while the gap for vertical CB$\rightarrow$VB transitions is around 3 eV at Γ, we see that there is no hope of going to a direct gap by local strain; instead, a modified wavefunction should be considered. The wavefunction of a very confined electron acquires broader k-components than the discrete $\mathbf{k}+\mathbf{K}$ components of Bloch states in bulk Si ($\mathbf{K}$'s being reciprocal lattice vectors). Therefore, there could be, for instance, substantial k-components of the CB state around $\mathbf{k}=0$, where holes do have the matching k-component. These modified wavefunctions could therefore result in a sizeable radiative recombination channel. Various calculations show that this requires very small sizes, on the order of five lattice constants (say, 2 nm).

Indeed silicon QDs show various photoluminescence phenomena, some reaching appreciable quantum yields. Efficient electrical excitation would attract the silicon industry to integrate sources on wafers, even if efficiencies are modest.

But insofar as scrambling wavefunctions is concerned, making the silicon shape irregular at the few-nanometre scale but still connected at large scale can produce much the same result. That is actually what PSi achieves. It is elaborated by means of photo-electrochemical etching of silicon (the etching in question is the anodic dissolution, preferably in the presence of light to photo-induce a flow of holes that feeds the electrochemical etching processes at the interface that would tend to stop otherwise, with depletion layers playing a key role to channel carriers in the forming of the porous skeleton). Doping and current densities should be in well-controlled margins to obtain the desired degree of porosity. Up to 60–70% of the matter can be removed, with the pores and distorted remaining 'struts' forming an intricate labyrinth. Using advanced drying techniques known as supercritical drying to avoid collapse caused by capillary forces, an even larger porosity can be made viable. The matter left between the pores stands somewhere between QDs and localized quantum wires, and it is possible that its characteristic size (transverse size) fits the range of a few nanometres signalled above. In this case, the specific surface of the porous system is very large ($500\,\mathrm{m}^2/\mathrm{g}$). Luminescence of PSi at visible wavelengths was discovered at the dawn of the 1990s and was investigated in depth during that decade (Bisi et al., 2000). The quantum confinement interpretation readily explains a transition energy $\hbar\omega \geq 2\,\mathrm{eV}$, lying $\sim 1\,\mathrm{eV}$ above the $1.1\,\mathrm{eV}$ bulk Si bandgap. But surface effects are certainly playing an important role as well. The system is not 'clean' enough to lend itself to a simplifying description. The large exchange surface and outer medium sensitivity (fluids or gas) obviously make it a good basis for many kinds of sensors.

Finally, the existence of 'macroporous silicon' must also be signalled. Using the same generic process in different conditions and initiating the etching at regularly organized pits on a face of a silicon wafer, holes can be made to propagate with long parallel shapes reaching a huge aspect ratio (a height/diameter ratio up to 100). Such systems were among the first used in 1995 to demonstrate the existence of a 2D photonic bandgap (see Chapter 16) at mid-infrared wavelength where Si is transparent (Grüning et al., 1996).

Exercises

Exercise 9.1 Fraction of outer atoms in a cubic dot.

(1) Calculate the fraction f of outer atoms in a cube of $N_{\mathrm{tot}} = N^3$ atoms. Base your reasoning on the smaller cube of the inner atoms, complementary to the outer ones.

(2) Using the formulation of f closest to the complementarity reasoning proposed above, find the sizes N and total number of atoms N_{tot} at which $f = 0.5, f = 0.1, f = 0.01$. Translate N into the cube size L using a simple cubic lattice parameter of $0.3\,\mathrm{nm}$.

Exercise 9.2 Size, energy, and monolayer fluctuations. We consider an ideal semiconductor with mass m^* in CB and reduced mass μ for photonic transitions (see Chapter 8). We assume infinite barriers for the confining QD or quantum wire or QW potential.

(1) Compare the confinement energies $E_{\rm conf} = E_1 - E_c$ for a QW, a quantum wire, and a QD, all with the same size parameter L, i.e. the quantum wire section is square and reads L^2 and the QD is a cube and its volume reads L^3. E_c is the bulk CB edge, and E_1 the first CB level ($E_1 \equiv E_{1,1}$ in the quantum wire and $E_1 \equiv E_{1,1,1}$ in the QD).

(2) Calculate the effect of a fluctuation of size $L + \Delta L$ on E_c in the three systems.

(3) Counting the dimensions in units of monolayer by defining $L = NL_{ML}$ with the monolayer thickness $L_{ML} = 0.26\,\text{nm}$, indicate at which thickness L a monolayer variation $\Delta L = L_{ML}$ results in a 5% variation of $E_{\rm conf}$ again in all three cases.

(4) If we are now interested in a 2% variation of the photon energy, which is $\hbar\omega = E_g + E_{\rm conf}(m^*/\mu)$ (i.e. replace m^* by μ in $E_{\rm conf}$), set up an equation for the thickness yielding the same monolayer tolerance criterion. In the approximation $E_{\rm conf} \ll E_g$, give numerical values obtained for $E_g = 1.7\,\text{eV}$ and $\mu = 0.16\,m_0$, about the values for CdSe ($m^* \simeq 0.2\,m_0$ in CB, not needed here). The assumption should be nearly valid. What can you conclude about CdSe QDs having emission in orange to blue bands in terms of size accuracy? Is the one-monolayer accuracy sufficient to get a narrow (2%) spectrum?

Exercise 9.3 Edges and corners of a QD We reuse the infinite barrier model in Exercise 9.2 of a cube, but we use a scaling of the energy with $N_{\rm tot}$ to capture the effect of edges and corners. The edge length is denoted by N and the lattice parameter is L_{ML}.

(1) Express the confinement energy of the QD as a function of $N_{\rm tot}$ using parameters of the referenced exercise.

(2) Deduce the naïve relative energy variation of $E_{\rm conf}$ associated with the removal of the edge atoms (their number is a linear function of N). Do the same for the eight corner atoms. What is the result for $N_{\rm tot} = 10$ and $N_{\rm tot} = 4$?

(3) For a model envelope wavefunction that would be the separable form $\psi(x, y, z) = \sin(\pi x/L)\sin(\pi y/L)\sin(\pi z/L)$ with $\psi = 0$ at the boundaries, based on the value of ψ the at a distance L_{ML} inside the edge and corner compared to the value of ψ at the same distance from a face centre, and reasoning in the spirit of perturbation theory, would the edge and corner atoms have more or less than the impact predicted above? Prolong the answer to the case of a more realistic wavefunction extending in the barriers around the QD.

Exercise 9.4 Statistics of separation of square QD levels. We consider the level quantization in two dimensions (x, y) for a relatively large and flat dot with dimensions $L \times L \times d$ and $d \ll L$, so that the third dimension is completely frozen in the first level that serves as an effective CB at energy E_c. The exercise is based on the fact that the integer numbers $\ell^2 + m^2$ are asymptotically uniformly distributed, as can be predicted using the 2D DOS results for electrons of mass $m^* \equiv m_c^*$ in Chapter 8. We assume $m^* = 0.1m_0$. Spin can be included or not, and a ~2% asymmetry added to lift the degeneracies $E_{m,\ell} = E_{\ell,m}$ of the exact square.

(1) Use mode counting in the spirit of 2D DOS to find the size $L = L_0$ such that $N_L = 201$ levels are taking place in the energy range of $[E_c, E_c + 500\,\text{meV}]$, so as to study the energy drop along a cascade of height $E_d = 500\,\text{meV}$. Their average spacing should thus be $E_0 = \frac{500}{200} = 2.5\,\text{meV}$. Maximum ℓ, m should be $\sim 16 \simeq \sqrt{\frac{4\times 200}{\pi}}$: note that an energy

bound amounts to delimiting discs in the ℓ, m plane, not squares. Check that 30 nm $< L_0 <$ 60 nm.

(2) Check numerically, by making a histogram of the spacing between ordered levels with a typical 1 meV bin size, that the statistics of *spacings*, $\Delta E = E_{q+1} - E_q$, where q is the ordering index, looks like a Poissonian $P(\Delta E) = \frac{1}{E_0}\exp(-\frac{\Delta E}{E_0})$. More numerous sets ($N_L \gg 201$) and/or a log scale can be explored to confirm the trend.

(3) The *largest* of a set of N Poissonian variables is normally given by the so-called Gumbel distribution, itself derived from the treatment of the cumulative distribution function $C_P(\Delta E) = \int_0^{\Delta E} P(\Delta E')d\Delta E'$. For our purpose here, the largest of N_L variables is reasonably approximated by $\Delta E_{\max} = \Delta E_N + E_0$, where ΔE_N is defined by $C_P(\Delta E_N) = 1 - 1/N_L$: if we cut $P(\Delta E)$ in N_L slices of equal weight, the largest slice starts at ΔE_N and the largest of N_L variables will have its expectation value around E_0 above the left boundary of this slice. Illustrate these assertions by a drawing of the Poissonian and the slice in question. Show that the largest spacing is on the order of $E_0 \ln(N_L) \equiv \frac{E_d}{N_L}\ln(N_L)$. Compare the result with *phonon* energies. Extrapolate to other L sizes: find when (size L_1), in a cascade of $E_d = 500$ meV, the largest spacing $\Delta E_{\max}$ exceeds the 36 meV LO phonon energy, and when (size $L_2 \gg L_1$) $\Delta E_{\max}$ exceeds a smaller threshold—identified for acoustic phonon scattering—of 1 meV.

Exercise 9.5 Layers of quantum dots. Determine how many layers of InAs-type quantum dots (QDs) bring a DOS equivalent to that of a QW of effective mass $m^* = 0.08m_0$, assuming a broadening of the first-level transition of $\Delta E = 10$ meV and neglecting the spin.

Assume that dots have a density of 400 per µm^2, a typical value for InAs on GaAs. Explain also, by comparing their confinement energy to ΔE, why their size does not matter (you can assume their size to be at most half their separation and use the above mass $m^* = 0.08$).

Exercise 9.6 Single-molecule detection. The fraction of photons emitted towards the lens in a standard microscopy experiment of a fluorophore on a glass substrate image through an air gap can be found from the solid angle Ω_{air} of the numerical aperture in air of the lens (use NA $= 0.9$): see how the corresponding angles are changed in glass and deduce the solid angle in glass of one cone Ω_{glass}. Then calculate the relative fraction $\eta_1 = \Omega_{\text{glass}}/(4\pi)$.

The various filters (dichroic, etc.) and glass interfaces take out about 75% of the light. A typical organic dye molecule can emit $3\,10^6$ molecules before bleaching. If a single-shot measurement requires 10^3 photons on a single (or a few) pixel(s) of the detector to get well above the noise, calculate how many measurements can be made using $\eta_2 = 0.25\,\eta_1$ and the above data. Conclude by discussing the possible benefits of immersion lenses for detecting single molecules.

Exercise 9.7 Absorption and surface conductance of an ideal graphene sheet. Let us consider the expression for the surface impedance obtained from $j_S = KE$, where the constant K is taken to be $K = \pi G_0/4$ with $G_0 = 2e^2/h$ the quantum of conductance. Note that this expression differs depending on theories, and agreement with experiment is only qualitative (Geim and Novoselov, 2007).

Calculate $(K)|E|^2$ and make the ratio with the flux $|E|^2\varepsilon_0 c$. You should be able to retrieve the absorption as $A = \pi\alpha$ with $\alpha \simeq 1/137$ the fine-structure constant within a few lines.

Exercise 9.8 Level spacing in the long dimension of a wire. Consider a wire of length L in its long dimension. Express the exact level spacing $E_{n+1} - E_n$ as a function of L, m^* and n. For large n, it becomes linear in n. Check that the resulting spacing is $\sqrt{E_n E_{\text{conf}}} = \sqrt{E_n E_1}$, taking care of the spin degeneracy. The important information is that if a large n is needed in order for E_n to attain some large confinement energy associated with the small wire dimensions, then the spacing will necessarily fall, in terms of order of magnitude, in between the confinement energy and E_n.

Part III

Advanced concepts in nanophotonics

10 Fundamental concepts of near-field optics

In this chapter, we will introduce the basic concepts of near-field optics. A most striking property is the possibility of confining light to subwavelength spots. This discussion will prove to be useful for clarifying the physical origin of the limitation of resolution in conventional optical systems. To this aim, in the first section, we introduce the concept of evanescent waves, which plays a key role in the description of the subwavelength structures of the electromagnetic field. The second part of the chapter will offer an alternative introduction to near-field optics by examining the radiation of an electric dipole. This point of view emphasizes the fact that to confine electromagnetic fields, we need to use charges. In other words, the confinement of the field simply comes from the fact that fields are large close to point-like charges, as seen in Coulomb's law. These two sections can also be viewed as a discussion of the concepts of near field, either in Fourier space (evanescent waves) or direct space (dipole radiation). In the third section, we establish a connection between the quasi-electrostatic and quasi-magnetostatic approximations familiar in the regime of low frequencies and the near-field regime used in optics for transverse-electric (TE) and transverse-magnetic (TM) polarization. Finally, in the last part of the chapter, we highlight some particular properties of the near field. It will be seen that the familiar concepts of phase, Fresnel reflection factor, field structure, and polarization need to be revisited in the near field.

10.1 Field confinement

10.1.1 Fundamentals

We introduced in Chapter 4 the angular spectrum of a field propagating in vacuum:

$$E(x, y, z) = \iint E(k_x, k_y, 0) \exp[i(k_x x + k_y y + \gamma z)] \frac{\mathrm{d}k_x}{2\pi} \frac{\mathrm{d}k_y}{2\pi} . \tag{10.1}$$

It has been shown that the field can be viewed as a sum of plane waves propagating along different directions specified by (k_x, k_y, γ) for real values of γ. We are going to show now that propagation is a low-pass filter. This property is the basis of the resolution limit of optical instruments.

For the sake of illustration, we consider two representative situations that exhibit large spatial frequencies. In both cases, the field varies rapidly on small length scales. We consider as a first example a beam blocked by a sharp edge or scattered by a small structure such as dust, a scratch, etc. A second example consists of a beam illuminating

Introduction to Nanophotonics. Henri Benisty, Jean-Jacques Greffet, and Philippe Lalanne, Oxford University Press.
 DOI: 10.1093/oso/9780198786139.003.0010

a tiny aperture with sharp edges in an opaque planar screen. These two examples illustrate a field that varies rapidly and therefore contains large spatial frequencies.

In order to examine the field behaviour during propagation, let us apply the propagation formula. We need to find the amplitude of each plane wave. We consider the case of a rectangular aperture with sharp edges illuminated at normal incidence. To compute the exact field in the aperture is a non-trivial question. Here, we assume that it is given by the incident field in the aperture and that the field is zero away from the aperture. The calculation of the integral (4.12) is elementary and yields the function $\frac{1}{k_x k_y} \sin\left(\frac{k_x a}{2}\right) \sin\left(\frac{k_y a}{2}\right)$. This expression gives the spectrum of the object. Note that the spectrum at high spatial frequencies $k_x \gg 1/a$ decays slowly as $1/k_x$. This slow decay is due to the presence of the field's discontinuities at the edges. It is known that by apodizing the edge discontinuity accelerates the decay. We have made the point that small structures always have large spatial frequencies. Now we address the following question: can these high spatial frequencies propagate in vacuum? We know that it is not possible to associate a propagation direction with a spatial frequency greater than ω/c. If we take a look at Eq. (4.9), we see that these spatial frequencies are associated with wavevectors such that γ is imaginary, namely to evanescent waves which decay exponentially away from the interface. Only plane waves such as:

$$k_x^2 + k_y^2 < \frac{\omega^2}{c^2} \tag{10.2}$$

have a real component of the wavevector along z and may therefore propagate. In short, only those waves satisfying Eq. (10.2) may propagate. The maximum wavevector in the plane (x, y) corresponding to a propagative wave is $\frac{\omega}{c} = \frac{2\pi}{\lambda}$ and the corresponding spatial frequency f_x is $1/\lambda$. This has a basic consequence for teledetection, microscopy, and, more generally, far-field imaging. Upon propagation *all the information carried by spatial frequencies* f_x *higher than* $1/\lambda$ *is lost.* In other words, the fine details, the structures of $E(x, y, 0)$ smaller than the radiation wavelength are lost during propagation. *This is nothing but the diffraction limit revisited in Fourier space.*

10.1.2 Near-field zone and evanescent waves

Let us imagine a surface on which structures smaller than the wavelength are etched. As the boundary conditions must be satisfied along the structures, boundaries, the fields must vary on length scales which depend on the geometric form of the structures. They can be much smaller than the wavelength. Hence, the field contains evanescent waves.

It is useful to note at this stage that the distance z along which the radiation is propagated defines a cutoff frequency. Let us consider a large wavevector, $k_x \gg \omega/c$. The z-component of the wavevector is given by $\gamma = \left(\frac{\omega^2}{c^2} - k_x^2\right)^{1/2} \simeq ik_x$. The decay is thus given by

$$\exp[i\gamma z] \simeq \exp[-k_x z].$$

By observing the field at a distance z, the field associated with the wavevector $k_x = 1/z$ is attenuated by a factor $1/e$. This wavevector corresponds to a period along the x-axis

$\Lambda = 2\pi/k_x = 2\pi z$. *It is useful to keep in mind that a field with period Λ along the x-axis is attenuated by a factor $1/e$ at a distance $z = \Lambda/2\pi$.*

We have just established a rule of thumb to specify the connection between Λ and the distance z. We can now use it to define the near-field zone as the zone where structures with Λ smaller than the wavelength λ can be detected. Hence, the distance z has to be smaller than $\lambda/2\pi$. *This condition provides a first definition of what is called the near field.*

10.1.3 Graphical representation of propagating and evanescent waves: Light cone

In this small section, we introduce three graphical representations of the domains corresponding to evanescent waves and propagating waves. In Fig. 10.1a we introduce the so-called light cone, which is the zone defined by $\omega > c\sqrt{k_x^2 + k_y^2}$. All wavevectors in the light cone correspond to propagating waves. In Fig. 10.1b, we consider waves at a fixed frequency, ω. All wavevectors within a disc with radius ω/c correspond to propagating waves, whereas all wavevectors with $k > \omega/c$ correspond to evanescent waves. This corresponds to a section of the cone at constant frequency. In Fig. 10.1c, we introduce the light line defined in the plane (ω, k_x) by the equation $\omega = ck_x$. This corresponds to the section of the cone at $k_y = 0$.

10.1.4 Physical origin of the evanescent waves

The purpose of this section is to give some insight into the origin of the evanescent waves. When studying waves propagating in vacuum, we know that the wavevector modulus must satisfy $k = \omega/c$, so that it seems that there is a cutoff. However, this is only valid in the absence of charges. A rapid variation of the field becomes possible in the vicinity of localized charges. This is a key point of nanophotonics: field confinement is always due to the existence of evanescent waves generated by nearby charges. These charges can be free electrons in metals or bounded polarization charges in dielectrics. The only constraint is that there must be some material nearby. Let us illustrate this

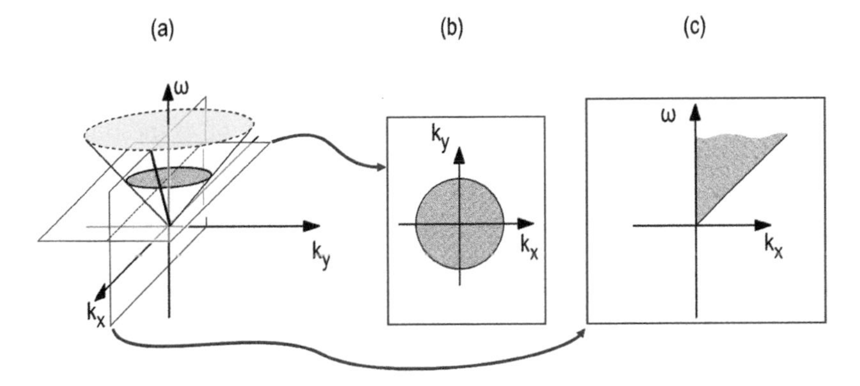

Fig. 10.1 Illustrating the decomposition of the field in plane waves. Large wavevectors correspond to evanescent waves. Only waves with a wavevector inside the circle $k_x^2 + k_y^2 < k_0^2$ are propagating.

idea with the simplest possible example: a scalar field produced by a monochromatic point-like source at the origin with amplitude $\delta(\mathbf{r})S(\omega)$. The Helmholtz equation is then given by:

$$\triangle E + k_0^2 E = \delta(\mathbf{r})S(\omega). \tag{10.3}$$

The term on the right-hand side of an otherwise homogeneous equation is a source term. If we introduce the spatial Fourier transform of the field, the Dirac $\delta(\mathbf{r})$ function transforms into a constant factor 1, and we find:

$$(-k^2 + k_0^2)E(k_x, k_y, k_z) = S(\omega). \tag{10.4}$$

This equation has a clear meaning. The term on the right-hand side describes the spectral content of the source both for frequency and spatial frequencies. Since we used a point-like source, it contains all the spatial frequencies. This term drives the field at all spatial frequencies. In other words, while vacuum contains only electromagnetic modes with $k = k_0 = \omega/c$, a point-like source can drive the field at different spatial wavevectors imposed by the spatial spectral content of the source. To summarize, the existence of a point-like source generates large spatial frequencies. Finally, we provide the explicit form of the solution of Eq. (10.3) taking $S(\omega) = 1$. The solution of this scalar problem is a spherical wave that can be represented as a superposition of plane waves,

$$\frac{\exp(ik_0 r)}{r} = 2i\pi \int \frac{1}{\gamma} \exp[i(k_x x + k_y y + \gamma|z|)] \, \frac{\mathrm{d}k_x}{2\pi} \frac{\mathrm{d}k_y}{2\pi}, \tag{10.5}$$

where γ is a function of k_x and k_y defined by Eq. (4.9) and $r = (x^2 + y^2 + z^2)^{1/2}$. This is the Weyl expansion derived in Complement 4.A (see also (Banos, 1966; Born, 1999). More generally, the origin of the large spatial frequencies lies in the spatial structure of induced currents in the materials. The basic idea put forward in this section is that evanescent waves are intimately related to charges. Hence, solving a radiation problem is another approach to studying the near fields. This is the point of view chosen in the next section.

10.2 Radiation in the near field

In the previous sections, we have introduced the concept of evanescent waves. The near-field zone has been defined as the zone where these evanescent waves contribute significantly to the field. Here, we adopt a different point of view and we show that near field effects can be discussed without invoking evanescent waves. We start by introducing the concept of near field by inspecting the field radiated by a dipole. We then generalize the discussion to any type of source. Of particular importance here is *the possibility of using electrostatics to study the spatial structure of the electromagnetic field in the near field.* A similar analysis could be done using the field radiated by a magnetic dipole as a starting point. We would then find that magnetostatics can be used to derive the spatial structure of the fields in the near field.

To do so, we first write the expression of the electrostatic field created by a static dipole. We will next discuss the field radiated by an oscillatory dipole. We show that the latter reduces to the former for distances such that $k_0 r \ll 1$, i.e. in the near field. We finally generalize the discussion of the applicability of electrostatics for computing electric fields in the near field.

10.2.1 Dipolar electrostatic field

We start by considering an electrostatic dipole with dipole moment $\mathbf{p}$. It is the source of an electrostatic field in vacuum given by the expression

$$\mathbf{E} = \frac{-1}{4\pi\varepsilon_0 r^3}[\mathbf{p} - 3(\mathbf{u}_r \cdot \mathbf{p})\mathbf{u}_r] ,$$

where $\mathbf{u}_r = \mathbf{r}/r$ and $\mathbf{r} = (x, y, z)$. Let us highlight two well-known aspects of the electrostatic field with regard to the radiated counterpart. Firstly, the electrostatic field is largest along the axis of $\mathbf{p}$, while we know, on the contrary, that the radiated field at a large distance is zero in this direction. Secondly, it decays as $1/r^3$ instead of $1/r$ for the radiated field at large distance. Next, let us study the polarization aspect. The static field has a variable orientation. This is why the relation between the dipole and the electric field cannot be given by a scalar linear function. Instead, we need a linear relation connecting two vectors. In a chosen basis, this is given by a matrix. It turns out to be practical to give an explicit form introducing the field $\mathbf{X}$ created by a dipole moment of unit amplitude carried by the unitary vector $\vec{e}_x$, the field $\mathbf{Y}$ created by a dipole moment of unit amplitude carried by the unitary vector $\vec{e}_y$ and the field $\mathbf{Z}$ created by a dipole moment of unit amplitude carried by the unitary vector $\vec{e}_z$. It is clear that the total field created by any dipole moment may be cast in the form

$$\mathbf{E} = p_x\mathbf{X} + p_y\mathbf{Y} + p_z\mathbf{Z}.$$

This form may be further condensed into the product of a matrix whose columns are the vectors $\mathbf{X}, \mathbf{Y}, \mathbf{Z}$ multiplied by the vector $\mathbf{p}$. Since the resulting relation should not depend on the basis chosen, the relation between the two vectors is given by a tensor. Note that the tensor may be written in the dyadic form

$$\mathbf{E} = \frac{-1}{4\pi\varepsilon_0 r^3}[\overleftrightarrow{\mathbf{I}} - 3\mathbf{u}_r\mathbf{u}_r]\mathbf{p} , \tag{10.6}$$

where $\overleftrightarrow{\mathbf{I}}$ stands for the unit tensor and $\mathbf{u}_r\mathbf{u}_r$ is the dyadic notation of a tensor defined by $(\mathbf{u}_r\mathbf{u}_r)\mathbf{X} = \mathbf{u}_r(\mathbf{u}_r \cdot \mathbf{X})$. To summarize, the tensor structure is necessary to account for this elementary property: the electrostatic field is not parallel to the dipole moment.

10.2.2 Electric field in the near field: Field radiated by a dipole

Let us turn now to the field $\mathbf{E}$ radiated by an oscillating dipole with circular frequency ω and dipole moment $\mathbf{p}$ (Eq. (2.42)):

$$\mathbf{E} = \frac{k_0^2}{4\pi\varepsilon_0}\frac{\exp(ik_0 r)}{r}[\overleftrightarrow{\mathbf{I}} - \mathbf{u}_r\mathbf{u}_r + (\frac{i}{k_0 r} - \frac{1}{k_0^2 r^2})(\overleftrightarrow{\mathbf{I}} - 3\mathbf{u}_r\mathbf{u}_r)]\mathbf{p} . \tag{10.7}$$

We first note that the radiated field is not parallel to the dipole moment, so that we again need a tensor. We also note that for distances much larger than the wavelength, the terms decaying as $1/r$ dominate. Conversely, when $k_0 r \ll 1$, the terms varying as $1/r^3$ dominate and the electrostatic field given in Eq. (10.6) is recovered. This is the electrostatic limit. It corresponds to the limit where retardation effects become negligible. There are different situations where this limit can be obtained. First of all, if we consider the limit where the light velocity becomes infinite, it is seen that k_0 becomes zero. In this limit, Maxwell equations are decoupled and the propagation equation becomes the Laplace equation. Interestingly, we clearly see in the form of the field radiated by the dipole that taking this limit amounts to considering that $k_0 r \ll 1$. This condition can be cast in different forms. The frequency should satisfy $\omega \ll c/r$ and, obviously, the electrostatic form is recovered at $\omega = 0$. The distance should satisfy $r \ll \lambda/2\pi$. These conditions ensure that the $1/r^3$ terms dominate, so that the field is well approximated by the electrostatic field.

This discussion may be used to obtain a second definition of the near-field zone: it is the zone where the $1/r^3$ and $1/r^2$ terms are no longer negligible in the radiated field. According to Eq. (10.7), this zone is defined by $k_0 r < 1$ or, equivalently, by $r < \lambda/2\pi$. Thus we recover the condition which appeared in our discussion of evanescent waves; we also note here again that *the length scale which characterizes the near field is not* λ, *but rather* $\lambda/2\pi$.

This behaviour is illustrated in Fig. 10.2, where we show the field above a dipolar scatterer illuminated by a plane wave. We have plotted the real part of the electric field in a plane at three different distances from the scatterer, $z = 0.05, 0.1$, and 1 in wavelength units. The incident field is polarized along the x-axis. There are many remarkable features in this figure. Let us first note the different length scales along the X and Y axis, demonstrating that the field is very confined in the plane $z = 0.05\lambda$. It is seen that for a distance $z = \lambda/10$, the width of the field distribution is already on the order of the wavelength, illustrating the previous discussion that pointed out that the limit of the near-field zone is given by $\lambda/2\pi$. The second remarkable feature is the polarization structure. Whereas the intensity distribution depends strongly on the polarization in the near field, it is seen that at a distance $z = \lambda$, the distribution of the intensity becomes almost isotropic.

From the previous discussion, we can draw a very important conclusion. The electric field has a strongly polarized structure in the near field. As is obvious from the analytic expression and from the figure, the scalar approximation is simply meaningless in the near field. We have seen that in the near field, the spatial structure of the electric field is given by the electrostatic field. This has been found for a particular case: the dipole field. However, this is a general result, as will be shown in the next section.

10.2.3 Electric field in the near field: Arbitrary charge distribution

Here, we generalize the previous discussion to any charge distribution. We start by considering the retarded scalar potential for an arbitrary time-dependent charge distribution $\rho(\mathbf{r}, t)$ in vacuum. The scalar potential produced by a charge density $\rho(\mathbf{r}, t)$ in a finite volume V is given by

$$\Phi(\mathbf{r}, t) = \frac{1}{4\pi\varepsilon_0} \int_V \frac{\rho(\mathbf{r}', t - |\mathbf{r} - \mathbf{r}'|/c)}{|\mathbf{r} - \mathbf{r}'|} \mathrm{d}x'\mathrm{d}y'\mathrm{d}z'.$$

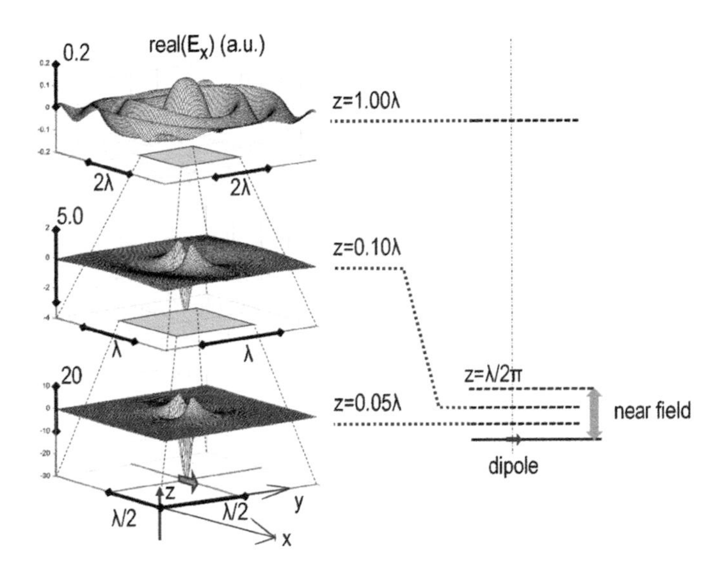

Fig. 10.2 Real part of the electric field generated by an electric dipole aligned along the x-axis located at $z = 0$.

It is seen that the retardation term $|\mathbf{r} - \mathbf{r}'|/c$ becomes negligible when light velocity tends to infinity. More precisely, we consider systems with a characteristic size L such that L/c is negligible when compared to a typical time scale of the charge distribution. For a monochromatic potential $\Phi(\mathbf{r}, t) = Re[\Phi(\mathbf{r})\exp(-i\omega t)]$ and $\rho(\mathbf{r}, t) = Re[\rho(\mathbf{r})\exp(-i\omega t)]$, the characteristic time scale is the period $2\pi/\omega$. The complex amplitude $\Phi(\mathbf{r})$ can be cast in the form:

$$\Phi(\mathbf{r}) = \frac{1}{4\pi\varepsilon_0}\int_V \rho(\mathbf{r})\frac{\exp[i\omega\frac{|\mathbf{r}-\mathbf{r}'|}{c}]}{|\mathbf{r}-\mathbf{r}'|}\mathrm{d}x'\mathrm{d}y'\mathrm{d}z'.$$

The retardation term becomes $\exp\left[i\omega\frac{|\mathbf{r}-\mathbf{r}'|}{c}\right]$. We now see that for monochromatic sources, this retardation term is nothing but the phase lag of the fields emitted at $\mathbf{r}$ by different points located at different positions $\mathbf{r}'$. We now consider that the retardation can be neglected. In this limit, the retarded potential becomes:

$$\Phi(\mathbf{r}, t) = \frac{1}{4\pi\varepsilon_0}\int_V \frac{\rho(\mathbf{r}', t)}{|\mathbf{r}-\mathbf{r}'|}\mathrm{d}x'\mathrm{d}y'\mathrm{d}z'.$$

It is seen that the scalar potential has the same structure than the electrostatic scalar potential. The time dependence of the potential reproduces the time dependence of the charge distribution without delay. This approximation is valid in three cases: (i) the circular frequency ω is zero; (ii) light velocity tends to infinity; (iii) $|\mathbf{r} - \mathbf{r}'| \ll \lambda/2\pi$. The third condition defines the near-field zone. It is the space zone where retardation effects can be neglected, so that it has an extension smaller than $\lambda/2\pi$. In this zone, the spatial structure of the electric field is given by the electrostatic form.

Let us give some numerical examples of $\lambda/2\pi$ in different frequency regimes. In the visible, the near-field zone corresponds to a length scale of 100 nm, so that near-field optics coincide with optics at the nanoscale. For microwaves at 1 GHz (cell phones are operated at 1.8 GHz), $\lambda = 30\,\text{cm}$, so that $\lambda/2\pi \approx 5\,\text{cm}$. For radiowaves at 200 kHz, $\lambda = 1.5\,\text{km}$, so that $\lambda/2\pi \approx 250\,\text{m}$, so that any ordinary building that has an antenna operating at this frequency is within its near field.

10.2.4 Magnetic field in the near field: Arbitrary current distribution

In this section, we extend the previous discussion to the case of a distribution of current $\mathbf{j}(\mathbf{r}, t)$ without any charge distribution. The retarded vector potential is given by the retarded potential solution Eq. (2.21). For a monochromatic current distribution, the vector potential can be cast in the form of Eq. (2.22). It is clearly seen that the retardation can be neglected if $|\mathbf{r} - \mathbf{r}'|\omega/c \ll 1$. In that case, $\exp(i|\mathbf{r} - \mathbf{r}'|\omega/c) \approx 1$, so that the retarded vector potential becomes identical to the magnetostatic solution.

10.3 The quasi-electrostatic and quasi-magnetostatic fields

In the previous section, we discussed the validity conditions of an electrostatic and magnetostatic approximation. This discussion is actually a standard discussion in electricity. When dealing with capacitors and coils, electric and magnetic fields are treated separately: a capacitor is considered to host electric energy, while a coil hosts only magnetic energy. These are approximations which are known as quasi-electrostatic approximations and quasi-magnetostatic approximations, respectively. The same discussion applies in near-field optics. While in electrostatics the magnetic field is strictly null and, conversely, for a static current, the electric field is strictly null, in the slowly varying regime, both electric and magnetic fields are non-zero but one of them dominates.

10.3.1 Quasi-electrostatic field

Here, we consider a system where the sources of the electromagnetic field are given by time-dependent charge densities. Examples are a capacitor in electricity and a polarized tiny particle with time-dependent surface charge in optics. The spatial structure of the electric field in the near field is given by $\mathbf{E}_C(\mathbf{r}, t) = -\nabla_{\mathbf{r}}\Phi(\mathbf{r}, t)$. In this expression, we have kept only the longitudinal part of the field hereafter denoted the Coulomb field $\mathbf{E}_C$ and neglected the contribution of the transverse part of the field $i\omega\mathbf{A}$, hereafter called the Faraday field $\mathbf{E}_F$. In other words, we have ignored the contribution of the potential vector to the electric field. This approximation is strictly valid in the static regime and is not correct in the dynamic regime. We now provide an estimate of the contribution of the vector potential to the electric field and derive a validity condition for this approximation.

We can make a dimensional estimate of the contributions of Φ and $\mathbf{A}$ to the electric field. Using the Lorenz gauge $\nabla \cdot \mathbf{A} = i\omega\Phi/c^2$, we find

$$A \sim \frac{\omega L}{c^2}\Phi,$$

where L is a characteristic length of the system. Similarly, the Coulomb field $\mathbf{E}_C(\mathbf{r}, t) = -\nabla_{\mathbf{r}}\Phi(\mathbf{r}, t)$ is given by $E_C \sim \Phi/L$, whereas the Faraday field is given by $E_F \sim \omega A$. We thus find:

$$E_F \sim E_C \left[\frac{\omega}{c} L\right]^2.$$

It follows that in the near field, the Faraday field is negligible. It can also be checked that the electric energy is much larger than the magnetic energy. This analysis is strictly equivalent to the slowly time-varying approximation or quasi-electrostatic approximation used in electricity.

10.3.2 Quasi-magnetostatic field

In the previous section, we considered that sources are electrical charges. An alternative situation corresponds to a current distribution with a spatial extension much smaller than the wavelength and no charge distribution. This is the case of a current loop smaller than the wavelength. We can repeat the same type of analysis. We start by assuming that the displacement current $\varepsilon_0 \frac{\partial \mathbf{E}}{\partial t}$ is negligible compared to $\mu_0 \mathbf{j}$, so that the magnetic field can be estimated to be

$$\frac{B}{L} \approx \mu_0 \mathbf{j}.$$

The Faraday electric field then follows from the Maxwell–Faraday equation,

$$\frac{E_F}{L} \approx \omega B.$$

It can be checked that the magnetic energy $U_M = B^2/2\mu_0$ is much larger than the electric energy $U_E = \varepsilon_0 E_F^2/2$:

$$\frac{U_M}{U_E} \approx \left[\frac{\omega}{c} L\right]^2.$$

This is equivalent to the quasi-magnetostatic approximation in electricity (Hauss and Melcher, 1989; Zangwill, 2013).

10.3.3 Quasi-static limit and polarization: A discussion in Fourier space

In the previous section we took as a starting point either a current distribution or a charge distribution, namely a quasi-magnetostatic system or a quasi-electrostatic system. Here, we change our perspective and start from an electromagnetic field in vacuum which can be either s-polarized (TE) or p-polarized (TM). Interestingly, we will find that the s-polarized field contains mostly magnetic energy, while the p-polarized field contains mostly electric energy. In other words, the splitting into quasi-electrostatics and quasi-magnetostatics in direct space becomes in Fourier space a splitting between TM (p) and TE (s) polarization.

Let us stress that the well-known relations between the electric and the magnetic fields for a plane wave are no longer valid for an evanescent wave characterized by k_x, k_y, γ. In particular, we will see that the fields are not necessarily orthogonal in the geometrical sense. We will also show that the ratio between the modulus of the electric and magnetic field in vacuum is no longer given by the familiar plane wave ratio $|B| = |E|/c$, so that the energy is no longer equally shared between electric and magnetic fields. Nevertheless, the starting point for finding the connection between the electric field and the magnetic field is still the Maxwell–Faraday equation $\nabla \times \mathbf{E} = i\omega \mathbf{B}$.

We start by considering an evanescent wave with a wavevector given by $\mathbf{k} = k_x\mathbf{e}_x + \gamma\mathbf{e}_z$. Here, we define a reference plane containing the propagation axis Ox and the decay axis Oz. This plane allows defining two linear polarizations: we define the s-polarized field when the *electric* field is perpendicular to this plane and the p-polarized field when the *magnetic* field is perpendicular to the plane. Hence, the electric field is given by $\mathbf{E} = (0, E, 0)$ for an s-polarized field. The magnetic field follows from the Maxwell–Faraday equation $\nabla \times \mathbf{E} = i\omega\mathbf{B}$ and is then given by $\mathbf{B} = (-\gamma, 0, k_x)E/\omega$. It can be checked that this is a transverse solution of Maxwell equations, as $\nabla \cdot \mathbf{E} = i\mathbf{k} \cdot \mathbf{E} = 0$. Note, however, that for an evanescent wave, $\gamma = i|\gamma|$ is a purely imaginary complex number, so that the equality $\mathbf{k} \cdot \mathbf{B} = 0$ and $\mathbf{k} \cdot \mathbf{E} = 0$ does not have the usual interpretation in terms of perpendicular vectors.

Let us now compare the modulus of the magnetic and electric field. $|\mathbf{B}|^2 = \mathbf{B} \cdot \mathbf{B}^* = (|\gamma|^2 + k_x^2)|\mathbf{E}|^2/\omega^2$ and not $(\gamma^2 + k_x^2)|\mathbf{E}|^2/\omega^2$. For a propagating wave, γ is real, so that $|\gamma|^2 + k_x^2 = \gamma^2 + k_x^2 = \omega^2/c^2$. Hence, the equality yields the usual result,

$$|\mathbf{B}| = \frac{|\mathbf{E}|}{c}.$$

For an evanescent wave, we now have $|\gamma|^2 = -\gamma^2$, so that

$$|\mathbf{B}| = \frac{|\mathbf{E}|}{c}\sqrt{2\frac{k_x^2}{k_0^2} - 1}.$$

It follows that for large wavevectors such that $k_x \gg k_0$, we have

$$\frac{cB}{E} \approx \sqrt{2}\frac{|k_x|}{k_0}.$$

Clearly, the magnetic field can be much larger than E/c. In other words, s-polarized evanescent waves contain mostly magnetic energy in the near field. A similar analysis (see Exercise 10.2) shows that for p-polarized evanescent waves, the electric field dominates and we find

$$\frac{E}{cB} \approx \sqrt{2}\frac{|k_x|}{k_0},$$

so that p-polarized evanescent waves can be used to represent subwavelength confined electric fields. In summary, the relation $\frac{E}{cB} = 1$ is no longer valid for evanescent waves. The electric and magnetic fields tend to decouple in the non-retarded (static) limit. Correspondingly, the *p-polarized evanescent waves can be associated with the electric field, whereas the s-polarized evanescent waves can be associated with the magnetic field.*

10.4 Revisiting simple concepts in the near field

10.4.1 Phase in the near field

At first glance, introducing phases in the near field seems contradictory to the statement that the near field is correctly described in the electrostatic limit. Indeed, in electrostatics, the fields are real and the phase is not a relevant concept. Note, however, that the dielectric constant is a relevant concept in electrostatics. Now, when dealing with fields in the optical regime, we need to describe the response of the materials using their dielectric properties at the optical frequencies. Hence, the dielectric constants are complex numbers in the frequency domain. For instance, a metallic nanoparticle will be polarized by an incident field. The induced dipole moment depends on the phase of the polarizability. It will be discussed in Chapter 14 that the electrons in the metallic nanoparticle have a collective motion which can be excited resonantly by the incident field. Depending on the frequency detuning between the driving incident field and the resonance frequency, the phase can change from zero to π. Clearly, the field scattered by the particle has a phase that depends on the induced dipole phase. This example shows that the phase of the fields is strongly related to the material properties. The phase may vary on a subwavelength scale if the material (e.g. a chain of nanoparticles) varies spatially. Other examples involve the familiar lumped elements (coils, capacitors) of electrical circuits which introduce phases in the response of subwavelength objects. We emphasized above the importance of the polarization in the structure of the electric field in the near field. Here, we emphasize that the phases are different for different terms of the electric field. This result is obvious upon inspection of the structure of the field radiated by a dipole. Indeed, it is seen in Eq. (10.7) that the terms varying as $1/r^2$ are in quadrature with the terms varying as $1/r$ or $1/r^3$.

10.4.2 Reflection factor in the near field

We now briefly consider the concept of reflection in the near field. We start by discussing the intuitive picture of a collimated beam impinging on a surface. We note that such a beam has a width which is much larger than the wavelength, so that the measurement of the reflected amplitude gives information averaged over the size of the beam. If we examine more closely the definition of the Fresnel reflection factor introduced in Chapter 4, we realize that it is defined as the response of a planar interface to an incident plane wave with a well-defined incident wavevector. From that perspective, reflection appears to be perfectly localized in k-space and poorly defined in direct space. Hence, the *reflection factor cannot be defined locally with arbitrary resolution.* However, it is always possible to decompose an incident beam into incident plane waves, apply a scattering operator to each of them, and sum the scattered fields (see Exercise 10.4). The smallest possible spatial resolution to define a local reflection factor is thus given by the smallest spot size on the order of the wavelength. This spatial resolution comes at the cost of losing angular dependence.

While Fresnel reflection factors are usually defined for real angles and therefore only for wavevectors with a modulus smaller than ω/c, it turns out that the Fresnel reflection factors can also be defined for incident evanescent waves with large wavevectors k_x. Care has to be taken when defining an incident evanescent wave. Such a wave decays exponentially as it approaches the interface and diverges at infinity, which seems to be unphysical. However, this evanescent wave can be used to represent the field within a *finite volume above the interface.* Consider, for instance, a point-like dipole located at $\mathbf{r} = (0, 0, z_0)$ above an interface located at $z = 0$. The field produced by this dipole contains evanescent waves, as discussed above. These have the form $\exp(i\gamma|z - z_0|)$, so that they decay away from the source. Since the evanescent waves growing away from the interface are only defined between 0 and z_0, there is no actual divergence issue.

We now ask the question: what is the reflection and transmission factor for an evanescent wave? It turns out that the same procedure can be applied to either propagating or evanescent waves when using a real wavevector k_x in order to describe the incident wave. With these notations (as opposed to using angles), there are no differences between the two cases $k_x > k_0$ and $k_x < k_0$. It follows that the usual forms of Fresnel reflection factors are valid. We find the following results for the two polarizations:

$$r_p = \frac{\varepsilon_1\gamma_2 - \varepsilon_2\gamma_1}{\varepsilon_1\gamma_2 + \varepsilon_2\gamma_1} \approx \frac{\varepsilon_1 - \varepsilon_2}{\varepsilon_1 + \varepsilon_2},$$

$$r_s = \frac{\gamma_1 - \gamma_2}{\gamma_1 + \gamma_2} \approx \frac{\varepsilon_2 - \varepsilon_1}{4}\frac{k_0^2}{k_x^2}. \tag{10.8}$$

In other words, the s-polarized reflectivity tends to zero, whereas the p-polarized reflection factor tends to the very factor that appears when computing the electrostatic image of a point-like charge above an interface. Hence, we recover the electrostatic and magnetostatic limits. In particular, we note that the magnetic field is continuous across a dielectric interface, so that no reflection is expected. By contrast, we found that the reflection factor for p-polarization tends to a limit which is consistent with the fact that the normal electric field is not continuous across the interface. It is important to note that this near-field reflectivity can be larger than unity. While this would contradict energy conservation in the far field, here the reflection does not have any energy conservation constraints. Indeed, we are dealing with evanescent waves, so that they do not contribute to the energy flux along the z-axis, as discussed in Chapter 4. A reflection factor larger than 1 indicates that the localized source has an image charge larger than the source. We know that the image charge is an effective concept to account for the surface charges that screen the electric field at an interface. A large surface charge may appear at a particular frequency if a surface charge oscillation mode is resonantly excited. The possibility of resonant surface waves is one of the most important aspects of nanophotonics. This resonance is associated with a microscopic resonance of the polarization of the material. Chapter 13 is devoted to the study of these excitations, called surface polaritons.

10.4.3 Polarization

When discussing polarization in isotropic media, it is often assumed that the electric field is perpendicular to a privileged direction given by the propagation direction of a collimated beam, so that only two components need to be specified to characterize the polarization. Obviously, there is no such privileged direction in the near field, so that one needs to consider all field components. Let us also point out that when dealing with an evanescent wave, the condition $\nabla \cdot \mathbf{E} = 0$ yields $k_x E_x + \gamma E_z = 0$, as for a propagating wave. Hence, we find

$$E_x = -\frac{i|\gamma|}{k_x} E_z,$$

indicating that the electric field is elliptically polarized in the (x, z) plane.

10.5 Energy confinement in the near field

10.5.1 Order of magnitude

One of the key issues of near-field optics is the possibility of confining electromagnetic energy and therefore drastically modifying light–matter interaction processes. We illustrate this statement with two simple examples. The first example is absorption and the second example is non-linear optics. In both cases, the phenomenon depends on the value of the electric field.

We now discuss how to generate large electric fields by confining the electromagnetic energy. The simplest approach is to focus a laser beam. Let us make a rough order-of-magnitude estimate. The electric field of a focused beam is typically given by

$$P = \frac{\varepsilon_0 c}{2} E^2 A,$$

where A is the beam cross section and P is the laser power. For $P = 1\,\text{mW}$ and $A = 10^{-12}\,\text{m}^2$, we find a field on the order of $E \approx 6\,10^5\,\text{Vm}$. We now consider a single photon confined in a cavity with typical size $L = 10\,\text{nm}$. The corresponding field is given by writing that the electric energy is the energy of a single photon:

$$\frac{\varepsilon_0 E^2}{2} V = \hbar\omega.$$

It follows that the typical field for a $3\,\text{eV}$ photon is on the order of $7\,10^6\,\text{Vm}$. In other words, confining the fields to nanometre-scale lengths provides a simple way of generating very large electric fields, even with few photons.

10.5.2 Confining electromagnetic energy

In this section, we explore the possibility of confining the electromagnetic energy in three dimensions on a single length scale a. We first consider the possibility of realizing an electromagnetic oscillator storing pure electromagnetic energy. We will show that such an oscillator cannot have a volume much smaller than the cube of the wavelength. By contrast, allowing the presence of charges, it becomes possible to design a subwavelength oscillator.

We assume that the oscillator has a typical size a along the three cartesian axes. Since the electromagnetic energy is the sum of electrical and magnetic energy, the oscillator energy density oscillates between a form $\varepsilon_0 E^2/2$ and a form $B^2/2\mu_0$. From the Maxwell equation $\nabla \times \mathbf{E} = i\omega\mathbf{B}$, we can estimate $B \approx E/a\omega$, assuming that a is the only relevant length scale of the problem. Since we are looking for a subwavelength resonance, the wavelength is not *a priori* considered to be a relevant length scale. The equality of the electric and magnetic energy yields:

$$\varepsilon_0 E^2 \approx E^2/[\mu_0 (a\omega)^2].$$

Hence, the typical size a of the oscillator is on the order of $c/\omega = \lambda/2\pi$. This order of magnitude analysis confirms what is known from cavities: the cavity size is on the order of $\lambda/2$.

However, if the system is not only made of fields in vacuum but includes material systems such as charges and wires, the energy can be stored in mechanical degrees of freedom of these systems. The simplest possible example of an oscillator able to store electromagnetic energy in a tiny volume is a hydrogen atom. The electron energy is the sum of electromagnetic interaction (Coulomb energy) and kinetic energy. Another example is a simple LC circuit. At low frequencies, a solenoid stores magnetic energy and a capacitor stores electrical energy in a subwavelength volume. We will see in Chapter 14 that electromagnetic energy at optical frequencies can be stored in nanoscale volumes by metallic nanostructures using plasmonic excitations. In that case, the energy oscillates between electrostatic energy and electron kinetic energy.

Exercises

Exercise 10.1 Diffraction by a subwavelength grating in the near field. We consider a grating in the plane $z = 0$, illuminated at normal incidence by a monochromatic plane wave with circular frequency ω. The field is described by a complex scalar amplitude $\Psi(x)$. The system is translationally invariant along the axis Oy. The amplitude transmission factor is given by:

$$t(x, y) = 1 + m\cos\left(\frac{2\pi x}{d}\right).$$

We consider two gratings with different periods $d = \lambda/10$ and $d = 1.5\,\lambda$. Show that the diffracted field is either propagating or evanescent. What is the cutoff spatial frequency between the two regimes? What would the cutoff spatial frequency be for a grazing incidence? What is the decay length of the evanescent field when $d = \lambda/10$?

Exercise 10.2 Relation between electric and magnetic energy for a p-polarized evanescent wave. Derive the relation between the amplitudes of electric and magnetic field in vacuum for a p-polarized evanescent wave with wavevector $k_x \gg k_0$:

$$\frac{E}{cB} \approx \sqrt{2}\frac{|k_x|}{k_0}.$$

Compare the electric and magnetic energy densities in vacuum.

Exercise 10.3 Electric field around a nanoparticle. We consider a spherical particle with radius $a = 10\,\text{nm}$ illuminated by a plane wave with an electric field linearly polarized along the x-axis. The particle is located at the origin. The field induces a dipole moment along the x-axis. Qualitatively sketch the electric field lines generated by the induced dipole in the near-field zone. Deduce qualitatively the intensity distribution in a plane $x = 50\,\text{nm}$ and in a plane $z = 50\,\text{nm}$. Assume that the intensity can be detected in the near field. Which polarization should be used to image the object?

Exercise 10.4 Local reflection of a focused propagating beam. In this exercise, we explore in more detail the concept of local reflection. In everyday life, we can define the reflectivity of a given object at a given position and for a given angle. When trying to define the local reflectivity of a subwavelength area, the situation is more subtle. To address this issue, we consider a beam illuminating a finite area.

(1) We consider a monochromatic 2D light beam depending only on x and z with TM polarization. Its amplitude is given by the magnetic-field $H_y(x,z)$ component along the (Oy) axis. The beam propagates towards negative z. Its amplitude is known in the plane $z = z_0 > 0$. Show that $H_y(x,z)$ can be cast in the form:

$$H_y(x,z) = \int \frac{dk_x}{2\pi} H_y(k_x, z_0) \exp[ik_x x - i\gamma(z - z_0)]. \tag{10.9}$$

(2) The beam impinges on a surface at $z = 0$ with reflection factor $r(k_x)$. Show that the reflected field is given by:

$$H_y(x,z) = \int \frac{dk_x}{2\pi} r(k_x)\, H_y(k_x) \exp(ik_x x + i\gamma(z + z_0)). \tag{10.10}$$

(3) In this question, we consider the case of a beam with uniform amplitude $H_y(k_x) = H_0$ for $|k_x| < \omega/c$ and null if $|k_x| > \omega/c$. Show that the beam waist of the incident field in the plane $z = z_0$ is on the order of λ.

(4) When dealing with an interface with a reflection factor depending on the incident angle, is it possible to define a reflection factor independent of the incident beam structure?

Exercise 10.5 Reflection of evanescent waves: The electrostatic limit. In this exercise, we consider the case an incident field generated by a point-like source in the near field of a flat interface. Thus, the incident field contains a broad spectrum of wavevectors, including evanescent waves. Here, we consider a 2D cylindrical incident wave invariant along the y-axis generated by a line source located at $(x = 0, z_0)$. We are interested in the very near-field case $z_0 \ll \lambda$, so that evanescent waves provide the leading contribution to the field. The incident field is given by a Hankel function,

$$\frac{-i}{4} H_0^1(k_0) = \frac{-i}{2} \int_{-\infty}^{\infty} \frac{dk_x}{2\pi} \exp[ik_x x] \frac{\exp[i\gamma|z - z_0|]}{\gamma},$$

where $\gamma = [k_0^2 - k_x^2]^{1/2}$ and $Im(\gamma) \geq 0$. Write the form of the reflected field $(z > 0)$ as a function of the reflection factor $r(k_x)$ defined for all real values of k_x. We now use the Fresnel reflection factor (which is valid for arbitrary large values of k_x),

$$r(k_x) = \frac{\varepsilon\gamma - \gamma_1}{\varepsilon\gamma + \gamma_1},$$

where $\gamma_1 = [\varepsilon k_0^2 - k_x^2]^{1/2}$. Show that in the limit $k_x \gg k_0$, the reflection factor *does not depend on* k_x. It follows that for very large wavevectors, there is no distortion upon reflection. What is the reflected field in this limit? Compare with the electrostatic image theory. Here, we have seen that the image of a point can be described with a simple reflection factor which is the form of the electrostatic reflection factor that is found when defining an electrostatic image to compute the electrostatic potential of charges above interfaces.

11 Introduction to superresolution optical imaging

Optical microscopy is a very successful technique for imaging objects with visible light down to the microscale. The purpose of this chapter is to explore the limits of optical imaging when dealing with small objects. As imaging is an everyday experience, we take for granted that we know what an image is. However, in order to explore its fundamental limits, we need a precise definition. The first goal of this chapter is to summarize the fundamental limits of optical imaging with conventional microscopes and to explain the physical origin of the so-called diffraction limit for resolution. We then introduce the key ideas that can be used to overcome this limit, with a special emphasis on near-field optical microscopy. We also introduce different techniques that allow us to produce images with a resolution better than the half-wavelength using far-field measurements. We will emphasize that different optical imaging techniques yield different kinds of information (e.g. local spectral reflectivities, topography, permittivity map).

11.1 Imaging and resolution

The goal of microscopy is to provide images of small features of an object. This should not be confused with detecting a small object. The difficulty when detecting a small object is the low signal, which is a sensitivity issue. Since it is possible to detect single photons emitted in the visible spectrum by single molecules using photomultipliers or avalanche photodiodes, there is no fundamental limit regarding the size of the object to be imaged. The key point is whether two small features or two small objects can be distinguished when the distance between them is reduced. The smallest distance that allows them to be distinguished is called the resolution of the imaging system.

11.1.1 What is optical imaging?

When thinking of an image, we first think of a picture (e.g. a portrait or a landscape) taken by a digital camera. The image consists of a number of pixels with different intensities and different colours. If we analyse what type of information is contained in the image, we come to the conclusion that a given spectral reflectivity is assigned to each pixel of the picture. This pixel reflectivity is supposed to be representative of the local reflectivity of the object itself. With this definition of imaging, the resolution depends on the smallest illuminating spot that we can produce.

Introduction to Nanophotonics. Henri Benisty, Jean-Jacques Greffet, and Philippe Lalanne, Oxford University Press.
 DOI: 10.1093/oso/9780198786139.003.0011

In the next section, we briefly remind the origin of the Abbe resolution limit of optical microscopes and emphasize the physical parameters that control the resolution. Sections 11.2 and 11.3 provide a short introduction to near-field microscopy. Section 11.4 is a brief discussion of the techniques used to achieve superresolution by solving the inverse problem using far-field data. Finally, we give a brief introduction to techniques based on imaging using localization of single emitters located on the object. Note that this chapter deals with optical imaging, namely imaging using optical waves. We do not consider here techniques based on electrical or mechanical probes such as scanning electron microscopy or atomic force microscopy.

11.1.2 What is the resolution limit of a confocal microscope?

Here, we estimate the spot size in the case of a confocal microscope. A confocal microscope produces a local illumination of the sample. This local illumination is produced by imaging a pinhole on the sample. The reflection or transmission is recorded. The image is formed by scanning the sample and recording the signal as a function of sample position. It is not our goal to give a detailed account of confocal microscopy. Instead, we introduce qualitatively the origin of the resolution limit. The so-called Rayleigh criterion is used to estimate the possibility of distinguishing between two different point-like objects. The two objects should be distinguishable if they are separated by a distance corresponding to a characteristic diameter Δx of the smallest spot. We consider an objective with diameter Φ and focal length f illuminated by an incident plane wave at normal incidence. The light is focused on the image focal plane. The largest incident angle θ_M is thus given by $\tan\theta_M = \Phi/2f$. Assuming that the medium has a real refractive index n, the corresponding wavevector along the x-axis is given by $k_{x,M} = n\frac{2\pi}{\lambda}\sin\theta_M$. Using the property of Fourier transforms, $\Delta x \Delta k_x \approx 2\pi$ and $\Delta k_x = 2k_{x,M}$, we find

$$\Delta x \approx \frac{\lambda}{2n\sin\theta_M}. \tag{11.1}$$

This result is usually derived in the framework of classical diffraction theory and assuming a circular diaphragm. The intensity distribution is then given by the so-called Airy function with a width $\Delta x = 0.61\lambda/(n\sin\theta_M)$. This is the well-known Abbe limit of resolution, also called the diffraction limit. It is due to the impossibility of producing a smaller spot using an objective. The origin of this limit can be traced to the fact that waves with very large wavevectors are evanescent, as discussed in Chapter 4. We now discuss how to optimize the resolution. The first option suggested by Eq. (11.1) is to use an objective with large numerical aperture $NA = n\sin\theta_M$. It is seen that by operating in a liquid with a large refractive index, it is possible to increase the NA. This is called an immersion objective. Another option is to place the sample on a lens made with a solid material to take advantage of a refractive index as large as n=4 for Si in the infrared. This is the so-called solid immersion lens (SIL). In the above discussion, we have shown that resolution stems from large wavevectors. Hence, working in a medium with refractive index n allows us to work with wavevectors n times larger. Another point of view comes from considering that for a given objective, working in a medium with refractive index n amounts to using a wavelength λ/n instead of λ. So far, we have discussed the fundamental limit of resolution. In practice, resolution is often limited by aberrations of the objective that may broaden the focal spot, so that the fundamental limit is attained with objectives specifically designed to minimize aberrations.

Finally, we examine the definition of imaging as a map or local reflectivity. We first note that when the microscope is operated with a large numerical aperture in order to increase the resolution, the sample is illuminated by a converging beam. We emphasize that the concept of local reflectivity is ambiguous. Indeed, the reflectivity is well defined as the response of a plane interface to a *plane wave.* Since the reflectivity for a plane wave depends on the angle of incidence, the reflected amplitude of a converging beam *is not an intrinsic property of the sample.* It does depend on the illumination conditions. To illustrate this discussion, we consider a 2D incident beam given by the superposition of plane waves:

$$E_i(x, z) = \int \frac{\mathrm{d}k_x}{2\pi} E_i(k_x) \exp(ik_x x - i\gamma z).$$

The reflected field is given by changing the sign of propagation along Oz and multiplying the amplitude of each plane wave by the reflectivity:

$$E_r(x, z) = \int \frac{\mathrm{d}k_x}{2\pi} r(k_x) E_i(k_x) \exp(ik_x x + i\gamma z).$$

Assuming that the reflectivity $r(k_x)$ is independent of the angle of incidence $k_x = \frac{2\pi}{\lambda} \sin\theta_i$, we obtain

$$E_r(x, z) = r \int \frac{\mathrm{d}k_x}{2\pi} E_i(k_x) \exp(ik_x x + i\gamma z) = rE_i(x, -z),$$

which is symmetrical to the incident beam with an amplitude rE_i. In general, the reflection factor depends on the angle, so that it cannot be factorized out of the integral. As different plane waves are reflected with different amplitudes and phases, the beam is distorted upon reflection on an interface. When distortion effects are limited, the beam can be described using angular, longitudinal (Imbert effect), and transverse (Goos–Hänchen effect; see Exercise 5.6) shifts.

In summary, the microscopy approach to imaging is based on the concept of local reflectivity. This leads to a fundamental limit given by the so-called diffraction limit, which is the smallest size of the spot that can be formed using an objective. This also introduces a difficulty when analysing the data, as the concept of local reflectivity is no longer an intrinsic property of the sample. More generally, determining the connection between the signal detected and the sample properties is not a trivial issue. This is often called the contrast mechanism. This question can be addressed by developing a procedure for computing the light scattered by the object. This question is discussed in Chapter 12.

11.2 Principles of near-field scanning optical microscopy (NSOM)

In this section, we introduce the principle of NSOM, also called scanning near-field optical microscopy (SNOM).

11.2.1 Beating diffraction limit using near-field localized illumination or detection

In the previous discussion, the ability to resolve fine structures was based on local illumination. We saw in Chapter 4 that the ability to produce a small spot is limited by the fact that waves with large wavevectors containing information on subwavelength details cannot propagate in vacuum. Hence, the idea is to bring a tiny source of light close to the object. The simplest method for realizing such a source is to use an opaque screen with a tiny aperture. It is then possible to form an image by scanning the object below the aperture and recording the scattered field, either in transmission or reflection. A contrast is expected as the object passes below the aperture. Of course, the object needs to be close to the aperture, because otherwise the light spot is no longer confined. In summary, the object needs to be scanned in the near field of a tiny aperture, hence the name: near-field scanning optical microscopy.

Ideally, a hole with diameter D in an opaque screen would produce a light source of a dimension given by D. This idea was first proposed by Synge in 1928 (Novotny, 2007*a*). It was impossible to implement such an idea at that time, so it was forgotten. The idea was discovered again by Ash and Nicholls and implemented in the microwave domain in 1972 (Ash and Nicholls, 1972). While the authors were aware of the importance of extending this concept to the optical regime, they concluded their paper indicating that it would probably be impossible to achieve in optics for practical reasons. Indeed, as discussed in the previous chapter, if we aim at observing an object with typical length scale $l = 50\,\text{nm}$, it is necessary to bring a spot with a diameter of 50 nm at a distance to the object of less than $l/2\pi \approx 8\,\text{nm}$. This was a formidable challenge before the invention of the scanning tunnelling microscope but became a routine task after the development of the scanning setups for near-field microscopy. The first optical image obtained using optical near-field microscopy was reported by D. Pohl in 1984 (Pohl et al., 1984). It was obtained using a small light source that was scanned on a sample. The signal was detected in transmission.

The principle of the technique is shown in Fig. 11.1. It consists of a tiny aperture with diameter Φ, usually drilled on an optical fibre coated with metal. Hence, the size of the illuminating spot is dictated by the aperture and roughly given by the aperture diameter denoted by Φ. However, in order to take advantage of this confinement, the aperture must be scanned close to the object. Indeed, we know that the field does not stay confined. More precisely, if the field is confined, it contains evanescent waves with wavevectors larger than $2\pi/\lambda$. We also know that such evanescent waves decay

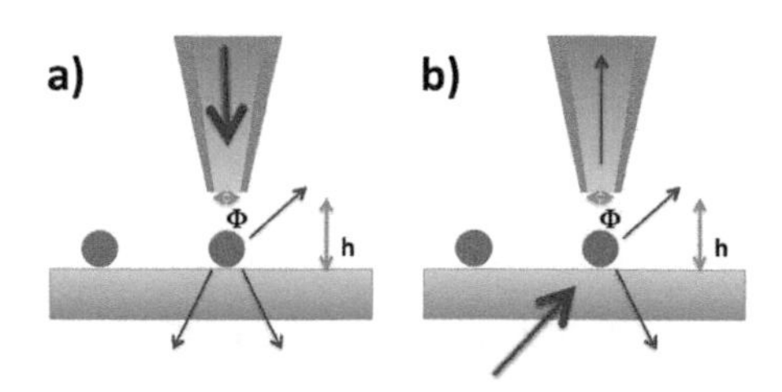

Fig. 11.1 Principle of NSOM: a) illumination mode; b) detection mode.

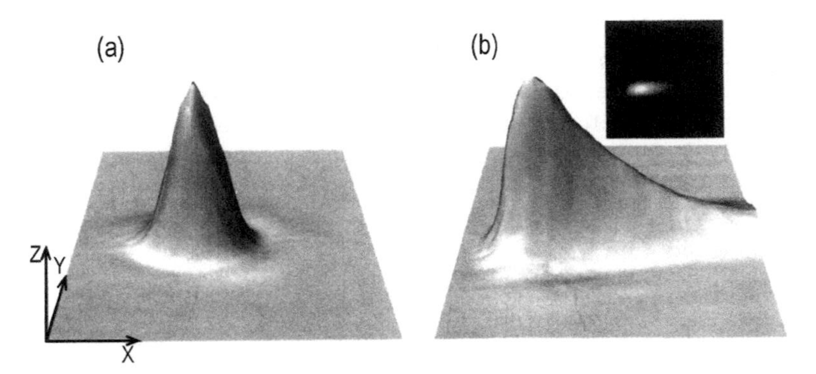

Fig. 11.2 Imaging evanescent electromagnetic fields. a) Image of the evanescent field intensity of a focused red beam internally incident on a bare prism face at greater than critical angle. The scan range is 40μm × 40μm. b) Image of the same red beam but with the beam impinging internally on a 53-nm-thick silver film exciting a surface plasmon. The exponentially decaying tail is due to surface plasmon propagation. Inset: 2D version of the image. Reprinted with permission from ref. (Dawson et al., 1994). © 1994 American Physical Society.

exponentially. At a distance h from the aperture, they decay as $\exp(-kh)$, so that there is a wavevector k cutoff given by $k_{\text{cutoff}} = 1/h$. The corresponding length scale is given by $2\pi/k = 2\pi h$. In other words, the resolution is roughly given by $max(\Phi, 2\pi h)$, where max denotes the largest value.

The first demonstration of an optical image with superresolution stimulated a lot of subsequent work. Other near-field imaging techniques are based on a wide illumination and a local detection. Here again, an optical fibre with a tiny aperture can be used to collect the signal locally. An example of the setup is shown in Fig. 11.1. There is no fundamental difference between the illuminating mode and the detection mode, as can be realized by invoking the reciprocity theorem. When the incident light illuminating the sample comes from the surface of a high-index material at an angle of incidence larger than the critical angle, the sample is only illuminated by evanescent waves. This particular setup is called photon scanning tunnelling microscope (PSTM). A useful property of this setup is to reduce the scattered light detected by the tip. Aperture microscopes are particularly useful to image field distributions. This potential is illustrated by the first image of a surface plasmon measured using a PSTM (see Fig. 11.2).

11.2.2 Apertureless mode

It is not necessary to use an aperture to obtain localized electromagnetic fields. An alternative possibility amounts to introducing a tiny scattering tip in the near field. This tip can be viewed as both a local source and a local detector, as seen in Fig. 11.3. As soon as an incident beam illuminates the tip, it scatters the light and produces a local illumination. The incident field also illuminates the sample directly and generates a local field which contains information on the subwavelength features of the sample. This local field is in turn scattered by the tip into the far field, where it can be collected. Hence, the tip also acts as a specific element that enables local-field detection. In most cases, a metallic tip is used. It is remarkable to realize that the tip also acts as a local polarizer. We know that the electromagnetic field is locally given by electrostatics, so that it is normal to a metal surface and particularly large where the radius of curvature is small. This is the so-called lightning rod effect. It follows that when illuminating a tip, a large field parallel to the axis of the tip is generated at the apex of the tip. In other words, the local illuminating field is linearly polarized. Conversely, when the tip is used as a detector, it also acts as a linear polarizer, as discussed in section 11.3.

11.2.3 Discussion

In practice, there are many aspects of producing and interpreting an image. We briefly address some of them in this section.

Controlling the tip position. An important practical issue is how to control the distance between the tip and the sample. It is important to avoid contact in order to avoid damaging the tip. A useful technique is the so-called shear-force technique. The tip is mounted on a cantilever which has an oscillatory movement parallel to the surface on resonance. As the tip approaches the interface, shear-force mechanisms perturb the mechanical movement when the tip–sample distance is on the order of a few nanometres. By detecting either a phase change or an amplitude modification of the oscillation, it is possible to control the tip–sample distance. A possible choice is to scan the sample by keeping the shear-force signal constant. It follows that the tip–sample distance is kept constant. Note that the physical origin of the shear force may vary, depending on the samples. However, it is found in practice that it has a range of a few nanometres.

Constant distance or constant height mode. The above procedure produces optical images that may be strongly correlated with the topography of the sample. If one is seeking to measure optical properties and not topography, this is a drawback. To understand the origin of the crosstalk between topography and optical signal, let us consider a simple flat interface between vacuum and a medium with refractive index n illuminated by a tip consisting of an optical fibre with a tiny aperture. The signal

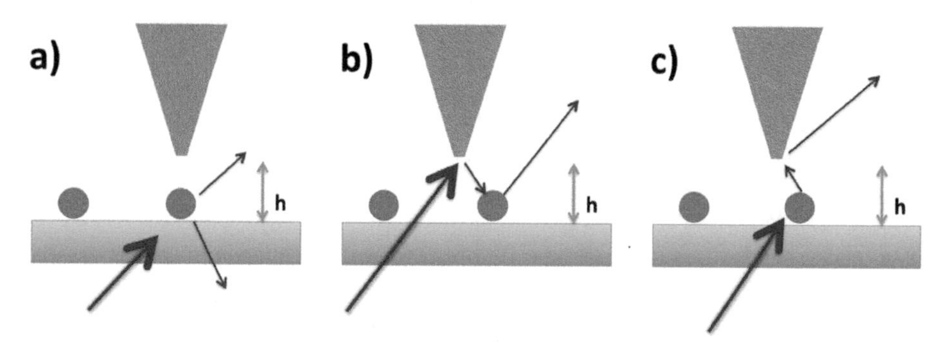

Fig. 11.3 Principle of apertureless near-field scanning optical microscopy. The sample is illuminated from the far field and scatters light towards the detector without superresolution (a). The incident field scattered by the tip provides a local illumination of the sample (b). The incident field scattered by the sample is locally detected by the tip and scattered towards the detector (c).

transmitted through the interface increases as the tip approaches the interface. This result is surprising at first glance, as the power emitted by the source is kept constant. Nevertheless, the mechanism is similar to the frustrated total internal reflection: as the distance between two prisms is reduced, the transmission is increased. In the case of an aperture, the electromagnetic waves with parallel wavevector in the interval $[k_0, nk_0]$ with $k_0 = \omega/c$ are evanescent in vacuum and propagating in the medium. As the tip approaches the interface, the evanescent waves decaying exponentially away from the tip can reach the interface and be coupled to the medium.

The above discussion shows that the signal $S(z)$ depends on the distance z to the interface. It is clear that if the tip height $Z_{tip}(x)$ varies as the tip is scanned over the sample, the signal will be $S[Z_{tip}(x)]$ and will also vary. In other words, the optical signal will carry information on the distance of the tip to the surface. This coupling between vertical position and optical signal is an example of an artefact if one aims to detect the optical properties of the sample.

Height modulation. We conclude this short discussion by pointing out the importance of a vertical modulation of the tip position. It is possible to enhance the near-field components of the signal by vertically modulating the tip height. To see this point, we assume that the signal is due to the detection of two evanescent waves, $S(x, z) = I_1 \cos(k_1 x) \exp(-k_1 z) + I_2 \cos(k_2 x) \exp(-k_2 z)$, with $k_1 \gg k_0$ and $k_2 \gg k_0$, so that $\gamma(k_1) \approx ik_1$ and $\gamma(k_2) \approx ik_2$. If we now modulate the tip vertical position using $Z_{tip}(t) = z_0 + h\cos(\Omega t)$, where h is typically an amplitude of 50 nm, we get

$$S[x, Z_{tip}(t)] \approx S(x, z_0) + \frac{\partial S}{\partial z} h \cos(\Omega t).$$

Using a lock-in amplifier, it is possible to measure the amplitudes of the different harmonics of the signal. The first term, $S(x, z_0)$, is constant and does not contribute to the signal. Only the second term is detected. It varies like the z-derivative of the signal. The ratio of the two contributions thus becomes $k_2 I_2 / k_1 I_1$ instead of I_2/I_1. Hence, the contribution of the large spatial frequencies (large k) is enhanced. This simplified discussion provides a first insight into the benefit of modulating the tip position to enhance the contribution of the large spatial frequencies to the signal and therefore increase the resolution of the image.

11.3 Modelling the signal of a near-field optical microscope: What is imaged by a scanning tip?

The purpose of this section is to establish a connection between the signal delivered by a scanning near-field microscope and the electromagnetic field. We assume that the tip is scanned in a plane at constant z and yields a map $S(x, y)$. Let us emphasize that in this section, we assume that the object of interest is the field itself. We do not attempt to image some structure. We simply address the question of the connection between the signal S delivered by the detector and the electromagnetic field in the plane.

We first note that the signal depends very strongly on the tip. An apertureless metallic conical tip and an aperture in a coated optical fibre do not produce the same images of the same field. Hence, understanding the image formation requires a connection to be established between the signal and the electromagnetic field, accounting for some properties of the tip. From a practical point of view, the question is whether the signal reproduces the intensity or rather some component of the electric or magnetic field.

The answer to that question can be derived from the reciprocity theorem (Porto et al., 2000). Let us consider two situations. The first one is the experimental situation. We aim to measure the fields denoted by $\mathbf{E}_{exp}, \mathbf{H}_{exp}$. The experimental situation is depicted in Fig. 11.4. The second situation is an auxiliary situation. We replace the detector by a point-like source which generates 'reciprocal' electromagnetic fields denoted by $\mathbf{E}_{rec}, \mathbf{H}_{rec}$. The amplitude of the current source is taken to be unity; its polarization is parallel to the orientation of an analyser in front of the detector. Two linear dipoles have to be used to model unpolarized detection. A sum over different points of the intensities is needed to model an extended detector. Using the reciprocity theorem, the electric field at the detector can be expressed as (Porto et al., 2000):

$$\mathbf{j}_{rec}\cdot\mathbf{E}_{exp}(\mathbf{r}_{tip}) = \int_{z=z_0} [\mathbf{E}_{exp}(\mathbf{r})\times\mathbf{H}_{rec}(\mathbf{r},\mathbf{r}_{tip}) - \mathbf{E}_{rec}(\mathbf{r},\mathbf{r}_{tip})\times\mathbf{H}_{exp}(\mathbf{r})]\cdot\mathbf{e}_z dxdy. \quad (11.2)$$

This expression shows that the signal is given by an overlap integral between the experimental field and the reciprocal fields, the latter of which play the role of filters. Indeed, the reciprocal fields have a spatial dependence, are polarized, and depend on frequency. Hence, they filter the fields spatially, in terms of polarization, and spectrally. The equation shows that the resolution of the instrument increases if the reciprocal fields are highly localized. It is interesting to note that *a priori*, both the magnetic field and the electric field can contribute to the signal. However, we note that Maxwell equations connect magnetic fields to electric fields. It is thus possible to derive a form of the signal in terms of only the electric or only the magnetic fields. We get:

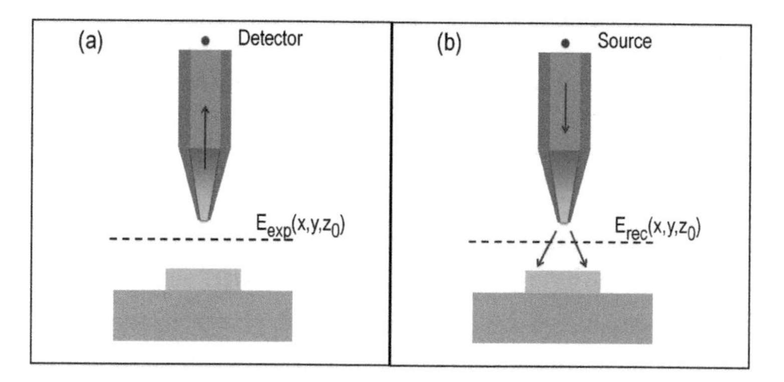

Fig. 11.4 Imaging near fields. We consider a tip used to detect the near field in the plane $z = z_0$, for example the field propagating in a waveguide denoted $\mathbf{E}_{exp}$. We assume that the mode in the fibre is coupled to a point-like detector linearly polarized perpendicularly to the figure. We introduce a reciprocal situation where the detector is replaced by a point-like dipolar source with current density $\mathbf{j}_{rec}$ also polarized along the axis perpendicular to the figure. The subscript rec stands for reciprocal. This source produces a field at the output of the tip denoted by $\mathbf{E}_{rec}$.

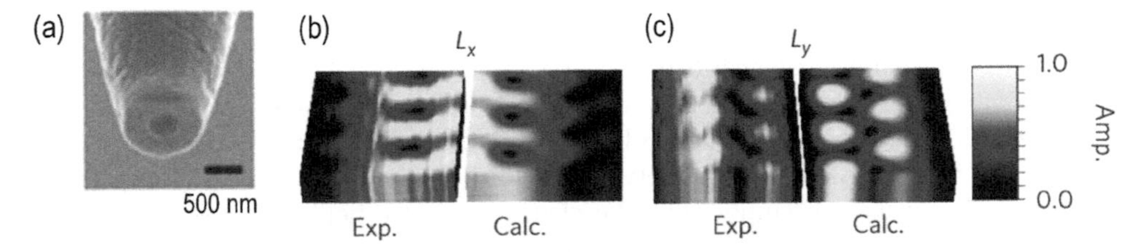

Fig. 11.5 Imaging near fields. (a) Tip; (b, c) comparison of measurement (left panels) with theory (right panels) of the amplitude of the fields measured 300 nm above a waveguide. Reprinted by permission from Springer Nature: (Le Feber, 2013) © 2014. L_x and L_y stand for the two polarization cases of the reciprocal dipole.

$$
\begin{aligned}
\mathbf{j}_{rec} \cdot \mathbf{E}_{exp}(\mathbf{r}_{tip}) &= \frac{-2i}{\omega\mu_0} \int_{z=z_0} \frac{\partial \mathbf{E}_{rec}(\mathbf{r}, \mathbf{r}_{tip})}{\partial z} \cdot \mathbf{E}_{exp}(\mathbf{r})\, \mathrm{d}x\mathrm{d}y, \\
&= \frac{2i}{\omega\varepsilon_0} \int_{z=z_0} \frac{\partial \mathbf{H}_{rec}(\mathbf{r}, \mathbf{r}_{tip})}{\partial z} \cdot \mathbf{H}_{exp}(\mathbf{r})\, \mathrm{d}x\mathrm{d}y.
\end{aligned} \tag{11.3}
$$

At first glance, these formulas suggest that any tip detects both the magnetic and the electric field. However, it is possible to detect a specific component of the electric field, for instance. To clarify this issue, let us take again the example of a metallic tip. When illuminated by a source at the detector position, the tip produces a localized electric field $\mathbf{E}_{rec}(\mathbf{r}, \mathbf{r}_{tip})$, owing to the lightning rod effect. The field is confined close to the apex and has a large z-component. A simple model of a field polarized along the z-axis and localized is given by $\mathbf{e}_z\delta(x - x_{tip})\delta(y - y_{tip})$. It can be considered to be a Dirac function $\mathbf{e}_z\delta(x - x_{tip})\delta(y - y_{tip})$. The tip also produces a magnetic field but it is an oscillating and less confined function. This explains why a metallic tip that produces images of the electric field with a good resolution fails to produce magneto-optical images with good resolution: the reciprocal electric field is confined but the reciprocal magnetic field is not, as shown in ref. (Walford et al., 2002). In order to detect the z-component of the magnetic field, one needs to use a tip with a localized magnetic dipole. This can be achieved using a metallic nanoring.

In Fig. 11.5 we show a comparison between measurement and theory using Eq. (11.3) for a symmetric aperture probe. This comparison has been performed by plotting the signal at 1570 nm on top of a photonic crystal with one row of missing holes playing the role of a waveguide, so that the electromagnetic field is known theoretically (see Chapter 17). The tip consisted of a glass cone coated with 200-nm-thick aluminium cladding.

11.4 Superresolution in the far field: Inverse problem and structured illumination

It was shown in Chapter 10 that propagation is a low-pass filter. We now discuss to what extent this is a limitation to superresolution imaging. We will point out that the loss of high spatial frequencies *of the field* does not necessarily mean that it is impossible to obtain maps of the *permittivity* with resolution better than half a wavelength, even though standard far-field linear optics is used.

11.4.1 Imaging fields and imaging structures

Let us stress that subwavelength features *of a propagating field* are lost because information contained in the evanescent waves is lost. However, this does not necessarily mean that information on an object to be imaged is lost. In order to analyse this issue, we need to clarify the connection between an object and the field scattered by the object. Imaging usually assumes that the electromagnetic field reflected by a surface conveys information on the local reflectivity of that surface. In a low-resolution context, the reflected field is the product of the incident field by the local reflectivity; therefore, there is no difference between reconstructing a field distribution and reconstructing a reflectivity map and therefore an object image. However, the concept of local reflectivity breaks down when the object under study varies on a wavelength scale. Hence, *we need a more fundamental definition of imaging* to be able to discuss spatial resolution at a subwavelength scale.

From a theoretical point of view, the light scattered by an object is entirely determined by the knowledge of the dielectric permittivity of the object $\varepsilon(\mathbf{r})$ and the illumination conditions.[1] This defines a well-posed scattering problem. Hence, a complete description of the object, *from the point of view of its interaction with light*, is the distribution of permittivity in space. Hence, imaging amounts to reconstructing a 3D *permittivity map* $\varepsilon(\mathbf{r})$ from a finite subset of *scattered-field measurements.* Here, we stress that *the link between the scattered field and the distribution of permittivity in a medium is not linear.* In Chapter 12, we will derive a general form of the scattered signal in the far field. It will be shown (see Eq. (12.17)) that the field amplitude scattered in a direction given by a unit vector $\mathbf{u}$ is proportional to

$$\int \chi(\mathbf{r})\mathbf{E}(\mathbf{r}) \exp(-ik_0\mathbf{u} \cdot \mathbf{r})d^3\mathbf{r},$$

which is the Fourier transform of the product of the dielectric contrast $\chi(\mathbf{r}) = \varepsilon_r(\mathbf{r}) - 1$ and the electric field in the object $\mathbf{E}(\mathbf{r})$. As the electric field $\mathbf{E}(\mathbf{r})$ strongly depends on the dielectric contrast, it is seen that the relation between the scattered field and the dielectric contrast is non-linear. We note that the scattered field in direction $\mathbf{u}$ may contain information on high spatial frequencies of the dielectric contrast $\chi(\mathbf{k})$. Using the fact that the Fourier transform of a product is a convolution product, we find that the scattered field is proportional to

$$\int \chi(\mathbf{k})\mathbf{E}(k_0\mathbf{u} - \mathbf{k})\mathrm{d}\mathbf{k}. \tag{11.4}$$

Hence, a large spatial frequency $\mathbf{k}$ of the function χ may contribute to the far field, provided that the field inside the object contains a spectral component at $k_0\mathbf{u} - \mathbf{k}$.

To summarize this discussion, the main idea is that the loss of information on the high spatial frequencies of the *electromagnetic field* does not mean that the information on *the details of the permittivity* $\varepsilon(\mathbf{r})$ is lost in the far field. In other words, the fundamental resolution limit does apply to imaging inasmuch as imaging is simply based on assuming that the field distribution has a one-to-one correspondence with a

[1] Note that magnetic permeability is μ_0 at optical frequencies.

reflectivity map. This limit does not apply to the permittivity reconstruction problem. It is thus possible to produce images with subwavelength resolution by *solving the inverse problem numerically using only far-field data.* This issue is further discussed in the next sections.

11.4.2 Spatial resolution in the single-scattering regime: First Born approximation

We now address the issue of the resolution limit when dealing with low dielectric contrast such that the Born approximation discussed in chapter 12 is valid. In that case, the field in the object can be approximated by the incident field and does not depend on the permittivity of the object. Using Eq. (12.32), it is seen that the scattered field is *linearly related* to the structure of the object. This paves the way to a simple reconstruction of the structure of the object directly from the knowledge of the field. Note, however, that the far field does not contain information on the high spatial frequencies of the object when using far-field illumination. Indeed, the amplitude of the field detected in the far field in a direction given by $k_0\mathbf{u}$ is proportional to the Fourier transform of $\chi(k_0\mathbf{u} - \mathbf{k}_{inc})$. The modulus of $\chi(k_0\mathbf{u} - \mathbf{k}_{inc})$ is bounded by $4\pi/\lambda$. Hence, by using several incident angles, it is possible to recover this information, but there is a cutoff at $4\pi/\lambda$. We can conclude that *in the Born approximation regime, namely in the single-scattering approximation*, the information on the object suffers from a fundamental limit. This is only twice as large as the resolution limit on the diffracted field.

11.4.3 Structured illumination

The second route that can be followed is to generate a structured illumination possessing high spatial frequencies on purpose. Let us assume that the dielectric contrast is low, so that the Born approximation is valid. Then we can replace the electric field in the scatterer by the incident electric field, so that the scattered amplitude is proportional to

$$\int \chi(\mathbf{k})\mathbf{E}_{inc}(k_0\mathbf{u} - \mathbf{k})d\mathbf{k}. \tag{11.5}$$

We can gain access to $\chi(\mathbf{k})$ for large values of $\mathbf{k}$ only if the incident field is non-zero for $k_0\mathbf{u} - \mathbf{k}$, which is a large wavevector. This is the principle of structured illumination: to generate an illuminating field with rapid spatial variations (Gustafsson, 2005). If far-field illumination is used, the maximum wavevector from the sample that can be detected is $2k_0$, so that there is a gain of a factor of 2 compared to the diffraction limit. An interesting alternative is to use a speckle pattern to illuminate (Mudry et al., 2012). However, it is possible to illuminate with larger spatial frequencies. In practice, one can deposit the sample to be imaged on a grating with a *subwavelength* period d. When illuminating this grating by a plane wave with wavevector $k_{x,inc}$, the grating generates high-order diffracted fields with wavevectors $k_{x,n} = k_{x,inc} + n2\pi/d$, which can be very large. These high-order terms are evanescent waves if $k_{x,n} > \omega/c$. It is seen from Eq. (11.5) that such a field will be able to translate a high frequency of the dielectric contrast into k-space. This approach is not, strictly speaking, a pure far-field technique, as a structured illumination containing very high spatial frequencies is necessarily an illumination with evanescent waves and therefore a near-field illumination. However, a textured substrate can be used, so that a pure far-field illumination and detection can still be used. The image then results from a numerical processing of the data (Röhrich and Koenderink, 2021).

11.4.4 Superresolution in the far field: Multiple scattering and the non-linear inverse problem

We now explore situations where the Born approximation is not valid. In that case, the electric field contains high spatial frequencies. It is thus possible to solve the inverse problem and recover subwavelength information on the permittivity from the far-field data. Some examples of subwavelength reconstruction with far-field data can be found in refs (Chen and Chew, 1998; Belkebir et al., 2006; Arhab et al., 2013). In Fig. 11.6, we show the result obtained in ref. (Chen and Chew, 1998) by reconstructing a 2D system consisting of a plastic pipe with permittivity $\varepsilon = 2.5$.

An inverse problem is an ill-posed problem, namely a problem with many different solutions. To find the right solution among the many possible solutions, some *a priori* information needs to be used. Here, it is known that the scatterer is inside a square box of 8 cm × 8 cm, so that the problem amounts to recovering the permittivity. The data consist of phase and amplitude of the scattered fields at five different antennas when illuminating with six different antennas at many different frequencies. The eleven antennas are located along a line placed at 40 cm from the center of the unknown object. Pulses with a broad spectrum between 3 and 9 GHz were used. The frequencies and distances were chosen such that there are no evanescent waves detected (10 GHz corresponds to $\lambda = 3$ cm). The key results shown in Fig. 11.6 are: (i) it is possible to reconstruct the object with an accuracy on the order of 7.5 mm, which is better than half the smallest wavelength $\approx$15 mm; (ii) if multiple scattering is not used, the algorithm does not retrieve the correct permittivity. This is a key issue which emphasizes that the resolution depends on the object shape and permittivity. This technique can also be implemented with visible light. Using a laser operating at 632.8 nm, it has been

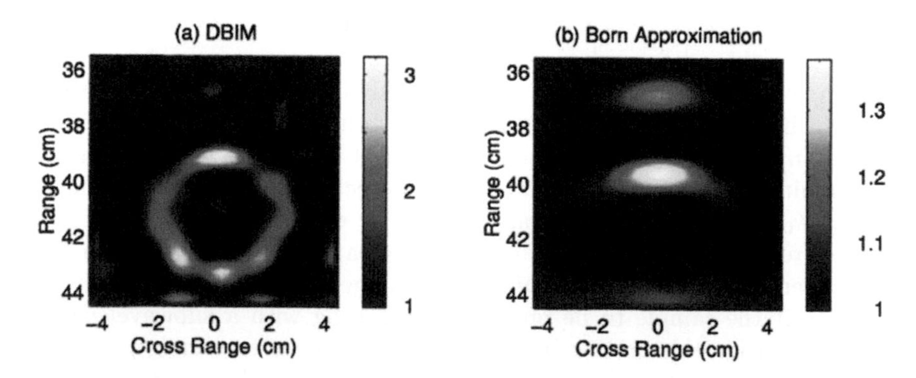

Fig. 11.6 Imaging by solving an inverse problem. Permittivity map of a plastic pipe with permittivity 2.5, inner diameter of 3.7 cm, and outer diameter of 4.8 cm. a) Reconstructed image using a multiple scattering model; b) reconstructed image using the first Born approximation. Figure taken from ref. (Chen and Chew, 1998).

possible to reconstruct the profile of a surface consisting of three triangular slits in an InP surface $\varepsilon = 13.6 + 1.5i$ with a width of 200 nm and a depth of 80 nm separated by a distance of 200 nm (Arhab et al., 2013). In this paper, the permittivity was assumed to be known.

In the previous discussion, we noted that superresolution cannot be achieved in the framework of Born approximation, which is a single-scattering approximation. We then saw that if a structured illumination is used, namely if we illuminate with evanescent waves, then superresolution can be achieved. Finally, we saw that superresolution can be obtained without illumination by evanescent waves. So a natural question arises: where are the high spatial frequencies of the field in the body coming from? The answer is multiple scattering. Let us think of a simple sample composed of two metallic nanospheres separated by a subwavelength distance. Each sphere is illuminated by the incident field and by the other nanosphere. The latter field contains evanescent waves. The origin of the superresolution is the presence of fields varying on a subwavelength scale inside an object. The measured scattered field depends on the product $\chi(\mathbf{r})\mathbf{E}(\mathbf{r})$. Solving the inverse problem amounts to disentangling the fine structure of the field $\mathbf{E}(\mathbf{r})$ and the fine structure of the dielectric contrast $\chi(\mathbf{r})$.

11.5 Superresolution in the far field: Localizing sparse markers with subwavelength accuracy

In this section, we discuss imaging techniques that aim to reconstruct the structure of a subwavelength object by detecting single molecules (or, more generally, single subwavelength objects) which are attached to the object. The molecules act as markers. By recording their position with subwavelength accuracy, it is possible to reconstruct the spatial structure of the object. Note that the techniques that we are going to discuss *do not provide information on optical properties or spectral properties of the objects.* In that sense, these techniques are very different from the far-field and near-field scattering techniques. The nature of extracted information is somehow comparable to the topography given by an atomic force microscope.

11.5.1 Localization and tracking of a single emitter

A single emitter can typically be a single molecule or a quantum dot (QD), like those seen in Chapter 9. They are detected by photoluminescence at λ_{emi} and excited at a shorter wavelength λ_{exc}. A filter removes the high-energy photons of the exciting light (a difficult task if the Stokes shift, essentially $\lambda_{\text{emi}} - \lambda_{\text{exc}}$, is small). One first issue to face in getting a signal from such emitters is to get a large enough flux. This entails ensuring a good collection efficiency, and hence a good extraction efficiency of the generated photons (see Fig. 11.7a: immersion has the benefits of both finer resolution and enhanced collection), and a detector with large quantum efficiencies, even for low fluxes such as avalanche photodiodes or intensified cameras.

The imaging optics transforms the point emitter into a point spread function (PSF), e.g. the classical Airy function for a diffraction-limited optics with perfect circular symmetry in the scalar approximation (see Fig. 11.7b). Because we have the *a priori* information that this function stems from a single emitter, we can get a much better idea of its position than the width of the Airy function (the Abbe limit). We

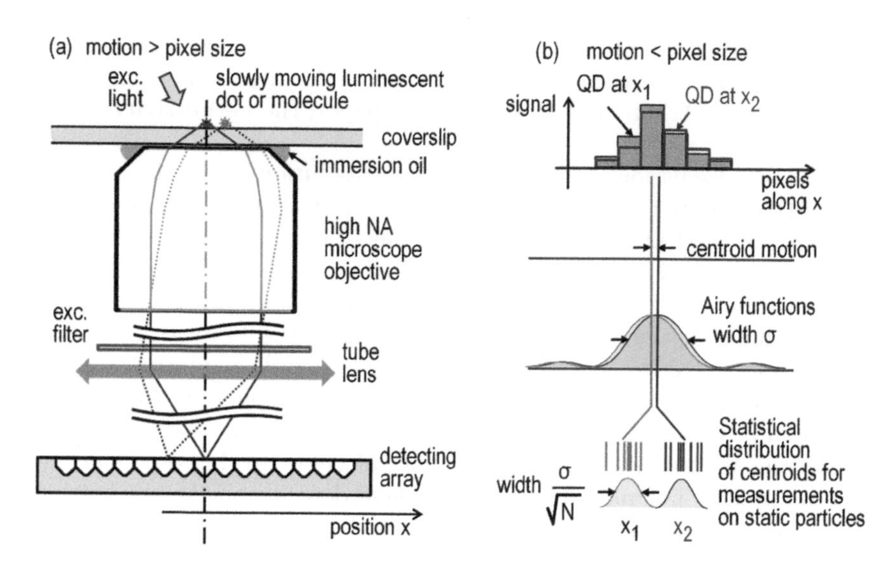

Fig. 11.7 (a) Microscope setup for localization of a single fluorescent quantum dot (QD) or molecule, with excitation, high-NA lens used in immersion mode, and geometric optics rays to the camera array. The pixel size of the array may typically be smaller than the diffraction-limited resolution. In the case shown, the source centre moves from x_1 to x_2, which is more (once magnified) than a pixel size. (b) Typical profile of photoluminescence counts from a single static QD, with width σ characteristic of the point spread function. For two close actual static positions x_1 and x_2, centroids are distinct and shall differ by an amount that is smaller than a pixel and smaller than the Airy limit. The noise on the centroid position (suggested by line distributions and their idealized histogram at the bottom) depends on the signal-to-noise ratio.

can assume that, practically, the pattern size is much larger than the pixel size of the camera, by a factor of 10 at least; only the object numerical aperture limits the resolution in a correctly designed microscopy setup. So the issue of localization boils down to asking how accurately one can determine the centre of an Airy pattern. The answer is that the accuracy of centre determination ultimately depends on the signal-to-noise ratio of the emission measurement, and thus on the photon flux. Specifically, the centre of the PSF can be localized with an accuracy $\sim \sigma/\sqrt{N}$, where σ is the PSF width and N is the number of detected photons (Betzig, 1995). Thus, with a signal-to-noise ratio as small as a few units, a neat subwavelength determination of, say, $\lambda/5$ can already be obtained. The accuracy can be much finer with a higher ratio, down to a few nanometres in the best cases in the visible range. By the same token, it is also possible to get an accuracy smaller than the pixel size, using the centroid motion of the discretized image. Both issues, the subpixel and subwavelength localization, are illustrated in Fig. 11.7, where, for simplicity, we use a 1D profile of the image. Here we solve an easy inverse problem because of the single emitter assumption. If two close emitters, say two molecules, lie together and emit together, we cannot know it by analysing the image shape. We could know it by comparison of intensities (we could have an intensity that is twice as large for instance, neglecting all chemical and electromagnetic interactions) or by more sophisticated quantum techniques using correlation and addressing photon bunching, such as the Hanbury Brown–Twiss measurements.

A conclusion can be drawn from the importance of a large signal-to-noise ratio: when it comes to the use of QDs as biological markers attached to some biological system, it is clear that the absence of bleaching compared to fluorescent molecules makes them crucial for biologists who need to track small objects in their path within or around cells or analyse the separation of two macromolecules such as DNA strands, etc. But, conversely, the blinking phenomenon that affects colloidal II-VI QDs, is a drawback for spatial tracking of moving biological nano-objects.

11.5.2 Volume sampling

In order to explain imaging techniques for less elementary objects, we start with a simple volume-sampling example based on the Brownian motion of gold nanoparticles in a solution. The goal is to image a porous medium with large pores and tiny pores much smaller than the wavelength naturally arising in an ensemble of latex beads sedimented in a water solution. An optical microscope cannot resolve the fine structures of the porous medium. In order to beat the diffraction limit, gold nanoparticles with a radius of 50 nm were added to the liquid. The gold nanoparticles move freely in the fluid due to Brownian motion. When illuminating this system with red light at 633 nm, the gold nanoparticles can be observed. The particles, concentration is chosen in such a way that the particles are well separated. Each of them yields an image that is a well-defined spot, as the exposure time is short enough that the particles do not move. Each spot is given by the PSF: it gives a diffraction-limited image. Assuming that $\sigma/\sqrt{N}$, as discussed in the above subsection, is small enough, it is thus possible to produce a map of the centres of the particles for each picture with a nanometre accuracy. As the system is sparse, each picture contains a small number of dots and produces an image which looks like a clear sky with stars and no very resolved global pattern, as seen in Fig. 11.8a. However, the procedure can be repeated many times. Due to Brownian motion, the gold nanoparticles explore the fluid space and can move in all volumes smaller than 50 nm. By summing all the pictures, we finally obtain a picture (Fig. 11.8b) with many dots in the pores and no dots in the rest of the sample. In summary, the key to this imaging technique is: (i) the possibility of having a large number of scatterers or emitters distributed over the object; (ii) the possibility of producing many different images with subwavelength accuracy of the positions of the emitters or scatterers. We again emphasize that this technique provides information on the spatial locations of the emitters or scatterers, not on the optical properties of the object. The key idea was proposed by E. Betzig in a seminal paper in 1995 (Betzig, 1995). The pictures shown in Fig. 11.8 were taken from the work of G. Moneron (Moneron et al., 2005). A remarkable step forward is the possibility of using single molecules as position markers. This is based on the ability to detect single molecules using both spatial and spectral selectivity, as first demonstrated by Moerner and Orrit (Moerner and Kador, 1989; Orrit and Bernard, 1990). Combining single-molecule detection and single-molecule localization led to new imaging techniques called STimulated Emission Depletion (STED) (Hell and Wichmann, 1994), PhotoActivated

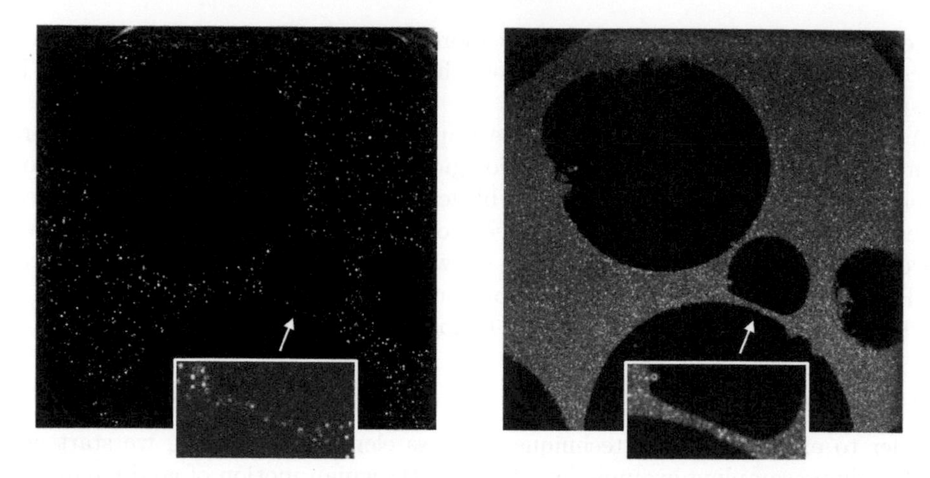

Fig. 11.8 Principle of superresolution imaging. Gold nanoparticles explore the available volume by Brownian motion. The lateral size of the field of view is 216 μm. Each image is processed by replacing the diffraction-limited spot with a point located at the centre of the point spread function. The left panel is a single processed image. A detail is enlarged and contrast-adjusted in the white box. The right image is the final result, the same detail is enlarged. Reproduced with permission from Moneron et al. 2005.

Localization Microscopy (PALM) (Betzig, 1995; Betzig et al., 2006), and STochastic Optical Reconstruction Microscopy (STORM) (Rust et al., 2006), all of which are briefly described hereafter.

11.5.3 PALM, STORM

This technique is based on the localization principle of single molecules which are attached to the structure to be imaged (e.g. a synapse, proteins). A typical density of 10^5 molecules per μm^2 is used. This number needs to be large in order to ensure that the object is densely mapped. In order to detect molecules one by one and accurately determine their centre, a small fraction of them have to emit fluorescent light simultaneously in order to avoid overlapping, as explained before. It is therefore necessary to find a way to activate a tiny fraction of these molecules. There are different techniques to achieve this. In PALM, photoactivatable fluorophores are used. Such fluorescent proteins become fluorescent or shift their excitation and emission spectra upon illumination with an ultraviolet laser, typically 405 nm. Since the activation of every molecule is independent from the rest, a low photoactivation intensity ensures a stochastic photoactivation of a subset of molecules. In the original STORM experiments, organic dyes Cy5 were switched between fluorescent and dark states in a controlled and reversible manner, but the recovery rate depends critically on the close proximity of a secondary dye, Cy3. Later, the use of an oxygen-reducing buffer allowed STORM to be performed with standard fluorescent probes and without the need of a secondary fluorophore (van de Linde et al., 2011).

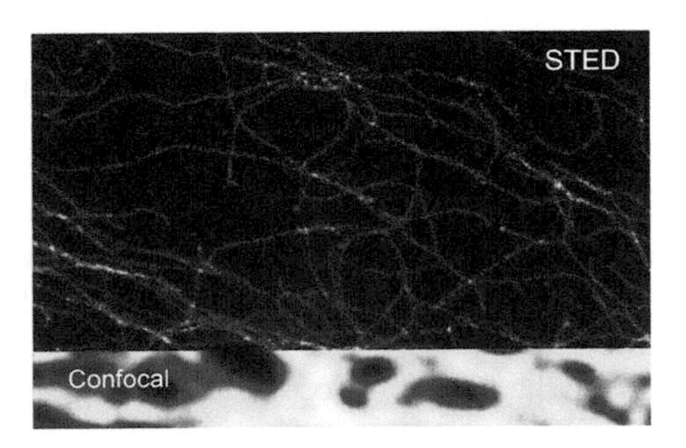

Fig. 11.9 Comparison of the image of a cellular cytoskeleton with confocal microscopy and with STED. This figure is part of the Nobel Lecture of Stefan W. Hell. ©The Nobel Foundation 2014 (Hell, 2015).

11.5.4 STED

The STED technique is also based on detecting molecules attached to a given structure to be imaged (see Fig. 11.9). Here, a large number of fluorescent molecules are excited with a first exciting laser pulse operating at λ_1. At this stage, the laser spot is diffraction-limited and excites a large number of molecules. These molecules have internal relaxation and emit fluorescence at a red-shifted wavelength λ_2 with a spontaneous time scale on the order of 1 ns. In order to be able to detect the fluorescence of a subwavelength region, a second laser pulse operating at the emission wavelength λ_2 is used to generate stimulated emission on a picosecond time scale, thereby inducing a fast radiative decay of the molecules. The key feature of the technique is its use of a doughnut-shaped beam such that the intensity is zero at the centre of the beam and therefore no stimulated emission takes place there. It follows that fluorescence observed well after the second pulse can only come from the very centre of the beam. In summary, using a second laser pulse amounts to turning off all the molecules except those which are in the centre of the doughnut beam. By increasing the doughnut pulse laser intensity, the volume where fluorescence is not suppressed is reduced to sizes much smaller than the wavelength. Here, the resolution is given by the value of the radius where the intensity reaches the threshold for stimulated emission: increasing the incident doughnut beam power increases the resolution.

11.6 Generating highly localized light spots with superoscillations

In this section, we briefly introduce the concept of superoscillation before discussing how it can be applied to imaging with superresolution. We also highlight the difficulties inherent in this type of technique. In our discussion of the resolution limit of the confocal microscopy technique, we have used the fact that the smallest spot that can be achieved has a size limited by the spatial frequency bandwidth—in other words, it is impossible to produce a spot with subwavelength localization when the maximum wavevector is given by $k_M = n\sin(\theta_M)\omega/c$. This statement is not mathematically correct. It is possible to exhibit functions with oscillations containing a very tiny fraction of the energy within a typical width that is much smaller than the inverse of the limited bandwidth. These features are called superoscillations. To illustrate this class of functions, we use the example given by M. Berry (Berry and Popescu, 2006):

$$f(x) = [\cos(kx) + ia\sin(kx)]^N \qquad (a > 1, N \gg 1). \tag{11.6}$$

$f(x)$ is periodic, with period $2\pi/k = \lambda$ for odd N. The function can be expanded in a Fourier series and the maximum spatial frequencies are clearly $\pm Nk$. Hence, we might expect that the smallest features would be on the order of $2\pi/Nk$. We now look carefully at the behaviour of the function close to the origin $x = 0$. It is seen that the function can be represented by:

$$f(x) = [\cos(kx) + ia\sin(kx)]^N \approx (1 + iakx)^N \approx \exp(iaNkx). \tag{11.7}$$

Hence, close to the origin, the function oscillates locally with a wavevector aNk, which is a times larger than the maximum wavevector, so that features with spatial extension on the order of $2\pi/aNk$ can be obtained. It is seen that the function has an amplitude on the order of $a^N = e^{Na}$ for $x = \lambda/2$, whereas it is on the order of unity for $x \ll 1/aNk$. In other words, the amplitude decays exponentially with the factor a in the region of superoscillations. There is no contradiction with the well-known property of Fourier transform $\Delta x \Delta k \geq 2\pi$ because the small features carry only an exponentially small fraction of the total energy.

The existence of this type of function raises the prospect of superresolution when using the superoscillation to locally illuminate an object. Practical implementation is extremely challenging due to the limited amount of energy contained in the localized oscillation, while the immediate vicinity is strongly illuminated.

Exercises

Exercise 11.1 A stethoscope is an acoustical near-field scanning microscope. The first experimental demonstration of superresolution using a near-field technique was reported by the group of Dieter Pohl. The group coined their technique 'optical stethoscopy' in reference to the medical doctor's device. The instrument consisted in a tiny aperture made on a silver coating on a dielectric prism. This aperture was scanned along a sample. By recording the signal as a function of tip position, it was possible to obtain an image with subwavelength features. The goal of this exercise is to further analyse this analogy.

(1) The pressure of an acoustic wave satisfies a scalar Helmholtz equation. Explain why the diffraction limit is the same for acoustic waves and optical waves.

(2) What is the order of magnitude of sound velocity in water?

(3) What is the order of magnitude of the wavelength in a body? To estimate the wavelength, we assume that sound velocity is the same as in water and we consider a 1 kHz acoustic wave.

(4) According to the Abbe resolution limit, is it possible to separate the sounds emitted by the heart and the belly of an adult? Explain why a medical doctor can perform this superresolution measurement.

Exercise 11.2 How near is near field? We consider the grating of period d described in Exercise 10.1 illuminated under normal incidence. In order to detect the diffracted field when $d \ll \lambda$, we use a tip which is scanned at a height h above the surface. Explain how to choose the height h as a function of d in order to detect the evanescent waves. Hint: use the rule of thumb introduced in section 10.1.2.

Exercise 11.3 Scanning near-field optical microscopy in illumination mode. We consider a sample locally illuminated through a tiny aperture in an aperture scanning tunnelling microscope. We assume that we detect the signal in transmission with a detector in the far field. By scanning the aperture at a distance h above the sample, it is possible to resolve subwavelength features of the sample, provided that they are separated by a distance larger than the size of the illuminating spot. Discuss whether there are evanescent waves involved in this setup.

Exercise 11.4 An effective dipole model: Local reflection mapping. It is possible to map the local permittivity $\varepsilon_m(x,y)$ of the material of a surface by detecting the field scattered by a tip scanned above the surface. In order to analyse the signal, the tip can be modelled by a tiny sphere with radius a and permittivity ε_p. Its dipole moment is $p = \varepsilon_0 \alpha E$, where the polarizability α is given by $\alpha = 4\pi a^3 \frac{\varepsilon_p - 1}{\varepsilon_p + 2}$. The field at a distance r along the axis of the dipole is given by $E = \frac{p}{2\pi r^3}$. The field of the dipole located at a distance h above the surface induces polarization charges on the interface. The field produced by these surface charges is equivalent to the field produced by an image dipole $p' = r(x,y)p$ located at a distance $2h$ from the dipole. The reflection factor is $r(x,y) = \frac{\varepsilon_m(x,y)-1}{\varepsilon_m(x,y)+1}$. Show that the system 'sphere on a surface' can be modelled by an effective polarizability,

$$\alpha_{\text{eff}} = \frac{\alpha(1+r)}{1 - \frac{\alpha r}{16\pi h^3}}.$$

12 Scattering, Green tensor, and local density of electromagnetic states

In the previous chapter, we saw that the electromagnetic field may vary on length scales smaller than the wavelength in the vicinity of matter. It is thus important to be able to compute the field in the vicinity of matter. When dealing with fields scattered in the far field, the exact form of the field at the scatterer position is often estimated using simple approximations. By contrast, in the near field, we are interested in the detailed form of the electromagnetic fields close to edges, tips, surfaces, etc. There are many theoretical ways to compute the electromagnetic field. A useful and powerful theoretical tool is the integral formulation of the field equations. A convenient feature of this formulation is that it allows some physical insight into the scattering mechanisms to be gained. In simple words, an integral formulation amounts to writing the electric field at any point as being the sum of the fields radiated by all the currents (either in the sources or induced in the matter). Thus, the basic quantity is the field radiated by an elementary volume. Since a small volume can be modelled by a dipole, it turns out that the key quantity is the field radiated by a dipole. It proves convenient to introduce the tensor describing the electric field radiated by a point-like dipole; this is known as the Green tensor. In the next section, we derive an integral equation for the electric field. We also derive the exact expression of the the Green tensor and carefully discuss its singularity as well as its physical meaning. For the sake of completeness, we also introduce the Green tensor as the solution of a partial differential equation whose source term is a point-like dipole. Then, we apply these concepts to the study of light scattering by arbitrary systems. This provides a framework for discussing the possibility of performing superresolution imaging by using far-field data and solving the inverse problem.

The second part of this chapter deals with the analysis of the link between the Green tensor and the local density of electromagnetic states. This quantity plays a key role in the study of spontaneous emission in the presence of nanostructures in the weak coupling regime.

12.1 The Green tensor and integral formulation of electromagnetism

The most usual formulation of the scattering of electromagnetic fields is based on partial differential equations. The propagation equation in a linear isotropic inhomogeneous medium characterized by the permittivity $\varepsilon_r(\mathbf{r})$ can be cast in the form.

Introduction to Nanophotonics. Henri Benisty, Jean-Jacques Greffet, and Philippe Lalanne, Oxford University Press.
 DOI: 10.1093/oso/9780198786139.003.0012

$$\nabla \times \nabla \times \mathbf{E} - \varepsilon_r(\mathbf{r})k_0^2\mathbf{E} = i\omega\mu_0\mathbf{j}_{inc}, \tag{12.1}$$

where $\mathbf{j}_{inc}$ stands for the current density in the sources generating an incident field. Here ε_r can be a function of both the considered point $\mathbf{r}$ and the frequency for inhomogeneous and dispersive media. Note also that its imaginary part accounts for losses. Once boundary conditions are specified, the problem is well posed. This type of formulation is very practical for simple shapes such as planes or spheres. For example, in Chapter 5, we used a general solution in terms of plane waves. The amplitudes of the plane waves were found by using boundary conditions. A similar procedure can be used for spheres using spherical functions. The solution is known as Mie theory (Bohren and Huffman, 1983). The problem becomes intractable for complex geometries, as general solutions are difficult to find. Instead, it is possible to reformulate the problem in terms of an integral equation which has a transparent physical meaning.

12.1.1 The Green tensor and the equivalence principle

The basic idea is to consider that the field propagates in vacuum[1] and to account for the presence of material media by introducing additional source terms. To proceed, let us add the term $k_0^2[\varepsilon_r - 1]\mathbf{E}$ on both sides of Eq. (12.1). We find:

$$\nabla \times \nabla \times \mathbf{E} - k_0^2\mathbf{E} = k_0^2[\varepsilon_r - 1]\mathbf{E} + i\omega\mu_0\mathbf{j}_{inc} = i\omega\mu_0\mathbf{j}_{ind} + i\omega\mu_0\mathbf{j}_{inc}, \tag{12.2}$$

where $\mathbf{j}_{ind} = -i\omega\varepsilon_0(\varepsilon_r - 1)\mathbf{E}$. As announced, this is a propagation equation in vacuum with a known source term, $\mathbf{j}_{inc}$, and an unknown source term due to the induced currents, $\mathbf{j}_{ind}$. The known source term generates the incident field and the (as yet unknown) source term generates the scattered field. In other words, *we have transformed a reflection/scattering problem into a radiation problem.* This is known as the *equivalence principle.* Eq. (12.1) considers that the field generated by the current density $\mathbf{j}_{inc}$ is scattered by the medium described by $\varepsilon_r(\mathbf{r})$, whereas Eq. (12.2) describes radiation in vacuum by two currents density distributions, $\mathbf{j}_{inc}$ and $\mathbf{j}_{ind}$. Yet, we still need to find the induced sources, $\mathbf{j}_{ind}$. As opposed to the radiation problem, where we assume that the currents are known, here we need to find the value of the currents induced by the incident fields in the medium.

We start from the partial differential equations for the electric field in Eq. (12.1). We seek inspiration from the Green function method introduced to solve the radiation problem in vacuum in Chapter 2. However, it is seen that we cannot solve for a scalar function because we look for a linear object that relates the current density (the source) to the electric field. If the equation would relate the x-component of the current density to the x-component of the electric field and so on, as is the case for the potential vector, a scalar function could be introduced. Yet, the operator $\nabla \times \nabla \times$ introduces relations between different components of the electric field to each component of the current density. A simpler way of expressing this fact is to point out that the electric field radiated in the far field is perpendicular to the propagation direction and therefore not always parallel to the current density. Thus, the object needed is a tensor connecting

[1] Exercise 12.8 deals with the case of scattering in a host medium with refractive index n_h.

the vector electric field to the vector sources. Therefore, we write the equation for the Green tensor in the form:

$$\nabla \times \nabla \times \overleftrightarrow{G}_0 - k_0^2 \overleftrightarrow{G}_0 = \overleftrightarrow{I} \, \delta(\mathbf{r} - \mathbf{r}'). \tag{12.3}$$

The source term is a point-like source. As compared to the scalar case, there is a unit tensor $\overleftrightarrow{I}$. This indicates that the source can be a dipole with three different possible orientations. This equation can be solved with radiation boundary conditions using techniques similar to the solution for the scalar case. The result can be cast in the following form:

$$\overleftrightarrow{G}_0 = (\overleftrightarrow{I} + \frac{1}{k_0^2} \nabla\nabla) \frac{\exp[ik_0 R]}{4\pi R}, \tag{12.4}$$

if $\mathbf{r} \neq \mathbf{r}'$. It is seen that this form contains terms that vary as $1/r, 1/r^2, 1/r^3$. It is consistent with the form of the electric field generated by an electric dipole given by Eq. (2.42). This form diverges when $\mathbf{r} = \mathbf{r}'$. This singularity can be handled as follows:

$$\overleftrightarrow{G}_0 = PV[\overleftrightarrow{I} + \frac{1}{k_0^2} \nabla\nabla] \frac{\exp[ik_0 R]}{4\pi R} - \frac{\overleftrightarrow{I}}{3k_0^2} \delta(\mathbf{r} - \mathbf{r}'), \tag{12.5}$$

where we have introduced a Dirac generalized function and where PV indicates that the integral is evaluated in the sense of a principal value. A detailed derivation of the Green tensor can be found in the paper by Yaghjian (Yaghjian, 1980) and in the books by Van Bladel (Van Bladel, 1991), Wang (Wang, 1990), and Tai (Tai, 1994). We present in Complement 12.A a derivation that gives some insight into the physical meaning of the Dirac singularity.

12.1.2 Integral relation and integral equation

It is now a simple matter of establishing the integral form of the electric field at any point $\mathbf{r}$. The general solution of the problem is obtained by a linear superposition of the contributions of the different sources:

$$\mathbf{E}(\mathbf{r}) = i\omega\mu_0 \int \overleftrightarrow{G}_0 (\mathbf{r}, \mathbf{r}') \, [\mathbf{j}_{inc}(\mathbf{r}') + \mathbf{j}_{ind}(\mathbf{r}')] d^3\mathbf{r}'. \tag{12.6}$$

The physical content of the equation is transparent. It shows that the field at any point is the superposition of the fields radiated by all current elements in the system. We now insert the explicit form of the induced current density and identify the first term as the incident field $\mathbf{E}_{inc}$ and the second term as the scattered field $\mathbf{E}_s$:

$$\begin{aligned} \mathbf{E}(\mathbf{r}) &= \mathbf{E}_{inc}(\mathbf{r}) + i\omega\mu_0 \int \overleftrightarrow{G}_0 (\mathbf{r}, \mathbf{r}') \, \mathbf{j}_{ind}(\mathbf{r}') d^3\mathbf{r}' \\ &= \mathbf{E}_{inc}(\mathbf{r}) + k_0^2 \int [\varepsilon_r(\mathbf{r}') - 1] \, \overleftrightarrow{G}_0 (\mathbf{r}, \mathbf{r}') \mathbf{E}(\mathbf{r}') d^3\mathbf{r}'. \end{aligned} \tag{12.7}$$

From Eq. (12.7), it is seen that the field can be computed everywhere outside the body, provided that the field is known inside the body.[2] Hence, the key issue is to find what the field inside the body is. To this aim, one restricts the integral relation (12.7) to the volume V, where $[\varepsilon_r(\mathbf{r}') - 1] \neq 0$. The resulting relation is an integral equation:

$$\text{If } \mathbf{r} \in V\ ; \quad \mathbf{E}(\mathbf{r}) = \mathbf{E}_{inc}(\mathbf{r}) + k_0^2 \int [\varepsilon_r(\mathbf{r}') - 1] \overleftrightarrow{G_0} (\mathbf{r}, \mathbf{r}')\mathbf{E}(\mathbf{r}')d^3\mathbf{r}'. \tag{12.8}$$

The unknown is the electric field. It appears both outside the integral and in the integrand. This equation is known as the Lippmann–Schwinger equation in physics. This formulation of the problem is a complete description of the scattering problem. It is equivalent to the set of partial differential equations with the boundary conditions. The integral formulation has many interesting properties that we briefly outline in what follows. It can be used to gain some physical insight into the scattering processes. In what follows, we discuss single scattering and multiple scattering. We derive a general form of the scattered field in the far field and introduce the scattering matrix. We use the integral formalism to derive approximate solutions for small scatterers or scatterers with low dielectric contrast.

12.1.3 Multiple scattering

The integral equation is extremely useful to provide a rigorous foundation for the concepts of single scattering or multiple scattering. From a mathematical point of view, it is possible to solve the equation by iteration. In order to simplify the notation, let us introduce a simplified notation:

$$\begin{aligned} [\varepsilon - 1]\mathbf{E} \quad &\rightarrow \quad V\mathbf{E}, \\ k_0^2 \int \mathrm{d}^3\mathbf{r}' \overleftrightarrow{G_0} (\mathbf{r}, \mathbf{r}')\mathbf{X}(\mathbf{r}') \quad &\rightarrow \quad G_0\mathbf{X}. \end{aligned} \tag{12.9}$$

It follows that

$$\begin{aligned} \mathbf{E} &= \mathbf{E}_{inc} + G_0 V\mathbf{E} = \mathbf{E}_{inc} + G_0 V\mathbf{E}_{inc} + G_0 V G_0 V\mathbf{E} \\ &= [1 + G_0 V + G_0 V G_0 V + G_0 V G_0 V G_0 V + ...]\mathbf{E}_{inc}. \end{aligned} \tag{12.10}$$

This expansion (known as the Liouville expansion in mathematics) can be interpreted as a multiple-scattering expansion. The term proportional to VG_0 is a single-scattering term, the term proportional to $(VG_0)^n$ corresponds to n scattering events. Fig. 12.1 represents this expansion graphically both within a given body and between different particles. The multiple-scattering terms exhibit the non-linear relation between the scattered field and V. This non-linear relation is at the heart of superresolution reconstruction of V from far-field scattering.

[2] Note that the body is defined by all the points where $[\varepsilon_r(\mathbf{r}') - 1] \neq 0$. This may be a single object like a bird or a plane. It can also be a set of different objects such as a collection of droplets or dust particles.

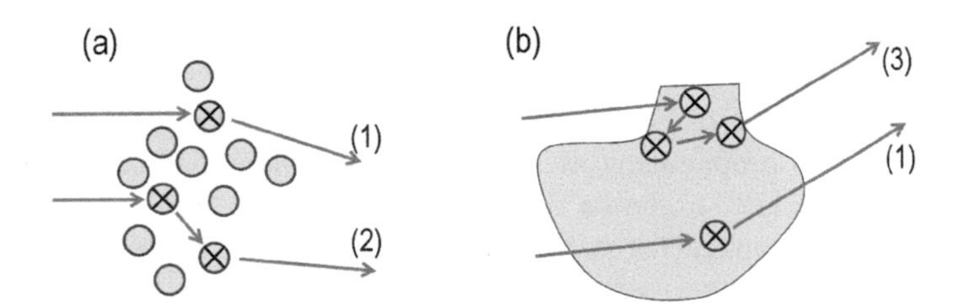

Fig. 12.1 Schematic representation of single, double, and triple scattering (a) in an ensemble of scatterers; and (b) inside a homogeneous body. The lines are not rays in the geometrical-optics sense. They are symbolic representations of the Green tensor G_0 accounting for propagation between two points, whereas the symbols account for the formal operator V.

12.1.4 The Green tensor: The Dyson equation

An alternative point of view is to introduce a Green tensor that accounts for the environment. Instead of defining the Green tensor $G_{nm,0}(\mathbf{r}, \mathbf{r}')$ as the electric field along axis n at point $\mathbf{r}$ produced by a dipole moment along the m-axis at position $\mathbf{r}'$ in vacuum, we consider the electric field produced by a dipole in an environment specified by $\varepsilon_r(\mathbf{r})$. To define this Green tensor, we start from Eq. (12.1):

$$\nabla \times \nabla \times \overleftrightarrow{G}(\mathbf{r}, \mathbf{r}') - \varepsilon_r(\mathbf{r}) k_0^2 \overleftrightarrow{G}(\mathbf{r}, \mathbf{r}') = \overleftrightarrow{I}\, \delta(\mathbf{r} - \mathbf{r}'). \tag{12.11}$$

It follows that the field is simply given by:

$$\mathbf{E}(\mathbf{r}) = \int \overleftrightarrow{G}(\mathbf{r}, \mathbf{r}')\, i\omega\mu_0\, \mathbf{j}_{inc}(\mathbf{r}') d^3\mathbf{r}'. \tag{12.12}$$

With this definition, it is seen that the field generated by a monochromatic dipole moment $\mathbf{p}$ corresponding to a current density $-i\omega\mathbf{p}\delta(\mathbf{r} - \mathbf{r}_0)$ is given by:

$$\mathbf{E}(\mathbf{r}) = \mu_0 \omega^2 \overleftrightarrow{G}(\mathbf{r}, \mathbf{r}_0)\, \mathbf{p}. \tag{12.13}$$

This new definition of the Green tensor is very important. Deriving the Green tensor for a complex geometry is a formidable challenge. However, for simple geometries, this can be done analytically. We give the form of the Green tensor for a system consisting of two media separated by a planar interface in Complement 12.B. It is derived in a simple manner by expanding the Green tensor in vacuum into plane waves and applying Fresnel factors and reflection/transmission laws to the plane waves.

Finally, we note that by inserting Eq. (12.12) into Eq. (12.7) and noting that this equation is valid for any distribution of currents, we get an integral equation for the Green tensor which can be written using the simplified notations:

$$G = G_0 + G_0 V G. \tag{12.14}$$

This equation is called the Dyson equation.

12.1.5 Extinction theorem

It is important to realize that the Green tensor $\overleftrightarrow{G_0}(\mathbf{r}, \mathbf{r}')$ used in Eq. (12.7) describes propagation in vacuum. Hence, all the fields propagated with the vacuum Green tensor throughout the system, including in metals or dielectrics, correspond to a phase velocity equal to c. At first glance, this may seem surprising. Let us, for instance, consider a system consisting of a slab of glass or gold. We are used to analysing this system in terms of incident fields propagating at velocity c and fields inside the medium with a phase velocity c/n. Eq. (12.7) offers a different point of view. The field is everywhere a superposition of fields propagating with velocity c. The Green tensor used in Eq. (12.7) is the vacuum Green tensor corresponding to waves propagating at speed c despite the fact that waves have a lower phase velocity in a dielectric or do not penetrate in a metal. Let us, for instance, consider a metallic body with volume V such that the field is zero inside the body. The integral equation yields:

$$\text{If } \mathbf{r} \in V; \quad \mathbf{E}_{inc}(\mathbf{r}) + k_0^2 \int [\varepsilon_r(\mathbf{r}') - 1] \overleftrightarrow{G}_0 (\mathbf{r}, \mathbf{r}')\mathbf{E}(\mathbf{r}')d^3\mathbf{r}' = 0. \tag{12.15}$$

This equation means that the incident field and the fields radiated by the induced currents in the skin depth interfere destructively inside the body in order to cancel fields propagating at velocity c inside the body. This result is a particular case of the extinction theorem also known as the Ewald–Oseen theorem (see ref. (Born, 1999)). It can be used to find the current density induced in the body $-i\omega[\varepsilon_r(\mathbf{r}') - 1]\mathbf{E}(\mathbf{r}')$. Once the current density is known, the integral can be used to compute the field at any other point outside the metal. Simple examples of this procedure are given in Exercise 2.4 to compute the field reflected by a perfectly conducting surface and in Exercise 12.1 to compute the reflected and transmitted field at an interface between vacuum and a dielectric. In this simple example, it is possible to check explicitly that the incident field is cancelled by the fields radiated by the induced currents.

At this point, the extinction theorem may appear a mathematical curiosity. However, it is possible to exhibit experimentally the interference between the incident field and the fields radiated by the induced currents by separating them in time, taking advantage of the time delay necessary to establish the polarization in the material. We show in Fig. 12.2 the transmission of monochromatic light on resonance with a cloud of cold rubidium atoms. The cloud is trapped at the focus of a laser beam. It has a cigar shape with a transverse waist of 230 nm and a longitudinal waist of 1.3 μm. The incident field can be turned on and off with an acousto-optic modulator on a subnanosecond time scale, generating a pulse with a 300 ns width (see dashed line). It is seen in the figure that the transmitted pulse has an intensity which is close to the incident intensity during the first nanoseconds. It decays within 30 ns, which is the lifetime of the atoms. This is also the time needed to polarize the atoms. Before being polarized, they do not scatter light and therefore do not modify the transmission. In other words, the beginning of the incident pulse propagates through the cloud of atoms as if it were a vacuum because the atoms are not yet polarized. After 30 ns, which corresponds to the time required to polarize the atoms, the transmitted field decays to a stationary value corresponding to the stationary transmission factor. This is due to the interference of the incident and scattered fields described by Eq. (12.7). Finally, when the incident pulse is turned off, light still emerges from the cloud due to the radiation by the atoms. This lasts for roughly 30 ns until all atoms have returned to their ground state. In summary, an experiment using short pulses allows us to separate in the time

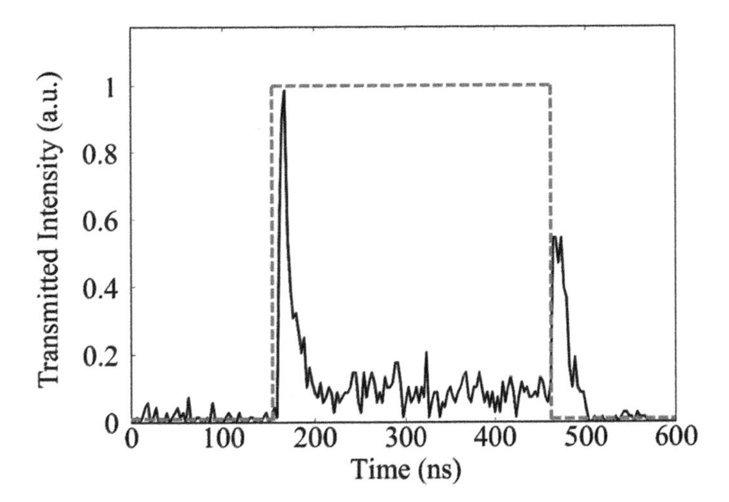

Fig. 12.2 Transmission of a focused incident resonant pulse by a cloud containing 100 rubidium atoms in a cigar-shaped trap with transverse waist of 230 nm and longitudinal waist of 1.3 μm. It is seen that between roughly 170 ns and 200 ns the cloud transparency changes from 1 to roughly 0.1. The reduced transmitted amplitude results from the interference between the scattered and incident light. Scattered light happens only after the atoms are polarized by the incident field. After the incident light is turned off, the scattered light is detected during roughly 30ns until all the atoms have radiatively relaxed. This time-resolved experiment reveals the interference of the incident and scattered fields in the forward direction by isolating the incident field at the beginning of the pulse and the radiated field just after the pulse. This interference is the physical mechanism behind the extinction theorem. Reprinted with permission from ref. (Jennewein et al., 2018). Copyright 2018 by the American Physical Society.

domain the incident and the scattered fields. In order to clearly observe this effect, it is necessary to use matter excitations with lifetimes longer than the modulation time of the incident field. This is why cold atoms are good candidates for observing this effect.

12.2 Far-field approximation: Scattering matrix

In this section, we consider a finite-sized object illuminated by a monochromatic plane wave with an electric field given by $\mathbf{E}_{inc} \exp(i\mathbf{k}_{inc} \cdot \mathbf{r} - i\omega t)$.

12.2.1 Far-field approximation

It is interesting to derive an explicit far-field form of the scattered field. We remind the reader that the far-field region was defined in section 2.3.1 of Chapter 2. To proceed, we start from Eq. (12.7), use the explicit form of the Green tensor, Eq. (12.4) and keep only the terms varying as $1/r$, which is a valid approximation if $R \gg \lambda/2\pi$. This amounts to taking the derivative of the exponential but not the derivative of $1/r$. Thus

$$\frac{\partial g(R)}{\partial x_j} \approx ik_0 \frac{R_j}{R} g(R),$$

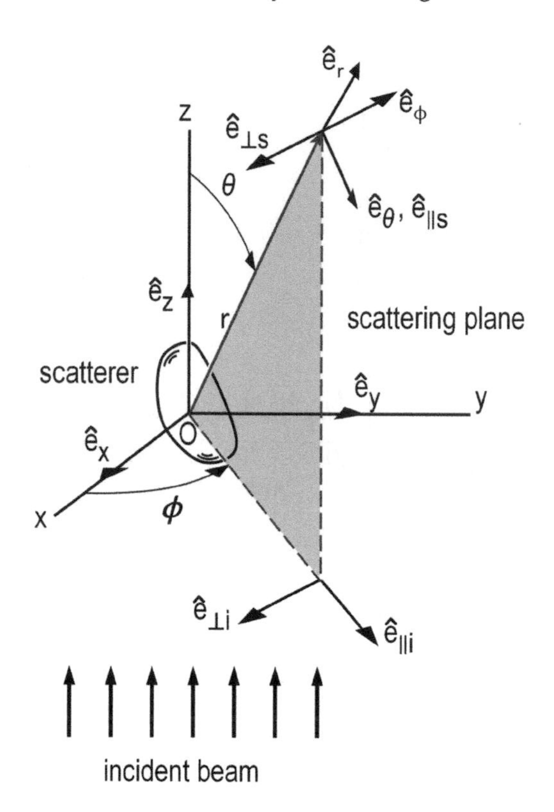

Fig. 12.3 Notations used for the scattering matrix.

where $g(R) = \exp(ik_0R)/R$ and $\mathbf{R} = \mathbf{r} - \mathbf{r}'$, Next, we assume that the scattering region has a maximum size L, and we evaluate the field in the so-called far-field region $r \gg L^2/\lambda$, so that $|\mathbf{r} - \mathbf{r}'| \approx r - \mathbf{u} \cdot \mathbf{r}'$ and $\mathbf{u} = \mathbf{r}/r \approx \mathbf{R}/R$, resulting in

$$\frac{R_j}{R} \approx \frac{r_j}{r} = u_j.$$

It follows that:

$$G_{0ij} = (\delta_{ij} + \frac{\partial_i \partial_j}{k_0^2})g(R) \approx (\delta_{ij} - u_i u_j)\frac{\exp(ik_0 r)}{4\pi r}\exp(-ik_0 \mathbf{u} \cdot \mathbf{r}'). \tag{12.16}$$

The tensor $\delta_{ij} - u_i u_j$ is the tensor that takes the transverse part of the field. Apart from this, the rest is essentially the usual far-field form of the potential vector. We finally find an explicit form of the scattered electric field:

$$\mathbf{E}_s(\mathbf{r}) \approx k_0^2 \frac{\exp(ik_0 r)}{4\pi r}[\overleftrightarrow{I} - \mathbf{uu}] \int [\varepsilon_r(\mathbf{r}') - 1] \exp(-ik_0 \mathbf{u} \cdot \mathbf{r}')\mathbf{E}(\mathbf{r}')d^3\mathbf{r}'. \tag{12.17}$$

This formula contains all the information connecting the scattered field and the permittivity of the object. The result is remarkably simple: it shows that the scattered-field amplitude is given by the Fourier transform of the polarization density in the scatterer.

We can use the discussion of the directionality of a radiating source (see section 2.4.2). Hence, for current distributions larger than the wavelength, the angular width of the scattered field may display lobes with an angular width equal to or larger than λ/L, where L is a typical transverse length of the scatterer. Thus, in order to get highly directional scattering, it is necessary to have large scatterers. For a subwavelength scatterer whose scattering is dominated by an induced electric dipole, scattering has a low directivity due to the dipolar scattering pattern. By increasing the scatterer size, other multipole contributions can interfere with the electric-dipole contribution and produce some directivity.

12.2.2 Scattering matrix

We have just seen that the scattered field in the direction specified by a wavevector $\mathbf{k}_s$ is perpendicular to the unit vector $\mathbf{u}_s = \mathbf{k}_s/k_0$. We can thus decompose the scattered electric field in two components. Using the coordinates defined in Fig. 12.3, we obtain:

$$\mathbf{E}_s = E_{\parallel,s}\mathbf{e}_{\parallel,s} + E_{\perp,s}\mathbf{e}_{\perp,s}, \tag{12.18}$$

where $\mathbf{e}_{\parallel,s} = \mathbf{e}_\theta$ and $\mathbf{e}_{\perp,s} = -\mathbf{e}_\phi$. Similarly, the incident electric field of a plane wave can be decomposed:

$$\mathbf{E}_{inc} = E_{\parallel,inc}\mathbf{e}_{\parallel,inc} + E_{\perp,inc}\mathbf{e}_{\perp,inc}, \tag{12.19}$$

where $\mathbf{e}_{\parallel,inc} = \cos\phi\mathbf{e}_x + \sin\phi\mathbf{e}_y$ and $\mathbf{e}_{\perp,inc} = \sin\phi\mathbf{e}_x - \cos\phi\mathbf{e}_y = -\mathbf{e}_\phi$.

Using these notations and Eq. (12.17), the scattered amplitudes are related to the incident amplitudes by

$$\begin{pmatrix} E_{\parallel,s} \\ E_{\perp,s} \end{pmatrix} = \frac{\exp(ik_0r)}{-ik_0r}\begin{pmatrix} S_2 & S_3 \\ S_4 & S_1 \end{pmatrix}\begin{pmatrix} E_{\parallel,inc} \\ E_{\perp,inc} \end{pmatrix}, \tag{12.20}$$

where the scattering matrix S is a function of frequency and incident and scattered directions.

12.2.3 Scattering cross section and absorption cross section

When an incident plane wave illuminates a scatterer, part of its energy may be dissipated in the particle and transformed into heat (absorbed power), and part of the incident power may be scattered. The absorbed and scattered power is not proportional to the incident power. It is proportional to the power per unit area at the scatterer location. Indeed, when illuminating a particle with an incident beam, the scattered and absorbed power increases if the beam is focused onto the particle. As the scattered and absorbed power is given in watts, the figure of merit of the particle characterizing how much power is scattered or absorbed must be given in m^2: the corresponding areas are called the scattering cross section, σ_s, and the absorption cross section, σ_{abs}.

Absorption cross section. Using the form of the absorption rate seen in Eq. (3.12) integrated over the volume of the scatterer, we get the absorption cross section by normalizing by the incident Poynting vector flux, $\varepsilon_0|\mathbf{E}_{inc}|^2/2$. It follows that the absorption cross section can be cast in the form

$$\sigma_{abs} = \frac{\omega}{c}\int_V \mathrm{d}^3\mathbf{r}\ \mathrm{Im}[\varepsilon_r(\mathbf{r})]\frac{|\mathbf{E}(\mathbf{r})|^2}{|\mathbf{E}_{inc}|^2}. \tag{12.21}$$

Scattering cross section. The scattering cross section is obtained by normalizing the total scattered Poynting vector integrated over a large sphere of radius R much larger than the scatterer by the incident Poynting vector. Hence, we get:

$$\sigma_s = \int_S \mathrm{d}A \frac{|\mathbf{E}_s(\mathbf{r})|^2}{|\mathbf{E}_{inc}|^2}. \tag{12.22}$$

12.3 Optical theorem

The aim of this section is to introduce the concept of the extinction cross section, which is the sum of the absorption cross section and the scattering cross section. To this aim, we apply energy conservation to a volume containing the scatterer. We then derive the optical theorem, which is an explicit form of the extinction cross section in terms of the amplitude of the field scattered in the forward direction.

12.3.1 Energy conservation

We consider a scatterer of arbitrary but finite volume V characterized by a permittivity $\varepsilon_r(\mathbf{r})$ (see Fig. 12.3). We assume that the scatterer is illuminated by a stationary monochromatic plane wave with incident electric field $\mathbf{E}_{inc}\exp(i\mathbf{k}_{inc}\cdot\mathbf{r})$. The total field is denoted by $\mathbf{E} = \mathbf{E}_{inc} + \mathbf{E}_s$. Note that this equation defines the scattered field everywhere, including inside the scatterer. The current density induced by the incident field in the scatterer is denoted by $\mathbf{j}_{ind}$. We start by writing the absorbed power in the scatterer:

$$P_{abs} = \frac{1}{2}\,\mathrm{Re}\left[\int_V \mathrm{d}^3\mathbf{r}\; \mathbf{j}_{ind}(\mathbf{r})\cdot[\mathbf{E}^*_{inc}(\mathbf{r}) + \mathbf{E}^*_s(\mathbf{r})]\right]. \tag{12.23}$$

The first term on the right-hand side can be interpreted as the power transferred from the incident field to the scatterer. This is called the extinction power, namely the power taken by the particle from the incident field. The second term is the power transferred from the scattered field to the induced currents, namely the opposite of the scattered power. To confirm this last statement, let us analyse a different problem without the incident field. We assume that we can create a current-density distribution equal to $\mathbf{j}_{ind}$ in a non-lossy system. The corresponding current-density distribution radiates the field

$$\mathbf{E}(\mathbf{r}) = \int \overleftrightarrow{G_0}(\mathbf{r},\mathbf{r}')\, i\omega\mu_0\, \mathbf{j}_{ind}(\mathbf{r}')d^3\mathbf{r}', \tag{12.24}$$

which is the scattered field $\mathbf{E}_s(\mathbf{r})$. Hence, we are dealing with a radiation problem. The energy conservation Eq. (3.13) applied to this stationary problem with no ohmic losses yields

$$0 = \frac{1}{2}\,\mathrm{Re}\left[\int_V \mathbf{j}_{ind}(\mathbf{r})\cdot\mathbf{E}^*_s\,\mathrm{d}^3\mathbf{r}\right] + P_s, \tag{12.25}$$

where P_s is the flux of the Poynting vector associated with the field $\mathbf{E}_s(\mathbf{r})$ radiated by $\mathbf{j}_{ind}$. It follows that Eq. (12.23) can be cast in the form

$$P_{abs} + P_s = P_{ext} = \frac{1}{2}\,\mathrm{Re}\left[\int_V \mathrm{d}^3\mathbf{r}\;\mathbf{j}_{ind}(\mathbf{r})\cdot\mathbf{E}^*_{inc}(\mathbf{r})\right]. \tag{12.26}$$

Hence, we see that the extinction power that we defined as the power transferred from the incident field to the particle appears to be equal to the sum of the absorbed power and the scattered power. This equation expresses energy conservation in the stationary regime. We now proceed to derive an explicit form of the extinction cross section known as the optical theorem.

12.3.2 The optical theorem

In order to define the extinction cross section, we normalize the extinction power by the incident Poynting vector. Since we consider an incident plane wave, $\mathbf{E}_{inc}(\mathbf{r}) = \mathbf{E}_{inc}\exp(i\mathbf{k}_{inc}\cdot\mathbf{r})$, the extinction power takes a particularly simple form:

$$\begin{aligned} P_{ext} &= \frac{1}{2}\,\mathrm{Re}\left\{\int_V \mathrm{d}^3\mathbf{r}\;\mathbf{j}_{ind}(\mathbf{r})\cdot\mathbf{E}^*_{inc}\exp(-i\mathbf{k}_{inc}\cdot\mathbf{r})\right\} \\ &= \frac{1}{2}\,\mathrm{Re}\left\{\mathbf{E}^*_{inc}\cdot\int_V \mathrm{d}^3\mathbf{r}\;\mathbf{j}_{ind}(\mathbf{r})\exp(-i\mathbf{k}_{inc}\cdot\mathbf{r})\right\}. \end{aligned} \tag{12.27}$$

We recognize the definition of the Fourier transform $\mathbf{j}_{ind}(\mathbf{k})$ of $\mathbf{j}_{ind}(\mathbf{r})$, so that we finally obtain:

$$P_{ext} = \frac{1}{2}\,\mathrm{Re}\left\{\mathbf{j}_{ind}(\mathbf{k}_{inc})\cdot\mathbf{E}^*_{inc}\right\}. \tag{12.28}$$

Hence, it is seen that the extinction power is proportional to the Fourier transform of the current density in the forward direction $\mathbf{k}_{inc}$. Using Eqs. (12.17) and (12.20), we can replace the Fourier transform of the current and introduce the scattered amplitude of the field in the forward direction. Finally, we find

$$\sigma_{ext} = \frac{2P_{ext}}{\varepsilon_0|\mathbf{E}_{inc}|^2} = \frac{4\pi}{k_0^2}\,\mathrm{Re}[\mathbf{u}_{inc}\cdot S(\mathbf{k}_{inc},\mathbf{k}_{inc})\mathbf{u}_{inc}], \tag{12.29}$$

where we have introduced $\mathbf{u}_{inc} = \mathbf{E}_{inc}/|\mathbf{E}_{inc}|$. The energy conservation Eq. (12.26) can be written in terms of scattering cross sections:

$$\sigma_{ext} = \sigma_{abs} + \sigma_s. \tag{12.30}$$

Attenuation length and the Beer–Lambert law. When a beam propagates in a suspension of particles containing n particles per unit length, the intensity decays exponentially following the so-called Beer–Lambert law, $I(x) = I(0)\exp(-n\sigma_{ext}x)$ (see Exercise 12.5).

12.4 Scattering by a particle

12.4.1 Scattering by a particle with low dielectric contrast: The first Born approximation

In this section, we show how the integral formulation of the electric field can be used to obtain in a very straightforward way an expression of the field scattered by an object of arbitrary shape. The problem that we address is the computation of the field scattered by a particle illuminated by a plane wave. The electric field can be derived everywhere using its integral form, provided that the field within the scatterer is known.

The first and more difficult problem to solve is finding the induced currents in the scatterer. To proceed, one can use the integral equation Eq. (12.7). The first term is the incident field and the second term represents the field produced by the induced currents. There are a number of cases where the scattered field is much smaller than the incident field, so that it is possible to simply assume that the field within the scatterer can be approximated by the incident field. This is known as the first Born approximation. It is difficult to give an accurate criterion to characterize the accuracy of this approximation. The general idea is that the scatterer should be small and the dielectric contrast should be small. A qualitative criterion is given by $|\mathbf{E} - \mathbf{E}_{inc}|/|\mathbf{E}_{inc}| \ll 1$ for any point within the scatterer. One can also derive sufficient conditions by deriving an upper bound of the integral that yields the electric field. For scattering of plane waves by more or less spherical objects, it can be shown that the set of conditions $|\varepsilon_r - 1| \ll 1$ and $L|n|/\lambda \ll 1$ are sufficient. The first one expresses the fact that the transmission factor is close to one, the second expresses the fact that the phase difference during propagation through the particle can be neglected. In case of an opaque material, it entails that the size of the scatterer should be much smaller than the skin depth. The scattered field $\mathbf{E}_s$ can thus be written

$$\mathbf{E}_s(\mathbf{r}) = k_0^2 \int [\varepsilon_r(\mathbf{r}') - 1] \overleftrightarrow{G}_0 (\mathbf{r}, \mathbf{r}') \mathbf{E}_{inc}(\mathbf{r}') d^3\mathbf{r}' \tag{12.31}$$

under the first Born approximation. In the case of an incident plane wave, the amplitude can be written as follows: $\mathbf{E}_{inc} \exp(i\mathbf{k}_{inc} \cdot \mathbf{r})$. It follows that the far-field scattered amplitude becomes

$$\mathbf{E}_s(\mathbf{r}) \approx k_0^2 \frac{\exp(ik_0 r)}{4\pi r} [\overleftrightarrow{I} - \mathbf{u}\mathbf{u}] \mathbf{E}_{inc} \int [\varepsilon_r(\mathbf{r}') - 1] \exp[-i\mathbf{q} \cdot \mathbf{r}'] d^3\mathbf{r}', \tag{12.32}$$

where $\mathbf{q} = k_0 \mathbf{u} - \mathbf{k}_{inc} = k_0(\mathbf{u} - \mathbf{u}_{inc})$ is the difference between the scattered and incident wavevectors.

12.4.2 Scattering by a small spherical particle with arbitrary dielectric contrast

In this section, we consider the case of a spherical particle with arbitrary dielectric contrast and radius $a \ll \lambda$. This condition entails that the field radiated by the induced currents can be described using the dipole approximation. According to Eqs (12.84) and (12.85), the field can be written as follows:

$$\mathbf{E}(\mathbf{r}) = \mathbf{E}_{inc}(\mathbf{r}) + k_0^2 \int [\varepsilon_r(\mathbf{r}') - 1] \overleftrightarrow{G}_{NS} (\mathbf{r}, \mathbf{r}') \mathbf{E}(\mathbf{r}') d^3\mathbf{r}' - \frac{[\varepsilon_r(\mathbf{r}) - 1]\mathbf{E}(\mathbf{r})}{3}. \tag{12.33}$$

We have explicitly separated the contribution of the singularity from the contribution of the non-singular Green tensor denoted by $\overleftrightarrow{G}_{NS}$. When the particle radius a becomes very small, the contribution of the principal value goes to zero. The contribution of the singularity is thus the leading term. Keeping only this term in Eq. (12.33) yields the field in the sphere:

$$\mathbf{E}(\mathbf{r}) + \frac{[\varepsilon_r(\mathbf{r}) - 1]\mathbf{E}(\mathbf{r})}{3} = \frac{[\varepsilon_r(\mathbf{r}) + 2]}{3}\mathbf{E}(\mathbf{r}) = \mathbf{E}_{inc}(\mathbf{r}). \tag{12.34}$$

This result is very interesting: it yields the relationship between the field inside the sphere and the field that illuminates the sphere. In other words, it accounts for the dielectric screening. It could have been found using an electrostatic approximation (see Exercise 14.4). Note that the explicit form of the field depends on the shape of the particle. Since the field inside a spherical particle is given by $\frac{3}{[\varepsilon_r(\mathbf{r})+2]}\mathbf{E}_{inc}(\mathbf{r})$, the induced dipole moment of the particle is

$$\mathbf{p} = 4\pi a^3 \frac{\varepsilon_r(\mathbf{r}) - 1}{\varepsilon_r(\mathbf{r}) + 2}\varepsilon_0 \mathbf{E}_{inc}(\mathbf{r}) = \alpha_{es}\varepsilon_0 \mathbf{E}_{inc}, \tag{12.35}$$

where the electrostatic polarizability of the particle is denoted by α_{es}. The field scattered by the particle can then be derived using the standard expression of the field radiated by a dipole.

Before moving to a more general approach to solve the integral equation, let us point out that the approximation made above has neglected higher-order terms in the expansion in powers of a. These terms are needed to account for radiative losses, which are, of course, not included in an electrostatic formula. Indeed, when using the form of the extinction cross section derived in Eq. (12.116) of Exercise 12.2 and the polarizability given by Eq. (12.35), we find that a particle with real permittivity has a zero-extinction cross section despite the fact that light is scattered. The reason is that the scattered power varies as a^6, so that we need to keep imaginary terms of the polarizability to order a^6. To proceed, we return to Eq. (12.33) and keep the contribution of the imaginary part of $\overleftrightarrow{G}_{NS}$, which is a constant equal to $ik_0/6\pi \overleftrightarrow{I}$. Noting that $\mathbf{p}/\varepsilon_0 = \int[\varepsilon_r(\mathbf{r}') - 1]\mathbf{E}(\mathbf{r}')d^3\mathbf{r}'$, Eq. (12.34) becomes

$$\frac{[\varepsilon_r(\mathbf{r}) + 2]}{3}\mathbf{E}(\mathbf{r}) = \mathbf{E}_{inc}(\mathbf{r}) + i\frac{k_0^3}{6\pi}\frac{\mathbf{p}}{\varepsilon_0}. \tag{12.36}$$

The physical content of this equation is clear: the field $\mathbf{E}(\mathbf{r})$ inside the particle is proportional to the sum of the incident field and the field produced by the particle on itself: the so-called radiation reaction. As we have seen, this term arises when accounting for the imaginary part of $\exp(ik_0 r)$, which is due to the retardation effects. The dipole moment is thus given by:

$$\mathbf{p} = \frac{4\pi a^3}{3}\varepsilon_0(\varepsilon_r - 1)\mathbf{E} = \alpha\varepsilon_0\mathbf{E}_{inc}. \tag{12.37}$$

From Eq. (12.36) and Eq. (12.37), we find that the polarizability can be cast in the form:

$$\alpha = \frac{\alpha_{es}}{1 - i\frac{\alpha_{es}k_0^3}{6\pi}}. \tag{12.38}$$

Hence, the polarizability has a non-zero imaginary part, despite the fact that the medium is non-lossy. The extinction cross section is no longer zero. The purpose of the above discussion is to highlight that accounting for retardation allows accounting for scattering losses. However, for the sake of simplicity, we have only kept the lowest-order terms in $k_0 a$ for the imaginary part of the field, and we have not included similar corrections for the real part. An exact form will be given in Chapter 14. A systematic derivation based on the Mie formalism can be found in ref. (Colom et al., 2016).

12.4.3 Scattering by a Rayleigh scatterer in a complex environment

In this section, we consider a small particle characterized by an electric polarizability α. The particle is not in vacuum, as in the previous section, but in an arbitrary environment described by a Green tensor $\overleftrightarrow{G}=\overleftrightarrow{G}_0 + \overleftrightarrow{S}$, where the tensor $\overleftrightarrow{S}$ accounts for the environment. Here, we aim to discuss the impact of the environment on the induced dipole. Due to the multiple scattering between the dipole and its environment, the field illuminating the dipole is modified as compared with a dipole illuminated in vacuum. We will show that this effect can be included in an effective polarizability often called 'dressed polarizability'. We start from the definition of the polarizability denoted by α^* and the definition of the field illuminating the dipole:

$$\mathbf{p} = \alpha^* \varepsilon_0 \mathbf{E}(\mathbf{r}_0)$$
$$\mathbf{E}(\mathbf{r}_0) = \mathbf{E}_{inc}(\mathbf{r}_0) + \mu_0 \omega^2 \overleftrightarrow{G} \mathbf{p}. \tag{12.39}$$

This equation is similar to Eq. (12.33). The difference is that here, we are assuming that the dipole is point-like. Let us first proceed by assuming that the point-like scatterer is in vacuum. We can formally solve the set of equations

$$\mathbf{p} = \alpha^* \varepsilon_0 [\mathbf{E}_{inc}(\mathbf{r}_0) + \mu_0 \omega^2 \overleftrightarrow{G}_0 (\mathbf{r}_0, \mathbf{r}_0)\mathbf{p}]. \tag{12.40}$$

The term proportional to the Green tensor in vacuum is the field produced by the dipole on itself. It is called radiation reaction. We can formally repeat the steps of the derivation of the previous section using $\overleftrightarrow{G}_0 (\mathbf{r}, \mathbf{r}') \approx \overleftrightarrow{I} [\delta(\mathbf{r} - \mathbf{r}')/3k_0^2 - ik_0/6\pi]$. Note that the vacuum Green tensor is a scalar for $\mathbf{r} = \mathbf{r}'$. With this form, it is seen that the singularity of the real part of the Green tensor produces a divergence. This non-physical divergence is a direct consequence of our crude assumption that the scatterer is point-like. This issue would not show up if we accounted properly for the spatial extension of the particle as we did in the previous section. We proceed formally:

$$\mathbf{p} = \alpha^* \varepsilon_0 [\mathbf{E}_{inc}(\mathbf{r}_0) + \mu_0 \omega^2 G_0'(\mathbf{r}_0, \mathbf{r}_0)\mathbf{p} + \mu_0 \omega^2 i G_0''(\mathbf{r}_0, \mathbf{r}_0)\mathbf{p}]. \tag{12.41}$$

The equation can be cast in the form:

$$[1 - \mu_0 \omega^2 G_0'(\mathbf{r}_0, \mathbf{r}_0)]\mathbf{p} = \alpha^* \varepsilon_0 [\mathbf{E}_{inc}(\mathbf{r}_0) + \mu_0 \omega^2 i G_0''(\mathbf{r}_0, \mathbf{r}_0)\mathbf{p}]. \tag{12.42}$$

Finally, we get:

$$\mathbf{p} = \alpha_{es} \varepsilon_0 [\mathbf{E}_{inc}(\mathbf{r}_0) + \mu_0 \omega^2 i G_0''(\mathbf{r}_0, \mathbf{r}_0)\mathbf{p}], \tag{12.43}$$

where we have introduced:

$$\alpha_{es} = \frac{\alpha^*}{1 - \alpha^* k_0^2 G_0'(\mathbf{r}_0, \mathbf{r}_0)}, \tag{12.44}$$

where α_{es} is the polarizability of the scatterer without accounting for the radiative losses. For instance, for a particle, it is the electrostatic polarizability. Once more the singularity in G_0' is an artefact, due to the fact that we have not properly accounted for the spatial extension of the scatterer as we did in the previous section. We now repeat the procedure to find:

$$\alpha = \frac{\alpha_{es}}{1 - i\frac{\alpha_{es} k_0^3}{6\pi}}, \tag{12.45}$$

where we have used $G''(\mathbf{r}_0, \mathbf{r}_0) = k_0/6\pi$. We emphasize that the observable quantity is not α^* but the polarizability α. The above discussion has the merit of showing how the radiation reaction impacts the polarizability. Let us illustrate its influence on a resonant polarizability of the form $\alpha^* = C/(\omega - \omega_0)$. Inserting this form into Eqs (12.44) and (12.45) yields

$$\alpha = \frac{C}{\omega - \omega_0 - Ck_0^2 G_0'(\mathbf{r}_0, \mathbf{r}_0) - i\frac{k_0^3}{6\pi}C}. \tag{12.46}$$

Here, it clearly appears that the real part of the Green tensor accounts for a spectral shift, whereas the imaginary part accounts for the radiative decay time. Let us now return to Eq. (12.39). Using $\overleftrightarrow{G} = \overleftrightarrow{G}_0 + \overleftrightarrow{S}$ and accounting for $\overleftrightarrow{G}_0$ yields

$$\mathbf{p} = \alpha\varepsilon_0[\mathbf{E}_{inc}(\mathbf{r}_0) + \mu_0\omega^2 \overleftrightarrow{S}\, \mathbf{p}]. \tag{12.47}$$

Solving this equation again yields

$$\mathbf{p} = \alpha_{\text{eff}}\varepsilon_0 \mathbf{E}_{inc}(\mathbf{r}_0), \tag{12.48}$$

where we have introduced the effective or dressed polarizability:

$$\alpha_{\text{eff}} = \frac{\alpha}{\overleftrightarrow{I} - \alpha k_0^2 \overleftrightarrow{S}}. \tag{12.49}$$

This is an important result that fully accounts for the interaction of the particle with its environment. If the particle is resonant and the environment is also resonant (e.g. a cavity or a plasmonic resonator), the particle may have a modified lifetime or even be strongly coupled to its environment (see chapters 7 and 19). Both behaviours are encoded in the resonances of the effective or dressed polarizability.

12.4.4 Scattering by an arbitrary particle: Numerical solutions

In this section, we briefly show how it is possible to numerically solve the problem of scattering by a particle of arbitrary shape and permittivity. The basic idea amounts to considering the particle as being a collection of small volume elements that can be modelled by dipoles. This is possible, provided that the volume elements are taken to be much smaller than the wavelength. To proceed, we start from the integral equation in the form given by Eq. (12.33), and we cast it in a slightly different form:

$$\frac{[\varepsilon_r(\mathbf{r}) + 2]}{3}\mathbf{E}(\mathbf{r}) = \mathbf{E}_{inc}(\mathbf{r}) + k_0^2 \int [\varepsilon_r(\mathbf{r}') - 1]\, \overleftrightarrow{G}_{NS}(\mathbf{r}, \mathbf{r}')\mathbf{E}(\mathbf{r}')d^3\mathbf{r}'. \tag{12.50}$$

The left term has a simple physical meaning: if one considers a small sphere, the field inside the sphere is given by $\mathbf{E}(\mathbf{r})$ and the field outside is given by $\frac{[\varepsilon_r(\mathbf{r})+2]}{3}\mathbf{E}(\mathbf{r})$, as shown in Eq. (12.34). Let us denote the illuminating field of a small volume element by $\mathbf{E}_{ill}$. We can write the following integral equation for this field:

$$\mathbf{E}_{ill}(\mathbf{r}) = \frac{[\varepsilon_r(\mathbf{r})+2]}{3}\mathbf{E}(\mathbf{r}) = \mathbf{E}_{inc}(\mathbf{r}) + k_0^2 \int 3\frac{\varepsilon_r(\mathbf{r}')-1}{\varepsilon_r(\mathbf{r}')+2} \overleftrightarrow{G}_{NS}(\mathbf{r},\mathbf{r}')\mathbf{E}_{ill}(\mathbf{r}')d^3\mathbf{r}'. \tag{12.51}$$

By extracting the singularity of the Green tensor, we have shifted from the internal field to the illuminating field. The next step to solve the problem is to introduce a mesh splitting the integral in several integrals over small volumes δV_n. Assuming that the electric field is constant over each volume, one can convert the integral equation into a linear system that can be solved using standard numerical routines. This yields the field for each cell. Once this is solved, the computation of the field at any point is a simple integral. Replacing the interaction between two neighbouring volume elements by the interaction between two dipoles is not very accurate. The method of moments (Wang, 1990) provides a more systematic procedure to solve the integral equation numerically.

12.5 Local density of states

So far, we have discussed how the scattered field is related to the permittivity of the object to be imaged. We have shown that the Green tensor is a very useful tool in that context. The purpose of this section is to briefly introduce the concept of the local density of electromagnetic states (LDOS). In particular, we will show how LDOS is related to the Green tensor. A generalization to open systems using a decomposition of the Green tensor in the basis of quasinormal modes is summarized in complement 7.A. A discussion of electric and magnetic density of states in lossy systems is provided in complement 12.D.

12.5.1 Introduction to the density of electromagnetic states (DOS) in vacuum

Before starting the discussion, we make a remark on semantics regarding the meaning of state, mode, and eigenfunction. The word 'state' is usually used in the context of quantum mechanics or statistical physics, whereas the word 'mode' is often used in the context of wave theory. A state is an eigenfunction of a Hermitian operator given a set of boundary conditions. Similarly, a mode of an acoustical or electromagnetic resonator is an eigenfunction of homogeneous wave equations for a given geometry. In the case of an open system or a lossy system, the time-dependent modes are exponentially decaying in time. They are the quasinormal modes already introduced in Chapter 4 and 7.

We have already discussed this concept for electrons in a semiconductor in section 8.2.1, where the term 'density of states'(DOS) was used for $g(E)$ such that $g(E)\mathrm{d}E$ is the number of electronic states (modes) with energy in the interval $E, E+\mathrm{d}E$. Here, we aim to find the number of *electromagnetic* modes such that $\rho(\omega)\mathrm{d}\omega$ is the number of electromagnetic states in the interval $[\omega, \omega+\mathrm{d}\omega]$. To begin with, we briefly remind the reader how to derive the density of electromagnetic states or, in other words, how to count the number of different solutions (plane waves) of Maxwell equations in vacuum. A difficulty immediately arises, as this number is infinite. However, it is possible to define the number of modes *per unit volume*. To proceed, it is useful to introduce a perfectly conducting virtual cubic box of size L and examine the limit when L tends to infinity. The modes are stationary waves with discrete wavevectors of the form $k_x = n\pi/L, k_y = m\pi/L, k_z = l\pi/L$, where n, k, and l are positive integers, so that the modes can easily be counted. In k-space, the volume occupied by a state is therefore $(\pi/L)^3$, so that the number of states in the volume element $\mathrm{d}k_x \mathrm{d}k_y \mathrm{d}k_z$ is $2L^3 \mathrm{d}k_x \mathrm{d}k_y \mathrm{d}k_z/(\pi)^3$, where the factor 2 accounts for the two possible polarizations of each state. It is seen that the number of modes increases as L^3, the volume of the box. We now let the size of the box increase and divide by the volume V in order to obtain the number of modes per unit volume. Here, the idea is that as the size of the box increases, the influence of the boundaries becomes negligible, and we end up finding an LDOS or DOS per unit volume which is an intrinsic property of vacuum. The DOS in k-space per unit volume is thus given by $2/\pi^3$. We can now easily find the DOS $g_v(\omega)$ using the dispersion relation $\omega = ck$. Indeed, the $\rho_v(\omega)\mathrm{d}\omega$ states per unit volume occupy the volume $\pi k^2 dk/2$ in k-space because we consider only positive values of k_x, k_y, k_z. Using the dispersion relation $k = \omega/c$, we find:

$$\rho_v(\omega)d\omega = \frac{\pi k^2 dk}{2}\frac{2}{\pi^3} = \frac{\omega^2}{\pi^2 c^3}d\omega. \tag{12.52}$$

So far, we have discussed stationary waves in a perfectly conducting box. An alternative procedure amounts to imposing periodic boundary conditions (the so-called Born–von Kármán condition); the modes are no longer stationary wave but propagating waves, and the number of modes is the same. From the periodic condition (e.g. $\exp(ik_x(x + L)) = \exp(ik_x x)$), it follows that the wavevector components are of the form $k_x = n2\pi/L, k_y = m2\pi/L, k_z = l2\pi/L$, where n, m, l are positive or negative integers. The DOS in k-space is now $1/4\pi^3$ and we have to consider positive and negative k_x, k_y, k_z, so that we recover the previous result for the DOS.

Let us use the DOS to count how many states $N(\omega)$ are available between 0 and ω in a volume V:

$$N(\omega) = V\int_0^{\omega} \rho_v(\omega')d\omega' = V\frac{\omega^3}{3\pi^2 c^3} = \frac{8\pi}{3}\frac{V}{\lambda^3}. \tag{12.53}$$

We find a very simple rule of thumb: the number of states with frequency smaller than ω is roughly given by the volume divided by $(\lambda/2)^3$.

We now illustrate the importance of this concept using two examples: blackbody radiation and spontaneous emission rate. Let us derive the form of the energy of the blackbody radiation. Each electromagnetic mode has a quantum of energy $\hbar\omega$ and the mean excitation number is given by Bose–Einstein statistics $n_{BE}(\omega) = 1/[\exp(\hbar\omega/k_B T) - 1]$. The product of these two terms by the DOS yields the black-body density of energy at temperature T:

$$u(\omega, T) = \left[\frac{\omega^2}{\pi^2 c^3}\right]\frac{\hbar\omega}{\exp(\hbar\omega/k_B T) - 1}. \tag{12.54}$$

As a consequence, the DOS has a strong influence on the thermodynamic properties of radiation. In particular, it plays an important role in energy and momentum transfer due to radiation.

The LDOS also plays a key role in the lifetime of the excited state in a two-level system. From the Fermi Golden Rule, it is known that the lifetime is proportional to the number of final states. When studying the rate of radiative relaxation, the radiative decay rate is therefore proportional to the number of electromagnetic states at the corresponding frequency. This can be seen by comparing the stimulated and spontaneous emission rates given by the Einstein coefficients. Their ratio is given by

$$\frac{A_{21}}{B_{21}\hbar\omega} = \frac{\omega^2}{\pi^2 c^3}, \tag{12.55}$$

which is nothing but the vacuum DOS. For stimulated emission, only the mode of the incident photon has to be considered, whereas for spontaneous emission, one has to sum over all possible electromagnetic states. Hence, the spontaneous emission coefficient is proportional to the DOS.

To summarize, the concept of DOS plays a key role when looking at the radiative decay of a two-level system and when looking at the thermodynamic properties of electromagnetic radiation. In what follows, we shall analyse why it is necessary to introduce a *local* DOS and how it can be related to the Green tensor.

12.5.2 Electron and phonon DOS

Let us first compare the DOS for light with the DOS for electrons or phonons. The DOS in k-space is also given by $1/4\pi^3$ for electrons. Here, we have accounted for the degeneracy due to the spin 1/2 of the electron. The total number of states in a crystal of volume V with N atoms is given by $2N$ for an s-band. Similarly, the total number of phonon states is given by 3N because this is simply the total number of degrees of freedom of the atoms. Hence, the number of states per unit volume is roughly N/V, which is the inverse of the volume a^3 of a unit cell of the crystal. If we now compare the number of states available in 1 m^3 for visible light in vacuum ($(2/\lambda)^3 \approx 10^{19}$) and for condensed matter excitations ($1/a^3 \approx 10^{30}$), we find a difference of 11 orders of magnitude.

This very crude estimate shows what is key to the efficiency of plasmons or optical phonons in increasing the energy density or in reducing the lifetime of quantum emitters. The DOS of polaritons benefits from the large number of states of condensed-matter excitations (electrons or phonons). However, many of these modes do not couple to light, so that a careful analysis is required to obtain an accurate description of the DOS. This is the purpose of the following sections.

12.5.3 Elementary introduction to the concept of LDOS

The goal of this section is to motivate the introduction of a *local* DOS and provide its general form. In this introductory discussion, we start the discussion with electrons. The number of states $dN(\omega)$ having a frequency comprising between ω and $\omega + \delta\omega$ is given by the DOS $g(\omega)$:

$$dN(\omega) = g(\omega)\delta\omega.$$

If the spectrum is discrete, then the DOS can be written in the form

$$g(\omega) = \sum_n \delta(\omega - \omega_n),$$

where n is a label of the states, so that they can be degenerate. We now introduce the concept of LDOS $g(\mathbf{r},\omega)$. The motivation is to account for the fact that not all the points are equivalent, except for translationally invariant media. For instance, at an interface between a metal or a semiconductor and vacuum, the electronic wavefunction modulus cannot be uniform and becomes progressively zero in the vacuum. The wavefunction is reflected, producing oscillations close to the interface. To account for this effect, we use the probability density to find an electron in state Ψ_n in the volume element $d\mathbf{r}$ given by $|\Psi_n(\mathbf{r})|^2 d\mathbf{r}$. This suggests the electronic LDOS should be defined by using the weighting factor, $|\Psi_n(\mathbf{r})|^2$:

$$g(\mathbf{r},\omega) = \sum_n |\Psi_n(\mathbf{r})|^2 \delta(\omega - \omega_n). \tag{12.56}$$

It is seen that upon integration of the LDOS over space, the DOS is recovered.

We now move on to electromagnetic states. We first motivate the need for an LDOS by taking two examples. We have seen that the LDOS is used to derive the spontaneous-emission decay rate of an atom. The decay rate is proportional to the number of final electronic and electromagnetic states. While in vacuum all points are equivalent, it turns out that the electromagnetic modes in a cavity, a waveguide, or even simply above a plane mirror present nodes and antinodes. As a result, an atom will be able to couple efficiently to the field if it is located in an antinode but will not be coupled if it is in a node. Similarly, if we are interested in the black-body energy density, it is seen that close to a perfectly conducting plane, the field is zero, so that the electromagnetic energy is zero. In summary, while the DOS is uniform in vacuum, it varies locally in the general case. The question is thus how to include this dependence.

The idea is to use the fact that the energy is proportional to the square of the wavefunction. Here, we use a scalar field for the sake of simplicity. We will give the vector form at the end of the section. Guided by Eq. (12.56), we introduce the definition

$$\rho(\mathbf{r},\omega) = \sum_n |\Psi_n(\mathbf{r})|^2 \delta(\omega - \omega_n), \tag{12.57}$$

where $\Psi_n(\mathbf{r})$ is a scalar mode satisfying the normalization condition.

$$\int_V |\Psi_n(\mathbf{r})|^2 dV = 1. \tag{12.58}$$

It follows that $\int_V \rho(\mathbf{r},\omega) dV = \rho(\omega)$.

12.5.4 Expansion of the Green function over the modes

This section has two goals. Firstly, we show that the LDOS is related to the Green function. Secondly, we derive an expansion of the field radiated by a source over the modes. To proceed, we first expand the Green function over a basis of modes. We start by defining the Green function as the solution of the equation

$$\Delta G(\mathbf{r},\mathbf{r}',\omega) + \frac{\omega^2}{c^2} G(\mathbf{r},\mathbf{r}',\omega) = -\delta(\mathbf{r}-\mathbf{r}'), \tag{12.59}$$

with the appropriate boundary conditions. We now consider the complete orthogonal set of modes $\{\Psi_n\}$ of the Laplacian operator

$$\nabla^2\Psi_n = -\frac{\omega_n^2}{c^2}\Psi_n, \tag{12.60}$$

with eigenvalues $-\frac{\omega_n^2}{c^2}$. We expand the Green function on this basis:

$$G(\mathbf{r},\mathbf{r}',\omega) = \sum_n c_n(\mathbf{r}',\omega)\Psi_n(\mathbf{r}). \tag{12.61}$$

Inserting this expansion in the equation that defines G, we obtain:

$$\sum_n c_n(\mathbf{r}',\omega)(\frac{\omega^2}{c^2} - \frac{\omega_n^2}{c^2})\Psi_n(\mathbf{r}) = -\delta(\mathbf{r}-\mathbf{r}'). \tag{12.62}$$

Since the $\{\Psi_n\}$ form an orthonormal basis, we have

$$\int \Psi_n\Psi_m^* d^3\mathbf{r} = \delta_{n,m}. \tag{12.63}$$

By multiplying the equation by $\Psi_m^*(\mathbf{r})$ and integrating, we obtain:

$$c_m(\mathbf{r}',\omega) = -\frac{\Psi_m^*(\mathbf{r}')}{\frac{\omega^2}{c^2} - \frac{\omega_m^2}{c^2}}. \tag{12.64}$$

Finally, the Green function can be cast in the form

$$G(\mathbf{r},\mathbf{r}',\omega) = -\sum_n c^2\frac{\Psi_n(\mathbf{r})\Psi_n^*(\mathbf{r}')}{\omega^2 - \omega_n^2}. \tag{12.65}$$

We may note in passing that this form is exactly the form obtained for the Green function in Eq. (2.66) provided that the sum over the label m is replaced by the integral over $\mathbf{k}$ with the DOS $1/(2\pi)^3$ and that the modes are taken to be plane waves, $\exp(i\mathbf{k}\cdot\mathbf{r})$.

12.5.5 Expansion of the fields radiated by a source over the modes

The expansion of the Green function over modes provides a systematic way of expanding the radiated fields produced by a source over the modes. Indeed, using the definition of the Green function, the solution of the equation

$$\nabla^2\Psi(\mathbf{r},\omega) + \frac{\omega^2}{c^2}\Psi(\mathbf{r},\omega) = S(\mathbf{r}) \tag{12.66}$$

is given by:

$$\Psi(\mathbf{r}) = -\int \mathrm{d}^3\mathbf{r}'\, G(\mathbf{r},\mathbf{r}',\omega)S(\mathbf{r}') = \sum_n C_n(\omega)\Psi_n(\mathbf{r}). \tag{12.67}$$

Hence, the radiated field can be decomposed as a sum over the modes. The amplitude of each mode is given by an overlap integral:

$$C_n(\omega) = \frac{c^2}{\omega^2 - \omega_n^2}\int \mathrm{d}^3\mathbf{r}'\, \Psi_n^*(\mathbf{r}')S(\mathbf{r}'). \tag{12.68}$$

12.5.6 Derivation of the local density of states

We can now obtain the expression of the LDOS. The key point is to note that[3]

$$\lim_{\eta\to 0} \mathrm{Im}\left[\frac{1}{(\omega+i\eta)^2-\omega_n^2}\right] = \frac{\pi}{2\omega_n}[\delta(\omega-\omega_n)-\delta(\omega+\omega_n)].$$

Using this result, it appears that we have

$$\sum_n |\Psi_n(\mathbf{r})|^2\delta(\omega-\omega_n)/2\omega_n = \lim_{\eta\to 0}\frac{1}{\pi c^2}\mathrm{Im}[G(\mathbf{r},\mathbf{r},\omega+i\eta)],$$

which is valid for positive values of ω. Finally, we get

$$\rho(\mathbf{r},\omega) = \sum_n |\Psi_n(\mathbf{r})|^2\delta(\omega-\omega_n) = \lim_{\eta\to 0}\frac{2\omega}{\pi c^2}\mathrm{Im}[G(\mathbf{r},\mathbf{r},\omega+i\eta)].$$

12.5.7 Vector case

Let us now consider briefly the vector case for non-lossy systems. The Green tensor satisfies Eq. (12.11). Assuming that the field can be expanded over a set of orthogonal modes $\mathbf{E}_m(\mathbf{r})$ satisfying

$$\int \mathbf{E}_m(\mathbf{r})\cdot\mathbf{E}_n^*(\mathbf{r})\mathrm{d}^3\mathbf{r} = \delta_{m,n}, \tag{12.69}$$

the Green tensor can be expanded as:

$$\overleftrightarrow{G}(\mathbf{r},\mathbf{r}',\omega) = \sum_m c^2\frac{\mathbf{E}_m(\mathbf{r})\mathbf{E}_m^*(\mathbf{r}')}{\omega_m^2-\omega^2}. \tag{12.70}$$

It follows that the field $\mathbf{E}(\mathbf{r})$ radiated by a current density $\mathbf{j}(\mathbf{r})$ is given by

$$\mathbf{E}(\mathbf{r}) = \sum_m C_m\,\mathbf{E}_m(\mathbf{r}), \tag{12.71}$$

where the mode amplitude C_m is given by an overlap integral:

$$C_m = \frac{i\omega}{\varepsilon_0}\frac{1}{\omega_m^2-\omega^2}\int \mathbf{E}_m^*(\mathbf{r}')\cdot\mathbf{j}(\mathbf{r}')\mathrm{d}^3\mathbf{r}'. \tag{12.72}$$

From Eq. (12.70), we can derive

$$\lim_{\eta\to 0}\mathrm{Im}\left[\overleftrightarrow{G}(\mathbf{r},\mathbf{r},\omega+i\eta)\right] = \sum_m \frac{\pi c^2}{2\omega}\mathbf{E}_m(\mathbf{r})\mathbf{E}_m^*(\mathbf{r})\delta(\omega-\omega_m). \tag{12.73}$$

It follows that

$$\rho(\mathbf{r},\omega) = \lim_{\eta\to 0}\frac{2\omega}{\pi c^2}\mathrm{Tr}\left\{\mathrm{Im}\left[\overleftrightarrow{G}(\mathbf{r},\mathbf{r},\omega+i\eta)\right]\right\}. \tag{12.74}$$

This definition gives a particular role to the electric field. This is legitimate when the topic of interest is the decay time of a system with an electric dipolar transition

[3] We use $\frac{1}{\frac{\omega^2}{c^2}-\frac{\omega_m^2}{c^2}} = \frac{1}{2\omega_m}\left[\frac{1}{\omega-\omega_m}-\frac{1}{\omega+\omega_m}\right]$ and $\lim_{\eta\to 0}\frac{1}{\omega-\omega_m+i\eta} = PV\frac{1}{\omega-\omega_m} - i\pi\delta(\omega-\omega_m)$.

which couples to the electric field only. In that case, only the projection of the LDOS on the axis of the dipole matters. Hence, the relevant LDOS is given by

$$\rho(\mathbf{r},\omega) = \lim_{\eta\to 0} \frac{2\omega}{\pi c^2} \mathrm{Tr}\left\{\mathrm{Im}\left[\mathbf{u}\cdot \overleftrightarrow{G}(\mathbf{r},\mathbf{r},\omega+i\eta)\cdot\mathbf{u}\right]\right\}, \tag{12.75}$$

where $\mathbf{u}$ denotes the dipole unit vector.

However, if one is interested in the electromagnetic energy in thermodynamic equilibrium, all modes are excited and the energy stored in the magnetic field needs to be computed specifically. In Eq. (12.75), the magnetic energy of the mode was included and assumed to be equal to the electric energy because of the mode normalization chosen. However, electric and magnetic energy are not necessarily the same in the near field, as discussed in Chapter 10. The derivation of a local electric and magnetic DOS is discussed in Complement 12.D. Furthermore, this discussion does not assume orthogonality of the modes, so that it can be used in the presence of lossy bodies. This approach is effective in deriving the LDOS at a given point and can be used to compute the consequences of the presence of the modes on the energy density or for the computation of a decay rate or a Casimir force. However, it does not exhibit the modes explicitly. Computing the LDOS for a lossy or open system can be achieved by searching the quasinormal modes, as discussed in Chapter 7, Complement 7.A.

12.A Complement: Derivation of the integral equation

In this complement, we derive the integral equation obeyed by the electric field. To proceed, we start from the general expression of the field in terms of the potential. The vector potential is

$$\mathbf{A}(\mathbf{r}) = \frac{\mu_0}{4\pi}\int \mathbf{j}(\mathbf{r}')\frac{\exp(ik_0R)}{R}d^3\mathbf{r}', \tag{12.76}$$

where $R = |\mathbf{r}-\mathbf{r}'|$. This solution assumes that the field decays at infinity. In this solution, the currents can be considered to be either produced by a source and called external currents, $\mathbf{j}_{ext}$, or induced in the material media $\mathbf{j} = -i\omega\varepsilon_0(\varepsilon_r - 1)\mathbf{E}$. The former are the source of the incident field, the latter are the source of the field scattered by the object. The vector potential can be cast in the form

$$\mathbf{A}(\mathbf{r}) = \mathbf{A}_{inc}(\mathbf{r}) + \mathbf{A}_{ind}(\mathbf{r}) = \mathbf{A}_{inc}(\mathbf{r}) - \frac{i\omega}{c^2}\int[\varepsilon_r(\mathbf{r}') - 1]\mathbf{E}(\mathbf{r}')g(R)d^3\mathbf{r}', \tag{12.77}$$

where $g(R) = \frac{\exp(ik_0R)}{4\pi R}$. We have denoted by $\mathbf{A}_{inc}$ the potential vector generated by the external currents. It corresponds to the field incident on the medium. Using the general definition of the field, we obtain

$$\mathbf{E}(\mathbf{r}) = -\frac{\partial\mathbf{A}(\mathbf{r})}{\partial t} - \nabla V(\mathbf{r}). \tag{12.78}$$

The scalar potential can be derived from the expression of the potential vector using the Lorenz gauge:

$$\nabla \cdot \mathbf{A}(\mathbf{r}) - \frac{i\omega V(\mathbf{r})}{c^2} = 0. \tag{12.79}$$

Inserting this expression of the scalar potential V into Eq. (12.78) yields

$$\mathbf{E}(\mathbf{r}) = i\omega \left[\mathbf{A}(\mathbf{r}) + \frac{1}{k_0^2} \nabla(\nabla \cdot \mathbf{A}(\mathbf{r})) \right]. \tag{12.80}$$

Using the explicit expression of the vector potential, we finally obtain:

$$\mathbf{E}(\mathbf{r}) = \mathbf{E}_{inc}(\mathbf{r}) + k_0^2 \left[\overleftrightarrow{I} + \frac{1}{k_0^2} \nabla\nabla \right] \int [\varepsilon_r(\mathbf{r}') - 1] \mathbf{E}(\mathbf{r}') g(R) d^3\mathbf{r}'. \tag{12.81}$$

The above formula yields the electric field both in and outside of the sources. Its physical meaning is transparent. It is seen that the field is the sum of two contributions: the field $\mathbf{E}_{inc}$ produced by the currents $\mathbf{j}_{inc}$ and the field radiated by the currents induced in the volume V. When evaluating the field in the sources, some care has to be taken because of the divergence of $g(R)$ when R tends to zero. This particular point will be discussed in the next section.

12.A.1 Singularity of the Green tensor and integral equation

Taking the derivative $\nabla\nabla$ of the integral does not introduce any difficulty, provided that the field is evaluated at a point $\mathbf{r}$ that does not belong to the distribution of currents. Indeed, in this case, $\mathbf{r} \neq \mathbf{r}'$, so that there is no singularity of the integrand. By contrast, if the point $\mathbf{r}$ belongs to the current sources, there is a logarithmic divergence of the integral. It is due to the second derivative, which yields a term that diverges as $1/r^3$. When combined with the r^2 dependence of the volume element, there is a $1/r$ term that produces a logarithmic divergence of the integral. In order to avoid this difficulty, it is possible to introduce an exclusion volume δV around the point $\mathbf{r}$ where the field is evaluated. In what follows, we will use a spherical exclusion volume and we will let its radius tend to zero. The field appears as the sum of two contributions,

$$\lim_{\delta V \to 0} \nabla\nabla \int_{V-\delta V} \mathbf{P}(\mathrm{r}') g(\mathbf{r} - \mathbf{r}') d^3\mathbf{r}' + \lim_{\delta V \to 0} \nabla\nabla \int_{\delta V} \mathbf{P}(\mathrm{r}') g(\mathbf{r} - \mathbf{r}') d^3\mathbf{r}' \tag{12.82}$$

where $\mathbf{P}$ denotes the polarization vector $\varepsilon_0(\varepsilon - 1)\mathbf{E}$. It is possible to exchange the order of derivation and integration for the first term, since the singularity has been removed. The evaluation of the second term is more involved. A derivation can be found in the books by Born and Wolf (Born, 1999) or in the book by Van Bladel (Van Bladel, 1991). However, the result is almost obvious from a physical analysis of the problem. Indeed, it is seen that the contribution of the small exclusion volume to the electric field is simply the field produced by a small sphere uniformly polarized. As we discussed previously, the electrostatic approximation is valid in the near field, when retardation effects are negligible. Thus, the contribution to the field of the exclusion volume is exactly the field produced by a sphere with uniform static polarization density $\mathbf{P}$. This is given by $-\mathbf{P}/3\varepsilon_0$. We finally obtain

$$\lim_{\delta V \to 0} \int_{V-\delta V} \nabla\nabla \mathbf{P}(\mathrm{r}') g(\mathbf{r} - \mathbf{r}') d^3\mathbf{r}' - \frac{\mathbf{P}}{3\varepsilon_0}. \tag{12.83}$$

It is worth noting that introducing an exclusion volume and letting it tend to zero is the definition of the principal value of the integral. Inserting this result into Eq. (12.81), we obtain a compact integral equation for the electric field,

$$\mathbf{E}(\mathbf{r}) = \mathbf{E}_{inc}(\mathbf{r}) + k_0^2 \int [\varepsilon_r(\mathbf{r}') - 1] \overleftrightarrow{G}_0 (\mathbf{r}, \mathbf{r}')\mathbf{E}(\mathbf{r}')d^3\mathbf{r}', \tag{12.84}$$

where the Green tensor $\overleftrightarrow{G}$ is given by:

$$\overleftrightarrow{G} = PV[\overleftrightarrow{I} + \frac{1}{k_0^2}\nabla\nabla]\frac{\exp[ik_0 R]}{4\pi R} - \frac{\overleftrightarrow{I}}{3k_0^2}\delta(\mathbf{r} - \mathbf{r}'). \tag{12.85}$$

In the above equation, PV indicates that the integral is evaluated in the sense of a principal value.[4] It can be checked that the Green dyadic coincides with the field radiated by a dipole, except that we have characterized the singularity at the origin and identified its physical meaning. The expression that we have derived for the electric field is an integral equation for the electric field: the unknown electric field appears in the integrand. The physical meaning of this equation is transparent. It is seen that the field at any point is the sum of two contributions: the incident field and the field radiated by the induced currents in the scattering object. Indeed, each volume element $d^3\mathbf{r}'$ of the scattering object has a dipole moment $\delta\mathbf{p} = \varepsilon_0[\varepsilon_r(\mathbf{r}') - 1]\mathbf{E}(\mathbf{r}')d^3\mathbf{r}'$ and radiates a field $k_0^2 \overleftrightarrow{G}_0 (\mathbf{r}, \mathbf{r}')[\varepsilon_r(\mathbf{r}') - 1]\mathbf{E}(\mathbf{r}')d^3\mathbf{r}'$. Hence, we see that the Green tensor is simply the operator that gives the field produced by an electric dipole. Since every point is illuminated by the fields produced by currents excited at all the other points of the medium, this is a multiple-scattering problem. Its solution generally involves taking into account all the interactions.

Let us stress here that the Green tensor is the Green tensor in vacuum. This may be surprising at first glance, as a volume element in a metal will emit a wave that will not be able to propagate through the metal. The decay of the wave emitted from a dipole in a bulk metal will emerge as the result of the coherent sum of all the scattered fields produced by all volume elements.

12.B Complement: Green tensor in layered systems

12.B.1 The Green tensor in vacuum

For layered systems, it is convenient to use the representation due to Sipe (Sipe, 1987), which consists in a decomposition over elementary plane waves. The first step is to decompose a field over a set of linearly polarized plane waves. We first note that a plane wave is characterized by a wavevector $\mathbf{k}^{\pm} = (\mathbf{k}_{\parallel}, \pm\gamma)$, where the sign $\pm$ depends on the propagation direction. The electric field of a plane wave can be written as the sum of an s-polarized component E_s and a p-component E_p in the form $\mathbf{E} = E_s\hat{s} + E_p\hat{p}^{\pm}$, where

[4] It is important to note that the shape of the exclusion volume does matter. If the exclusion volume is chosen to be a cube or a sphere, the electrostatic contribution to the electric field of the exclusion volume is $-\frac{\mathbf{P}}{3\varepsilon_0}$. If the shape changes, the value of the contribution also changes. However, the sum of the principal value and the singularity contributions does not depend on the choice of the shape of the exclusion volume.

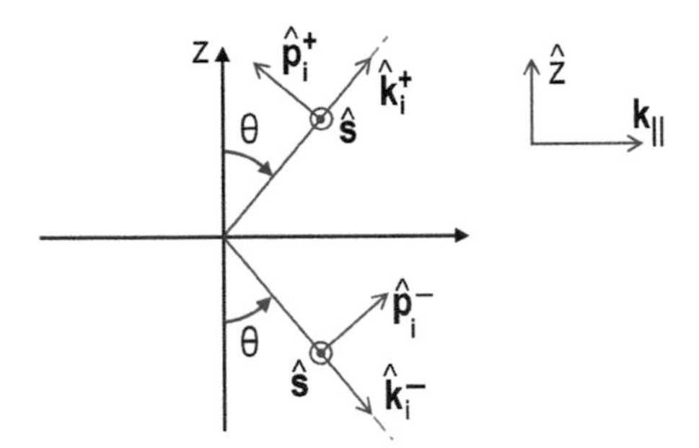

Fig. 12.4 Vectors for Green tensor definition. The horizontal axis is along $\mathbf{k}_{\|}$.

$\hat{s} = \mathbf{k}_{\|} \times \hat{\mathbf{z}}/|K|$, and $\hat{p}_i^{\pm} = [-\gamma_i \mathbf{k}_{\|}/|K| \pm K\hat{\mathbf{z}}]/(n_i k_0)$, where $\gamma_i = [\varepsilon_i \omega^2/c^2 - K^2]^{1/2}$, and we choose the solution with positive real part and positive imaginary part. These unit vectors are orthogonal and perpendicular to the wavevector (see Fig. 12.4). They can be used to project the field onto its two polarized components.

Using the dyadic notation for the tensors, the projection of the field onto the s-polarized unit vector is given by $\hat{s}\hat{s}\mathbf{E} = \hat{s}(\hat{s} \cdot \mathbf{E})$. The Green tensor in vacuum is defined by the relation

$$\mathbf{E}(\mathbf{r}) = \mu_0 \omega^2 \overleftrightarrow{\mathbf{G}}_0(\mathbf{r}, \mathbf{r}')\mathbf{p}. \tag{12.86}$$

Its explicit expression is given by

$$\begin{aligned}\overleftrightarrow{\mathbf{G}}_0(\mathbf{r}, \mathbf{r}', \omega) = \frac{k_0 e^{ik_0 R}}{4\pi} &\left[\left(\frac{1}{k_0 R} + \frac{i}{(k_0 R)^2} - \frac{1}{(k_0 R)^3}\right)\overleftrightarrow{\mathbf{I}}\right. \\ &+ \left.(\mathbf{u}_r\ \mathbf{u}_r)\left(\frac{3}{(k_0 R)^3} - \frac{3i}{(k_0 R)^2} - \frac{1}{k_0 R}\right)\right],\end{aligned} \tag{12.87}$$

where $R = |\mathbf{r} - \mathbf{r}'|$, $\overleftrightarrow{\mathbf{I}}$ is the identity tensor and $\mathbf{u}_r \mathbf{u}_r$ is a tensor in dyadic notation such that $(\mathbf{u}_r\ \mathbf{u}_r)\mathbf{A} = \mathbf{u}_r(\mathbf{u}_r \cdot \mathbf{A})$. In order to derive an expansion of the Green tensor over plane waves, we use the Weyl formula, Eq. (4.57) and insert it into Eq. (12.4). The integration and derivation can be exchanged if $\mathbf{r} \neq \mathbf{r}'$ so that the Green tensor can be written

$$\overleftrightarrow{\mathbf{G}}_0(\mathbf{r}, \mathbf{r}', \omega) = \frac{i}{2} \int \frac{d^2\mathbf{k}_{\|}}{4\pi^2} \frac{1}{\gamma} \left[\hat{s}\hat{s} + \hat{p}_1^+ \hat{p}_1^+\right] e^{i\mathbf{k}_{\|}\cdot(\mathbf{r}-\mathbf{r}')} e^{i\gamma(z-z')}, \tag{12.88}$$

where the transverse projector $1 - \mathbf{u}\mathbf{u} = \hat{s}\hat{s} + \hat{p}_1^+ \hat{p}_1^+$ is written in terms of s-polarized $\hat{s}$ and p-polarized $\hat{p}_1^+$ (respectively, $\hat{p}^-$) unit vectors for $z > z'$ (respectively, $z < z'$).

12.B.2 The Green tensor above an interface

To account for the interface, we simply decompose the field in plane waves that are either s-polarized or p-polarized, multiply their amplitude by the relevant Fresnel factor using the dyadic tensor $\left[\hat{s} t_{21}^s \hat{s} + \hat{p}_1^+ t_{21}^p \hat{p}_2^+\right]$, and then recombine the resulting (reflected or transmitted) plane waves with the appropriate z-component of the wavevector.

In the case of a Green tensor relating currents in the lower half-space ($\mathbf{r}'$ in medium 2) to a field in the upper half-space ($\mathbf{r}$ in medium 1), one has:

$$\overleftrightarrow{\mathbf{G}}(\mathbf{r},\mathbf{r}',\omega) = \frac{i}{2}\int \frac{d^2\mathbf{k}_\parallel}{4\pi^2}\frac{1}{\gamma_2}\left[\hat{s}t^s_{21}\hat{s} + \hat{p}^+_1 t^p_{21}\hat{p}^+_2\right] e^{i\mathbf{k}_\parallel\cdot(\mathbf{r}-\mathbf{r}')}e^{i\gamma_1 z - i\gamma_2 z'}. \tag{12.89}$$

In the expression, $\hat{s} = \mathbf{k}_\parallel \times \hat{\mathbf{z}}/|\mathbf{k}_\parallel|$ and $\hat{p}^\pm_i = -[\gamma_i \mathbf{k}_\parallel/|\mathbf{k}_\parallel| \mp |\mathbf{k}_\parallel|\hat{\mathbf{z}}]/(n_i k_0)$. The transmission factors are defined by:

$$t^p_{21} = \frac{2n_1 n_2 \gamma_2}{\varepsilon_1\gamma_2 + \varepsilon_2\gamma_1} \quad ; \quad t^s_{21} = \frac{2\gamma_2}{\gamma_1 + \gamma_2} \tag{12.90}$$

Note that the transmission factor for p-polarization has a pole that corresponds to the surface wave (see section 13.3.2). Thus the Green tensor contains all the information on surface waves.

In the case of two points lying above the interface in medium 1, the tensor can be cast in the form.

$$\overleftrightarrow{\mathbf{G}}(\mathbf{r},\mathbf{r}',\omega) = i\int \frac{d^2\mathbf{k}_\parallel}{4\pi^2}\left[\frac{\hat{s}r^s_{12}\hat{s} + \hat{p}^+_1 r^p_{12}\hat{p}^-_1}{2\gamma_1}e^{i\gamma_1(z+z')}\right.$$
$$\left. + \frac{\hat{s}\hat{s} + \hat{p}\hat{p}}{2\gamma_1}e^{i\gamma_1|z-z'|} + \frac{i\delta(z-z')}{k_0^2\varepsilon_1}\hat{z}\hat{z}\right] e^{i\mathbf{k}_\parallel\cdot(\mathbf{r}-\mathbf{r}')}, \tag{12.91}$$

where the reflection factors are

$$r^p_{12} = \frac{-\varepsilon_1\gamma_2 + \varepsilon_2\gamma_1}{\varepsilon_1\gamma_2 + \varepsilon_2\gamma_1} \quad ; \quad r^s_{12} = \frac{\gamma_1 - \gamma_2}{\gamma_1 + \gamma_2}, \tag{12.92}$$

and where we have defined $\hat{p} = [|\mathbf{k}_\parallel|\hat{z} - \gamma_3\hat{z}(z-z')/|z-z'|]/k_0 n_3$.

12.B.3 The Green tensor for a two-interfaces system

We now consider a two-layer system. The upper medium is denoted by 1 and lies above $z = d$. The lower medium ($z < 0$) is medium 2. Medium 3 is defined by $d > z > 0$. The Green tensor relating the currents in medium 2 to the field in medium 1 with a film of medium 3 between media 1 and 2 is given by

$$\overleftrightarrow{\mathbf{G}}(\mathbf{r},\mathbf{r}',\omega) = \frac{i}{2}\int \frac{d^2\mathbf{k}_\parallel}{4\pi^2}\frac{1}{\gamma_2}\left(\hat{s}t^s_{12}\hat{s} + \hat{p}^+_1 t^p_{12}\hat{p}^+_2\right) e^{i[\mathbf{k}_\parallel\cdot(\mathbf{r}-\mathbf{r}')}e^{i[\gamma_1(z-d)-\gamma_2 z']}, \tag{12.93}$$

where

$$t^{s,p}_{21} = \frac{t^{s,p}_{31}t^{s,p}_{23}e^{i\gamma_3 d}}{1 - r^{s,p}_{31}r^{s,p}_{32}e^{2i\gamma_3 d}}. \tag{12.94}$$

Note that the two-interface Green tensor is very similar to the single-interface one, except that the single-interface transmission coefficient has to be replaced by a generalized transmission coefficient taking into account the multiple reflections. When $\mathbf{r}$ and $\mathbf{r}'$ are in the film (medium 3), the Green tensor can be cast in the form:

$$\overleftrightarrow{\mathbf{G}}(\mathbf{r},\mathbf{r}',\omega) = \int \frac{d^2\mathbf{k}_{\|}}{4\pi^2}\overleftrightarrow{\mathbf{g}}(\mathbf{k}_{\|},z,z')\exp[i\mathbf{k}_{\|}.(\mathbf{r}-\mathbf{r}')], \tag{12.95}$$

where $\overleftrightarrow{\mathbf{g}}(\mathbf{k}_{\|},z,z')$ is the sum of three contributions:

$$\begin{aligned}\overleftrightarrow{\mathbf{g}}(\mathbf{k}_{\|},z,z') =& \frac{i}{2\gamma_3}[\hat{s}\hat{s}+\hat{p}\hat{p}]\exp[i\gamma_3|z-z'|] - \frac{1}{k_0^2\varepsilon_3}\delta(z-z')\hat{\mathbf{z}}\hat{\mathbf{z}} \\ &+ \frac{i}{2\gamma_3}[\hat{s}\rho_{32}^s\hat{s}+\hat{p}_3^+\rho_{32}^p\hat{p}_3^-]\exp[i\gamma_3(z+z')] \\ &+ \frac{i}{2\gamma_3}[\hat{s}\rho_{31}^s\hat{s}+\hat{p}_3^-\rho_{31}^p\hat{p}_3^+]\exp[i\gamma_3(d-z+d-z')],\end{aligned} \tag{12.96}$$

where

$$\rho_{31}^{s,p} = \frac{r_{31}^{s,p}}{1-r_{31}^{s,p}r_{32}^{s,p}e^{i2\gamma_3 d}} \quad ; \quad \rho_{32}^{s,p} = \frac{r_{32}^{s,p}}{1-r_{31}^{s,p}r_{32}^{s,p}e^{i2\gamma_3 d}} \tag{12.97}$$

and

$$r_{ij}^s = \frac{\gamma_i-\gamma_j}{\gamma_i+\gamma_j}; \quad r_{ij}^p = \frac{\varepsilon_j\gamma_i-\varepsilon_i\gamma_j}{\varepsilon_j\gamma_i+\varepsilon_i\gamma_j}. \tag{12.98}$$

12.C Complement: Density of states, optical etendue, energy flux, second principle, and spatial coherence

12.C.1 Optical etendue and number of modes

An important quantity in optics is the optical etendue, dG. It is a characteristic of the size of a beam both in terms of transverse size and angular aperture. It is defined as

$$dG = n^2 dS\ \cos\theta\ d\Omega,$$

where n is the real refractive index (we consider non-lossy media), θ is the angle between the beam propagation direction and the normal of the surface dS, and $d\Omega$ is the solid angle $\sin\theta d\theta d\phi$. At thermodynamic equilibrium, the energy flux propagating through the area dS in the direction $d\Omega$ is given by the product of the black-body radiance by the optical etendue. This equality suggests that the optical etendue is related to the DOS at a given frequency. This is indeed what we are going to show now.

We choose to define the z-axis parallel to the normal of dS. We now characterize the electromagnetic states by their wavevector components k_x and k_y. We have fixed the frequency ω. The z-component of the wavevector γ is then given by the dispersion relation. We find $\gamma = \sqrt{n^2\omega^2/c^2 - k_x^2 - k_y^2}$. We choose a positive root, as light propagates towards positive frequencies. In order to derive the DOS, we use the Born–von Kármán periodic boundary conditions with period L along the x- and y-axis, so that the number of states is given by $\frac{dk_x dk_y}{(2\pi/L)^2} = L^2\frac{dk_x dk_y}{4\pi^2}$ or $\frac{dk_x dk_y}{4\pi^2}$ per unit area. It follows that the DOS is given by $1/4\pi^2$ in the phase space (x, y, k_x, k_y). We now introduce a unit vector, $\mathbf{s} = \frac{c\mathbf{k}}{n\omega}$. Its coordinates can be expressed as follows: $s_x = \sin\theta\cos\phi$, $s_y = \sin\theta\sin\phi$, $s_z = \cos\theta$. The elementary area $ds_x ds_y$ can be expressed as $\cos\theta\sin\theta d\theta d\phi = \cos(\theta)d\Omega$, where $\cos\theta\sin\theta$ is the Jacobian of the coordinate transformation connecting (s_x, s_y) to (θ, ϕ).

We thus finally obtain the number of states dN propagating through an area L^2 in a given solid angle,

$$\mathrm{d}N = \frac{\mathrm{d}k_x \mathrm{d}k_y}{(2\pi)^2/L^2} = \frac{n^2 L^2 \cos\theta \, \mathrm{d}\Omega}{\lambda^2} = \frac{\mathrm{d}G}{\lambda^2}, \tag{12.99}$$

where we have introduced the wavelength in vacuum $\lambda = 2\pi c/\omega$. In other words, the optical etendue divided by λ^2 measures the number of modes propagating through dS in the solid angle dΩ. This is an important quantity for assessing an energy flux in a homogeneous medium.

12.C.2 Optical etendue conservation at an interface and the second principle of thermodynamics

We now consider the case of an interface between two homogeneous media with different refractive indices. It is important to examine how energy can be transferred at the interface. According to Eq. (12.53), the number of modes is proportional to n^3. One may imagine that if the high-index material at equilibrium at temperature T is put in contact with the medium with a lower index, the energy flux will flow to a medium with fewer modes, so that the energy per mode will increase. Since the energy per mode is given at equilibrium by $\hbar\omega/[\exp(\hbar\omega/k_B T) - 1]$, increasing the energy per mode amounts to increasing the temperature. In other words, energy goes from a cold medium to a hot medium. We know that this would violate the second principle. This does not happen, because several modes are totally reflected. This raises the question of the transformation of the optical etendue (namely the number of modes) through refraction. It was shown in Chapter 5 that the tranverse wavevectors k_x, k_y are conserved when light propagates through an interface normal to the z-axis separating homogeneous media. This is nothing but the Fresnel refraction laws. Indeed, the quantity dk_xdk_y, and therefore the optical etendue, is conserved as light propagates through interfaces. It follows that the number of modes is conserved through refraction, and therefore the second principle is not violated. This relation also gives a fundamental upper bound to the concentration that can be achieved when collecting light in an area dS_1 and concentrating it in a smaller area dS_2 using refraction or reflection:

$$\frac{\mathrm{d}S_1}{\mathrm{d}S_2} = \frac{n_2^2 \cos\theta_2 \mathrm{d}\Omega_2}{n_1^2 \cos\theta_1 \mathrm{d}\Omega_1}. \tag{12.100}$$

12.C.3 Optical etendue and spatial coherence

We finally use the connection between the optical etendue and the number of modes to characterize the spatial coherence of a beam. The basic idea is that the field associated with a single mode is spatially coherent. Indeed, the loss of coherence comes from the superposition of different modes with random and uncorrelated amplitudes. Our discussion shows that a beam containing a single mode ($dN = 1$) has an optical etendue dG given by λ^2. Hence, by imposing d$N = 1$, we can find a relation between the area and the solid angle d$S = \frac{\lambda^2}{n^2 \cos\theta \mathrm{d}\Omega}$ of a beam containing a single mode, namely a spatially coherent beam.

12.D Complement: Local electric and magnetic density of electromagnetic states

The derivation given in the chapter is based on the explicit form of the LDOS given by Eq. (12.57). It is based on the existence of an orthogonal set of eigenmodes. This approach cannot be used in the presence of losses, because modes are no longer orthogonal. In this section, we follow ref. (Joulain et al., 2005) to sketch a derivation of the LDOS in the presence of losses and accounting for the vector nature of electromagnetic fields: we write the electromagnetic energy density at equilibrium either in terms of the LDOS or by directly computing the fields at equilibrium. The comparison yields the explicit form of the LDOS in terms of the Green tensor.

We consider a system made of arbitrary lossy bodies in vacuum. We start by noting that the energy density at thermal equilibrium in vacuum (i.e. black-body radiation) can be written in the form

$$U(\omega) = \rho_v(\omega)\hbar\omega\left[\frac{1}{2} + \frac{1}{\exp(\hbar\omega/k_BT) - 1}\right], \tag{12.101}$$

where $\rho_v(\omega)$ is the vacuum DOS. Now, in the presence of lossy bodies, such as an interface separating a lossy half-space from vacuum, the energy density is no longer uniform. Hence, we can write it as

$$U(\mathbf{r},\omega) = \rho(\mathbf{r},\omega)\hbar\omega\left[\frac{1}{2} + \frac{1}{\exp(\hbar\omega/k_BT) - 1}\right], \tag{12.102}$$

by definition of the local density of electromagnetic states $\rho(\mathbf{r},\omega)$. On the other hand, the electromagnetic energy density in vacuum between the lossy bodies can be expressed in terms of the electric and magnetic fields

$$\langle U\rangle = \frac{\varepsilon_0}{2}\left\langle|\mathbf{E}(\mathbf{r},t)|^2\right\rangle + \frac{\mu_0}{2}\left\langle|\mathbf{H}(\mathbf{r},t)|^2\right\rangle. \tag{12.103}$$

In thermodynamic equilibrium, the field correlation functions are stationary and depend only on $t - t'$. It follows that they can be expressed in terms of cross-spectral densities:

$$\langle E_k(\mathbf{r},t)E_l(\mathbf{r}',t')\rangle = \mathrm{Re}\left[\int_0^\infty \frac{d\omega}{2\pi}\mathcal{E}_{kl}^{(N)}(\mathbf{r},\mathbf{r}',\omega)e^{-i\omega(t-t')}\right], \tag{12.104}$$

$$\langle H_k(\mathbf{r},t)H_l(\mathbf{r}',t')\rangle = \mathrm{Re}\left[\int_0^\infty \frac{d\omega}{2\pi}\mathcal{H}_{kl}^{(N)}(\mathbf{r},\mathbf{r}',\omega)e^{-i\omega(t-t')}\right]. \tag{12.105}$$

The fluctuation–dissipation theorem (Agarwal, 1975) establishes a connection between these cross-spectral densities, $\mathcal{E}_{kl}^{(N)}(\mathbf{r},\mathbf{r}',\omega)$, $\mathcal{H}_{kl}^{(N)}(\mathbf{r},\mathbf{r}',\omega)$, and the Green tensor. It follows (Joulain et al., 2005) that the energy can be cast in the form

$$U(\mathbf{r},\omega) = \frac{\hbar\omega}{[\exp(\hbar\omega/k_BT) - 1]}\frac{\omega}{\pi c^2}\mathrm{Im}\mathrm{Tr}\left[\overleftrightarrow{\mathbf{G}}^E(\mathbf{r},\mathbf{r},\omega) + \overleftrightarrow{\mathbf{G}}^H(\mathbf{r},\mathbf{r},\omega)\right]. \tag{12.106}$$

The electric Green tensor was already introduced by $\mathbf{E}(\mathbf{r}) = i\omega\mu_0\int\overleftrightarrow{\mathbf{G}}^E(\mathbf{r},\mathbf{r}',\omega)\mathbf{j}(\mathbf{r}')\mathrm{d}^3r'$.

Similarly, the magnetic field is related to the magnetization by $\mathbf{H}(\mathbf{r}) = \int \overleftrightarrow{\mathbf{G}}^{H}(\mathbf{r},\mathbf{r},\omega)$ $\mathbf{M}(\mathbf{r}')\,\mathrm{d}^3\mathbf{r}'$ (Joulain et al., 2005). A comparison of Eqs (12.102) and (12.106) shows that the LDOS is the sum of an electric contribution ρ^E and a magnetic contribution ρ^H:

$$\rho(\mathbf{r},\omega) = \frac{\omega}{\pi c^2}\mathrm{ImTr}\left[\overleftrightarrow{\mathbf{G}}^{E}(\mathbf{r},\mathbf{r},\omega) + \overleftrightarrow{\mathbf{G}}^{H}(\mathbf{r},\mathbf{r},\omega)\right] = \rho^E(\mathbf{r},\omega) + \rho^H(\mathbf{r},\omega). \tag{12.107}$$

In what follows, we shall discuss a few examples to illustrate the modification of the LDOS due to the presence of a lossy half-space.

Vacuum. In the vacuum, the Green tensors $\overleftrightarrow{\mathbf{G}}^{E}$ and $\overleftrightarrow{\mathbf{G}}^{H}$ obey the same equation and have the same boundary conditions. Therefore, their contributions to the electromagnetic energy density are equal:

$$\mathrm{Im}[\overleftrightarrow{\mathbf{G}}^{E}(\mathbf{r},\mathbf{r},\omega)] = \mathrm{Im}[\overleftrightarrow{\mathbf{G}}^{H}(\mathbf{r},\mathbf{r},\omega)] = \frac{\omega}{6\pi c}\overleftrightarrow{\mathbf{I}}. \tag{12.108}$$

The LDOS is thus obtained by multiplying the electric-field contribution by two. After taking the trace, the usual result for vacuum is retrieved:

$$\rho_v(\mathbf{r},\omega) = \rho_v(\omega) = \frac{\omega^2}{\pi^2 c^3}. \tag{12.109}$$

As expected, we note that the LDOS is homogeneous and isotropic.

12.D.1 Plane interface

We now consider a plane interface separating vacuum (medium 1, in the upper half-space) from a semi-infinite material (medium 2, in the lower half-space) characterized by its complex dielectric constant $\varepsilon_2(\omega)$. The material is assumed to be homogeneous, linear, isotropic, and non-magnetic. The expression of the LDOS at a given frequency and at a given height z above the interface in vacuum is obtained by inserting the expressions of the electric and magnetic Green tensors for this geometry (Sipe, 1987) into Eq. (12.107). Note that in the presence of an interface, the magnetic and electric Green tensors are no longer the same. Indeed, the boundary conditions at the interface are different for the electric and magnetic fields.

Let us consider some specific examples for real materials like metals and dielectrics. We first calculate $\rho(z,\omega)$ for aluminium at different heights. Aluminium is a metal whose dielectric constant is well described by a Drude model for near-ultraviolet, visible, and near-infrared frequencies (Palik, 1997):

$$\varepsilon(\omega) = 1 - \frac{\omega_p^2}{\omega(\omega + i\Gamma)}, \tag{12.110}$$

with $\omega_p = 1.747\ 10^{16}$ rad s^{-1} and $\Gamma = 7.596\ 10^{13}$ rad s^{-1}. In Fig.12.5 we have plotted the LDOS $\rho(\mathbf{r},\omega)$ in vacuum in the near-UV–near-IR frequency domain at four different distances to the interface. We first note that the LDOS increases drastically when the distance to the material is reduced. As discussed in the previous paragraph, at larger distances from the material, one retrieves the vacuum DOS. Note that at a given distance, it is always possible to find a sufficiently high frequency for which the corresponding wavelength is small compared to the distance, so that a far-field situation is retrieved. This is clearly seen when looking at the curve for $z = 1\,\mu\text{m}$, which coincides with the vacuum LDOS. When the distance to the material is decreased, additional modes are present: these are the evanescent modes which are confined close to the interface and which cannot be seen in the far field. Moreover, aluminium exhibits a resonance around $\omega = \omega_p/\sqrt{2}$. We anticipate the next chapter by noting that below this frequency, the material supports surface-plasmon polaritons, so that there are additional modes contributing to the LDOS in the near field, close to the interface. The enhancement is particularly important at a resonant frequency which corresponds to $\text{Re}[\varepsilon_2(\omega)] = -1$. Also note that in the low-frequency regime, the LDOS increases. Finally, Fig. 12.5 shows that it is possible to have a LDOS smaller than that of vacuum at some particular distances and frequencies. Figure 12.6 shows the propagating and evanescent waves' contributions to the LDOS above an aluminium sample at a distance of 10 nm. The propagating contribution is very similar to that of the vacuum LDOS. As expected, the evanescent contribution dominates at low frequency. It also dominates close to the surface-plasmon polariton resonance, where pure near-field contributions dominate.

Let us now turn to the comparison of $\rho(z,\omega)$, with the usual definition often encountered in the literature (Wijnands et al., 1997), which corresponds to $\rho^E(z,\omega)$. In Fig. 12.7 we plot, ρ, ρ^E, and ρ^H above an aluminium surface at a distance $z = 10\,\text{nm}$. In this figure, it is possible to identify three different domains for the LDOS behaviour. We note again that in the far-field situation (corresponding here to high frequencies, i.e. $\lambda/2\pi \ll z$), the LDOS reduces to the vacuum situation. In this case, $\rho(z,\omega) = 2\rho^E(z,\omega) = 2\rho^H(z,\omega)$. Around the resonance, the LDOS is dominated by the electric contribution ρ^E. Conversely, at low frequencies, $\rho^H(z,\omega)$ dominates.

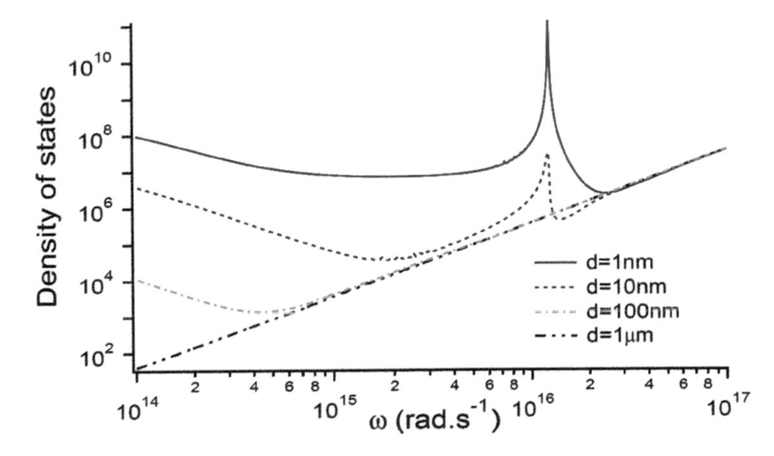

Fig. 12.5 Local density of states versus frequency at different heights above a semi-infinite sample of aluminium. From ref. (Joulain et al., 2003).

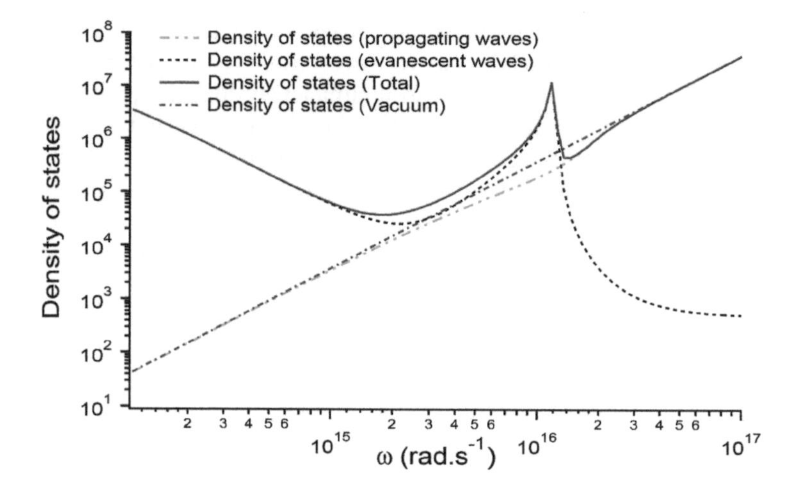

Fig. 12.6 Density of states (DOS) contributions due to the propagating and evanescent waves compared to the total DOS and the vacuum DOS. These quantities are calculated above an aluminium sample at a distance of 10nm. From ref. (Joulain et al., 2003).

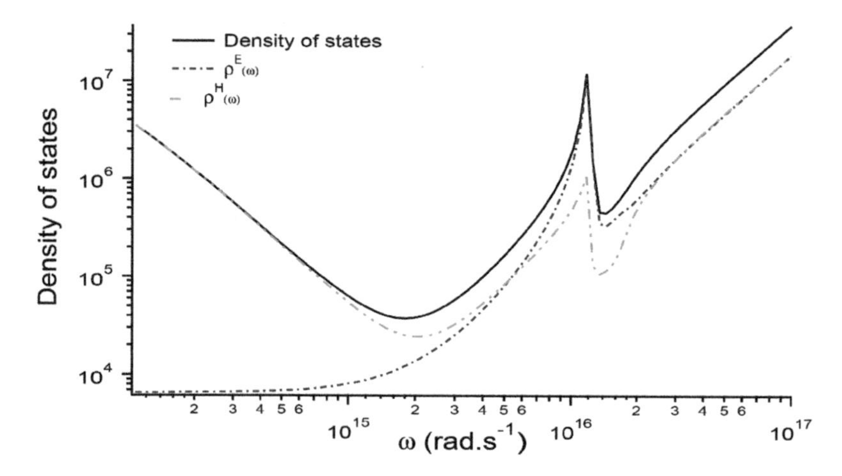

Fig. 12.7 Local density of states at a distance z = 10nm above a semi-infinite aluminium sample. Comparison with $\rho^E(z,\omega)$ and $\rho^H(z,\omega)$. From ref. (Joulain et al., 2003).

Thus, Fig. 12.7 shows that we have to be very careful when using the approximation $\rho(z,\omega) = \rho^E(z,\omega)$. Above aluminium and at a distance z = 10 nm, it is only valid in a narrow range between $\omega = 10^{16}$ rad s^{-1} and $\omega = 1.5 \times 10^{16}$ rad s^{-1}, i.e. around the frequency where the surface wave exists.

12.D.2 Asymptotic form of the LDOS in the near field

In order to get more physical insight, we have calculated the asymptotic LDOS behaviour in the three regimes mentioned above. As we have already seen, the far-field regime ($\lambda/2\pi \ll z$) corresponds to the vacuum case. To study the near-field situation

($\lambda/2\pi \gg z$), we focus on the evanescent contribution due to the large wavevectors $k_\parallel$, as suggested by the results in Fig. 12.6. In this (quasi-static) limit, the Fresnel reflection factors reduce to

$$\lim_{K\to\infty} r^s_{12} = \frac{\varepsilon_2 - 1}{4(k_\parallel/k_0)^2}, \tag{12.111}$$

$$\lim_{K\to\infty} r^p_{12} = \frac{\varepsilon_2 - 1}{\varepsilon_2 + 1}. \tag{12.112}$$

Asymptotically, the expressions of $\rho^E(z,\omega)$ and $\rho^H(z,\omega)$ are (Joulain et al., 2003):

$$\rho^E(z,\omega) = \frac{\rho_v}{|\varepsilon_2+1|^2}\frac{\varepsilon_2''}{4k_0^3z^3}, \tag{12.113}$$

$$\rho^H(z,\omega) = \rho_v\left[\frac{\varepsilon_2''}{8k_0z} + \frac{\varepsilon_2''}{2|\varepsilon_2+1|^2k_0z}\right]. \tag{12.114}$$

At a distance $z = 10$ nm above an aluminium surface, these asymptotic expressions match almost perfectly with the evanescent contributions ($k_\parallel > k_0$) of ρ^E and ρ^H. These expressions also show that for a given frequency, one can always find a distance z to the interface below which the dominant contribution to the LDOS will be the one due to the imaginary part of the electric-field Green function that varies like $(k_0z)^{-3}$. But for aluminium at a distance $z = 10\,\text{nm}$, this is not the case for all frequencies. As we mentioned before, this is only true around the resonance. For example, at low frequencies, and for $z = 10\,\text{nm}$, the LDOS is actually dominated by the magnetic contribution $\rho_v\varepsilon_2''/(8k_0z)$. This is due to the fact that the magnetic field is continuous at the interface, while the electric field has a very large dielectric screening.

12.D.3 Spatial oscillations of the LDOS

Let us now focus on the LDOS variations at a given frequency ω vs the distance z to the interface. There are essentially three regimes. First, for distances much larger than the wavelength, the LDOS is given by the vacuum expression ρ_v. The second regime is observed close to the interface where oscillations are observed. Indeed, at a given frequency, each incident plane wave on the interface can interfere with its reflected counterpart. This generates an interference pattern with a fringe spacing that depends on the angle and the frequency. Upon adding the contributions of all the plane waves over angles, the oscillating structure disappears, except when close to the interface. This leads to oscillations around distances on the order of the wavelength. This phenomenon is the electromagnetic analogue of Friedel oscillations, which can be observed in the electronic DOS near interfaces (Ashcroft and Mermin, 1976). For a highly reflecting material, the real parts of the reflection coefficients are negative, so that the LDOS decreases while approaching the surface. These two regimes are clearly observed for aluminium in Fig. 12.8. The third regime is observed at small distances, as seen in Fig. 12.8. It is seen that the evanescent contribution dominates and ultimately the LDOS always increases as $1/z^3$, following the behaviour found in Eq. (12.113).

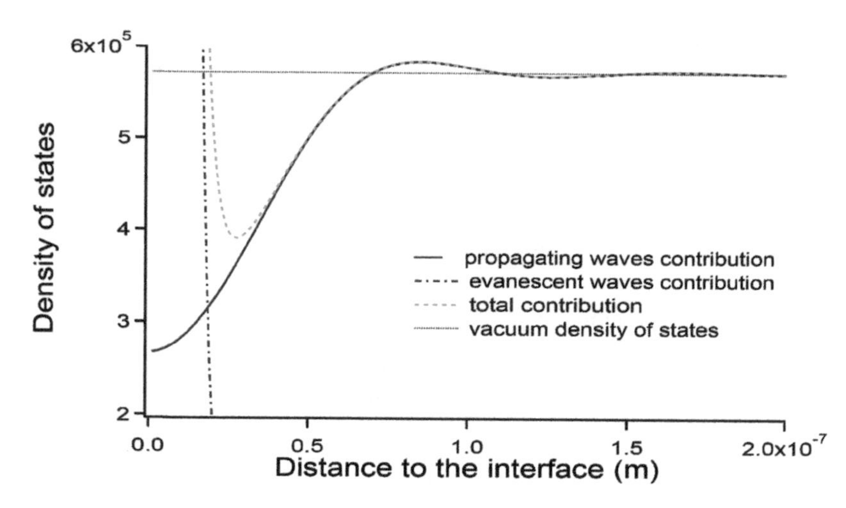

Fig. 12.8 Local density of states vs the distance z from an aluminium–vacuum interface at the aluminium resonant frequency. From ref. (Joulain et al., 2003).

Exercises

Exercise 12.1 Computing Fresnel factors using the integral formalism. Extinction theorem. The purpose of this exercise is to derive the Fresnel reflection factor by solving the integral equation for an interface-separating vacuum from a homogeneous medium. It will be seen that the field radiated by the induced polarization cancels the incident field inside the medium (extinction theorem) on the one hand and generates both the transmitted and the reflected field on the other. The half-space $z < 0$ is filled with a non-magnetic medium with complex permittivity ε_r describing either a dielectric or a metal. It is illuminated by a plane wave with the electric field polarized along Oy (see Fig. 12.9). We want to derive the reflection and transmission amplitudes by performing a radiation calculation. We first need to find the field radiated by the current density induced in the medium. The current density is thus $\mathbf{j} = -i\omega\varepsilon_0(\varepsilon_r - 1)\mathbf{E}$. The permittivity $\varepsilon_r = N^2$ is the square of the refractive index N for a non-magnetic medium.

(1) Using the results of Chapter 5, what are the reflection and transmission amplitudes for an interface between a vacuum and a medium with refractive index N?

(2) Derive the form of the transmitted field in the medium at a depth z, noting that it is the sum of the fields radiated by three types of sources:

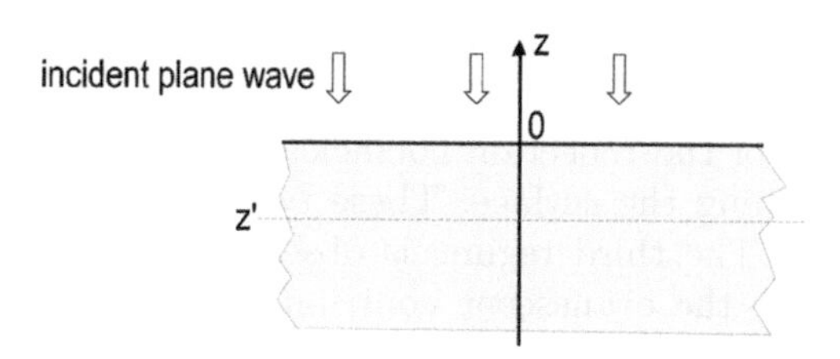

Fig. 12.9 Scheme of the studied system.

(a) the current density in the source which is generating the incident field;
(b) the induced current density with $z' > z$, i.e. from above the studied plane;
(c) the induced current density with $z' < z$.

Using Eq. (4.76), derive contributions (b) and (c) in the form of an integral depending on the unknown electric field in the medium $E(z')$.

(3) The previous equation is an integral equation for the field $E(z')$. We seek a solution in the form of a plane wave:

$$E(z') = A\ \exp(-iKz') \quad \text{with } \mathrm{Im}(K) > 0.$$

Denoting $K = N\omega/c$, determine the form of A and N as a function of ε_r. Compare with the transmission factor given by the Fresnel factor.

Exercise 12.2 Rayleigh-scatterer-scattering matrix. A Rayleigh scatterer is a scatterer much smaller than the wavelength, so that it can be described by its electric dipole moment $\mathbf{p} = \alpha\varepsilon_0\mathbf{E}_{inc}$, where α is the polarizability. Using Eqs (12.17) and (12.20), prove that the scattering matrix can be cast in the form

$$S = \frac{-i\alpha k_0^3}{4\pi}\begin{pmatrix}\cos(\theta) & 0\\ 0 & 1\end{pmatrix}. \tag{12.115}$$

Use this form to discuss the polarization of the scattered field in the forward direction ($\theta = 0°$) and for $\theta = 90°$ for an incident monochromatic depolarized light (i.e. incoherent superposition of two linear polarizations with the same amplitude and random phase). Using the optical theorem, derive the extinction cross section:

$$\sigma_{ext} = \frac{\omega}{c}\mathrm{Im}(\alpha). \tag{12.116}$$

Exercise 12.3 Symmetry properties of the scattering matrix of a sphere. Use the Curie principle (the symmetries of the sphere are to be found in the scattering matrix) to prove that the non-diagonal elements of the scattering matrix of a sphere are null. Use the reciprocity theorem to prove that $S(\mathbf{u}_s, \mathbf{u}_{inc}) = S(-\mathbf{u}_{inc}, -\mathbf{u}_s)$.

Exercise 12.4 Scattering matrix in the Born approximation (Rayleigh–Gans approximation). The purpose of this exercise is to derive the scattering matrix of a particle with permittivity ε_r and with arbitrary shape and volume V in the first Born approximation. The particle is in vacuum.

(1) Using Eqs (12.20) and (12.32), prove that the scattering matrix can be cast in the form

$$S = \frac{-i\alpha k_0^3}{4\pi}\begin{pmatrix}\cos(\theta) & 0\\ 0 & 1\end{pmatrix}, \tag{12.117}$$

where the effective polarizability α depends on the directions and is given by $\alpha = [\varepsilon_r - 1]Vf(\mathbf{k}_{inc} - k_0\mathbf{u}_s)$, where the form factor $f(\mathbf{q})$ is given by

$$f(\mathbf{q}) = \frac{1}{V}\int_V \exp(i\mathbf{q}\cdot\mathbf{r})d^3\mathbf{r}. \tag{12.118}$$

(2) Use the form factor to explain qualitatively the connection between the size of a particle and its scattering angular pattern.

Exercise 12.5 Decay length. Beer–Lambert law. We consider a colloidal suspension containing n scatterers per unit volume with an extinction cross section, σ. A collimated beam with area A and intensity $I(x)$ propagates in the system. Write the intensity difference between x and $x + dx$ using the definition of the extinction cross section. Show that the beam intensity decays exponentially with a decay length given by $1/(n\sigma)$.

Exercise 12.6 Local density of states (LDOS) close to an interface. It will be seen in Chapter 19 that the lifetime of a two-level system with an electric dipole moment is inversely proportional to the LDOS ρ^E. We consider a two-level system emitting at 700 nm located above a vacuum–metal interface such that the imaginary part of the metal is non-zero at 700 nm. Eq. (12.114) provides an asymptotic form of the LDOS valid in the near field. Assuming that it can be used at a distance of 100 nm, derive the ratio of the lifetimes of the emitter when it is at two different distances, 100 nm and 5 nm. The coupling to more evanescent modes does not increase the rate of light emission. Do you expect to observe light emission as the emitter approaches the surface?

Exercise 12.7 Coherence volume. In this exercise, we introduce the coherence volume in vacuum with two different points of view. We first estimate the volume where two photons are indistinguishable. Two photons with different frequencies will stay in phase during a time given by the coherence time $1/\Delta\nu$. In other words, during this time window, they cannot be distinguished. If these two photons are propagating along the same axis, it is possible to introduce a longitudinal length given by $c/\Delta\nu$, where they stay in phase. We consider a mode characterized by a solid angle $d\Omega$ and a spectral width $\Delta\nu$. We define the coherence volume ΔV as the product of the transverse coherence area (see Complement 12.C) and the longitudinal coherence length. Derive $\Delta V = \frac{1}{\Delta\Omega}\frac{\lambda}{\Delta\lambda}\lambda^3 = \frac{Q\lambda^3}{\Delta\Omega}$, where we have introduced a quality factor Q to characterize the spectral width.

We now seek the smallest volume, called the coherence volume V_{coh}, which is required to have a single mode in a given frequency interval $d\omega$. Note that we consider a single mode and therefore a single polarization, so that the density of states is given by $\frac{\omega^2}{2\pi^2 c^3}$. Show that the coherence volume V_{coh} is given by $V_{coh} = \frac{\lambda^3 Q}{4\pi}$. Check that it agrees with the previous form when assuming $\Delta\Omega = 4\pi$.

Exercise 12.8 Propagation equation in a host medium. We consider that sources and the scattering medium with permittivity ε_r are in a host medium with refractive index n_h. Derive the equations:

$$\nabla \times \nabla \times \mathbf{E} - \varepsilon_r(\mathbf{r}) n_h^2 k_0^2 \mathbf{E} = i\omega\mu_0 \mathbf{j}_{inc},$$

$$\nabla \times \nabla \times \overleftrightarrow{G}_h - n_h^2 k_0^2 \overleftrightarrow{G}_h = \overleftrightarrow{I}\, \delta(\mathbf{r} - \mathbf{r}'),$$

$$\overleftrightarrow{G}_0 = PV[\overleftrightarrow{I} + \frac{1}{n_h^2 k_0^2}\nabla\nabla]\frac{\exp[i n_h k_0 R]}{4\pi R} - \frac{\overleftrightarrow{I}}{3 n_h^2 k_0^2}\delta(\mathbf{r} - \mathbf{r}'),$$

$$\mathbf{E}(\mathbf{r}) = \mathbf{E}_{inc}(\mathbf{r}) + k_0^2 \int [\varepsilon_r(\mathbf{r}') - n_h^2]\, \overleftrightarrow{G}_h(\mathbf{r}, \mathbf{r}')\mathbf{E}(\mathbf{r}')d^3\mathbf{r}'.$$

Part IV

Plasmonics

13
Propagating surface plasmons

The term 'surface plasmon' is often used for both polarization oscillation in metallic nanoparticles called localized surface plasmons (LSPs) and surface waves propagating along a metallic interface. This chapter covers the latter case. LSPs will be considered in more detail in Chapter 14.

From the point of view of electrodynamics, surface plasmons propagating along interfaces are a particular case of surface waves, a topic that was extensively covered in the early days of radiowave propagation along the earth (Brekhovskikh, 1960; Baños, 1966; Felsen and Marcuvitz, 1996; King et al., 1992). From the point of view of optics, surface plasmons are a particular case of surface polaritons, hence the name surface plasmon polariton (SPP). Bulk polaritons were introduced by Hopfield (Hopfield, 1958) to describe the modes resulting from the strong coupling between the electromagnetic field and the polarization in a bulk medium. Surface polaritons are modes of an interface resulting from the strong coupling (see section 7.8.4) of electromagnetic waves and matter polarization. This vision was developed in the 1970s and 1980s. Several excellent monographs are available (Boardman, 1982; Agranovich and Mills, 1982; Raether, 1988). Subsequent works are summarized in (Zayats et al., 2005; Barnes et al., 2003; Schuller et al., 2010).

Finally, from the point of view of a quantum-mechanical description of the energy levels of electrons in a metal, a bulk plasmon is a collective excitation of electrons that emerges when going beyond the independent-electron approximation and including the Coulomb electron–electron interaction. This electronic excitation is not included in a band-structure picture. Indeed, band structure is based on a mean-field approximation which assumes non-interacting electrons. Excellent introductions can be found in well-known textbooks (Kittel, 2004; Ashcroft and Mermin, 1976) and in more advanced classic texts (Ziman, 1972; Pines, 1964; Kittel, 1966). The goal of this chapter is to provide an introduction to the different points of view and emphasize the key physical properties of surface plasmons.

As seen from the cited references, surface plasmons have been known for many years. There has been a revival of interest in surface plasmons with the advent of nanophotonics and the ability to observe them in the near field and control them. *Plasmons play an important role in nanophotonics because they allow electromagnetic energy to be confined at optical frequencies both in space at the nanometre scale and in time at a scale of a few fs.* This is a unique property of plasmons: the possibility to build an optical resonator (Chapter 7) significantly smaller than the wavelength with deep-subwavelength mode volumes. With the deployment of nanotechnology, the first two decades of the millennium have seen a rapid expansion of plasmonic circuitry, towards highly integrated

Introduction to Nanophotonics. Henri Benisty, Jean-Jacques Greffet, and Philippe Lalanne, Oxford University Press.
 DOI: 10.1093/oso/9780198786139.003.0013

photonic signal-processing systems with subwavelength plasmonic waveguides and active end devices for detection and emission of surface plasmons (Maier et al., 2001; Gramotnev and Bozhevolnyi, 2010; Schuller et al., 2010). Completely novel and impressive results have been obtained; they will be covered in chapters 18 and 19. Hereafter, we focus more on the physics behind the applications (Stockman, 2011) while keeping an ever-watchful eye on applications.

In the first section, we will show how the plasma frequency can be seen as the natural oscillation frequency of electrons in a thin film. This (solid-state) point of view will be generalized to bulk plasmons in an electron gas in the second section using a hydrodynamic model that assimilates the free electrons to fluids in motion. This analysis will allow us to illustrate the concept of the polariton: an electromagnetic wave coupled to a polarization excitation in the material.[1] We will then adopt the macroscopic electrodynamics point of view and derive the dispersion relation of a surface wave. In this approach, the material properties are accounted for by using a dielectric constant without any microscopic model. Hence, this discussion will be equally valid for radiowaves or optical waves, for metals or dielectrics. We will then focus on the case of SPPs propagating at an interface between a metal and a dielectric. Gap SPPs propagating between two parallel metallic planes with subwavelength separation will then be studied. We finally briefly introduce plasmons on graphene and surface phonon polaritons and discuss the contribution of surface polaritons to the local density of states (LDOS).

13.1 Surface and particle electron oscillation modes: Introductory examples

To introduce the concept of plasma oscillation, we consider a thin metallic film. The metal is described by a simple model: we assume that there are n free and independent electrons per unit volume. The crystal is modelled by a uniform positively charged background. This is the so-called jellium model (Pitarke et al., 2006). The purpose of this section is to illustrate the essence of a plasmon: it is an oscillatory collective mode of the electrons. To proceed, we work in the framework of classical mechanics.

Let us now assume that a static homogeneous electric field $E_{ext}\,\hat{\mathbf{x}}$ is applied normally to the film along the x-axis with $E_{ext} < 0$ (see Fig. 13.1). A force $-eE_{ext}\,\hat{\mathbf{x}}$ is exerted on the electrons, so that they will be displaced by x, with $x > 0$. A negative static surface charge $-nex$ will appear on the right interface and a positive surface charge on the left side. These surface charges produce a static field that cancels the external field in the metal. Let us now assume that the external electrostatic field is turned off at time $t = 0$, so that the system has been driven out of equilibrium and is free to move. The electrons in the film are accelerated by the electric field generated by the surface charges. When they return to their initial position, they have acquired a momentum, so that they keep moving along the x-axis and therefore generate an electric field of opposite sign. This process will be repeated and the electrons will oscillate back and

[1] Depending on the nature of the excitation (plasmon, phonon, exciton), there are different types of polaritons called plasmon polaritons, phonon polaritons, etc. These polaritons can be bulk polaritons or surface polaritons at interfaces. Hence the name 'surface plasmon polariton', often simplified as plasmon or surface plasmon for the sake of brevity.

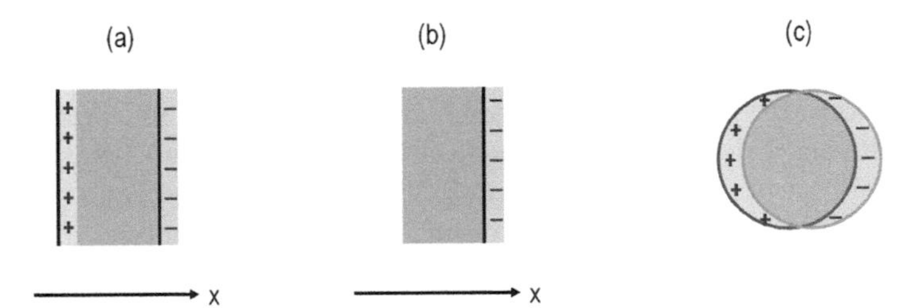

Fig. 13.1 Oscillation of the electron gas for: (a) a thin film; (b) vacuum–metal interface; (c) a nanosphere.

forth. It is easy to quantitatively describe this phenomenon by using Newton's law applied to a single electron,

$$m\frac{\mathrm{d}^2x}{\mathrm{d}t^2} = -eE_x, \tag{13.1}$$

where we have neglected the magnetic force and dissipative mechanisms. Using Gauss's theorem, it can be shown that the field generated by a sheet carrying a surface charge nex is $(nex/2\varepsilon_0)\mathbf{u}$, where $\mathbf{u}$ is an outward unit vector. Accounting for the two interfaces, we find $E_x = nex/\varepsilon_0$. It follows from Newton's equation that the movement of one electron is given by

$$\frac{\mathrm{d}^2x}{\mathrm{d}t^2} + \frac{ne^2}{m\varepsilon_0}x = 0. \tag{13.2}$$

This simple argument allows us to introduce in a simple way the plasma frequency ω_p:

$$\omega_p^2 = \frac{ne^2}{m\varepsilon_0}, \tag{13.3}$$

which appears to be the natural frequency of the collective oscillation of the electrons in the bulk of the film. To summarize, the oscillation is produced by an electric field due to all the electrons. This is why it is called a collective oscillation. With this simple argument, we have captured the essence of the plasmon: *it is the natural collective oscillation of the electrons characterized by the plasma frequency.* As pointed out in Chapter 7, a resonator corresponds to a periodic exchange of energy between two forms of energy: kinetic and potential energy for a mass attached to a spring, electric and magnetic energy for an LC circuit or an optical microcavity. It is shown in Exercise 13.1 that this system corresponds to a periodic exchange of energy between electric energy (the system is a capacitor) and the *kinetic energy* of the electrons. It is seen that the inertia of the system is not due to magnetic energy as in a photonic cavity but to the kinetic energy of the electrons. This is the key to confining electromagnetic energy in subwavelength volumes beyond the diffraction limit as derived in section 10.5.2.

It turns out that for noble metals, the electronic density is such that the plasma frequency is in the near ultraviolet or visible, while highly doped semiconductors can sustain surface plasmons in the mid-infrared. Hence, plasmonic oscillations can be used to fabricate nanoscale resonators at optical frequencies.

We now consider the case of a single interface (see Fig. 13.1b). In other words, we consider that the thickness of the film goes to infinity, so that the force is due to the charge density at a single interface. It follows that the electric field takes the value $nex/2\varepsilon_0$. The oscillation frequency is therefore $\omega_p/\sqrt{2}$.

We finally consider the case of a nanosphere. For a sphere much smaller than the wavelength, retardation effects can be neglected, so that we can use the electrostatic form of the field generated by a uniform polarization field P_x and the field $E_x = -P_x/3\varepsilon_0$; see ref. (Jackson, 1975) for the derivation of this useful relationship. Inserting this expression into Eq. (13.1) yields

$$m\frac{\mathrm{d}^2x}{\mathrm{d}t^2} = -e\frac{-P_x}{3\varepsilon_0}. \tag{13.4}$$

The polarization P_x is due to the displacement of the electrons, so that we have $P_x = -nex$. Upon inserting this expression in Eq. (13.4), we find

$$\frac{\mathrm{d}^2x}{\mathrm{d}t^2} + \frac{ne^2}{3m\varepsilon_0}x = 0, \tag{13.5}$$

so that the resonance frequency of the plasmon in a nanosphere is given by $\omega_p/\sqrt{3}$. The previous discussion has illustrated the fact that the resonance frequency depends on the geometry, as well known in mechanics for tuning forks or organ pipes.

In Fig. 13.2, we show an experiment reporting the energy lost by accelerated electrons scattered by a surface. After interacting with either bulk plasmons or surface plasmons, they can lose an energy which is given by $n\hbar\omega_p + m\hbar\omega_p/\sqrt{2}$, where n, m are positive integers. The values measured are 15.3 eV and 10.3 eV. The ratio is 1.49,

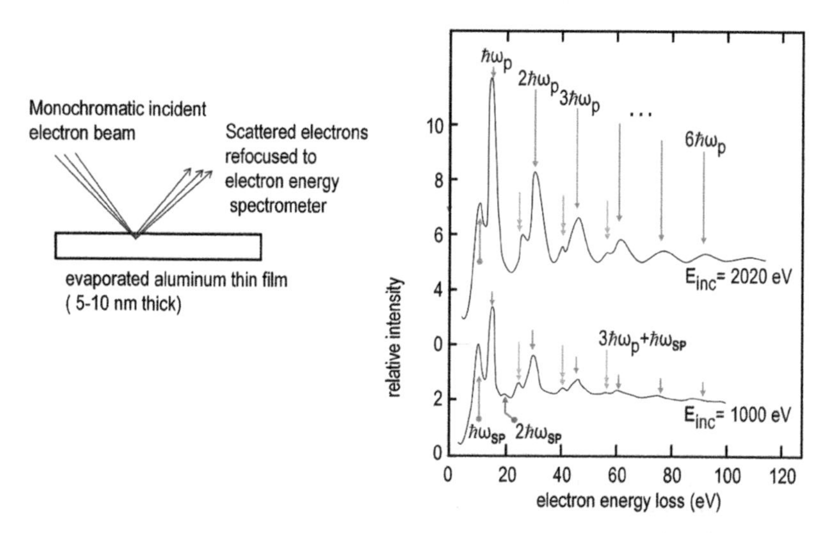

Fig. 13.2 Electron energy-loss spectroscopy (EELS) showing the quantization of the energy of plasmons and surface plasmons. Electrons accelerated with different energies (2020, 1520, and 1000 eV) are incident at 45° and scattered by a thin aluminium film (with thickness estimated to be in the range of 5–10 nm). The curve displays the number of reflected electrons at 45° versus the lost energy. Adapted from ref. (Powell and Swan, 1959).

slightly greater than the predicted 1.41. This experiment illustrates the quantization of the exchange of energy between electrons and plasmons. The experiment can be simply interpreted by considering that the electrons emit bulk plasmons or surface plasmons whose energies are quantized.

13.2 Bulk plasmon

13.2.1 Hydrodynamic model: The concept of the polariton

We now introduce a more general analysis of the concept of plasmon. We do not consider a specific geometry. Instead, we look for a general equation describing a *charge-density wave* in an infinite homogeneous electron gas. Such a wave is called a bulk plasmon. It can be understood in very simple terms using the concept of acoustic wave in a fluid. An acoustic wave in a gas is a density wave. When considering an acoustic wave in an electron gas, where the positive ions are treated as a uniform positively charged background (the jellium), it appears clearly that the regions with large electronic density carry a negative charge, whereas the regions where the density is reduced leave a positive charge due to the jellium. Such a charge distribution generates an electrostatic field in the frame of the acoustic wave. This electric field is transported by the acoustic wave, so that a *longitudinal* electric field is carried by the acoustic wave. This shows that the plasmon is both a mechanical and an electromagnetic wave.

The charge-density oscillation can also be viewed as a polarization wave, hence the name polariton, first introduced by Hopfield (Hopfield, 1958). The purpose of this section is to provide a simple and explicit model of a polarization wave in the case of a metal within the jellium model (Pitarke et al., 2006). In this section, we will use a hydrodynamic model to derive the equation of the charge-density wave, adopting a mechanical point of view. In the following section, we study the same polarization wave, adopting an electromagnetic point of view. To begin, we write Euler's equation and the mass conservation and Gauss–Maxwell equations,

$$
\begin{aligned}
nm\left[\frac{\partial \mathbf{v}}{\partial t} + \mathbf{v}\cdot\nabla\mathbf{v}\right] &= -ne\mathbf{E} - \nabla P_e, \\
\nabla\cdot(n\mathbf{v}) &= -\frac{\partial n}{\partial t}, \\
\nabla\cdot\mathbf{E} &= \frac{n-n_0}{\varepsilon_0}(-e),
\end{aligned}
\tag{13.6}
$$

where P_e is the electronic pressure, $\mathbf{v}$ is the velocity of electrons, and $n(x,t)$ is the number of electrons per unit volume. We finally introduce the compressibility of the electron gas $\partial P_e/\partial n = m\beta^2$, where β^2 is on the order of v_F^2, v_F being the Fermi velocity (Ashcroft and Mermin, 1976). A detailed discussion of the value of β is given in ref. (Halevi, 1995). For large frequencies $\omega\tau \gg 1$, where τ is a typical relaxation time, we have $\beta^2 = 3v_F^2/5$. When looking for a small amplitude perturbation $n_1(x,t) = n(x,t) - n_0$ and $P_{e1} = P_e - P_{e0}$, where the subscript 0 indicates the equilibrium value, the non-linear term $\mathbf{v}\cdot\nabla\mathbf{v}$ can be neglected.

Let us comment on this set of equations. For a neutral gas, the electric force $-ne\mathbf{E}$ in Euler's equation would be suppressed. One would then find the usual propagation equation for acoustic waves. Here, after linearizing, we find a set of two coupled linear equations:

$$\nabla^2 n_1 - \frac{5}{3v_F^2}\frac{\partial^2 n_1}{\partial t^2} = -\frac{5n_0 e}{3mv_F^2}\nabla\cdot\mathbf{E},$$
$$\nabla\cdot\mathbf{E} = \frac{n_1}{\varepsilon_0}(-e). \tag{13.7}$$

This system clearly exhibits the coupling between the acoustic wave and the electric field. It is seen that the electron density satisfies a propagation equation with a source term given by the divergence of the electric field. Similarly, the equation describing the longitudinal component of the electric field is driven by the electron-density modulation $n - n_0$. The key idea here is that the acoustic and electromagnetic fields are no longer modes of the system. The mode of the system is a hybrid mode called the polariton. It is half a photon and half a phonon.

It is now a simple matter of eliminating the electric field and finding the propagation equation for the electron-density wave that accounts for both the pressure force and the electric force. When searching for a solution of the form $\exp(ikx - i\omega t)$, we find the dispersion relation

$$\omega^2 = \omega_p^2 + \frac{3v_F^2}{5}k^2. \tag{13.8}$$

It turns out that the electric force yields the ω_p^2 contribution, which is much larger than the pressure contribution $v_F^2 k^2$ for wavevectors in the optical regime (i.e. $k \ll \omega/v_F$). It follows that in the optical regime, the dependence of ω on the wavevector can be neglected.

To summarize this section, it was shown that the plasmon appears to be an acoustic mode in an electron gas. As the particles are charged, an additional electric force has to be accounted for. It turns out that this electric force yields the dominant contribution, so that the modes are non-dispersive inasmuch as $k \ll \omega_p/v_F$.

13.2.2 Bulk plasmon: An electromagnetic model

When studying the propagation of electromagnetic waves in vacuum, we always focus on transverse waves, as longitudinal solutions do not exist. This is no longer the case in a material medium. Plasmons are longitudinal solutions of Maxwell equations. In the previous section, we studied the propagation of coupled mechanical and electromagnetic waves using a hydrodynamic model of metals. We found that the electromagnetic solution has an electric field which is parallel to the wavevector. In other words, it is a longitudinal electromagnetic field. We saw in Chapter 4 that longitudinal fields are defined by $\nabla \times \mathbf{E} = 0$ and $\nabla \cdot \mathbf{E} \neq 0$. The dispersion relation is given by Eq. (4.35). In the particular case of a non-magnetic and non-local medium such as a plasma, the dispersion relation becomes $\varepsilon_{\parallel}(\mathbf{k}, \omega) = 0$. The hydrodynamic model has been used to derive the longitudinal non-local permittivity for an electron gas, Eq. (3.61). Neglecting losses ($\nu = 0$), we recover the dispersion relation, Eq. (13.8). This result was expected, as we had used the same model to derive the permittivity. Nonetheless, it is instructive to find the same dispersion relation either by starting from hydrodynamic equations or Maxwell equations. This is a consequence of the dual nature of the polariton.

If we also neglect the dependence on the wavevector, the permittivity is given by $\varepsilon_r(\omega) = 1 - \omega_p^2/\omega^2$, so that we find $\omega = \omega_p$ in agreement with the local approximation of Eq. (13.8). Non-local models more elaborate than the hydrodynamic model, which account for the k-dependence of the dielectric constant (for instance, with the Feibelman d-parameters), are discussed in refs (Ashcroft and Mermin, 1976; Ziman, 1972; Pines, 1964; Ford and Weber, 1984).

13.3 Surface electromagnetic wave

So far, we have introduced the concept of the polariton using the particular case of a plasmon polariton in a bulk medium, focusing on longitudinal electromagnetic modes. We now consider modes which propagate along an interface and are confined close to the interface. More precisely, we look for a solution that decays exponentially away from the interface. In other words, the single interface acts as a waveguide. At this stage, we do not make any particular assumptions regarding the specific properties of the medium. Hence, the surface-mode dispersion relation that will be found can be applied to any material (e.g. metals, dielectrics) and any frequency range (e.g. radiowaves, infrared, visible). When dealing with surface waves propagating at interfaces between metals and dielectrics at infrared or optical frequencies, the surface waves are called SPPs.

However, other surface waves exist, such as surface phonon polaritons in the infrared, THz between dielectrics, or surface radiowaves between metals and dielectrics. The only assumptions made in what follows are that the media are local, isotropic, and non-magnetic. Hence, they are characterized by complex frequency-dependent dielectric constants ε_r. We refer to the upper medium ($z > 0$) with the index 1 and the lower medium ($z < 0$) with the index 2, as indicated in Fig. 13.3. We denote by $\mathbf{k}$ the wavevector, by (k_x, k_y, γ) its Cartesian components, and by k its modulus.

13.3.1 Dispersion relation for the non-magnetic case

Just like in Chapter 6 on waveguides, we look for source-free solutions of the Maxwell equations. The electric field obeys the Helmholtz equation in both media $i = 1, 2$:

$$\nabla^2 \mathbf{E}_i + \varepsilon_i \frac{\omega^2}{c^2} \mathbf{E}_i = 0. \tag{13.9}$$

For a p-polarized solution (also called TM for transverse magnetic), we seek a solution of the form (k_x was denoted by β in the context of guided waves in Chapter 6)

$$\begin{aligned} z > 0 : \quad & E_{x1} = E_0 \exp[ik_x x + i\gamma_1 z], \\ z < 0 : \quad & E_{x2} = E_0 \exp[ik_x x - i\gamma_2 z], \end{aligned} \tag{13.10}$$

which satisfies the continuity condition along the interface. Here,

$$\gamma_1 = [\varepsilon_1 \omega^2/c^2 - k_x^2]^{1/2} \tag{13.11}$$

with $\text{Im}(\gamma_1) > 0$ and

$$\gamma_2 = [\varepsilon_2 \omega^2/c^2 - k_x^2]^{1/2} \tag{13.12}$$

with $\text{Im}(\gamma_2) > 0$, so that the waves decay exponentially far from the interface. We look for transverse waves, so that, by definition, $\nabla \cdot \mathbf{E} = 0$ in each medium. In Fourier space,

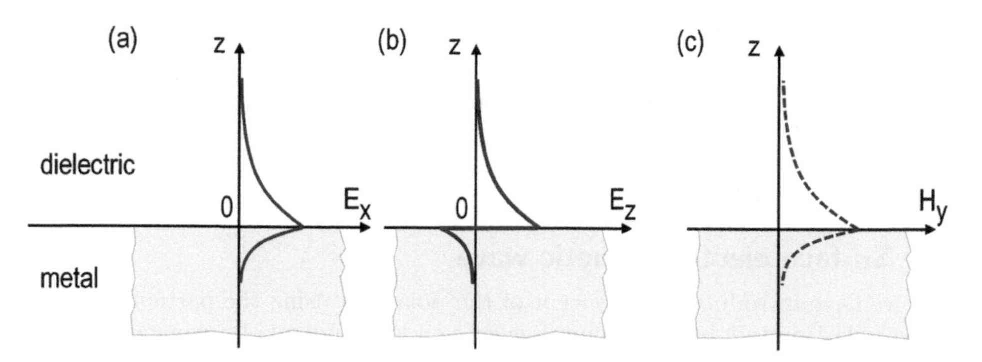

Fig. 13.3 Schematic representation of the transversal exponential decay along z of the three electromagnetic field components of a surface mode propagating along the x-axis.

this relation becomes $\mathbf{k} \cdot \mathbf{E} = 0$, where $\mathbf{k} = (k_x, 0, \gamma)$. We stress that this equation does not have the usual geometrical meaning of two perpendicular real vectors because $\mathbf{k}$ is complex. In other words, transverse (i.e. $\nabla \cdot \mathbf{E} = 0$) should not be confused with perpendicular. It follows that

$$\begin{aligned} z > 0 : \quad & E_{z1} = -\frac{k_x E_0}{\gamma_1} \exp[ik_x x + i\gamma_1 z], \\ z < 0 : \quad & E_{z2} = \frac{k_x E_0}{\gamma_2} \exp[ik_x x - i\gamma_2 z]. \end{aligned} \tag{13.13}$$

If we now enforce the continuity condition of the z-component of $\varepsilon\mathbf{E}$ at the interface, we have

$$\varepsilon_1 \gamma_2 = -\varepsilon_2 \gamma_1. \tag{13.14}$$

This equation is the dispersion relation of the surface mode. To obtain an explicit form connecting the wavevector and the frequency, we take the square of both terms. Note that in doing so, we lose the sign at this point, and we will find the solution of two equations $\varepsilon_1 \gamma_2 = \pm \varepsilon_2 \gamma_1$. For TM polarization, the solution for k_x, denoted by $k_{sp}(\omega)$, is

$$k_{sp}^2(\omega) = \frac{\omega^2}{c^2} \frac{\varepsilon_1 \varepsilon_2}{\varepsilon_1 + \varepsilon_2}. \tag{13.15}$$

We remember that Eq. (13.15) has two branches corresponding to the solution of $\varepsilon_1 \gamma_2 = -\varepsilon_2 \gamma_1$ and $\varepsilon_1 \gamma_2 = \varepsilon_2 \gamma_1$. Only the first is of interest to us.

At this point, it is not obvious that this dispersion relation provides a surface wave. We need to make sure that the wave decays exponentially away from the interface. Let us assume that medium 1 is a non-lossy dielectric, so that ε_1 is a real positive number, and that medium 2 is lossless ($\mathrm{Im}(\varepsilon_2) = 0$). In order to ensure that we obtain a surface mode, we need to check that $k_{sp} > n_1 \omega / c$, so that the wave decays exponentially in medium 1. Eq. (13.15) entails that

$$\frac{\varepsilon_2}{\varepsilon_1 + \varepsilon_2} > 1. \tag{13.16}$$

This condition can be satisfied only if $\varepsilon_1 + \varepsilon_2 < 0$. It follows that the mode is necessarily exponentially decaying in medium 2, as $\gamma_2 = [\varepsilon_2\omega^2/c^2 - k_x^2]^{1/2}$ is purely imaginary. We have found a surface wave for the TM polarization. A similar surface wave can be found for transverse-electric (TE) polarization for magnetic media with permeability μ_i. The dispersion relation is then given by

$$\mu_1\gamma_2 = -\mu_2\gamma_1. \tag{13.17}$$

13.3.2 Link with resonances of the reflection factor

An alternative approach for finding the dispersion relation consists in looking for the poles of the Fresnel reflection coefficient. The reason for looking at the Fresnel factors is simple. Since we can write $E_r^{s,p} = r_{s,p}E_{inc}^{s,p}$, it is seen that the reflection factor $r_{s,p}$ can be considered to be a linear response factor to the incident field $E_{inc}^{s,p}$, viewed as an external excitation. As for any linear system, a resonant response is the signature of the excitation of a mode of the system. Hence, writing that the denominator of $r_{s,p}$ is zero yields the pole which is the signature of a mode of the interface. It can be checked that, alongside Eq. (13.14), $\varepsilon_1\gamma_2 + \varepsilon_2\gamma_1$ is indeed the denominator of the Fresnel reflection factor for TM polarization for a non-magnetic material.

We can thus easily generalize the approach to a magnetic material for both polarizations. Since the Fresnel reflection coefficients can be cast in the form

$$r_s = \frac{\mu_2\gamma_1 - \mu_1\gamma_2}{\mu_2\gamma_1 + \mu_1\gamma_2} \quad ; \quad r_p = \frac{\mu_1\varepsilon_2\gamma_1 - \mu_2\varepsilon_1\gamma_2}{\mu_1\varepsilon_2\gamma_1 + \mu_2\varepsilon_1\gamma_2}, \tag{13.18}$$

it follows that the corresponding dispersion relations are:

$$s\text{-polarization}: \quad \mu_2\gamma_1 + \mu_1\gamma_2 = 0 \quad ; \quad p\text{-polarization}: \quad \mu_1\varepsilon_2\gamma_1 + \mu_2\varepsilon_1\gamma_2 = 0. \tag{13.19}$$

This approach is of particular interest when dealing with more complex systems such as thin-film stacks. It accounts for guided modes, interface modes, and the coupling between these modes. A technical remark might be useful at this stage. The reader may be familiar with a presentation of the Fresnel reflection coefficient using the incident angle θ_i as a variable instead of the parallel component k_x of the wavevector. For the case of a propagating incident wave in a lossless dielectric medium with refractive index n_1, it is essentially a matter of taste to use k_x or $n_1(\omega/c)\sin\theta_i$. For the case of a surface wave, we seek a real value of k_x which is larger than $n_1\omega/c$, so that the notation $n_1(\omega/c)\sin\theta_i$ is not well suited.

13.3.3 Polarization of the surface wave

We have seen that the electric field of a surface mode propagating along the x-axis has two components along the x- and z-axes. Moreover, the z-component of the electric field is complex. Hence, the electric field has an elliptic polarization in the (x, z)-plane. This peculiar polarization is easily understood. According to Eq. (1.52) the existence of a z-component of the electric field entails the presence of a surface charge $P_{z2} - P_{z1}$ along the interface. Hence, the surface mode can be viewed as a surface-charge-density wave propagating along the x-axis, as depicted in Fig. 13.4. Since in the vacuum above

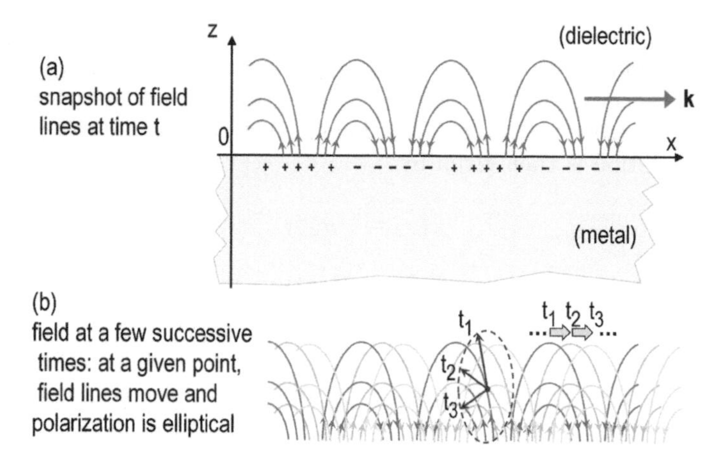

Fig. 13.4 Polarization of the surface plasmon polariton. The figure illustrates the surface-charge-density wave. It follows that the electric field has a normal component at the interface that oscillates. The figure shows that the continuity of the field in the vacuum requires a curvature of the field lines. The electric field is thus elliptically polarized in the plane (x, z).

the interface the field lines must be continuous, there must be an x-component of the field to close the line fields (see Fig. 13.4).

It is worth emphasizing the difference between the charges participating in the current density in the skin depth and the *surface* charge associated with the field discontinuity. While the current density $\mathbf{j} = -i\omega\mathbf{P} = -i\omega\varepsilon_0(\varepsilon_2 - 1)\mathbf{E}$ penetrates the metal over the skin depth, which is typically on the order of 15 nm for gold, the surface charge does not penetrate the metal, as $\nabla\cdot\mathbf{P} = \varepsilon_0(\varepsilon_2 - 1)\nabla\cdot\mathbf{E} = 0$ below the interface. The contribution to the surface charge is a pure surface term given by $P_{z2} - P_{z1}$. For a metal–vacuum interface, it is simply given by P_{z2}. From a physical point of view, a surface charge must have some finite extension along the z-axis. One has to account for screening (see Complement 3.D.3) to introduce the relevant length scale. It is the Thomas–Fermi length scale, which is on the order of 0.1 nm.

13.4 SPPs

In this section, we consider the specific case of surface modes propagating at the interface between a metal and a dielectric in the optical regime. As mentioned in the introduction, these surface waves are called SPPs. For the sake of brevity, they are often called 'surface plasmons'. The typical metals supporting surface plasmons are noble metals such as gold and silver. However, it is important to stress that there is a large family of materials supporting surface plasmons (Boltasseva and Atwater, 2011; Naik et al., 2013).

13.4.1 Dispersion relation of an SPP for a Drude model

In this section, we discuss the dispersion relation of a surface plasmon at an interface between a dielectric medium with a real dielectric constant ε_1 and a Drude metal to account for the metal dispersion. As already discussed, this is a crude model for noble metals when the frequency approaches the plasma frequency. Nevertheless, it simplifies the mathematics. Inserting the Drude form $\varepsilon_2(\omega) = 1 - \omega_p^2/(\omega^2 + i\nu\omega)$ of the dielectric constant in the dispersion relation given by Eq. (13.15), we obtain

$$k_{sp} = \frac{n_1\omega}{c}\left[\frac{\omega^2 - \omega_p^2 + i\omega\nu}{(1+\varepsilon_1)(\omega^2 + i\omega\nu) - \omega_p^2}\right]^{1/2}. \tag{13.20}$$

Since the permittivity is complex-valued owing to absorption, it is impossible to find a solution of the dispersion equation for a real wavevector and a real frequency. We have already encountered this issue when dealing with the dispersion equation of a plane wave propagating in a homogeneous lossy medium in section 4.7 and Complement 4.C. Two approaches discussed in chapters 4, 6, and 7 are possible. Either we may fix ω real and find solutions with a complex-valued k_{sp} (these are the guided or radiation modes of Chapter 6), or we may fix k_{sp} real and find solutions with a complex-valued frequency; these are the quasinormal modes of Chapter 7 which form the frequency spectrum of the system. The two dispersion relations are shown in Fig. 13.5 for both choices. The first striking feature is that they are different.

Let us analyse the physical content of these dispersion relations. First of all, we note the existence of an upper branch existing for $\omega > \omega_p$. In this frequency regime, a non-lossy metal has a permittivity $1 - \omega_p^2/\omega^2$ which is positive, so that it behaves as a dielectric. Hence, this branch does not correspond to a surface mode. It is not the solution of $\varepsilon_1\gamma_2 + \varepsilon_2\gamma_1 = 0$ but the solution of $\varepsilon_1\gamma_2 - \varepsilon_2\gamma_1 = 0$. In other words, it corresponds not to the pole of the reflection factor but instead to the zero of the reflection factor, namely the locus of the Brewster angle in the plane (k_x, ω). This is connected to the Brewster angle θ_B by the relation $k_x = n_2(\omega/c)\sin(\theta_B)$. In what follows, we ignore this branch. The lower branch is seen to be below the light line, as expected for a surface mode.

Quasinormal surface-plasmon modes: complex-valued frequency. When we impose a real value for k_{sp} in solving the dispersion relation, we obtain a complex-valued frequency $\tilde{\omega}$. When plotting the real part of $\tilde{\omega}$ as a function of k_{sp}, we observe a horizontal asymptote. Upon examination of the dispersion relation of Eq. (13.15), it is seen that k_{sp} may diverge if $\varepsilon_1 + \varepsilon_2(\omega) = 0$. This equation defines the frequency of the asymptote. For a lossy metal, the equation has a solution for only a complex-valued frequency $\tilde{\omega}_{sp}$. In the limit of large k_{sp}, the complex-frequency solution of Eq. (13.20) is given by

$$\tilde{\omega}_{sp} = \frac{\omega_p}{\sqrt{\varepsilon_1 + 1}}\sqrt{1 - \frac{(\varepsilon_1+1)\nu^2}{4\omega_p^2}} - i\frac{\nu}{2}, \tag{13.21}$$

with an amplitude decay time given by $2/\nu$. The existence of a horizontal asymptote has two major consequences. It indicates that at that particular frequency, the density of quasinormal modes presents a peak. The second consequence is the existence of surface modes with very large wavevectors and highly confined fields. Note that these modes are delocalized along x (they are exponentially damped in z but vary as $\exp(ik_{sp}x)$) and have very low group velocity. Their fast decay rate is only dictated by ν, a parameter of the permittivity.

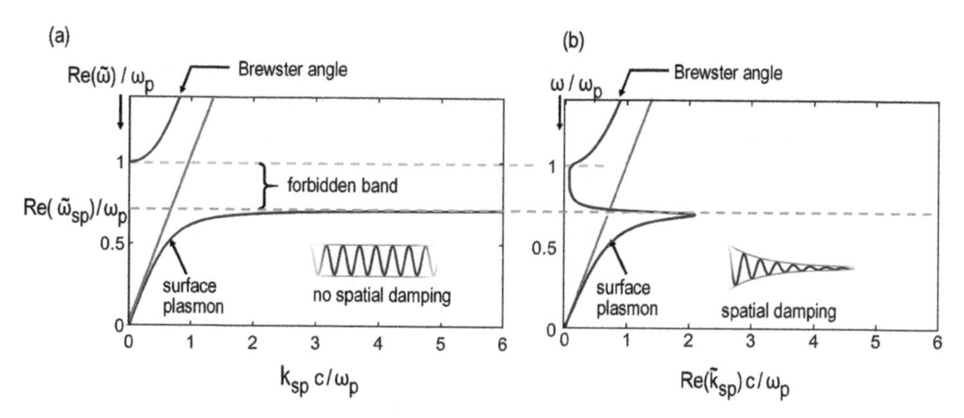

Fig. 13.5 Dispersion relations of a surface plasmon propagating at an interface between a lossy metal (here aluminium is described by a Drude model with $\omega_p/(2\pi c) = 1.25\,10^5\text{cm}^{-1}$, $\nu/(2\pi c) = 4.24\,10^3\text{cm}^{-1}$) and vacuum. (a) Dispersion relation in the $\text{Re}(\tilde{\omega}) - k$ quasinormal-mode representation. Slow plasmons with large k show up with a lifetime limited by the metal damping rate. The horizontal asymptote (dashed line) occurs for the complex-valued frequency $\tilde{\omega}_{sp}$, see Eq. (13.21). (b) Dispersion relation in the $\omega - \text{Re}(\tilde{k})$ guided-mode representation. The relation exhibits an expected backbending at modest wavevector values, preventing the existence of slow surface plasmons with large k's at real frequencies. The grey solid line is the light line.

Guided surface-plasmon modes: complex-valued $\tilde{k}$. The $\omega - \tilde{k}$ guided-mode representation is straightforwardly obtained by plotting Eq. (13.20) for a real-valued frequency to obtain complex-valued k_{sp}, denoted by a tilde to emphasize their complex nature. Below the light line, the imaginary part of k_{sp} is only due to ohmic losses. The latter are supplemented with radiation losses above the light line. $\text{Im}(k_{sp})$ is related to the inverse of the attenuation length of the plasmon mode (see Section 13.4.2). This plasmon mode is the analogue of the guided photonic modes of Chapter 6 with a spatial damping.

The $\omega - \tilde{k}$ representation offers a perspective that is radically contrasted with the $\tilde{\omega} - k$ representation. The dispersion relation exhibits a backbending for a relatively low value of $\text{Re}(\tilde{k}) = k_M = \frac{\omega_p}{c}\,(\frac{\varepsilon_1/2}{\varepsilon_1+1})^{1/2}$. In this real frequency representation, no slow plasmons with large k are launched by a radiating dipole, even when the emission frequency is closed to $\omega_p(\varepsilon_1+1)^{-1/2}$ and the dipole is very close to the surface. This result is in contrast to the previous analysis with quasinormal modes. It also contrasts with metal–insulator–metal (MIM) waveguides with small gaps, presented below, for which we will show that guided *gap* plasmons with large k's exist at any frequency.

Discussion. The absence of a gap at frequencies slightly larger than $Re(\tilde{\omega}_{sp})$ in the $\omega - \tilde{k}$ representation is puzzling. It suggests that there are plasmonic modes in this frequency interval, which does not exist in the $\tilde{\omega} - k$ representation. This is similar to what happens in a metal with real negative permittivity. There is no propagating mode, but a solution can be found with a complex wavevector. Similarly, in a Bragg mirror, there is a frequency gap without propagating modes. Nevertheless, modes are found when searching for solutions with a complex wavevector. These will be carefully analysed in chapters 15 and 16.

If we are interested in deriving the density of states (DOS) from the dispersion relation, we get different results from the two dispersion relations. Which one should be used when evaluating the decay rate of quantum emitters using the Fermi Golden Rule or the thermal energy at equilibrium? In this context, we aim to find the spectrum of the system, namely the ensemble of eigenfrequencies of the system with the corresponding modes. This is what was done in Chapter 7 for resonators and corresponds to the quasinormal modes. Hence, the relevant DOS is the density of quasinormal modes (see Complement 7.A). As for any resonator, the spatial structure of the mode is imposed by the geometry of the system and the losses are accounted for by the imaginary part of the frequency. When studying the LDOS using the Green tensor in Chapter 12, we observed a peak at the surface-plasmon frequency in Fig. 12.5. The peak cannot emerge from the dispersion relation with backbending.

We now discuss the excitation of the delocalized modes with large k's (Lalanne et al., 2019). We consider the radiation of an oscillating monochromatic (frequency ω) dipole placed just above the metal surface, (see Complement 7.A for a related discussion on nanoresonators). Intuitively, we anticipate that all quasinormal modes are excited in phase at the dipole position, so that their respective contributions add up constructively just beneath the dipole. Away from the dipole on the surface, all the sinusoidal quasinormal modes are not phased and the sum of their field is no longer constructive, so that the contribution of high-k modes to the reconstructed field rapidly decreases away from the dipole on the metal surface. Moreover, very high-k modes have a very fast transverse decay ($\propto k^{-1}$) and are thus excited only when the dipole is very close to the surface in the evanescent tail of the mode. The higher the k, the closer the dipole should be for efficient excitation. This implies that, as the separation distance between the oscillating dipole and the surface decreases, a hot spot just beneath the dipole takes shape and becomes more intense and localized. The hot-spot-formation process is well known as 'quenching' in the literature, a detrimental process that has been observed in many experiments with fluorescent molecules. Since the quasinormal-mode excitation scales as $\propto 1/(\omega - \tilde{\omega}_{sp})$ (section 7.6.2), we see that quenching is strongly enhanced when the emitting frequency is matched with slow plasmon frequency.

Slow plasmons are often regarded as localized plasmons, as will be discussed in the next subsection. To avoid any confusion, we emphasize that, in the $\tilde{\omega} - k$ representation, the high-k plasmons are spatially delocalized over the whole surface; it is their coherent excitation by near-field 'sources' that leads to an extremely localized field. In the subwavelength regime, this field is nothing but the field produced by the electrostatic image of the dipole with a resonant amplitude proportional to the electrostatic reflection factor $(\varepsilon - 1)/(\varepsilon + 1)$, as discussed in section 10.4.2.

13.4.2 Typical length scales for a propagating surface plasmon

There are three different lengths characteristic of a surface plasmon driven by a monochromatic source with real frequency. It is seen from Eq. (13.15) that the wavevector is complex if there are losses. The imaginary part of k_{sp} accounts for the decay of the surface mode upon propagation along the interface. A characteristic decay length can be defined by $\delta_x = 1/\text{Im}(k_{sp})$. There are two other characteristic lengths accounting for the exponential decay of the surface mode away from the interface. They are given by $\delta_{zi} = 1/\text{Im}(\gamma_i)$ in medium i and are found by inserting Eq. (13.15) into Eqs (13.11) and (13.12):

$$\frac{1}{\delta_{zi}} = \text{Im}[\gamma_i] = \frac{\omega}{c}\text{Im}\left[\frac{\varepsilon_i^2}{\varepsilon_1 + \varepsilon_2}\right]^{1/2}. \tag{13.22}$$

Typical orders of magnitude are given in Table 13.1 for the case of a surface plasmon. It is clearly seen that the surface plasmon has a decay length along the x-axis on the order of a few micrometres in the visible range but considerably larger (up to half a metre) in the far-infrared. Regarding the confinement of the wave close to the interface, it is seen that the decay length δ_{z2} in the metal almost does not depend on the wavelength. It is mainly given by the skin depth in the metal and is on the order of 12 nm for gold. By contrast, the decay length δ_{z1} of the surface wave in medium 1 changes dramatically. For a vacuum–gold interface, the decay length in a vacuum varies between 165 nm in the visible (633 nm) and 700 μm in the infrared at 36 μm. These results can be interpreted qualitatively using the non-lossy Drude model $\varepsilon_2 = 1 - \omega_p^2/\omega^2$. Let us first analyse the decay length in a metal. In the limit of low frequencies $\omega \ll \omega_p$, we can approximate the permittivity $\varepsilon_2 \approx -\omega_p^2/\omega^2$. From Eq. (13.22), we find

$$\delta_{z2} \approx c/\omega_p.$$

Hence, the decay length of the plasmon in a metal given by the Drude model is non-dispersive and given by the plasma wavelength. Its typical value is on the order of a few tens of nanometres. By contrast, the decay length in the dielectric is much larger. By denoting the wavelength in the dielectric by $\lambda_1 = c/(\sqrt{\varepsilon_1}\omega)$, we find that $\delta_z 1 \approx \lambda_1 \omega_p/\omega$, so that the decay length increases significantly in the infrared.

Let us now discuss the physical origin of the long propagation length in the far-infrared. Since the surface mode is loosely confined close to the interface in the infrared, most of its electromagnetic field is in the vacuum and only a tiny fraction is in the (lossy) metal (see Fig. 13.6). Hence, Joule losses become negligible when the decay length in vacuum increases. This remark can be generalized: *surface-plasmon losses are reduced when most of the field of the mode is in a non-lossy dielectric and not in the metal.* It is often useful to design *hybrid metallo-dielectric structures* to reduce losses by taking advantage of this idea. We finally note that the confinement in a dielectric is due to

Table 13.1 Decay length for a surface plasmon propagating along a gold–vacuum interface. Data taken from ref. (Etchegoin et al., 2006) for 633 nm and 1 μm and from ref. (Ordal et al., 1985) for 10 and 34 μm.

λ_i (μm)	0.633	1	10	36
δ_x (μm)	9.8	91.6	38,880	504,243
δ_{z1} (μm)	0.165	0.51	57.3	702.67
δ_{z2} (μm)	0.014	0.012	0.011	0.013

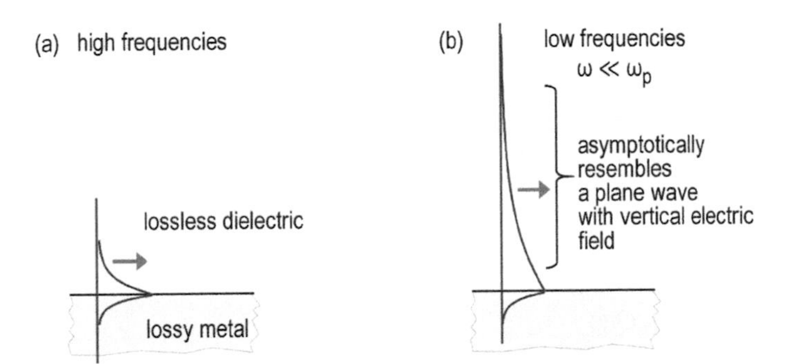

Fig. 13.6 Why does a surface plasmon propagate over large distances in the infrared although the imaginary part of the metal dielectric constant is very large in the infrared? The explanation is found in the distribution of the modal field between the lossless dielectric and the lossy metal regions. In the far-infrared, the modal field is dominantly localized in the lossless dielectric, while in the visible, a larger fraction is in the metal.

the fact that $k_x^2 > \varepsilon_1 \omega^2/c^2$. We note, in particular, that for very large wavevectors k_x, $\delta_{zi} = 1/\mathrm{Im}(\gamma_i) \approx 1/k_x$, so that large vectors are strongly confined.

13.4.3 How to excite a surface plasmon

It is clearly seen in Fig. 13.7 that propagating waves and surface plasmons are on different sides of the light line. At a given frequency, a propagating wave has a wavevector smaller than ω/c, whereas a surface plasmon always has a wavevector larger than ω/c. Hence, the phase cannot be the same along a plane interface and the two fields cannot satisfy boundary conditions. As a consequence, an incident plane wave on a plane interface cannot excite a surface plasmon. In this section, we address the question of the excitation of a surface mode and review different schemes. These are depicted in Fig. 13.6. All of them need to fulfil the same condition: to generate a real parallel wavevector that matches the parallel surface-plasmon wavevector.

A rigorous discussion requires an accurate distinction between surface plasmons of the two representations. However, for simplicity reasons, we will not distinguish the two cases.

Point-like source. It is possible to use a point-like source. According to Weyl's expansion, the spherical wave generated by a point source contains evanescent waves with a continuum spectrum of wavevectors. Of course, these components are evanescent waves localized close to the source. Hence, the source has to be close to the surface. More generally, any subwavelength source located in close proximity to the surface can excite surface modes. It can be an atom, a tiny particle scattering an incident beam, the aperture of an elongated optical fibre, as used in near-field microscopy, a scattering tip, as used in an atomic force microscope or scanning tunnelling microscope, a tiny scratch in the surface, etc. In order to illustrate this excitation, we give the amplitude of a guided surface plasmon excited by a point-like dipole located above a metallic interface in Complement 13.A and by a tiny slit in the metal in Complement 13.B.

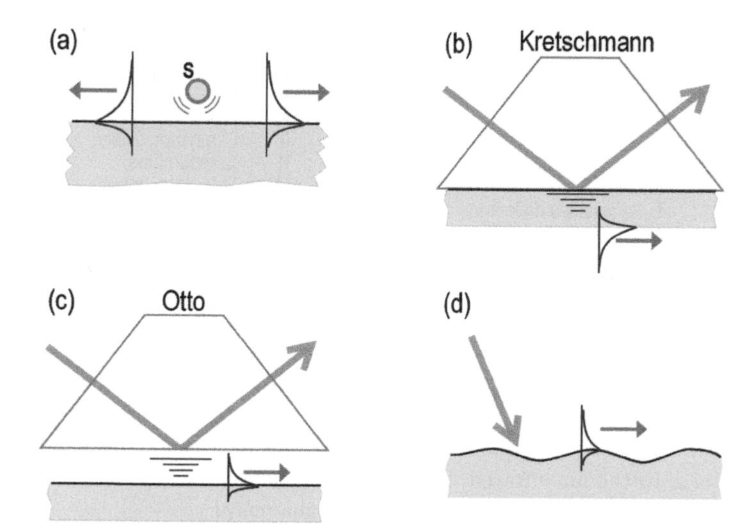

Fig. 13.7 Surface-plasmon mode excitation. (a) Localized source: a subwavelength scatterer or a subwavelength aperture generate confined fields which contain large-wavevector components; (b) Kretschmann configuration; (c) Otto configuration: the wave in the prism has a wavevector with modulus $n\omega/c$; (d) the waves diffracted by the grating have a wavevector of the form $\mathbf{k}_{inc} + \mathbf{G}$.

Kretschmann configuration. One can use a metal film with a thickness smaller than the skin depth separating two dielectric media with different dielectric constant n_1 and $n_2 > n_1$. Here, the key idea is to take advantage of a large refractive index to increase the modulus of the wavevector from $n_1\omega/c$ to $n_2\omega/c$. By illuminating from the side of the high-refractive-index medium with a plane wave with wavevector $k_x = n_2(\omega/c)\sin\theta_i$, it is possible to excite the surface mode on the other side by taking advantage of the fact that the incidence angle can be chosen so that $k_x > n_1\omega/c$. This technique is called the Kretschmann configuration. Strictly speaking, with an incident plane wave of infinite spatial extent, it is the quasinormal plasmon mode that is excited; however, experimentally, beams have finite sizes and guided surface plasmons are launched. The first image taken with near-field optical microscopy of a plasmon launched in this configuration can be seen in Fig. 11.2.

Otto configuration. A similar but less practical technique consists in using the evanescent part of the field totally internally reflected in a prism. This is known as the Otto configuration. It requires accurate control of the spacing between the prism and the metal surface.

Grating excitation. A grating with a period d can be used, so that the nth order of the grating has a wavevector $k_{x,n} = n_1(\omega/c)\sin\theta_i + n2\pi/d$. If the phase-matching condition $k_{sp} = n_1(\omega/c)\sin\theta_i + n2\pi/d$ is *approximately* satisfied[2], an incident plane

[2] Just like the presence of the prism perturbs the surface plasmon of the isolated interface in the Otto and Kretschmann setups, the presence of the grating perturbs all the quasinormal modes for all

wave may effectively excite a surface mode. The efficiency of the coupling depends on the parameters of the surface profile of the grating. It can be optimized to allow a total absorption, as was first reported by using a gold grating operating in the visible at a frequency where a flat surface is a good mirror (Hutley and Maystre, 1976). This is a very counterintuitive result, since the modulation depth of the sinusoidal grating is very low, so that the surface appears almost flat. This is another example of critical coupling (see Chapter 7).

13.4.4 Electrostatic limit

In this paragraph, we extend the previous discussion to the non-retarded limit. Let us consider a dipole source oscillating at a frequency $\omega = 2\pi c/\lambda$ at a distance d above an air–metal interface such that $d \ll \lambda$. As the distance is much smaller than the wavelength, the interaction between the source and the interface can be described within the electrostatic approximation, as discussed in Chapter 10. It follows that the interface can be modelled by introducing an image charge. The standard electrostatic formalism (Jackson, 1975) allows the introduction of an electrostatic reflection factor given by $\frac{\varepsilon-1}{\varepsilon+1}$. This yields a resonance for $\varepsilon+1=0$ and hence for ω_{sp}, given in Eq. (13.21), when using a Drude model. As already discussed in Chapter 10, the resonant reflection factor is larger than unity. It simply means that the resonant localized field produced by the electrostatic image can be viewed as a coherent superposition of non-radiative surface-plasmon modes with large wavevectors coherently and resonantly driven by the dipolar source. This result can also be derived by searching a mode of the Laplace equation for the scalar potential for a system with one interface separating two homogeneous media (Zayats et al., 2005).

13.4.5 Surface mode on a lossy Drude metal: Is it a surface plasmon?

We now compare the case of low frequencies with the case of optical frequencies. Surface modes propagating along metallic surfaces have been known since the early days of radiowaves. The question arises if there is a difference with a surface plasmon. We start by analysing the Drude model in the low and high frequency regimes. It is easily seen that we can approximate the dielectric constant by

$$\begin{aligned} \omega \gg \nu, \quad & \varepsilon(\omega) \approx 1 - \frac{\omega_p^2}{\omega^2}, \\ \omega \ll \nu, \quad & \varepsilon(\omega) \approx 1 + i\frac{\omega_p^2}{\omega\nu} = 1 + \frac{i\sigma}{\omega\varepsilon_0}, \end{aligned} \tag{13.23}$$

where $\sigma = ne^2/m\nu$ is the DC conductivity. This form is enlightening, as it shows that the optical properties of the metal are dominated by the plasmon response in the optical regime, whereas they are dominated by the drag force in the low-frequency regime. In the former case, the dielectric constant is almost a negative real number, whereas in the latter case, it is almost a purely imaginary number. It follows that the

k's. Actually, it is these perturbed modes (which are leaky modes, see Chapter 17) that are excited. These modes have field distributions and complex-valued frequencies that are slightly different from those of the flat interface. This is why the phase-matching condition is only approximately fulfilled.

plasmonic (oscillatory) character of the surface wave is only meaningful in the regime $\omega \gg \nu$. Indeed, if we rewrite the equations of motion of the electrons including the friction term as in Eq. (13.2), we obtain

$$-\omega^2 x = i\omega\nu x - \frac{ne^2}{m\varepsilon_0} x. \tag{13.24}$$

For large and small frequencies, we have different approximate expressions:

$$\begin{aligned} \omega \gg \nu, \quad & -\omega^2 x + \frac{ne^2}{m\varepsilon_0} x = 0, \\ \omega \ll \nu, \quad & -i\omega\nu x + \frac{ne^2}{m\varepsilon_0} x = 0. \end{aligned} \tag{13.25}$$

It is clearly seen that the mechanical oscillation regime which is the essence of a plasmon corresponds only to the case $\omega \gg \nu$. Instead, for frequencies smaller than ν, the electronic response of the medium is dominated by the viscous term. Since the typical value of ν is 10^{14} Hz, we note that a surface wave is hardly a surface plasmon for frequencies in the infrared or smaller. On the other hand, the plasmonic behaviour (i.e. oscillatory behaviour) dominates the metal response in the case of femtosecond pulses.

In summary, from a macroscopic electrodynamics point of view, metal interfaces support surface modes for all frequencies. However, from a microscopic point of view, the underlying behaviour of the electrons is very different in the low and large frequency regimes. The essence of the plasmon is the collective oscillation which takes place only in the regime where $\omega \gg \nu$. Beyond this semantics remark, this distinction is important, because the detailed form of the dispersion relation is different for low frequencies and high frequencies.

13.4.6 Dispersion relation of surface waves in the radiofrequency range

Let us analyse in more detail the nature of the surface waves propagating along a conductive surface at low frequencies. We can give a more explicit form of the dispersion relation of Eq. (13.15) in this regime. At radiofrequency, the dielectric constant of a metal is $\varepsilon = 1 + \frac{i\sigma}{\omega\varepsilon_0} \approx \frac{i\sigma}{\omega\varepsilon_0}$; the modulus of the dielectric constant is much larger than 1. A Taylor expansion of the wavevector of the surface wave can thus be obtained:

$$k_{sp} = \frac{\omega}{c}\left[\frac{1}{1+\frac{1}{\varepsilon}}\right]^{1/2} \approx \frac{\omega}{c}\left(1 + \frac{i\omega\varepsilon_0}{2\sigma}\right). \tag{13.26}$$

It is seen that in the limit of the perfect conductor, σ becomes infinite, so that the wavevector becomes $k = \omega/c$. This requires that the mode not be entirely confined at the interface. Here, we recover the concept of surface waves used in the context of radiowaves propagating along perfectly conducting wires or perfectly conducting surfaces. Note, in particular, that in this limit, there is no more damping, as the wavevector becomes real. Let us now check that Eq. (13.26) is a valid solution of the surface-wave-dispersion relation $\varepsilon_1\gamma_2 + \varepsilon_2\gamma_1$. In the radio regime, the analysis is very different from the case of the surface plasmon in the optical regime. We consider the case of an interface separating a metal (medium 2) from vacuum (medium 1) and use the notation $\varepsilon = |\varepsilon| \exp(i\phi_\varepsilon)$ and Eq. (13.22):

$$\gamma_2 \approx \frac{\omega}{c}[\varepsilon]^{1/2} \approx \frac{\omega}{c}|\varepsilon|^{1/2} \exp(i\phi_\varepsilon/2),$$
$$\gamma_1 \approx \frac{\omega}{c}\left[\frac{1}{\varepsilon}\right]^{1/2} \approx -\frac{\omega}{c}\left[\frac{1}{|\varepsilon|}\right]^{1/2} \exp(-i\phi_\varepsilon/2). \tag{13.27}$$

The choice of the determination of the square root is imposed by the condition $\mathrm{Re}(\gamma) + \mathrm{Im}(\gamma) > 0$ (see Complement 4.B), so that we have to include a minus sign for γ_1. It clearly appears, then, that the condition $\varepsilon\gamma_1 + \gamma_2 = 0$ is satisfied.

To summarize, inserting the Drude model in the dispersion relation of a surface wave yields two limits: the surface plasmon for $\omega > \nu$ and the surface wave with $k = \omega/c$ for $\omega \ll \nu$. The latter is the so-called Zenneck mode (also called the Sommerfeld surface wave) (Baños, 1966), see section 18.6 for historical details and section 18.6.3 for an overview of the related concept of spoof surface plasmon. Note that when dealing with THz waves, the nature of the mode is closer to a radio surface mode (Zenneck mode) than to a surface plasmon.

13.5 Gap SPPs

As we have seen, owing to the backbending, the propagation constant of guided bound plasmon modes at a single metal/dielectric interface is not significantly larger than the free-space wavevector. Consequently, the decay length of the plasmon evanescent field is not much smaller than the wavelength. The mode is not extremely confined.

This observation has inspired a new class of plasmon waveguides that have the potential to confine the electromagnetic fields at deep-subwavelength scales. They consist of an insulating core and conducting cladding. These MIM heterostructures guide light via the refractive-index differential between the core and cladding. They support both TE and TM guided modes. TE modes require wavelength-scale gap thicknesses; they may propagate over long propagation distances of several tens or hundreds of wavelengths at optical frequencies. They are not extremely confined and will not be further discussed.

Among the TM modes, one with a plasmonic character exists even for ultra-thin dielectric layers. It is the so-called gap-plasmon mode. The latter is well known at microwave frequencies with almost perfectly conducting metals, where it exhibits a normalized propagation constant equal to the refractive index of the insulator, even for gaps much smaller than the wavelength (see the introductory part of Chapter 6 for more details). At optical frequencies it is exactly the same, except that the metal absorption is much larger and the mode propagates only over short distances. However, since its 'effective wavelength' vanishes, gap-plasmon modes remain quite interesting. They even play a key role in plasmonic circuitry, e.g. for energy transport through subwavelength metal slits or grooves in metallic thin films, artificial magnetism at optical frequency, and anomalous dispersion with hyperbolic media (Chapter 18). They are also responsible for the huge field enhancement of plasmonic nanoantennas formed

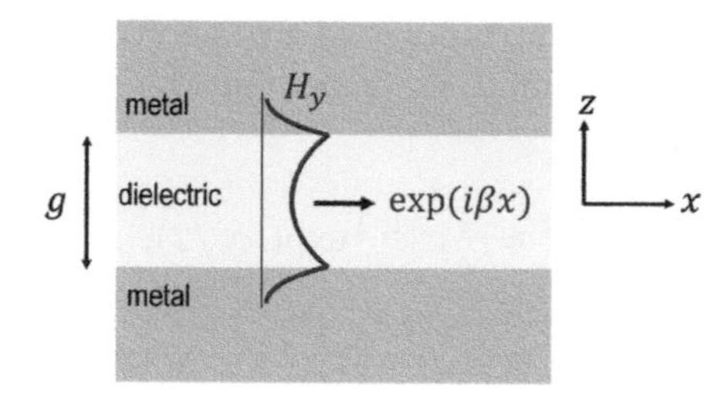

Fig. 13.8 Metal–insulator–metal heterostructure with its symmetric gap-plasmon mode (the H_y field component is symmetric with respect to the z-axis). The gap width is g.

with two almost-touching nanoparticles (Chapter 19) and for slow-light effects, as we will discuss hereafter.

13.5.1 Geometry and dispersion relation

The general study of the gap-plasmon (guided) modes is performed like for all guided waves, for a real frequency and a complex wavevector in the $(\omega, \tilde{k})$ representation. In this section, we consider propagation along axis x and denote by β the propagation constant, following the notation of Chapter 6. For reasons of simplicity, we will assume that the metal relative permittivity ε_m is real, negative, and varies with the frequency. In a first step, neglecting ohmic losses is simpler mathematically[3] and physically for providing energetic interpretations. However, we will systematically discuss the validity of the main formulas in the presence of absorption, and will also emphasize the limitations due to absorption. In fact, the following derivation remains essentially valid for complex-valued ε_m, resulting in complex values for the propagation constant β and $n_g = c\frac{\mathrm{d}\beta}{\mathrm{d}\omega}$.

The MIM parameters are sketched in Fig. 13.8, along with the mode profile of the gap-plasmon mode, which results from the coupling of the surface plasmons supported by each individual interface. In the limit of small thicknesses, two new modes are created from this coupling, one being symmetric $H_y(-z) = H_y(z)$ and the other antisymmetric. Hereafter, we consider only the symmetric plasmon, which exists for an arbitrarily small value of the gap width g if surface quantum correction is not considered (Pitarke et al., 2006). Since absorption is neglected, the gap-plasmon propagation constant β along the waveguide x-axis is real.

Solving Maxwell equations for this two-boundary system is similar to the planar waveguide case analysed in Chapter 6. Hereafter, we summarize the main steps of the derivation. The gap-plasmon mode may be solved via the vector Helmholtz equation under the constraint of tangential E- and H-field continuities at the interface $z = \pm g/2$. Due to symmetries, the general solution for the magnetic field is

$$H_y(z)\exp(i\beta x), \tag{13.28}$$

[3] There is a small additional complexity if complex-valued ε_m are considered by properly defining the square-root function in the complex plane, see Complement 4.B.

with

$$H_y(z) = h_0 \left[\exp[\gamma_d(z - \frac{g}{2}) + \exp[-\gamma_d(z + \frac{g}{2})]\right] \quad \text{for } |z| \leq \frac{g}{2},$$
$$H_y(z) = h_0[1 + \exp(-\gamma_d g)] \exp[-\gamma_m(|z| - \frac{g}{2})] \quad \text{for } |z| > \frac{g}{2}, \tag{13.29}$$

where h_0 is a complex magnetic-field amplitude that will be normalized afterwards. Since the magnetic field satisfies the Helmholtz equation in the metal and in the dielectric, we have $\gamma_j = [\varepsilon_j \frac{\omega^2}{c^2} - \beta^2]^{1/2}$, with $j = d, m$. Note that γ_m and γ_d are *both positive*, as is usually the case for the square root of a real number. The electric-field components, $E_z(z)\exp(i\beta x)$ and $E_x(z)\exp(i\beta x)$, can be straightforwardly deduced using Maxwell equations:

$$-i\omega\varepsilon_0\varepsilon_j E_z(z) = \frac{\partial H_y}{\partial x}, \quad ; \quad i\omega\varepsilon_0\varepsilon_j E_x(z) = \frac{\partial H_y}{\partial z}. \tag{13.30}$$

Using the continuity of the tangential electric field for E_x at $z = \pm g/2$, we obtain the transcendental dispersion relation

$$-\frac{\gamma_m}{\varepsilon_m}[1 + \exp(-\gamma_d g)] = \frac{\gamma_d}{\varepsilon_d}[1 - \exp(-\gamma_d g)]. \tag{13.31}$$

Eq. (13.31) can be solved numerically by computing $\beta(\omega)$. For nanophotonics applications, an important geometry is the case of a gap thickness g much smaller than the wavelength $\lambda = 2\pi c/\omega$. In the regime $\gamma_d g \ll 1$, γ_m and γ_d become comparable and equal to β and the gap-plasmon fields become

$$H_y(z) = 2h_0,$$
$$E_z(z) = 2h_0 \frac{\beta}{\omega\varepsilon_0\varepsilon_d},$$
$$E_x(z) = h_0 \frac{i\beta}{\omega\varepsilon_0\varepsilon_d}[\exp[\gamma_d(z - g/2)] - \exp[-\gamma_d(z + g/2)]] \tag{13.32}$$

in the gap and

$$H_y(z) = 2h_0 \exp[-\gamma_m(|z| - g/2)],$$
$$E_z(z) = 2h_0 \frac{\beta}{\omega\varepsilon_0\varepsilon_m} \exp[-\gamma_m(|z| - g/2)],$$
$$E_x(z) = h_0 \frac{x}{|x|} \frac{2i\gamma_m}{\omega\varepsilon_0\varepsilon_m}[\exp[-\gamma_m(|z| - g/2)] \tag{13.33}$$

in the metal. The dispersion relation of Eq. (13.31) simplifies considerably, and we get

$$\beta = \frac{\omega}{c} n_{neff} = -\frac{1}{g}\frac{2\varepsilon_d}{\varepsilon_m}. \tag{13.34}$$

13.5.2 Slow gap plasmon

Slow modes for photonic or plasmonic structures have many interesting properties including miniaturization and field enhancement (see section 16.4). In nanophotonics, two standard schemes are commonly studied. It is important to contrast their differences, advantages, and drawbacks (Lalanne et al., 2019). In the first scheme, slow light is implemented with periodic photonic waveguides. The energy is purely electromagnetic and the slowness originates from photons that bounce forwards and backwards as they propagate in the periodic structure. It immediately appears that photonic-crystal slow light operates in a narrow frequency range, satisfying a phase-matching condition, and the characteristic dimension of the structure, i.e. the waveguide period a, therefore scales with the wavelength λ of interest, $a \approx \lambda/2$.

The second scheme with MIM waveguides that we study in this section is radically different. Slowness is achieved in a translational invariant waveguide and originates from a subwavelength confinement that is automatically accompanied by a coupling of electromagnetic fields with the free electrons of metals (Bozhevolnyi and Khurgin, 2016), so that the energy is stored both in the electromagnetic field and in the kinetic energy of the electrons (see section 10.5.2 and exercises 13.1 and 14.1). The fact that MIM media support slow modes can be immediately seen by derivation of Eq. (13.34) to compute the group index, defined as $n_g = c\frac{\partial \beta}{\partial \omega}$, of the gap plasmon mode. We find that, in the limit of small gaps, the group index is

$$n_g = n_{\text{eff}}[1 - \frac{\omega}{\varepsilon_m}\frac{\partial \varepsilon_m}{\partial \omega}], \tag{13.35}$$

and diverges as $1/g$ like n_{eff} (see Eq. (13.34)). In other words, the group velocity v_g is proportional to g.

Our main results, Eqs (13.34) and (13.35), are quite simple. They remain valid for complex-valued ε_m, leading to complex-valued β's or effective indices n_{eff} or group indices n_g (see Section 13.5.5). We emphasize that their simplicity is due to our asymptotic treatment for small gaps; see (Lalanne et al., 2019) for a test of their accuracy with fully numerical data.

13.5.3 Physics of wave deceleration

There are different ways to understand the wave deceleration for small gaps. Perhaps the most unified way, which will also be used in Chapter 16 for photonic-crystal waveguides, consists in considering slowness as resulting from antiparallel power flows. For instance, slow light in photonic waveguides is due to two counterpropagating waves; each carries some energy, but together they constitute a quasistationary pattern. Similarly, in the present case, because the permittivities of metallic and dielectric layers are opposite, the x-component of the Poynting vector $\mathbf{S}$ is positive in the dielectric layer, whereas it is negative in the metallic layers. Hence, the total time-averaged power is reduced, owing to the negative contribution.

In this picture, stopped light at zero group velocity is often seen as resulting from a perfect balance, for which the energy flows in the metal and the dielectrics are opposite. This point of view, albeit intuitive and widespread, carries some flaws. This may be realized by considering the asymptotic expressions of the electromagnetic fields in the gap and in the metal at small group velocities. From Eqs (13.32) and (13.33), we infer that the power flows in the gap and in the two surrounding metal layers are respectively $-4h_0^2/(\omega\varepsilon_0\varepsilon_m)$ and $2h_0^2/(\omega\varepsilon_0\varepsilon_m)$, where h_0 is the magnetic field at the MI interface. In particular, this implies that, no matter how small v_g is and no matter the convention for normalization, the power flow in the dielectric gap is twice the total power flow in the metal layers, as soon as the gap plasmon enters the slow regime for tiny gaps.

13.5.4 Field enhancement

As will be shown in Chapter 17, the electric and magnetic fields of slow Bloch modes of photonic-crystal waveguides both scale proportionally to the square root of the group index $\sqrt{n_g}$. Slow plasmons offer different perspectives. This can be immediately realized by referring to the waveguide-mode formalism developed in Chapter 6. Applying Eq. (17.11) to the present gap-plasmon case (remember that metal absorption is neglected), we get

$$v_g \frac{1}{4} \int \mathrm{d}y \int \mathrm{d}z \left(\varepsilon_0 \frac{\partial \omega \varepsilon}{\partial \omega} |E|^2 + \mu_0 |H|^2 \right) = 1. \tag{13.36}$$

In Eq. (13.36), the integral runs over any transverse cross section of the waveguide and unity on the right-hand side comes from the fact that we consider normalized modes with unitary power flows. We recall that, since the integrand on the left-hand side is reminiscent of the electromagnetic energy density in dispersive and lossless media, Eq. (13.36) readily interprets the group velocity as the energy velocity (defined by the ratio of the spatial average of the time-averaged Poynting vector to the spatial average of the time-averaged energy density).

Great care should be taken when inferring from Eq. (13.36) how the electric and magnetic fields scale with the group index (Lalanne et al., 2019). The reason for this is that spectral dispersion matters a lot. Actually, as the plasmonic waveguide dimensions decreases, the field spatial profile changes. Self-sustained oscillations between magnetic and electric energies no longer hold, and the energy balance is restored by considering the kinetic energy of the free carriers involved in the subwavelength SPPs. The precise analysis depends on the exact geometry, but the general trend is that the kinetic and electric energies balance (both scale with n_g^2), while the magnetic energy becomes negligible for normalized modes with unitary power flows. The scaling is easily derived from Eq. (13.32) when h_0 is chosen, so that $4\omega h_0^2/(\varepsilon_0 k^2 c^2) = 1$ to have a unitary power flow of

$$|\mathbf{E}| \propto n_g \quad ; \quad |\mathbf{H}| \propto \text{constant}. \tag{13.37}$$

Noticeably, the electric-field increase rate for small group velocities is much larger for plasmons than for photons: n_g vs $\sqrt{n_g}$. The reason comes from the additional transverse-geometry change (gap-width reduction) that is accompanying plasmon slowdowns.

It is easy now to understand why the power flow of a slow mode is always different from zero, even as $v_g \to 0$. If we refer to Eq. (13.36), we see that the left-side term of the equation does not vanish as $v_g \to 0$, despite the v_g prefactor, because the surface integral of the electromagnetic energy density diverges with n_g.

13.5.5 Attenuation due to absorption

Attenuation of the gap modes is a very important factor that one should consider for real applications. In reality, plasmons (or photons) cannot propagate indefinitely at low speed, because of inevitable fabrication imperfections and ohmic loss. What happens is that as the group velocity decreases, the energy density increases, and not only optical non-linearities are boosted but also scattering and absorption in the linear regime. The dominant loss mechanism for slow plasmons is absorption in the metal. Simple reasoning leads us to expect that the absorbed power $P_{abs} = \omega/2 \iint \mathrm{Im}[\varepsilon_m]|\mathbf{E}|^2 \mathrm{d}S$ scales as $|\mathbf{E}|^2$, i.e. as n_g^2 according to Eq. (13.37). Fortunately, this prediction is pessimistic. Indeed, in view of Eq. (13.36), which is valid for any geometry, the integral of $|\mathbf{E}|^2$ over a cross section scales as n_g, so that the correct behaviour is $P_{abs} \propto n_g$.

The reason why the simple argument from perturbation theory fails is that the slowdown in plasmonic waveguides is always accompanied by a strong modification of the transverse mode profile. This mode-profile modification is accounted for in Eq. (13.36), but not in the simple reasoning. As the gap width is shrinking, the gap-plasmon effective index n_{eff} increases (see Eq. (13.34)), so that the penetration depth of the gap-plasmon mode into the metal $\lambda/[2\pi(n_{\text{eff}}^2 - \varepsilon_m)^{1/2}]$ shrinks too. We move from a spatial extension given by the skin depth $\lambda/[2\pi(-\varepsilon_m)^{1/2}]$ to a regime dominated by the large wavevector of the mode $\lambda/[2\pi n_{\text{eff}}]$, so that the surface integral of P_{abs} is effective over a reduced surface close to the interfaces.

Actually, the absorption in the slow-light regime is drastic. For instance, for an $Ag/SiO_2/Ag$ waveguide, $\varepsilon_{Ag} = -27 + i1.5$ at $\lambda = 800\,\text{nm}$, the attenuation length at $1/e$ is $L_p \approx 300\,\text{nm}$ for $n_g = 10$, whereas it is barely $100\,\text{nm}$ for $n_g = 30$.

13.5.6 Bandwidth

Eqs (13.34) and (13.35) evidences that slowness is guaranteed at any frequency, provided that $g \ll c/\omega$. This result markedly contrasts with the one previously obtained for slow (guided) plasmons at single metal–insulator interfaces, for which slowness is achieved only for a certain frequency achieved at the backbending. Another important characteristic of slow gap plasmons is the spectral bandwidth of this interaction. A major limitation of any slow-light structure is dispersion. Indeed, the operating slowness is fixed at a central frequency, but, in general, the transported signal possesses a bandwidth and a severe limitation occurs if the slowness changes over the bandwidth.

Imagine a small variation $\Delta\omega$ of the driving frequency ω. It results in a small variation, Δn_g of n_g. To quantify the limiting impact of group-velocity dispersion, in Chapter 17 we define a dimensionless coefficient B such that $\Delta\omega/\omega = B\Delta n_g/n_g$, which quantifies the slow-mode bandwidth, i.e. the variation $\Delta\omega$ of ω due to a small variation Δn_g at a fixed value n_g. Large bandwidths are achieved for large B's and, conversely, small bandwidths for small values. By using a Drude model $\varepsilon_m = \varepsilon_\infty(1 - \omega_p^2/\omega^2)$, it is easily found that

$$n_g = \frac{2c\varepsilon_d}{g}\frac{1}{\varepsilon_m^2}\frac{\mathrm{d}\varepsilon_m}{\mathrm{d}\omega} \tag{13.38}$$

for the slow gap plasmon and that the coefficient B is independent of n_g. Actually, $B = -1/3$, whereas the same coefficient for slow photonic waveguides varies as n_g^2 (see Chapter 17). Slow plasmons offer a spectral bandwidth considerably larger than that of photonic-crystal waveguides.

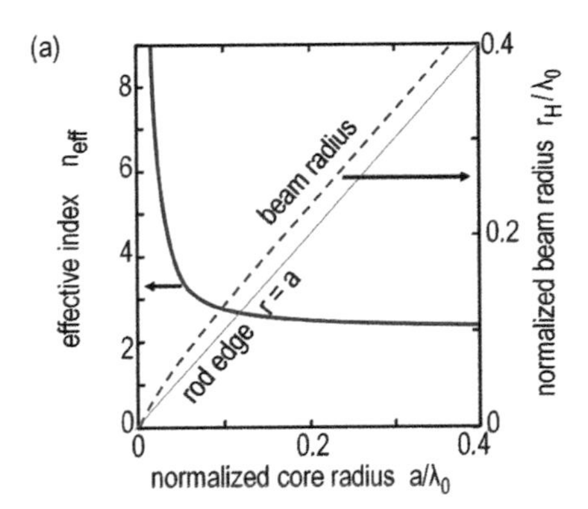

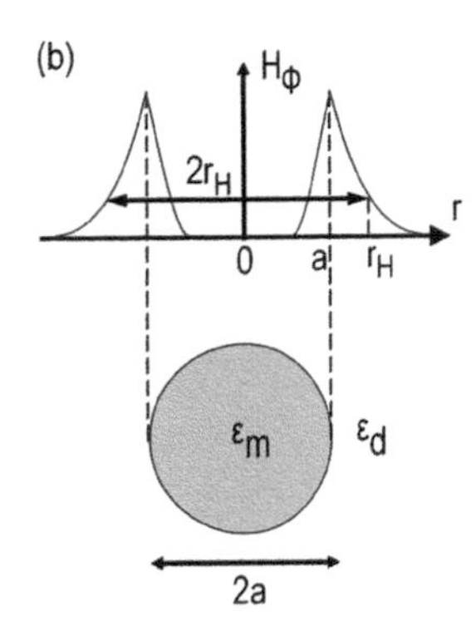

Fig. 13.9 (a) Effective index n_{eff} (solid line) and beam radius r_H (dashed line) of a metallic ($\varepsilon_m = -19$) rod in a background dielectric $\varepsilon_d = 4$ with respect to the normalized core radius a/λ_0. (b) The beam radius is defined at $1/e^2$ in intensity. Adapted with permission from ref. (Takahara et al., 1997) ©(1997) Optical Society of America.

13.5.7 Insulator–metal–insulator plasmonic waveguides

As shown in Chapter 6, decreasing the core thickness of symmetric dielectric slabs reduces the number of supported guided modes, and in the limit of vanishing thicknesses d the fundamental mode becomes the only mode with no cutoff with a field distribution that penetrates far into the cladding and eventually (at $d = 0$) becomes a bulk plane wave. As a result, the mode size of a dielectric waveguide decreases when the waveguide diameter decreases to a certain value, and then increases to infinity when the diameter is reduced further. Thus, decreasing the diameter of an optical fibre or the thickness of a guiding slab to zero cannot lead to subwavelength light confinement of the guided mode. This feature constitutes the main hurdle in achieving higher degrees of miniaturization and integration of photonic devices.

The situation for surface-plasmon guided modes in insulator-metal-insulator (IMI) waveguides and metal nanowires (see Fig. 13.9) is completely different.[4] When the diameter d of a cylindrical metal nanowire is reduced, the fundamental SPP mode (whose magnetic field has axial symmetry and is perpendicular to the nanowire axis like for the MIM waveguide) experiences a strong monotonic increase in localization and a significant reduction in its phase and group velocities (Takahara et al., 1997). As a result, the diameter of the guided-plasmon mode can be decreased to just a few nanometres, limited only by the increased dissipative losses, the atomic structure of matter, and spatial dispersion (Pitarke et al., 2006). This feature and the associated physics are formally identical to those of the MIM situation carefully analysed in section 13.5, and the derivation of the $1/d$ scaling of the phase and group velocities, as well as the attenuation length due to dissipative losses, is not repeated here.

[4] Long-range surface plasmons that propagate over many wavelengths with a very weak attenuation and a field profile that extends far into the cladding also exist. They will not be documented in this chapter: see (Berini, 2009) for an in-depth overview.

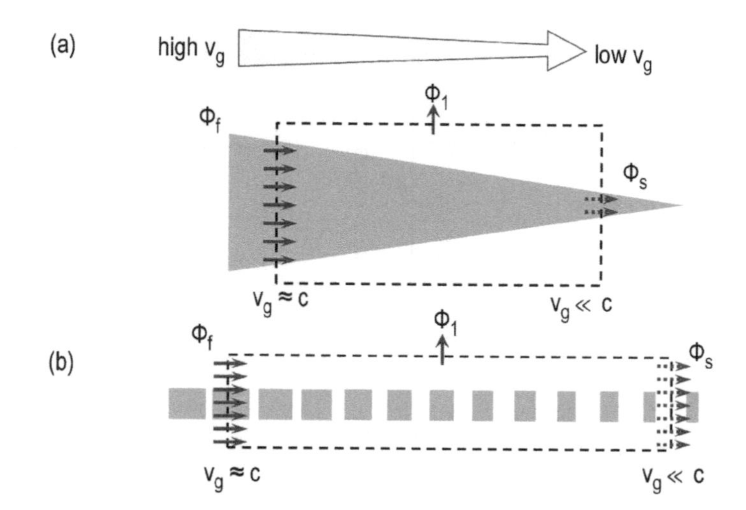

Fig. 13.10 Adiabatic slowdown of plasmon (a) or light (b) with tapered structures. Perfect focusing ($\phi_s = \phi_f$) is possible for structures that are progressively tapered (no back-reflection), do not convert the guided mode into free-space photons in the cladding ($\phi_1 = 0$), and do not absorb. Note that, for these structures, the application of the Poynting theorem to the dashed rectangle implies that the slow modes, with arbitrarily small group velocities, transport energy, since their transverse power flow ϕ_s is equal to ϕ_f and is thus non-null. This is yet another intuitive way to realize that a null group velocity does not imply a null power flow (see Section 13.5.3).

13.5.8 Plasmon nanofocusing

One of the most tantalizing prospects of plasmonic subwavelength waveguides is their ability to concentrate light energy into nanoscale regions as small as a few nanometres. This plasmon nanofocusing is typically achieved using tapered metallic guiding nanostructures such as tapered metal rods, metal wedges (see Fig. 13.10), or tapered nanogaps between two metallic media (Gramotnev and Bozhevolnyi, 2010). Theoretically, plasmon nanofocusing can occur adiabatically with nanofocusing structures that are weakly tapered (i.e. the taper angle is sufficiently small), so that the propagating plasmon mode does not 'feel' the taper and thus does not experience any significant reflections from it, even when reaching arbitrarily small speeds. As illustrated in Fig. 13.10, adiabaticity is a direct consequence of the possibility of normalizing the power flow of slow plasmons, no matter how slow they are (Lalanne et al., 2019). There is nothing specific there to slow plasmons, and adiabaticity can also be achieved with slow light in periodic waveguides (see Fig. 13.10b and the discussion on slow light in Chapter 16).

Thus, ideally, plasmons can be stopped; i.e. their energy velocity can be reduced to zero. Note, however, that this is only possible if one neglects the absorption loss. The energy velocity is never zero when the propagation losses are extremely high (Reza et al., 2008), as is the case for slow plasmons.

13.6 Graphene surface plasmons

Surface plasmons correspond to a surface charge at the interface between a metal and a dielectric. Another interesting system supporting surface plasmons is graphene deposited at the interface between two non-magnetic dielectrics (Luo et al., 2013). We introduced the local and dispersive conductivity of graphene in the optical regime as $j_{x,S} = K(\omega)E_x$. We saw in Chapter 9 that $K = e^2/4\hbar$ for pristine graphene. When modifying the electronic density either by doping or by applying a gate voltage, the conductivity can be modified significantly, owing to intraband and interband transitions. It can be modelled as follows:

$$\begin{aligned} K &= K_{intra} + K_{inter}, \\ K_{intra} &= \frac{ie^2 E_F}{\pi\hbar^2(\omega + i\nu)}, \\ K_{inter} &= \frac{e^2}{4\hbar}\left[\theta(\hbar\omega - 2E_F) + \frac{1}{\pi}\ln\left|\frac{2E_F - \hbar\omega}{2E_F + \hbar\omega}\right|\right]. \end{aligned} \tag{13.39}$$

The first term of the interband transition is responsible for graphene's absorptivity of 2.3%, observed in the visible (see Exercise 9.7). The intraband contribution has the form of a Drude model and dominates for $E_F \gg \hbar\omega$.

The derivation of the dispersion relation and the spatial structure of the plasmons is very similar to what has been done in section 13.3.1; see ref. (Luo et al., 2013), for instance. The general form of the x- and z-components of the electric fields in both media is the same. The only difference is due to the continuity condition of εE_z at the interface. A discontinuity has to be included to account for the presence of the surface charges in graphene. It is shown in Exercise 13.6 that the dispersion relation is given by

$$\frac{\varepsilon_1}{\gamma_1} + \frac{\varepsilon_2}{\gamma_2} = \frac{K(\omega)}{\omega\varepsilon_0}. \tag{13.40}$$

In the limit of large wavevectors $k_x \gg n_1 k_0, n_2 k_0$ and large doping $2E_F > \hbar\omega$, the Drude contribution to the conductivity dominates and the dispersion relation takes the simple form

$$k_x = \frac{\pi\varepsilon_0\hbar^2}{e^2 E_F}(\varepsilon_1 + \varepsilon_2)(\omega^2 + i\omega\nu).$$

A key feature of graphene plasmonics is the existence of large wavevectors $k_x \gg \omega/c$. Indeed, from the dispersion relation, it is seen that

$$\frac{k_0}{k_x} \approx \frac{4\alpha}{\varepsilon_1 + \varepsilon_2}\frac{E_F}{\hbar\omega}, \tag{13.41}$$

where alpha is the fine-structure constant, so that the surface-plasmon wavevector is much larger than the wavevector in vacuum. This implies that the field is extremely confined close to the graphene sheet in the z-direction. Using $\alpha \approx 1/137$, we see that $k_x \approx 100k_0$. It follows that the volume of a mode of frequency ω is typically λ^3 in vacuum but $10^{-6}\lambda^3$ for a graphene plasmonic mode at the same frequency. It is thus possible to achieve a large electric field with few plasmons.

Another key feature of graphene plasmonics is the ability to modulate the Fermi energy by applying a gate voltage and therefore to modulate the dispersion relation of the plasmons. This paves the way to the development of optoelectronic applications.

A limitation of graphene plasmonics is absorption due to interband transition. If the plasmon frequency is larger than $2E_F$, an electron–hole pair can be created by interband transition. This very efficient loss mechanism impedes the existence of surface plasmons. As a consequence, it is easier to operate in the infrared and THz range.

In the next section, we introduce surface phonon polaritons propagating at an interface between two dielectrics in the infrared spectral range. It turns out that graphene is often deposited on h-BN or on SiC. Both materials support surface phonon polaritons, which can hybridize with graphene plasmons.

13.7 Surface phonon polaritons

In the case of a metal and a frequency in the range $[\nu, \omega_p]$, we have seen that the dielectric constant has a negative real part. In this regime, the surface wave has the character of a surface plasmon. There are other materials with a negative dielectric constant. In agreement with the Kramers–Kronig relations, the frequency has to be close to a resonant excitation of the medium. In the infrared, crystals can absorb light due to the coupling to optical phonons. There is a frequency range called the Reststrahlen band where the dielectric constant has a negative real part. According to simple models (see Exercise 3.6), the dielectric constant can be written

$$\varepsilon(\omega) = \varepsilon_\infty \frac{\omega_L^2 - \omega^2 - i\nu\omega}{\omega_T^2 - \omega^2 - i\nu\omega}, \tag{13.42}$$

where ω_L is the longitudinal frequency and ω_T is the transverse optical frequency. These frequencies are due to the presence of optical phonons. Like for electrons, a longitudinal solution exists at ω_L. It corresponds to a charge-density wave. Here, it is a polarization charge density. A detailed discussion can be found in textbooks, e.g. ref. (Ziman, 1972) or ref. (Ashcroft and Mermin, 1976).

There are some differences with the plasmon case. The dielectric constant is negative only in the spectral range $\omega_T < \omega < \omega_L$. The corresponding wavelength is typically between $10\,\mu$m and $40\,\mu$m, depending on the mass of the atoms. Hence, a surface phonon polariton can exist only in the infrared or near-THz.

Fig. 13.11 is an example of a dispersion relation of a surface-phonon polariton. It corresponds to the case of a wave propagating at the interface between GaAs and vacuum. A very important similarity to the surface-plasmon case is the existence of a horizontal asymptote in the dispersion relation. Note that the dispersion relations of guided and quasinormal surface phonon modes coincide when loss is neglected; however, we again emphasize that the asymptote, which corresponds to a peak in the LDOS close to the interface, only exists in the $\tilde{\omega} - k$ representation. The importance of the peak for phonon polaritons in the context of near-field thermal radiation will be further discussed in Chapter 21.

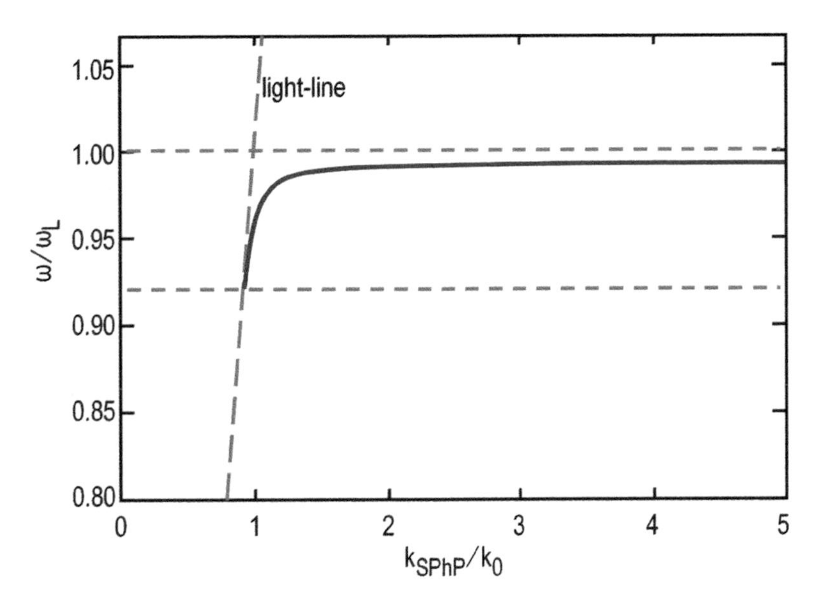

Fig. 13.11 Dispersion relation of a surface phonon polariton propagating along a GaAs–vacuum interface with dielectric constant given by a non-lossy Lorentz model $\varepsilon(\omega) = \varepsilon_\infty \frac{\omega_L^2-\omega^2}{\omega_T^2-\omega^2}$. $\omega_L = 292.1\,\mathrm{cm}^{-1}, \omega_T = 267.8\,\mathrm{cm}^{-1}, \varepsilon_\infty = 11$. Note that the dispersion relations $\tilde{\omega} - k$ and $\omega - \tilde{k}$ coincide when loss is neglected.

13.8 Surface plasmon contributions to the LDOS

The concept of LDOS was introduced in Chapter 12. It plays a key role when studying the lifetime of an emitter or the electromagnetic energy at equilibrium in the near field. Close to metallic surfaces, the LDOS is increased by orders of magnitude due to the electronic contributions. An explicit asymptotic form of the LDOS in the vicinity of an interface was given in Eq. (12.114) and the spatial and frequency dependence is illustrated in Figs 12.6 and 12.7. It is seen that the plasmon contribution results in a peak at the frequency of the asymptote. Hereafter, we compare the increase of the LDOS obtained with photonic crystals, surface plasmons on gold, and surface phonon polaritons on GaAs in the infrared. We then mention that the plasmonic LDOS plays a key role in the Casimir force, a pure quantum electrodynamics phenomenon. The application for tailoring light emission will be discussed in Chapter 19 and the application to heat transfer at the nanoscale will be discussed in Chapter 21.

13.8.1 Increasing the DOS: Surface waves and slow light

It is known that slow-velocity systems can be used to increase the DOS. The mechanism is depicted in Fig. 13.12. As the dispersion relation becomes flat close to the band edge, the number of states (represented by dots) with a frequency in the interval $\Delta\omega$ increases. This behaviour is known as Van Hove singularity. A major advantage of photonic crystals as compared to plasmonic systems is that there are almost no losses in dielectric media. Since the DOS diverges (the group velocity becomes zero), these

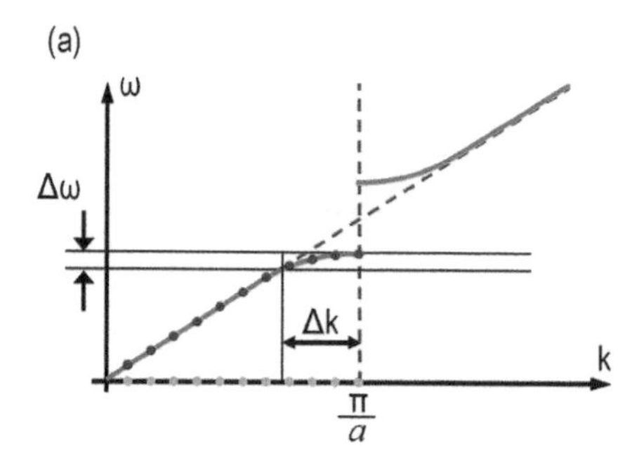

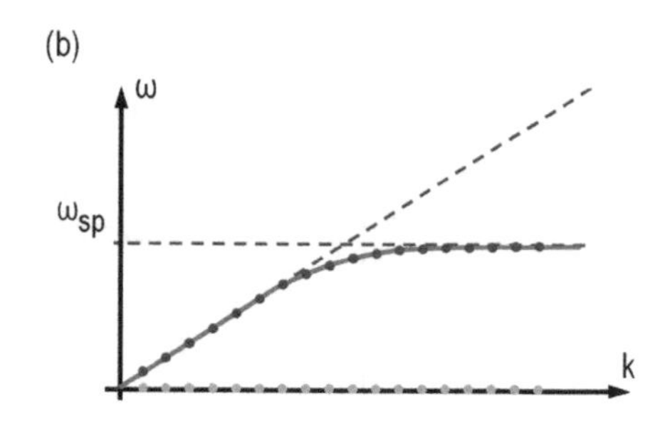

Fig. 13.12 Increasing the density of states. States are characterized by k and ω. They correspond to a point located on the dispersion relation. In k-space, the DOS per unit length is constant and takes the value $1/2\pi$. Fig. (a) shows that in ω-space, the DOS increases close to the gap edge and Fig. (b) shows that the increase close to the asymptote is considerably larger, as the asymptote is not limited along the k-axis.

systems seem to be the perfect solution to increase the DOS. Yet, it is important to emphasize that the number of states available when using a waveguide is always finite. In order to understand this apparent paradox, let us first remind the derivation of the DOS for a 1D system. We consider a waveguide branch characterized by a dispersion relation $k(\omega)$ for a mode propagating in a periodic waveguide with period a. In order to count the number of modes, we again consider that the system has a finite length L and we introduce the so-called Born–von Kármán (or periodic) boundary conditions, which stipulate that the system is periodic with period L. It follows that $k = p\,2\pi/L$. The number of modes in the interval $\mathrm{d}\omega$ corresponding to an interval $\mathrm{d}k$ is given by

$$g(\omega)\mathrm{d}\omega = \frac{L}{2\pi}\mathrm{d}k = \frac{L}{2\pi}\frac{\mathrm{d}k}{\mathrm{d}\omega}\mathrm{d}\omega = \frac{L}{2\pi}\frac{1}{v_g}\mathrm{d}\omega. \tag{13.43}$$

It is seen that the DOS diverges as the group velocity goes to zero. However, one should keep in mind that *this divergence is integrable*, so that the number of states in a finite interval $[\omega_1, \omega_2]$ always remains finite.

This is clearly seen in Fig. 13.12 where we represent schematically the dispersion relation of a guided wave close to a band edge. It is seen that the modes are simply displaced. They are removed from the gap and their density increases close to the band edge. An upper value of the number of modes involved is clearly the size of the Brillouin zone $2\pi/a$ divided by the interval between two modes $2\pi/L$. We find L/a. The period of a photonic crystal is on the order of the wavelength, so that we obtain an estimate of an upper bound of the number of modes in a 1D photonic crystal given by L/λ.

We now analyse the contribution of surface plasmons and surface phonon polaritons to the DOS. It is seen in Fig. 13.12 that the number of states provided by a surface plasmon at resonance is infinite, as the dispersion relation seems flat and unbounded. This is not correct and is a consequence of the model of the dielectric constant that does not account for the non-locality. A non-local dielectric constant introduces a cutoff (Ford and Weber, 1984) in the dispersion relation given by ω_p/v_F, where v_F is the Fermi velocity. We can now easily compare the DOS due to surface plasmons to the vacuum

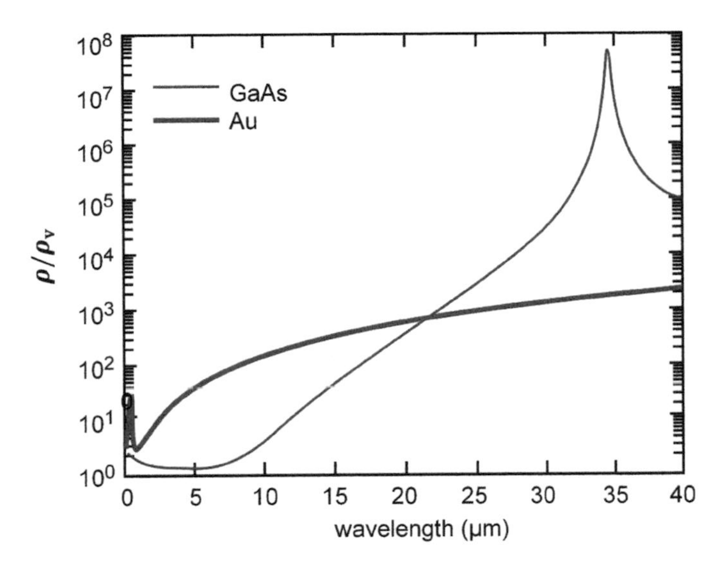

Fig. 13.13 Normalized local density of states ρ/ρ_v at a distance of 10 nm above an interface separating vacuum from gold or GaAs.

DOS. A rough estimate of the DOS is given by dividing the area of a disc with a radius $\pi k_{sp,max}^2$ by the area per state $4\pi^2$,

$$\frac{\pi k_{sp,max}^2}{4\pi^2} \approx \frac{\omega_p^2}{4\pi v_F^2}, \tag{13.44}$$

which is clearly much larger than what we found for dielectrics, where the order of magnitude is $1/\lambda^2 \approx \omega^2/c^2$. We remark that c/v_F is typically on the order of 300.

The above discussion has not taken into account the dependence on z of the LDOS. This is a critical issue, as a mode with wavevector k parallel to the interface decays exponentially as $\exp(-kz)$ for values of $k \gg k_0$. We now turn to a quantitative analysis. An explicit form can be derived from the imaginary part of the Green tensor. Of particular interest is the asymptotic expression of the normalized LDOS at a distance z from the interface such that $z \ll \lambda$:

$$\frac{\rho(z,\omega)}{\rho_v(\omega)} = \frac{\mathrm{Im}[\varepsilon]}{|\varepsilon+1|^2}\frac{1}{4(k_0 z)^3}, \tag{13.45}$$

where $g_v(\omega)$ stands for the LDOS in vacuum. The surface-plasmon resonant contribution is clearly given by the term $1/|\varepsilon+1|^2$, which is responsible for the peaks observed in the visible for gold and in the infrared for GaAs in Fig. 13.13 at a distance of 10 nm from the surface. It is seen that the higher quality factor of the surface phonon polariton results in a much larger enhancement. Note also that the LDOS is enhanced over the whole spectral range where there are losses. These contributions are due to non-radiative modes with large wavevectors which are not surface plasmons.

In both cases, the presence of the surface wave results in a peak in the LDOS enhancement. It clearly appears that the enhancement is orders of magnitude larger for surface phonon polaritons than for surface plasmons.

Local density of states and Casimir force. Another consequence of the contribution of surface plasmon to the LDOS is the Casimir force between two metallic parallel plates. Casimir force is a pure quantum electrodynamics effect that manifests itself at macroscopic scale. Casimir proposed (Casimir, 1948) that there is an attractive force between two parallel perfectly conducting surfaces at zero temperature separated by a gap of width d. Since then, his remarkable prediction has been measured experimentally with great accuracy (Lamoreaux, 1997; Decca et al., 2003; Jourdan et al., 2009). However, when comparing the measurements with the data, the assumption of a perfect conductor cannot be used any longer (Genet et al., 2000). A careful analysis shows that the surface plasmons are responsible for the forces actually observed (Henkel et al., 2004; Intravaia and Lambrecht, 2005). We give here a brief qualitative discussion of this effect. We refer the reader to ref. (Henkel et al., 2004) for a further discussion. The gap behaves as a waveguide with a set of modes. From the quantum electrodynamics point of view, each mode (k, ω) has a harmonic-oscillator Hamiltonian and its energy is $(n + 1/2)\hbar\omega$. It follows that the total minimal electromagnetic energy when $n = 0$ (the so-called vacuum energy) is given by $\sum_n \hbar\omega_n/2$, where the sum is over all modes of the gap. The origin of the force is now easily understood by realizing that if the gap is reduced, the number of available modes is reduced, so that the electromagnetic energy is reduced. This means that the vacuum electromagnetic energy acts as an attractive potential. In the original derivation of Casimir, modes of a planar waveguide with perfectly conducting surfaces were used. When accounting properly for the optical properties of metals, it turns out that the DOS is dominated by the surface-plasmon contribution (Henkel et al., 2004).

13.A Complement: Surface plasmon excited by an electric dipole

The purpose of this complement is to provide the explicit form of the surface plasmon emitted by a dipole. We consider an interface separating a metal with permittivity ε_2 in the lower half-space ($z < 0$) from a dielectric with permittivity ε_1 in the upper half-space. An electric dipole moment $\mathbf{p}_0 \exp(-i\omega t)$ is located at a distance d from the interface in medium 1. The emitted field is given by the Green tensor and can be accurately computed numerically using the integral form (Complement 12.B.2). The total field radiated by the monochromatic dipole close to the surface contains two contributions, a quasi-spherical wave often called direct wave and a surface plasmon. The characteristics of the two contributions and their relative weight with the metal conductivity are discussed in section 18.6. The detailed form can be found in ref. (Baños, 1966). Here, we give the surface-plasmon expression, which is found by extracting the contribution of the pole of the reflection factor to the Green tensor (see Complement 12.B.2). We only give an asymptotic form valid for a radial distance larger than a few wavelengths (Archambault et al., 2009). This simplified form has the structure of a cylindrical wave in the horizontal plane with a complex wavevector $k_{sp}(\omega)$ given by the surface-plasmon dispersion relation, Eq. 13.15; it is exponentially decaying along the

z-axis with the usual dependence $\exp(i\gamma z)$ and it is elliptically polarized in the plane $(\hat{\rho}, \hat{\mathbf{z}})$, where the unit vector $\hat{\rho}$ is defined by $\hat{\rho} = \boldsymbol{\rho}/\rho$. For a dipole along the z-axis $\mathbf{p} = p\hat{\mathbf{z}}$, the electric field in medium $n = 1, 2$ is given by

$$\mathbf{E}_{sp,n}(\rho, z) = \frac{\exp(ik_{sp}\rho + i\gamma_n|z|)}{\sqrt{k_{sp}\rho}} \left(\hat{\rho} - \frac{k_{sp}}{\gamma_n}\hat{\mathbf{n}}_m\right) p_0 \exp(i\gamma_1 d) M, \tag{13.46}$$

where $\hat{\mathbf{n}}_1 = \hat{\mathbf{z}}$, $\hat{\mathbf{n}}_2 = -\hat{\mathbf{z}}$ and

$$M = \mathbf{e}^{-i\frac{\pi}{4}} \frac{k_{sp}\gamma_2}{\varepsilon_0\sqrt{8\pi}} \frac{\varepsilon_2\gamma_1 - \varepsilon_1\gamma_2}{\varepsilon_1^2 - \varepsilon_2^2}.$$

A similar expression can be derived for a dipole oriented along the x-axis:

$$\mathbf{E}_{sp,n}(\rho, z) = \frac{\exp(ik_{sp}\rho + i\gamma_n|z|)}{\sqrt{k_{sp}\rho}} \left(\hat{\rho} - \frac{k_{sp}}{\gamma_n}\hat{\mathbf{n}}_m\right) p_0 \exp(i\gamma_1 d) \cos(\theta) M', \tag{13.47}$$

where $\cos(\theta) = \hat{\mathbf{x}} \cdot \hat{\rho}$ and

$$M' = -\mathbf{e}^{-i\frac{\pi}{4}} \frac{\gamma_1\gamma_2}{\varepsilon_0\sqrt{8\pi}} \frac{\varepsilon_2\gamma_1 - \varepsilon_1\gamma_2}{\varepsilon_1^2 - \varepsilon_2^2}.$$

It is seen that the amplitude of the plasmon emitted in a direction making an angle θ with the dipole orientation decays as $\cos(\theta)$.

13.B Complement: Surface plasmon launched by an isolated slit

Metallic surfaces patterned with subwavelength indentations possess a variety of interesting optical properties, with applications ranging from sensing and plasmonic devices to integrated circuits mixing photonics and electronics (Stockman, 2011; Maier et al., 2001; Bozhevolnyi et al., 2006). In these structures, SPPs launched at the input and output sides of the indentation play a key role (see exercise 18.6). In this section, we summarize the main results known for an iconic problem, the launching of guided surface plasmon modes by an isolated subwavelength slit illuminated by an incident plane wave (see Fig. 13.14).

We introduce the permittivities of the dielectric and metal materials, ε_d and ε_m. The slit is assumed to be infinitely long (no energy is reflected from the bottom of the slit). The normally incident plane wave is polarized perpendicularly to the slit. We define the plasmon excitation efficiency η as the ratio between the plasmon power flux (launched on both sides of the slit) and the flux of the Poynting vector of the incident field integrated over the slit entrance.

Fig. 13.14 summarizes the most important results for an air–gold interface and for four different wavelengths: $0.6, 1, 3,$ and $1\,\mu$m. Other noble metals used at frequencies close to the plasma frequency exhibit a similar behaviour. The computational data,

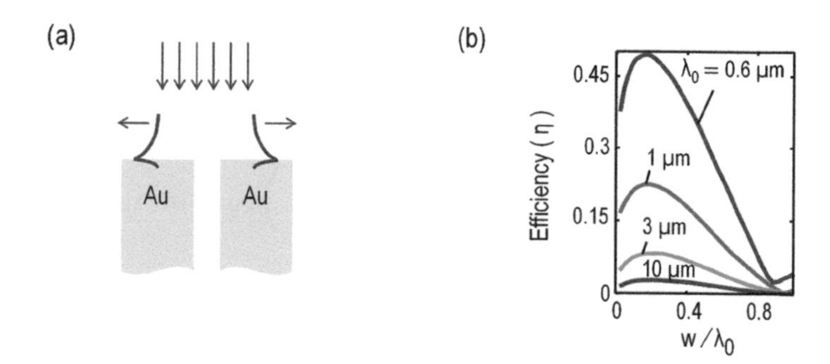

Fig. 13.14 Plasmon launching efficiency by a subwavelength semi-infinite slit perforated in a gold substrate as a function of the normalized slit width w/λ_0 for different wavelengths. The slit is illuminated by a normally incident plane wave linearly polarized with the electric field perpendicular to the grooves (TM). The total efficiency corresponds to the efficiency of the two plasmon modes launched in the two opposite directions on the air–gold interface. The predictions are obtained for the gold frequency-dependent permittivity tabulated in ref. (Palik, 1997). Fig. (b) is adapted with permission from ref. (Lalanne et al., 2005) © 2005 American Physical Society.

obtained with a fully vectorial approach, can be quantitatively explained with an approximate model that provides a closed-form expression for the efficiency

$$\eta = \sqrt{\frac{\varepsilon_d}{|\varepsilon_m|}} f(w/\lambda_0), \tag{13.48}$$

where $f(x)$ is a sinc-like function which weakly depends on the dielectric constants of the problem (Lalanne et al., 2005).

Most noteworthy are the following observations:

- η strongly depends on the slit width, with a maximum value reached for a width approximately equal to $\lambda_0/4$. The existence of this maximum and a minimum for $w = \lambda_0$ has been verified experimentally with far-field measurements performed on a slit doublet, and the sinc-like behaviour has been qualitatively verified by detecting the guided surface plasmons launched at the transmission facet of a single slit with a near-field probe (Kihm et al., 2008).
- η is fairly large (≈50%) at visible wavelength. However, it rapidly decreases in the near-infrared, as it scales as $|\varepsilon_m|^{-1/2}$. This scaling law, which is valid for any subwavelength indentation, is due to the dispersion of the distribution of the plasmon modal field (see Fig. 13.6). It indicates that the visible part of the electromagnetic spectrum is the most exciting region for surface-plasmon circuitry on metal films (Bozhevolnyi et al., 2006; Gramotnev and Bozhevolnyi, 2010) and that surface phonon polaritons are better suited in the infrared.
- η can be enhanced by immersing the sample in a dielectric material; it increases proportionally to the refractive index of the material.

There are several ways to increase the plasmon-generation efficiency, either by using highly oblique incidences or grooves instead of slits. For grooves, the light that is coupled to the gap-plasmon mode in the slit is back-reflected at the bottom of the slit, and when it reaches the aperture it scatters into additional plasmons. Under constructive interference, the groove behaves as a resonator and the efficiency may exceed unity ($\eta \approx 2 - 3$), implying that the plasmon scattering cross section is larger than the geometrical area of the slit. Indeed, the plasmon-generation efficiency can be further increased by arraying grooves or ridges, like in a grating plasmon coupler, at the expense of miniaturization.

Exercises

Exercise 13.1 Bulk plasmon resonance frequency. Importance of electron kinetic energy. The purpose of this exercise is to derive the bulk plasmon frequency $\omega_p^2 = ne^2/m\varepsilon_0$ of a thin metallic film with area A and thickness d containing n electrons per unit volume using an energetic approach. We use the notations of section 13.1. For an RLC oscillator, the energy oscillates between electric energy stored in the capacitor and magnetic energy stored in the inductance. For a resonant electromagnetic cavity, the energy oscillates between the energy of the electric field $\varepsilon_0 E^2/2$ and the energy of the magnetic field $B^2/2\mu_0$. The key difference between plasmons and electromagnetic waves is the fact that the magnetic energy is negligible as compared to the kinetic energy of the electrons.

(1) Prove that the capacitance C of the thin film is $\varepsilon_0 A/d$. The electric energy is $CV^2/2$, where V is the applied voltage. As in section 13.1, we denote by x the amplitude of the electron displacement. Show that the electric energy U_E and the kinetic energy U_K of electrons oscillating at frequency can be cast in the form:

$$U_E = \frac{n^2 e^2 x^2 Ad}{2\varepsilon_0} \quad ; \quad U_K = nAd\,\frac{m\omega^2 x^2}{2}.$$

(2) By equating the electric and kinetic energies, derive the oscillator frequency. The possibility of storing the energy as either electrostatic energy or electron kinetic energy can be viewed as the fundamental reason for the possibility of reducing the size of the resonator to well below half a wavelength. This is a key difference from THz and radiowaves, where most metals can be considered to be perfect conductors, so that the field does not penetrate in the media and energy is not stored in the mechanical degrees of freedom of the charge carriers.

Exercise 13.2 Surface-plasmon resonance frequency. Derive the resonance frequency of a plasmon oscillation at a single interface between a bulk metal and vacuum using the elementary model of the first section. Show that the resonance frequency of the surface plasmon is given by $\omega_p^2 = ne^2/2m\varepsilon_0$.

Exercise 13.3 Surface-plasmon resonance and electrostatic reflection factor. Derive the limiting form of the reflection factor for p-polarization and non-magnetic materials in the limit $k_x \gg \omega/c$. Prove that it is independent of k_x. Compare with the reflection factor $(\varepsilon_1 - 1)/(\varepsilon_1 + 1)$ used to evaluate the amplitude of the image charge generated by an interface in electrostatics. Using the non-lossy Drude model for ε_1 and the idea that a surface resonance corresponds to a linear response divergence, derive the resonance frequency of the interface.

Show that the resonance frequency is complex for a lossy Drude model. What is the physical meaning of the imaginary part of the frequency? What is the quality factor of the oscillation?

Exercise 13.4 s-polarized surface wave. Derive the dispersion relation of an s-polarized surface wave propagating along an interface-separating vacuum and a medium with a permittivity ε and a permeability μ. Show that the dispersion relation is given by $\mu_2\gamma_1 + \mu_0\gamma_2 = 0$.

Exercise 13.5 Wood anomaly. Fig. 13.15 shows the absorptivity map of a gold grating in the plane $(k_{x,inc}, k_{y,inc})$. Show that the dispersion relation of a surface plasmon propagating along a metal–vacuum interface in the plane (k_x, k_y) for a given frequency is given by a circle with radius $k_{sp}(\omega)$. An incident plane wave with incident wavevector $(k_{x,inc}, k_{y,inc})$ can excite

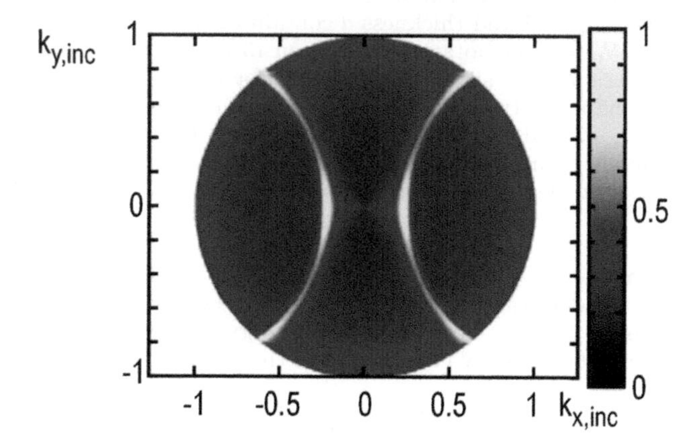

Fig. 13.15 p-polarized absorptivity of a gold grating with period d along the x-axis in the plane $(k_{x,inc}, k_{y,inc})$.

a surface plasmon in the presence of a periodic grating with period d along the x-axis. What is the condition that needs to be satisfied by $(k_{x,inc}, k_{y,inc})$ to excite a plasmon? Plot in the plane (k_x, k_y) the locations of the solutions. Given that light is absorbed when a surface plasmon can be excited, discuss the absorptivity map in the plane $(k_{x,inc}, k_{y,inc})$. Wood anomaly is also discussed in Exercise 18.6.

Exercise 13.6 SPP polarization. We consider a surface plasmon propagating along a metal-vacuum interface. The metal is described by a non-lossy Drude model. Derive the value of k_x such that the surface plasmon is circularly polarized in the plane (x, z). What is the polarization for lower values of k_x?

Exercise 13.7 Graphene surface plasmon. We consider a surface-charge-density wave propagating in graphene between medium 1 and 2 along the axis x. The goal of the exercise is to adapt the derivation of section 13.3.1 to derive the dispersion relation of the surface plasmon on graphene. The general form of the fields in media 1 and 2 are the same. The continuity of $E_x = E_0$ is still valid. The continuity relation of the z-component of the electric field has to be modified by including a surface charge.

(1) Show that the continuity condition (see section 5.2) of εE_z is replaced by

$$\varepsilon_1 E_{z1} - \varepsilon_2 E_{z2} = \rho_S/\varepsilon_0.$$

(2) Using charge conservation, show that $\rho_S = -k_x K E_0/\omega$.

(3) Derive the dispersion relation $\frac{\varepsilon_1}{\gamma_1} + \frac{\varepsilon_2}{\gamma_2} = \frac{K}{\omega\varepsilon_0}$.

(4) In the limit $k_x \gg n_1 k_0, n_2 k_0$, we have $\gamma_1 \approx \gamma_2 \approx i k_x$. Show that $k_x = i(\varepsilon_1 + \varepsilon_2)\omega/K(\omega)$.

(5) If $E_F \gg \hbar\omega$, $K \approx K_{intra}$. Show that

$$k_x = \frac{\pi\varepsilon_0\hbar^2}{e^2 E_F}(\varepsilon_1 + \varepsilon_2)(\omega^2 + i\omega\nu).$$

Neglecting the loss term ν, this relation shows a characteristic quadratic behaviour of the dispersion relation $k_x \propto \omega^2$.

14
Localized surface plasmons

14.1 Introduction

Bounded geometries, such as metallic particles of various shapes or voids in metals (see Fig. 14.1), can support surface-plasmon modes. These modes are called localized surface plasmons (LSPs). They are oscillations of the electronic density. The physical origin of these modes is a periodic exchange of energy between the kinetic energy of the electrons and potential electrostatic energy. This is the key ingredient for achieving electromagnetic oscillations at a subwavelength scale (see sections 7.1 and 10.5.2 and Exercise 14.1).

Plasmonic-based nanostructures are of considerable interest because of their unique ability to confine light at a deep-subwavelength scale, something that cannot be achieved with conventional lenses in 2D and photonic microcavities in 3D, which are diffraction-limited. This feature can be used to perform near-field imaging beyond the diffraction limit, as discussed in Chapter 11, to implement nanoscale light sources, enhance the capacity of data storage devices, or guide light at the nanoscale to name a few

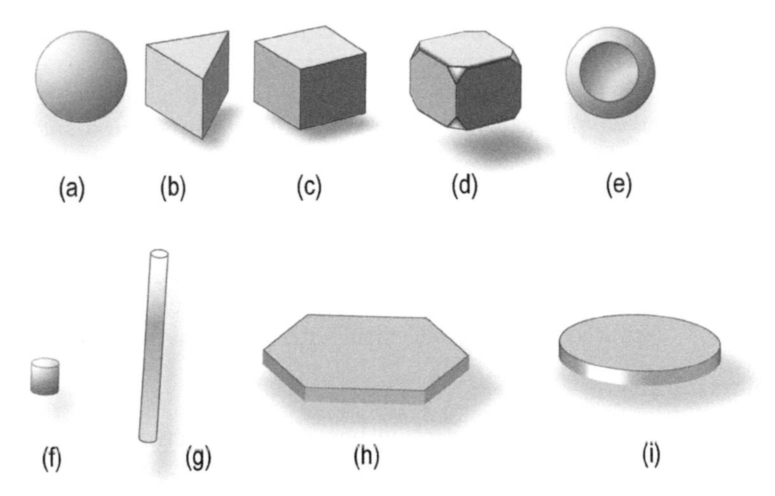

Fig. 14.1 Schematic illustration of configurations of single-metal nanostructures for localized plasmon resonance. The sizes are at the nanometre scale (from tens to hundreds of nm): (a) nanosphere, (b) nanoprism, (c) nanocube, (d) nanocage, (e) nanoshell, (f) nanorod, (g) nanowire, (h) nanosheet or flake, and (i) nanodisc.

Introduction to Nanophotonics. Henri Benisty, Jean-Jacques Greffet, and Philippe Lalanne, Oxford University Press.
 DOI: 10.1093/oso/9780198786139.003.0014

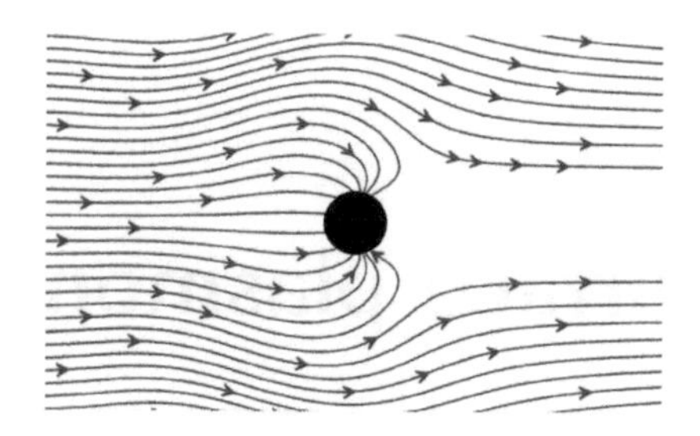

Fig. 14.2 Resonant excitation of the surface phonon–polariton mode of a SiC nanoparticle illuminated by a plane wave coming from the left. The figure shows the bending of energy flow towards the particle, which absorbs and scatters the light beyond its actual geometrical cross section. The concentration of the energy flow illustrates the field-enhancement mechanism.

examples (Maier et al., 2001; Stockman, 2011). The confinement of light comes with another important property: the enhancement of the field. To illustrate how this field enhancement results from the resonant behaviour of the particle, we show in Fig. 14.2 the lines of the Poynting vector of a plane wave illuminating a nanosphere at resonance. It is seen that the energy is funnelled into the particle. The energy enhancement is clearly visible, as the flux captured by the particle corresponds to the flux over an area with an order of magnitude given by $\lambda^2/4$ and is concentrated in an area which is the particle area on the order of a^2, where a is the nanosphere radius. The field enhancement can be used for many applications and, in particular, to increase non-linear effects. A major application is the giant enhancement of the Raman effect known as surface-enhanced Raman scattering (SERS) or tip-enhanced Raman scattering). Another application is local heating, which can be used for, e.g., local destruction of tumour cells or heat-assisted magnetic data storage.

Nanoparticles sustaining plasmon resonances can be used as nanoantennas to tailor the coupling between propagating plane waves and tiny sources such as atoms, molecules, or quantum dots (QDs), as will be discussed in Chapter 19.

14.2 A tutorial example: Metallic nanosphere

The goal of this section is to introduce the key concepts of metallic nanoparticles using the simplest possible example: a nanosphere. The particle has a radius a and a permittivity ε_p. It is located in a host dielectric with permittivity ε_h.

14.2.1 Polarizability

As discussed in section 2.3.2, the interaction between the electromagnetic field and a current distribution $\mathbf{j}(\mathbf{r})$ in a volume V much smaller than a wavelength cube can be characterized by a dipole moment, $\mathbf{p}$, given by

$$-i\omega\mathbf{p} = \int_V \mathbf{j}(\mathbf{r})\mathrm{d}\mathbf{r}, \tag{14.1}$$

unless this quantity is zero (as is sometimes the case for symmetry reasons, e.g. for a ring). The induced dipole is proportional to the incident electric field $\mathbf{E}_{inc}$. It was shown in Chapter 12 that the dipole moment is related to the incident electric field using the polarizability, which is a matrix in the general case. For a metallic sphere, the polarizability is a scalar. In this introductory section, we assume that we can neglect retardation effects, so that the electrostatic form of the field inside a sphere can be used. The derivation is given in Exercise 14.4. Within this approximation, the field inside the particle is uniform:

$$\mathbf{E}_{int} = \frac{3\varepsilon_h}{\varepsilon_p + 2\varepsilon_h}\mathbf{E}_{inc}. \tag{14.2}$$

The dipole moment is found to be:

$$\mathbf{p} = \varepsilon_0\varepsilon_h\alpha_{es}\mathbf{E}_{inc}, \quad \text{with} \quad \alpha_{es} = 4\pi a^3\frac{\varepsilon_p - \varepsilon_h}{\varepsilon_p + 2\varepsilon_h}. \tag{14.3}$$

This form is the leading term valid in the limit of a small radius a. We use the subscript es because it coincides with the form derived using an electrostatic derivation (Exercise 14.4). Let us stress that neglecting retardation effects leads to a polarizability that predicts the resonance but fails to account for radiative losses and therefore does not fulfil energy conservation. A corrected approximate polarizability that accounts for the radiation reaction and satisfies energy conservation was given in Eq. (12.38). An exact form of the polarizability of a sphere illuminated by a plane wave can be derived using the so-called Mie theory (Bohren and Huffman, 1983). This will be discussed in section 14.2.5.

14.2.2 Resonance and quality factor (Q-factor)

We continue the presentation with the modes. Given that a metallic particle has ohmic and radiative losses, we seek quasinormal modes with a complex-valued frequency. They are given by the poles of the polarizability as a function of frequency.

The resonance frequency corresponds to the pole of the electrostatic polarizability in Eq. (12.38) and is given by

$$\varepsilon_p(\omega) + 2\varepsilon_h = 0. \tag{14.4}$$

Because of absorption, ε_p is complex-valued, and so is ω. The meaning of the imaginary part of the frequency has already been introduced in Chapter 7. It corresponds to the decay time of the plasma oscillation. We will denote the complex-valued frequency by a tilde: $\tilde{\omega}_{sp} = \omega'_{sp} + i\omega''_{sp}$ with $\omega''_{sp} < 0$, owing to our convention $\exp(-i\tilde{\omega}t)$.

For the sake of illustration, we use the Drude model $\varepsilon_p(\omega) = 1 - \omega_p^2/(\omega^2 + i\nu\omega)$. Inserting this model in Eq. (14.4) and solving for ω, we find a complex frequency denoted by $\tilde{\omega}_{sp}$

$$\tilde{\omega}_{sp} = \frac{\omega_p}{\sqrt{2\varepsilon_h + 1}}\sqrt{1 - \frac{(2\varepsilon_h + 1)\nu^2}{4\omega_p^2}} - i\frac{\nu}{2}. \tag{14.5}$$

We now discuss the response of the particle to a monochromatic excitation at frequency ω and explore the existence of a resonance of the dipole amplitude as the real

Table 14.1 Resonance wavelength and Q-factor for different metals and semiconductors. Data taken from ref. (Palik, 1997).

Material	gold	silver	aluminium	SiC
λ_{sp}(nm)	505.4	353.6	138.2	10 735
Q	6.6	18.8	19.9	195.5

frequency ω approaches the real part of the pole of the system. We expand the permittivity around the real frequency ω_{sp} defined by $\varepsilon'(\omega_{sp}) + 2\varepsilon_h = 0$, where $\varepsilon' = \mathrm{Re}(\varepsilon)$:

$$\varepsilon(\omega) = -2\varepsilon_h + \left(\frac{\mathrm{d}\varepsilon'(\omega)}{\mathrm{d}\omega}\right)_{\omega_{sp}} (\omega - \omega_{sp}) + i\varepsilon''(\omega_{sp}). \tag{14.6}$$

Using a Drude model, we find

$$\omega_{sp} = \frac{\omega_p}{\sqrt{2\varepsilon_h + 1}}\sqrt{1 - \frac{(2\varepsilon_h + 1)\gamma^2}{\omega_p^2}}.$$

It is seen that this result slightly differs from the real part of $\tilde{\omega}_{sp}$, so that the resonant response of a system does not occur at the exact frequency of the mode. This is an important feature of all lossy resonators. The resonant response associated with the excitation of a physical quantity such as dipole amplitude is resonant, but the resonant frequency usually differs from the real part of the mode frequency. Note also that the resonant response to an external excitation depends on the physical quantity studied. Absorption and scattering maximums occur close to the resonance but at different frequencies, given that the functions to be optimized are different.

Close to resonance, we insert the expansion Eq. (14.6) in the polarizability Eq. (14.3) and obtain a resonant form:

$$\alpha_{es}(\omega) = \frac{-12\pi a^3}{\left(\frac{\mathrm{d}\varepsilon'}{\mathrm{d}\omega}\right)} \frac{1}{\omega - \omega_{sp} + i\frac{\varepsilon''}{\left(\frac{\mathrm{d}\varepsilon'}{\mathrm{d}\omega}\right)}}. \tag{14.7}$$

The square modulus of the polarizability is thus a Lorentzian function with a full width at half-maximum (FWHM) given by $2\varepsilon''/(\mathrm{d}\varepsilon'/\mathrm{d}\omega)$ and a Q-factor:

$$Q = \frac{\omega'_{sp}\left(\frac{\mathrm{d}\varepsilon'_p}{\mathrm{d}\omega}\right)_{\omega_{sp}}}{2\varepsilon''_p}. \tag{14.8}$$

With our hypothesis (static limit and spherical particle), the Q-factor depends only on the permittivity of the material. In fact, this form of the Q-factor is valid for deep-subwavelength particles of any shape, as shown in (Wang and Shen, 2006).

In Table 14.1, we give a variety of Q-factors that can sustain a phonon–polariton resonance for different metals and for SiC. It is seen that silver has a resonance much narrower than gold.

14.2.3 Absorption and scattering cross section

We introduced the concept of cross section in section 12.2.3 to characterize the properties of a particle. In order to find the scattering cross section of a particle in vacuum ($\varepsilon_h = 1$), we simply compute the power radiated by the induced dipole using Eq. (2.48) and the electrostatic polarizability, Eq. (14.3). We find (see Exercise 14.5)

$$\sigma_{sc} = \frac{|\alpha|^2}{6\pi}\frac{\omega^4}{c^4} = \frac{8\pi}{3}a^6\frac{\omega^4}{c^4}\left|\frac{\varepsilon_p - 1}{\varepsilon_p + 2}\right|^2. \tag{14.9}$$

In order to find the absorption by a particle in vacuum, we can also perform a direct calculation and integrate the absorption rate $< \mathbf{j}\cdot\mathbf{E}_{int} >= \omega\varepsilon_0 \mathrm{Im}[\varepsilon_p]|\mathbf{E}_{int}|^2/2$ over the volume of the particle. As the field is uniform in the particle, we get (see Exercise 14.6) the following using Eq. (14.2) with $\varepsilon_h = 1$:

$$\sigma_{\mathrm{abs}} = 12\pi a^3 \frac{\omega}{c}\frac{\mathrm{Im}(\varepsilon_p)}{|\varepsilon_p + 2|^2} = \frac{\omega}{c}\mathrm{Im}(\alpha_{es}). \tag{14.10}$$

A particle sustaining a plasmonic resonance can produce both absorption and scattering. It is important to know which process dominates. It turns out that small particles mostly absorb, whereas larger particles mostly scatter. The difference stems from the fact that absorption is proportional to the volume, while scattering is proportional to the square of the dipole moment and therefore to the square of the volume. Hence, as the particle radius a goes towards zero, the scattering cross section decays faster, so that small particles are essentially absorbing particles. Conversely, as the size of the particles increases, scattering becomes the dominant mechanism and the particles become scatterers with low losses. A very good example is silver nanoparticles. A silver sphere much larger than the wavelength is highly reflecting. By contrast, glass containing silver nanospheres appears as a coloured absorbing medium.

These very different behaviours raise the question of the transition between the two regimes. The size of the particle corresponding to the transition between the two regimes can be estimated by equating the two cross sections. We find:

$$a_{transition} = \frac{\lambda}{2\pi}\left[\frac{9\varepsilon_p''}{2|\varepsilon_p - 1|^2}\right]^{1/3}. \tag{14.11}$$

At resonance, $\varepsilon' = -2$, so that:

$$a_{transition} \approx \frac{\lambda_{sp}}{2\pi}\left[\frac{\varepsilon_p''}{2}\right]^{1/3}. \tag{14.12}$$

For silver particles, the transition between absorption and scattering typically takes place around 35 nm. For SiC, it takes place at 70 nm. As a rule of thumb, the transition between dominant absorption and dominant scattering takes place around 50 nm for metallic nanoparticles.

14.2.4 Colour of an ensemble of metallic nanoparticles

We now discuss colour mechanisms due to either scattering or absorption. Colour due to absorption is widely used in pigments in paints and inks. Similarly, an ensemble of tiny gold particles absorbs at 520 nm in the green, so that the colour of the non-absorbed light, either in transmission or in reflection, is reddish.

For larger particles, the scattering cross section becomes comparable or larger than the absorption cross section, and both mechanisms contribute to the appearance. The effect of scattering is well known in relation to the propagation of sunlight in the atmosphere. In transmission, at sunset when the atmosphere thickness is large, the sun appears red, while the light scattered in the atmosphere is blue. For gold nanoparticles larger than 50 nm, scattering dominates, so that, like for the atmosphere, the colour in transmission and in scattering differs. The scattered light is green, corresponding to the plasmon resonance, and the transmitted light has the complementary reddish colour. These considerations can be used to understand the beautiful colours of the Lycurgus Cup in the British Museum and the stained glass of the Sainte-Chapelle in Paris (see Exercise 14.7).

14.2.5 Beyond the quasistatic approximation: Radiative losses

Previously, we assumed that the spherical particles were sufficiently small and used the electrostatic approximation to derive the Q-factor or the cross section. If the size of the particle is larger than, let us say, $\lambda/20$, this approximation breaks down and retardation effects need to be included. Furthermore, even for very small particles, the electrostatic approximation leads to an inconsistency when analysing the energy conservation of non-lossy particles. The inconsistency is removed when accounting for retardation effects (see section 12.4.2).

An exact expression of the electric polarizability of a spherical particle with radius a and refractive index $n_p = \sqrt{\varepsilon_p}$ in a host medium of refractive index $n_h = \sqrt{\varepsilon_h}$ can be found in ref. (Colom et al., 2016). A convenient Taylor expansion to lowest order of $k_0 a$ is given by

$$\alpha \approx \frac{\alpha_{es}}{1 - \frac{3k_0^2 n_h^2 a^2}{5}\frac{\varepsilon_p - 2\varepsilon_h}{\varepsilon_p + 2\varepsilon_h} - i\frac{k_0^3 n_h}{6\pi}\alpha_{es}}. \tag{14.13}$$

A remarkable property of this formula is its ability to introduce a term with an imaginary part, even for a nanoparticle made with a non-lossy material (ε_p and α_{es} are both real-valued in this case). This imaginary part accounts for the radiative losses.

When gradually increasing the radius of the particle, the resonance wavelengths red shift and the resonance linewidths are broadened. This can be seen directly by inserting the Drude formula in Eq. (14.13) and calculating the poles. The trends are illustrated in Fig. 14.3 for three different radii. Note that the scattering and absorption cross sections are normalized by a universal value, $3\lambda^2/8\pi$, which will be introduced in the upcoming sections. The broadening is a direct consequence of the increase of the radiative losses. The red-shift can be understood using the simple model of plasmon resonance given by Eq. (13.1). The resonance frequency is proportional to the restoring electrostatic force, which is proportional to the charge density at the interfaces.

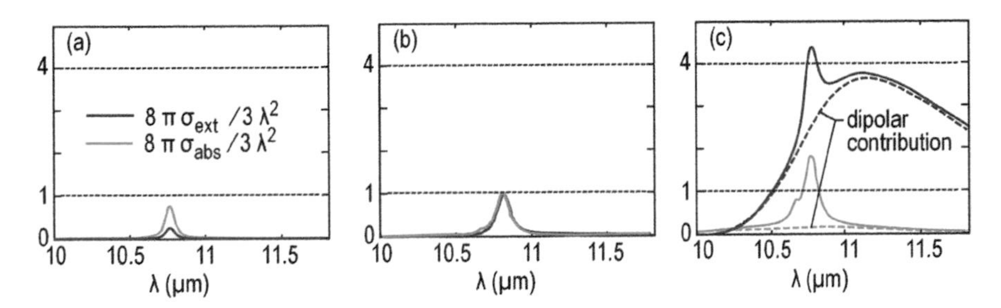

Fig. 14.3 Normalized spectra of the scattering and absorption cross section of SiC nanoparticles for three different sizes. (a) For small particles ($a = 51\,\text{nm}$), a peak due to the electric dipolar mode is observed and the absorption cross section dominates. (b) At critical coupling ($a = 69\,\text{nm}$), the absorption and scattering cross sections are equal to $3\lambda^2/8\pi$. (c) For a larger particle ($a = 189\,\text{nm}$), new modes appear. The dipolar contribution is shown with dotted lines, while the full Mie calculation is shown with plain lines. The dipolar absorption cross section decreases as expected beyond the critical coupling. The scattering cross section has a red-shifted peak which reaches the maximum value, $3\lambda^2/2\pi$. Spectral features due to higher-order modes also appear.

This model neglects retardation effects which introduce a dephasing and hence a reduction of the restoring force.

We now discuss the absorption cross section beyond the electrostatic approximation. Eq. 14.10 has been derived in the context of a small dipolar particle with no retardation. For larger dipolar particles, we need an exact form, one that is derived in Exercise 14.5 from the energy conservation and the optical theorem:

$$\sigma_{abs} = k_0 \text{Im}(\alpha) - \frac{k_0^4 |\alpha|^2}{6\pi}. \tag{14.14}$$

We use this exact form to derive universal bounds of the absorption and extinction cross section valid for both lossy and non-lossy dipolar particles in Complement 14.A. Remarkably, these forms can be applied to radiowave antennas, plasmonic nanoparticles, and atoms.

14.2.6 Higher-order modes of nanospheres

So far, we have only discussed the electric dipolar modes of the nanospheres. However, the Mie solution involves an expansion over spherical harmonics of order n, l instead of the simple ansatz used in Exercise 14.2. When including an angular dependence ($l \neq 0$), the resonances are given by

$$\omega_l = \sqrt{\frac{l}{2l+1}}\omega_p, \tag{14.15}$$

when using a Drude model and assuming that the particles are in vacuum. The general resonance condition for a sphere in vacuum is given by $\varepsilon_p(\omega) + \frac{l+1}{l} = 0$. An l-order mode corresponds to a surface-charge oscillation with l nodes.

The modes with even l cannot be excited by a plane wave, as their dipole moment is zero. The modes with odd l have a non-zero dipole moment, but in practice, the dipole value is negligible for very small spheres, so that they are hardly observed. However, high-order modes can be efficiently excited by localized sources such as molecules located in the vicinity of the particle or by collimated electron beams passing close to the particle.

When exploring higher values of n, which is needed for large particles, multipolar modes appear. Their maximum extinction cross section for non-lossy particles varies as $(2n+1)\lambda^2/2\pi$ and the maximum absorption cross section varies as $(2n+1)\lambda^2/8\pi$. We note that these formulas are in agreement with the dipolar case ($n=1$) previously discussed.

14.2.7 Confinement effects

For deep-subwavelength particles, radiative losses are negligible and the Q-factor does not depend on the radius (or on the exact form): see Eq. (14.8). However, the experiments (Kreibig and Genzel, 1985) show a broadening of the spectra when the radius decreases. This broadening can be modelled using the following form of the damping factor ν:

$$\nu(R) = \nu + A\frac{v_F}{a}, \tag{14.16}$$

where A is a dimensionless factor, v_F is the Fermi velocity, and a is the sphere radius. The correction term, which is size-dependent, fits the experimental data (Kreibig and Genzel, 1985). In a classical picture, this term accounts for the collisions of the electrons with the surface. In this picture, v_F/a is a typical time between two surface-scattering events. In a quantum picture, the electronic states of the free electrons inside a sphere are no longer plane waves. Indeed, the electronic quantum states are obtained by solving the Schrodinger equation in the particle, so that the energy spectrum depends on the specific shape of the particle. Accounting for the boundary conditions leads to a modified energy spectrum and therefore a modified polarizability (Hache et al., 1986). Both approaches lead to a polarizability that is consistent with a damping term described by Eq. (14.16). Note that this quantum model neglects electronic collisions, so that it is valid, provided that the particle is smaller than the electronic mean free path, which is typically on the order of tens of nanometres for metals at an ambient temperature.

14.3 Controlling particle modes

This section highlights different strategies to tune the resonant frequency of a plasmonic resonator.

14.3.1 Influence of the dielectric environment

The first technique consists in modifying the dielectric material around the plasmonic resonator. Increasing the host permittivity $\varepsilon_h > 1$ results in a red-shift. The reason is quite simple. By introducing a dielectric, the surface charge is reduced by a factor ε_h. Hence, the electric field acting on the electrons inside the particle is also reduced, resulting in a lower resonance frequency. This sensitivity to changes of the external permittivity is mostly exploited for sensing applications (see Chapter 20).

14.3.2 Influence of the shape

An alternative route to modify the resonant frequency is a modification of the shape of the particle. Here, we consider three examples.

(1) *Ellipsoid.* A simple model with an analytical solution is given by an ellipsoid with the equation

$$\frac{x^2}{a_1^2} + \frac{y^2}{a_2^2} + \frac{z^2}{a_3^2} = 1.$$

It can be shown (Bohren and Huffman, 1983) that the electrostatic polarizability can be cast in the form

$$\alpha_i = \frac{4\pi}{3} a_1 a_2 a_3 \frac{\varepsilon_p - \varepsilon_h}{\varepsilon_h + L_i(\varepsilon_p - \varepsilon_h)}, \tag{14.17}$$

with L_i given by

$$L_i = \frac{a_1 a_2 a_3}{2} \int_0^\infty \frac{\mathrm{d}q}{(q + a_i^2)\sqrt{(q + a_1^2)(q + a_2^2)(q + a_3^2)}}, \tag{14.18}$$

so that $L_1 + L_2 + L_3 = 1$.

It is possible to derive a closed-form expression for spheroids which are ellipsoids with two identical axes. For cigar-shaped spheroids with $a_2 = a_3$, $L_2 = L_3 = (1 - L_1)/2$, and the geometrical factor L_1 is given as a function of the eccentricity e:

$$L_1 = \frac{1 - e^2}{e^2}\left(-1 + \frac{1}{2e}\ln\frac{1 + e}{1 - e}\right); \; e^2 = 1 - \frac{a_2^2}{a_1^2}. \tag{14.19}$$

For oblate (pancake-shaped) spheroids, $(a_1 = a_2)$, $L_1 = L_2 = (1 - L_3)/2$, and the geometrical factor L_1 is given as a function of the eccentricity e:

$$L_1 = \frac{g(e)}{2e^2}\left(\frac{\pi}{2} - \tan^{-1} g(e)\right) - \frac{g^2(e)}{2}; \; g(e) = \left(\frac{1 - e^2}{e^2}\right)^{1/2}; \; e^2 = 1 - \frac{a_3^2}{a_1^2}. \tag{14.20}$$

The resonance takes place for $\varepsilon_p + \varepsilon_h L_i/(1 + L_i) = 0$, so that modifying the eccentricity allows the resonance frequency to be modified.

(2) *Nanorod.* Another class of nanoparticles is the nanorod. It can be viewed as a finite-length wire supporting cylindrical surface plasmons propagating along the axis. At first glance, this system is equivalent to a metallic line used for radiowaves. As discussed in section 13.5.7, the field can be confined close to the wire because the electric energy at this stage is confined not in the magnetic field but in the kinetic energy of the electrons. The dispersion relation of the modes along a metallic wire differs significantly in optics from what is obtained in radiowaves. A rigorous treatment based on Maxwell equations is given in ref. (Novotny, 2007*b*), showing that the effective wavelength of the mode can be cast in the form

$$\lambda_{\text{eff}} = n_1 + n_2 \frac{\lambda}{\lambda_p}, \tag{14.21}$$

where λ_p is the plasma wavelength $2\pi c/\omega_p$ and n_1, n_2 are coefficients with dimensions of length dependent on the antenna geometry and static dielectric properties. When the nanorod radius a satisfies $a \ll \lambda_p$, the coefficient n_2 is proportional to a. This form is valid, provided that the metal can be described by a Drude model. In order to get some physical insight, we introduce a very simple model in Exercise 14.3. The key idea is that the skin depth for a Drude model is given by λ_p, so that the field penetrates in the nanorod when the radius a is smaller than λ_p. For such a system, the dominant contribution to the inductance becomes the kinetic inductance, i.e. the inertia of the electrons.

A nanorod with finite length L can be viewed as a Fabry–Perot cavity. The modes propagating along the rod are reflected at the extremities with a reflectivity denoted by $|r|e^{i\phi}$. The resonance condition is thus given by $4\pi L/\lambda_{\text{eff}} + 2\phi = p2\pi$, where p is an integer. It turns out that this condition can still be used as L decreases, even when the nanorod becomes a nanosphere. *Hence, the Fabry–Perot picture is still valid for simple particles.*

(3) *Core–shell structure.* An interesting structure that can be synthesized chemically consists in a gold shell of thickness t grown on a core dielectric sphere with permittivity ε_c and placed in a host dielectric with permittivity ε_h (Prodan et al., 2003). Such a system can be viewed as the assembly of a sphere and a cavity, both of which can sustain plasmon resonances. The core–shell structure sustains plasmon modes resulting from the hybridization of these modes. The coupling strength depends on the thickness of the shell. Indeed, if the thickness is much larger than the skin depth, the two modes cannot interact. Controlling the radii and the thickness thus offers degrees of freedom to control the resonances.

14.3.3 Experimental investigation of nanoparticle modes using electron energy-loss spectroscopy

Imaging the field structures of nanoparticles with sizes on the order of 50 nm is beyond the capability of scanning near-field optical microscopy. It is possible to map the field distribution of the plasmonic modes of a nanoparticle using electron energy-loss spectroscopy (EELS). The technique consists in measuring the energy loss of a high-energy electron (typically 100 keV) which passes near or through a metallic nanoparticle. Upon interaction with a plasmonic mode with frequency ω_{sp}, the electron loses an energy equal to $\hbar\omega_{sp}$. A map of the mode can be performed by focusing the electron beam down to a nanometre-scale size and scanning the beam over the particle (see Fig. 14.4).

As the energy transferred by the electron is given by $\mathbf{j} \cdot \mathbf{E}$ and given that the current density associated with an electron with velocity $v\mathbf{e}_z$ is $\mathbf{j} = -ev\mathbf{e}_z\delta(x)\delta(y)\delta(z - vt)$, it is seen that the energy loss is sensitive to the z-component of the electric field. More precisely, it can be shown (Garcia de Abajo and Kociak, 2008) that the energy loss is proportional to $\text{Im}[G_{zz}(x, y, k_z = \omega/v; x, y, k'_z = -\omega/v, \omega)]$. This quantity is the Fourier transform along z, evaluated at $k_z = \omega/v$ of the local density of states (LDOS) at frequency ω and position (x, y). Indeed, as the electron travels along the z-axis at a constant speed, it loses the spatial resolution along this axis. This technique benefits from the nanometre-scale resolution of electron optics to produce a spatial map of the plasmonic mode. By contrast, energy resolution of the LDOS is limited by the resolution of electron energy measurement, which is usually on the order of 0.1 eV and down to slightly less than 10 meV on most advanced setups.

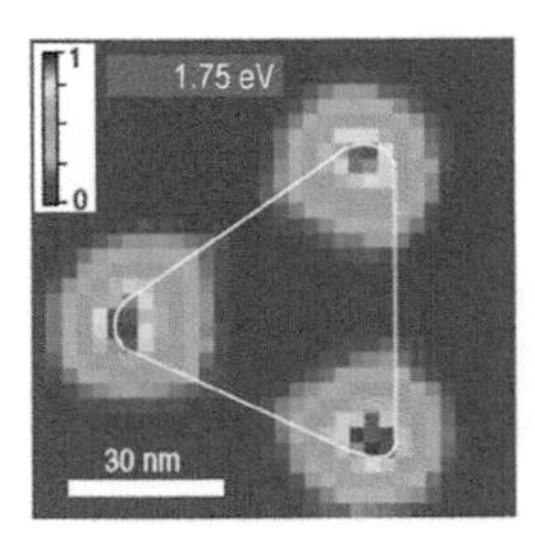

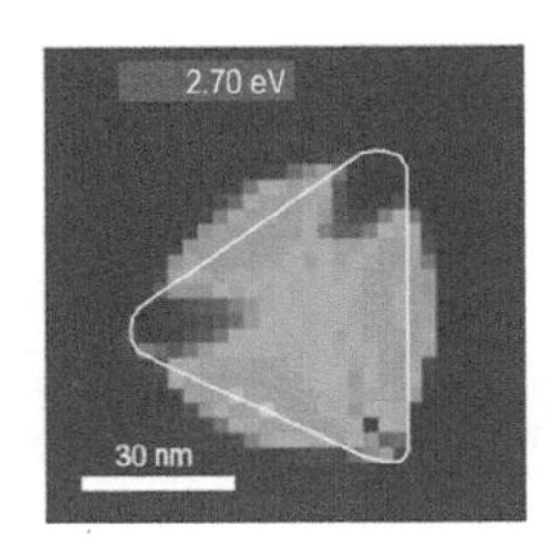

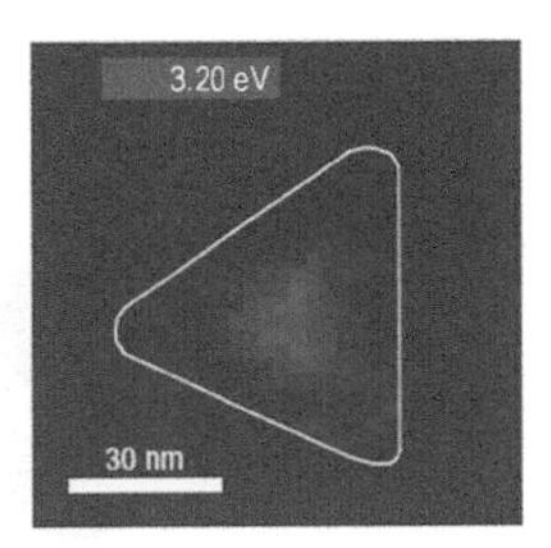

Fig. 14.4 Mapping the plasmonic modes of silver nanoparticles. The three figures represent the electron energy-loss spectroscopy signal obtained at three different frequencies (photon energy given in eV) for a silver nanoprism with a side of 78 nm. Reprinted by permission from Springer Nature: ref. (Nelayah et al., 2007) ©2007.

14.4 Field enhancement

In the introduction, we pointed out that localized resonances can enhance the electromagnetic field in tiny volumes. In this section, we use simple arguments to derive an explicit expression of the enhancement factor, pointing out two different mechanisms: field confinement in a subwavelength resonator and the lightning rod effect.

14.4.1 Nanocavity

In this section, we consider a localized resonator. For the sake of clarity, we consider the example of a plasmonic cavity also called the metal–insulator–metal (MIM) cavity, as depicted Fig. 14.5, but the exact type of resonator is not used in the argument. We only assume that there is a resonant electromagnetic mode. In order to estimate the enhancement factor, we write that in the stationary regime, the power transferred from the illuminating beam to the particle (extinction power) is equal to the power lost by absorption or radiation. By definition of the extinction cross section $\sigma_{ext}(\theta,\phi)$ (or the effective area in antenna terms), which is a function of the incident direction, the extinction power is

$$P_{ext} = \sigma_{ext}(\theta,\phi)\frac{\varepsilon_0\, c\,|E_{inc}|^2}{2}. \tag{14.22}$$

The losses of the resonator mode can be cast in the form κU, where κ is the decay rate due to radiative and non-radiative losses and U is the energy which can be cast in the form $\frac{\varepsilon_0|\mathbf{E}|^2}{2}V$, where V is the mode volume and $\mathbf{E}$ is the amplitude of the field in the cavity.[1] Introducing the quality factor $Q=\omega/\kappa$, the balance in the stationary regime yields:

$$P_{ext} = P_{abs} + P_{sc} = \frac{\omega}{Q}\frac{\varepsilon_0|\mathbf{E}|^2}{2}V. \tag{14.23}$$

It follows that the enhancement factor is given by

$$|K|^2 = \frac{|\mathbf{E}|^2}{|\mathbf{E}_{inc}|^2} = \frac{\sigma_{ext}(\theta,\phi)\, c\, Q}{\omega\, V}. \tag{14.24}$$

[1] The definition of the mode volume is discussed in section 7.5.2.

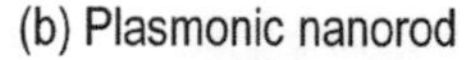

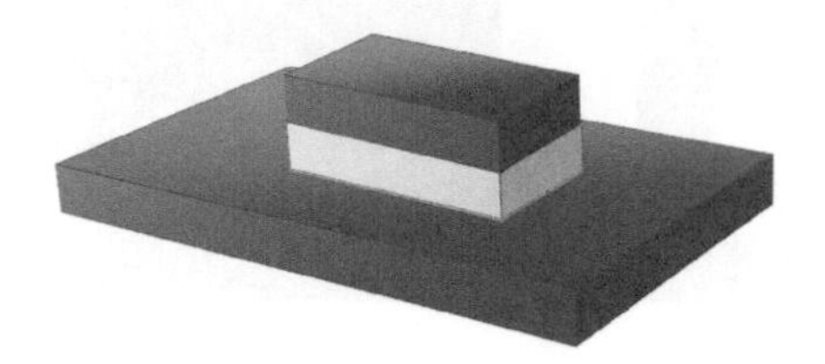

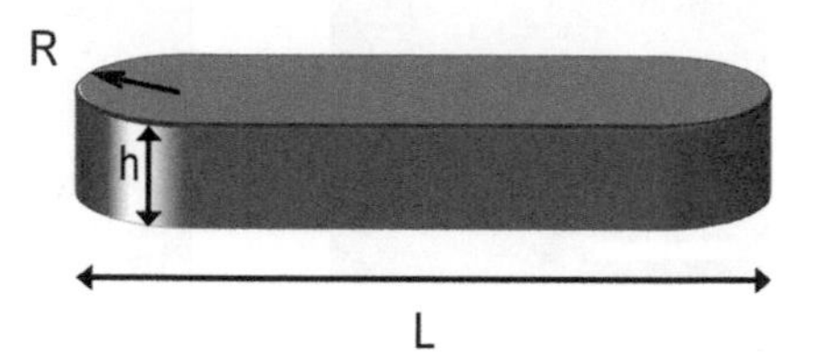

Fig. 14.5 Field enhancement at the nanoscale. (a) A Metal-insulator-metal plasmonic cavity can be resonantly excited. The electric field confined in the gap is enhanced. (b) A nanorod enhances the field due to the plasmonic dipolar resonance and also due to the lightning rod effect at the tip.

We now replace the extinction cross section by its expression in terms of the gain given by Eq. (2.60), which characterizes the directivity of the resonator

$$|K|^2 = \left[\frac{3}{4\pi^2}Q\frac{\lambda^3}{V}\right]\frac{G(\theta,\phi)}{6} = F_p\frac{G(\theta,\phi)}{6}. \tag{14.25}$$

We see that the enhancement factor provided by the resonator is given by the product of the Purcell factor (see section 19.3.4) and the resonator gain (see section 2.4.2), which is a function of the angle of emission.

Let us now discuss the physical meaning of these two terms. We assume that the resonator sustains a single resonant mode. The Purcell factor characterizes the enhancement of the LDOS; it increases when the field is confined in a subwavelength volume V. As for any resonator, the mode field is enhanced if there is a large quality factor or, in other words, if losses are small. In particular, radiative losses should be suppressed.

However, if radiative losses are suppressed in all directions, owing to reciprocity, the mode cannot be excited by an incident plane wave. Hence, to optimize the field enhancement, radiative losses should be suppressed in all directions except one, in order to permit efficient excitation of the system from that direction. The gain, which is proportional to the normalized emission angular pattern, quantitatively measures the resonator's ability to couple efficiently to one direction and suppress coupling to other directions.

14.4.2 Lightning rod effect

An alternative mechanism to enhance the field locally is the lightning rod effect. This is not a plasmonic effect but is often associated with a plasmonic resonance to produce a field enhancement. In this paragraph, we derive an approximate formula for the enhancement at the apex of a nanorod sustaining a plasmonic resonance. Our derivation is valid for any dipolar resonance, assuming that all the charge is concentrated at some apex of the structure. In electrostatics, a large electric field can be produced at the apex of a tip due to the large local charge density. Owing to the Coulomb theorem, the normal component of the electric field on a conductor is given by σ/ε_0, where σ is the surface charge density. This surface charge density can be roughly estimated as $Q/2\pi R^2$, where Q is the charge and R is the radius of curvature of the apex. If we consider, e.g., a nanowire with length L with radius of curvature at the extremities, the field can be estimated as follows. On resonance and at critical coupling, the nanowire is a dipole and has a polarizability given by $\alpha = i3\pi/k_0^3$, as shown in Complement 14.A.1. It follows that the dipole moment is given by $p = i3\pi\varepsilon_0 E_{inc}/k_0^3$. As the dipole moment can be cast in the form QL, we can estimate the charge Q. The external field is then given by $Q/(2\pi\varepsilon_0 R^2)$. Finally, we obtain:

$$|K| = \frac{|\mathbf{E}|}{|\mathbf{E}_{inc}|} = \frac{3}{k_0^3 R^2 L}. \tag{14.26}$$

This reasoning can be extended to other geometries such as Fig. 14.4 or Fig. 14.5b. For instance, a triangle such as the one in Fig. 14.4 sustains a dipolar mode. If the field is along the midline of the triangle, all the charge is confined in one apex of the triangle with radius of curvature R and thickness h. The charge density is then $\sigma \approx Q/\pi h R$ and the enhancement factor is roughly given by:

$$|K| = \frac{|\mathbf{E}|}{|\mathbf{E}_{inc}|} = \frac{3}{k_0^3 R h L}. \tag{14.27}$$

This concept of field enhancement due to a combination of geometrical effects and plasmonic resonance has been put forward by Li, Stockman, and Bergman, who proposed to use a chain of nanospheres with decreasing radius to realize a system that can efficiently confine the energy (Li et al., 2003). This system was coined a 'nanolens' because it can focus light to the nanoscale. Note, however, that the chain of nanoparticles is a non-imaging system, so that it would be more appropriate to use the term nanoconcentrator.

14.5 Ultrafast dynamics of nanoparticles

An important feature of localized plasmons is their ability to manipulate light on a femtosecond scale. In order to understand the dynamics of localized plasmons, it is important to review the different processes that can take place. To study these processes, a convenient experiment consists in using a pump-probe technique (see ref. (Voisin et al., 2001) for a review). A femtosecond laser with a typical pulse width on the order of 10–100 fs is used to illuminate a thin metallic film or a nanoscale particle. The pulse excites a plasmon. This plasmonic excitation decays by exciting an electron to a state above the Fermi level. This is commonly called a hot electron. This plasmon–electron energy transfer process occurs on a time scale on the order of 10 fs. After the pulse, the distribution of electrons contains hot electrons and is no longer given by a Fermi–Dirac distribution. It is called 'athermal distribution'. This electronic distribution can be probed using a second femtosecond laser and measuring the reflectivity or extinction as a function of the time delay between the two pulses. By following the modification of the optical properties as a function of time, it is possible to observe the relaxation dynamics of the hot electrons in the conduction band.

The experiments show that the excited electrons thermalize over a typical time on the order of 300–500 fs. Thermalization means that the distribution returns to a Fermi–Dirac distribution but with a much larger temperature called the electronic temperature. This thermalization is due to a fast energy exchange process between electrons due to Coulomb electron–electron interaction. This first relaxation step is followed by a second relaxation process whereby the electrons transfer their energy to the phonons of the lattice. This electron–phonon decay time is on the order of 1 ps (Voisin et al., 2001), so that the electrons return to the lattice temperature after a few picoseconds. Finally, the particle lattice releases its energy in the environment through conduction, convection, and radiation.

Hence, illuminating a plasmonic particle allows hot electrons to be generated. If the particle is electrically connected to a semiconductor (e.g. a nanoparticle embedded in a semiconductor or a metallic particle deposited on a semiconductor), the electrons can be transferred to the semiconductor within this picosecond before they relax by electron–phonon coupling. This technique to generate hot electrons paves the way to light-detection strategies or photo-induced chemical reactions.

14.A Complement: Universal bounds of the absorption and scattering cross section of lossy and non-lossy dipolar scatterers

14.A.1 Resonant absorption, critical coupling

In this complement, we show that the absorption cross section of a lossy dipolar particle has an upper bound which is universal and given by $3\lambda^2/8\pi$. We also show that this value is attained when the radiative losses are equal to the absorption losses, a condition called critical coupling (see section 7.7).

The existence of a maximum is a consequence of the existence of a trade-off between two competing effects: absorption losses that vary with the volume and radiative losses which vary with the square of the volume. On one hand, absorption increases if the volume increases. On the other hand, radiative losses, which are negligible for small particles, increase even faster, so that the fraction of energy which is absorbed decays for large volumes. This trade-off cannot be addressed when using the electrostatic approximation of the polarizability, which ignores radiative loss. To proceed, we use the notation $\alpha = \alpha' + i\alpha''$ in Eq. (14.14):

$$\sigma_{abs} = k_0\alpha'' - \frac{k_0^4}{6\pi}(\alpha'^2 + \alpha''^2). \tag{14.28}$$

Note that we have made no assumption regarding the form of the polarizability, so that this derivation is valid for any dipolar emitter, be it a plasmonic nanoparticle or a small radio antenna. To maximize the absorption, we need to choose $\alpha' = 0$. It is then seen that $\partial\sigma_{abs}/\partial\alpha'' = 0$ for $\alpha'' = 3\pi/k_0^3$. This extremum corresponds to the trade-off between absorption and radiation called critical coupling. In this regime, we have

$$\sigma_{ext} = \frac{3\lambda^2}{4\pi}, \ \sigma_{sc} = \frac{3\lambda^2}{8\pi}, \tag{14.29}$$

so that

$$\sigma_{abs} = \sigma_{ext} - \sigma_{sc} = \frac{3\lambda^2}{8\pi}. \tag{14.30}$$

Let us repeat that this derivation only accounts for the electric-dipole contribution. An actual particle may have other multipolar contributions to the absorption, as seen in Fig. 14.3.

14.A.2 Upper limit of the scattering cross section of a non-lossy dipolar particle

It is known in atomic physics that the maximum extinction cross section of a two-level atom on resonance is given by $3\lambda^2/2\pi$, a result which is twice as large as the extinction cross section that we found in the previous section for a dipolar particle. This result may be surprising at first glance, as one would expect, naïvely, that a resonant particle scatters more than a single atom. This naïve expectation is indeed correct. A particle scatters more, but in the sense that *the resonance spectrum is much broader*, so that it scatters more *if it is illuminated by white light* (see Complement 14.B for a discussion of the physical origin of this broadening, which can be associated with a classical superradiance effect). If illuminated by a laser tuned on resonance, an atom scatters more than a metallic nanoparticle! This is due to the fact that there are no internal losses in an atom, so that the amplitude of the dipole moment on resonance is larger and it radiates more.

Let us now deal with a non-lossy particle. Can we recover the extinction cross section of a resonant atom[2]? We now seek the maximum value of the extinction cross section $\sigma_{ext} = k_0 \mathrm{Im}(\alpha)$. To proceed, we use the form of the polarizability that accounts for radiative losses given in Eq. (12.38). A simple calculation shows that the extinction cross section is given by:

$$\sigma_{ext} = \frac{6\pi}{k_0^2} \frac{\left(\frac{\alpha' k_0^3}{6\pi}\right)^2}{1 + \left(\frac{\alpha' k_0^3}{6\pi}\right)^2}. \tag{14.31}$$

This function of α' has an upper bound given by $6\pi/k_0^2 = 3\lambda^2/2\pi$. Let us emphasize that the existence of a maximum is a consequence of an initial assumption: that we are discussing the contribution to the scattering cross section of the dipolar mode. However, as the size of a particle increases, the dipolar approximation breaks down and multipolar terms have to be included, yielding a much larger scattering cross section, as seen in Fig. 14.3.

[2] Note also that the scattering cross section is often called the absorption cross section in the context of atomic physics. When dealing with condensed-matter particles, absorption is only used to characterize energy transfer to other degrees of freedom, resulting finally in heat generation. In the case of an atom, the energy is not transformed into heat. In atomic physics, the word 'absorption' is used to indicate that a photon energy has been used to promote an electron to a higher energy state. The only relaxation pathway is radiation.

14.B Complement: Radiative decay time, superradiance, and the Chu–Wheeler limit

To reconcile the idea that a resonant nanoparticle scatters more than a single atom despite the fact that a single atom has a larger monochromatic extinction cross section, we have pointed out that the spectral width is much larger, so that it scatters more than an atom when illuminated by white light. Here, we aim to explore the physical mechanism responsible for this spectral broadening. We will show that it is due to the constructive interference of the field radiated by all the electrons of the particle. This effect is known as classical superradiance. To proceed, we use a Drude model, $\varepsilon = 1-\omega_p^2/(\omega^2+i\nu\omega)$, where ν accounts for the ohmic losses. Inserting this permittivity into Eq. (12.35), we find the electrostatic polarizability:

$$\alpha_{es} = -\frac{4\pi a^3}{3}\frac{\omega_p^2}{\omega^2 - \omega_p^2/3 + i\omega\nu}. \tag{14.32}$$

Inserting this expression into Eq. (12.38) to account for radiative losses and expanding the expression close to the resonance, we obtain a Lorentzian form:

$$\alpha = -2\pi a^3 \frac{\omega_p/\sqrt{3}}{\omega - \omega_p/\sqrt{3} + i(\nu + N\nu_{cl})}, \tag{14.33}$$

where $\nu_{cl} = e^2\omega^2/(6\pi\varepsilon_0 mc^3)$ is the classical radiation decay rate for an electron and $N = n4\pi a^3/3$ is the number of electrons in the particle. This Lorentzian form shows that the decay rate of the plasmon is given by $\nu + N\nu_{cl}$. The dependence of the decay rate with N is a clear manifestation of the collective radiation of the electrons. The decay rate increases linearly with the number N of electrons, i.e. with the particle size. Given that the number of electrons is roughly 1000 per nm^3, it is seen that a particle with a radius of 10 nm is expected to radiate a million times faster than a single electron. This is indeed what is observed. A typical decay time is 1 ns for an atom and 1 fs for a localized plasmon in a particle where the decay rate is on the order of the ohmic decay rate. It is seen that as the size increases, the radiative width $N\nu_{cl}$ may dominate the natural width ν due to metallic losses.

The linear dependence of the decay rate with N is often called classical superradiance. It can be understood using a simple argument, assuming that each electron is a classical dipolar antenna. With that picture, the constructive interference of the fields emitted by each electron produces an emitted field enhanced by a factor N, so that the emitted power increases by a factor N^2. The energy initially stored in the system is proportional to the number of excited electrons N, so that the decay rate varies as N (see also Chapter 19).

Finally, we note that a decay rate proportional to the number of electrons $N = n4\pi a^3/3)$ is also proportional to the particle volume. Hence, we find a Q-factor in agreement with the Chu–Wheeler limit, which was established in the context of small antennas (see Exercise (14.9)).

Exercises

Exercise 14.1 A simple model of the plasmon mode of a nanosphere. We aim to derive the dynamic equation describing the movement of a single electron in a metallic nanosphere containing n electrons per unit volume with mass m and charge $-e$. We neglect ohmic loss. We further assume that all electrons move together along the x-axis with a displacement denoted by $x(t)$. This displacement produces a polarization $P_x = -nex$ that generates an electric field, $E_x = -P_x/3\varepsilon_0$. We have assumed that retardation effects are negligible (electrostatic approximation) because the size of the nanosphere is supposed to be much smaller than the wavelength in vacuum corresponding to the oscillation frequency, $\lambda = 2\pi c/\omega$.

(1) Prove that $\ddot{x} + \frac{ne^2}{3m\varepsilon_0}x = 0$.

(2) By multiplying the dynamic equation by $\dot{x}$, show that the energy per unit volume can be cast in the form

$$U = \frac{1}{2}n\,m\,\dot{x}^2 + \frac{n^2e^2x^2}{6\,\varepsilon_0}. \tag{14.34}$$

(3) The electrostatic energy per unit volume of a uniformly polarized sphere is given by $\frac{1}{2}PE$. Identify the two terms of the energy and show that the plasmon mode corresponds to a periodic oscillation of the energy between kinetic energy and electrostatic energy of the electrons.

Exercise 14.2 An equivalent LC model of the plasmonic nanosphere resonator. We consider the metallic nanosphere with radius a described in Exercise 14.1 and aim to introduce an equivalent capacitance, C, and an equivalent kinetic inductance, L. Show that the intensity flowing through the plane perpendicular to the electric field passing through the centre of the nanosphere is given by $I = -n\,e\,\dot{x}\,\pi\,a^2$. We now use Eq. (14.34) and identify the kinetic energy with $LI^2/2$ and the electrostatic energy with $Q^2/2C$. By writing the kinetic energy of all the electrons of the sphere in the form $LI^2/2$, show that $L = 4m/(3ne^2\pi a)$. The inductance L is called kinetic inductance, as it is due to kinetic energy rather than magnetic energy. Using $I = \dot{Q}$ and identifying the electrostatic energy with $Q^2/2C$, show that $C = 9\pi\varepsilon_0 a/4$. Check that $LC\omega_0^2 = 1$ with $\omega_0^2 = \omega_p^2/3$. Compare with the solution in the case of a non-lossy metal, Eq. 14.5.

Exercise 14.3 Simple model of propagation along a metallic nanowire. Propagation along a wire results from an exchange of energy between electric energy and magnetic energy. A simple model of electrical line can be derived by computing the capacitance Γ and inductance Λ per unit length. In this exercise, we will show that for a nanorod, the mass of the electrons gives a leading contribution to the inductance. This is the origin of the differences between the radiowave regime and the optical regime. We consider, for the sake of simplicity, a coaxial wire with a core wire of radius a_1 and an external wire with radius a_2. The space between the two wires is a dielectric with permittivity ε_h. It can be shown that:

$$\Gamma = \frac{2\pi\varepsilon_0\varepsilon_h}{\ln(\frac{a_2}{a_1})};\quad \Lambda = \frac{\mu_0}{2\pi}\ln\frac{a_2}{a_1}. \tag{14.35}$$

We now compare the kinetic energy with the magnetic energy per unit length $\Lambda I^2/2$. We assume that the nanowire is small enough that the current density can be considered to be uniform in the wire. Show that the kinetic energy K per unit length of the electrons is given by:

$$K = \frac{1}{2}\Lambda' I^2, \text{where } \Lambda' = \frac{\mu_0}{2\pi}\frac{\lambda_p^2}{2\pi a_1^2}. \tag{14.36}$$

(Hint: $I = j\pi a_1^2 = (nev)\pi a_1^2$.) Compare the kinetic and magnetic contributions. Using the formula $n_{\text{eff}} = c\sqrt{\Gamma\Lambda'}$, derive the effective wavelength $\lambda_{\text{eff}} = \lambda/n_{\text{eff}}$ in the limit $a_1 \ll \lambda_p$:

$$\lambda_{\text{eff}} \approx \frac{\lambda}{\lambda_p}\frac{a_1}{n_h}\sqrt{2\pi \ln(a_2/a_1)}. \tag{14.37}$$

Exercise 14.4 Polarizability of a nanosphere. We consider a nanosphere with radius a smaller than 30 nm, illuminated by a monochromatic plane wave with a wavelength in vacuum of $\lambda_0 = 633$ nm. Explain why an electrostatic approximation can be used to derive the electromagnetic field in the particle. Within this approximation, we introduce the scalar potential $V(\mathbf{r})$. The nanoparticle is described by an isotropic dielectric relative permittivity ε_m and the surrounding medium by ε_h. The particle is in a uniform electric field $\mathbf{E}_{ext}$ oriented along the axis Oz.

Write the Poisson equation satisfied by $V(\mathbf{r})$ and the boundary conditions.

Using the spherical coordinates (r, θ, ϕ), where θ is the angle formed by axis Oz and vector $\mathbf{r}$, we seek a solution $V(\mathbf{r})$ in the form

$$r < a : V_m = C_m z = C_m r\cos\theta,$$

$$r > a : V_h = -E_{ext} r\cos\theta + C_h\frac{\cos\theta}{r^2}.$$

Prove that $C_h = a^3\frac{\varepsilon_m - \varepsilon_h}{\varepsilon_m + 2\varepsilon_h}E_{ext}$ and show that the field $\mathbf{E}_{int}$ inside the particle is given by:

$$\mathbf{E}_{int} = \frac{3\varepsilon_h}{\varepsilon_m + 2\varepsilon_h}\mathbf{E}_{ext}.$$

The potential $C_h\cos\theta/r^2$ has the structure of an electrostatic dipolar potential $\frac{p\cos\theta}{4\pi\varepsilon_0 r^2}$. Show that the polarizability α defined by $\mathbf{p} = \alpha\varepsilon_0\varepsilon_h\mathbf{E}_{ext}$ is given by:

$$\alpha = 4\pi a^3\frac{\varepsilon_m - \varepsilon_h}{\varepsilon_m + 2\varepsilon_h}.$$

Exercise 14.5 Scattering cross section and absorption cross section of a Rayleigh scatterer. The time-averaged power radiated by a particle with dipole moment $\mathbf{p}$ is given by $P_{\text{rad}} = \frac{1}{4\pi\varepsilon_0}\frac{\omega^4|\mathbf{p}|^2}{3c^3}$.

Derive the scattered power P_{sc} and show that the scattering cross section of a Rayleigh scatterer with polarizability α is $\sigma_{sc} = \frac{k_0^4|\alpha|^2}{6\pi}$.

Using the form of the extinction cross section in Eq. (12.108) and the energy conservation, derive the absorption cross section $\sigma_{abs} = k_0\text{Im}(\alpha) - \frac{k_0^4|\alpha|^2}{6\pi}$.

Exercise 14.6 Absorption cross section. In this exercise, we consider a direct calculation of the absorption cross section σ_{abs} of a nanosphere with permittivity ε_p in vacuum. We compute the absorbed power P_{abs} and normalize it by the incident Poynting vector of a plane wave $\varepsilon_0 c|E_{\rm inc}|^2/2$, $\sigma_{\rm abs} = \frac{2P_{\rm abs}}{\varepsilon_0 c|E_{\rm inc}|^2}$.

Using Eq. (3.12) and the form of the field inside the particle given by Eq. (14.2), show that $\sigma_{\rm abs} = 12\pi a^3 \frac{\omega}{c} \frac{{\rm Im}(\varepsilon_p)}{|\varepsilon_p+2|^2}$.

Check that this result can be cast in the form $\sigma_{\rm abs} = \frac{\omega}{c}{\rm Im}(\alpha_{es})$. This calculation is based on an approximate form of the polarizability derived in the electrostatic limit. Compare with the result derived in the previous exercise and show that the two results are in agreement in the limit of small lossy particles.

Exercise 14.7 The Lycurgus Cup. A Roman cup called the *Lycurgus Cup* can be seen at the British Museum in London. Its colour differs in transmission and in reflection: it is reddish in transmission and green in reflection. The colour is attributed to silver nanoparticles. The colour of parts of the stained glass at the Sainte-Chapelle in Paris is red in both reflection and transmission and is also attributed to the presence of silver nanoparticles. In both cases, the fact that colours have survived over centuries is due to the chemical stability of metallic nanoparticles embedded in glass. Explain these seemingly contradictory observations and make a guess on the size of the nanoparticles.

Exercise 14.8 Local heating using metallic nanoparticles. It is possible to deposit heat very locally in a medium by inserting metallic nanoparticles. When illuminating them on resonance, the energy absorbed is converted into heat and can be used to locally heat. A possible application is cancer treatment. By functionalizing metallic nanoparticles, they can be selectively attached to tumours. It is then possible to illuminate them with a wavelength chosen to be in the near-infrared, so that the body does not absorb much but the nanoparticle absorbs resonantly. Hence, the tumour temperature may increase by a few degrees. This can be enough to cause destruction of the tumour.

What is the energy absorbed by a nanoparticle with absorption cross section $\sigma_{\rm abs}$ illuminated by a local flux per unit area ϕ during a time Δt?

This energy is converted into heat. We consider that the particles are in a fluid. How many particles per unit volume are needed to increase the fluid temperature by 10 K? Give an expression as a function of $\sigma_{\rm abs}$, ϕ, and Δt and of the heat capacity per unit volume denoted by ρc_p, where ρ is the fluid mass per unit volume. We take $\Delta t = 200\,{\rm s}$, $\phi = 4\,{\rm W.cm^{-2}}$, $\sigma_{\rm abs} = 4\,10^{-14}\,{\rm m}^2$, and the water value $\rho\, c_p = 4.18\,10^6\,{\rm J/(K\,m^3)}$.

Exercise 14.9 Classical superradiance. Derive Eq. (14.33). The linear dependence on N of the radiative decay rate is a signature of the superradiance. Derive the Q-factor, assuming that the radiative losses are the leading mechanism (use $N = n4\pi a^3/3$). Compare with the Chu–Wheeler limit discussed in section 2.4.4.

Part V

Artificial media: Photonic crystals and metamaterials

15

Propagation in periodic media (I): Bloch modes and homogenization

The theory of electronic transport in periodic potentials has led to the concept of band structures in semiconductors. The solution of the Schrödinger equation requires wavefunctions called Bloch functions (after the Swiss physicist Felix Bloch, who established the quantum theory of solids during his thesis in 1928), which are the product of $\exp(i\mathbf{k}\cdot\mathbf{r})$ by periodic functions, $\mathbf{k}$ being the Bloch wavevector. Bands are obtained by computing the energy (or frequency) $\omega(\mathbf{k})$ and the eigenvectors (or quasinormal modes). In photonics, this solid-state approach is well described in ref. (Joannopoulos et al., 2008), for instance.

However, it provides only a partial description that does not address the fate of impinging beams. Electron beams are indeed rarely manipulated in solids at the relevant band energies of a few eV. For instance, the band structure does not explain how many periods are required to attain a near-unity reflection. These issues additionally require us to consider the dispersion diagram obtained by computing the real and complex Bloch wavevectors $k(\omega)$ for a range of frequencies. Thus a synthetic view of the properties of photonic structures relying at least locally on a periodic pattern, e.g. the periodic waveguides, photonic-crystal cavities, and metal hole arrays in chapters 17 and 18, involves considering both the $k(\omega)$ and $\omega(k)$ dispersion diagrams (Yariv and Yeh, 1984).

The eigenvectors obtained from the $k(\omega)$ and $\omega(k)$ calculations are identically referred to as Bloch modes in the literature. This is confusing. The eigenvectors obtained by calculating the frequency for fixed k's should be called quasinormal Bloch modes (like the quasinormal modes of cavities, see Chapter 7), since they represent the states in the sense of solid-state physics. For non-absorbing materials, $\omega(k)$ is real and modes have an infinite lifetime. However, for absorbing materials, $\omega(k)$ becomes complex and the quasinormal Bloch mode becomes exponentially damped in time while remaining pseudo-periodic and completely delocalized in space (k is real). Conversely, eigenvectors obtained by calculating k's for a fixed real frequency ω should be called normal Bloch modes (like the guided modes of waveguides presented in Chapter 6). For absorbing materials, $k(\omega)$ becomes complex and the normal Bloch mode becomes exponentially damped (or localized) in space while remaining completely periodic in time (ω is real).

Albeit inevitably twice as long, the dual view is adopted hereafter for the very important textbook example of 1D photonic crystals, a simple system that illustrates most of the physical features of the more complex 2D and 3D photonic-crystal systems. We will consider non-absorbing materials only, so that the quasinormal and normal

Introduction to Nanophotonics. Henri Benisty, Jean-Jacques Greffet, and Philippe Lalanne, Oxford University Press.

DOI: 10.1093/oso/9780198786139.003.0015

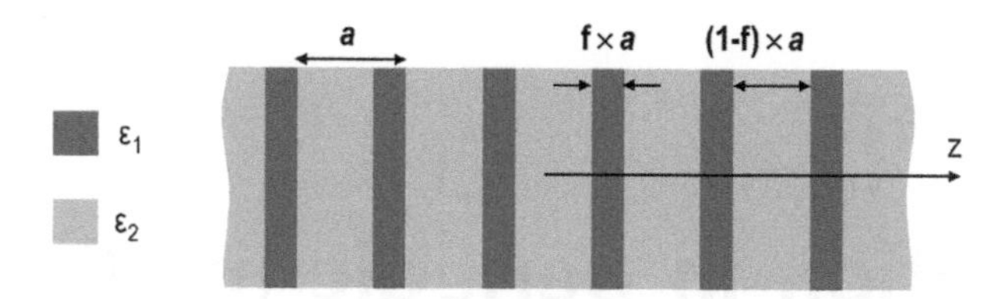

Fig. 15.1 Periodic thin-film stack composed of alternate uniform layers of relative permittivities ε_1 and ε_2 with period a. The parameter f such that the thickness of the ε_1 layers is $f \times a$ is defined as the fill factor. We assume that the materials are not magnetic: $\mu = \mu_0$ everywhere.

Bloch modes are identical in the bands but not in the gap. To calculate the dispersion diagrams and the corresponding Bloch modes, we rely on the classical plane-wave expansion method, essentially the Fourier decomposition of the field periodic part. For 1D systems, this method has many good competitors (Yariv and Yeh, 1984), but it is worth working it out, as it is almost systematically used for 2D and 3D systems. The study of the properties of the gap and the band edge are left for the next chapter, and we focus on the first allowed band and homogenization theory, as it leads to the concept of artificial materials, metasurfaces, and subwavelength optics in general.

15.1 Thin-film stack: Effective index

15.1.1 Pseudo-periodicity

Let us start with the simplest structure, a periodic stack of period a in the spatial direction z composed of thin uniform layers with relative permittivity ε_1 and ε_2 and respective thickness fa and $(1-f)a$ (see Fig. 15.1). We are looking for the electromagnetic modes (i.e. the solutions of the source-free Maxwell equations) that may propagate in the periodic stack. Because the structure is invariant in the x- and y-directions, the x- and y-dependence of the fields can be written $\exp(ik_x x + ik_y y)$. The propagation constants k_x and k_y are arbitrary *a priori*. Since the continuity conditions are satisfied at the interfaces for any x and y, note that k_x and k_y are the same in every layer: they are invariants. As such, we often say that the parallel wavevector components, k_x and k_y in the present case, are conserved. This is nothing other than Snell's law. For the sake of simplicity, hereafter, we take $k_x = k_y = 0$.

The solution is fully analytical and can even be solved with closed-form expressions (Yariv and Yeh, 1984), but the calculations are rather long, even for a simple bi-layer 1D case. The approach used hereafter is much less analytical, but will be applicable to the more general cases of structures that are periodic in two or three dimensions. It is referred to as the plane-wave expansion method in the literature. Because $k_x = k_y = 0$, the modes linearly polarized along the x-direction can be written

$$E_x(z,t) = \mathsf{E}(z)\exp(-i\omega t), \qquad (15.1)$$
$$H_y(z,t) = \mathsf{H}(z)\exp(-i\omega t).$$

The time dependence $\exp(-i\omega t)$ will be omitted in the following. In addition, the Bloch theorem (also called the Floquet theorem in the mathematics community), which is easily demonstrated by the physicist under the assumption that the solution is unique, stipulates that the solution is pseudo-periodic; the electromagnetic fields $\mathsf{E}(z)$ and $\mathsf{H}(z)$ can be written as the product of a periodic function by $\exp(ik_z z)$, with k_z a complex number in general. Expanding in Fourier series the periodic contribution, one obtains

$$\mathsf{H}(z) = \sum_m \mathsf{H}_m \exp(\jmath(k_z + mK)z), \tag{15.2}$$
$$\mathsf{E}(z) = \sum_m \mathsf{E}_m \exp(\jmath(k_z + mK)z),$$

where $K = 2\pi/a$ is the reciprocal-vector modulus. We define the effective index of the Bloch mode by

$$k_z = (\omega/c)n_{\text{eff}}, \tag{15.3}$$

c being the speed of light in vacuum (the analogy with the effective index of guided modes is mainly mathematical: both describe a phase behaviour along a specific axis through k). Because of the periodicity, k_z is defined modulo K and the definition of Eq. (15.3) raises a difficulty that will be discussed later through homogenization. In the absence of charges and current densities, the divergence equations, Eq. (1.2), among the Maxwell equations are satisfied, since the magnetic and electric fields are then curls, Eq. (1.3). Injecting Eq. (15.1) into the Maxwell curl equations, we have

$$-i\omega\mu_0\ \mathsf{H}(z) = \partial\mathsf{E}/\partial z, \tag{15.4}$$
$$-i\omega\varepsilon_0\varepsilon(z)\ \mathsf{E}(z) = \partial\mathsf{H}/\partial z, \tag{15.5}$$

and by introducing the Fourier series expansion of the relative permittivity

$$\varepsilon(z) = \sum_p \hat{\varepsilon}_p \exp(i\,pK\,z), \tag{15.6}$$

we obtain

$$-i\omega\mu_0 \sum_m \mathsf{H}_m \exp(i(k_z + mK)z) = \sum_m i(k_z + mK)\mathsf{E}_m \exp(i(k_z + mK)z), \tag{15.7}$$
$$-i\omega\varepsilon_0 \sum_{p,q} \hat{\varepsilon}_p \exp(ipKz)\mathsf{E}_q \exp(i(k_z + qK)z) =$$
$$\sum_m i(k_z + mK)\ \mathsf{H}_m \exp(i(k_z + mK)z). \tag{15.8}$$

Identifying each term of the Fourier series, and using the change of variable $p+q=m$ in Eq. (15.8), we obtain two sets of equations satisfied for any relative integer m:

$$\omega\mu_0\ \mathsf{H}_m = -(k_z + mK)\mathsf{E}_m, \tag{15.9}$$
$$\omega\varepsilon_0 \sum_q \hat{\varepsilon}_{m-q}\mathsf{E}_q = -(k_z + mK)\mathsf{H}_m. \tag{15.10}$$

15.1.2 Frequency or wavevector eigenvalue approaches

There are two important approaches to looking for a solution of the above system of equations. Within a first approach, the wave frequency ω is real-valued and fixed, and the dispersion $k_z(\omega)$ or $n_{\text{eff}}(\omega)$ is computed. We are calculating the 'guided' modes at frequency ω that can be excited by a source emitting at the same frequency. The modes are the analogue of the waveguide modes encountered in Chapter 6 (we could call them normal Bloch modes). Within the second approach, the frequencies ω are calculated for a given k_z. We then consider the natural resonance modes of the stack, i.e. the quasinormal modes with possibly complex frequencies, see the Chapter 7 on optical resonators. They are connected to the local density of electromagnetic states. In the present case, since there is no leakage, the computed spectrum shall be real if we neglect absorption by considering that ε is a real number.

Adopting the first approach, Eqs (15.9) and (15.10) lead to an eigenvalue problem that can be written in a matrix format (infinite matrices to be truncated later),

$$\begin{bmatrix} \mathbf{D}_K & \omega\mu_0\mathbf{I} \\ \varepsilon_0\omega\,\mathbf{T}_\varepsilon & \mathbf{D}_K \end{bmatrix} \begin{bmatrix} \mathsf{E} \\ \mathsf{H} \end{bmatrix} + k_z \begin{bmatrix} \mathsf{E} \\ \mathsf{H} \end{bmatrix} = \begin{bmatrix} 0 \\ 0 \end{bmatrix}, \tag{15.11}$$

where $\mathbf{I}$ is the identity matrix, $\mathbf{T}_\varepsilon$ is a Toeplitz matrix built from the Fourier coefficients of $\varepsilon(z)$, $(\mathbf{T}_\varepsilon)_{m,n} = \hat{\varepsilon}_{m-n}$, $\mathbf{D}_K$ is a diagonal matrix with successive spatial harmonics $(\mathbf{D}_K)_{m,n} = mK\,\delta_{m,n}$, and E and H are column vectors formed by the coefficients E_m and H_m. Eq. (15.11) is a dispersion relation which links the mode momentum k_z (or effective index n_{eff}) to the mode frequency ω. The k_z's are obtained as the eigenvalues and the Bloch modes are the associated eigenstates. Material dispersion can be trivially accounted for by the parametric dependence of $\mathbf{T}_\varepsilon$ on the wave frequency.

Alternatively, with the second approach, we may fix k_z and calculate ω. Because of the pseudo-periodicity (Floquet's theorem, dictating Eq. (15.2)), it is sufficient to fix k_z in the first Brillouin zone from $-K/2$ to $K/2$. With this approach classically used in solid-state physics to get band diagrams of crystalline solids (Chapter 8), the eigenvalue equation reads:

$$\begin{bmatrix} \mathbf{0} & \mathbf{T}_\varepsilon^{-1}(k_z\mathbf{I}+\mathbf{D}_K) \\ (k_z\mathbf{I}+\mathbf{D}_K) & \mathbf{0} \end{bmatrix} \begin{bmatrix} \mathsf{E} \\ \mathsf{H} \end{bmatrix} + \frac{\omega}{c} \begin{bmatrix} \mathsf{E} \\ \mathsf{H} \end{bmatrix} = \begin{bmatrix} 0 \\ 0 \end{bmatrix}. \tag{15.12}$$

In contrast with the first approach, taking into account material dispersion is not trivial, since $\mathbf{T}_\varepsilon$ depends on ω and Eq. (15.12) becomes a non-linear eigenvalue problem. A key property of the eigenvalue equation is that if $\varepsilon(z)$ is real (no Joule loss), the eigenvalues of the matrix $(k_z\mathbf{I}+\mathbf{D}_K)\mathbf{T}_\varepsilon^{-1}(k_z\mathbf{I}+\mathbf{K}_D)$ are real for real values of k_z. This can be seen by first considering that for real $\varepsilon(z)$, $\hat{\varepsilon}_p = (\hat{\varepsilon}_{-p})^*$ and thus $\mathbf{T} = \mathbf{T}_\varepsilon^\perp$, where $\perp$ denotes the transpose-conjugate operation. $\mathbf{T}_\varepsilon$ and, therefore, $\mathbf{T}_\varepsilon^{-1}$ are Hermitian matrices. Noting that $(k_z\mathbf{I}+\mathbf{D}_K)^\perp = (k_z\mathbf{I}+\mathbf{D}_K)$, it follows that $(k_z\mathbf{I}+\mathbf{D}_K)\mathbf{T}_\varepsilon^{-1}(k_z\mathbf{I}+\mathbf{D}_K)$ is Hermitian too. Thus ω is real if $\varepsilon(z)$ is real for this special 'closed' case without leakage.

15.1.3 Main features of dispersion relations: Bands and gaps

Illustrative results calculated with a computer are shown in Fig. 15.2. In such practical implementations, the matrices of infinite dimension are truncated but retain enough

Fourier harmonics mK to achieve a good accuracy. For the usual dielectric materials, it is enough to consider $|m| < M = 5$ ($2M+1 = 11$ Fourier harmonics are retained) for each vector **E** and **H**. Note that $2(2M+1) = 22$ eigenvalues are thus calculated, but they are all twice degenerate. As shown in ref. (Yariv and Yeh, 1984), the Bloch mode is composed of two counterpropagating local plane waves, and the total number of Bloch modes is only two at a given frequency: one with a positive propagation constant, k_z, and the other with an opposite constant, $-k_z$ (not shown in Fig. 15.2). Note that the Fourier expansion approach adopted here is not describing the Bloch modes as two sets of counterpropagating plane waves. In what follows, we shed more light on the properties and the physical meaning of the dispersion curves in Fig. 15.2.

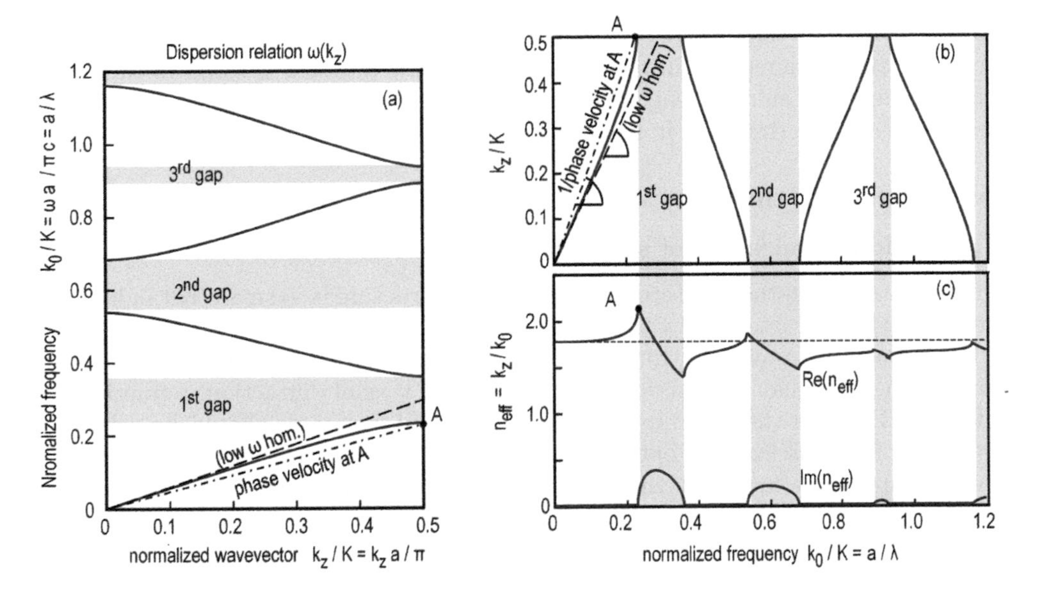

Fig. 15.2 Dispersion relations of Bloch modes in 1D periodic media. Eq. (15.12) is used for (a) and (b) and Eq. (15.11) is used for (c). The results are obtained for a typical stack $\sqrt{\varepsilon_1} = n_1 = 1.6$, $\sqrt{\varepsilon_2} = n_2 = 2.3$ and the thicknesses are given by the quarter-wave condition $\lambda_0/4n$, thus $f = n_2/(n_1+n_2) \simeq 0.59$ is the fill fraction of material 1. (a) Dispersion relation in the conventional 'solid-state' form $\omega = \omega(k_z)$, with the first few gaps as grey areas. (b) Same dispersion curve represented by exchanging the horizontal and vertical axes. (c) Spectrum of real and imaginary parts of effective index in 'guided wave' form $k_z(\omega)$. Note that the curves in (b) and (c) relate to the same underlying $k_z(\omega)$ (just the vertical axis changes), except in the gaps shown with grey bands. In either (a), (b), or (c), the dashed black or blue lines represent the dispersion of the bulk homogenized material with $n_{\text{eff}} = \sqrt{\langle\varepsilon\rangle}$; note also the point A at $k_z = \pi/a$ in the three plots and the dash-dotted line that is related to the phase velocity v_φ and effective index $n_{\text{eff}} = c/v_\varphi$ at the sole point A.

Two different zones appear:

- *Photonic bandgaps:*
  In specific frequency ranges, no mode is found in the $\omega(k_z)$ representation (a), or, equivalently, a complex propagation constant is obtained in the $k_z(\omega)$ representation (c), although the materials are lossless, $\varepsilon(z)$ being real.
- *Allowed photonic bands:*
  Outside the bandgaps, the periodic stack supports truly propagative modes with a real propagation constant (k_z is real). Let us emphasize that the wavevector k_z (or n_{eff}) strongly depends on ω, whereas the two constituent materials were assumed to be dispersionless.

In Chapter 16, we will consider bandgaps; we will show that the imaginary part of the propagation constant k_z is due to the evanescent character of the mode. In the following, we focus on the allowed bands, especially on the first band at small energies. We will introduce the 'homogenized' (effective) material in the subwavelength and deep-subwavelength regimes in relation to the effective index. Then we will turn to the homogenization of subwavelength gratings (i.e. photonic crystals with a finite depth) which play an important role in modern photonics, as discussed in Chapters 17 and 18.

15.2 Periodic thin-film stacks with small periods

15.2.1 Homogenization of subwavelength materials

In most optical materials, the atomic or molecular structure is so small that light propagation may be characterized by their refractive indices and the materials are considered as homogeneous. Conversely, for structures substantially larger than the wavelength, light propagation may be described by the classical laws of diffraction, refraction, and reflection. Between these two extremes, the Maxwell equations should be solved rigorously to fully disclose the optical properties of the structure. But still there is an intermediate region for which the structure is too fine to give rise to diffraction in the usual sense but too coarse for the medium to be strictly considered as homogeneous. In this region, of great interest in nanophotonics and often referred to as the *subwavelength domain*, simple homogenization techniques can most often give a good physical understanding of the medium properties, even if they are not strictly exact. Homogenization of electromagnetic waves in composite media in the subwavelength domain is an old but still very active topic.

Initially, various effective medium approaches were used to determine the dielectric constant of composite disordered materials: the Maxwell–Garnett or Bruggeman ones are the most popular (see Complement 15.A). When it comes to periodic structures with some underlying order, e.g. arrays of needles or simply the kind of layered stack considered above, those approaches turn out to be inadequate even in the static limit, i.e. when the period is infinitely smaller than the wavelength ($a/\lambda \to 0$). The theory of composite materials is made easier for periodic structures because of the Bloch theorem, dictating that in periodic systems, spatial frequencies are discretized. For a subwavelength period, only a single spatial frequency is a progressive wave, all the others being evanescent. The far field of the composite then becomes strictly governed by specular light. In contrast, in a disordered subwavelength composite, a continuous

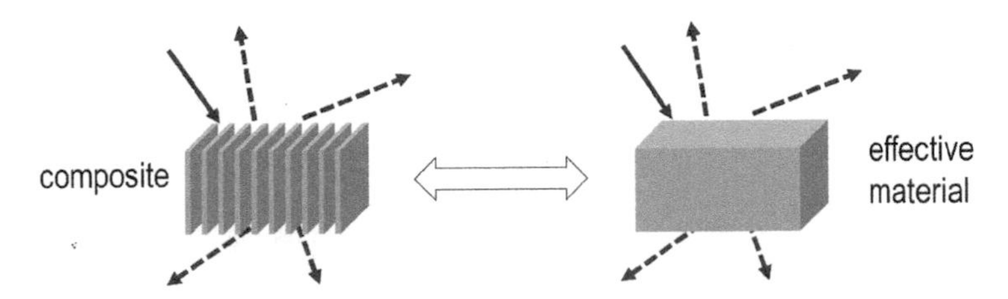

Fig. 15.3 The equivalence between a periodic composite and a homogeneous medium in the static limit implies that the field scattered by the finite-size composite on the left is strictly the same as that scattered by the uniform material with the same shape on the right.

spectrum of spatial frequencies is present (think of the Fourier transform of $\varepsilon(z)$, for instance), so that the diffuse light can only be ignored at deep-subwavelength scales in the static limit.

Finally it should be noted that the analysis that we will derive for periodic systems can be applied without change to a number of different physical situations, so that the formulas quoted hereafter for effective permittivities and permeabilities also apply to electrical conductivity, thermal conductivity, diffusivity, fluid permeability, and the shear matrix of antiplane elasticity (Milton, 2002).

15.2.2 Static limit of 1D periodic structures

The important result of this chapter, which is justified in section 15.2.3, is the following:

> In the static limit ($a/\lambda \to 0$), *a periodic structure is rigorously equivalent to a homogeneous material.* 1D periodic composites are equivalent to *uniaxial* materials, while 2D or 3D periodic composites are in general equivalent to *biaxial* materials.

The equivalence implies that if one considers a material with a periodicity a and an arbitrary closed-surface boundary (a parallelepiped in Fig. 15.3), and if it is illuminated by an incident wave with an infinitely large wavelength compared to the period ($a/\lambda \to 0$), the fields scattered outside the composite structure are exactly those that would be obtained if the composite was homogeneous, while the same statement for fields scattered inside holds if we take the field average over a unit cell. For 1D composites, simple closed-form expressions exist for the ordinary (n_o) and extraordinary (n_e) effective indices of the artificial material in the static limit

$$n_o = \langle \varepsilon(z) \rangle^{1/2} \text{ and } n_e = \langle 1/\varepsilon(z) \rangle^{-1/2}, \tag{15.13}$$

where $\varepsilon(z)$ could be either a continuous or discontinuous function of z and $\langle \ldots \rangle$ represents spatial averaging over one period. Thus a 1D composite, periodic in the z-direction and invariant in the two others, has an effective permittivity tensor in the canonical (x, y, z) basis whose diagonal terms are the harmonic and arithmetic averages of $\varepsilon(z)$

$$\varepsilon_{\text{eff}} = \begin{pmatrix} \langle \varepsilon \rangle & 0 & 0 \\ 0 & \langle \varepsilon \rangle & 0 \\ 0 & 0 & \langle \frac{1}{\varepsilon} \rangle^{-1} \end{pmatrix}. \tag{15.14}$$

For the lamellar two-material composite of Fig. 15.4 and the definition of filling factor f of material 1, we have the following indices: $n_o = (f\varepsilon_1 + (1-f)\varepsilon_2)^{1/2}$ and $n_e =$

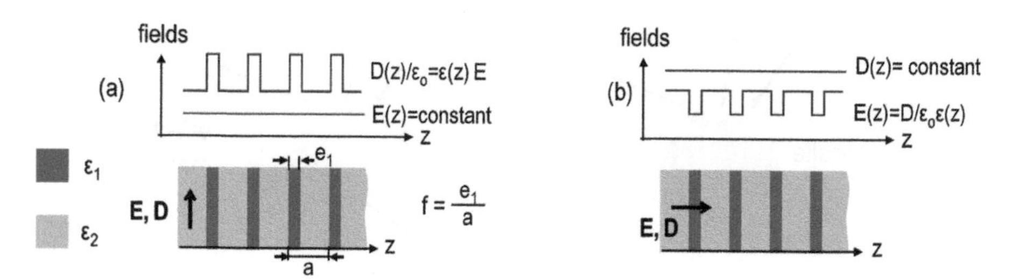

Fig. 15.4 Sketch of the dielectric and field profiles when $a/\lambda \to 0$ (a) for an electric field parallel to the interfaces, and (b) for fields normal to the interfaces. The filling factor f of material 1 is defined as indicated.

$(f/\varepsilon_1 + (1-f)/\varepsilon_2)^{-1/2}$. In general, by precisely controlling the fill factor, it is possible to synthesize artificial materials whose refractive indices continuously span the range $[(\varepsilon_1)^{1/2}, (\varepsilon_2)^{1/2}]$. Note in relation to Chapter 18 that $\langle\varepsilon\rangle$ may have a negative real part, whereas $\langle 1/\varepsilon\rangle^{-1}$ may, nevertheless, have a positive real part, so that the usual ellipsoidal isofrequency surface becomes an open hyperboloid (having two manifolds).

It is important to realize that such so-called *form birefringence*, available through artificial layered stacks, is an order of magnitude larger than natural birefringence available in existing materials. For $\varepsilon_2 = 2.3^2 = 5.29$ (titania TiO_2) and $\varepsilon_1 = 1$, birefringence can be as large as $\delta n = n_o - n_e \simeq 0.5$ for $f = 0.5$. This value has to be compared with the natural birefringence, $\delta n = 0.17$ for the calcite crystal (also known as Iceland spar, one of the crystals offering the highest δn), and $\delta n \simeq 0.01$ for quartz at visible frequencies.

15.2.3 Microscopic polarization and fields

This subsection is dedicated to simply explaining the physical origins of the harmonic and arithmetic averages appearing in the dielectric tensor in the long-wavelength limit. The critical dimension over which a wave significantly varies is its wavelength in the material. In the static limit $a/\lambda \ll 1$, the field cannot vary spatially at the scale of any of the individual layers of the composite stack, because the layer thicknesses become much smaller than the wavelength. However, it may vary at the interfaces between the layers. These two simple statements are sufficient for the following.

First, let us consider waves that are propagating with an electric field (E_x or E_y) that is parallel to the interfaces. Since the field is constant in every layer and is continuous at the interface, it is actually constant throughout the stack, $E(z) = E_0$ (see Fig. 15.4a), so that, on spatial averaging, we may write: $\langle E(z)\rangle = E_0$. We can now use the dielectric displacement field $D = \varepsilon_0 E + P$ that formally defines ε in $D = \varepsilon_0\varepsilon E$. By averaging $D(z)$ for any of those two constant components E_x or E_y, we obtain

$$\langle D(z)\rangle = \langle\varepsilon(z)E(z)\rangle = \langle\varepsilon(z)E_0\rangle = \langle\varepsilon(z)\rangle E_0 = \langle\varepsilon(z)\rangle\langle E(z)\rangle. \tag{15.15}$$

By intuitively defining the effective permittivity ε_{eff} as the average response over the material,

$$\langle D \rangle = \varepsilon_{\text{eff}} \langle E \rangle \,, \tag{15.16}$$

we obtain $\varepsilon_{xx} = \varepsilon_{yy} = \langle \varepsilon(z) \rangle$, which justifies the arithmetic mean made above.

Second, let us consider waves that are propagating with an electric field (E_z) that are perpendicular to the interfaces. We now have to use the field component that is continuous across the interfaces—that is, the component D_z, for which $\langle D_z(z) \rangle = D_0$ holds everywhere throughout the stack. We then obtain

$$\langle E_z(z) \rangle = \langle D_z(z)/\varepsilon(z) \rangle = \langle D_0/\varepsilon(z) \rangle = D_0 \langle 1/\varepsilon(z) \rangle = \langle D_z(z) \rangle \langle 1/\varepsilon(z) \rangle \,, \tag{15.17}$$

implying that $\varepsilon_{zz} = \langle 1/\varepsilon(z) \rangle^{-1}$, which now justifies the harmonic mean.

15.2.4 Brief discussion for 2D and 3D structures

The simple rule expressed for the layered stack, with a well-defined diagonal anisotropic tensor, does not hold for higher-dimensional composites. We may easily guess that the presence of spatially dependent field components, even in the static limit, seriously compromises a simple treatment. But, for 2D composites that are invariant along z and periodic along x and y (note that the z-direction was the periodic direction chosen for 1D structures), while there are no closed-form rigorous expressions in general, the ε_{zz} component of the permittivity tensor can be simply expressed as $\langle \varepsilon \rangle$, since E_z is continuous for all interfaces. Within the xy-plane, anisotropy is also present in general. The two dielectric axes (where $\mathbf{P}$ and $\mathbf{E}$ are parallel) must be chosen as x and y in order to get a diagonal tensor

$$\varepsilon_{\text{eff}} = \begin{pmatrix} \varepsilon_{xx} & 0 & 0 \\ 0 & \varepsilon_{yy} & 0 \\ 0 & 0 & \langle \varepsilon \rangle \end{pmatrix} . \tag{15.18}$$

Approximate expressions for its ε_{xx} and ε_{yy} elements exist, as well as upper and lower bounds (Jackson, 1975); see Exercise 15.2, which tackles the issue of rectangle-shaped disconnected inclusions on a rectangular lattice. Generally speaking, 2D structures behave as biaxial materials, except if some specific symmetry conditions are fulfilled. For 3D periodic structures, no closed-form expression exists, even for ε_{zz}. Of course, in electrostatics, a number of closed-form expressions have been developed for simple shapes, and they can also be profitably used to get closed-form expressions for the effective properties of the composite. But this remains rare and is often quite involved.

To give a complementary insight, it is instructive to remember that, after all, the static dielectric constant is physically understood *ab initio* as the consequence of microscopic polarizability (defined by $\mathbf{p} = \alpha \mathbf{E}_{\text{local}}$). So, provided this latter polarizability is well defined (for an atom, it depends on its orbitals, which, in turn, depend on the chemical/physico-chemical bonds to surrounding atoms), the issue of the dielectric response of, say, alloys (ordered or amorphous) bears much resemblance to our composite response problem. A tenet of the dielectric theory of bulk materials in this respect is the Clausius–Mossotti relationship, which is fully based on the premise that the Lorentz local field is $\mathbf{E}_{\text{local}} = \mathbf{E} + \mathbf{P}/3\varepsilon_0$. Interested readers will profitably consult

Aspnes's view on this (Aspnes, 1982), which we use in Complement 15.A on homogenization methods for disordered composites when approaching the Maxwell–Garnett and Bruggeman methods. Aspnes proposes to circumvent the less intuitive usual presentation of the Lorentz local field and urges us not to be afraid to confront directly the issue of averaging the microscopic field. Such ideas are a helpful entry point to tackle many non-trivial issues of shape-related dielectric response. We now turn our attention to the subwavelength regime of large (but not infinitely large) wavelengths to guess what expansions can be made in this case of great practical interest in nanophotonics.

15.2.5 Taylor expansion for small periods

The static limit (or the deep-subwavelength regime) discussed above is of moderate importance in practice, because it is rarely reached in deterministic man-made devices at optical wavelengths. It could be thought that nature has different options, but it often chooses structures that are also operating in the subwavelength regime when the period is only slightly smaller than the wavelength (periods of a few tenths of a wavelength), as can be seen for some insect wing scales, feathers, leaf coatings, or for mother-of-pearl made of aragonite platelets and biopolymers.

In order to have quantitative predictions in the subwavelength regime, numerical tools are requested. The problem is treated with the greatest generality in the framework of metamaterials (see Chapter 18) because of the opportunity to get non-trivial properties such as artificial magnetism and chirality, even if constituent materials are neither magnetic ($\mu \equiv \mu_0$ everywhere) nor chiral. What we do here is easier and of practical importance. We start from the fundamental Bloch mode of the periodic medium and study how its propagation changes upon departing from the static limit. To do that, let us admit that the effective index (or the propagation constant) accepts a Taylor expansion in the variable (a/λ), and let us go back to 1D z-periodic systems and introduce the scaling parameter α defined by

$$\alpha = \frac{K}{k_0} = \frac{\lambda}{a} \,. \tag{15.19}$$

We start again with Eqs (15.9) and (15.10), valid for all m

$$\mu_0 c \, \mathsf{H}_m = -(n_{\text{eff}} + m\alpha) \, \mathsf{E}_m, \tag{15.20}$$

$$\frac{1}{\mu_0 c} \sum_q \hat{\varepsilon}_{m-q} \, \mathsf{E}_q = -(n_{\text{eff}} + m\alpha) \, \mathsf{H}_m, \tag{15.21}$$

and now look for a solution with a Taylor expansion in the limit $\alpha^{-1} \to 0$ (in the limit of large α's)

$$n_{\text{eff}} = n^{(0)} + n^{(1)}\alpha^{-1} + n^{(2)}\alpha^{-2} + \ldots, \tag{15.22}$$

$$\mathsf{H}_m = \mathsf{H}_m^{(0)} + \mathsf{H}_m^{(1)}\alpha^{-1} + \mathsf{H}_m^{(2)}\alpha^{-2} + \ldots, \tag{15.23}$$

$$\mathsf{E}_m = \mathsf{E}_m^{(0)} + \mathsf{E}_m^{(1)}\alpha^{-1} + \mathsf{E}_m^{(2)}\alpha^{-2} + \ldots. \tag{15.24}$$

We then insert Eqs (15.22), (15.23) and (15.24) into Eqs (15.20) and (15.21), and identify terms with a α-dependency of order $O(\alpha), O(1), O(\alpha^{-1}), O(\alpha^{-2})$, etc. Terms in $O(\alpha)$ for $m \neq 0$ and in $O(1)$ for $m = 0$ give us the following identities:

$$0 = \mathsf{E}_m^{(0)} = \mathsf{H}_m^{(0)}, \quad \text{for all } m \neq 0, \tag{15.25}$$
$$\mu_0 c\, \mathsf{H}_0^{(0)} = -n^{(0)}\, \mathsf{E}_0^{(0)}, \tag{15.26}$$
$$(\mu_0 c)^{-1} \hat{\varepsilon}_0\, \mathsf{E}_0^{(0)} = -n^{(0)}\, \mathsf{H}_0^{(0)}, \tag{15.27}$$

showing that in the limit $\frac{a}{\lambda} \to 0$, the zeroth-order term of the expansion of n_{eff} is given by

$$n^{(0)} = \sqrt{\hat{\varepsilon}_0}, \tag{15.28}$$

in agreement with Eq. (15.13). Moreover, we find that the Bloch mode is simply a plane wave, since solely $\mathsf{H}_0^{(0)}$ and $\mathsf{E}_0^{(0)}$ do not vanish and $\mathsf{H}_p^{(0)} = \mathsf{E}_p^{(0)} = 0$ for all $p \neq 0$: *the 1D periodic composite strictly behaves as a homogeneous material in the static limit.*

Let us now abandon the asymptotic limit. Analysing terms in $O(1)$ for $m \neq 0$, we get

$$\mu_0 c\, \mathsf{H}_m^{(0)} = -m\, \mathsf{E}_m^{(1)}, \quad \text{(which is identically 0 for } m \neq 0), \tag{15.29}$$
$$(\mu_0 c)^{-1} \hat{\varepsilon}_m\, \mathsf{E}_0^{(0)} = -m\, \mathsf{H}_m^{(1)}, \quad \text{for all } \; m \neq 0, \tag{15.30}$$

showing that the Bloch mode is no longer a plane wave as soon as we escape the static limit (the first-order terms are non-zero). Terms in $O(\alpha^{-1})$ for $m = 0$ give us

$$\mu_0 c\, \mathsf{H}_0^{(1)} = -n^{(1)} \mathsf{E}_0^{(0)} - n^{(0)} \mathsf{E}_0^{(1)}, \tag{15.31}$$
$$(\mu_0 c)^{-1} \hat{\varepsilon}_0\, \mathsf{E}_0^{(1)} = -n^{(1)} \mathsf{H}_0^{(0)} - n^{(0)} \mathsf{H}_0^{(1)}, \tag{15.32}$$

from which we infer, using Eqs (15.26) and (15.27), that

$$n^{(1)} = 0. \tag{15.33}$$

Finally, by considering terms in $O(\alpha^{-1})$ in Eq. (15.20), we obtain $\mu_0 c\, \mathsf{H}_0^{(1)} = -m\, \mathsf{E}_m^{(2)}$, which allows us to establish, after injecting into Eq. (15.30), that

$$\mathsf{E}_m^{(2)} = \frac{\hat{\varepsilon}_m}{m^2}\, \mathsf{E}_0^{(0)}. \tag{15.34}$$

The second-order expansion term of n_{eff} is provided by terms in $O(\alpha^{-2})$ for $m = 0$:

$$\mu_0 c\, \mathsf{H}_0^{(2)} = -n^{(2)} \mathsf{E}_0^{(0)} - n^{(0)} \mathsf{E}_0^{(2)}, \tag{15.35}$$
$$(\mu_0 c)^{-1} \sum_m \hat{\varepsilon}_m\, \mathsf{E}_m^{(2)} = -n^{(2)} \mathsf{H}_0^{(0)} - n^{(0)} \mathsf{H}_0^{(2)}, \tag{15.36}$$

which allows us to obtain with Eq. (15.34) the relation

$$\sum_{p \neq 0} \frac{1}{p^2}\, \hat{\varepsilon}_p \hat{\varepsilon}_{-p} = 2 n^{(0)} n^{(2)}. \tag{15.37}$$

Finally, and since it may likewise be shown that the third-order $n^{(3)}$ term vanishes, by collecting the two first non-vanishing orders, one gets

$$n_{\text{eff}}^2 = \hat{\varepsilon}_0 + \sum_{p \neq 0} \frac{1}{p^2} \hat{\varepsilon}_p \hat{\varepsilon}_{-p} \left(\frac{a}{\lambda}\right)^2 + O\left(\left(\frac{a}{\lambda}\right)^4\right) . \tag{15.38}$$

The result is superimposed on the exact calculation on Fig. 15.5. A similar expression holds for the extraordinary effective index for transverse-magnetic (TM) polarization (see Exercise 15.9).

The quadratic expansion of n_{eff} in Eq. (15.38) is valid for any periodic 1D structure (lamellar or continuous, lossy or transparent), the $\hat{\varepsilon}_p$'s being the Fourier coefficients of the periodic relative permittivity $\varepsilon(z)$ and $\hat{\varepsilon}_0 = \langle \varepsilon(z) \rangle$ being the mean value.

For transparent media ($\varepsilon(z)$ is real), $\hat{\varepsilon}_{-p} = (\hat{\varepsilon}_p)^*$ and n_{eff} decrease as the wavelength increases, just like most real materials. This is unfortunate, because it implies that we cannot create negative dispersion with artificial dielectrics. Note, however, that n_{eff} varies quickly with the wavelength (as is the case in a waveguide) and the artificial material is highly dispersive, i.e. much more dispersive than natural non-resonant materials.

Initially, we defined the effective index as $k_z = (\omega/c)n_{\text{eff}}$ in Eq. (15.3) and noted that the definition was equivocal and even unclear because k_z is defined modulo K. Now, with Eq. (15.38) (and, for TM, Eq. (15.45) in Exercise 15.9), we even lift the ambiguity out of the static limit.

15.2.6 Effective permittivity and effective index

Constitutive relations describe the properties of a linear material by linking **D** to **E** and **H** to **B**. When injecting the constitutive relations into the Maxwell equations, one studies wave properties and describes how waves propagate in a material. We note that the term 'refractive index' is ambiguously used in optics, since we often talk of the refractive index n of a material, although we intend to say that the waves propagating in the material take the form $\exp(i(\omega/c)nz)$. Thus, the microscopic polarization approach of section 15.2.3 and the effective index approach are completely different in nature. The former relies on a property of the composite material, its effective permittivity,

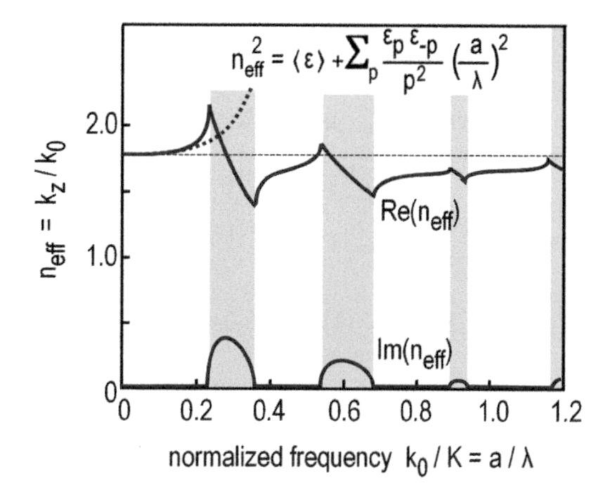

Fig. 15.5 Second-order approximation (dotted line) superimposed on the $n_{\text{eff}}(\omega)$ relation.

whereas the latter relies on a property of the wave that propagates in the material, the effective index.

The presentation was somewhat ambiguous at this level; for instance, when describing the behaviour of 1D periodic structures, one first explores the effective index n_{eff} of the fundamental Bloch mode (a wave property) and then assumes that $n_{\text{eff}}^2 = \varepsilon_{\text{eff}}$, but on the other hand, one also tries to average the response of the material to light by writing relations on averaged macroscopic quantities $\langle D \rangle = \varepsilon_{\text{eff}} \langle E \rangle$. Thus two different concepts are linked artificially and too rapidly. Actually, the quadratic expansion for n_{eff} describes how the propagation constant changes as one departs from the static limit, but not how the effective permittivity changes; one cannot simply identify n_{eff}^2 with an effective permittivity ε_{eff} that would account for all properties of the composite, e.g. reflection and transmission at the interfaces. The reason is deeply rooted in the nature of the Bloch modes: out of the static limit, the Bloch mode is no longer a plane wave, as it has a periodic spatial dependence and the material cannot be replaced by a homogeneous material. As is well known, this is only possible in the static limit, where we found that $\varepsilon_{\text{eff}} = n_{\text{eff}}^2$ (see Eq. (15.37) for $a/\lambda \to 0$ and Eq. (15.15)).

The procedure which consists in inferring effective material properties from the measurement of wave properties, such as the complex values of the transmission and reflection coefficients of a thin film of artificial material assumed to be homogeneous, is referred to as 'retrieval' in the literature. Indeed, there are now a significant number of theoretical and numerical works on that matter, but this is out of the scope of this book. What comes immediately into play, and this is not intuitive, is spatial dispersion (see section 3.2.4), the analogue of the usual temporal dispersion, which results in a dependence of the permittivity and permeability with the direction of the propagation. Normally, such a dependence is significant only at nanoscopic length scales and is assumed to be absent for the sake of simplicity. However, it does come into play for artificial materials with periods that are only slightly smaller than the wavelength when one tentatively homogenizes them. For instance, in Exercise 15.5, we study the homogenization of gratings that are periodic in the z-direction. The frequency ω and the perpendicular wavevector component k_z being fixed, the exercise proposes a derivation of a Taylor expansion of the propagation constant, k_x or k_y, which is perpendicular to the grating interfaces. It is found that the coefficient of the Taylor expansion actually depends on k_z in a complicated manner, which is an additional manifestation of spatial dispersion on the effective properties of composites operating out of the quasi-static limit.

An interesting consequence of the spatial dispersion, which will be studied in Chapter 18, is artificial magnetism at optical frequencies, i.e. the possibility of creating a magnetic like artificial response without magnetic materials, just by considering that the total induced magnetic moment of a homogenized medium has two contributions, one directly coming from the average magnetization $\int (\mathbf{B}/\mu_0 - \mathbf{H}) dV$ and the other from the average microscopic current density, $\frac{1}{2} \int \left(\mathbf{r} \times \frac{\partial \mathbf{P}}{\partial t}\right) dV$, coming from the velocity of charged particles in the medium (Agranovich and Gartstein, 2006).

15.3 Grating equation

A common periodic object of optics is the diffraction grating. Its usual role is to change the direction of light either in transmission (deviation of a beam) or reflection (reflection other than specular). Hereafter, we focus on the simple case of a grating that is periodic along the z-direction (period a) and invariant along the y-direction (see Fig. 15.6a). When the grating is illuminated by a plane wave with frequency ω and wavevector perpendicular to the y-axis (the so-called classical diffraction by opposition to the conical diffraction, hence we can treat it in 2D rather than 3D), two independent solutions have to be considered: one for TE polarization (electric field E_y parallel to the grooves) and the other for TM polarization (magnetic field H_y parallel to the grooves). The Bloch theorem or the Floquet theorem both stipulate that the main field component (E_y or H_y) takes the same generic form over all space, $E_y(x,z) = \exp(ik_z z)\mathsf{e}_y(x,z)$, where k_z is the parallel wavevector of the incident plane wave (whatever the other k-component k_x is) and $\mathsf{e}_y(x,z)$ is a periodic function of z of period a. Thus $\mathsf{e}_y(x,z)$ can be written as a Fourier series (with coefficients that depend on x) and, in total, $E_y(x,z)$ features all the harmonics $k_z + mK$, m being a relative integer (the diffraction order), and takes the form $E_y(x,z) = \Sigma_m[c_m(x)\exp(i(k_z + mK)z)]$. The coefficients $c_m(x)$ are in general complicated functions of the x-coordinate.

However, in the substrate and cover/superstrate regions below and above the grating, $E_y(x,z)$ takes a particularly simple form. Since these regions are homogeneous, $E_y(x,z)$ satisfies the Helmholtz equation, implying that each element of the sum also satisfies the Helmholtz equation. It directly follows that $c_m(x) \propto \exp(i\gamma_{m(\mathrm{i,s})}x)$, with $\gamma^2_{m(\mathrm{i,s})} + (k_z + mK)^2 = \varepsilon_{(\mathrm{i,s})}(\omega/c)^2$, where subscripts or superscripts (i,s) refer to the incident medium (cover/superstrate) for ‘i’, and to the substrate for ‘s’. Some $\gamma_{m(\mathrm{i,s})}$ are real and correspond to progressive plane waves, while others are purely imaginary (or complex if $\varepsilon_{(\mathrm{i,s})}$ is complex). This is the so-called plane-wave Rayleigh expansion that is valid above and below the grating—not inside, of course. We are going to see that this expansion corresponds to the usual grating law. The latter provides the diffraction-order directions but not the distribution of the incident energy among these orders.

Let us first present the grating law in vector form, and let us consider the general case of conical incidence for which the incident wavevector has a component along the y-axis. We name $\mathbf{k}_{\mathrm{inc},||} = k_z\mathbf{u}_z + k_y\mathbf{u}_y$, $\mathbf{k}_{\mathrm{diff},||} = k_{\mathrm{diff},z}\mathbf{u}_z + k_y\mathbf{u}_y$ and $\mathbf{K} = K\mathbf{u}_z$. Then, due to the Bloch–Floquet theorem and the grating invariance along the y-direction, we have

$$\mathbf{k}_{\mathrm{diff},||} = \mathbf{k}_{\mathrm{inc},||} + m\mathbf{K}\,, \tag{15.39}$$

with m an integer. Note that we omit the m-dependence of $\mathbf{k}_{\mathrm{diff},||}$ for simplicity. This dictates the two in-plane components of the diffracted-order wavevectors (the grating equation is a momentum-conservation law in the parallel plane, that of the grating interfaces with substrate or cover). The other rule to find the out-of-plane x-component of the wavevectors (which dictates the direction of the diffracted orders in the substrate and/or superstrate) is to apply the conservation of energy: the frequency of the wave is unchanged, and thus the modulus of the diffracted wavevector is known and must be $n_i\omega/c$ in the cover (incident medium) and $n_s\omega/c$ in the substrate.

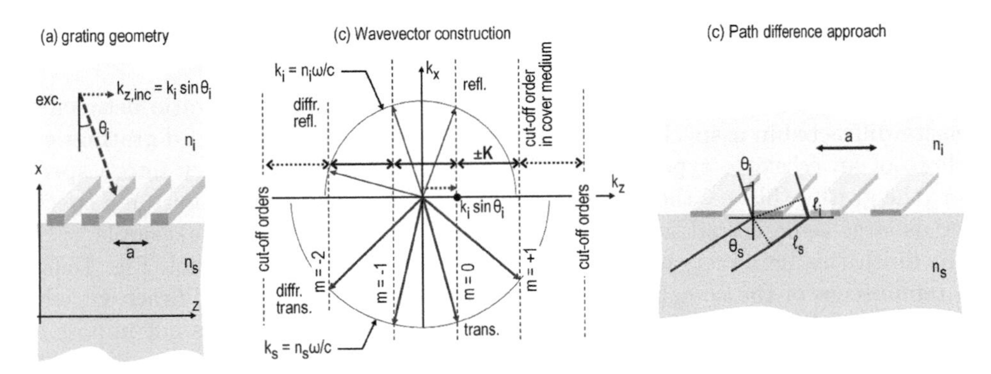

Fig. 15.6 Classical grating diffraction between two media: the incident medium (superstrate/cover) and the substrate with refractive indices n_i and n_s. (a) Diffraction geometry: period a, incidence angle θ_i, parallel component of the incident wave $k_z = n_\mathrm{i}\frac{\omega}{c}\sin\theta_\mathrm{i} = k_\mathrm{i}\sin\theta_\mathrm{i}$). (b) The component $k_{\mathrm{inc},z}$ gets shifted through diffraction by multiples of $K = \frac{2\pi}{a}$ from the starting one ($m = 0$, black dot). The circles used to construct diffraction directions have radii $|k_i| = n_\mathrm{i}\frac{\omega}{c}$ and $|k_\mathrm{s}| = n_\mathrm{s}\frac{\omega}{c}$. In the substrate, only the orders $m = -2, -1, 0$ and $+1$ are propagating. All the other orders are evanescent or cutoff. (c) Sketch, in the transmission case, of the path difference $\ell_\mathrm{i} + \ell_\mathrm{s}$ that must be a multiple of λ.

It is then clear that the rule $|\mathbf{k}_{\mathrm{diff},||}|^2 + |\mathbf{k}_{\mathrm{diff},x}|^2 = (n_{\mathrm{i,s}}\omega/c)^2$ for progressive and evanescent waves in media (i, s) also involves $k^2_{\mathrm{diff},||} \leq (n_{\mathrm{i,s}}\omega/c)^2$ for progressive waves only. Then the finite set of allowed values of m, also known as the *allowed diffraction orders*, are dictated by $|k_{\mathrm{diff},z} + mK| < n_{\mathrm{i,s}}\omega/c$. In the simple case of identical media above and below the gratings, the allowed orders are the same. The limit imposed on m by the inequality is known as a *diffraction cutoff.* For classical mountings ($k_y = 0$), Eq. (15.39) becomes $k_{\mathrm{diff},z} - k_{\mathrm{inc},z} = mK$ and the directions of the diffracted orders are found by targeting the intersection of lines at these $k_{\mathrm{diff},z}$ values with the circles of radii $n_{\mathrm{i,s}}\omega/c$, as shown in Fig. 15.6. Here a case of different cutoff regimes for $m = 1$ is shown: this order is allowed only in the larger-index substrate.

Now we are reminded of another very classical way of describing the diffraction directions, as in Fig. 15.6c: if we count the optical paths of two rays impinging at a distance of one period from each other and request a constructive interference condition, i.e. an optical path difference integer multiple of the wavelength λ, we obtain the same rule.

Denoting the angles of the two plane waves by θ_i (incident wave) and θ_s (diffracted wave in substrate), we write the path difference as the sum of the two contributions shown; thus $\ell_\mathrm{i} + \ell_\mathrm{s} = m\lambda$. Under such a condition, the field scattered by any period will be in phase in the θ_s-direction, so they will add coherently. Of course, such simple paths are more intuitively valid in the vanishingly thin grating limit (a screen with openings, for instance), but with the coherent scattering being dictated in the far field, the reasoning holds for any grating but for propagating (not evanescent) diffracted orders only. Simple geometry relates these paths to angles, taking care of angle conventions (see Fig. 15.6 caption). We get $m\lambda = \ell_\mathrm{i} + \ell_\mathrm{s} = an_\mathrm{i}\sin(\theta_\mathrm{i}) - an_\mathrm{s}\sin(\theta_\mathrm{s}) \rightarrow m\frac{\lambda}{a} = n_\mathrm{i}\sin(\theta_\mathrm{i}) - n_\mathrm{s}\sin(\theta_\mathrm{s})$, which gives the same $k_{z,\mathrm{i}} - k_{z,\mathrm{s}} = mK$ rule if we write $\lambda = \frac{2\pi c}{\omega}$:

$$m\frac{2\pi}{a} = \frac{\omega}{c}(\sin(\theta_\mathrm{i}) - \sin(\theta_\mathrm{s})) = k_{z,\mathrm{i}} - k_{z,\mathrm{s}}. \tag{15.40}$$

The usual issue of grating optics is the diffraction efficiency, i.e. the ratio between the intensity diffracted in a specific order and the incident intensity. Blazed gratings such as those of an échelette type are designed to maximize the efficiency into a specific order (the grating 'blazes' the incident energy into one specific direction) in a spectral range as large as design allows around a privileged (wavelength, angle) couple.

To finish this reminder, we mention that the graphical construction Fig. 15.6b is also reminiscent of the so-called Ewald construction used in X-ray diffraction, which is, however, basically a fully 3D construction, unlike ours, which does not impose any periodicity-related restriction on k_x.

15.4 Homogenization of subwavelength gratings

Consider the very simple task of designing an anti-reflection coating for operation at a single wavelength at normal incidence. Theory tells us that a single-layer coating may achieve perfect anti-reflection, provided its refractive index n_coat is the square root of that of the substrate and the layer is a quarter-wave thick ($\lambda/4n_\mathrm{coat}$; see Exercise 15.11). Unfortunately, most optical glasses have refractive indices around 1.55, so $n_\mathrm{coat} \simeq 1.245$ is desirable. Sadly, no such material is commonly available (MgF_2 with $n_\mathrm{coat} \simeq 1.38$ is about the best at visible frequencies), so more complex solutions have to be found, such as porous sol–gels or subwavelength diffraction grating, often referred to as metasurfaces in the modern literature and in Chapter 18.

We shall use the knowledge of nanophotonics and Bloch modes in particular to deepen the understanding and technology of subwavelength gratings and periodic metasurfaces. Hereafter, we consider under what conditions a subwavelength grating can be substituted by an artificial homogeneous layer in a formal way. In Chapter 18, through a selection of real examples, we will illustrate how this substitution is useful for initial designs and the limits of the simplification provided by the substitution.

15.4.1 Statement of the problem and static limit

Let us consider the diffraction problem shown on Fig. 15.7a, where a subwavelength lamellar grating of depth h is illuminated at oblique incidence by a linearly polarized plane wave with a free-space wavelength λ. The refractive index of the incident medium (cover, superstrate) is n_i and that of the substrate is n_s. 'Lamellar' means that the permittivity is assumed to be invariant along the direction normal to the substrate within the grating depth h.

In this section, we seek to answer the following question: under what conditions may the complex diffraction problem of Fig. 15.7a be approximated by a simple refraction–reflection problem on a homogeneous thin film?

To answer this question, it is instructive to start by considering that the grating period is much smaller than the wavelength of the incident illumination. We know that there is a genuine equivalence between periodic artificial media and homogeneous material in the static limit. The grating is equivalent to a uniaxial or biaxial thin film with a thickness equal to the grating depth. For instance, for 1D periodic structures, the ordinary n_o and extraordinary n_e indices of refraction are simply $\langle\varepsilon\rangle^{1/2}$ and $\langle\frac{1}{\varepsilon}\rangle^{-1/2}$.

The static limit is rather an academic case. With state-of-the-art nanofabrication, it is only possible to manufacture structures with in-plane features two to four times smaller than the 'inner' optical wavelength, $\lambda_{\mathrm{mat}} = \lambda\langle\varepsilon\rangle^{-1/2}$. For these actual artificial media, there is no such thing as a theorem of equivalence between periodic structures and homogeneous media. On the contrary, the physical properties of real composite materials may sometimes strongly differ from those of homogeneous media.

15.4.2 Gratings with subwavelength periods

For real gratings with a finite period, it is important to bear in mind that the substitution proposed in Fig. 15.7 can at best be a well-converged approximation for some aspects of the problem but not all of them. We will be concerned here with an approximate physical equivalence, especially valid for the waves propagating in the far field.

Consider the diffraction shown in Fig. 15.7a. As the incident plane wave scatters at the upper interface of the grating, it excites an infinite number of reflected diffraction orders and downward-propagating guided-like Bloch modes. The latter reach the substrate interface and excite upward-propagating Bloch modes and transmitted diffraction orders. The dynamical diffraction can be seen as resulting from the bouncing back and forth of the upward- and downward-propagating Bloch modes that may absorb some energy (if the grating material is absorbing) and distribute the incident energy between the different diffraction orders. In general, three conditions are sufficient to achieve an approximate physical equivalence between the two heuristic models.

Condition (I). The first condition requires that only the zeroth diffraction orders propagate in the substrate and in the incident medium; all the other higher diffracted orders have to be evanescent. Whether a diffraction order propagates or not is given by the grating equation. If only the zeroth transmitted and reflected orders are to propagate, it is immediately deduced from the grating equation that the following condition holds: $|\mathbf{k}_{\mathrm{inc},\|} + m\mathbf{K}| > \max(n_{\mathrm{i}}, n_{\mathrm{s}})\omega/c$ for all $m \neq 0$, where $\max(\ldots)$ is the maximum of the arguments. We also remind the reader that $\mathbf{k}_{\mathrm{inc},\|}$ is the component of the incident wavevector parallel to the interface. For instance, for a 1D grating illuminated with an angle of incidence θ, the grating period a should be smaller than a geometrical cutoff value a_g

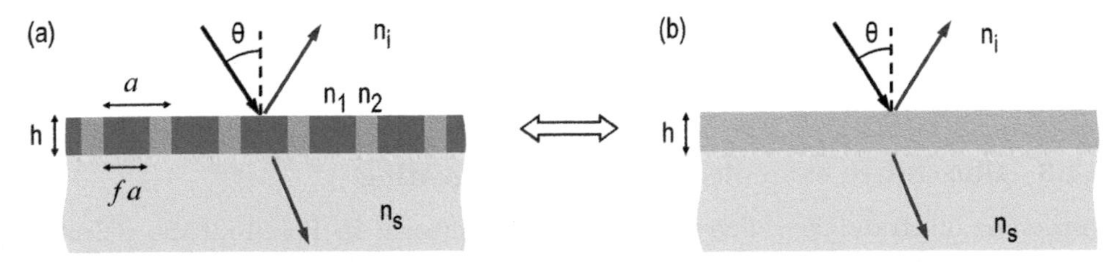

Fig. 15.7 Substitution of a subwavelength grating (a) of period a by a homogeneous thin film (b) of identical thickness h. The grating permittivity is assumed to be independent of the x-direction that is perpendicular to the substrate. For the sake of illustration, a lamellar grating with fill factor f and permittivities ε_1 and ε_2 is shown.

$$a < a_g = \frac{\lambda}{\max(n_\mathrm{i}, n_\mathrm{s}) + n_\mathrm{i}\sin(|\theta|)}, \tag{15.41}$$

which does not depend on the grating profile and materials; it is just deduced from the grating equation.

Condition (II). The second condition is related to the number of propagating Bloch modes that are able to propagate in the grating layer. One requires that only two pairs of Bloch modes (a single pair if polarization degeneracy is lifted) travel backwards and forwards between the two grating interfaces in the same way that multiple-beam interference of local plane waves occurs in the equivalent thin film in Fig. 15.7b. In sharp contrast with the previous condition, the fact that only two Bloch modes propagate in the grating (all the others are evanescent) for a given frequency depends mainly on the grating geometry (not on the depth) and weakly on the diffraction geometry. It is actually independent of n_i and n_s and weakly depends on $\mathbf{k}_{\mathrm{inc}\|}$. This defines a second cutoff that could be called a structural cutoff a_s, which is roughly proportional to the wavelength with a proportionality factor that depends on the grating materials and geometry.

Condition (III). The last condition is to avoid tunnelling of evanescent Bloch waves and is related to the grating depth. If the latter is large enough, all the evanescent Bloch modes that are excited at the upper and lower grating interfaces and are exponentially damped as they propagate through the grating do not tunnel through the grating region and do not participate in the multiple-beam interference. Typically, for dielectric gratings, the impact of evanescent modes on the grating's effective properties is critically driven by the grating depth and is significant for grating depths smaller than a quarter of the wavelength.

Fig. 15.8 sketches the field distribution of a grating that satisfies all three conditions. The far-field diffraction pattern is only composed of the specularly reflected and transmitted plane waves (dashed arrow). The grey areas, just above and below the grating interfaces, represent the grating near-field zones, which are composed, in addition to the specular zeroth order, of an infinite number of evanescent diffraction orders that exponentially decay away from the grating interfaces. In the middle of the grating, the electromagnetic field is solely composed of two counterpropagative Bloch modes which are bouncing back and forth. In the light grey zones inside the grating near the interfaces, the electromagnetic field is much more intricate, as it additionally supports many evanescent Bloch modes.

15.4.3 Illustrative example: High-contrast grating

In order to illustrate the different regimes, we represent in Fig. 15.9 the reflectivity of a particular version of the generic system depicted in Fig 15.8 as a function of the normalized wavelength $\frac{\lambda}{a}$ and the normalized thickness $\frac{h}{a}$. We deal here with silicon ridges ($\varepsilon_1 = 10, \varepsilon_2 = 1$) suspended in air ($n_i = n_s = 1, f = 0.4$). These structures are also called high-contrast gratings (HCG) (Mateus et al., 2004). The grating is illuminated under TE polarization at normal incidence $\theta = 0$.

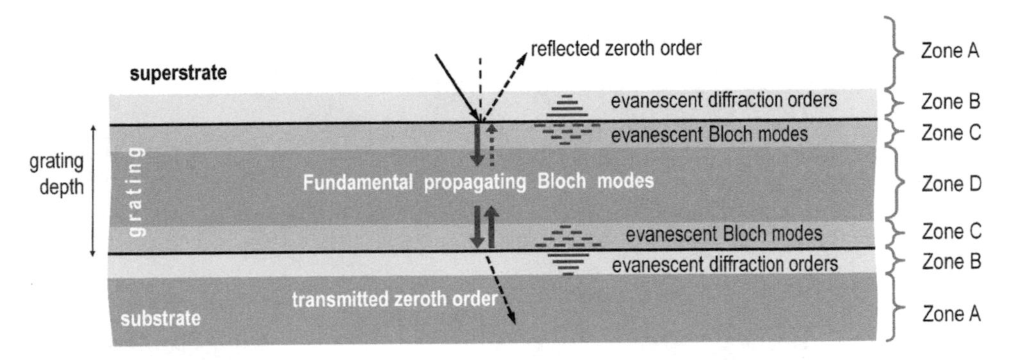

Fig. 15.8 Sketch of lamellar subwavelength grating that satisfies the three conditions. **Zone A**. *Condition (I)*: The zeroth transmitted and reflected orders are shown with dashed black arrows.

Zone B. The field is altered by evanescent plane waves; zone B thickness diminishes when the period is reduced.

Zone C. The field is altered by evanescent Bloch modes; zone C thickness diminishes when the period is reduced.

Zone D. *Condition (II)*. A single Bloch mode (in blue) is bouncing back and forth and dictates the phase difference of the reflected and transmitted planes waves. *Condition (III)*. The evanescent Bloch-mode field is negligible if the grating depth is larger than zone C thickness.

For $\frac{\lambda}{a} = 1$, the reflectance pattern exhibits a rapid spectral variation for all depths. This variation corresponds to a redistribution of the energy between the different diffracted orders when the positive- and negative-first diffraction orders in air are cut off. It is known as Rayleigh anomaly for $a = a_g$, predicted by Eq. (15.41) for $\sin(|\theta|) = 0$.

The second marked behaviour is the regular reflectance pattern obtained for $\frac{\lambda}{a} > 1.7$. The pattern is a remnant of Fabry–Perot fringes observed in thin-film stacks, thereby implying that a single fundamental propagative Bloch mode is excited and is bouncing back and forth in the grating region. The structural cutoff is likely to be $a_s = \frac{\lambda}{1.7}$ for this geometry.

Thus, for $\frac{\lambda}{a} < 1.7$, the grating supports at least two propagative Bloch modes. These modes propagate without coupling in the grating, but when they reach the grating interfaces, they transmit some energy into the diffraction orders and back-reflect into the counterpropagative Bloch modes. The reflection fringe pattern, a sort of chessboard, is much less intuitive, but still regular. In particular, we note that the apertured semiconductor membrane in air exhibits an unexpected broadband high reflectance (> 0.97) for some specific values of the grating depth ($h \approx 0.8a$, or $1.45a$) and for any wavelengths larger than a and smaller than $1.7a$.

For $\frac{\lambda}{a} = 1.7$, another cutoff is visible in the reflectance pattern. Note that for this limit, already pointed out above, the cutoff, in contrast to the Rayleigh cut-off, is less conspicuous for small values of h, being again abrupt for larger ones. The phenomenon, while similar to the Rayleigh cutoff, now corresponds to the passing-off of a propagative Bloch mode at large periods, which becomes evanescent at $\frac{\lambda}{a} = 1.7$. Although it is evanescent, the Bloch mode participates in the energy transfer between the two grating

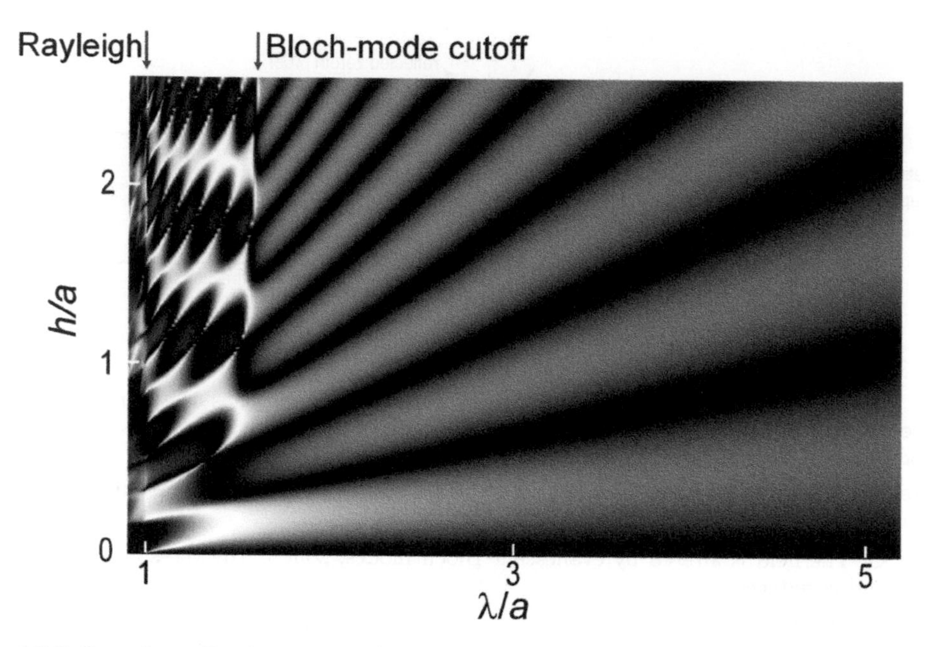

Fig. 15.9 Specular reflectivity map of a 1D lamellar grating suspended in air (parameters: $\varepsilon_1 = 10, \varepsilon_2 = 1, n_i = n_s = 1, f = 0.4$) illuminated by a normally incident plane wave under TE polarization (**E** is parallel to the ridges). The scale is linear, and the maximum reflectance (bright zone) is 100%. Adapted with permission from (Lalanne et al., 2006) ©(2006) IEEE.

interfaces by tunnelling, the issue related to condition (III). Hence, there is a progressive variation. For small grating depths, many evanescent Bloch modes participate in the transfer (even those with strong decay) and the reflectance pattern presents a smooth variation with the period-to-wavelength ratio. Conversely, as the tunnelling distance h increases, the energy transfer becomes less efficient, and the reflectance variations become more rapid (Lalanne et al., 2006).

In the present case, the refractive index of the grating ridges is larger than the substrate one and the Rayleigh anomaly is observed for a period a_g that is smaller than a_s. This holds in practice when a layer of high refractive index is patterned on a substrate of low refractive index (TiO_2/SiO_2 at visible wavelengths or silicon-on-insulator at telecom wavelengths). Conversely, when the grating is directly etched into the substrate, the geometrical (Rayleigh) cutoff is always larger than the structural (Bloch wave) one, $a_g > a_s$.

The substitution of a grating by a homogeneous film may serve as a guideline for an initial design of subwavelength optical components, as it ensures a drastic reduction of complexity. Even if the three conditions are fulfilled, the substitution remains qualitative, because the propagating Bloch modes generally differ from plane waves as well as their excitation and scattering. For dielectric materials, such as polymers or even semiconductors, the substitution is approximately valid for $\lambda/a >\sim 2-3$. For metallo-dielectric mixtures, it is additionally required that the Bloch modes penetrate the metal almost uniformly. Another scale then plays a key role, the *skin depth*; this is $\simeq 25\,\text{nm}$ for noble metals in the visible and near-infrared (see Exercise 4.2). Thus, for the attainment of a genuine Bloch-mode homogenization regime with metals, the metal layer widths should not be larger than ~ 10 nm.

A set of applicative examples of smart designs of subwavelength surfaces with homogenization are: (i) form birefringence with binary gratings for manufacturing wave plates; (ii) wire-grid polarizers and filters; (iii) subwavelength optical surfaces for broadband anti-reflection coating; (iv) hyperbolic metamaterials; and (v) metasurfaces for lensing and beam-shaping applications. Their principle of operation will be reviewed in Chapter 18.

15.A Complement: Homogenization of disordered composites

15.A.1 Clausius–Mossotti vs interacting dipole approach

Let us consider a disordered composite composed of inclusions of polarizable particles (material A with dielectric constant $\varepsilon_{\mathrm{A,abs}} = \varepsilon_A\varepsilon_0$) in a matrix of material B ($\varepsilon_{\mathrm{B,abs}} = \varepsilon_B\varepsilon_0$).The most tractable approach towards homogenization assumes that the inclusions can be viewed as tiny dipoles that get polarized under the excitation by the so-called Lorentz local field $\mathbf{E}_{\mathrm{local}}$. In particular, for identical particles uniformly distributed, we have a single-valued distribution of α's, and we can derive the overall response of a composite in exactly the same way as when establishing the Clausius-Mossotti relation for a solid of polarizable atoms or molecules. In locally homogeneous media, we recall that the basic definition of polarizability $\mathbf{P} = \bar{n}\mathbf{p}$ is the product of the local density of objects $\bar{n}$ by their common local dipole $\mathbf{p}$.

Before we proceed, let us signal the clarifications proposed by Aspnes (Aspnes, 1982). The textbook derivation of Clausius–Mossotti is what Aspnes calls an 'average/solve' approach. The construction of the local field from the macroscopic field $\mathbf{E}_{\mathrm{local}} = \mathbf{E} + \mathbf{P}/3\varepsilon_0$, while being the key point, does not lend itself to much physical understanding. One carves a fictitious sphere large enough to lend its surface to continuous physical descriptions. One then subtracts its negative contribution, stemming from its surface-charge density $\sigma = \mathbf{P}\cdot\mathbf{n}$ ($\mathbf{n}$ the outward normal). Hence, the second term $\mathbf{P}/3\varepsilon_0$, adds positively to $\mathbf{E}$, since in doing so we removed matter that screened the considered local point from the outer imposed field. As for the 'atomically near' field produced by close atoms at the very centre of this fictitious sphere, it is tediously derived as zero from an evaluation of the field of successive dipole shells around the considered point, e.g. in a cubic lattice whose symmetries are clearly helpful. This two-step reasoning conveniently gets to the right result, but clearly the fictitious sphere hampers physical understanding, wrongly suggesting that 'atomically far' atoms differ in some respect from 'atomically near' ones. Aspnes cleverly proposes an explicit 'solve/average' approach, where the macroscopic field is quite unknown at the start. One first solves for dipoles interacting together (in which each dipole is submitted to an external field and the fields of other dipoles), which is doable. The solution is a direct relation between the field applied initially from the outside (just screened by existing physical interfaces and thus called $\mathbf{E}_{\mathrm{int}}$ in ref. (Aspnes, 1982)) and $\mathbf{P}$, free from any use of the macroscopic field.

The last step is to *explicitly calculate* the macroscopic field as an average over the microscopic one, whereby simple identities of integral calculus for a small volume around each dipole yield a fixed contribution per dipole, $\langle \mathbf{E} \rangle_{\text{small volume}} = -\mathbf{p}/3\varepsilon_0$. Proceeding in this way, the macroscopic field is explicitly found to be $\mathbf{E} = \mathbf{E}_{\text{local}} - \mathbf{P}/3\varepsilon_0$. We believe that this treatment has good heuristic value, notably in grasping the effects of different geometries and how the summation on dipoles, leading to cancellation in an infinite medium, will differ in finite geometries (slabs, shells, rods, etc.). In all such situations, the trick of the Lorentz fictitious sphere rather obscures thinking.

We will not comment on this any further, and we now proceed with the adapted version of the Clausius–Mossotti result for our 'A in B' composite. We have, for the resulting ε_{eff} dielectric response,

$$\frac{\varepsilon_{\text{eff}} - \varepsilon_B}{\varepsilon_{\text{eff}} + 2\varepsilon_B} = \frac{\bar{n}_A\,\alpha}{3\,\epsilon_0}, \tag{15.42}$$

where $\bar{n}_A$ is the density (m^{-3}) of A inclusions. From this point on, one should further relate α to the field with a relation that is shape-dependent. This will lead us to the Maxwell–Garnett case.

15.A.2 Maxwell–Garnett

The simplest case is that of spherical inclusions of radius r_A. Then electrostatics shows that $\alpha = 4\pi\epsilon_0 \frac{\varepsilon_A - \varepsilon_B}{\varepsilon_A + 2\varepsilon_B} \times r_A^3$ (this derives from the study of the outer field of a single embedded sphere when the $r \to \infty$ limit is an imposed field $\mathbf{E}_0$ that plays the role of $\mathbf{E}_{\text{local}}$). Then we have

$$\frac{\varepsilon_{\text{eff}} - \varepsilon_B}{\varepsilon_{\text{eff}} + 2\varepsilon_B} = \frac{\bar{n}_A}{3\,\epsilon_0}\,4\pi\epsilon_0 \frac{\varepsilon_A - \varepsilon_B}{\varepsilon_A + 2\varepsilon_B} \times r_A^3 = f_A \frac{\varepsilon_A - \varepsilon_B}{\varepsilon_A + 2\varepsilon_B}, \tag{15.43}$$

with the last equality defining the volume fraction f_A (or 'fill fraction', also used in photonic-crystal descriptions) as the $\frac{4\pi}{3} r_A^3 \bar{n}_A$ term, the product of the individual volume by $\bar{n}_A$. This is the Maxwell–Garnett result, and simple algebra yields ε_{eff} as a rational function of f_A.

It can thus be seen that we are summing the polarizabilities rather than the dielectric constants. This is a physically meaningful outcome. The result can thus be generalized for several kinds of inclusions of different fill fractions and dielectric constants—(f_A, ε_A), $(f_{A'}, \varepsilon_{A'})$—etc. provided that we remain in a dilute regime, hence the sum must obey $f_A + f_{A'} + \ldots \ll 1$ (and the distances among constituents should be much larger than their size).

15.A.3 Bruggeman

An issue with the Maxwell–Garnett obviously arises if we raise the volume fraction of A in B. At some point there will be a percolation of A particles (or they will be very close to each other) and if we traverse the non-obvious zone where $f_A \sim f_B$ to reach $f_B \ll \boldsymbol{f}_A$, we must reverse the roles of A and B. It is easily seen that the formula is not symmetric (see the discussion around Eq. (24) in ref. (Aspnes, 1982)). So its lack of validity is rather obvious as soon as one leaves the dilute regime, which can be defined by, say, $f_A < 0.15$ for practical purposes.

Bruggeman proposed a theory that does not make assumptions about 'who is the matrix and who is the inclusion'. He proposed instead that the permittivity of the unknown effective medium has to be found in a self-consistent manner. The corresponding formula is often given in the original terms of Bruggeman, who worked on conductivities (the 'A in B or B in A' conundrum was all the more obvious for metal-containing mixtures, where percolation changes properties more dramatically than in insulating dielectrics). Again, Aspnes (Aspnes, 1982) gives a dielectric version that is worth reproducing:

$$0 = f_A \frac{\varepsilon_{\text{eff}} - \varepsilon_A}{\varepsilon_{\text{eff}} + 2\varepsilon_A} + f_B \frac{\varepsilon_{\text{eff}} - \varepsilon_B}{\varepsilon_{\text{eff}} + 2\varepsilon_B}. \tag{15.44}$$

Note that the left-hand term becomes zero just because of the self-consistency assumption. By the same token, this restores the A/B symmetry.

We do not review the developments on various shapes of inclusions, which are well documented in the literature and in some textbooks. The above steps must essentially be revisited when the polarizability α becomes a tensor, $\overline{\overline{\alpha}}$: for a cigar-shaped ellipsoid, there would be, for instance, in the appropriate frame, the long-axis component differing from the two short-axis ones. A term often used to name the numerical parameters that describe these polarizabilities is 'depolarization': basically, the field inside such inclusions lies somewhere between the $\mathbf{P}/3\varepsilon_0$ value for a sphere and the $\mathbf{P}/\varepsilon_0$ value for a slab (furthermore, if a simple scalar coefficient can apply, as in this brief account, it implies that we consider the field along the eigenvectors of the polarizability tensor).

Exercises

Exercise 15.1 Simple application of homogenization formulas.

(1) Find the result of the homogenization of the following thin-film stacks along the stack axis z for either an electric field along x (ε_{xx}) or an electric field along z (ε_{zz}), with the assumption that both media have the same thickness: → silicon ($\varepsilon_1 = 3.5^2$) and silica ($\varepsilon_2 = 1.5^2$); → silicon ($\varepsilon_1 = 3.5^2$) and silicon nitride ($\varepsilon_2 = 2.0^2$); → silica ($\varepsilon_1 = 1.5^2$) and some polymer ($\varepsilon_2 = 1.44^2$).

(2) Find the filling factor of the larger-index stack constituent, $f = \frac{\text{thickness 1}}{\text{thickness 1+thickness 2}}$, which causes the effective value of ε_{xx} to coincide exactly with the arithmetic mean $\frac{1}{2}(\varepsilon_1 + \varepsilon_2)$. Solve the same question for ε_{zz}

Exercise 15.2 Homogenization of 2D photonic crystals. We tackle the issue of 2D periodic rectangular (but not dilute) inclusions of a medium 2 ($\varepsilon_2 = 12$) in medium 1 ($\varepsilon_1 = 2$) (see the generic Fig. 15.10). A particular case is, for instance, that the rectangle dimensions are one-half of the period in each of the two dimensions x and y but do not need to be identical. We look for effective values of ε_{xx} and ε_{yy}. Proceed as follows.

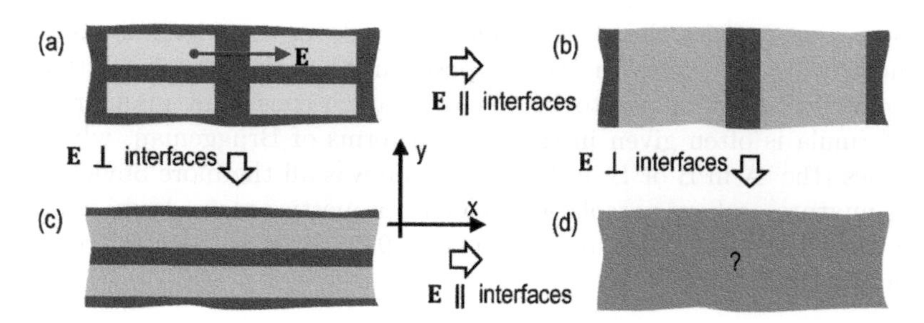

Fig. 15.10 Treatment of a 2D periodic structure made of rectangular inclusions and invariant along z. (a) Initial structure and field component considered, E_x here; (b) averaging along y; (c) averaging along x; (d) 2D averaging gives two differing approximations ('x then y' and 'y then x'), bounding the exact result.

(1) A first field direction and a first Cartesian direction are chosen, e.g. E_x, D_x, and x (see Fig. 15.10a). Then the structure is averaged along this direction as stripes, *as though each stripe were an infinitely extended layered medium*, (see Fig. 15.10c), with the correct choice (here the harmonic average, since the field crosses the interfaces).

(2) Get a first bound of ε_{xx} by averaging along y, thus from (c) to (d) in the figure.

(3) Get a second bound of ε_{xx} by averaging along y first, thus from (a) to (b), and then along x, thus from (b) to (d).

(4) Do you need to treat the other field direction with the identical ratio assumption made above? Treat an example with a different ratio of 5 to 2 along x but still 1 to 1 along y.

Exercise 15.3 Homogenization of anisotropic photonic crystals. Consider the same exercise as above, but with different fill factor along each direction: along x, the medium 2 rectangles have a size of one-half of the x-period (50%), while along y, these rectangles occupy 80% of the y-period. Depending on whether you start by averaging along x or along y, your final values of ε_{xx} differ (the same is also true for the two ways of approximating ε_{yy}, of course).

Find the two approximate average values and their difference.

What can you then state physically about the actual value of the homogenized dielectric constant with respect to the two values obtained by the distinct averaging choices?

Exercise 15.4 Dispersion relations of 1D photonic crystals. Let us consider a two-layer 1D periodic medium with a relative permittivity, $\varepsilon(z)$. The layer indices are $\sqrt{\varepsilon_1} = n_1 = 1.6$, $\sqrt{\varepsilon_2} = n_2 = 2.3$ and the thicknesses are given by the quarter-wave condition $\lambda_0/4n$. The fill fraction of material 1 is thus $f = n_2/(n_1 + n_2) \simeq 0.59$.

(1) Calculate the Fourier coefficients of $\varepsilon(z)$.

(2) Write a program to solve for the eigenproblems of Eqs (15.11) and (15.12) to find the dispersion relations $\omega = \omega(k_z)$ (and $k_z = k_z(\omega)$) shown in Figs. 15.2b and c. A Matlab program is about 30 lines. Study the convergence of the dispersion as a function of the truncation order M.

Exercise 15.5 Classical diffraction. We consider a 1D grating of period a illuminated by a plane wave at an angle θ with respect to the normal axis. The medium of incidence has a refractive index n_i and the substrate a refractive index n_s.

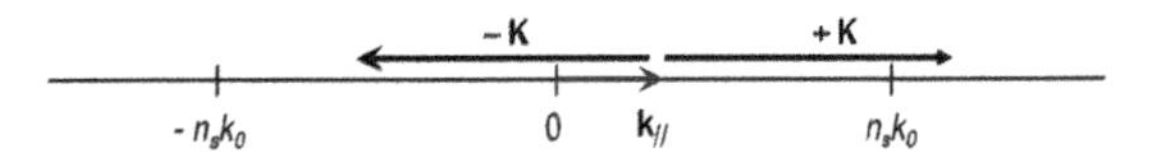

Fig. 15.11 Help specifying a single allowed order in the classical diffraction condition.

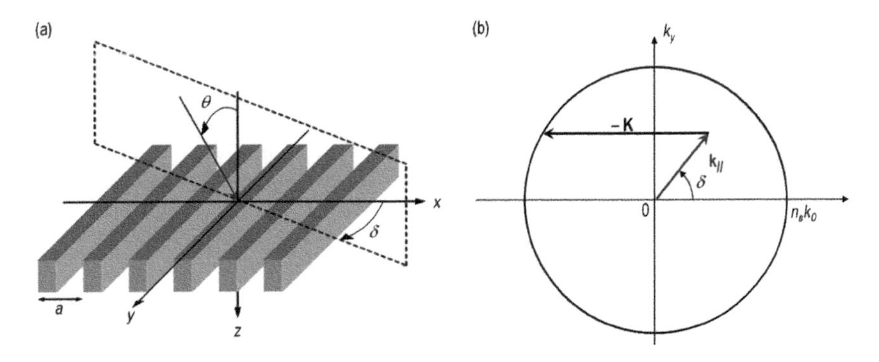

Fig. 15.12 Conical diffraction geometry. (a) Perspective view; (b) top view of wavevectors.

Under which condition (on the wavelength λ and the period a) is the zeroth order in transmission the only *propagating* order? You might get help from Fig. 15.11. How is this condition evolving with the angle of incidence θ?

Conical diffraction. We consider the same 1D grating as above, but illuminated in a conical diffraction configuration: the incidence plane is no longer normal to the invariance direction of the grating (see Fig. 15.12). Calculate, as a function of the angle δ, the condition under which the zeroth order in transmission is the only propagating order. For $\delta = \pi/2$, how is this condition evolving with the angle of incidence θ?

Exercise 15.6 Evanescent fields in the substrate and superstrate. We assume that the period of a subwavelength grating is 140 nm and the wavelength of an impinging radiation from vacuum is $\lambda = 1200$ nm, the beam being at normal incidence.

(1) Find the wavevectors of the different harmonics mK.

(2) Find the decay length in vacuum of these harmonics, thanks to the Helmholtz equation (the in-plane wavevector of these harmonics is mK).

(3) Find the harmonic number $m_{40} > 0$ at which the decay length is smaller than 30 nm, i.e. $\lambda/40$, and the value $m_{2400} > 0$ for which it is smaller than 0.5 nm. What physical/technological issue is met if considering even smaller decay lengths?

Exercise 15.7 Metal composite. We consider a disordered composite consisting of inclusions of an idealized metal with a relative dielectric constant $\varepsilon_A = -30$ in a silica matrix, $\varepsilon_B = 2.2$ (at a near-infrared frequency). Using the Maxwell–Garnett or the Bruggeman method, find the average dielectric constant of a disordered system with a volumic fraction of 1% metal. Redo the calculation for 5% metal.

Find the formal metal volumic fraction f_{zero} that cancels the Maxwell–Garnett average. If the inclusions are metal spheres of radius $a = 30\,\text{nm}$, find the average distance of metal spheres in this case. Conclude on the validity of considering independent spheres for this f_{zero} value and for the first $f = 1\,\%$ value.

Exercise 15.8 Zeroth-order cutoff condition in the substrate. The cutoff in the case of gratings was established for the case of identical indices of outer media, $n_{\text{cover}} = n_{\text{substrate}} = n$.

(1) Discuss what happens to the cutoff for a wave impinging from medium 1 (cover) on a grating with a substrate such that $n_{\text{substrate}} > n_{\text{cover}}$, notably the possibility of orders in the substrate far field that are absent in the cover far field. Provide the construction based on semicircles that generalizes that of Fig. 15.6b.

(2) Next, discuss the converse case.

(3) For normal incidence, $n_{\text{substrate}} = 3.5 > n_{\text{cover}} = 1$, and $\lambda = 1000\,\text{nm}$, find the range of period a such that there are only $m = \pm 1$ diffracted orders in the substrate, and show that the interference of these two orders causes a basic interference pattern whose pitch (along z) is very directly related to a.

Exercise 15.9 Effective index for TM polarization and normal incidence. In section 15.2.5, we derived a quadratic expansion for the effective index of a periodic 1D structure (lamellar or continuous, lossy or transparent) (see Eq. (15.38)) for TE polarization (electric field is parallel to the grating grooves). By adopting the same approach, derive the quadratic expansion for TM polarization (magnetic field is parallel to the grating grooves) and show that the extraordinary effective index reads as

$$n_{\text{eff}}^2 = \frac{1}{\hat{a}_0} + \frac{1}{\hat{a}_0^3} \sum_{p \neq 0} \frac{\hat{a}_{-p} b_p}{p} \left(\frac{a}{\lambda}\right)^2 + O\left(\left(\frac{a}{\lambda}\right)^4\right). \qquad (15.45)$$

Inspiring help may be found in J. Mod. Opt. **43**, 2063 (1996). In Eq. (15.45), the $\hat{a}_p$ are the Fourier coefficients of the inverse of the relative permittivity, $\frac{1}{\varepsilon(z)} = \sum \hat{a}_p \exp(ipKz)$, and $b_p = \sum_{q \neq 0} \frac{\hat{\varepsilon}_{p-q} \hat{a}_q}{q}$. Similar expressions hold for 2D-periodic composites.

Exercise 15.10 Effective index for oblique incidence: Spatial dispersion. Let us consider a 1D lamellar subwavelength grating with period a in the z-direction and a finite depth in the x-direction. The grating relative permittivity $\varepsilon(z)$ is assumed to be independent of the x- and y-directions and is illuminated by a plane wave (frequency ω) at oblique incidence (angle θ) in classical mounting; see, for instance, Fig. 15.6. We denote by n_{i} the refractive index of the incident medium. We assume that the period is small, and would like to know the effective index to be considered for the substitution by a homogeneous thin film. In the chapter main text, only the normal incidence case was derived (see Eq. (15.38)). Inspiring help may be found in J. Mod. Opt. **66**, 850 (2019).

For that purpose, let us calculate the x-component, $k_x = n\omega/c$ (we denote by n the effective index for notational simplicity), of the propagation constant of the fundamental Bloch mode of the periodic medium $\varepsilon(z)$ for a fixed value of its z-component, $k_z = \omega/c\ n_{\mathrm{i}} \sin(\theta)$, and let us look at how k_x changes as one departs from the static limit. We are going to find that k_x (and thus the equivalent index n) has a complex dependence with k_z, which is a manifestation of spatial dispersion. We assume TE polarization. The electric field E is along the y-direction and can be written $E = \exp(j(\omega/c)nx)\sum_p s_p \exp(j(pK + k_z)z)$. We further introduce two dimensionless parameters: $\beta = \frac{k_z}{\omega/c} = n_{\mathrm{i}} \sin(\theta)$ and the scaling parameter $\alpha = \frac{K}{\omega/c} = \frac{\lambda}{a}$.

We now look for a solution with a Taylor expansion in the limit $\alpha^{-1} \to 0$:

$$n^2 = N_0 + N_1\alpha^{-1} + N_2\alpha^{-2} + \ldots, \tag{15.46}$$

$$s_m = s_m^{(0)} + s_m^{(1)}\alpha^{-1} + s_m^{(2)}\alpha^{-2} + \ldots. \tag{15.47}$$

(1) Justify the form chosen for the electric field.

(2) Show that the electric field satisfies $\Delta E + (\omega/c)^2\varepsilon(z)E = 0$. Remember that the divergence of the displacement vector (not that of the electric field) is null.

(3) By introducing the Fourier series expansion of the relative permittivity, namely $\varepsilon(z) = \sum_p \hat{\varepsilon}_p \exp(i\,pK\,z)$, show that

$$\forall\, m, \qquad -[n^2 + (m\alpha + \beta)^2]s_m + \sum_p \hat{\varepsilon}_{m-p}s_p = 0. \tag{15.48}$$

(4) We now solve Eq. (15.48) from $\alpha = 2$ down to $\alpha = -2$ in a recursive manner using the Taylor series, remembering that α diverges.

- Show that the largest term in α^2 of the series is $\forall\, m, m^2 s_m^{(0)}$ and deduce that $s_m^{(0)} = 0$, $\forall\, m \neq 0$.
- The largest terms all being null, let us consider the term of degree 1 in α. Show that the term is $-2m\alpha\beta s_m^{(1)}$ for all $m \neq 0$. Then deduce that $s_m^{(1)} = 0$, $\forall\, m \neq 0$.
- We now consider the terms of degree 0, successively for $m = 0$ and $m \neq 0$ in Eq. (15.48). Show that this leads to:

$$N_0 = \hat{\varepsilon}_0 - \beta^2, \tag{15.49}$$

$$s_m^{(2)} = \hat{\varepsilon}_m/m^2\ s_0^{(0)}, \quad m \neq 0. \tag{15.50}$$

Note that the dependence of N_0 with β^2 is just expected from the static limit of the tensor.

- Now we iterate with the terms of degree -1 in α^{-1}. Strictly following the same approach, show that

$$N_1 = 0, \tag{15.51}$$

$$s_m^{(3)} = \hat{\varepsilon}_m/m^2\ s_0^{(1)} - 2\beta\hat{\varepsilon}_m/m^3\ s_0^{(0)}, \quad m \neq 0\,. \tag{15.52}$$

- By considering the terms of degree -2 in α^{-2}, show that

$$N_2 = \sum_{p\neq 0} \hat{\varepsilon}_p\hat{\varepsilon}_{-p}/p^2, \tag{15.53}$$

$$s_m^{(4)} = \left(\frac{\beta^2\hat{\varepsilon}_m}{m^4/4} - \frac{\hat{\varepsilon}_m\hat{\varepsilon}_0}{m^4} + \sum_{p\neq 0}\frac{\hat{\varepsilon}_p\hat{\varepsilon}_{m-p}}{p^2m^2}\right)s_0^{(0)} - \frac{\beta\hat{\varepsilon}_m}{m^3/2}s_0^{(1)} + \frac{\hat{\varepsilon}_m}{m^2}s_0^{(2)}, \quad m \neq 0\,. \tag{15.54}$$

- For the degree -2, we do not find any dependence of N_2 on β. By continuing the iterative procedure, show that

$$N_3 = -2\beta \sum_{p\neq 0} \frac{\hat{\varepsilon}_p \hat{\varepsilon}_{-p}}{p^3} = 0, \tag{15.55}$$

$$N_4 = (4\beta^2 - \hat{\varepsilon}_0) \sum_{p\neq 0} \frac{\hat{\varepsilon}_p \hat{\varepsilon}_{-p}}{p^4} + \sum_{p,q\neq 0} \frac{\hat{\varepsilon}_{-p} \hat{\varepsilon}_{p-q} \hat{\varepsilon}_q}{p^2 q^2} . \tag{15.56}$$

Exercise 15.11 Anti-reflection with a thin film. In this classical exercise, we derive the condition fulfilled by a thin film for antireflection coating at normal incidence. We denote by h the film thickness and by n its refractive index. The film is assumed to be deposited on a substrate of refractive index n_s and the incident medium has a refractive index n_i.

(1) We first consider the interface between two media labelled by '1' and '2' and the layer illuminated by a plane wave impinging on the interface from medium '1'. The incident field generates a transmitted wave and a reflected one. We denote by t_{12} and r_{12} the (in general) complex transmission and reflection coefficients. Importantly, the coefficients are defined for normalized plane waves with a Poynting vector equal to one. Using the continuity of the parallel components of the electric and magnetic field, calculate t_{12} and r_{12}. Verify that $t_{12} = t_{21}$. This reciprocity condition is due to the normalization.

(2) We now consider the entire system with the anti-reflection layer and a normally incident plane wave from the incident medium. In the classical Fabry-Perot picture, the solution of this scattering problem is composed of a reflected plane wave with a reflection coefficient r, a transmitted plane wave with a transmission coefficient t, and two counterpropagative plane waves bouncing back and forth in the deposited layer (see chapter 7). We denote by f and b the excitation coefficients of the waves propagating forwards and backwards, respectively. Write the system of two equations verified by f, b, r, and t, eliminate b and f, and calculate r and t.

(3) Show that $r = 0$ is attained if $n = (n_i n_s)^{1/2}$ and $h = \lambda/(4n)$.

16
Propagation in periodic media (II): Photonic crystals

In this chapter, we focus our attention on the inhibition of wave propagation in non-absorbing periodic media for which photonic gaps are formed. We saw in Chapter 15 that in order to find the Bloch modes in a medium with 1D periodicity along z, one has to look for a propagation constant k_z that is calculated as an eigenvalue of a system of equations parametrized by the arbitrary frequency ω. We mostly dealt with solutions that had a real k_z, and with the relevance of defining an effective index in the subwavelength regime.

In regimes where solutions have a non-zero imaginary k_z, we first get an idea of the main features of inhibited propagation by looking in section 16.1 at a finite quarter-wave stack made of alternating materials.

In section 16.2, we describe a simpler two-wave approach that gives most of the essential features of bandgaps in one dimension. We are then ready to examine more closely how to deal with finite or semi-infinite systems in section 16.3. We complement this with a discussion in section 16.4 of the various flavours of slow-light propagation, which occurs typically but not exclusively near bandgaps.

Section 16.5 deals with 2D periodicity, a case where the naming 'photonic crystal' is entirely justified. The concepts of bands and gaps were indeed matured in solid-state physics to explain electron propagation/inhibition in bulk conductors, semiconductors, or metals. The discussion is centred around the attainment of an omnidirectional bandgap.

A large part of the interest in photonic crystals is due to their ability to control the emission rate in spontaneous emission processes, as articulated by Yablonovitch in 1987 for periodic structures: the density of states (DOS) being the main factor in the spontaneous emission rate, an absence of states in the gap translates into an inhibited spontaneous emission. The 2D case is best suited to grasping the issue of omnidirectional bandgaps and is also the basis for many applications, hence its importance. Finally, Complement 16.A is dedicated to the photonic-crystal fibre and Complement 16.B addresses 3D photonic bandgaps, which were seen at their inception as a holy grail of light control.

Introduction to Nanophotonics. Henri Benisty, Jean-Jacques Greffet, and Philippe Lalanne, Oxford University Press.
 DOI: 10.1093/oso/9780198786139.003.0016

16.1 Bandgap opening and bands: Qualitative discussion

16.1.1 Transmission spectrum of the quarter-wave stack

To quickly get an idea of the physical meaning of the complex constant k_z found when solving the eigenvalue equation in the gap (see chapter 15), we consider the reflection of a normally incident plane wave on a periodic thin-film stack with a finite thickness. Fig. 16.1 represents the transmission spectrum as a function of normalized frequency $T(\lambda_0/\lambda)$ of a dielectric stack composed of six pairs with alternating layers of indices $n_1 = 1.6$ and $n_2 = 2.3$ of thicknesses equal to $t_1 = \lambda_0/(4n_1)$ and $t_2 = \lambda_0/(4n_2)$, respectively, deposited onto a glass substrate (see inset of Fig. 16.1). The calculation is performed with the classical 2×2 matrix approach described in textbooks such as Yeh's one (Yeh, 1988). This stack is called a quarter-wave stack for the central wavelength λ_0, and acts as a Bragg mirror with a broad, flat-bottomed transmission dip around λ_0. In the absence of absorption, $1 - T(\lambda/\lambda_0)$ represents the reflection $R(\lambda/\lambda_0)$; around the central wavelength, almost all the incident energy is reflected, with only a tiny fraction transmitted.

The reason is that reflections at the successive interfaces tend to interfere constructively in the reflected beam, redistributing the energy of the incident beam in favour of the reflected one.

16.1.2 Intensity profile inside the quarter-wave stack

Let us add more insight into the Bragg reflection by considering the field intensity distribution inside the stack for the central wavelength λ_0: as shown in Fig. 16.2, it is composed of a stationary wave in the incident medium, an exponentially damped oscillatory wave in the stack, and a weak transmitted plane wave in the substrate. Because the field intensity at the bottom interface is very weak, the reflected field at this interface is also weak; the total field inside the stack is dominated by the gap Bloch mode propa-

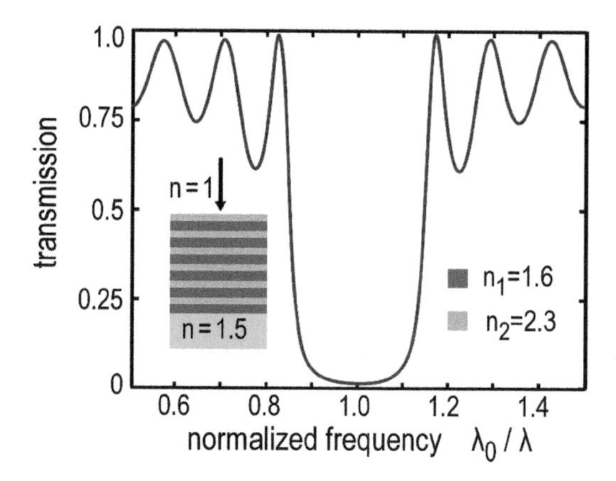

Fig. 16.1 Transmission spectrum of a quarter-wave thin-film stack designed for the central wavelength λ_0, as a function of normalized frequency λ_0/λ. The stack is deposited on a glass substrate (refractive index $n = 1.5$) and illuminated from air at normal incidence.

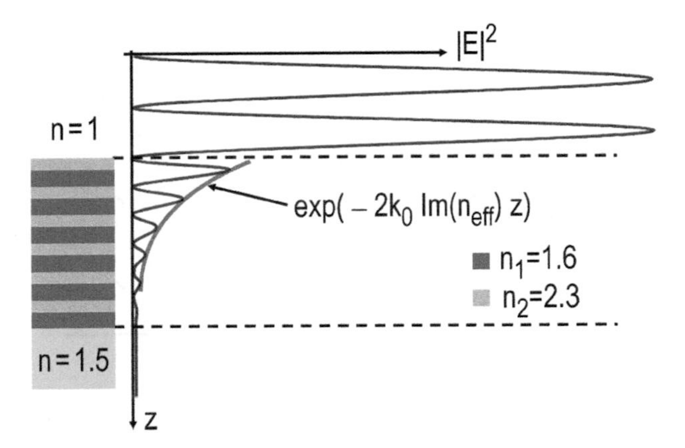

Fig. 16.2 Sketch of the intensity $|E|^2$ inside a stack of finite thickness for the central wavelength ($\equiv$ Bragg frequency) $\lambda = \lambda_0$. The envelope of the oscillating field in the stack exponentially decreases and the penetration depth is given by $\lambda_0 / (2\pi \mathrm{Im}(n_{\mathrm{eff}}))$.

gating downwards in the stack and the Bloch mode propagating upwards from the rear interface is not seen in the exponentially damped oscillatory wave. Therefore, the field-intensity envelope (thick line) is proportional to $\exp(-2k_0 \mathrm{Im}(n_{\mathrm{eff}})z) \equiv \exp(-2\mathrm{Im}(k_z)z)$ (z is downward). The imaginary part of k_z, or, even more strikingly, of the effective index, is related not to any absorption (materials are lossless), but to the evanescent character of the Bloch modes associated with the photonic-gap frequencies.

16.1.3 Intuitive insight into bandgap opening: Phase-matching condition

We study again a 1D periodic stack and pay attention to its Bloch modes. In Chapter 15, we considered the source-free Maxwell curl equations for such a stack, derived coupling equations between field harmonics $\mathsf{E}_m, \mathsf{H}_m$ and dielectric constant Fourier components $\hat{\varepsilon}_q$ (see Eqs (15.9) and (15.10)), and finally solved the system of coupled equations numerically to obtain the Bloch-mode propagation constant k_z as a function of the frequency ω. In order to have an intuitive understanding of what is happening in the gap, we eliminate the magnetic harmonics H_m in Eqs (15.9) and (15.10) to obtain the Helmholtz-like equation for the electric-field harmonics

$$-(k_z + mK)^2 \mathsf{E}_m + (\omega/c)^2 \sum_q \hat{\varepsilon}_{m-q} \mathsf{E}_q = 0 \, . \tag{16.1}$$

Let us start by considering the trivial asymptotic case without any modulation, $\varepsilon(z) = \hat{\varepsilon}_0$. Bloch modes are then pairs of counterpropagative plane waves, $E^{(-)} = \exp(-ik_{PW}z)$ and $E^{(+)} = \exp(+ik_{PW}z)$, with $k_{PW} = k_0\sqrt{\hat{\varepsilon}_0}$, obeying the dispersion relation $\omega/c = \pm k_z/\sqrt{\hat{\varepsilon}_0}$. The latter are shown in Fig. 16.3a as two straight solid lines.

As soon as the modulation is non-zero, the two plane waves become pseudo-periodic Bloch modes and we have to solve Eq. (16.1). By considering infinitely small modulations and thus neglecting all the Fourier coefficients $\hat{\varepsilon}_m$ for $m \neq 0$ (they are proportional to the refractive index modulation), Eq. (16.1) becomes $-(k_z + mK)^2 \mathsf{E}_m + (\omega/c)^2 \hat{\varepsilon}_0 \mathsf{E}_m = 0$, whose solutions are

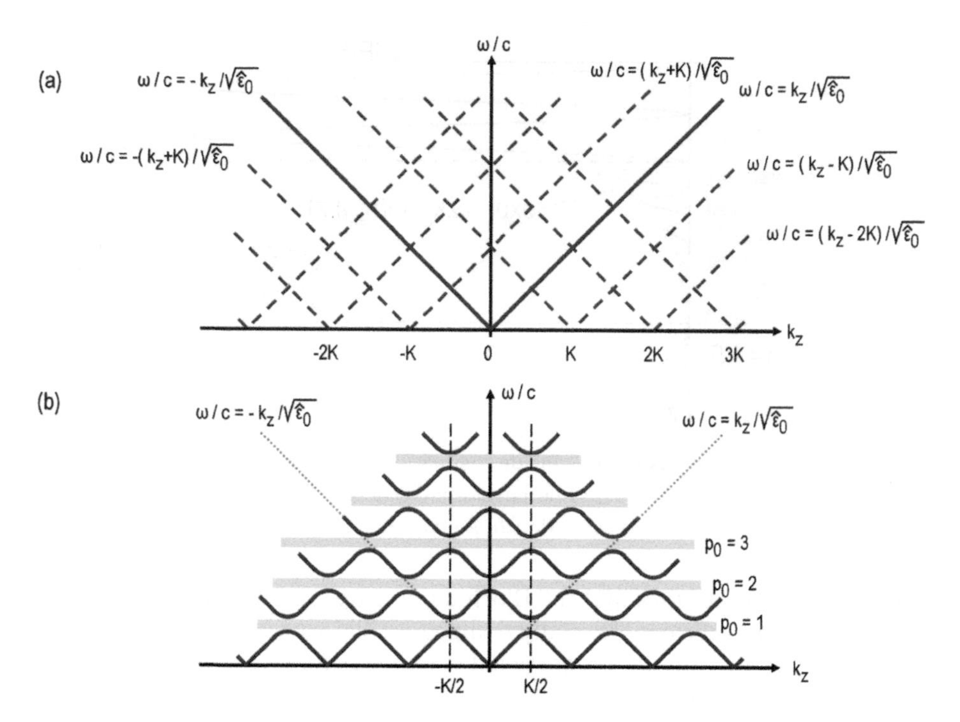

Fig. 16.3 Bandgap opening. (a) Solid lines: dispersion relation of the plane waves propagating in a homogeneous material with a refractive index $\hat{\varepsilon}_0^{1/2}$; dashed lines (empty lattice limit): K-periodic dispersion relations of a Bloch mode in a periodic stack with an infinitely small refractive index modulation Δn compared to $\hat{\varepsilon}_0^{1/2}$. (b) Dispersion curves $\omega(k_z)$ including the bandgaps (true lattice). Since the relation dispersion is K-periodic, it is sufficient to consider it on an interval of length K. The conventional one, ensuring also a minimal $|k_z|$, is the first Brillouin zone ($-K/2 < k_z < K/2$), as indicated.

$$\omega/c = \pm(k_z + mK)/\sqrt{\hat{\varepsilon}_0}\,. \tag{16.2}$$

We thus identify a first basic effect of periodicity: the dispersion relation of Bloch modes in the weak-modulation limit are identical to that of plane waves, except for periodic shifts by K multiples : $\omega(k_z + mK) = \omega(k_z)$. They are shown with dashed lines in Fig. 16.3a. Now remains the important step of bandgap opening.

In order to intuitively understand it, let us go back to the Helmholtz equation $\partial^2\mathsf{E}/\partial z^2 + k_0^2 \sum_p \hat{\varepsilon}_p \exp(i\,pK\,z)\mathsf{E} = 0$ and again consider the weak-modulation limit. In this limit, we expect that the two plane waves, $\mathsf{E}^{(-)} = A\exp(-ik_{PW}z)$ and $\mathsf{E}^{(+)} = A\exp(+ik_{PW}z)$, are nearly solutions. We inject these solutions into the Helmholtz equation to obtain

$$\partial^2\mathsf{E}^{(-)}/\partial z^2 + k_0^2\,\hat{\varepsilon}_0\,\mathsf{E}^{(-)} + k_0^2 A \sum_{p\neq 0} \hat{\varepsilon}_p \exp(i\,pK\,z - i\,k_{PW}\,z) \approx 0\,, \tag{16.3}$$

$$\partial^2\mathsf{E}^{(+)}/\partial z^2 + k_0^2\,\hat{\varepsilon}_0\,\mathsf{E}^{(+)} + k_0^2 A \sum_{p\neq 0} \hat{\varepsilon}_p \exp(i\,pK\,z + i\,k_{PW}\,z) \approx 0\,, \tag{16.4}$$

where '≈ 0' means that the error made is $\ll A$. Indeed, since $|\hat{\varepsilon}_p| \ll |\hat{\varepsilon}_0|$ for $p \neq 0$, we could neglect the latter summation terms to again obtain the dispersion relations shown as straight dashed lines in Fig. 16.3a. However, such a neglect holds if upon integration of the differential equations, all the sum's terms oscillate without matching the phase of any of the main terms of both equations $\mathsf{E}^{(\pm)}$. But if one of the terms does match, it should not be neglected. This is exactly what happens for gap opening: matching occurs in each of the two equations with the main term of the other one, causing coupling, and thus scattering, between $E^{(+)}$ and $E^{(-)}$. Mathematically, this happens if there is an integer p_0 such that $-p_0 K + k_{PW} = -k_{PW}$. If we neglect the sum terms that do not match, this leads to the appearance of crossed terms

$$\partial^2 \mathsf{E}^{(-)}/\partial z^2 + k_0^2\, \hat{\varepsilon}_0\, \mathsf{E}^{(-)} + k_0^2\, \hat{\varepsilon}_{-p_0}\, \mathsf{E}^{(+)} \approx 0\,, \tag{16.5}$$

$$\partial^2 \mathsf{E}^{(+)}/\partial z^2 + k_0^2\, \hat{\varepsilon}_0\, \mathsf{E}^{(+)} + k_0^2\, \hat{\varepsilon}_{p_0}\, \mathsf{E}^{(-)} \approx 0\,. \tag{16.6}$$

In other words, the contrapropagative plane waves $\mathsf{E}^{(+)}$ and $\mathsf{E}^{(-)}$ are coupled completely symmetrically at the exact condition $-p_0 K + k_{PW} = -k_{PW}$, which can be called a Bragg condition. The Bloch modes that result from these coupled modes are stationary patterns that do not carry any energy in the lossless case.

We will admit without proof that, as long as a coupling exists (i.e. as long as we are in the bandgap), the actual Bloch-mode propagation constant k_z (not k_{PW}) satisfies the same phase matching as $\mathsf{E}^{(+)}$ and $\mathsf{E}^{(-)}$, i.e.

$$\mathrm{Re}(k_z) = p_0\, \frac{K}{2}, \quad \rightarrow \quad \mathrm{Re}(n_{\text{eff}}) = p_0\, \frac{\lambda}{2a}, \tag{16.7}$$

where Re(...) has been introduced to take into account that, at variance with plane waves, the propagation constant of the gap Bloch mode is complex. In the $\omega(k_z)$ diagram, the coupling between the counterpropagative waves results in a gap opening and an absence of Bloch mode, as shown in Fig. 16.3b. This parallels the textbook introduction of energy gaps between energy bands in solid-state physics when starting from free electrons. We emphasize that all the k_z-periodic bandgap openings in Fig. 16.3b, which correspond to a given band label $p_0 = 1, 2, 3, \ldots$, result from the coupling between two dashed dispersion lines that are offset horizontally by increasing momenta $p_0 K$, with the two coupling coefficients given by $\hat{\varepsilon}_{p_0}$ and $\hat{\varepsilon}_{-p_0}$. Conversely, in the $k_z(\omega)$ representation, the Bloch mode in the gap is a stationary pattern. Its complex propagation constant reflects its evanescent character, and its real part is strictly given by the phase-matching condition of Eq. (16.7). Although this equation has been derived in an approximate manner in the limit of small perturbation, it is exact (see Yeh's detailed account (Yeh, 1988), for instance, for lamellar periodic structures). It indeed holds in much more complex systems, such as ridge waveguides with periodic holes or the segmented waveguides presented later. The gap–Bloch-mode dispersion relations of the real part of the effective index, $n_{\text{eff}} = \lambda/a, 2\lambda/a$, and $3\lambda/a$, are represented as blue curves in Fig. 16.4a.

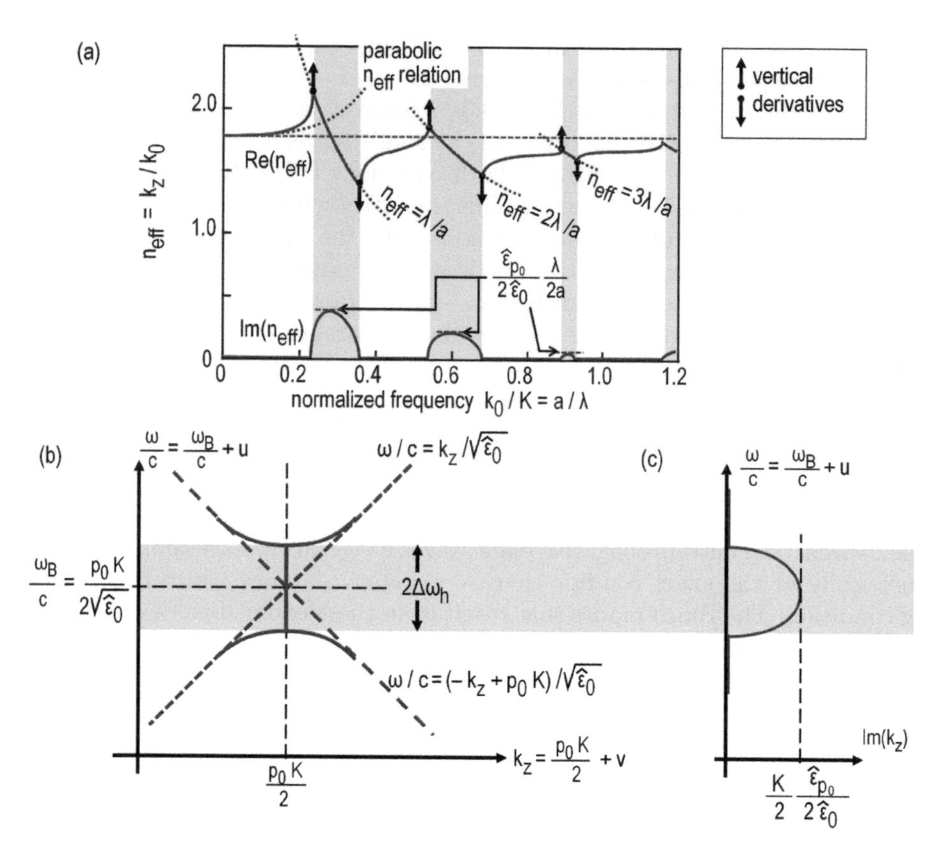

Fig. 16.4 Evanescent Bloch modes in the gap. (a) Same as in Fig. 15.5. We simply added the phase-matching condition, i.e. Eq. (16.7), which provides the real part of the effective index of evanescent gap Bloch modes with the dashed hyperbolic curves. The latter perfectly match the computed dispersion relation in the gap. We also added the black arrows to indicate that the slope of the dispersion curve is infinite at the boundary of the gaps (the group velocity is zero). (b) Hyperbolic dispersion obtained with the coupled-mode theory at the gap of order p_0 at the boundary of Brillouin zones. (c) Imaginary part of the effective index in the gap.

16.2 Coupled-wave method

16.2.1 Coupled equation for field harmonics

The phase-matching Bragg condition of Eq. (16.7) provides the value of the effective index in the gap, but does not tell the gap spectral range over which it occurs. To know this, one should solve the coupled equations for the electric-field harmonics and for full analyticity, some approximations have to be made: we are going to retain only two harmonics in the coupled equations, in the same way as most textbooks to date.

Let us consider the gap of order p_0 that opens up at the crossing of the central dispersion branch with another one separated by a momentum difference p_0K. The Bragg point defined by the intersection of the solid line $\omega/c = k_z\sqrt{\hat{\varepsilon}_0}$ and the dotted line $\omega/c = (p_0K - k_z)/\sqrt{\hat{\varepsilon}_0}d$ in Fig. 16.3a is situated at $k_z = p_0K/2 = p_0\pi/a$. Its (Bragg) frequency ω_B is then easily shown to be $\frac{\omega_B}{c} = \frac{p_0\pi\sqrt{\hat{\varepsilon}_0}}{a}$.

To derive closed-form expressions for the main gap properties, we assume that in the vicinity of the Bragg point ($\omega = \omega_B, k_z = p_0\pi/a$), the Bloch mode only has two predominant Fourier components, S_{p_0} and S_{-p_0}. Then, the system of Eq. (16.1) above, $-(k_z + mK)^2\mathsf{E}_m + (\omega/c)^2\sum_q \hat{\varepsilon}_{m-q}E_q = 0$ for any m, simplifies, and we get two equations with two unknowns:

$$\begin{aligned} \left[(\omega/c)^2\hat{\varepsilon}_0 - k_z^2\right]\mathsf{E}_0 \quad &+ \quad (\omega/c)^2\hat{\varepsilon}_{p_0}\mathsf{E}_{-p_0} = 0 \;\; (m=0), \\ (\omega/c)^2\hat{\varepsilon}_{-p_0}\mathsf{E}_0 + \left[(\omega/c)^2\hat{\varepsilon}_0 - (k_z - p_0K)^2\right]&\mathsf{E}_{-p_0} = 0 \;\; (m=-p_0)\,. \end{aligned} \tag{16.8}$$

To make the treatment algebraically lighter, we use local variables u, v centred around the Bragg point $(\omega_B, p_0\pi/a)$, reminding the reader that our approach is appropriate only for weak modulations $|\hat{\varepsilon}_{\pm p_0}| \ll |\hat{\varepsilon}_0|$. Both u and v have the dimensions of wavevectors. Specifically, we write

$$k_z = p_0\frac{K}{2} + u, \tag{16.9}$$

$$\frac{\omega}{c} = \frac{\omega_B}{c} + v\,. \tag{16.10}$$

The system of Eq. (16.8) can then be worked out to retain only linear u, v terms in the square brackets:

$$\left[v\sqrt{\hat{\varepsilon}_0} - u\right]\mathsf{E}_0 \quad + \quad \hat{\varepsilon}_{p_0}\frac{K}{4\,\hat{\varepsilon}_0}\mathsf{E}_{-p_0} = 0 \;\; (m=0), \tag{16.11}$$

$$\hat{\varepsilon}_{-p_0}\frac{K}{4\,\hat{\varepsilon}_0}\mathsf{E}_0 + \left[v\sqrt{\hat{\varepsilon}_0} + u\right]\mathsf{E}_{-p_0} = 0 \;\; (m=-p_0)\,. \tag{16.12}$$

16.2.2 Dispersion relation

For this linear homogeneous system, non-zero solutions exist for E_0 and E_{-p_0} if the determinant is zero. We thus obtain a dispersion relation of hyperbolic nature

$$v^2 - \frac{u^2}{\hat{\varepsilon}_0} = \hat{\varepsilon}_{p_0}\,\hat{\varepsilon}_{-p_0}\,\frac{K^2}{16\,\hat{\varepsilon}_0^3} = |\hat{\varepsilon}_{p_0}|^2\,\frac{K^2}{16\,\hat{\varepsilon}_0^3}\,, \tag{16.13}$$

the second equality being valid if $\varepsilon(z)$ is real. The geometry of these hyperbolas is depicted in Fig. 16.4b.

The width of the photonic gap is the difference between the ordinates of the hyperbola apex aligned at the Bragg wavevector or Brillouin zone edge $u = 0$. They read $v = \pm|\hat{\varepsilon}_{p_0}|\,(K/4)\,\hat{\varepsilon}_0^{-3/2}$. Noting $\Delta\omega_h$, the gap half-width in frequency, we thus obtain for the normalized gap width $2\Delta\omega_h/\omega_B$ the expression

$$2\,\frac{\Delta\omega_h}{\omega_B} = \frac{1}{p_0}\,\frac{|\hat{\varepsilon}_{p_0}|}{\hat{\varepsilon}_0}\,. \tag{16.14}$$

The main result is that the gap width is proportional to the amplitude of the p_0-th Fourier component of $\varepsilon(z)$. This result is similar to the perturbation results of quantum mechanics, whereby energy levels are affected at first order by the perturbation strength (usually denoted by $\langle\psi\,|V|\,\psi\rangle$). Here, the $1/p_0$ scaling is due to the normalization by ω_B: the absolute size $\Delta\omega_h/c = |\hat{\varepsilon}_{p_0}|\,(K/2)\;\hat{\varepsilon}_0^{-3/2}$ is independent of p_0 (except for the Fourier coefficient).

16.2.3 Propagation constant in the photonic gap

Let us now calculate the complex propagation constant in the gap, and let us do this, in particular, at the centre frequency ($v = 0$). Prolonging the algebraic hyperbolic dispersion of Eq. (16.13) in the gap, we obtain $u^2 = -\,|\hat{\varepsilon}_{p_0}|^2\;K^2/(16\,\hat{\varepsilon}_0^3)$, hence a purely imaginary u. Thus, in the gap centre, the complex propagation constant $k_z(\omega_\mathrm{B}) = n_\mathrm{eff}(\omega_\mathrm{B})k_0$ is given by

$$k_z(\omega_\mathrm{B}) = \frac{K}{2}\left[p_0 + i\frac{|\hat{\varepsilon}_{p_0}|}{2\hat{\varepsilon}_0}\right] \tag{16.15}$$

(see also Fig. 16.4c). So, we find that the spatial damping rate $\mathrm{Im}(k_z)$ is also directly proportional to the relative strength of the specific Fourier component $|\hat{\varepsilon}_{p_0}|\,/\hat{\varepsilon}_0$. Overall, large gaps and strong spatial decays are both promoted by large modulations.

16.2.4 Bloch-mode localization around band edges

We now examine the Bloch modes themselves (the eigenfunctions, in the quantum mechanics analogy)—that is, the two components that we have retained, E_0 and E_{-p_0}. They give the electric field according to $E_y(z) = \mathsf{E}_0\exp(ik_zz)+\mathsf{E}_{-p_0}\exp(ik_zz-ip_0Kz)$. At the band edge, for $k_z = p_0K/2$ or $u = 0$, the Bloch modes become $E_y(z) = \mathsf{E}_0\exp(ip_0Kz/2) + \mathsf{E}_{-p_0}\exp(-ip_0Kz/2)$. At the ‘valence’ band edge (lower frequency, $v < 0$), the degenerate linear system gives $\mathsf{E}_{-p_0} = \mathsf{E}_0$, provided $\varepsilon(z)$ is real. The field is then a stationary wave:

$$E_y(z) = 2\mathsf{E}_0\cos(p_0Kz/2). \tag{16.16}$$

For the first (fundamental) gap $p_0 = 1$, the period of $|E_y(z)|^2$ is $\pi/K = a/2$: half the stack period. The antinodes (intense field) are located in the high refractive index regions. The effective index is also large, consistent with Fig. 15.2a, c, where it is obviously larger than at the Bragg central point discussed above. Conversely, at the ‘conduction’ band edge (higher frequency, $v > 0$), we have with identical assumptions $\mathsf{E}_{-p_0} = -\mathsf{E}_0$, and therefore

$$E_y(z) = 2\mathsf{E}_0\sin(p_0Kz/2). \tag{16.17}$$

Now for the first (fundamental) gap, $p_0 = 1$, the antinodes of the field coincide with the low refractive index regions. Consistently, the frequency being higher, the effective index k_z/k_0 is lower.

A good knowledge of these localization properties is vital for explaining a number of subsequent phenomena. For instance, if there is some residual absorption in one of the two materials of the stack and not in the other, the corresponding losses will be experienced by the field much more strongly if the antinodes are in that medium. More generally, knowledge of a system's wavefunction is always useful for predicting how it behaves under specific perturbations (disorder, losses, defects ...).

16.3 Semi-infinite and finite cases: Penetration depth and Bloch-mode resonances

16.3.1 Penetration depth of a semi-infinite stack

The stack presented above is nothing but a dielectric mirror, often called an interference mirror or a distributed Bragg reflector (DBR—see Chapter 20). Besides the modulus of its reflectivity, the phase of the reflectivity coefficient varies rapidly with the wavelength. It is often convenient, for instance if such mirrors are used to bound a cavity, to introduce an equivalent mirror with a phase independent of the wavelength but with the same reflection amplitude. Such an idealized metal-type mirror has a wavelength-independent reflectivity amplitude ρ, letting $0 < |\rho|^2 \leq 1$, and a given phase, not necessarily 0 or π.

Let us first denote by $r(k) = |r| \exp(i\varphi)$ the reflection coefficient of a plane wave of wavelength $\lambda = 2\pi/k$ impinging from a semi-infinite medium with refractive index n onto a semi-infinite periodic composite operating in the bandgap (see Fig. 16.5a). At the infinitesimally larger frequency $k + dk$, assuming that $|r|$ is independent of λ, we can write $r(k+dk) = |r| \exp(i\varphi + i\,(d\varphi/dk)\,dk)$. If we locate the metal-type mirror at a distance L below the interface $z = 0$ (see Fig. 16.5b), its reflection seen from the plane $z = 0$ can be written $r_{\mathrm{met}}(k) = \rho \exp(2iknL)$, to take into account the phase round trip over a distance L. And at the slightly larger frequency $k + dk$, the metal reflection becomes $r_{\mathrm{met}} = \rho \exp(2i\,nL\,k + 2i\,nL\,dk)$. By equating in the two expressions obtained at k and at $k + dk$ the first-order increment in dk, we obtain:

$$|r| \exp\left(i\frac{d\varphi}{dk}dk\right) = \rho \exp(2i\,nL\,dk). \tag{16.18}$$

Hence, for this to be at least locally true on a finite range of dk, we can equate the phases themselves and get an expression for what is often called a penetration depth:

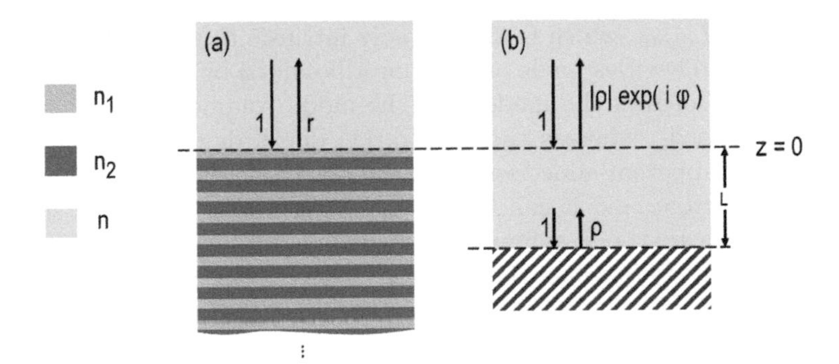

Fig. 16.5 Definition of the penetration length of a dielectric mirror by relating the reflection problem of a semi-infinite periodic stack (a) to a much simpler problem (b), with a 'metal' mirror reflectance that is independent of the wavelength. If r in (a) is given by $r(k) = |r| \exp(i\varphi)$ for a given wavelength ($k = 2\pi/\lambda$), then ρ and L in (b) are given by $L = c(d\varphi/d\omega)/(2n)$ and $\rho = |r| \exp\left[i(\varphi - ck(d\varphi/d\omega))\right]$.

$$L = L_{\text{pen}} = \frac{1}{2n} \frac{d\varphi}{dk} \,. \tag{16.19}$$

16.3.2 Extensions and caveat

Most often, the variation of L_{pen} is dictated by the position of the frequency inside the bandgap rather than by, e.g., the material's own dispersion. However, it can be shown that φ has an inflection point around the Bragg central frequency; hence, it only has a third non-zero derivative there, the zero second derivative of φ meaning that the derived penetration depth does not vary much in the central region of the photonic gap. Nevertheless, if the outer material is substantially dispersive, the round-trip phase associated to L should be modified exactly as is done for Fabry–Perot cavities, by using the group index $n_g \equiv n + \omega dn/d\omega$ instead of the phase index n. We can now re-express ρ by substituting L in the trivially obtained expression $\rho = |r| \exp(i\varphi - 2iknL)$, leading to a more intrinsic expression:

$$\rho = |r| \exp\left(i\varphi - k\frac{d\varphi}{dk}\right) \,. \tag{16.20}$$

This expression is analogous to the relation between group and phase index; it rightly suggests that we are playing with a group delay of the stack. The concept of group delay is more familiar in the context of Bragg structures with small index modulations and long penetration depths that are embedded in long optical fibres, devices known as Fibre Bragg gratings (FBG), because the fibre's guidance is secondary for the current purpose.

A caveat must be made, however, concerning the ambiguity of the term 'penetration depth'. For metallic reflection, it refers to the spatial damping length corresponding to a $1/e$ attenuation of the wave that penetrates the metal, referred to as the skin depth. The analogue quantity in our case of dielectric Bragg mirrors is not Eq. (16.19). It is the decay length of the Bloch mode dictated by the inverse of the imaginary part of the propagation constant; hence, $L_{\text{damp}} = \lambda / [2\pi \text{Im}(n_{\text{eff}})]$. For instance, we see that the outer medium value n intervenes in the definition of L_{pen}, while it should by no means intervene in L_{damp}, which is undoubtedly intrinsic to the mirror.

When it comes to cavities made of medium n bounded by such mirrors, the mirror decay length L_{damp} defines the mode size (the mode volume defined in chapter 7 is the concept often used), whereas the penetration depth describes, for the particular cavity index n, the apparent added length in the cavity medium due to the distributed reflection in the mirrors, and it can be shown to essentially define the cavity quality factor (much as in a textbook symmetric Fabry–Perot interferometer, cf. Chapter 7: the mirrors reflectivity determines the finesse F, whereas the cavity order p determines the quality factor $Q = pF$).

16.3.3 Bloch modes for the reflection of a semi-infinite stack

Bloch modes are modes of periodic media, but actual devices are never mathematically periodic, having some finite (or semi-finite) extent. How much, then, can we use the concept of Bloch modes to analyse actual finite photonic devices? Probably, a completeness theorem for normal Bloch modes, such as was discussed for optical waveguides (Vassallo, 1991), would be a welcome mathematical answer. Rather, let us try to intuitively approach the problem and consider the scattering problem shown in Fig. 16.6. A plane

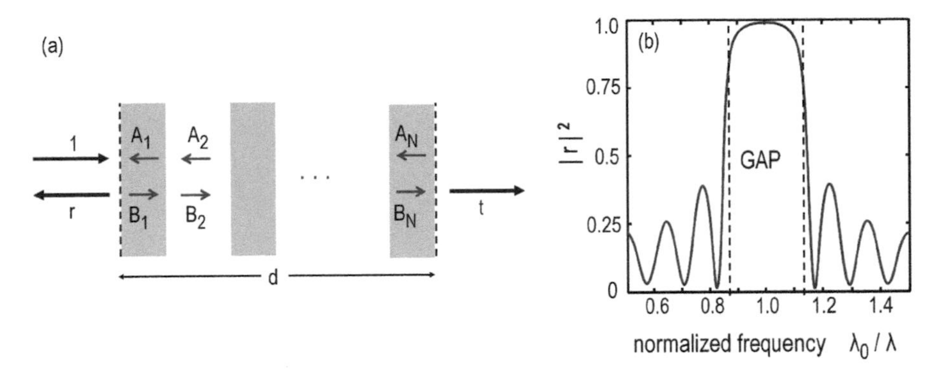

Fig. 16.6 (a) Scattering of a normally incident plane wave on a thin-film stack with a finite thickness d. The surrounding semi-infinite media have the same refractive index n; (b) reflection spectrum $|r|^2$ with Bragg peak, complementary of $|t|^2$ (stack as in Fig. 16.1).

wave with unit amplitude launched in a semi-infinite medium (refractive index n) impinges on a periodic system with thickness d. We assume that the half-space on the other side, the transmission region, has the same index, n, for the sake of simplicity. This classical layout generates back-reflected and transmitted plane waves with reflection and transmission coefficients r and t. The reflected-intensity spectrum, shown in the right panel, can be calculated with textbook 2×2 transfer-matrix methods throughout all the layers. The computation essentially consists of calculating all the plane-wave coefficients A_p and B_p (forward and backward plane waves, $p = 0, 1, \ldots, N$) and matching the boundary conditions at the interfaces. The spectrum exhibits an oscillatory behaviour with a flat top around the central wavelength of the gap. In the absence of absorption, we have $|r|^2 + |t|^2 = 1$.

Let us now consider the related problem shown in Fig. 16.7a. The same plane wave is now impinging on the same periodic stack, but now the stack is semi-infinite (its thickness is infinite). The classical 2×2 transfer-matrix method becomes inappropriate: we do not know how to end it, i.e. how to express an outgoing wave condition in the periodic half-space. We need to incorporate the Bloch modes, i.e. the natural entities that are extended and well defined. To enhance our intuition, it is useful to remark that a modal decomposition is exactly what is done in the incident semi-infinite half-space, since the incident and the reflected plane waves are nothing but the modes of the uniform medium. The common point that we can grasp is that in the first Brillouin zone of this 1D periodic system, $[-\pi/a, \pi/a]$, we also find a pair of counterpropagating Bloch modes at a given frequency, as mentioned in section 16.1. So our task consists essentially in identifying the two coefficients of these Bloch modes. We do this with a slightly more synthetic formalism: we note that at the interface shown with a dashed line in Fig. 16.7a, the continuity of the tangential (x and y) electric and magnetic fields (x, y components) leads to two non-degenerate equations that prolong the reflection/transmission picture:

$$1 \times \langle x|\mathsf{PW}^+\rangle + r_\infty\langle x|\mathsf{PW}^-\rangle = t_\infty\langle x|\mathsf{BM}^+\rangle, \tag{16.21}$$

$$1 \times \langle y|\mathsf{PW}^+\rangle + r_\infty\langle y|\mathsf{PW}^-\rangle = t_\infty\langle y|\mathsf{BM}^+\rangle\ , \tag{16.22}$$

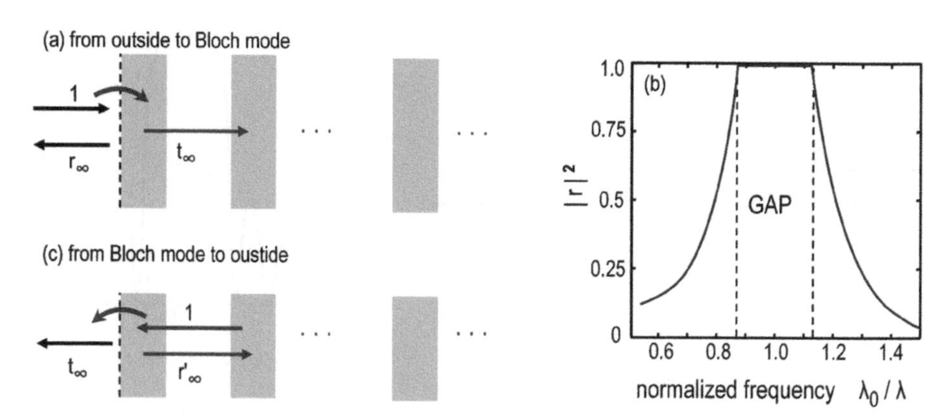

Fig. 16.7 Scattering of a normally incident plane wave on a semi-infinite thin-film stack. (a) Sketch of the geometry. (b) Characteristic reflection spectrum $|r_\infty|^2$. In the gap, $|r_\infty|^2 = 1$, and in the bands there are no oscillations at all. (c) Sketch of the reciprocal scattering problem that defines a new scattering coefficient r'_∞. Note that $|r_\infty|^2 = |r'_\infty|^2$ in the bands, since energy conservation stipulates that $|r_\infty|^2 + |t_\infty|^2 = 1 = |r'_\infty|^2 + |t_\infty|^2$.

where $|\mathsf{PW}^+\rangle$ and $|\mathsf{PW}^-\rangle$ are column vectors formed by the electromagnetic components of the right- and left-propagating plane waves in the incident medium; similarly, $|\mathsf{BM}^+\rangle$ is a column vector formed by the electromagnetic components of the right-propagating Bloch mode at the interface (taking into account the Bloch mode's z-dependence). These two equations are easily solved in practice for the unknown reflection and transmission coefficients, which are denoted by r_∞ and t_∞ to emphasize that they specifically apply to the interface problem between two semi-infinite half-spaces.

A typical reflectance spectrum is shown in Fig. 16.7b. As expected, we find that $|r_\infty|^2 = 1$ in the gap, since there is no absorption loss. We also find that there is no oscillation in the bands. Note that upon approaching the band edge, we have $|r_\infty|^2 \to 1$, implying that coupling to a slow Bloch mode is very inefficient.

Similarly, we may also consider the scattering of a left-propagating Bloch mode (see Fig. 16.7c). This defines a new reflection coefficient r'_∞, since reciprocity arguments imply that the transmission coefficients in (a) and (c) are identical under appropriate normalization.

16.3.4 Bloch modes of a finite-length stack

The approach developed in Fig. 16.7 shows that Bloch modes can be used without infinitely extended periodicity. The next logical step after a semi-infinite case is to treat the finite case. The minimal length to sensibly describe the field as a Bloch mode is the unit period a. Let us consider that inside the finite stack of Fig. 16.8, of total size d, the field is not described as a set of local plane waves scattering at each interface, but rather as two counterpropagating Bloch modes $|\mathsf{BM}^+\rangle$, and $|\mathsf{BM}^-\rangle$, scattered only at the two boundaries of the stack. With reference to Fig. 16.8a, feeding the stack

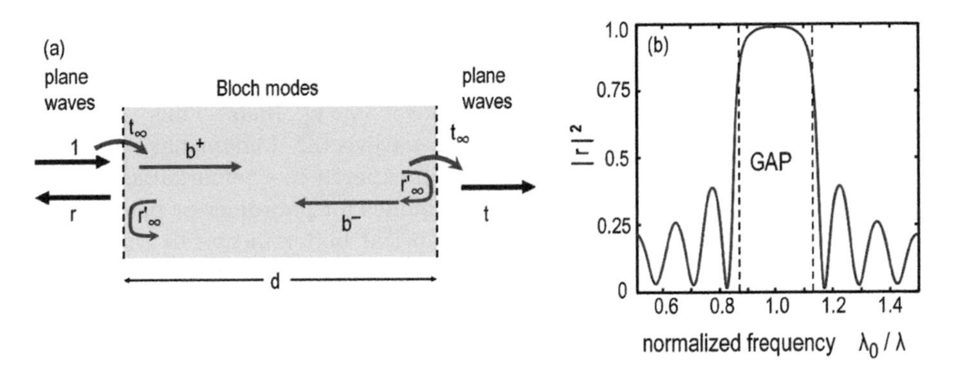

Fig. 16.8 Homogenized version of the finite-stack scattering of Fig. 16.6. (a) Bloch modes are generated and undergo round trips in the stack. (b) The reflection and transmission coefficients calculated thanks to these Bloch modes are exactly those of Fig. 16.6a, as shown in (b).

from the left with unity amplitude (plane wave), we can describe the two scattering processes at each boundary with the two amplitudes b^+ and b^- as follows:

$$b^+ = t_\infty + r'_\infty b^- \exp(i\,k_0\,n_{\text{eff}}\,d), \tag{16.23}$$

$$b^- = r'_\infty b^+ \exp(i\,k_0\,n_{\text{eff}}\,d), \tag{16.24}$$

$$t = t_\infty b^+ \exp(i\,k_0\,n_{\text{eff}}\,d)\;. \tag{16.25}$$

By eliminating b^+ and b^- from these three equations, we finally get the classical Fabry–Perot transmission for the finite stack of thickness d:

$$t = \frac{t_\infty^2 \exp(ik_0 n_{\text{eff}} d)}{1 - {r'_\infty}^2 \exp(2ik_0 n_{\text{eff}} d)}\;. \tag{16.26}$$

Eq. (16.26) shows that the entire stack can be seen as a Fabry–Perot resonator, forcing a Bloch mode to bounce back and forth over a distance d between two semi-infinite half-spaces that act as two mirrors with reflection coefficient r'_∞. It is this Fabry–Perot response that is responsible for the oscillating behaviour in the 'valence' and 'conduction' bands, with a fringe spacing dictated by the inverse of the Bloch-mode group index $n_g = c/v_g = n_{\text{eff}} + \omega(dn_{\text{eff}}/d\omega)$. In the gap, the Bloch mode is a standing wave with an evanescent character; as such, it does not carry any energy. However, because of the finite thickness of the homogenized material, the homogenized field is a superposition of two evanescent waves $(b^+|\mathsf{BM}^+\rangle + b^-|\mathsf{BM}^-\rangle)$, and this superposition carries some net power that flows by tunnelling through the stack. Said in another language, that of dichroic filters, in the filter stopband, the residual transmission (rejection is never infinite) is the flow carried by two counterpropagating evanescent Bloch modes (tunnelling effect).

This point of view, and the simple explanation it provides for the fringes (for instance, why reflectivity fringes eventually peak at unity or why they are so regular), shows that depicting the stack as a single system supporting two counterpropagating Bloch modes represents a considerable conceptual simplification.

16.4 Structural slow light

Near the edges of the bandgaps, the slopes of the dispersion relations $d\omega/dk_z$ are zero (conversely, those of $n_{\text{eff}}(\omega)$ are infinite), as shown in Fig. 16.4a. Thus, a wavepacket impinging on the stack within a frequency or wavevector domain narrow enough to match a locally linear region of the diagram will experience a remarkable slowdown.

There are several reasons for studying slow modes for photonics or plasmonics (Notomi, 2010). First, functionalities such as all-optical buffer memories that may alter or release on-demand delayed wavepackets can be envisioned. This is an important component for all-optical processing devices such as photonic routers. The second reason is that various light–matter interactions can be enhanced in slow-light structures, because the interaction time becomes longer and the spatial length of an optical pulse is compressed in proportion to the group velocity (if the group velocity dispersion is negligible), leading to enhancement of the field intensity per unit input power. The last reason is simply compactness (spatial compression). With slowness, more pulses can be stored per unit length. As readers may notice, the last two reasons are similar to the advantages of high-Q nanocavities. In fact, slow-light media and high-Q micro and nanocavities have a lot in common.

We have already encountered slow waves in Chapter 13 when studying the propagation of electromagnetic fields in tiny metal–dielectric–metal gaps. In these structures, slowness originates from the free electrons of the metal that couple with the electromagnetic fields. Slowness may also originate from material dispersion, i.e. from situations in which the velocity of light pulses can be fully described in terms of a spatially uniform but frequency-dependent refractive index of a material. A typical class of slow-light material dispersion is associated with the occurrence of a sharp dip in an absorption or gain feature. Simple saturation effects can lead to such behaviour, as well as more advanced effects such as electromagnetically induced transparency or coherent population oscillations (Khurgin, 2010). In general the effect takes place in a very narrow spectral region around the absorption line and the slowness is not accompanied by an electromagnetic field enhancement.

In contrast to material dispersion or slow plasmons for which extraneous degrees of freedom (manifested, e.g., by the spectroscopic properties of atoms or electrons) are needed to store the energy, with structural slow light, the energy is purely electromagnetic and slowness arises from multiple scattering: as it propagates through the periodic medium, light bounces forwards and backwards. It immediately follows that a phase-matching condition, namely Eq. (16.7), should be satisfied and that the characteristic dimension of slow-light structures (the present stack or periodic waveguides in practice) is much larger than that of slow-plasmon structures. The characteristic dimension is the period a and is comparable to $\approx \lambda/(2n)$. This phase-matching condition is also responsible for the fact that structural slow light operates in quite a narrow frequency range, with a bandwidth that is much smaller (about 100 times smaller) than that of slow plasmons, remaining, however, orders of magnitude larger than slowness encountered with material dispersion.

To understand what is happening at the gap edges in Fig. 16.4a, let us express the group velocity $v_g = d\omega/dk$ of the Bloch mode in the reduced units u, v used above (see Eq. (16.10)). Their definition indicates that $v_g = dv/du$. In the region of locally parabolic dispersion, close enough to the band edges ($u/K \ll |\hat{\varepsilon}_{p0}|/\hat{\varepsilon}_0$), we easily find that

$$\frac{v_g}{v_\varphi} = \frac{\sqrt{\hat{\varepsilon}_0}\, v_g}{c} = \pm \frac{4\, u}{K\, |\hat{\varepsilon}_{p0}|} \,. \tag{16.27}$$

With the same approximations, it can be additionally shown that the Bloch-mode coefficients E_0 and E_{-p0} obey the following ratio:

$$\frac{\mathsf{E}_{-p0}}{\mathsf{E}_0} = \pm \frac{|\hat{\varepsilon}_{p0}|}{\hat{\varepsilon}_0} \left[v_g/v_\varphi - 1 \right] \,. \tag{16.28}$$

Eq. (16.28) essentially says that a slow Bloch mode is a standing wave built with two counterpropagating travelling waves both propagating at the speed of light. Note that normalizing the Bloch-mode energy flow ($\propto |\mathsf{E}_0|^2 - |\mathsf{E}_{-p0}|^2$) of a slow Bloch mode propagating towards the z-direction to a given value, one readily obtains that the Bloch-mode electromagnetic fields ($\propto \mathsf{E}_0$ or E_{-p0}) scale as $(v_g/v_\varphi)^{-1/2}$, since $|\mathsf{E}_0|^2 - |\mathsf{E}_{-p0}|^2 = \mathsf{E}_0{}^2 \left[1 - |\mathsf{E}_{-p0}/\mathsf{E}_0|^2\right] = 2\,|\mathsf{E}_0|^2\; v_g/v_\varphi$.

The behaviour of slow-light devices, mainly based on photonic-crystal waveguides or coupled-resonator optical waveguides, is essentially identical to the previous two-wave picture. These devices have received much attention because of their promise for optical signal processing applications and enhanced non-linear light–matter interaction. Details are provided in Chapter 17.

16.5 2D photonic crystals

16.5.1 Oblique incidence in a 1D stack

To grasp the interest of higher-dimensional periodic photonic structures, we start by illustrating the role of oblique incidence in a standard stack, using the tools of bands and wavevectors. A lot of the thrust to study higher-dimensional structures was triggered by the possibility of controlling the DOS. The most striking target is to cancel this DOS in bandgaps, for the control of spontaneous emission (Yablonovitch, 1987; John, 1987), as said in the introduction; then it should be cancelled for all possible modes of a structure, or at least for large subsets. We explain here why, when including the role of oblique directions of incidence, a 1D structure cannot achieve such a cancellation. The geometry of a typical case is pictured in Fig. 16.9a, b. The periodicity is along z, and we consider that $k_x = 0$. Snell's law says that k_y, the tangential momentum, is a conserved quantity in (b). The main result is that at oblique incidence, the z-components are reduced compared to normal incidence. Since these components are those involved in the periodicity and Bragg effects, this has the important consequence that a *higher* frequency is needed to reach the Bragg condition than at normal incidence. The trend is dictated by the degree of k_z reduction, which is approximately quadratic in k_y for elementary wavevectors of both media, shown in (b).

As a result, see how the bands of normal incidence in Fig. 16.9c (blue lines in the (ω, k_z)-plane) are prolonged in the k_y-directions (grey lines, solid for the first gap, dotted for the second gap, dashed for the third). Let us remind the reader that this graph is the band structure of Bloch modes, whose k_z are not those of local plane waves.

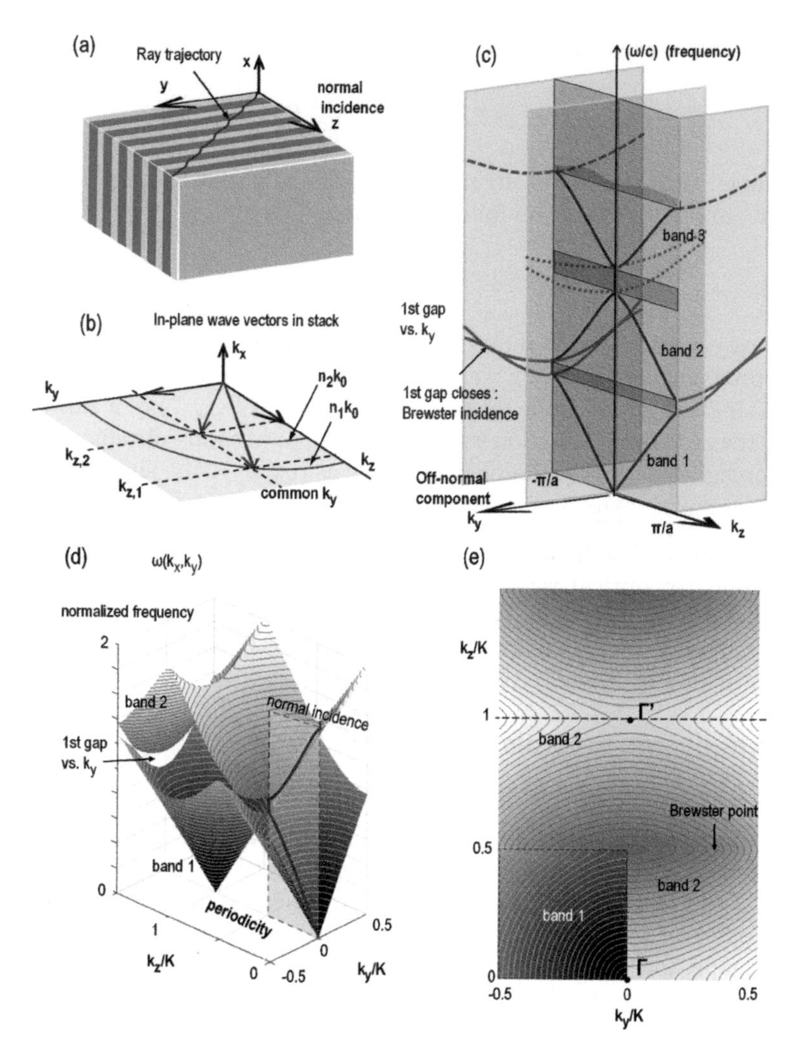

Fig. 16.9 Oblique incidence in a stack. (a) Real space; (b) wavevectors in the yz-plane; (c) selected loci of the $\omega(k_y, k_y)$ surfaces, with the normal incidence case in the (ω, k_z) plane and the gap evolution with oblique angle in the (ω, k_z)-planes; (d) split view of full surface for the two first bands; (e) top view of (d) evidencing constant-energy surfaces and their distortion around band edges and at specific points.

Here, note that an example with a smaller bandgap than in the above sections was chosen, so as to better depict its k_y-related evolution. We also add that the TM polarization was chosen (E-field in the y, z-plane). As a consequence, there is a *reduction* of Fresnel reflection down to the Brewster internal incidence, where this reflection vanishes. This is manifested by the indicated closing of the first gap at a largely oblique incidence (here it is less than 45° but we are dealing with internal wavevectors and angles, unlike the usual Brewster windows).

Next, we illustrate the two first bands in the form of cuts of their surfaces in Fig. 16.9d, with added constant frequency contours: this is in our opinion too rare an illustration of the relationship between several concepts. The periodicity along k_z is evidenced by the repetition of the low-frequency cone. The didactic 1D band structure is shown in the vertical Cartesian cut surrounded by a dashed-line boundary. The gap and its evolution with k_y and its Brewster closure are most visible on the left-side replica. We also see that there is a 'pass' between the two cones of band 1, and that the second band has its trough just above that pass.

In Fig. 16.9e, a top view provides what is called a wavevector diagram, or constant-energy surface (akin to Fermi surfaces of solid-state physics). In the lower left part, a cut in the map of band 2 allows band 1 to be seen, with its concentric contours that are distorted at the Brillouin zone edge $k_z = \pi/a = 0.5K$. The topology of band 2 also features 'passes' located at all Γ points. The trough of band 2 at the Brillouin zone edge is locally elliptical in shape, but the Brewster closure point gives a singular local topology whereby surfaces can be prolonged from band 2 to band 1.

A very important element that we get from this picture is that there is no limit to the upward motion of the bandgap ('blue shift') with oblique incidence, as k_z is bound to diminish when k_y increases at a given frequency. In our case, we started from a rather narrow gap, so, not very far from the minima, there are k_y values where band 1 has raised enough to provide modes at the normal incidence Bragg frequency: this is general, and destroys the possibility of drastically affecting the photonic DOS in 1D dielectric stacks.

16.5.2 2D dielectric photonic crystals: The role of the Brillouin zone

For a 2D photonic crystal, the lattice geometry can be square, rectangular, or triangular (hexagonal). There is no other 2D Bravais lattice. If we stick to the most symmetrical cases, we have two lattices (see Fig. 16.10): (a) the square and (b) the triangular lattices. Both are defined with two vectors $\mathbf{a}_1$ and $\mathbf{a}_2$ of norm $|a_1| = |a_2| = a$, making an angle of 90° (square) or 60° (triangular).

Then the Fourier decomposition can be made on reciprocal lattice vectors, denoted by $\mathbf{K}_1, \mathbf{K}_2$ for this 2D case instead of K in 1D. Using a unit vector normal to the periodicity plane $\mathbf{u}_x$, we have, using the vector product,

$$\mathbf{K}_1 = (\mathbf{a}_2 \times \mathbf{u}_x)/S_{\text{cell}}, \quad \mathbf{K}_2 = (\mathbf{a}_1 \times \mathbf{u}_x)/S_{\text{cell}}, \quad \text{where } S_{\text{cell}} \equiv |(\mathbf{a}_1, \mathbf{a}_2, \mathbf{u}_x)|, \tag{16.29}$$

with S_{cell} as the unit cell area and (.,.,.) denoting the mixed product (volume). All quantities such as ε now have Fourier coefficients $\hat{\varepsilon}_{p,q}$.

To describe waves in such 2D photonic crystals with invariance along x, we use a 2D space position $\mathbf{r}$ in the y, z-plane, and we have Bloch modes that can take two polarizations, E or H. We focus here on the so-called H-polarization, with $\mathbf{E}$ in the y, z-plane and $\mathbf{H} = H\mathbf{u}_x$. (In the other case, the E-polarization, the $\mathbf{E}$-field would be the scalar field along the invariant x-direction and $\mathbf{H}$ would be in the y, z-plane.) $\mathbf{H}$ reads as follows:

$$\mathbf{H} = \sum_{p,q} \mathsf{H}_{p,q} \exp(i[p\mathbf{K}_1 + q\mathbf{K}_2 + \mathbf{k}].\mathbf{r})\ \mathbf{u}_x = \sum_{\mathbf{K}} \mathsf{H}_{\mathbf{K}} \exp\left(i(\mathbf{K} + \mathbf{k}).\mathbf{r}\right)\ \mathbf{u}_x, \tag{16.30}$$

where it suffices that $\mathbf{k}$ spans the first Brillouin zone, as sketched in Fig. 16.10a, b for the two lattices of interest (it is the cell bounded by lines drawn at half-distance of each (p, q) reciprocal lattice node, minimizing the modulus in the $(\mathbf{k} + \mathbf{K})$ periodic family of vectors).

Prior to solving Maxwell equations (in section 16.5.4), we can intuitively generalize what we learned in 1D. The fundamental gap came from the interaction of two waves coupled by the largest Fourier components $\hat{\varepsilon}_{\pm 1}$ (with vector K) and occurred at the Bragg frequency that verified $|k| = \hat{\varepsilon}_{\pm 1}^{1/2}\,\omega_{\mathrm{B}}/c$ for $k = \frac{1}{2}K$, because then $-k = k - K$. We must go on a similar quest, but a direction-dependent one. For the square lattice and for 'atoms' of simple shape, the largest components are $\hat{\varepsilon}_{\pm 1,0}$ and $\hat{\varepsilon}_{0,\pm 1}$. For the triangular one, the six Fourier components that provide the first coupling opportunity have indices $(\pm 1, 0)$, $(0, \pm 1)$, and $\pm(1, 1)$ (see Fig. 16.10b). In the specific directions of these four or six vectors, we have exactly the same ingredients as in 1D. The Bragg frequency is thus expected to stem from the same relation $\left|\frac{1}{2}K_1\right| = \hat{\varepsilon}_{\pm 1}^{1/2}\,\omega_{\mathrm{B,short}}/c$. But these are the directions with shortest $|\mathbf{k}|$ at the zone edge, where the coupling is thus first found, whereas the directions of largest $|\mathbf{k}|$ at the zone edge are the *vertices* of the Brillouin zone. For the square lattice, the vertex lies at $|\mathbf{k_v}| = \frac{1}{2}(\mathbf{K}_1 + \mathbf{K}_2)$, a vector $\sqrt{2}$ longer than $\frac{1}{2}\mathbf{K}_1$. At this point, coupling is provided by $\hat{\varepsilon}_{\pm(1,1)}$, with a two-wave interaction fully analogous to the 1D one. For the triangular lattice, vertices are so-called K points situated at $\mathbf{k}_v = \frac{1}{3}(2\mathbf{K}_1 + \mathbf{K}_2)$, and coupling is now provided not so intuitively by two vector pairs *not collinear* to $\mathbf{k}_v$, specifically $\pm\mathbf{K}_1$ and $\pm(\mathbf{K}_1 + \mathbf{K}_2)$ in Figure 16.10b.

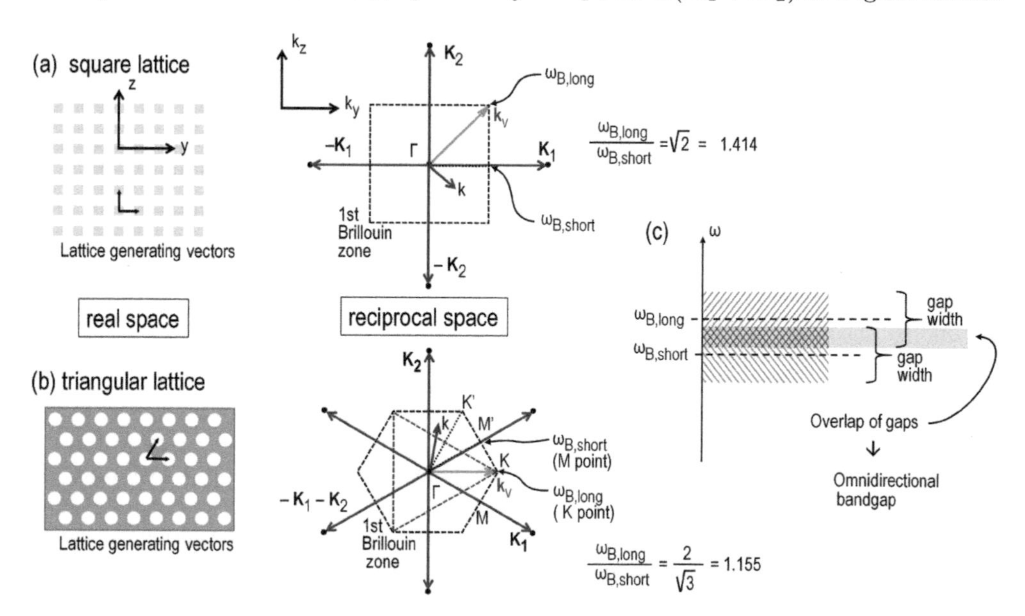

Fig. 16.10 (a) Square lattice in real space and its first Brillouin zone in reciprocal space, with the longest and shortest distances from its centre to symmetry points of its edges indicated; (b) same for the triangular lattice, which is the 'roundest'; (c) principle of the overlap of bandgaps (the two hatched ranges) centred at $\omega_{\mathrm{B,short}}$ and $\omega_{\mathrm{B,long}}$ to form an omnidirectional bandgap in the periodicity plane.

At variance with the square lattice, they do not involve Fourier components from beyond the six nodes already considered, $(\pm 1, 0)$, $(0, \pm 1)$, and $\pm(1,1)$ (moreover, they give identical Fourier components for simple atoms).

Compared to the 1D oblique case, where the Bragg frequency was continuously shifting up when away from the periodicity axis, we now have a different scenario where $\mathbf{k}$ deviates off a shortest K_1 direction and travels the Brillouin zone edge sideways to a longest one. Starting from the short-axis frequency $\omega_{\mathrm{B,short}}$, dictated by $|\mathbf{k}| = |\mathbf{K}_1|/2$, the Bragg frequency should first rise when scanning angles off that axis, but when approaching the vertices $\mathbf{k}_v$ of the Brillouin zone, the Bragg frequency should stall around the larger value $\omega_{\mathrm{B,long}}$, dictated by $|\mathbf{k}_v|$, the longest wavevector of the Brillouin zone, as detailed above. If the in-plane angle is further increased, symmetry implies that we should go back to the short one.

Hence, in the lattice with the roundest Brillouin zone, the hexagonal one, since $\left|\frac{1}{2}\mathbf{K}_1\right| = \cos(\frac{\pi}{6})\,|\mathbf{k}_v| \simeq 0.866\,|\mathbf{k}_v|$, there is only a $\frac{2}{\sqrt{3}} \sim 1.15$ ratio for $\omega_{\mathrm{B,long}}/\omega_{\mathrm{B,short}}$. Conversely, for the more 'pointy' square Brillouin zone, $\left|\frac{1}{2}\mathbf{K}_1\right| = \cos(\frac{\pi}{4})\,|\mathbf{k}_v| \simeq 0.707\,|\mathbf{k}_v|$, there is a larger ratio $\sqrt{2} \sim 1.41$ for $\omega_{\mathrm{B,long}}/\omega_{\mathrm{B,short}}$.

16.5.3 Omnidirectional bandgap: Two antagonist factors

We now ask what is the likelihood of obtaining an omnidirectional bandgap in 2D: a range of frequencies such that there is no band in any of the in-plane y, z-directions (Villeneuve and Piché, 1994; Padjen et al., 1994)(if a z-propagation were added, we would be in a completely different situation, and would recover more modes; this is discussed in Complement 16.B, and the issue of out-of-plane modes is discussed in Chapter 17 on periodic waveguides).

The fundamental property expressed by Eq. (16.14) (the gap width $2\Delta\omega_h$ scales with the $|\hat{\varepsilon}_{p_0}|$ Fourier component, here $\hat{\varepsilon}_1$) essentially holds here as well, except for some cases arising from possible cancellations due to the very dielectric pattern, i.e. the photonic 'atoms' of the photonic crystal. Hence, we need the Fourier components of both the short and long vectors along the symmetry axes of the Brillouin zone to be large enough for individual gaps to overlap, as shown in Fig. 16.10c, in spite of the given ratio, $\omega_{\mathrm{B,short}}/\omega_{\mathrm{B,long}}$. Obviously, the triangular lattice offers a much easier target in this respect: a relative gap width of 10% is enough to grant an overlap with a 15% centre frequency difference, whereas it requires roughly a 25% gap width to cope with the $\sqrt{2} \simeq 1.41$ factor of the square lattice, a substantially larger difference to mitigate.

Another important factor that should be borne in mind in order for an omnidirectional band gap to survive in higher-dimensionality photonic crystals (man-made or natural) is the need to share the available dielectric contrast among several Fourier components. Consider, for simplicity, the following continuous $\varepsilon(\mathbf{r})$ maps also illustrated in Fig. 16.11a, b rather than some piecewise constant versions:

$$\text{square lattice} \quad \hat{\varepsilon}_{(1,0)} = \hat{\varepsilon}_{(-1,0)} = \hat{\varepsilon}_{(0,1)} = \hat{\varepsilon}_{(0,-1)} = A,$$

$$\varepsilon = \hat{\varepsilon}_0 + 2A\cos(\mathbf{K}_1.\mathbf{r}) + 2A\cos(\mathbf{K}_2.\mathbf{r}) = \hat{\varepsilon}_0 + 2A\cos\left(2\pi\frac{y}{a}\right) + 2A\cos\left(2\pi\frac{z}{a}\right); \tag{16.31}$$

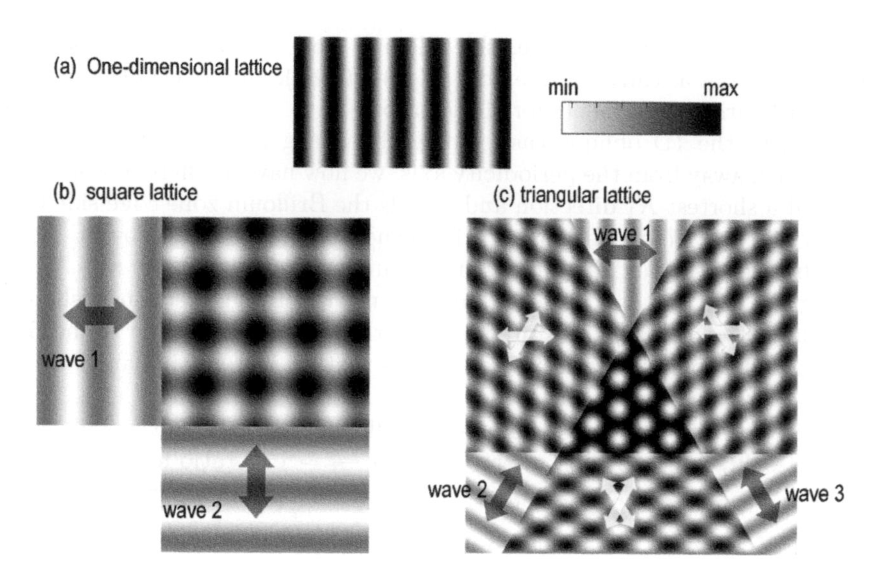

Fig. 16.11 (a) 1D modulation between min and max values (black and white, see colour bar on the top right), serving as a reference; (b) square lattice, superposition of two modulations of half the reference amplitude to allow the sum to fit in the [min,max] range; (c) same idea for the triangular lattice with three modulations, denoted by wave 1, 2, 3. Their pairwise combinations are indicated by pairs of arrows. These modulations now have less than half the reference amplitude to allow their sum to fit into the desired range.

$$\text{triangular lattice} \quad \hat{\varepsilon}_{(1,0)} = \hat{\varepsilon}_{(-1,0)} = \hat{\varepsilon}_{(0,1)} = \hat{\varepsilon}_{(0,-1)} = \hat{\varepsilon}_{(1,-1)} = \hat{\varepsilon}_{(-1,1)} = A,$$
$$\varepsilon = \hat{\varepsilon}_0 + 2A\cos(\mathbf{K}_1.\mathbf{r}) + 2A\cos(\mathbf{K}_2.\mathbf{r}) + 2A\cos((-\mathbf{K}_1 - \mathbf{K}_2).\mathbf{r})\,. \qquad (16.32)$$

The extreme values reached for the square case, Eq. (16.31), are $\hat{\varepsilon}_0 \pm 4A$, thus spanning $8A$. Hence, if we have a given bound due to material limitations $\Delta\varepsilon = \varepsilon_{\max} - \varepsilon_{\min}$, we find that $A = |\hat{\varepsilon}_{(0,1)}| \leq \Delta\varepsilon/8$, instead of the larger modulation capability $|\hat{\varepsilon}_{\pm 1}| \leq \Delta\varepsilon/4$ in the equivalent 1D problem for the same $\Delta\varepsilon$ material-based limitation.

In the triangular case, we used the three symmetric vectors $\mathbf{K}_1, \mathbf{K}_2, -\mathbf{K}_1 - \mathbf{K}_2$ at 120° of each other. The extreme values of distributions such as that of Eq. (16.32) now span from $9\,A$ to $10.4\,A$ (the actual value depends on relative phases), so that the penalty of modulation limitation is now *worse*, being typically $|\hat{\varepsilon}_{\pm 1}| \leq \Delta\varepsilon/10$.

This means that there is a large cost in terms of bandgap width, from the need to 'spread' the modulation among all Fourier components of ε. Subsequently, the target of the omnidirectional bandgap, which could be perceived as a somewhat benign evolution from the 1D bandgap demands, appears to be very demanding. For the case of the square lattice, we had found that the difference between Brillouin zone extrema (short and long segments) implied that at least a 25% relative width of the gap in a given direction must be achieved, a target made much harder by the halved modulation capability.

The situation is better for the triangular lattice: the modulation capability, now diluted along three directions, is more affected than in the case of the square lattice by a ratio 5/4, but it more than compensates for this penalty by offering a gap opening for a 10% width instead of a 25% width, hence recovering a 5/2 factor and alleviating by roughly a factor of 2 the demand on $\Delta\varepsilon$ to attain an omnidirectional bandgap. Hence, this triangular lattice is the preferred one for studying the omnidirectional bandgap.

16.5.4 2D dielectric photonic crystals: Band calculation for the H-case

We now proceed towards rigorous band structures, confining ourselves to the 'H'-case, whereby the field $\mathbf{H} = H_x \mathbf{u}_x$ is scalar along the invariant axis. Textbooks on photonic crystals document other polarizations, as well as higher dimensionality (vector fields are required in 3D photonic crystals) and convergence issues (Joannopoulos et al., 2008; Plihal and Maradudin, 1991).

We start from Maxwell equations in a source-free, non-magnetic medium with locally isotropic media:

$$\nabla \times \mathbf{E} - i\omega\mu_0 \mathbf{H} = 0, \qquad \nabla \times \mathbf{H} + i\omega\varepsilon_0\varepsilon(y,z)\mathbf{E} = 0, \tag{16.33}$$

which we re-express with $\mathbf{H}$ only by eliminating $\mathbf{E}$:

$$\nabla \times [\eta(y,z)\, \nabla \times \mathbf{H}] - \frac{\omega^2}{c^2}\mathbf{H} = 0, \tag{16.34}$$

$$\eta(y,z) = \frac{1}{\varepsilon(y,z)}. \tag{16.35}$$

The natural quantity in this equation is thus the *inverse* of the relative dielectric constant, denoted by η. We label its Fourier coefficients with vectors $\mathbf{K}''$ in anticipation of the convolution that will ensue from the product:

$$\eta(y,z) = \eta(\mathbf{r}) = \sum_{\mathbf{K}''} \hat{\eta}_{\mathbf{K}''} \exp(i\mathbf{K}''.\mathbf{r}), \tag{16.36}$$

with the summation running on all reciprocal wavevectors of Eq. (16.30). Denoting by $\mathbf{k}+\mathbf{K}$ the wavevector $p\mathbf{K}_1+q\mathbf{K}_2+\mathbf{k}$ in the exponential Eq. (16.30) of a given (p,q)-field harmonic, the curl product results in the form:

$$\sum_{\mathbf{K}}\sum_{\mathbf{K}'} \hat{\eta}_{\mathbf{K}-\mathbf{K}'}(\mathbf{k}+\mathbf{K}) \times [(\mathbf{k}+\mathbf{K}') \times \mathsf{H}_{\mathbf{K}'}] \exp\left(i(\mathbf{k}+\mathbf{K}).\mathbf{r}\right) =$$

$$-\frac{\omega^2}{c^2}\sum_{\mathbf{K}} \mathsf{H}_{\mathbf{K}} \exp\left(i(\mathbf{k}+\mathbf{K}).\mathbf{r}\right), \tag{16.37}$$

whereby we replaced $\mathbf{K}'' = \mathbf{K} - \mathbf{K}'$ in $\hat{\eta}$ to assemble exponential terms. We can now identify each spatial harmonic $\mathbf{K}$ or, in more functional terms, project on the $\exp\left(i(\mathbf{k}+\mathbf{K}).\mathbf{r}\right)$ basis to obtain a matrix relation between harmonics:

$$\sum_{\mathbf{K}'} \hat{\eta}_{\mathbf{K}-\mathbf{K}'}(\mathbf{k}+\mathbf{K}) \times [(\mathbf{k}+\mathbf{K}') \times \mathsf{H}_{\mathbf{K}'}] = -\frac{\omega^2}{c^2}\sum_{\mathbf{K}} \mathsf{H}_{\mathbf{K}}. \tag{16.38}$$

In our case, the innermost vector product gives a vector in the y,z-plane perpendicular to $\mathbf{k}+\mathbf{K}'$. Hence, its vector product with $\mathbf{k}+\mathbf{K}$ involves the scalar product of the two vectors, $(\mathbf{k}+\mathbf{K}').(\mathbf{k}+\mathbf{K}')$, as can also be deduced from rules on double vector products. We can also write that the field is scalar along x, and project the whole equation onto the x-axis. We then get:

$$\sum_{\mathbf{K}'} \hat{\eta}_{\mathbf{K}-\mathbf{K}'} \, (\mathbf{k}+\mathbf{K}).(\mathbf{k}+\mathbf{K}') \; \mathsf{H}_{\mathbf{K}'} = \frac{\omega^2}{c^2} \sum_{\mathbf{K}} \mathsf{H}_{\mathbf{K}}. \tag{16.39}$$

This is an eigenvalue equation providing $(\omega/c)^2$ as a function of any 2D $\mathbf{k}$-parameter on an infinite basis (in practice, convergence is acceptable for a large enough subset).

In principle, a scan of an irreducible part of the first Brillouin zone provides full knowledge of the dispersion, thanks to symmetries. It is, for instance, the triangle ΓMK in the triangular case Fig. 16.10b. For the gentle 'atoms' with rather simple shapes that are feasible in nanophotonics at the (sub)wavelength scale, there is not even a strong need to scan the inner part of this triangle, as it does not contain any extrema or singularity: notably, all the knowledge relative to band edges can be obtained along the symmetry lines. There is a similar idea in solid-state physics whereby, even for 3D Brillouin zones, only a circuit is scanned. We are adopting the same approach and therefore limiting ourselves in the next subsections to displaying band structures calculated on the three segments of the archetypal ΓMK *circuit*.

16.5.5 Triangular arrays of round holes in high-index matrices

It is now time to examine the fruits of the above approach by focusing on the example of round holes in a high-index matrix (rods of vacuum in a matrix, in a 3D view). This example is related to feasible photonic crystals, provided an x-guiding mechanism is added (the topic of Chapter 17), while here we remain purely 2D for both the system and the wavevectors.

We must now express $\hat{\eta}_{\mathbf{K}}$ for our landscape $\eta(y,z)$ of round holes. We start from

$$\eta(y,z) = \eta_b + (\eta_a - \eta_b)\theta_R(y,z) \; , \tag{16.40}$$

where $\theta_R(y,z)$ takes the value one in the rod of radius R and zero outside it (and we set $2R \leq a$ to avoid overlapping the holes), $\eta_a = 1/\varepsilon_a$ is the value in the rod of dielectric constant ε_a, and $\eta_b = 1/\varepsilon_b$ is the value in the matrix of dielectric constant ε_b. The Fourier transform has to be made on a unit cell, $\hat{\eta}_{\mathbf{K}} = S_{\text{cell}}^{-1} \int \eta(\mathbf{r}) \exp(-i\mathbf{K}.\mathbf{r}) \, dy \, dz$. The constant contribution η_b gives only a $\mathbf{K} = 0$ component, whereas the integral of $\theta_R(y,z)$ can be expressed by a first-order Bessel function $J_1(\xi)$:

$$\hat{\eta}_{\mathbf{K}} = \eta_b \delta_{\mathbf{K},0} + (\eta_a - \eta_b) \frac{\pi R^2}{S_{\text{cell}}} \frac{2J_1(|\mathbf{K}| \, R)}{|\mathbf{K}| \, R} \; . \tag{16.41}$$

The Bessel function appearing here roughly resembles the sine function, so the right-hand fraction is an oscillating function of $|\mathbf{K}| \, R$ akin to the familiar $\sin(\xi)/\xi$ sinus cardinal function appearing in elementary diffraction treatments of apertures. In the established language of X-ray diffraction, it is the 'form factor' of the round hole.

A calculation result for a typical case of a matrix of constant $\varepsilon_b = 12$ perforated by a triangular array of air holes $\varepsilon_a = 1$ is shown in Fig. 16.12. The hole radius R

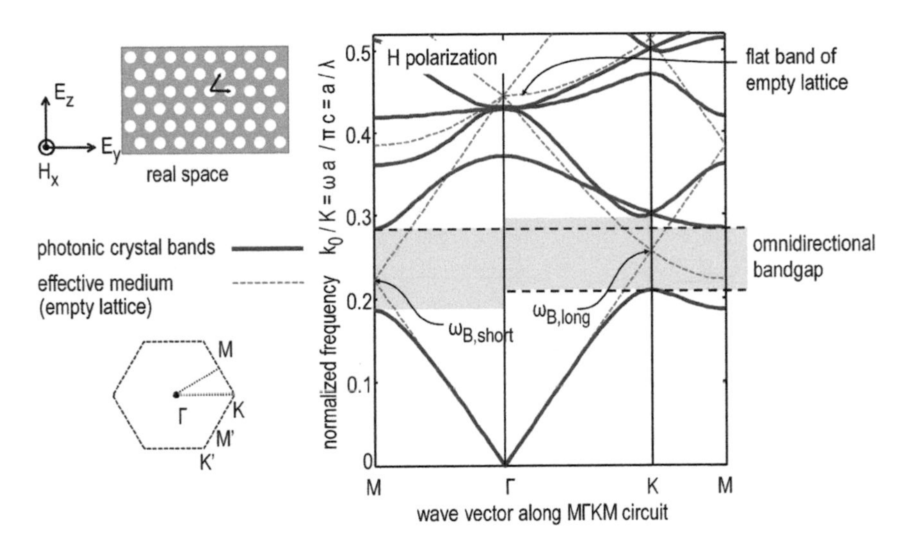

Fig. 16.12 Band structure for H-polarization of a triangular 2D photonic crystal made of round air rods $\varepsilon_a = 1$ in a matrix $\varepsilon_b = 12$ with an air filling fraction $f = 0.35$ (top left scheme), with the Brillouin zone in the bottom left scheme. The dashed grey lines are the band structure of an empty lattice (uniform media) with the same low-frequency dispersion. The Bragg frequencies and the appearance of flat bands on the latter are indicated.

is conveniently parametrized through the filling fraction $f \equiv f_a$ of air here. Since $S_{\text{cell}} = a^2\sqrt{3}/2$, we have $f = \pi R^2/S_{\text{cell}} = (2\pi/\sqrt{3})(R/a)^2$. The $f = 0.35$ value chosen here results in the size of the holes being similar to the size of the veins that separate them.

We can now check the tools and trends provided above on this particular case. We have added as dashed grey lines the empty lattice case, which fits well at low frequency with the exact dispersion along both ΓM and ΓK. They stem from an empty lattice made of a uniform $\varepsilon = 6.75$ medium, which is the homogenized system for this H polarization ($\varepsilon = 6.75$ is logically less than the value $\langle\varepsilon(y,z)\rangle = 8.15$ for E-polarization; see section 15.2.2).

We also see that the gaps at M and K build up around the Bragg frequency of this empty lattice. The two distinct bandgaps along ΓM and ΓK and their overlap, the *omnidirectional photonic bandgap*, are apparent. The width of this latter, here $\Delta u \sim 0.08$ in units of $u = a/\lambda$, is the key quantity for spontaneous emission control.

At higher frequencies, going to the second gap that lies around the Γ point, the band structure gets more complex. Notice that even the gap-free empty lattice displays very flat bands with apparently low group velocity. This is an *artefact of the band structure representation*: eigenmodes with all possible symmetries comprise the superposition of two plane waves in the same way that modes are constituted in a dielectric slab with perfect-conductor boundaries (see Complement 6.D) for which the group velocity is similarly diminished.

The quest for the bandgap is conveniently illustrated by the *gap map* of Fig. 16.13: this map describes the bandgap edges as a function of f for a constant $\varepsilon_b = 12$ host

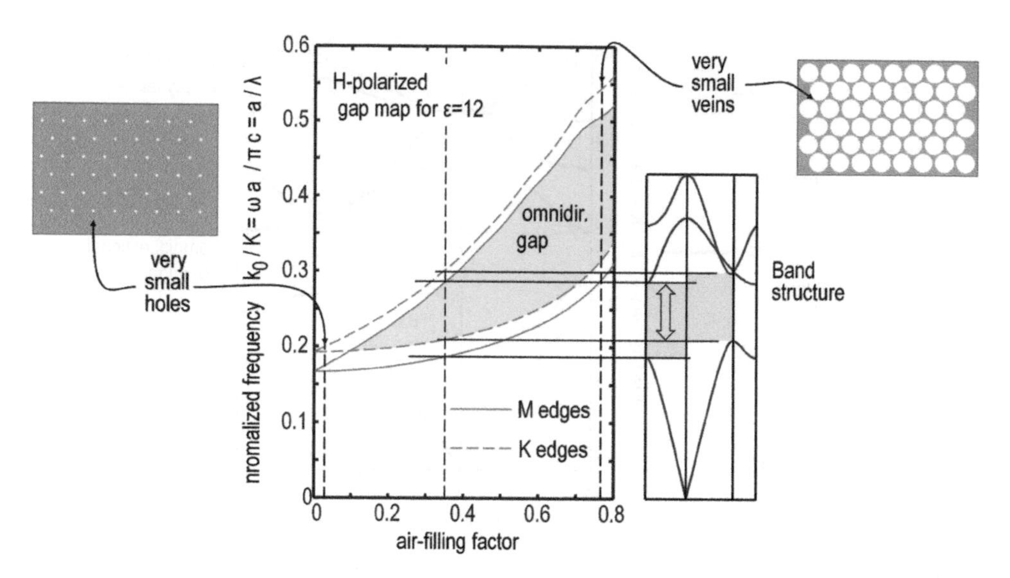

Fig. 16.13 Gap map for the H-polarization, and for air holes in a $\varepsilon_b = 12$ matrix. The four grey lines represent the four band-edge positions, as indicated by the horizontal lines linking to the sketched band structure on the right. Illustration of the lattice appearance in the limit cases of small and large f is provided, with very small holes in one case and very large holes and tiny veins in the other case.

material and variable hole sizes. As suggested by the aligned band structure on the right, the relevant edges are the upper bandgap edge at M-point (top of omnidirectional gap, bottom of 'conduction band') and the lower bandgap edge at K-point (bottom of omnidirectional gap, top of 'valence' or 'dielectric' band): the omnidirectional bandgap is achieved only in the grey area delimited by these values. Hence, there is no omnidirectional bandgap below $f \sim 0.1$. This is a relatively small threshold value due to the large matrix index $\varepsilon_b^{1/2} = \sqrt{12} \simeq 3.46$, hence a large dielectric contrast. The same map in a lower-index medium, e.g. $\varepsilon_b^{1/2} = 2.5$ typical of gallium nitride or some forms of titania TiO_2, would demand larger f-values to get this omnidirectional bandgap. The variations with f of the two edges have different slopes, which can be ascribed to the distinct degree of overlap of the relevant band-edge Bloch modes with the air holes.

The plot could be prolonged to around $f = 0.9$. But already at the present limit, $f = 0.8$, the holes are not far from touching, and the veins are very thin. In this area, accuracy (the number of plane waves in band calculation, for instance) becomes demanding. This limit of thin veins is also technologically challenging. At this point, we can additionally mention that a comparatively quite tiny bandgap region exists for the same air-cylinder structure in the E-polarization. So it is formally possible to have a bandgap frequency range common to both polarizations. However, in realistic situations, the presence of the third dimension and the peculiarities of emitters are extra factors that make this issue of the common polarization bandgap less important.

Nevertheless, a number of different lattices (square, rectangular, honeycomb) have been devised with variable inserted 'atoms' (tubes, prismatic shapes, ...). They have been helpful to understand symmetries and light–matter interaction among other things (Joannopoulos et al., 2008; Padjen et al., 1994).

16.5.6 2D crystalline optics: Superprism and supercollimator

Bloch modes of a photonic crystal may behave very differently from usual plane waves. In the 2D context, some angular and spectral behaviours that differ strikingly are instructive. They help us to take steps towards photonic-crystal devices based on '2D crystalline optics'. Two such cases are described below, namely the superprism and the supercollimator. An issue of both fundamental and applied interest is how to 'contact' the specific Bloch modes acting in these structures. It concerns, for instance, exploring which directions a 2D photonic crystal must be cut along, and even at what exact position in the crystal lattice. Indeed, the mode-matching problem with outer waves has no trivial solution, and can benefit from modified crystal edge 'atoms', or from a short tapering. We leave these issues aside to focus on the two devices.

The superprism. In the superprism, pictured in Fig. 16.14a, a region close to a band edge, or a Brillouin zone corner, is targeted. In particular, the quickly evolving shape of the iso-frequency curves $\mathbf{k}(\omega)$ is exploited.

The main idea is to address specific wavevectors $\mathbf{k}$ in the crystal bands at specific frequencies, by using wavevector conservation at an interface, which holds for the component parallel to this interface (within a reciprocal wavevector of the interface, $K = 2\pi/a_{\text{interface}}$, just as occurs for gratings; see section 15.3). Hence, we set one component, k_y in Fig. 16.14a, and the generated Bloch mode that goes away from the interface has a group velocity dictated by the intersect of the k_y = constant line and the iso-frequency curve $\mathbf{k}(\omega)$, with $\omega = 2\pi/\lambda$. It is then clear that operating near a saddle point of the landscape $\omega = \omega(k_y, k_z)$ of the targeted band (such as those of Fig. 16.9d) results in an extreme dependence of the Bloch-mode direction on frequency (and obviously on the impinging direction as well).

By designing a device with a long Bloch-mode propagation, and good coupling (no unwanted reflections, no unwanted transmission orders, etc.), as shown in Fig. 16.14b, a huge spatial separation of different wavelengths can be realized, even for frequencies that differ by, say, 1% or so. Separation in common glass prisms is usually modest, even if a dispersive glass is chosen: a few degrees of angular variation for the whole visible spectrum (400–750 nm), for instance. But with the superprism Bloch modes, we can obtain nearly the opposite: tens of degrees of angular shift for a 1% variation in wavelength. Usual gratings perform somewhere in between the two extremes.

The supercollimator. The supercollimator exploits a situation of Bloch waves that is nearly opposite to the superprism. Here, for a large range of input wavevectors at the device interface (and with a given frequency), the group velocity has a nearly constant direction and modulus. The shape of iso-frequency curves liable to provide this effect is pictured in Fig. 16.14c.

The bands inside a photonic crystal can adopt a variety of shapes, but remember that in the empty lattice limit, the iso-frequency surfaces are circles. Moving away from this limit, a more general situation of a photonic crystal is one that features elliptical iso-frequency lines and has Bloch waves with a pair of effective indices, as

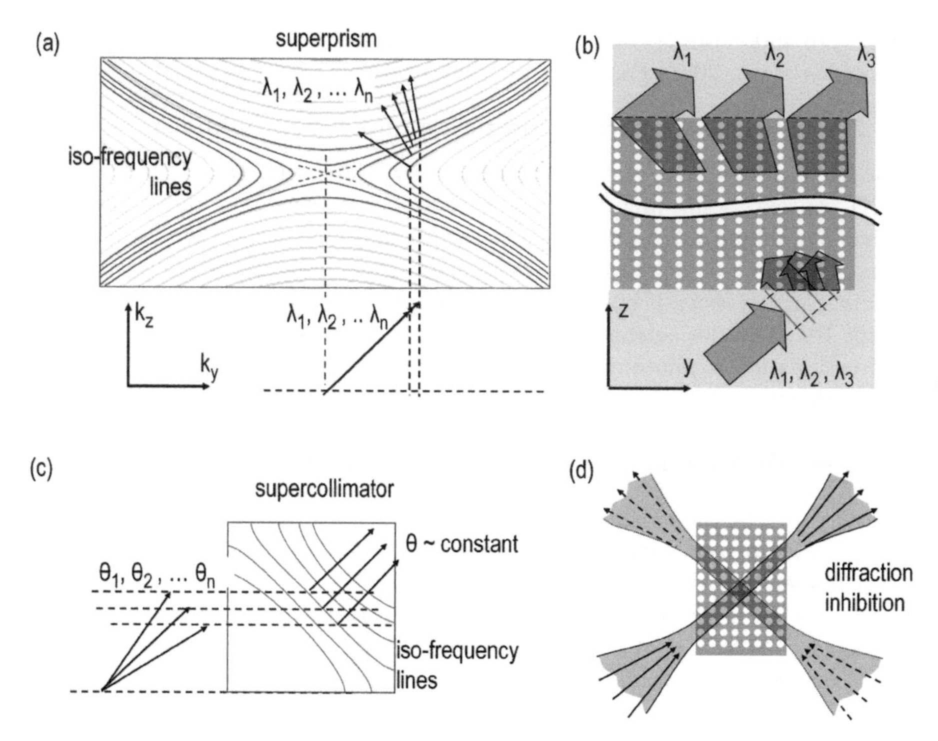

Fig. 16.14 Superprism and supercollimator. (a) Band structure for the superprism effect: in the vicinity of a singular point, there is a quick variation of group velocity direction, $\mathbf{v}_g$ with wavelength λ; the conserved quantity when entering the crystal is a single wavevector component, in this case k_y. (b) Sketch of the beams (composed of Bloch modes) in a superprism device. (c) Band structure for the supercollimator effect: around a given frequency, bands flatten out. At a given frequency, different k-values propagate with parallel group velocities $\mathbf{v}_g$; the conserved quantity is again a single wavevector component, in this case k_z. (d) A focused beam enters a supercollimator and the apparent beam diffraction is inhibited. Dashed arrows show a crossing beam.

in birefringent materials. But there are also situations such as the one pictured here, whereby the curvature of bands locally vanishes. This indeed also happened at some points in the superprism band structure of Fig. 16.14c, but there are cases where, additionally, the k-domain over which the curvature vanishes is much more extended than in standard band structures.

In such a case, the coupling of a variety of incident directions nevertheless results in a Bloch wave in a well-defined, collimated direction. The consequence can be seen in the implementation of Fig. 16.14d: a clearly convergent beam focused on one edge of a photonic crystal defines a beam inside the crystal whose convergence or divergence is negligible at first order. One interesting use of such a system is a possible beam-crossing device. If the waves on the edges are divergent waves coming from a waveguide, their divergence is inhibited over a substantial physical path, allowing the positioning of a collecting waveguide far enough from the input, hence with much more design freedom for photonic circuitry of an integrated device.

16.A Complement: 3D photonic crystals

The full control of the photonic DOS, and specifically its 100% inhibition, is not achieved in a 2D photonic crystal if we consider the third direction. Much in the same way as was done for oblique incidence (see section 16.5.1), the 2D bandgaps evolve and their frequency increases if a **k**-component along the invariant direction, namely k_x, is included. Hence, if we consider photonic structures with periodicity and/or invariance, the achievement of a zero photonic DOS $\rho(\omega) \equiv 0$ in a given range requires 3D crystals.

The plane-wave method of Eq. (16.38) essentially holds in 3D. The **H**-field is, of course, a general vector and no longer a scalar, and no specific polarization can be defined. The eigenvalue equation is not much more complex, but the vector basis to handle the double vector product should be chosen with care (Joannopoulos et al., 2008).

The main issue of the 3D photonic crystal is how to choose the lattice and atoms to get a large and fabrication-tolerant omnidirectional bandgap. It turns out that among the cubic lattice family, the first Brillouin zone of the *fcc* (face-centred cubic) lattice is reasonably 'round', being a truncated octahedron (it is the same Brillouin zone as that of diamond or silicon crystals, easily found in solid-state physics textbooks).

If we stick to the spirit of simple 'atoms' discussed in 2D, we may replace rods with spheres. It was shown that a diamond lattice of spheres offers quite a broad omnidirectional (3D) bandgap (Ho et al., 1990). Electromagnetic reasons have been invoked to explain this trend, pertaining to the need to ensure closed loops for lines of both electric and magnetic fields (Joannopoulos et al., 2008). But the feasibility of the diamond lattice in photonics is quite awkward; it is a very 'empty' lattice that is not easily amenable to common nanophotonic manufacturing/etching.

Three other crystal geometries are worth mentioning, as they could lead to practical 3D photonic crystals, albeit not always easily in the (nano)photonic domain.

- **Yablonovite lattice:** This crystal, named after its proponent, E. Yablonovitch, consists of a lattice of galleries. At a surface of the crystal, a triangular lattice defines points where galleries are initiated. At each lattice point, three straight round holes are made that mimic the (111) oriented galleries of silicon crystals: they have to be oriented at 35° off the surface normal, and three of them at azimuths $0°, 120°$, and 240° are defined at each hole. The air fraction must be well above 50% to get large bandgaps. This generalizes the teaching of the gap map made in 2D (see Fig. 16.13).

- **Inverse opal lattice:** Opals are stacks of dielectric spheres. Silica in nature, for instance, can form opals, as sketched in Fig. 16.15a. Opals themselves do not yield omnidirectional bandgaps, not least because the air-filling fraction of a compact sphere assembly is far too low. Artificial opals are obtained by letting a solution of spheres slowly or forcibly sediment to form often imperfect crystals with many local faults. These man-made opals then serve as a template for the synthesis of *inverse* opals: by means such as chemical vapour deposition (CVD), another material is 'infiltrated' across a couple of layers. It is then possible to selectively dissolve the silica template to retain only the deposited material, e.g. silicon. The resulting structure is hard to picture (see Fig. 16.15b), but can possess a full 3D

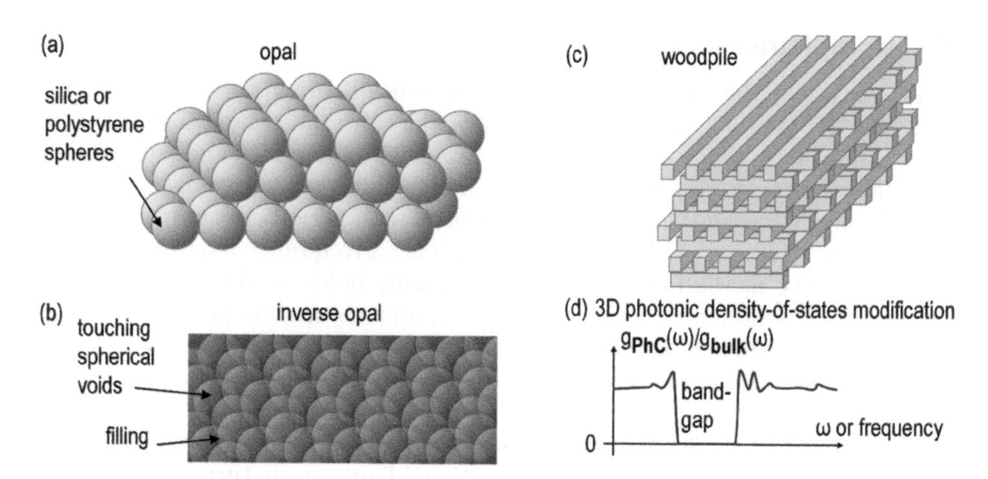

Fig. 16.15 3D photonic crystals. (a) Opal ideal structure; (b) sketch of an inverse opal, obtained by removing the spheres after the opal has been infiltrated by a material that withstands the removal procedure (a dissolution, for instance); (c) woodpile structure. Note the four-stack period in the vertical direction; (d) scheme of 3D DOS modification vs frequency, $g_{\mathrm{PhC}}(\omega)/g_{\mathrm{bulk}}(\omega)$, in a 3D photonic crystal with omnidirectional bandgap.

photonic bandgap, i.e. a modification of the 3D photonic DOS with a genuine zero, as sketched in Fig. 16.15d, where a simplified flat background is dented by the bandgap.

- **Woodpile lattice:** The wood pile lattice is a geometrically simpler proposal, not least because it requires not etching, but simpler methods, to get a reasonable structure. It consists of a stack of (often square) rods with staggered directions, 0° and 90°, and alternating in-layer rod positions, so that the actual unit cell measures four layers in height, as can be seen in Fig. 16.15d.

Overall, the poor feasibility of 3D photonic crystals has remained their Achilles' heel, but the self-assembled ones turned out to be remarkable photonic periodic systems, either for themselves or as templates for inverted structures.

16.B Complement: Photonic-crystal fibres

Photonic crystal fibres (PCFs) became the first object incorporating the 'photonic crystal' concept into commercial items. They make use of some of the concepts seen in this chapter, albeit in a modified fashion.

The geometrical basis of PCFs can be related to the addition of a wavevector in the invariant dimension, as was considered for the 1D oblique incidence case in section 16.5.1. For PCFs, we add a vertical wavevector along k_x to a periodic structure in the y, z-plane. Hence, the dielectric map is still $\varepsilon \equiv \varepsilon(y, z)$, but the modes have a much more complex structure due to the addition in all the fields of a propagation term, $\exp(ik_x x)$.

The reader will also recognize the analogy with the developments on waveguides in Chapter 6, whereby the field had a profile in a plane and a propagation constant $k_x = n_{\text{eff}}\omega/c$ along the invariant direction. And, indeed, PCFs can mostly be viewed as guides with a (transversally) periodic cladding.

In the spirit of guidance and its many light-control applications, PCFs are of particular interest as fibres and guides due to: (i) their ability to engineer the dispersion far more than in traditional fibres, as in these latter, the dispersion is bounded by a structure made of two glasses of very close indices (Ge-doped silica/silica, for instance: a 1% index step typically, used between core and cladding); and (ii) their ability to widely change the mode size and scan a number of non-linear regimes.

It is instructive here to provide the main physical grounds that, in hindsight, have made PCFs attractive fibres, with reference to Fig. 16.16a, b:

- Grazing incidence: The Fresnel reflection becomes large ($R \sim 1$) at grazing incidence, so even a modest index contrast such as silica/air can provide strong multiple reflections and gap opening around certain directions.

- Larger period: Expected photonic effects such as bandgaps occur at a much larger period than the $\frac{\lambda}{2\langle n\rangle}$ standard Bragg formula due to the involvement of the *transverse wavevector* only, i.e. the two smaller wavevector components, while the large k_x component along the invariant direction plays no Bragg-type role.

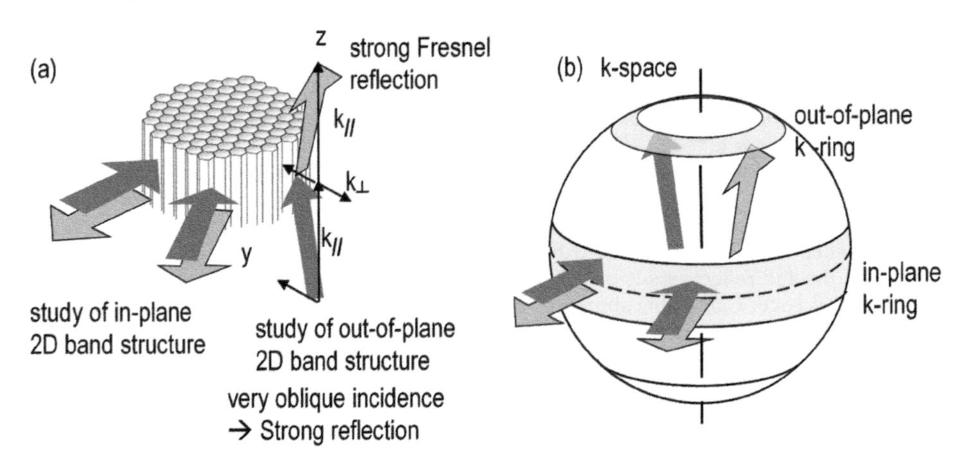

Fig. 16.16 Light behaviour in photonic crystal fibres. (a) 2D crystalline structure with near-perfect invariance along one axis and configuration of wavevectors around this for its practical use; (b) the strong reflection at very oblique incidence, even with a modest air–glass index contrast, builds up into a nearly perfect bandgap in a ring of k-space near the 'Arctic area' of the sphere (the actual ring may be distorted by the six-fold lattice symmetry). This is at variance with the classical 2D bandgap that corresponds to a belt of controlled propagation around the sphere's equator. Hence, for photonic crystal fibres, since only the in-plane component of the wavevector, $k_\perp$, matters to define the bandgap, a large period and an easier fabrication results for photonic crystal fibre structures.

Exercises

Exercise 16.1 Bandgap of a periodic stack with usual materials. Use the formula $\lambda_{\text{Bragg}} = 2an_{\text{eff}}$ to locate the approximate wavelength of the first bandgap for the two following periodic stack cases made of two materials by using a static effective-index approximation:

→ 100 nm silica ($n_1 = 1.5$) and 150 nm titania ($n = 2.2$).

→ 100 nm air ($n_1 = 1.0$) and 150 nm silicon ($n = 3.5$).

For which case do you expect the result to be most accurate?

Exercise 16.2 Inhibited transmission in a bandgap for a finite stack. A stack similar to that of Fig. 16.2 comprises 10 pairs. Assume that the index $n_1 = 1.6$ is the same but the index n_2 is such that the minimum power transmission inside the first bandgap is $T_{10} = 0.01414$.

Propose a first-order approximation for the transmission of a similar stack with 20 or 30 pairs (T_{20} and T_{30}). Discuss whether these approximations overestimate or underestimate the transmission by distinguishing Bloch-mode decay and interfaces. Propose a way of bracketing the correct upper and lower bound values.

Exercise 16.3 Bandgap width intuition based on simple phase analysis. We propose an intuitive method to qualitatively find the bandgap width of Eq. (16.14).

We consider a semi-infinite stack with indices $n_1 \simeq n_2$ as in Fig. 16.5, with quarter-wave thicknesses at a given λ_0. We consider only normal incidence. Check that the reflection coefficient of one period (two interfaces) at the Bragg wavelength λ_0 reads $r = \frac{\Delta n}{\langle n \rangle}$ with $\Delta n = n_2 - n_1$ and $\langle n \rangle = \frac{1}{2}(n_1 + n_2)$ with good accuracy.

Justify the fact that most of the Bragg reflection is built up after a useful number of periods N such that $N = 1/r$.

Calculate the round-trip phase change $\Delta\varphi = \varphi - 2\pi$ for a one-period slice resulting from a frequency shift $\Delta\omega = \omega - \omega_0$.

Drawing inspiration from the 'phasor diagram' (addition of complex amplitudes viewed in complex plane) of the various reflections, and from the fact that the reflected waves at Bragg wavelength are in-phase by successive multiples of 2π for the first gap studied here, the phase of the Nth reflection is naïvely shifted by $N\Delta\varphi$. Thus, we argue that when $N\Delta\varphi \sim \pi/2$, the last of the N 'useful' reflections is on the verge of contributing out-of-phase, so it corresponds to the stopband edge. Using this simple picture, deduce the half gap $\Delta\omega_h/2$ from the value of $\Delta\varphi$ determined above. Check for consistency with Eq. (16.14).

Exercise 16.4 Gap calculation in 2D photonic crystal.

(1) Assume that in a 2D photonic crystal, the *relative* bandgap opening in a given in-plane direction $\Delta\omega_h/\omega_0$ is a constant, the same in any direction. What should it be in the case of the square lattice to start opening an omnidirectional bandgap?

(2) Same question for a triangular lattice.

(3) Same question if an omnidirectional bandgap opening of 15% relative width is desired.

(4) In the triangular lattice, the reciprocal lattice vectors are of the form $n_1\mathbf{b}_1 + n_2\mathbf{b}_2$. There are six equivalent vectors around the origin. Show that in one of the next rings of vectors, there are 12 equivalent vectors with the same norm. It is possible to modulate at

the spatial frequency imposed by these 12 vectors. If we assume only these modulations exist in the Fourier spatial decomposition, we have a flavour of 12-fold symmetry, reminiscent of quasicrystal issues or Archimedean tiling issues. Reconsider the issue of attaining an omnidirectional bandgap when the angular separation between adjacent vectors is now only $\pi/6$: what is the largest off-angle around one of the 12 vector directions? Using the evolution at oblique incidence (Exercise 16.6), show that the bandgap is much less demanding in terms of modulation strength around one direction. Run a small simulation program to guess the relationship between the individual modulation carried by each of the 12 vectors and the total $\varepsilon(y, z)$ in-plane modulation, probing random modulation phases.

Exercise 16.5 Supercollimator and superprism. We borrow a simple dispersion relation from that of elementary phonon models in solid-state physics: $\omega(k_y, k_z) = \omega_0\,(\sin^2(k_y a/2) + \sin^2(k_z a/2)$, with a the lattice constant of a square lattice.

What is the first Brillouin zone?

Study the equifrequency curves for the three domains (1) $\omega \ll \omega_0$, (2) $(\omega - \omega_0) \ll \omega_0$, (3) $|\omega - \sqrt{2}\omega_0| \ll \sqrt{2}\omega_0$.

Show that domain (2) is well adapted to the supercollimator effect (much like Fig. 16.14) and the superprism effect in different regions of the k-space.

Show that domain (3) presents an isotropic (in-plane) but locally parabolic dispersion. Contrast with domain (1) and discuss the orientation of the group velocity.

Exercise 16.6 Evolution of the bandgap at oblique incidence. Consider a quarter-wave stack made of GaAs ($n_2 = 3.5$) and GaAlAs ($n_1 = 3.17$), at $\lambda_0 = 1000\,\text{nm} = 2\pi c/\omega_0$.

Based on the first Fourier component of dielectric constant modulation, find the relative width of the bandgap at normal incidence.

Describe the approximate evolution of the bandgap as a function of the *internal off-normal angle* θ_2 in GaAs in the form $\omega_c = \omega_0 \times f(\theta_2)$. At what approximate angle $\theta_{2,c}$ will the centre of the bandgap ω_c move at the high-frequency edge ω_+ of the normal incidence bandgap? At which larger angle $\theta_{2,l} > \theta_{2,c}$ will the lower bandgap edge reach the same point, ω_+?

What are the corresponding angles in the air, outside the stack? What happens for the same situations with slightly stronger index modulations, such as 3.0/3.5?

17 Periodic waveguide

In this chapter, we describe an important building block of modern photonics: combining waveguides and gratings. We are concerned with longitudinal periodicity, whereby the periodicity axis and the waveguiding axis are common. Compared to translation-invariant waveguides, the main concept that we need to introduce is the light line, in section 17.1, as well as its generalization to the light cone when considering photonic-crystal slabs. In section 17.2, we describe various periodic guides, mostly in view of integrated optics applications: modulated ridge waveguides for lasers and related devices, segmented waveguides, photonic-crystal waveguides using 'missing rows'. In section 17.3, we go to a regime where outside waves are coupled and study resonant waveguide gratings that serve as filters or couplers to fibres. The last section, 17.4, discusses an elegant way to build high-Q cavities based on modulated photonic-crystal waveguides. Such structures provide several perspectives in signal processing or light–matter interaction enhancement.

Complement 17.A addresses the issue of light extraction by periodic waveguides, notably for InGaN-based light-emitting-diode (LED) structures, and Complement 17.B summarizes the advanced theory of periodic waveguides (orthogonality, Bloch-mode excitation by dipolar sources, etc.).

17.1 Waveguide and periodicity

17.1.1 Waveguide with finite or vanishing modulation

We consider the dielectric lossless geometries of period a presented in Fig. 17.1, restricted to the Oxz-plane. Two typical cases are a modulation by etching grooves, Figs. 17.1a, b, and modulation of the index of the whole core, Fig. 17.1c. Case (a) implies that the cover index is naturally one of the two indices of the truly periodic region, unlike case (c).

The vanishing modulation of the Fig. 17.1b limit helps with setting first-order rules for guidance: in case (a), the modulation may vanish either due to a large duty-cycle f of one constituent (f is the ratio of the darker ridge region along x over period a), i.e. a vanishing groove width, or due to a vanishing depth (depth $\ll$ thickness d). In case (c), it can be due notably to a small index contrast between the two core constituents. This is what happens, for instance, in Bragg fibres due to minute material modifications by localized periodic irradiations. Within this limit, guidance is physically ensured by assuming the average index of the core to be larger than the single index n_1 of the two semi-infinite cladding regions ($n_1 \equiv n_3$ in the notation of section 6.2).

Introduction to Nanophotonics. Henri Benisty, Jean-Jacques Greffet, and Philippe Lalanne, Oxford University Press.
 DOI: 10.1093/oso/9780198786139.003.0017

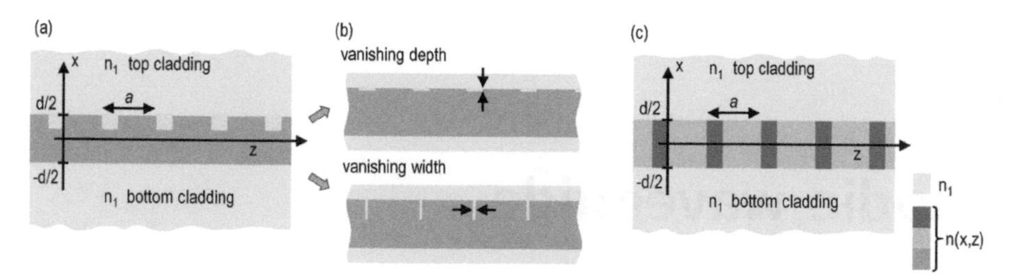

Fig. 17.1 Typical periodic waveguides with identical cladding media of index n_1: (a) the modulation is typically made by etching grooves away; (b) two paths for vanishing modulation of etched grooves: vanishing depth or width; (c) the modulation affects the whole core, made of two different materials, with an average index generally larger than n_1.

17.1.2 Dispersion relation of grating waveguides

The Floquet–Bloch theorem of fully periodic structures holds for the electromagnetic modes of periodic waveguides (Peng et al., 1975), either for normal Bloch modes obtained from $k(\omega)$ calculations or for quasinormal Bloch modes obtained from $\omega(k)$ calculations: omitting the $\exp(-j\omega t)$ dependence, the field components can be decomposed in Fourier series along the z-axis for each x:

$$\mathsf{E}_y(x,z) = \sum_m \mathsf{e}_{y,m}(x)\exp(j(k_z + mK)z), \tag{17.1}$$

where we have taken a y electric-field component, as it leads to the simplest example for a dielectric structure (the transverse-electric (TE)-polarized mode in the sense of guided waves and gratings). The integer index m labels the spatial harmonics and $K = \frac{2\pi}{a}$. Note that each harmonic varies independently along the x-axis. In the most general case, both ω and k_z might be complex. However, such considerations would lead to complicated treatments, and in the intuitive picture developed in this section, we will consider lossless materials and focus our attention on truly guided Bloch modes with real ω and k_z.

To qualitatively understand the main features of the dispersion relation in the (ω, k_z)-plane, it is advisable to start from the unperturbed waveguide modes (without periodicity), then follow the same approach as in chapters 15 and 16: first add the periodicity in the regime of vanishing modulation, and then move to the finite modulation regime, especially to open gaps at the crossing branches.

We follow these three steps in Fig. 17.2, with the reminder of the uniform waveguide dispersion in panel (a). At low frequencies, the fundamental TE-polarized guided mode effective index verifies $n_{\text{eff}} \to n_1$, bordering the cladding light line, whereas the same fundamental mode is confined inside the core at large frequencies; thus, $n_{\text{eff}} \to n_{\text{core}}$, see section 6.2.3. We consider a single guided mode for simplicity, but the following reasoning applies to every guided mode.

In Fig. 17.2b, we add the periodicity to (a) *in the limit of vanishing modulation.* Thus the dispersion $\omega(k_z)$ is indistinguishable from that of the uniform waveguide, and we simply need to add the two Brillouin zone edges at $k_z = \pm\frac{\pi}{a} = \pm\frac{K}{2}$ with vertical lines to further repeat the dispersion for all mK multiples, as for the 1D Bloch mode

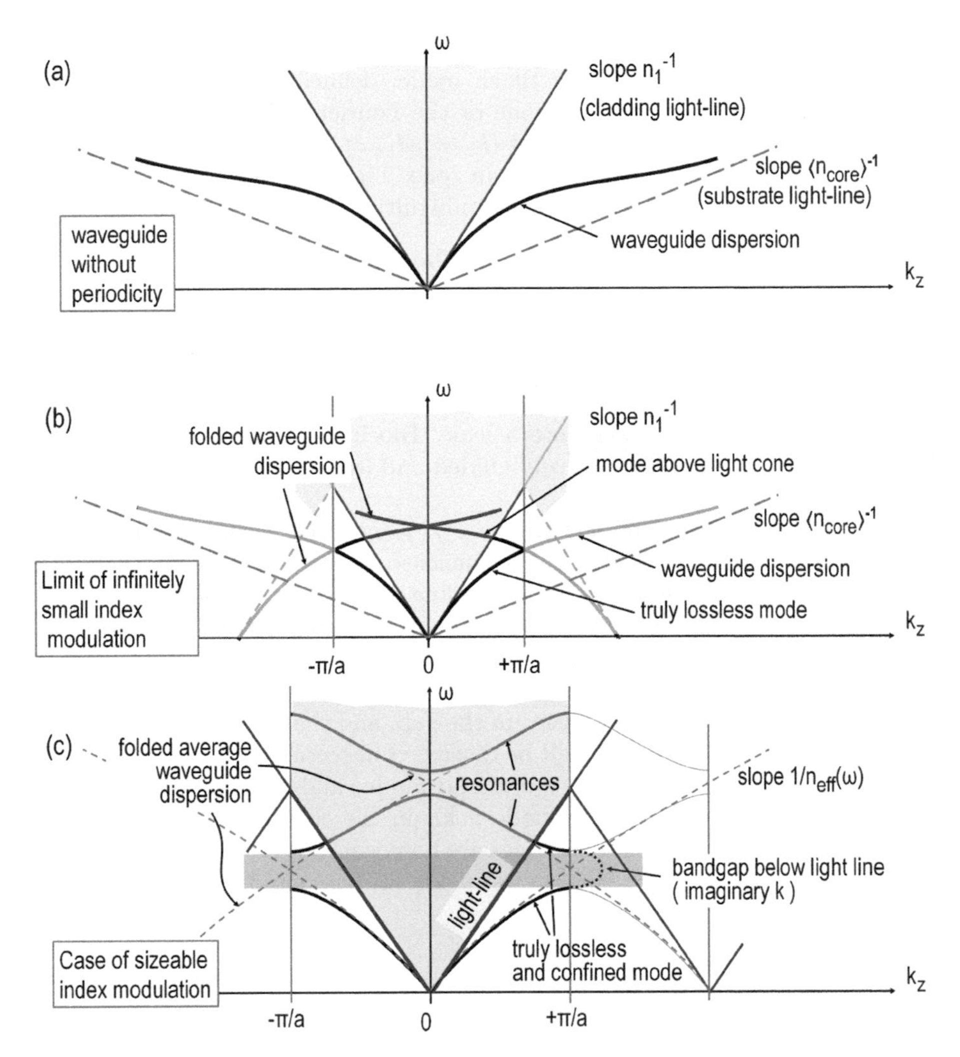

Fig. 17.2 (a) Waveguide dispersion with the two limit lines: the core line $\omega = ck_z/n_{\text{core}}$ and the cladding light line $\omega = ck_z/n_1$; (b) folding of a waveguide dispersion in the limit of vanishing index modulation; (c) bands of a system with sizeable index modulation. The bandgap may survive below the light cone inside the first Brillouin zone, at the top of the two symmetric triangular areas bounded by the zone edges at $\pm\pi/a$ and by the light lines.

in Chapter 15. We also know that this is equivalent to folding the branches at $\pm\frac{K}{2}$, hence the drawing of the folded waveguide dispersion.

The last step shown in Fig. 17.2c consists of going to finite modulation. Compared to the vanishing modulation case, important changes of the dispersion curve dominantly occur at the intersection of the initial band with its folded version. The intersection implies a coupling, as explained in section 17.1, and results in the opening of bandgaps.

We also note that it is only in the vanishing modulation limit that we can safely introduce an effective index for the Bloch mode, defined as a ratio $k_z c/\omega$. This is the only case that ensures that only one of the Fourier terms in Eq. (17.1) is much larger than the others, the one such that $(k_z + mK, \omega)$ falls on the uniform waveguide dispersion, even if k_z is in the first Brillouin zone. The resulting effective index is that of the guided mode, and is thus free from ambiguity.

17.1.3 Guided and resonance modes: The light line

The previous construction outlines the main scope but neglects an important effect: due to periodicity, Bloch modes possess many spatial harmonics, and some of them may leak their energy into the claddings. If just one of them leaked, this would completely transform the nature of the Bloch mode from a truly guided one with both ω and k_z real (in the $\omega(k)$ formulation) into a leaky Bloch mode with a finite lifetime (ω complex). Discriminating between truly guided and leaky Bloch modes will lead us to the concept of the light line.

Leaky Bloch modes are different in the $\tilde{\omega}(k_z)$ and $\tilde{k}_z(\omega)$ formulations, and so are their dispersion curves, especially for frequencies around which leakage occurs. Although a complete introduction of the light line requires us to consider both formulations, we restrict ourselves only to discriminating between truly guided and leaky quasinormal Bloch modes hereafter, which is much easier. We note that quasinormal Bloch modes ($\tilde{\omega}(k_z)$ formulation) and normal Bloch modes ($\tilde{k}_z(\omega)$ formulation) are identical for truly guided modes, except in the gap, and that the existence of a gap for truly guided normal Bloch modes will be discussed in section 17.1.4.

The question of discriminating between truly guided and leaky Bloch modes amounts to assessing the behaviour of Bloch-mode fields in the cladding. The Fourier components simplify this task: since the claddings are homogeneous media, each of the Fourier components $\mathsf{E}_{y,m}(x)$ of the Fourier series in Eq. (17.1) should satisfy the Helmholtz equation in these regions,

$$\left(\frac{\partial^2}{\partial x^2} + \frac{\partial^2}{\partial z^2} + \frac{n_1^2\omega^2}{c^2}\right)(\mathsf{e}_{y,m}(x)\exp\left[j(k_z + mK)z\right]) = 0, \forall\, m, \tag{17.2}$$

so that applying the z-derivative, we get:

$$\left[(k_z + mK)^2 - \frac{\partial^2}{\partial x^2}\right]\mathsf{e}_{y,m}(x) = \frac{n_1^2\omega^2}{c^2}\mathsf{e}_{y,m}(x). \tag{17.3}$$

In the $\tilde{\omega}(k_z)$ formulation, truly guided modes have real ω's. Let us assume that ω is real and consistently ask what this means for k_z. It is clear from Eq. (17.3) that the choice between progressive solutions along x (negative $\frac{\partial^2}{\partial x^2}$, thus leaky waves) and evanescent ones (positive $\frac{\partial^2}{\partial x^2}$, thus confined waves) now depends on the respective values of k_z and m:

$$|(k_z + mK)| < \frac{n_1\omega}{c} \quad \rightarrow \quad \text{propagation along } x, \tag{17.4}$$

$$|(k_z + mK)| > \frac{n_1\omega}{c} \quad \rightarrow \quad \text{evanescent along } x. \tag{17.5}$$

Now, we can consider all harmonics altogether: if, for any integer m, Eq. (17.5) is satisfied, every spatial harmonic is evanescent in the cladding (the Bloch mode is confined transversally), ω and k_z are real, and all is consistent to promote the fact that the quasi-normal Bloch mode is truly guided; it lives forever and propagates along the z-direction without attenuation, like the guided modes of Chapter 6. Since the dispersion relation is periodic, all the modes show up in the first Brillouin zone. It is easy to see that a necessary and sufficient condition for a mode to be truly guided, Eq. (17.5), is very simple: its k_z value in the first Brillouin zone (the smallest value of the $|(k_z + mK)|$ discrete set) should satisfy

$$k_z \in \left[-\frac{\pi}{a}, \frac{\pi}{a}\right[\quad \text{and} \quad |k_z| > \frac{n_1\omega}{c}. \tag{17.6}$$

Thus the boundaries

$$|k_z| = \frac{n_1\omega}{c} \tag{17.7}$$

define the so-called *light line*. Note that we have already met the light line in Chapter 6, it marked an upper bound of the guided-mode dispersion, as outlined in Fig. 17.2a. Its slope in the (k_z, ω)-axis is dictated by $\frac{1}{n_1}$.

The light line separates two different physics.

Below the light line, we have *truly guided Bloch modes* that cannot be excited by illuminating the waveguide from the far field away from the z-axis. The modes have exactly the same properties as the guided modes of translation-invariant waveguides: their fields exponentially decay away from the waveguide, they obey orthogonality relations, the transverse-mode profile can be finely tuned by changing geometrical parameters (the groove depth or width, for instance), the group velocity is equal to the energy velocity outside the gaps and is a real quantity defined by the slope of the dispersion curve; see more details in Complement 17.B. Perhaps the only difference with the translation-invariant case is the appearance of gap modes and slow modes with vanishing group velocity; see section 17.1.6.

Above the light line, at least one of the Fourier terms in the series corresponds to a leaky wave in the cladding, propagating towards $x \to \infty$ and offering a channel for extracting energy from, or coupling energy, to the waveguide. By crossing the light line, we thus move from a perfect waveguide picture to a much more complex picture. Resonant Bloch modes may be represented in at least two ways. Strictly following the procedure adopted in Chapter 15, either ω in the $\tilde{\omega}(k)$ representation or k_z in the $\tilde{k}(\omega)$ representation becomes complex. Above the light line, Bloch modes are either damped in space along the periodicity axis or damped in time. We will call them resonant Bloch modes (other names are found in the literature for these modes: 'guided-mode resonance', 'waveguide-mode resonance', etc.). The first approach leads to the concept of quasinormal Bloch modes with a finite lifetime and an infinite spatial extent along z. The picture is similar to that presented in Chapter 7 with a mode whose physics closely mimics that of the Fabry–Perot quasinormal modes of slabs. The field diverges away from the waveguide and the imaginary part of the frequency $\tilde{\omega}$ is inversely proportional to the mode lifetime. In the second approach, it is $\tilde{k}_z$ that becomes a complex quantity, indicating that the Bloch mode is exponentially damped in the propagation direction and leaks its energy from the core to the cladding as it propagates along the z-direction. We will develop this for resonant waveguide gratings and couplers (see section 17.3). Note that $\text{Im}(\tilde{k}_z)$, like $\text{Im}(\tilde{\omega})$, is the same for all harmonics.

In general, the dispersion relations of resonant Bloch modes can be tracked by exciting the grating waveguide from the far field with collimated plane waves. We have two options in practice. First, we may fix the angle of incidence θ and tune the wavelength to measure spectra that depend on θ. The field inside the grating—or any scattering coefficient—then exhibits a series of peaks (or dips), and the amplitude response $s(\omega)$ around each peak varies as (Petit, 1980)

$$s(\omega) \propto \frac{1}{\omega - \tilde{\omega}}, \tag{17.8}$$

which gives rise to a Lorentzian-type (Fano in general) response for $|s(\omega)|^2$. The peak characteristics are determined by the resonance frequency $\mathrm{Re}(\tilde{\omega})$ and the linewidth, which is proportional to $\mathrm{Im}(\tilde{\omega})$. The 'real dispersion' relation of the quasinormal Bloch modes is then obtained by plotting each peak position $\mathrm{Re}(\tilde{\omega})$ as a function of the wavevectors $\mathrm{Re}(\tilde{\omega})\,\frac{n_1}{c}\sin\theta$ for the set of angles used.

Second, the 'real dispersion' of the leaky Bloch mode can be recovered by fixing the incident plane-wave frequency ω while varying the angle of incidence. The response then also exhibits a series of peaks, whose linewidth is related to $\mathrm{Im}(\tilde{\mathbf{k}}_z)$ and position to $\mathrm{Re}(\tilde{\mathbf{k}}_z)$. The amplitude response $s(\theta)$ around each peak varies as (Petit, 1980)

$$s(\theta) \propto \frac{1}{k_z - \tilde{k}_z} \tag{17.9}$$

for every fixed frequency. We note that the two sets of dispersion curves, $\mathrm{Re}(\tilde{\omega})$ as a function of (real) k_z and $\mathrm{Re}(\tilde{k}_z)$ as a function of (real) ω, are identical below the light line (and outside the gaps—in fact, when the imaginary part vanishes), but may be very different above it, especially when close to near the light line.

In practice, both approaches are helpful. Quasinormal Bloch modes help with designing filters in the spectral domain and leaky Bloch modes help with designing grating waveguides with a finite length in the z-direction, such as grating couplers that couple light from outside beams to classical translation-invariant waveguides, as will be discussed later in this chapter.

An important feature of the light line is the frequency at which it crosses the first Brillouin zone. An easy calculation, using the normalized frequency unit $u = \frac{\omega a}{2\pi c}$, shows that this crossing frequency is simply $u_{\max} = \frac{1}{2n_1}$. Above $u > u_{\max}$, at least one of the harmonics $(k_z + mK)$ leaks into the cladding, ruling out the existence of truly guided modes. Hence the subscript 'max': $u_{\max}$ is the limit frequency below which guided modes have a chance to exist.

17.1.4 Bandgap below the light cone

We consider the realistic situation of a waveguide with sizeable refractive index modulation along z and large index contrasts between core and cladding. The clue we can get from the dispersion relation of Fig. 17.2c is that if these indices greatly differ, the waveguide dispersion can be rather close to the n_{core} light line at $|k_z| = \frac{K}{2}$. Then there is more room for the folded dispersion, before it meets the light line on its way back to $k_z = 0$, to open up a bandgap entirely below the light line, as depicted in Fig. 17.2c (in this typical example, however, the second branch, above the gap, has a limited span before it meets the light line). For extreme cases such as a thick silicon slab in air, a small part of the third folded branch may re-emerge below the light line before the $|k_z| = \frac{K}{2}$ boundary. But then we would be in a multimode guiding regime.

The first bandgap, or more precisely all those bandgaps that fall below the light line, are not full bandgaps: if a radiating dipole is placed in the waveguide and emits with a frequency in the gap, the emission will not be forbidden and light will escape in the cladding. We have a bandgap for guided modes only.

There is also an additional difference with the simple 1D-stack problem analysed in Chapter 15. The inspection of the dispersion curve cannot guarantee that the gap mode is a truly guided and evanescent mode. The reason is that the dotted curve in Fig. 17.2c lies in the complex plane and not in the real plane $\omega(k_z)$. Thus it may happen that the gap mode features some leakage, implying that when it is excited, it only partly reflects light, since some light leaks out into the cladding.

17.1.5 2D photonic-crystal slab: The light cone

We are interested here in generalizing to waveguides that possess periodicity not only along one dimension but more generally along two dimensions: say, z and y. Apart from complexities related to polarization, we must essentially replace the light line of a (k_y, ω) diagram by a light cone of a 3D (k_y, k_z, ω) representation. In this latter, the 2D bands are the manifolds seen in Chapter 16. The intersection of the cone with the multi-folded dispersion curve should ideally be visualized in 3D. This issue relates to that of guided light extraction tackled in Complement 17.A.

We emphasize here the case of 2D photonic crystals with a triangular lattice of period a, which exhibit a strong omnidirectional bandgap, as it is the pioneer embodiment of a structure with omnidirectional bandgap in 2D. So we are considering the perforated slab shown in Fig. 17.3 (top left), often termed a 'photonic-crystal membrane'. Rather than a 3D representation, we plot the normalized frequency $u = a/\lambda$ of branches along a circuit ΓMK of the first Brillouin zone.

Full calculations of this 3D problem are, of course, possible (Johnson et al., 1999; Chutinan and Noda, 2000), but we take a heuristic view in order to assess an essential question: how easy is it to get the light cone above the full 2D bandgap? And how much does the teaching of the 1D periodic case, Fig. 17.2, helps?

We can take as a first approximation for the bandgaps a system of air holes with infinite x-extension perforated in a single matrix index equal to the effective index $n_{\text{eff}}^{\text{slab}}$ of the unperforated slab. Using the very approximate dispersion manifolds deduced from this system, we can examine the light-cone position vs the bandgap between the relevant manifolds. A perforated slab has less 'guidance' than its unperforated counterpart, thus we expect bands with $n_{\text{eff}} < n_{\text{eff}}^{\text{slab}}$. With silicon as the slab material, $n = 3.4$ around $\lambda = 1550\,\text{nm}$, and a typical value of $n_{\text{eff}}^{\text{slab}} \simeq 2.4$, bands would behave far from the gap much as in slabs with, say, $n_{\text{eff}} \sim 2.0$, due to the air holes. Such values correspond to a bandgap centred around $u \sim 0.3$, as shown in Fig. 17.3.

The ΓM and ΓK parts of the circuit are vertical planes of the (k_y, k_z, ω) space where we can reason as in the 1D , k_z, ω case above. At point M, since $k_{\text{M}} = \pi/a$, the vacuum light-cone frequency is $\omega_{\text{M}} = k_{\text{M}}c/1 = \pi c/a$; thus, it is exactly at $u_{\text{M}} = \omega_{\text{M}}a/2\pi c = 0.5$. So, the air band starting at $u \simeq 0.35$ at point M only has a tiny span below the light

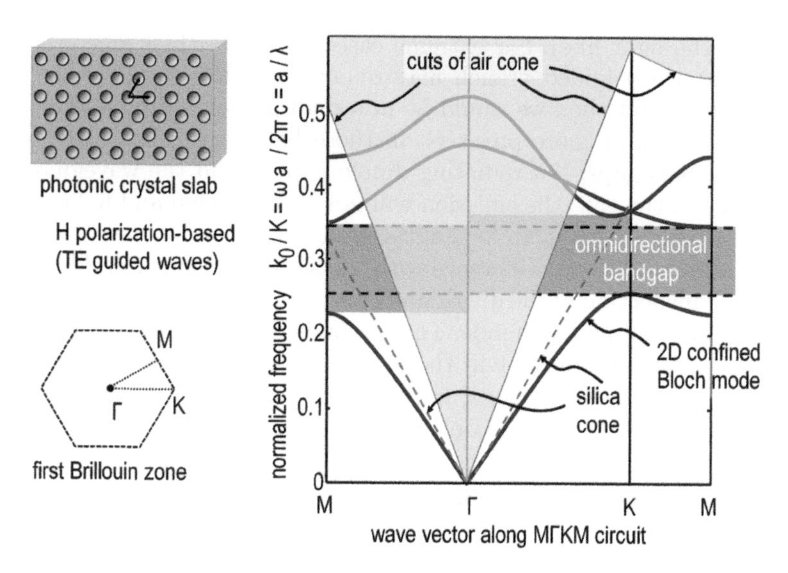

Fig. 17.3 Dispersion relation of a triangular photonic crystal etched in a semiconductor slab, sketched on the top left with its Brillouin zone on the bottom left. Similarly to the 1D case, the air cone shown with the light grey shaded area separates the truly guided Bloch modes from the resonant ones. The omnidirectional gap for guided modes is shown with a darker shaded band. For a silica cladding ($n = 1.48$), the light cone has a larger apex angle (shown as dashed grey lines) and the bandgap does not entirely fit below the cone.

cone before turning into a resonance. We let the reader examine the slightly better ΓK situation.

We note that, as indicated by the dashed grey line, if silica is used to fill the membrane, for instance to support or passivate it, the cone opens more, so that part or all of the bandgap is no longer below but above it. When moving from the air case to the silica case, this situation shall affect the K edges in spite of its larger u_K ($\Gamma K/\Gamma M = 2/\sqrt{3} \simeq 1.15$).

This regime, with a light cone so low that it transforms all branches above the bandgap into resonant Bloch modes, may be exploited for light extraction from guided modes in emitting devices such as LEDs (see Complement 17.A). Through patterning, the initial truly guided modes of the unpatterned stack that usually funnel a large fraction of the spontaneously emitted light are transformed into leaky Bloch modes that decouple part of that light towards the cladding, the other part being most often absorbed.

17.1.6 Slow-light periodic waveguides

The main motivation (Krauss, 2007) for using slow light on a chip is the space compression. Whether the slowing-down process is due to material or structural dispersion, delays are increased and more pulses can be stored per unit length. Therefore, one immediately sees the 'unique' potential offered by small group velocities for integration and miniaturization, especially if one realizes that, in contrast to electronic devices, scaling down photonic devices by reducing their dimensions is virtually impossible because of the diffraction limit. The latter can be overcome only with surface plasmons, but then one faces ohmic losses. Since the pioneer observation of slow light in periodic waveguides with a photonic-crystal waveguide (Notomi et al., 2001), many successful demonstrations underpinning the miniaturization promise have been made, such as ultra-compact modulators.

The master equation that relates the modal field intensity and the group velocity can be simply established by repeating the analysis done for z-invariant waveguides in Chapter 6. For an arbitrary 3D periodic waveguide (e.g. the segmented waveguide in Fig. 17.4b or the photonic-crystal waveguide in Fig. 17.5), the modal field of a truly guided mode at frequency ω can be written

$$[\mathsf{E}(\mathbf{r}), \mathsf{H}(\mathbf{r})] = [\mathsf{e}(\mathbf{r}), \mathsf{h}(\mathbf{r})] \exp(jk_z z), \tag{17.10}$$

where $\mathsf{e}(\mathbf{r})$ and $\mathsf{h}(\mathbf{r})$ are periodic functions of z, e.g. $\mathsf{e}(x, y, z + a) = \mathsf{e}(x, y, z)$, which are exponentially decaying away in the cladding. By applying the Lorentz reciprocity theorem to a lossless waveguide ($\mathrm{Im}(\varepsilon) = 0$), it is easy to derive (Lecamp et al., 2007)

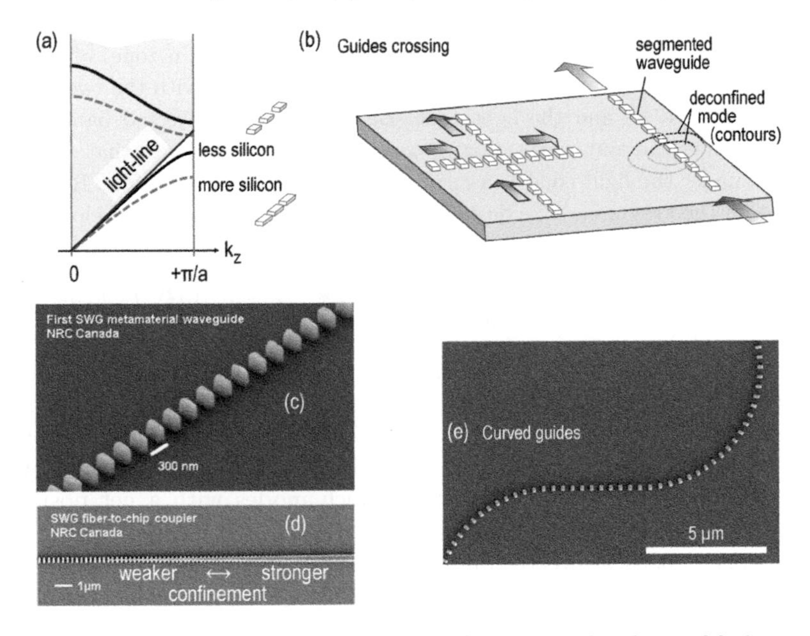

Fig. 17.4 Segmented waveguide channel composed of a periodic chain of Si boxes on silica. (a) Dispersion relation with less silicon (dashed line) or more silicon (solid line). (b) The segment chains, seen as waveguides. A deconfined mode profile is shown on the right side, with dashed contours in a cross section plane. On the left of the same substrate, the guides are used to implement a crossing of independent signals with very low crosstalk. (c) Close view of 300-nm-period waveguide with ~60% air. (d) Guide with graded effective index decreasing to the left to deconfine the mode in that region and ensure good coupling in or out to a fibre (further to the left, not shown); (e) Top view of a realization with curved guide and small silicon fraction. (c–e) Courtesy of Pavel Cheben, NRC.

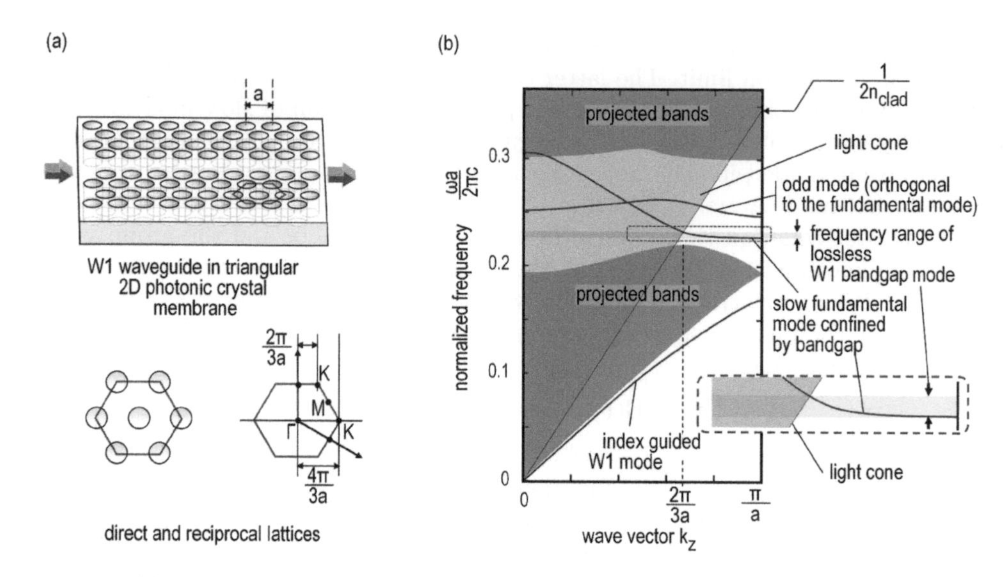

Fig. 17.5 (a) Sketch of a W1 waveguide in a 2D photonic-crystal membrane, i.e. a single missing row of holes. The real space and reciprocal space (Brillouin zone) of the 2D crystal are outlined below. (b) Dispersion relation of the W1 waveguide, with the two first projected bands (see Exercise 17.8) and the light cone. Below the first projected band lies an index guided branch. Note the tiny range of truly-guided bandgap guidance that survives for the second branch under the light cone as a result from the various constraints. Resonance Bloch modes appear in the grey area. The dashed box is magnified in the inset below.

$$\frac{1}{2}\iint \mathrm{Re}(\mathbf{e}\times\mathbf{h}^*)\cdot\hat{\mathbf{z}}\,dx\,dy = \frac{1}{4a}\frac{d\omega}{dk_z}\iiint(\mathbf{e}^*\cdot\varepsilon\mathbf{e}+\mu_0|\mathbf{h}|^2)\,dx\,dy\,dz, \tag{17.11}$$

where the triple integral runs over any unit cell of the periodic waveguide delimited by two infinite cross sections separated by the period. Just like for translation-invariant waveguides, the energetic interpretation of Eq. (17.11) evidences that the group velocity $d\omega/dk_z$ of the wavepacket is equal to the velocity at which energy is conveyed by the mode along the waveguide. For normalized Bloch modes with a net positive flow of unity, the left-hand side of Eq. (17.11) equals one. Then the equality of the electric and magnetic energy densities directly implies that the electric and magnetic fields both scale as

$$|\mathbf{E}|,|\mathbf{H}| \propto \sqrt{n_g}, \tag{17.12}$$

provided that one additionally neglects the spatial mode profile dispersion as the group velocity vanishes (Lalanne et al., 2019). This well-known scaling that boosts light–matter interactions in the slow-light regime is the main property leveraged for slow-light applications (besides delay lines and buffers), e.g. sensing, lasing, or switching.

Theoretically, there is no fundamental limit to how much light may be slowed down by structural dispersion, but in practice, this is different. Ironically, the promising scaling of Eq. (17.12) is also the Achilles' heel of slow light. This can be immediately realized by considering the impact of a tiny perturbation—a fabrication imperfection, for instance, modelled as a Dirac local permittivity change $\Delta\varepsilon\delta(\mathbf{r})$—on the transport of light in a waveguide. In the Born approximation, this defect acts as a local induced current, $\mathbf{J} = -j\omega\Delta\varepsilon\mathbf{E}\delta(\mathbf{r})$, which is proportional to $\Delta\varepsilon$ and the driving field (the incident slow mode $\mathbf{E}$).

This current may radiate into the cladding, induce absorption if the defect absorbs, or back-scatter light into the waveguide by exciting the counterpropagating slow mode. The power P_{rad} radiated into the cladding is proportional to $|\mathbf{J}|^2$. Since $\mathbf{J} \propto \mathbf{E}$ and $|\mathbf{E}| \propto \sqrt{n_g}$, we have $P_{rad} \propto n_g$. Similarly, since the power P_{abs} absorbed by the defect is proportional to $\mathbf{J} \cdot \mathbf{E}$, P_{abs} also scales with n_g. These scalings are acceptable per se, but unfortunately the back-scattered power P_{back} that is proportional to $|\mathbf{E} \cdot \mathbf{J}|^2$ (the excitation modal coefficient of the normalized back-propagating mode is $\frac{1}{4}\mathbf{E} \cdot \mathbf{J}$: see Complement 17.B) is scaling with n_g^2. This tells us that the influence of fabrication imperfection is strong at small group velocities, and that at extremely small group velocities, light cannot propagate; it is immediately back-scattered and localized in microcavities that are randomly formed by imperfections in the waveguide. In fact, the very notion of group velocity loses all meaning at those velocities for both photonic and plasmonic waves (Le Thomas et al., 2009; Lalanne et al., 2019).

Many geometries have been studied, from coupled-ring resonators and large arrays of coupled photonic-crystal microcavities to monomode single-missing-row photonic-crystal waveguides (Notomi, 2010; Scullion et al., 2011; Jean et al., 2020). Invariably, the main limitation is the fabrication imperfection that eventually distorts the propagating pulses more than the group-velocity dispersion, and prevents propagation over long distances. Depending on geometries and fabrication accuracy, the propagation length varies between 50 and 500 μm for $n_g = 100$, and between 1000 and 10,000 μm for $n_g = 30$.

Below the light line, the group velocity, v_g, is always well defined, being the derivative, i.e. the dispersion slope. When it comes to the resonances above the light line, especially in the vanishing modulation regime where resonances have high Qs, we can still use physical intuition to connect the spatial and temporal scales through the value of v_g that can be deduced by tracking small variations δk_z vs $\delta\omega$ along the dispersion curve (see Fig. 17.9): if the characteristic decay time of the resonance is τ, it shall be accompanied by a spatial decay of the guided wave given by $L_{\mathrm{decay}} = v_g\tau$, implying that the imaginary parts of ω and k_z are simply proportional, with a real proportionality factor given by v_g, to first order in small imaginary parts.

17.2 Periodic waveguiding geometries and effective index descriptions

17.2.1 The segmented waveguide channel

The segmented waveguide geometry is schematically depicted in Fig. 17.4. It is formed by a 1D periodic array of small pieces of high-index material—say, silicon—often disconnected from each other and laying on a low-index substrate—say, silica—a typical case being silicon in SOI (silicon-on-insulator), the workhorse of silicon photonics. The segmented waveguide machinery (Cheben et al., 2018) at telecom wavelengths offers many degrees of freedom to transport, bend, or tapper light, or even to reduce its speed (Jean et al., 2020) by using the properties seen in chapters 16 and 17 on artificial dielectrics and bandgaps of periodic media.

The dashed and solid curves in Fig. 17.4a represent two generic dispersion relations, with the silica light line plotted in blue. We see that part of the dispersion relation is below the light line. Depending on the relative fractions of silicon and air gaps in each period, the mode goes upwards if there is less silicon, or downwards if there is more. Pictorially, the mode hops across the air gaps between the silicon segments and by enlarging or reducing the air gaps, the phase velocity or effective index may be largely controlled. In particular, this implies that various Bloch mode profiles can be built up on demand and adapted by controlling the fraction of etched silicon. Note that the system is often made more symmetrical by burying it in silica on the top as well.

This situation can be pushed to an interesting extreme by adding as much air as possible to approach the cladding light line. Then, we encounter a phenomenon that we have seen for translation-invariant waveguides in Chapter 6: the mode profile tends to swell in the claddings just as in slab waveguides (Halir et al., 2018). We are in the so-called deconfinement regime, where the mode transverse size can be many times larger than the physical cross section.

Interesting perspectives on deconfined modes usually concern sensing or coupling to monomode fibres (Cheben et al., 2018), as shown in Fig. 17.4d. A more unexpected application is waveguide crossing. In general, a deconfined mode is less sensitive to a 90° crossing's perturbation, because at the crossing locus there is only little more high-index material seen in the swollen mode profile than elsewhere. The traditional recipe for crossings in guided optics rests on deconfinement by an enlarged guide cross-section, not a thinned one, making the guided mode close to a gentle plane wave inside the core. However, the crossing is still felt strongly by the mode tails on the waveguide sides. Another option is to shrink the waveguide in order to get a swollen mode and a very tiny *continuous* core. But this turns out to be much more delicate than the segmented waveguide fabrication.

17.2.2 The photonic-crystal slab waveguide with defects

Next, consider the photonic-crystal slab or membrane made of a triangular array of circular holes (see section 17.1.5 and Fig. 17.3a).

We then elaborate a *channel* guide by introducing a line defect in the 2D array of holes. Very often, such defects are made by removing or modifying one or a few rows of holes. Thus the smallest possible period is generally the lattice parameter a itself. In general, guided Bloch modes are used at frequencies that fall in the gaps of the 2D photonic-crystal band structure, so that any defect of the periodic structure result not in a leakage in the 2D plane but in a leakage in the third out-of-plane dimension.

For the typical case of a single missing row, popularly known as the 'W1' waveguide and shown in Fig. 17.5a, what is the resulting dispersion of the channelled mode? In Fig. 17.5b, we first see that, in agreement with the period of the W1 waveguide which is also a, the first Brillouin zone is the usual interval $[-\frac{\pi}{a}, \frac{\pi}{a}[$. This size is smaller than ΓM by a factor of $\sqrt{3}/2$. Following refs (Johnson et al., 1999; Chutinan and Noda, 2000), we draw the 'projected bands' of the periodic 2D photonic-crystal membrane with a blue area: they result from a 'supercell' analysis with a period a in z and an artificial period along y with perfectly matched layers. These projected bands depict the fact that for the associated (k_z, ω) pair, at least one Bloch mode of the 2D lattice membrane exists.

With this in mind, we can interpret all the branches below the first projected band as being localized in the higher-index W1 core by the lower effective index of the surrounding 2D photonic crystal, a situation specific to our choice of a defect that adds a dielectric into former holes and thus draws bands downwards in frequency.

As for the folded branch that emerges in the 2D photonic gap, it is not index-guided, or only partly so, depending on the parameters and position in the band. Due to the specific interactions between 'projected bands' and the W1 guided mode, we observe that this band is particularly flat in the region adjacent to the zone edge, and turns upwards rather sharply when k_z reaches the value of one of the K-point wavevectors (see Exercise 17.8 on the origin of projected bands and the vertical line at $k_z = 2\pi/(3a)$).

But then, almost immediately after this upward turn, the light line shows up and the W1 mode becomes a resonant Bloch mode. This scenario depends on the index but often holds for common membrane materials such as silicon or III-V compounds. So the range left for true bandgap guidance reduces to a small value of, say, less than 0.01 in normalized units of $u = \omega a/2\pi c$ or 2% in relative frequency terms, or typically 40 nm at a telecom wavelength of 1550 nm.

Indeed, the light line can even dent a sizeable part of the bandgap when we have the popular situation of W1 fabricated in silicon-on-silica. Due to the index of $n_{\text{Silica}} \simeq 1.46$, the bandgap of Fig. 17.5b can be mostly lost to the light cone or, at best, only a minute part of the W1 waveguide branch survives. Hence, variants of these canonical but imperfect W1 waveguides have been investigated to mitigate this issue.

The W1 waveguide is much more complex to fabricate than the segmented waveguide; thus one may wonder: why is it so popular? Initially, it was believed that the existence of a 2D gap surrounding the waveguide would be advantageous when implementing high-efficiency 90° waveguide turns or 90° waveguide crossing, or that the waveguide attenuation would be smaller, since any defect such as roughness would not scatter in plane. Actually, none of these expectations were correct. The attenuations of W1, segmented, and ridge waveguides are typically all equal to a few tens of dB/cm for similar technology efforts.

So what are the advantages of the W1? The most important one is dispersion engineering, as the in-plane gap guidance leaves room for greatly varying the size and position of the first rows surrounding the defect row and therein monitoring the dispersion. Another unique feature is the flatness of the dispersion of the first folded band, unrivalled by the segmented waveguide. More extended flatness (in k-space) implies larger bandwidth for slow light, so that slow-light experiments performed with segmented waveguides are rare and recent (Jean et al., 2020).

17.2.3 Effective media and waveguide design

Waveguide dispersion design for passive or active devices often entails delicate demands for the refractive indices involved in the vertical stack or lateral cladding. Small differences in cladding indices can result in large changes to the mode profile. This is notable for guidance with tiny index contrasts $\Delta n \sim 0.02$, doped silica in planar or fibre forms, lithium-niobate guided electro-optics, and indium-phosphide laser-diode stacks.

Another case of small index differences is the shallow ridge channel, even with high-index contrast constituents (see Fig. 17.6b, right panel). The small difference is then between the local vertical effective indices in the core and in the lateral channel's claddings; it dictates the in-plane confinement and the characteristic length of the transverse profile decay.

Since growth and overgrowth are heavily constrained in terms of material quality, epitaxy requirements, etc., it is better to open the design space by changing some of the apparent stack indices through subwavelength structuring. The benefit is the ability to operate on a more optimized guided-optics basis in terms of either effective index, mode profile, or confinement factor (see section 6.2), to ease various subsequent optical functions. An example of an array of tiny holes—in other words, a nanostructured cladding rather than just a shallow etch—is shown in the left part of Fig. 17.6b. By varying the hole depth and diameter, a broad range of effective indices can be synthesized while leaving an essentially flat surface, easing further technology steps for integration requirements.

Taking a more general look, we may classify such attempts by placing in two categories: those based on cladding engineering and those based on core engineering. We use the latter to illustrate the issues at stake. Consider the geometry of Fig. 17.6a, whereby the whole core is now periodic for simplicity. We operate below the light line (the period a is small enough, namely $a < \frac{\lambda}{2n_{\text{eff}}}$). There are exact tools to solve for the modes of such periodic guides, but it is often desirable to get some intuition. There are two natural ways to average the system: either horizontally (h) first and vertically (v) next, a process that we denote (h→v), or the converse, (v→h) (Bougot-Robin et al., 2014*a*; Bougot-Robin et al., 2014*b*).

- In the (h→v) process, the core medium is first homogenized along z as a single stack of average index n_{core} or average dielectric constant ε_{core}. This is easily done taking into account the considerations of Chapter 15 for TE polarization, with the electric field oriented along the interface. Then the resulting stack $\{n_1, n_{core}, n_1\}$ is solved along x to find its TE-guided modes.
- In the (v→h) process, each vertical slice is treated as a guide and solved analytically (a rigorous approach here would be to solve for its Bloch modes and to match them properly next). The two effective indices thus obtained, $n_{\text{eff},1}$ and $n_{\text{eff},2}$, now form a 1D periodic photonic-crystal sequence along z, whose dispersion can be found as outlined in Chapter 15. This approach has been commonly practised to find the bandgap of distributed feedback (DFB) laser structures described later in this chapter.

It turns out that these approaches do not yield the same result for strong index contrast. This is not surprising, and we already encountered a similar situation when homogenizing any 2D array of, say, rectangles, by averaging along one and the other direction (see Chapter 15). The reason can be ascribed to the poor treatment of high spatial harmonics, in agreement with the idea that field discontinuities are polarization dependent and occur differently at the corners of the structure.

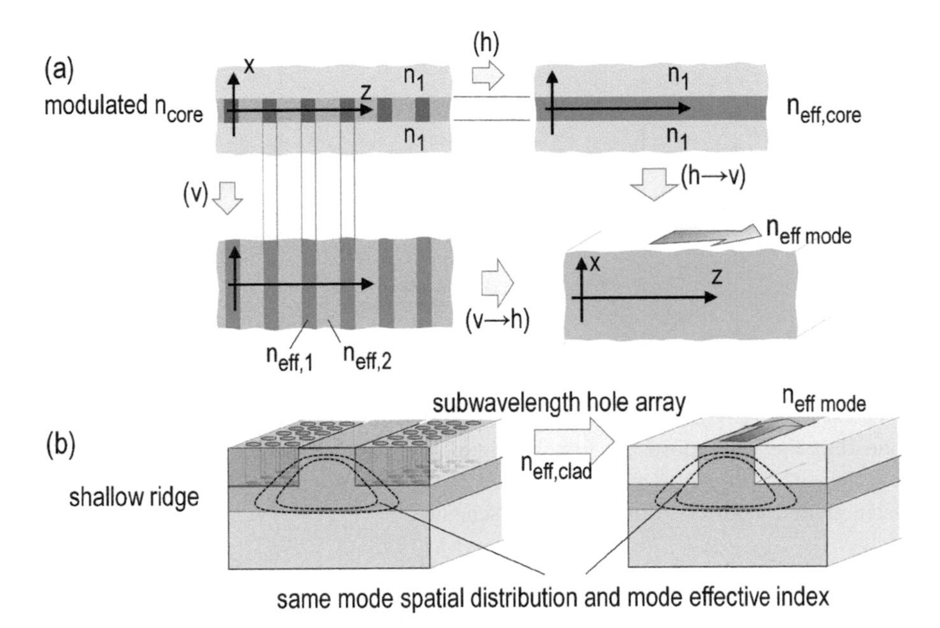

Fig. 17.6 (a) Scheme of the homogenization processes to get a modal index from a periodic waveguide using first either the (h)orizontal (z-axis) homogenization or the (v)ertical (x-axis) homogenization. (b) Example of subwavelength array of holes used to attain some effective index in the top cladding of a shallow ridge, a geometry of full 3D character. The dotted contours sketch a fundamental mode profile.

17.2.4 The modern distributed feedback laser diode

Now we are more familiar with guided geometries, it is easier to describe a very successful embodiment of periodic waveguides: the DFB laser diode. It became a workhorse of optical telecommunications in the early 1980s, to produce coherent beams from otherwise multimode laser diodes due to the large-gain bandwidth of semiconductors and the lengthy $\sim 400\,\mu$m diode cavity.

As we can see in Fig. 17.7, in a DFB, a grating is carved into a channel waveguide such that the fundamental guided mode lies below the substrate light line. A shallow-ridge waveguide is often preferred (see Fig. 17.6b and Chapter 20) to accommodate electrical and thermal requirements. Periodicity is achieved mainly by etching away a part of the high-index core, and generally stopping the process well before eroding the gain region itself. Alternatively, only the sides of the channel can be etched: the guide width appears modulated, the ridge being shaped with wavy walls. A similar method is to add periodic pieces of high-index material to the channel sides.

Insofar as lasers are concerned, there is a benefit to the slow light appearing on the modal dispersion near the Brillouin zone edge: from laser physics, the longer a photon stays in the system, the more it is likely to trigger stimulated emission and reach a gain threshold towards a lasing action. However, the exact band-edge solutions cannot in general cope with boundary conditions at the facets. Rather, a stationary wave forms

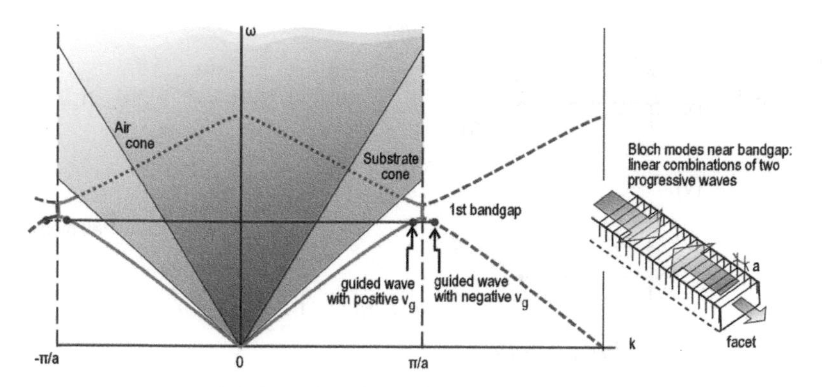

Fig. 17.7 Dispersion relation of a typical distributed feedback laser diode, with a substrate light cone that lies just above the first bandgap. Lasing typically occurs on the two typical Bloch modes with low group velocity outlined in the figure, which are linear combination of progressive guided modes shown in the side scheme.

up at the device length scale, e.g. $L_{\text{laser}} \sim 400\,\mu\text{m}$ as the sum of two counterpropagating slow progressive Bloch modes. Simplified zero-field boundary conditions at the facets provide an approximate way to predict that the departure Δk from the exact edge has to be on the order of π/L_{laser}, showing that the longer the laser, the slower the Bloch modes involved in the stationary pattern. Thus a DFB laser is nothing but a Fabry–Perot cavity formed by two counterpropagative slow modes. This operation principle privileges the lasing of a single mode on the fundamental branch (instead of the many similar modes all within the centre of the gain band for a Fabry–Perot of the same length), even though another mode, which must be averted, shows up at the upper band edge as well (see Chapter 20).

17.3 The resonant waveguide grating

17.3.1 Using resonant waveguide gratings as filters

Until now, we discussed mostly truly guided modes. We shall now focus on the resonant modes that can be excited by waves incident from the cladding.

The name 'resonant waveguide grating' (RWG) (or 'resonant grating waveguide') is commonly used to address this situation (Peng et al., 1975; Vincent and Nevière, 1979; Wang and Magnusson, 1994; Rosenblatt et al., 1997). One reason for the popularity of RWGs is their great feasibility with modern lithographies, offering high quality factors without the pain of multi-layer deposition control required for low-loss Fabry–Perot cavities, the canonical resonant structure of optics. Another reason lies in another shortcoming of the Fabry–Perot: its non-resonant signals *must be* on the reflected side. Consider the typical task of the wavelength-multiplexing context at nodes of optical networks that consists of extracting/routing M selected wavelengths among the N ones of a large comb, as shown in Fig. 17.8a, b for $M = 3$ among $N = 6$ spectral lines.

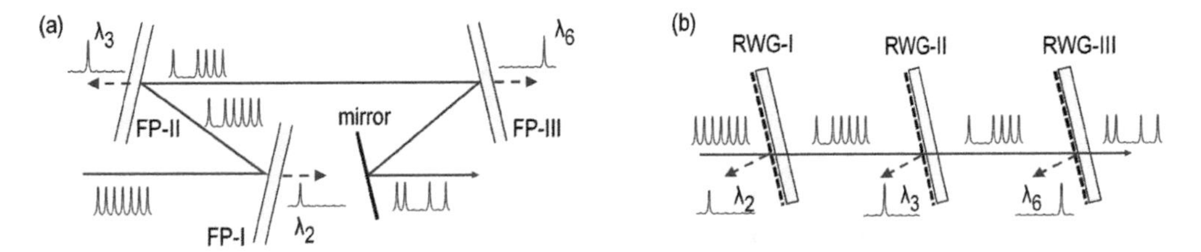

Fig. 17.8 (a) Geometry needed to extract three spectral channels, $\lambda_2, \lambda_3, \lambda_6$, with three Fabry–Perots (FP-I, FP-II, FP-III) in series: each resonates and transmits a narrow line at the specific desired frequency; (b) in-line geometry for three RWGs performing the same task, the narrow resonance now being on the reflected side.

The Fabry–Perot solution (a) results in a complex path, whereas the RWG solution (b) is 'in-line'. This is fundamentally due to the confinement at a different transverse momentum from the impinging wave in the RWG, the non-resonant through-channel having an unaffected transverse momentum.

We shall give clues to the RWG operation below by using a simple graphical extension of the classical grating equation seen in Chapter 15. Non-resonant gratings deflect incident beams for all wavelengths $\lambda = 2\pi c/\omega$, with a smooth spectral variation of diffraction efficiency. The converse situation, a highly wavelength-dependent diffraction efficiency at a given angle, must involve a longer residence time of the resonant photon in the structure to get the resolving power. Remember that the same is well established for spectrometers, whether Fabry–Perot-based or grating-based: their resolving power is proportional to the photon residence time τ_{res} (a textbook diffraction exercise is to establish that grating spectrometers' resolution relates to the very number of grating lines rather than the τ_{res}; but both views are reconciled if the instrument aperture and focal length are duly considered). Hence, any grating with sharp spectral features stores photons resonantly for some time, and must belong to the RWG family.

17.3.2 Resonance excitation condition: Visual representation

In Fig. 17.9a we illustrate a wave that is impinging on the grating with some momentum $k_{z,1} = k_1 \sin\theta_1$ from the cladding of index n_1 and start the construction as previously, with a circle of index n_1 and the various harmonics at $k_{z,1}+mK$. For simplicity, we take identical cover and substrate media. It suffices to change the radius of the semicircle to adapt the construction. Now, we pay attention to those values of $k_{z,1} + mK$ that approximately match the propagation constant of the resonant Bloch mode. At a given real frequency, we must consider matching the complex $\tilde{\mathbf{k}}_z$ of the leaky Bloch mode. This matching condition is in essence different from the grating equation condition, because the propagation constant is complex-valued. In particular, this implies that the matching does not involve an infinite number of grating periods, but only a finite number $N_{\text{dec}} = [a \times \text{Im}(\tilde{\mathbf{k}}_z)]^{-1}$ linked to the finite decay length $(\text{Im}(\tilde{\mathbf{k}}_z))^{-1}$ of the leaky mode. As suggested by the picture of Fig. 17.9b of a 'blurred' mode dispersion

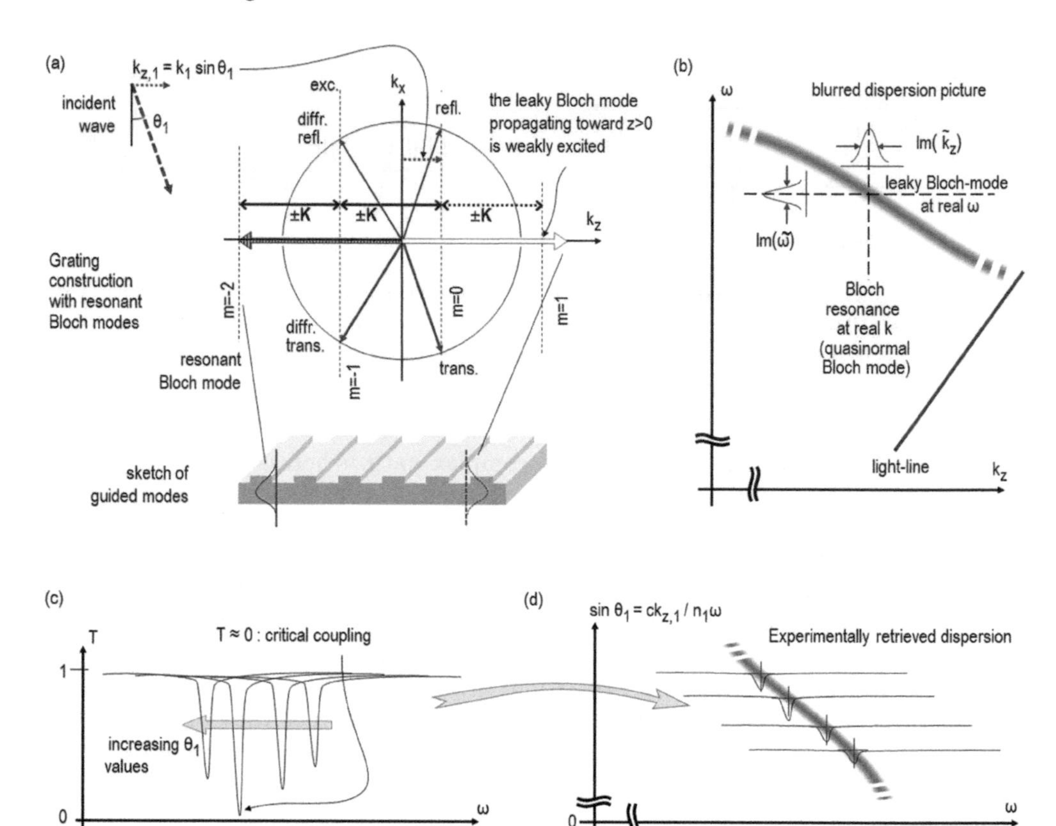

Fig. 17.9 Resonant waveguide grating: extension of the wavevector construction (see the basic grating construction in Fig. 15.6). (a) The wave is incident from the cladding of index n_1 with an incident wavevector of z-component $k_{z,1} = n_1 \frac{\omega}{c} \sin\theta_1 = k_1 \sin\theta_1$. Multiples of $K = \frac{2\pi}{a}$ may add to $k_{z,1}$. The circle has radius $|k| = n_1 \frac{\omega}{c}$. Order $m = -2$, outside the circle, feeds the left-going mode, but the right-going mode is poorly excited because $|k_{z,1} + 1 \times K - \mathrm{Re}(\tilde{\mathbf{k}}_z)| \gg |\mathrm{Im}(\tilde{\mathbf{k}}_z)|$ for order $m = 1$. (b) Sketch of a 'blurred' dispersion relation. A cut at a given real k_z is a *quasinormal Bloch mode* with complex frequency (thus a spread real-frequency spectrum). A cut at a given real frequency ω is a *leaky Bloch mode* whose k_z width is associated with a spatial decay. (c) Schematic set of RWG transmission spectra at variable angles θ_1. When the transmission dip goes to 0, we have a critical coupling between the incident wave and resonant Bloch mode. (d) Retrieval of the dispersion of the resonant branch from the set of recorded spectra.

in (k_z, ω) space, the leaky Bloch mode (at real ω) is well excited if

$$|k_{z,1} + mK - \mathrm{Re}(\tilde{\mathbf{k}}_z)| < |\mathrm{Im}(\tilde{\mathbf{k}}_z)|. \tag{17.13}$$

For simplicity, this condition is often replaced by $k_{z,1} + mK - k_{\text{guided}} \approx 0$, where k_{guided} is usually understood as the propagation constant of the guided mode that would exist in the guide in the absence of grating (without etched grooves, for instance) or, more suitably, in a homogenized version of the grating waveguide.

In the traditional temporal-domain picture, the momentum-matching condition is replaced by a frequency one (see the horizontal dashed line in Fig. 17.9b). The incident plane wave stores its energy in the quasinormal Bloch mode during a sizeable residence time τ_{res} related to the imaginary part of the complex frequency. Hence, much as in any resonant Fabry–Perot cavity, the resonant Bloch mode is fed by the incident field during the residence time. The resonance builds up and the field enhancement is proportional to $\tau_{\mathrm{res}} = 1/\mathrm{Im}(\tilde{\omega})$.

Similarly, in the momentum picture, the field enhancement results from the accumulation in space of all the incident energy that is impinging on the N_{dec} periods. Note that the two pictures are consistent, since $N_{\mathrm{dec}}a = \tau_{\mathrm{res}} \times v_g$, insofar as there is a sufficient degree of resonance to track the resonance and v_g can be seen as the derivative of the 'blurred' dispersion relation.

In practice, a set of transmission spectra at variable angles θ_1, such as sketched in Fig. 17.9c, is measured. The spectra exhibit Fano resonances (Pietroy et al., 2007) and the spectral positions of the dips are then directly exploited to reconstruct the dispersion diagram (see Fig. 17.9d). When a dip reaches $T \simeq 0$ at its lowest point, we have a situation of critical coupling. For a lossless system, everything is reflected (as desired for filtering). For a lossy system, we have optimized the transfer of energy to the resonant mode, and reflection may vanish as well.

To conclude, even if it does not account for the imaginary part effects, the simple picture of Fig. 17.9a is a very useful tool for designing RWGs. For instance, by varying the frequency, one can guess how the circles swell or shrink, and estimate one's chances to hit, say, a forward guided mode instead of a backward one, or a ray incident from a substrate of different index, etc. Of course, we must know the resonance (complex) dispersion $\omega(k_z)$ for quantitative design, which ideally has better accuracy than the one deduced from the homogenized (non-leaky) limit, but many qualitative aspects are also well derived within this limit.

Zooming out, now, it must be said that the RWG phenomenology has an old record: see the history of the so-called Wood's anomalies of gratings in Chapter 18. This suggests that the concept has a large future and can be extended. An interesting subset of extensions relies on exploiting slow resonant Bloch modes at, for instance, normal incidence. Potentially one may then couple to two degenerate modes that are counter-propagating. It is clear from Fig. 17.9c that in order to exploit both directions, we are pushed to operate with $k_{z1} \simeq 0$. The in-plane gap that results from this interaction flattens the guided/resonance bands $\omega(k)$; thus ω can remain unchanged even when k varies, implying that a slightly focused beam can be used to excite the resonance or, equivalently, that compact devices can be implemented. Such flattened resonance bands clearly point out a design avenue for diminishing the basic angular sensitivity of RWGs, which is often not desired (Ding and Magnusson, 2004; Fehrembach et al., 2007). Indeed, significant progress on this path is possible at the expense of a more delicate design to control the role of the bandgap.

Note that our reasoning also applies to a 2D in-plane patterning, although the graphical 3D effort is clearly harder (see Complement 17.A on light extraction).

17.3.3 Surface-emitting laser-diode structures

It is amazing to note that, in the completely disconnected field of DFB lasers, slow resonant modes (not rigorous slow Bloch modes) were exploited in the 1980s. At that time, the dearth of high-resolution grating at the required first-order period $a = \lambda/2n_{\text{eff}} \simeq 220\,\text{nm}$ for 1550 nm InP-based lasers led to the use of second- or third-order DFB lasers, with periods $a \sim 440$ or $\sim 660\,\text{nm}$, more amenable to optical lithography fabrication. The second-order case was naturally behaving as a surface-emitting laser through its first order, while the presence of a bandgap at the second order (to ensure wavelength-selective feedback) meant that the regime of slow modes was readily attained. The context of miniature lasers and photonic crystals led to a rebirth of surface-emitting lasers of this kind, which rely on Bloch modes and not on vertical cavities, some 30 years after the first wave of such devices. Still, the depth of the conceptual analysis of such DFB lasers around 1990 was very good and is worth reading (Henry, 1986; Kazarinov and Henry, 1985); of particular note is the radiative loss analysis at the different band edges, but progress in photonics in the 2010s has also made available 2D in-plane modulations that effectively control and transfer the coherent in-plane oscillation into a net surface emission.

17.3.4 Waveguide couplers for fibre coupling

The grating coupler is perhaps the first device relying on a leaky Bloch mode and used in integrated optics. These devices were initially conceived as perturbations of classical waveguides with tiny grooves. Resonant light, obtained for specific combinations of incident angles and light frequency, is exciting the leaky Bloch mode and is coupled efficiently into a guided mode of the waveguide. Since the grating grooves are tiny, the transverse mode profiles of the leaky and guided modes are almost the same, and the coupling efficiency is remarkably high (Petit, 1980). However, in general, the coupling length is long, typically a few hundred wavelengths.

With the development of silicon photonics in the 2000s, grating couplers became the preferred link between conventional fibre optics and highly integrated photonic chips. In that context, the realization of remarkably miniaturized devices capable of directly coupling a fibre to a ridge waveguide has become routine (Roelkens et al., 2010). As sketched in Fig. 17.10, such couplers typically aim at collecting the field radiated by a cleaved monomode glass fibre, with a $\sim 8\,\mu\text{m}$ diameter, to funnel it into a silicon channel waveguide standing on silica (say, 200×400 nm cross section, hence $\sim 1,000$ times more tightly confined). With strong and aperiodic modulations, tilted illuminations that favour unidirectional couplings, and on-chip adiabatic tapers that funnel light in the waveguide, remarkably high efficiencies above 80% are routinely achieved for large bandwidths of a few tens of nanometres (Roelkens et al., 2010). Note that essentially the same physics and design rules hold for compact couplers aimed at launching surface plasmon polaritons.

17.4 Longitudinal waveguide confinement: Cavities

The line-defect photonic-crystal 'W1' waveguide studied above gave us the opportunity to grasp how the hole array periodicity defines a 2D bandgap and how to further use

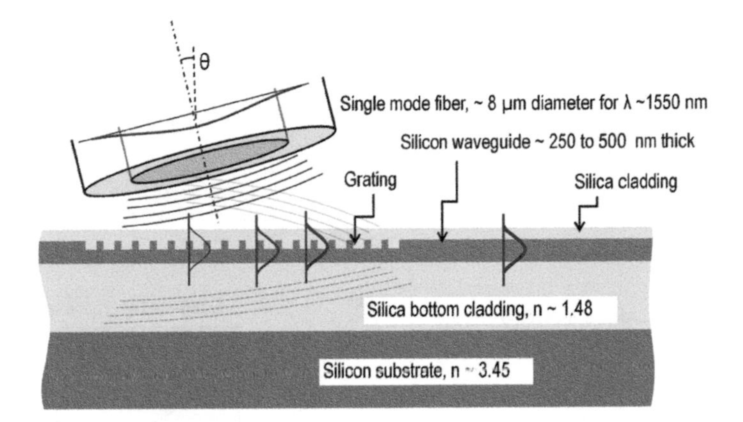

Fig. 17.10 Scheme of a fibre SOI-chip grating coupler (SOI: Silicon-On-Insulator). Note the angle $\theta \neq 0$. Thinner wavefronts are drawn to suggest uncoupled light that is firstly reflected or transmitted above and below the coupler.

this bandgap to confine light in this so-called W1 channel. We can now wonder whether the same basis can provide even more confinement in the longitudinal direction to form cavities.

The first thought when localizing light in a short segment of a waveguide is to simply close its ends, or in other words, to consider a finite-length defect of a few contiguous missing holes in line, instead of an infinite line: the full 2D in-plane photonic crystal is then used to stop light longitudinally as well as transversally. Technically, a set of n contiguous missing holes has been termed an 'Ln cavity' (see Fig. 17.11a).

For a mode confined in such a cavity, the W1 waveguide's translation invariance is now fully lost, so there is no possibility of defining a wavevector, and in principle the light-line concept is of no use. Light leakage to the continuum (air above and below the membrane) is unavoidable, but the amount of leakage can still be impacted by design: we expect that the mode in the central part of the Ln cavity is similar in most respects to a standing-wave Fabry-Perot pattern formed by two counterpropagative W1 waveguide modes, so the issue is how much reflection we get from the closed end and how much unwanted scattering arises there. Alternatively, in the spirit of antenna radiation, leakage is mitigated when an in-plane Fourier decomposition computed in a plane just above or below the slab (valid as a far-field plane-wave decomposition) has minimal components within the air light cone and manages to push as many modal field components outside the cone as possible.

This issue has been tackled by two approaches, both equally instructive:

- It can first be thought as a mode-matching problem. The mode in the perfect periodic slab (in the gap below the light cone) and that of the local W1 section do not match well; hence, modes of the continuum are excited, resulting in radiation to air. Following this approach, a careful tapering of the structure with local Bloch modes evolving from one type (W1) to the other (photonic-crystal slab) can drastically mitigate outer radiation and help reach reflectivities close to one (Lalanne et al.,

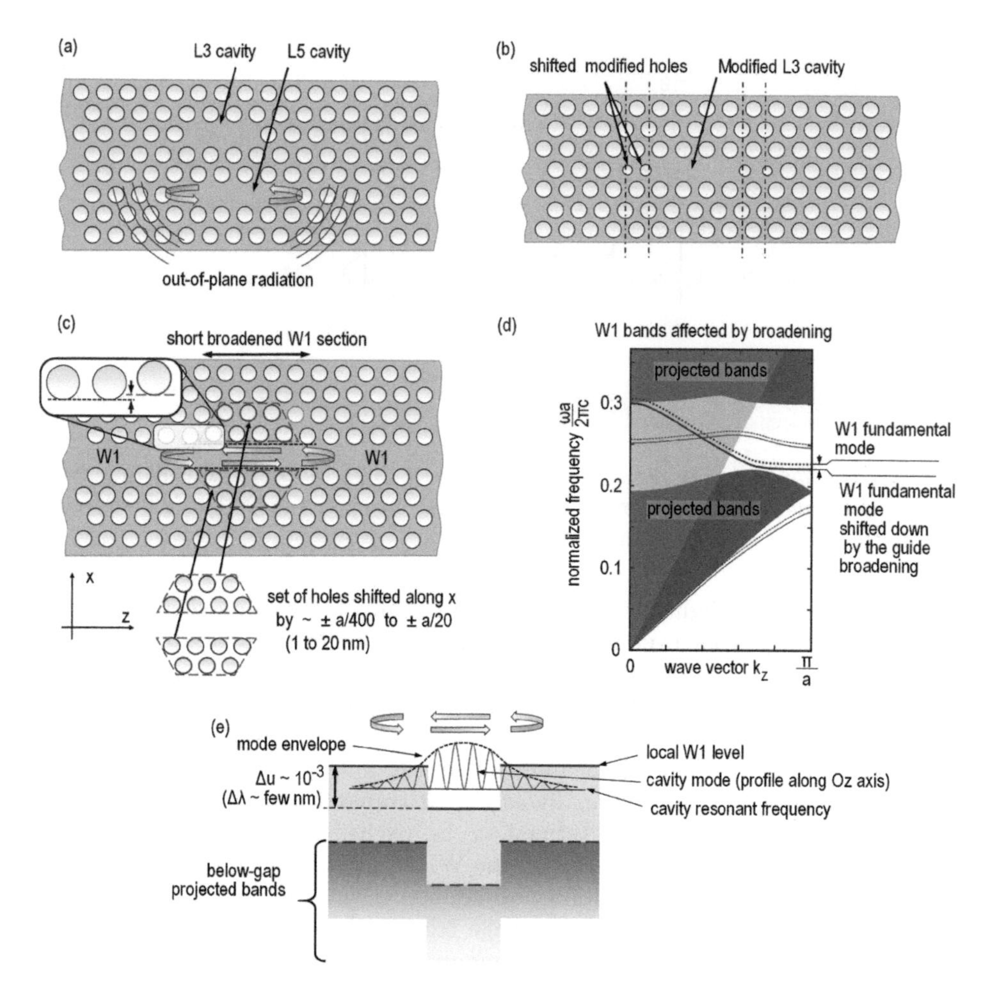

Fig. 17.11 (a) Cavities Ln (n=3 and 5) formed by removing n aligned holes in a 2D photonic crystal, with a sketch emphasizing a Fabry–Perot view of the cavity mode and its radiation losses at the end mirrors. (b) A modified L3 cavity that features fewer losses and a higher Q-factor. (c) Ultra-high-Q cavity made by modifying the width of a waveguide in a short section (trapezoidal set of holes indicated, with the local displacement shown in the inset of three holes). (d) Effect of the guide broadening on the fundamental mode branch in the 2D bandgap is a shift to lower frequencies. (e) Structure of the cavity mode viewed as a quantum well; the similar nature of the Bloch wavefunction along the structure ensures minimal radiation losses. After ref. (Song et al., 2005).

2008). The elegance here is that Bloch modes at every local period are known, therein helping to develop an analytical and intuitive view.

- Another way of mastering the radiation is to consider that holes are of subwavelength size; therefore, each hole should radiate much like an individual induced dipole. So mastering the overall outer emission at resonance means finding a destructive interference of these local emitters, much as in a stealth vehicle approach with designed vanishing radar scattering. Typically, by moving the holes that close the Ln cavity longitudinally by fractional amounts and by tuning their diameter by a couple of tens of per cent (Akahane et al., 2003), see Fig. 17.11b, various optimized designs can be produced to significantly increase increase the Q-factor of such cavities by factors of 10 to 100. The inconvenience is that all these modifications interact, so less insight is gained.

Yet another elegant way to localize light was found by simply modulating the geometry of a W1-type waveguide (Song et al., 2005). In Fig. 17.11c, we show an exemplary spatial arrangement where the width is locally enlarged (the two first rows of holes adjacent to the guide are shifted) and in Fig. 17.11d, we show the corresponding band diagram. The idea is to lower the frequency of the guided mode associated with the central section locally, and thus bracket it by standard sections of comparatively larger frequency to thoroughly taper propagating Bloch modes into gap ones.

The situation thus created is very comparable to the physics of quantum wells (QWs—see Fig. 17.11e and Chapter 8). Now, the Bloch-mode profiles of the periodic section with broadened core guide (defect) and the local W1 section (mirror) have a good mode matching. Thus, the cavity quasinormal-mode field-distribution has a local structure dictated by the surrounding 'atoms', but its envelope decays gradually in the W1 waveguide sections with narrower core. To maintain adiabaticity, only a small parametric change is permitted between the broader centre and the side barriers. It could be feared that, in turn, the two barriers define a shallow well, so that the decay in the barriers is quite extended. In general, the change in wavevector is, however, one order larger than the change in frequency in a perturbation analysis, so the decay would not be very long either. But here, because of the peculiar nature of the W1 band, there is a much flatter dispersion (lower group velocity) than the standard 1D bandgap theory (seen in section 16.2.2) would entail. Hence, with some extra tuning around the W1 geometry and associated parameters, a rather large region near the Brillouin zone edge behaves with a remarkably flat $k(\omega)$ relation, a strong 'photonic mass' if one can compare a boson band and a fermion band. This means, in turn, that for a rather shallow well and a modest shift of the frequency of the band, by analytical continuation, a relatively large Im(k) arises in the evanescent part of the mode. This means in practice that the mode volume $\tilde{V}$ (see section 7.5.2) is typically no more than twice as large as the central constant section volume, even though such a section is typically only $2a$ to $3a$ long. As the mode volume and the quality factors are the primary ingredients to obtain large Purcell factors (see Complement 7.A), this type of cavity realizes an extremely good performance in this respect. The attainment of Q-factors around 10^6 was routinely achieved by teams tackling these systems in around 2010, while the volume $\tilde{V}$ could be maintained around the volume $(\frac{\lambda}{n})^3$ (where $n = n_{\text{matrix}} \sim 3.4$ is the index of the matrix semiconductor), which is often used as a gauge in these matters.

For the sake of completeness, let us mention that similar high performances in terms of Qs and Vs can be achieved in ridge waveguides with tapered mirrors (Lalanne et al., 2008), so that the 2D bandgap confinement does not appear to be a critical factor in this respect.

17.A Complement: Light extraction from guided modes in various LEDs

We address the issue of light extraction from a guided mode by a grating. In many light-emitting devices (most notably LEDs), the spontaneous emission from excited carriers uniformly feeds all the in-plane guided modes supported by the semiconductor high-index layer. In the absence of corrugation, that light remains trapped. As nitride-based LEDs (GaN/InN) are key to modern lighting (a feat worthy of the 2014 Nobel Prize for its Japanese inventors) their efficiency widely benefits from the extraction of this light, which otherwise undergoes an absorption in one of the LED constituents.

Here, we emphasize the issue that a given reciprocal lattice wavevector $\mathbf{K}$ is not able to extract all the guided light propagating in all azimuthal in-plane directions, but only a fraction thereof.

The reason why is schematically depicted in Fig. 17.12a, b. The geometric construction simply requires us to consider the set of possible guided directions as a circle of radius $n_{\text{eff}}\frac{\omega}{c}$ in the horizontal k-space, and to remark that the addition of a given

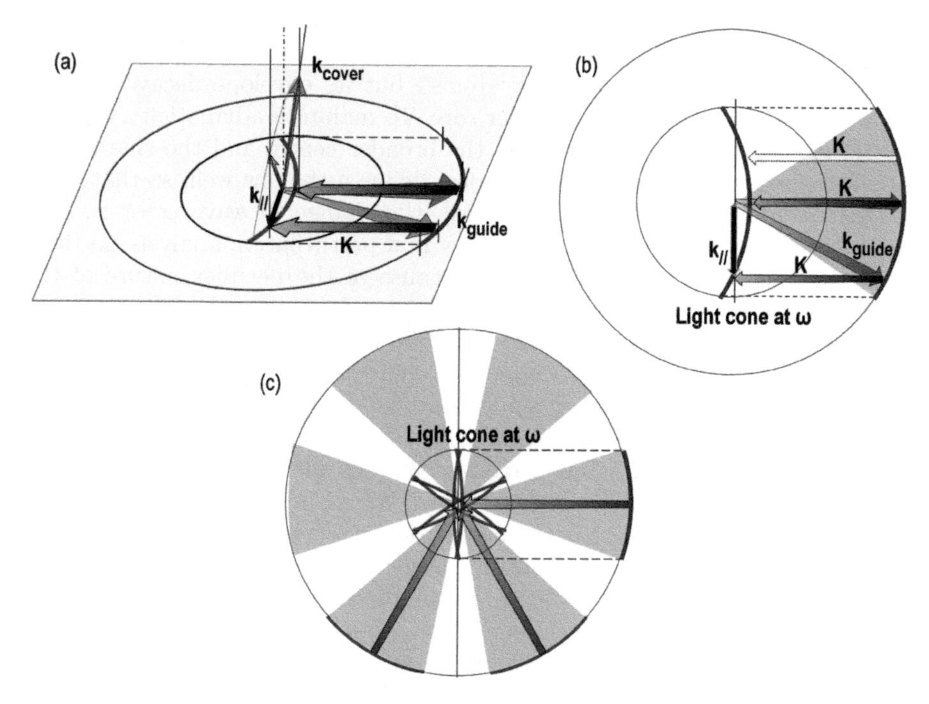

Fig. 17.12 Extraction of guided modes. (a) A 3D view of a guided mode of given frequency ω (outer in-plane circle) being scattered to a channel in air, with a given $\mathbf{k}_{\|} = \mathbf{k}_{\text{guide}} - K$. (b) In-plane view: Light is extracted only inside the light cone, hence defining a truncated arc (in blue) on the guided mode, which represents a modest fraction of the full circle. (c) In-plane view for a threefold symmetric structure and a larger ratio n_{eff}/n_1, showing the six truncated arcs (in blue) now extracting altogether. However, six non-extracted angular sectors (in white) remain.

K translates this circle to an off-centre position. It can then be easily concluded that only the arc falling inside the cladding disc of radius $n_1\frac{\omega}{c}$ matches the conditions of energy (frequency) and momentum conservation. The corresponding directions of extraction $\mathbf{k}_{\text{out}}$ are deduced in Fig. 17.12a, b for the 'cover' case, from the knowledge of the in-plane projection and the vector magnitude.

If the mismatch between n_{eff} and n_1 is sizeable, the arc of intersection with the cladding disc represents a moderate angular sector, amounting asymptotically to $2\frac{n_1}{n_{\text{eff}}}$ radians—that is, a fraction $\frac{n_1}{\pi n_{\text{eff}}}$ of the 2π azimuthal degree of freedom.

Sure enough, with a 2D grating, there are several wavevectors. For the square and hexagonal (also called triangular) lattices, the numbers of distinct **K**-values are four and six, respectively. We get the superposition of four or six arcs (see Fig. 17.12c), and thus an extra factor of four or six. The total extracted fraction becomes $\frac{4n_1}{\pi n_{\text{eff}}}$ or $\frac{6n_1}{\pi n_{\text{eff}}}$. This fraction thus reaches $\sim$35% (square lattice) or 55% (hexagonal lattice) for the case of high-index LED materials ($n_{\text{eff}} \simeq 3.5$) with air cladding.

Numerous extra considerations come into play. Often, an epoxy dome encapsulates the LED chip, so that a more relaxed condition, $n_1 \sim 1.5$, applies. But if light is eventually collected in air, the brightness (irradiance in Watts/m^2/sr) does not increase because the dome is a lens magnifying the emitting area viewed from outside (a consequence of Abbe's law, or an étendue conservation law for two different media).

Another issue arises in a nanophotonic context if we start from modes of a strongly textured emitting system where anisotropy strongly affects the in-plane emission pattern itself: for instance, in-plane emission does not occur in all angles due to directional bandgaps, be it in a singly or doubly periodic structure, an effect eased by a strong index modulation. In other words, the Bloch-wave dispersion $\mathbf{k}_{\parallel}(\omega)$ for the photons trapped in the structure is no longer a circle.

It can seem strange to start from such a complex system, but if, for instance, the modified isofrequency curves (locus of $\mathbf{k}_{\parallel}(\omega)$ at a given ω) are six isolated ring-like shapes, each situated at one of the six corners of a hexagonal Brillouin zone, the same strategy for extracting as in Fig. 17.12c can then prove much more successful: the addition of a weak grating that brings $\mathbf{k}_{\parallel}+\mathbf{K}$ inside the light cone can now lead to full extraction, because only a reduced isofrequency curve is concerned along each of the 60°-spaced equivalent directions. However, no simple prediction can be made: full 3D calculations are needed to truly quantitatively address such systems. Besides, not all LED materials/wavelengths are equivalent, and various constraints related to carrier management can rule out deep in-plane texturing (David et al., 2012).

We now discuss the specific case of blue-emitting InGaN LEDs, which contain a thick GaN ($n_2 \sim 2.5$) largely multimode waveguide grown on a sapphire substrate ($n \simeq 1.7$) and thin QWs, as shown in Fig. 17.13. An extracting grating is etched on top (David et al., 2012).

The etched layer has two consequences: in the spirit of the average effective index discussed in section 17.2.3, it appears as a layer of average index around 1.7 (air and GaN average), which is comparable to sapphire. So, the guided modes that are evanescent in sapphire will also be evanescent in the grating. This means that only the tail of the guided modes interacts with the grating and scatters efficiently.

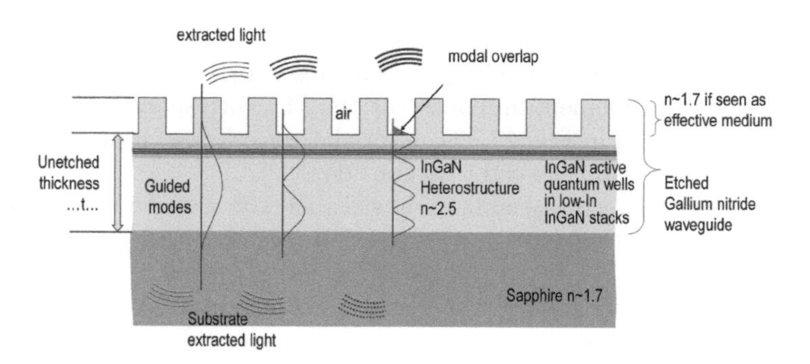

Fig. 17.13 Structure of a GaN-based LED with an extracting grating. Note the variable t that measures the remaining core (GaN) waveguide thickness left unetched.

Hence, the fundamental mode and perhaps the first few high-order modes have a weak interaction with the grating, so that decoupling photons launched in these modes requires a long propagation distance. This is not desired in practice, since on the other hand, the launched photons are also absorbed or scattered in the substrate. Instead, a rapid extraction is required in LEDs with couplers that are not working as perturbations, which contrasts with the comparatively weak gratings of DFB lasers.

Of course, the scattered field induced in the grating layer undergoes upward and downward reflections that translate into interference effects and a noticeable modulation of the field (David et al., 2012). This modulation oscillates with, e.g., core thickness, on a fast $\lambda/(2n_2)$ scale, much as the emission of a source in a planar Fabry–Perot cavity. Even though the way the extracted power is split vertically is important, it is not as important as the extraction length itself, which should be short enough, as hinted above, with the provision that a very short extraction length means a strong change of the environment that goes beyond a perturbation approach.

In Exercise 17.9, we propose an account of the trend of extraction length in this situation based on the guided-field profile. The extraction length (inverse of $\mathrm{Im}(k_z)$) is found to scale like t^{-3} with t the *unetched remaining thickness* of the waveguide. In a nutshell, this is due to the combination of a t^{-3} trend from the value of $|\mathsf{E}(x)|^2$ in the grating and a t^{-1} factor from normalization.

17.B Complement: Fundamentals of the guided Bloch mode

In Chapter 6, we provided a set of general properties of guided modes in translation-invariant waveguides, e.g. the mode orthogonality for lossless or lossy waveguides, the existence of pairs of modes with opposite propagation constants, and the equivalence of the group and energy velocities for lossless waveguides. To demonstrate these properties, we used integral forms of Maxwell equations with the divergence theorem.

A similar approach can be followed for periodic waveguides. Fig. 17.14 shows an arbitrary, non-symmetric (i.e. no mirror symmetry by a z-plane) waveguide that is periodic along the z-direction (periodicity a) with arbitrary transverse cross sections. We assume non-magnetic materials and denote by ε the relative permittivity tensor, $\varepsilon(x, y, z) = \varepsilon(x, y, z + a)$. In this complement, we do not provide the derivation but

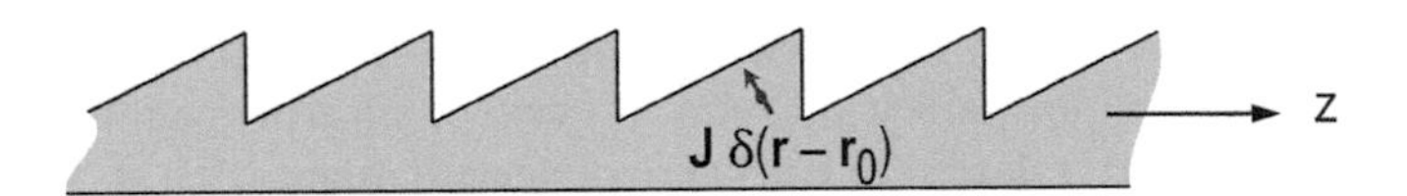

Fig. 17.14 Periodic waveguide without mirror-symmetry plane.

instead simply give the results; the interested reader may find demonstrations and more details in ref. (Lecamp et al., 2007). For a given frequency ω, there exist many guided and leaky Bloch modes. Forward Bloch modes potentially encompass leaky Bloch modes ($\mathrm{Im}(\beta) > 0$) and also a finite number of truly guided Bloch modes, either progressive ($\mathrm{Im}(\beta) = 0$ in non-absorbing materials) or exponentially damped ($\mathrm{Im}(\beta) > 0$). The forward modes are labelled by positive p's,

$$[\mathsf{E}_p(\mathbf{r}), \mathsf{H}_p(\mathbf{r})] = [\mathsf{e}_p(\mathbf{r}), \mathsf{h}_p(\mathbf{r})] \exp(j\beta_p z), \tag{17.14}$$

where $\mathsf{e}_p(\mathbf{r})$ and $\mathsf{h}_p(\mathbf{r})$ are periodic functions of z. The wavevector component k_z is denoted by β in this complement to focus on the useful modal subscripts β_p.

Existence of pairs of modes with opposite propagation constants. For translation-invariant waveguides, the existence is guaranteed by the invariance. For periodic waveguides, the existence is not guaranteed in general. Pairs of modes with opposite propagation constants and an identical frequency exist under certain conditions:

- (C1) for reciprocal materials (the permittivity tensor and its transpose are equal);
- (C2) for truly guided propagating Bloch modes ($\mathrm{Im}(\beta) = 0$) in non-absorbing media ($\mathrm{Im}(\varepsilon) = 0$), even if the materials are non-reciprocal. For this important case of practical interest, the counterpropagating Bloch mode takes a simple expression $[\mathsf{E}^*(\mathbf{r}), -\mathsf{H}^*(\mathbf{r})] = [\mathsf{e}^*(\mathbf{r}), -\mathsf{h}^*(\mathbf{r})] \exp(-j\beta z)$, directly derived from the forward Bloch mode $[\mathsf{e}(\mathbf{r}), \mathsf{h}(\mathbf{r})] \exp(j\beta z)$ by conjugating the curl Maxwell equations.

Note that if the periodic waveguide possesses a mirror-symmetry z-plane, all the field components of a Bloch mode with a propagation constant $-\beta$ can be directly deduced from those of the Bloch mode with a propagation constant $+\beta$. Otherwise, two Bloch-mode computations need to be performed: one for $+\beta$ and the other for $-\beta$.

Thus, whenever conditions (C1) or (C2) are verified, Bloch modes are pairwise; the backward Bloch mode will be labelled by the index $-p$ with a propagation constant $\beta_{-p} = -\beta_p$. Hereafter, we assume that condition (C1) is fulfilled.

Equivalence of the group and energy velocities for lossless waveguides.. The equivalence between the group velocity $d\omega/d\beta$ of the wavepacket and the velocity at which energy is conveyed by a truly guided progressive Bloch mode along the waveguide is given in Eq. (17.11).

Bloch-mode orthogonality. We consider two Bloch modes at the same frequency ω. They are labelled by the indices q and p. We have the orthogonality relation

$$\iint [\mathsf{e}_{-q}(\mathbf{r}) \times \mathsf{h}_p(\mathbf{r}) - \mathsf{e}_p(\mathbf{r}) \times \mathsf{h}_{-q}(\mathbf{r})] \cdot \hat{\mathbf{z}}\, dx\, dy = N_p\, \delta_{p,q}, \tag{17.15}$$

where $\delta_{p,q} = 1$ for $p = q$ and 0 otherwise, and N_p is a complex constant used for normalization. Note that normalization relies on two counterpropagative Bloch modes and that the orthogonality relation of Eq. (17.15) is not defined with the usual Poynting $\mathrm{E} \times \mathrm{H}^*$ products.

For truly guided progressive Bloch modes, it is possible to replace e_{-q} by e^*_q and h_{-q} by $-\mathsf{h}^*_q$ to obtain

$$\frac{1}{2}\iint [\mathsf{e}_q(\mathbf{r}) \times \mathsf{h}^*_p(\mathbf{r})] \cdot \hat{\mathbf{z}}\, dx\, dy = N_p \delta_{p,q}, \tag{17.16}$$

which reintroduces an orthogonality in the energy/power sense, implying that the total power carried in any periodic waveguide by a set of truly guided progressive Bloch modes can be decomposed as a sum of the powers carried by each of them.

Light emission in periodic waveguide. We consider an electric-dipole source $\mathbf{J}\,\delta(\mathbf{r}-\mathbf{r}_0)$ located at $\mathbf{r}_0$ with an arbitrary polarization (see Fig. 17.14). For $z > z_0$, the field radiated can be expanded as a sum over the forward Bloch modes

$$\sum_p \alpha_p\, [\mathsf{e}_p(\mathbf{r}), \mathsf{h}_p(\mathbf{r})] \exp(j\beta_p z) \tag{17.17}$$

and, similarly, for $z < z_0$, it can be expanded as a sum over the backward Bloch modes

$$\sum_p \alpha_{-p}\, [\mathsf{e}_{-p}(\mathbf{r}), \mathsf{h}_{-p}(\mathbf{r})] \exp(-j\beta_p z). \tag{17.18}$$

In Eqs (17.17) and (17.18), the α's are the modal amplitude coefficients. Then, for normalized Bloch modes ($N_p = 1$), we have

$$\alpha_p = -\frac{1}{4}\mathsf{e}_{-p}(\mathbf{r}_0) \cdot \mathbf{J} \exp(-j\beta_p z_0), \tag{17.19}$$

$$\alpha_{-p} = \frac{1}{4}\mathsf{e}_p(\mathbf{r}_0) \cdot \mathbf{J} \exp(j\beta_p z_0). \tag{17.20}$$

Note that the coupling coefficient α_p of the forward Bloch mode p depends on the backward Bloch mode, and vice versa. The emission into truly guided Bloch modes ($[\mathsf{e}_{-p}, \mathsf{h}_{-p}] = [\mathsf{e}^*_p, \text{-}\mathsf{h}^*_p]$) deserves particular attention. For a linearly polarized source $\mathbf{J} = I\mathbf{u}$, I being a complex number and $\mathbf{u}$ a 3×1 real unitary vector, we find that the powers P_p and P_{-p} coupled into the forward and backward Bloch modes are equal:

$$P_p = P_{-p} = \frac{1}{16}|I|^2\, |\mathsf{e}_p(\mathbf{r}_0) \cdot \mathbf{u}|^2 \cdot \hat{\mathbf{z}}. \tag{17.21}$$

Perhaps counterintuitively, the power equality of Eq. (17.21) holds independent of the source location $\mathbf{r}_0$ in the unit cell of the periodic waveguide and of the existence or non-existence of a mirror-symmetry z-plane. Additionally, since the electric fields of a normalized truly guided progressive Bloch mode scale as $v_g^{-1/2}$, the radiated power scales as $1/v_g$ in the slow-light regime, possibly diverging. Note that this divergence holds for lossless and infinitely long periodic waveguides, and that any waveguide termination or residual fabrication defect would prevent it.

Exercises

Exercise 17.1 Generalized light lines for given cladding angles. The issue of the light line may be generalized to sort among far-field waves in the cladding. Let us define useful photons (modes) as those that verify $\theta < \theta_{\max}$ with $\theta_{\max}$ some arbitrary angle (with respect to the x-axis in our notation). Show that in a dispersion diagram, the corresponding modes lie above an adapted light line whose slope is related to $n_1 \sin\theta$.

Exercise 17.2 Light line and 1D periodicity. Consider a system whose guided mode is described by a constant effective index $\beta \equiv k_z = n_{\text{eff}} \frac{\omega}{c}$, and for which the two claddings have the same index n_1, much as in Fig. 17.2b, but with a straight dispersion.

(1) Re-express the light line and the dispersion using the normalized frequency $u = \frac{\omega a}{2\pi c}$. Express $u_{\max}$, the ordinate at which the light line meets the Brillouin zone edge $k_z = \frac{\pi}{a}$.

(2) What is the normalized frequency u_{leak} of the crossing of the waveguide dispersion with the light line, assuming it happens after the first fold?

(3) Express this limit as a diffraction cutoff for a wave from outside to be coupled to the guided Bloch mode. You may use the above exercise to express the cutoff condition as $\theta = \frac{\pi}{2}$.

Exercise 17.3 Light line, multiple modes, and different cladding.

(1) Starting from the same frame as the above exercise, we now consider that the guide may support multiple modes and treat them within the same assumption, $\beta_q \equiv k_{z,q} = n_{\text{eff},q} \frac{\omega}{c}$. Since the fundamental mode has the higher effective index, predict in which frequency and wavevector order the different modes q reach the light line.

(2) Consider a single-mode waveguide with two different claddings, $n_{\text{cover}} = n_1 < n_{\text{substrate}} = n_3$. Depict how the two light lines of these substrates are crossed by a Bloch-mode dispersion. Depict what waves are in the substrate when the cover light line is crossed.

Exercise 17.4 Period and coupler or resonant waveguide grating. We consider the silicon-on-insulator grating-coupler situation shown in Fig. 17.10b. Relate the angle θ to the period and the grating Bloch-mode effective index (use $n_{\text{eff}} = 2.8$ for calculation).

Find the angular dispersion by relating it to the frequency dispersion and relating this latter to the size of the system, which is about $Na = 15\,\mu\text{m}$ at $\lambda = 1.55\,\mu\text{m}$ to fit to cleaved single-mode fibre beams that have not undergone much diffraction in air.

Explain why it is not favourable to operate close to $\theta = 0$ to couple in a given waveguide. Find the range around the exact condition $\theta = 0$ by exploiting the same rule as above for the tolerance.

Exercise 17.5 Silicon nitride resonant waveguide grating. We consider a silicon nitride waveguide with a typical effective index of 2.0. The waveguide stands over a silica cladding of index 1.5.

(1) Targeting the coupling from air to a resonant guided mode for various photonic classical wavelengths such as 532 nm, 633 nm, 1064 nm, propose designs in terms of (angle, period) pairs that provide this coupling without diffracted beams in air.

(2) Tackle the same issue but without diffracted beams in the silica substrate as well.

Exercise 17.6 The light line and 2D photonic crystals. Using the same notation as in Exercise 17.2, $u = \frac{\omega a}{2\pi c}$ but now considering 2D photonic crystals of period a with square or triangular lattices, find the different frequencies u_{max} and u_{leak} in the directions of high symmetry (X and J in the square lattice, M and K in the triangular lattice).

Exercise 17.7 Integrated optics with channel and planar regions. Assume that a channel waveguide is defined by a thick silicon ridge of about 400 nm and effective index $n_{\text{eff}} \simeq 3.5$ at around $\lambda = 1500$ nm, standing on a silica ($n = 1.5$) substrate. This waveguide is laterally surrounded by a rather thin silicon sheet of $\sim$150 nm thickness whose effective index lies between the two values above at, say, $n_{\text{eff}} \simeq 2.5$.

Describe what happens when a weak corrugation of period a is made on the channel waveguide: proceed by defining the 'in-plane' light line, which enables one to determine the cutoff period that allows leakage from the channel to the planar region.

Can you further arrange a variation of the period to focus light at a given point of the planar region using geometric optics and the grating equation? What would be the issue regarding the wavefront amplitude thus generated?

Exercise 17.8 Projected bands. Consider a supercell in a triangular photonic lattice of round holes or 'atoms': a rectangular elongated cell of dimensions $(a \times Ma)$ that contains $Ma\sqrt{3}$ holes.

Study its first Brillouin zone: it should be rectangular, and smaller than the hexagon at the bottom of Fig. 17.5a. Consider now multiple replicas of this rectangular cell tiling the hexagon.

Deduce from this tiling how the projected bands appear along the long axis $[0, a]$ of the small Brillouin zone.

In principle, you can also identify that the K-point at the top right of the hexagon now appears at two-thirds of the $[0, a]$ span, as shown in Fig. 17.5b.

Exercise 17.9 Coupling of fundamental mode from a thick GaN waveguide. Study the situation of the fundamental mode in the configuration of Fig. 17.13. Replace the top grating by an index $n_{\text{clad}} = 1.7$ the same as the substrate one, and assume it is thick enough to allow adoption of the approximation of a symmetric slab with a mode-effective index close to the core index $n_{\text{core}} = 2.5$. You can use $k_x \simeq \pi/t$ to find a first approximation of the effective index as a function of n_{core} and $\lambda^2 t^{-2}$, and deduce the exponential decay along x in the cladding vs t. Show that the squared field at $x = t/2$ has a scaling t^{-2} with respect to the maximum field at $x = 0$. Using modal normalization, show that the mode's squared field at the targeted interface scales as t^{-3} and that the same rule holds for the fractional intensity in all the top cladding (thus the extracting grating). The consequence is that for thick waveguides, the fundamental mode is so weakly extracted by the grating that the photons are lost through all other loss mechanisms.

Exercise 17.10 Anisotropies.

(1) Consider a ridge waveguide with a core made of a composite medium structured normally to the propagation axis. The effective medium approach for the core of the waveguide may lead to different averaging according to the polarization (field direction). Using the naïve homogenization formulas (see section 15.2.2), find the two indices to be implemented for TE-like and TM-like behaviour.

Study whether a cladding index exists that would exclude guidance in one of the two polarizations.

(2) In the cladding proposed for the shallow ridge-waveguide (Fig. 17.6b), discuss the anistropy of the cladding's own effective index. Discuss whether the TE-like polarization in the shallow-ridge waveguide is more or less confined than the naïve index-average expectation in this case.

Exercise 17.10: Anisotro[illegible]

[illegible]

18 Metamaterials and metasurfaces

18.1 Introduction: The ideas behind 'meta'

The basic idea behind the word 'metamaterial', and spin-off concepts such as 'metasurfaces' is to artificially synthesize new materials that have a property that is not found in naturally occurring materials. Metamaterials are often composite systems that mix existing materials in a smart way to extend the range of available properties already offered by nature. The design is often a challenge.

Thus, making a metamaterial amounts in a restricted sense to producing a slab of material that behaves with respect to impinging waves in a reasonable range of frequency and wavevector (angle/direction) as though it would have some desired values of permittivity $\varepsilon(\omega)$ and permeability $\mu(\omega)$. In particular, the possibility of getting any positive or negative (or complex) values for these quantities gives food for thought. It is obviously a nice bridge between engineers with demands such as steering, filtering, and funnelling beams for given needs and, say, electromagnetism and optics scientists.

Historically, radar and microwave communities considered 'frequency-selective surfaces' made of alternating layers of metals and dielectrics (Munk, 2000) and metallo-dielectric composites (Collin, 1960), and made great strides towards the general idea suggested above. About 40 years after these efforts, the accumulated knowledge on nearby fields (such as photonic crystals and plasmonics) and the efforts to bridge the terahertz frequency gap fostered a more unified vision of electromagnetism which resulted in the term 'metamaterial' or 'metasurface' being the successful one.

Also, in 1968, Victor Veselago considered what the electromagnetic waves in a general medium with any signs of both $\varepsilon(\omega)$ and $\mu(\omega)$ should be (Veselago, 1968). He arrived at counterintuitive conclusions on what should be taken as the index of refraction (reminding that $n = \sqrt{\epsilon}$ for non-magnetic materials). These considerations spurred rich investigations at the turn of the millennium. We shall thus start this chapter with a class of artificial materials in which the EM properties, as represented by the permittivity and permeability, can be controlled in such a way that it is possible to achieve negative permittivity and negative permeability simultaneously, but whose anisotropy is at most a side effect. We will next examine the converse case of a strong effect of anisotropy, focusing on hyperbolic metamaterials. Finally, we will present some basics of the field of 'metasurfaces', which boomed after the decline of 3D metamaterials, much as 2D photonic crystals emerged after the initial works on 3D photonic crystals.

Introduction to Nanophotonics. Henri Benisty, Jean-Jacques Greffet, and Philippe Lalanne, Oxford University Press.
 DOI: 10.1093/oso/9780198786139.003.0018

18.2 Metamaterials with a negative index

18.2.1 The dream of a perfect lens

In Chapter 4, we saw that the radiation of a microscopic dipole in a uniform medium is split between propagative plane waves and decaying evanescent waves, making it impossible to retrieve high-spatial-frequency information outside the near-field region. Can we pick up those evanescent waves and exactly compensate for their decay, so that no information is lost? This task could be assigned to a material whose 'refraction' properties have to be uncommon to achieve this job. We are going to detail below how a hypothetical material, often referred to as a 'perfect lens', having negative permittivity and permeability, can help achieve this task.

18.2.2 Electromagnetism of negative permittivity and permeability materials

Let us study a uniform material that possesses a negative relative permittivity and a negative relative permeability, $\varepsilon < 0$ and $\mu < 0$. This choice does not cause mathematical contradictions; in particular, it does not change the fact that since the material is translation invariant, the EM modes are plane waves $\propto \exp(i\mathbf{k}.\mathbf{r} - i\omega t)$ with a propagation constant $\mathbf{k}$. We omit the ω-dependence hereafter.

Since the Maxwell equations for such a mathematically clear situation take the form

$$\mathbf{k} \times \mathbf{H} = \omega\varepsilon\mathbf{E} \qquad \text{and} \qquad \mathbf{k} \times \mathbf{E} = -\omega\mu\mathbf{H}, \tag{18.1}$$

one immediately sees that $\{\mathbf{E}, \mathbf{H}, \mathbf{k}\}$ constitutes a *left-handed* set when both ε and μ are negative, whereas they usually constitute a *right-handed* set for positive values. For this reason, materials with $\varepsilon < 0$ and $\mu < 0$ are frequently called *left-handed materials*. However, the Poynting vector

$$\mathbf{S} = \mathbf{E} \times \mathbf{H} \tag{18.2}$$

always constitutes a right-handed set with $\mathbf{E}$ and $\mathbf{H}$. Therefore, $\mathbf{S}$ and $\mathbf{k}$ are opposite, or, equivalently, the energy (which is identical to the group velocity, since the same demonstration as in Chapter 6 is valid for uniform materials) and phase velocities are opposite. Solving Eq. (18.1) is easy and the positive $\varepsilon\mu$ product naturally appears. Retaining the vector aspect, we can write down the plane-wave dispersion relation that includes a unit vector in the direction of energy flow $\mathbf{S}/|\mathbf{S}|$,

$$\mathbf{k} = -\sqrt{\varepsilon\mu}\,\frac{\omega}{c}\,\frac{\mathbf{S}}{|\mathbf{S}|}, \tag{18.3}$$

where the minus sign is the direct consequence of the opposite directions of $\mathbf{k}$ and $\mathbf{S}$. We just recall that for a standard material with $\varepsilon > 0$ and $\mu > 0$, the very same relation holds but without the minus sign. It is appropriate to recall that the refractive index n is not a material property, but a *wave* property, which tells us how plane waves, i.e. the modes here for a uniform medium, are propagating. Thus, the classical expression 'a material with a refractive index' is an inappropriate wording, and conveniently referring to the minus sign in Eq. (18.3), one may refer to materials with $\varepsilon < 0$ and $\mu < 0$ as negative-index metamaterials with

$$n = -\sqrt{\varepsilon\mu}. \tag{18.4}$$

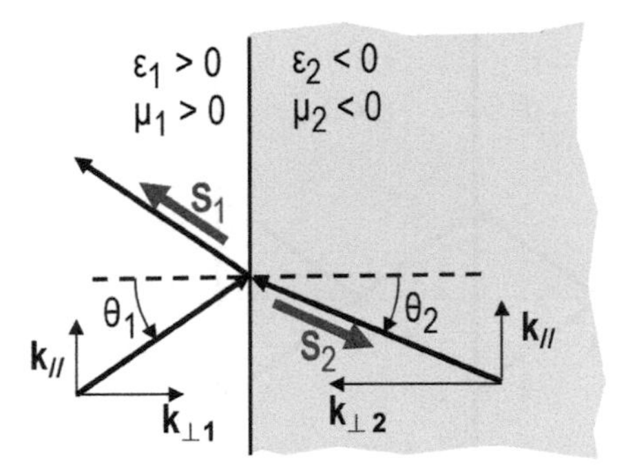

Fig. 18.1 Negative refraction at the interface between a positive-index material and a negative-index material. Thick arrows represent the direction of the Poynting vector **S**, whereas thin arrows depict the direction of wavevector **k**, and thus the direction of phase velocity v_φ.

18.2.3 Snell's law at negative-index material interfaces

Antiparallel phase and group velocities immediately affect Snell's law. Let us now consider an interface between a right-handed material (ε_1, μ_1) and a left-handed material (ε_2, μ_2). To match both field evolutions along the interface, the parallel momentum $\mathbf{k}_{//}$ must be conserved. Now, we must consider what the *outgoing* waves are in the left-handed material. To take energy out, the Poynting vector must be directed outwards (see Fig. 18.1). The change of the relative signs of the $\mathbf{k}, \mathbf{S}$ pair now implies a specific change in the Poynting vector: the refracted beam must be 'mirror-imaged' about the normal to the surface; i.e. there is a reversal of the component of **S** along the surface, compared to the usual case with a material with $\varepsilon > 0$ and $\mu > 0$. We can say that Snell's law still applies with the negative index on side 2,

$$\sqrt{\varepsilon_1 \mu_1} \sin(\theta_1) = -\sqrt{\varepsilon_2 \mu_2} \sin(\theta_2), \tag{18.5}$$

yielding in the case of Fig. 18.1 a negative sign for θ_2.

18.2.4 Flat lens with negative-index metamaterial

The most striking example of what we can do with an ideal negative-index material is a flat lens, i.e. a thin slab with $\varepsilon < 0$ and $\mu < 0$ and with a thickness d, immersed in a background with positive permittivity and permeability, $-\varepsilon$ and $-\mu$ (see Fig. 18.2).

According to Snell's law, the refracted angle is equal to the opposite of the angle of incidence, hence all the rays emanating from a point source A located at a distance z_1 from the front side of the lens are deviated so as to converge on a point A'' in the lens before diverging again. Released from the medium, the light reaches a focus for a second time at a point A' located at a distance $z_2 = d - z_1$ from the right side of the lens. In this traditional ray picture (we are concerned with progressive waves here, not with evanescent ones), we already have a number of important properties:

- The optical path from the external focus to the internal focus is zero; it is extremal as in the classical lens, according to the Fermat principle.
- The lens has no optical axis; the magnification is always unity.

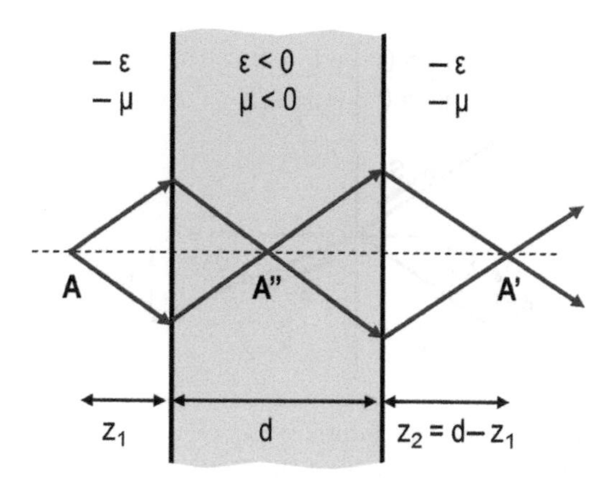

Fig. 18.2 Veselago's flat lens (thickness d). Light rays initially diverging from a point source A located at a distance z_1 from the lens are set in reverse and converge to a point A'' inside the lens, before diverging and finally focusing for a second time at a point A' located at a distance $z_2 = d - z_1$ from the right side of the lens.

- The geometrical aberrations are null, the point-to-point correspondence is perfect.
- The other relevant quantity, the impedance of the medium $Z = Z_0\sqrt{\mu/\varepsilon}$, is equal to the vacuum impedance Z_0 for $\mu = \varepsilon$, so that the lens is perfectly matched to the background and the interfaces exhibit zero reflection.

An entirely new exciting prospect arises from the realization that the flat lenses also perfectly 'transmit' evanescent plane waves with arbitrarily large parallel wavevectors $\mathbf{k}_{//}$ (Pendry, 2000), providing an infinitely sharp resolution.

There are two ways to be convinced that perfect imaging occurs. In the first way, we may consider a 2×2 transfer-matrix approach, just like in classical thin-film stack calculation, and incorporate evanescent fields with parallel momenta larger than $k_0\sqrt{\varepsilon\mu}$—see details in refs (Pendry, 2000; Solymar and Shamonina, 2009). The second way is to consider that, by virtue of the correspondence between material properties and coordinate transforms (Nicolet et al., 1994; Leonhardt, 2006), the slab with negative ε and μ implements a coordinate transform $z' = -z$, so that the field (e.g. the parallel components E_x, E_z, H_y for transverse-magnetic (TM) waves) located at a distance z from the front side of the slab is perfectly reproduced at a distance z from the front side of the slab inside the slab with a sign change for the transverse components H_z, E_z. In this vision, after two sign changes for the transverse components, the lens implements two real coordinate transforms that perfectly reproduce the field created by the source on the other side of the slab. This perspective is developed in Exercise 18.1.

The proposition of a perfect imaging system has raised many expectations, followed by the realization of how daunting a challenge it was, as it rested on the peculiar resonant behaviour of interfaces involving matched materials with positive and negative indices (Milton et al., 2007). It is not that electromagnetism is becoming inconsistent for such interfaces, but that the behaviour becomes singular, since even a minute addition of ohmic loss in the lens drastically impacts the performance (Garcia and Nieto-Vesperinas, 2002). Anyhow, the concept has led to conceptually new works that are interesting in themselves. The following sections will illustrate direct spin-offs, known as superlensing devices.

18.2.5 Seminal negative-index metamaterial demonstrations

In essence, the first demonstration of negative-index material results from the association of two resonances, one with electric dipoles (EDs) and the other with magnetic dipoles (MDs). To understand it at least qualitatively, it is important to first consider the essence of negative permittivity, and to refer to the simplest classical picture of atom–field interactions, the popular Lorentz oscillator model. Within this model, the electron is a light mass connected to the much heavier nucleus by a spring, and the effective dielectric constant ε_{eff} is given by

$$\varepsilon_{\text{eff}} = 1 + \frac{\omega_p^2}{\omega_0^2 - \omega^2 + 2i\gamma\omega}, \tag{18.6}$$

where γ is the damping constant, ω_p the analogue of the plasma frequency, and ω_0 the resonance frequency of the atomic transition. We immediately see that, on the low-frequency side of a resonance, where, unfortunately, the loss is maximal, the relative permittivity is negative, suggesting that any resonant contribution, $\propto 1/(\omega - \omega_0)$, is unavoidably inducing a sign change of the response. In general, the higher the quality factor (Q-factor), the narrower the spectral region with negative permittivity and the stronger the 'negativity'.

The idea of implementing artificial magnetism with a negative permeability directly follows on from the Lorentz dipole model, replacing the ED resonance with an MD one. See ref. (Solymar and Shamonina, 2009) for a nice historical overview of the idea. As is well known from elementary electromagnetics, MDs can be implemented by a circulating current on a closed loop, leading to a magnetic moment with a magnitude given by the product of the current and the area of the loop and with a direction perpendicular to the plane of the loop. Figure 18.3 depicts the famous so-called split-ring resonator (SRR), a metal loop interrupted to form a resonant LC circuit with a resonance frequency $\omega_{LC} = (LC)^{-1/2}$ completely controlled by the dimensions of the 'cut loop' (split ring). It then becomes intuitively apparent that a periodic array, a crystal, formed by a collection of subwavelength SRRs, playing the role of meta-atoms, may present a Lorentz-oscillator magnetic response, with a negative effective permeability μ_{eff} on the low-frequency side of its resonance, provided the artificial MD may be considered as electromagnetically independent from its neighbours (see Exercise 18.2).

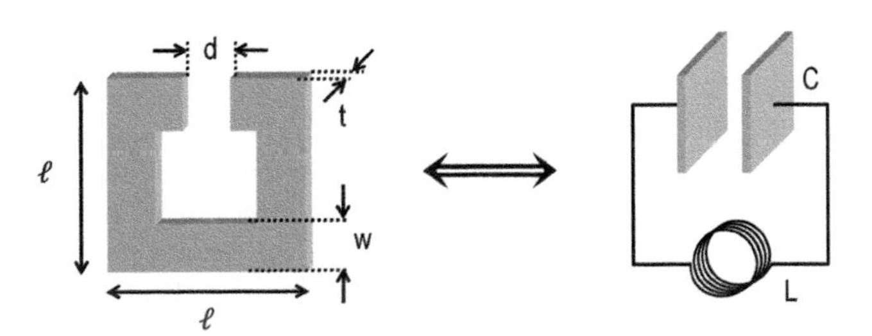

Fig. 18.3 Illustration of the analogy between a usual LC circuit and a cut metallic loop, also called a split-ring resonator formed with a plate capacitor $C = \varepsilon_0 wt/d$ (w, d, and t being the ring width, the capacitor gap width, and the ring thickness) and an inductance $L = \mu_0 \ell^2/t$, (ℓ being the side size of the ring); $\omega_{LC} = (LC)^{-1/2}$. See Exercise 18.2 for details.

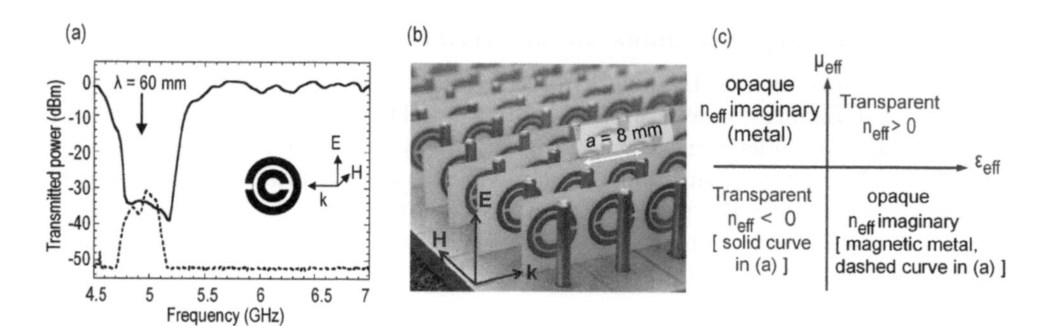

Fig. 18.4 (a) First observation of a negative-index material at microwave frequencies. The solid curve represents the transmission of the split-ring-resonator (SRR) array only (no metallic post). When the posts are added (dotted curve), the transmission is drastically reduced at all frequencies except in the band for which the split rings resonate; (b) the metamaterial consists of a SRR, created lithographically on a circuit board, and an array of metallic posts. The period is $a = 8.0\,\text{mm}$, which corresponds to 1/8 of the resonance wavelength of the split rings; (c) the four-quadrants of operation for a metamaterial, with either positive or negative ε_{eff} and μ_{eff}. Reprinted with permission from ref. (Smith et al., 2000) ©(2000) American Physical Society.

Figure 18.4 shows the first negative-index metamaterials demonstrated at microwave frequencies. The basic idea consists in superimposing in a single structure two metamaterials known for providing negative permittivity and negative permeability. Specifically, the former is a metallic-wire structure, which provides a predominantly free-electron-type response to EM fields, and the latter is an SRR structure, which provides a predominantly magnetic response. Thus, at the heart of the negative-index concept is a fundamental physics approach in which a continuous material, described by the relatively simple EM parameters ε_{eff} and μ_{eff}, conceptually replaces an inhomogeneous collection of scattering objects. In general, the continuous material parameters are of a tensor nature and are also frequency-dependent, but they nevertheless represent a considerable reduction in complexity for describing wave-propagation behaviour. The effective parameter response stems from the universal resonant response of a harmonic oscillator to an external frequency-dependent perturbation, in which the details of a material are replaced conceptually by a collection of harmonically bound charges that are either electric or fictitious magnetic.

The microwave experiments of Fig. 18.4 were performed with an incident wave confined between two parallel metallic plates (the upper one is removed in the photograph). In order to couple to the metal rings, the magnetic field has to be perpendicular to the plane of the rings, and in order to couple to the rods, the electric field has to be parallel to the rods, and of course the propagation direction is perpendicular to both the electric and magnetic fields. The main experimental results are shown in Fig. 18.4a. If only the split-ring resonators are present (i.e. all metal rods are removed), the transmission spectrum shown with a solid curve reveals the existence of a stopband between the frequencies 4.7 and 5.2 GHz. Following the interpretation in ref. (Smith et al., 2000), the ring array acts as a material with a negative permeability in the stopband and the propagation is forbidden, since $n_{\text{eff}} = \sqrt{\mu_{\text{eff}}\varepsilon_{\text{eff}}}$ with $\varepsilon_{\text{eff}} \simeq 1$ and $\mu_{\text{eff}} < 0$ is imaginary. The metamaterial behaves as a magnetic metal; see the lower-right quadrant in Fig. 18.4c.

When the rods are included (dotted curve), the stopband turns into a passband and the attenuation diminishes from 50 dBm to about 32 dBm, corresponding to a 18 dBm increase (the remaining imaginary parts of ε are still large in these pioneer realizations). Again, the interpretation is simple: the rod array acts as a material with a negative effective permittivity over the entire spectral range of (see the upper-left quadrant in Fig. 18.4c), and when the effective permittivities and permeabilities are both negative, $n_{\text{eff}} = \sqrt{\mu_{\text{eff}}\varepsilon_{\text{eff}}}$ becomes real. The stopband turning into a passband appears to provide convincing evidence that a material with negative permittivity and permeability can propagate EM waves (Smith et al., 2000), despite the considerable amount of absorption/attenuation observed in the experimental data. As discussed in Fig. 18.1, if a metamaterial has negative permittivity and permeability, one should observe negative refraction.

This beautiful initial demonstration was followed by a negative-refraction Snell's law experiment performed on a wedge-shaped (prism-like) metamaterial with a structure similar to that in Fig. 18.4b (see ref. (Shelby et al., 2001) for details), and by other experiments where material losses were reduced and the structure presented a better impedance match to free space. These additional measurements were enough to demonstrate that materials with negative refractive index are not a 'myth'.

18.2.6 Negative-index metamaterial at optical frequencies

In the late 2000s, a sustained effort was made by the community to push the operating frequency of the metamaterials deeper and deeper into the terahertz region, to ultimately reach optical frequencies. However, in practice this proved difficult. While it is easy to find negative-permittivity materials (e.g. ordinary metals), the same is not true for negative permeability. Natural materials do not exhibit any significant magnetic response at such high frequencies (and even at frequencies well below the optical range (see the section entitled 'The dispersion of the magnetic permeability' on p. 269 in ref. (Landau et al., 1984)), and the creation of artificial magnetism at optical frequencies remains a daunting challenge.

Although the lump-element model in Exercise 18.2 shows that the magnetic resonance wavelength λ_{LC} scales with the structural size, $\lambda_{LC} = 2\pi\sqrt{\varepsilon}\,\ell\,(w/d)^{1/2}$, this linear scaling breaks down at high frequencies, notably towards visible ones. The main reason comes from the scaling of the other typical characteristic length that should be considered, the skin depth, $\delta = \lambda/(2\pi\sqrt{\varepsilon_m})$. Inserting the expression of the Drude metal relative permittivity $\varepsilon_m(\omega) = 1-\omega_p^2/\omega^2$ at high frequencies with $\gamma(\omega) \ll \omega < \omega_p$, we find that the skin depth becomes a constant, independent of the wavelength (see Chapter 13). Thus, as we scale all the dimensions of metallic-loop resonators down to the visible, the EM field largely penetrates in the metal and the resonator no longer works as in the long wavelength regime, with a field essentially located outside the metal. We will discuss this issue later in relation to the operating principle of wire-grid polarizers.

Perhaps the most convincing demonstration of negative-index metamaterials operating in the visible and telecom ranges of the spectrum with a thick sample (a real metamaterial, not an optically thin sample) is the one achieved in 2008 at Berkeley

(Valentine et al., 2008); see Fig. 18.5. Abandoning the former SRR geometry, the demonstration is based on a 3D optical metamaterial made of stacked 'fishnet' structures obtained by etching a hole array in alternating layers of silver and magnesium fluoride. Tested in a prism-like experiment (see Fig. 18.5b), negative refraction is unambiguously observed over a broad spectral range beyond $\lambda = 1500\,\text{nm}$ (see Fig. 18.5c). What is remarkable in this experiment is that, in sharp contrast with earlier experiments that were mainly limited to optically thin samples, this observation is made from transmission measurements with a thick 21-layer sample. As shown in the right panel of Fig. 18.5c, the measurements of the negative index n (as deduced from Snell's law) match the full-vector computational results of the normalized propagation constant $n_{\text{eff}} = k_z/k_0$ ($k_x = k_y = 0$) of the fundamental Bloch-mode of the fishnet

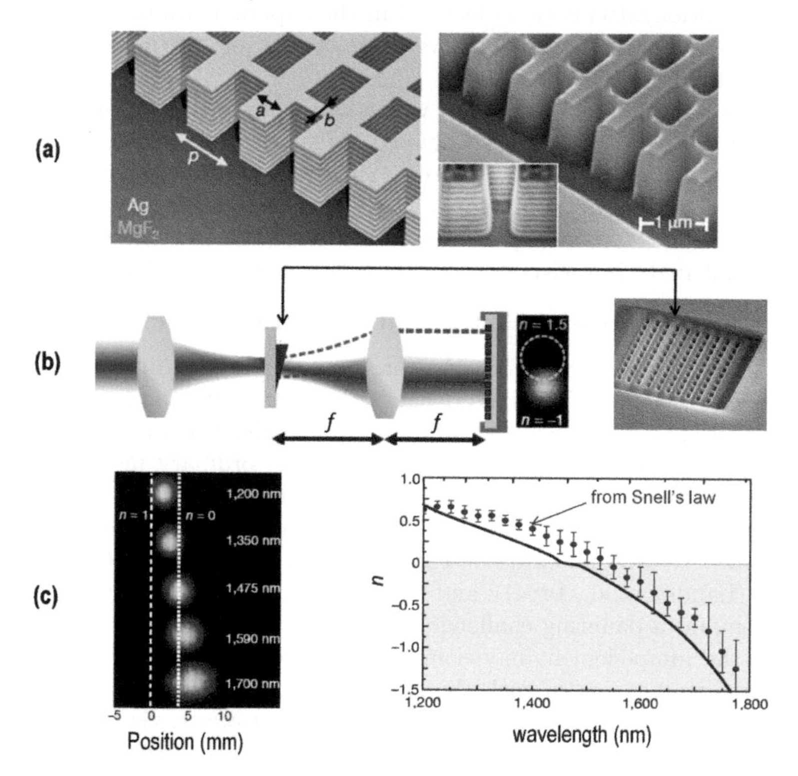

Fig. 18.5 Demonstration of negative refraction at telecom frequencies with fishnet metamaterials. (a) Diagram and SEM image of the 21-layer fishnet structure fabricated in ref. (Valentine et al., 2008): $p = 860\,\text{nm}$, $a = 565\,\text{nm}$, and $b = 365\,\text{nm}$. The structure consists of alternating layers of 30 nm silver (Ag) and 50 nm magnesium fluoride (MgF_2); (b) experimental setup used to observe negative refractive results obtained with a wedged, prism-like fishnet. (c) Left: Images of the beam on the camera for various wavelengths. The abscissa gives the beam shift, and lines for $n = 1$ and $n = 0$ are indicated. Right: Measurements (circles) and simulation (solid curve) of the fishnet refractive index. The solid curve shows $\text{Re}(n_{\text{eff}})$, the normalized propagation constant normal to the layers, for the $k_{||} = 0$ fundamental Bloch mode. Reprinted by permission from ref. (Valentine et al., 2008) ©(2008) Springer Nature.

rather well. Another important feature is the remarkably high figure of merit (FOM), FOM $= 2\text{Re}(n_{\text{eff}})/\text{Im}(n_{\text{eff}}) = 3.5$ (a large FOM corresponds to low loss), which is predicted numerically.

It is instructive to capture the physical origin of the negative effective permittivity in the experiment. The fishnet is composed of two intersecting subwavelength channels that couple and exchange energy. The negative effective permittivity originates from the longitudinal (z-direction) channel that consists of air holes in a metal film, which transport energy through resonant tunnelling. Conversely, the negative effective permeability originates from the transverse (x- or y-direction) channel formed by metal-insulator-metal (MIM) waveguides that support the propagation of gap-plasmon modes. This mode, which is studied in detail in Chapter 13, plays a crucial role in nanophotonics; we will meet it again at the end of the chapter when studying wire-grid polarizers. In Fig. 18.6, we show an enlarged view of an MIM stack. We also depict the field profile of the y-component of the magnetic field of the gap-plasmon mode. It follows directly from Maxwell curl equations that the x-component of the electric field is antisymmetric, so that the corresponding electric current J_x in the metal is also antisymmetric (arrows). Either using a Fabry–Perot picture in which the gap-plasmon mode is bouncing back and forth along x between the facets, or with a lumped-elements model associating a capacitance with the two facets and an inductance with the cut loop, it becomes intuitively clear that the transverse channels of the fishnet structure behave as SRRs.

However great and convincing they may be, the demonstrations of negative-index metamaterials at telecom frequencies have not had a genuine impact on applications. The literature extensively mentioned the critical issue of excessive ohmic losses in metals at high frequencies to explain a feeling of disappointment regarding the immense hopes. To mitigate the loss, interesting new materials for metamaterial components and telecommunication devices have been developed (Boltasseva and Atwater, 2011; Naik et al., 2013), often offering a better FOM than metals with the highest conductivities, e.g. gold and silver. Attempts at doping metamaterials with gain have even been made to mitigate absorption by amplification.

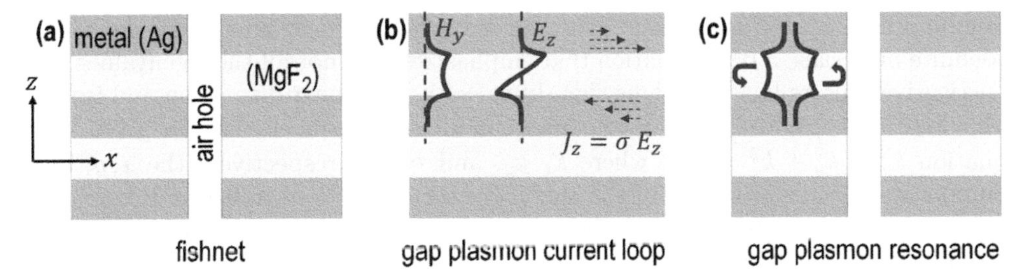

Fig. 18.6 Implementation of current-loop resonators with thin metal–insulator–metal (MIM) stacks. (a) Sketch of the fishnet stack. (b) The blue curve shows the symmetric y-component of the magnetic field of the gap-plasmon mode formed by the coupling of the surface plasmons of the two metal-dielectric interfaces. The dotted arrows show the associated x-component $J_x \propto \frac{\partial H_y}{\partial z}$ of the induced current, which is antisymmetric with respect to the middle of the MgF_2 layers. (c) Fabry–Perot resonance formed by the bouncing of the gap plasmon modes reflected by the holes. Finite-length MIMs are also called 'cut-wire pairs' in the related literature.

Ohmic losses are not the only issue. Perhaps more fundamentally, it should be noted that the design of an isotropic negative-index metamaterial has to fulfil challenging requirements. For example, strong resonances are needed inside the unit cell to build a large magnetic response from materials that are non-magnetic in essence (Landau et al., 1984) and the interaction between every individual atom should remain weak, so that *isotropic* effective properties can be simply inferred by spatially averaging at a characteristic length much larger than the meta-atom size $\lambda/\mathrm{Re}(\sqrt{\varepsilon_m})$ and yet significantly smaller than the wavelength λ (Merlin, 2009). *These daunting requirements have somehow never been realized: not at any visible wavelength, and not even in the microwave region.*

For instance, even for the beautiful fishnet structure of Fig. 18.5a, the magnetic resonance arises from a transverse MIM resonance that spreads across four or five unit cells. If doping is provided to mitigate absorption, it is not four or five cells that are involved, but rather 20. In other words, the fishnet structure is far from what can be described as an artificial medium. Its optical properties are more related to those of gratings, with a delocalized resonance and a huge spatial dispersion, rather than a collection of meta-atoms that reproduce the properties of every individual atom by mere ensemble averaging.[1]

18.3 Hyperbolic metamaterials and superresolution

In this section, we consider highly anisotropic metamaterials known as hyperbolic metamaterials. They display hyperbolic dispersion, which originates from one of the principal components of their electric or magnetic effective tensors having the opposite sign to the other two principal components. We will consider the special case of non-magnetic metamaterials for which all the diagonal components of the effective permeability tensor are μ_0 and the effective permittivity tensor is thus

$$\varepsilon_{\text{eff}} = \begin{bmatrix} \varepsilon_{||} & 0 & 0 \\ 0 & \varepsilon_{||} & 0 \\ 0 & 0 & \varepsilon_{\perp} \end{bmatrix}, \tag{18.7}$$

with $\varepsilon_{\perp} > 0$ and $\varepsilon_{||} < 0$, for instance. Here, the subscripts $||$ and $\perp$ indicate components parallel and perpendicular to the anisotropy axis. These media have been called indefinite materials, a denomination that emphasizes that not all the eigenvalues of the constitutive tensors have the same sign. In vacuum, the linear dispersion and isotropic behaviour of propagating modes implies a spherical isofrequency surface given by the equation $k_x^2 + k_y^2 + k_z^2 = \frac{\omega^2}{c^2}$, where k_x, k_y, and k_z are respectively the x, y, and z components of the wavevector of the mode. For usual textbook anisotropic cases, with positive diagonal tensor elements, the spherical isofrequency surface of vacuum is distorted into an ellipsoid. The interesting properties of indefinite materials stem from the isofrequency surface of extraordinary (TM-polarized, E_z) modes,

$$\frac{k_x^2 + k_y^2}{\varepsilon_{||}} + \frac{k_z^2}{\varepsilon_{\perp}} = \frac{\omega^2}{c^2}. \tag{18.8}$$

[1] Spatial dispersion has profound implications for many applications of metamaterials (Shapiro et al., 2006). The difficulty is exacerbated by the dielectric meta-atoms that also exhibit artificial magnetism (see section 18.4), for which the implementation of magnetism requires resonators with a size that is typically equal to the wavelength in the particle ($\approx \lambda/\sqrt{\epsilon_d}$, with $\epsilon_d \approx 3$).

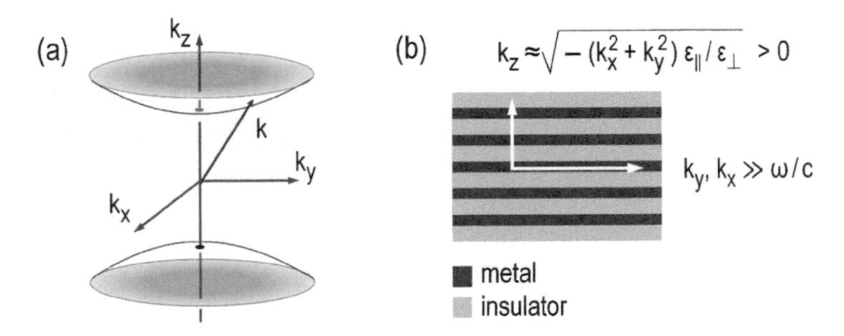

Fig. 18.7 Hyperbolic metamaterials. (a) Hyperboloid isofrequency surface for a uniaxial medium with an extremely anisotropic dielectric response, $\varepsilon_{\perp} < 0$ and $\varepsilon_{||} > 0$. Hyperboloid surfaces are also obtained for the reciprocal case of $\varepsilon_{\perp} > 0$ and $\varepsilon_{||} < 0$. (b) Simple and classical example of hyperbolic metamaterials implemented with alternating layers of metals and insulators. For arbitrarily large in-plane wavevectors, k_z remains positive. Remember that, in this figure, *we assume an idealized case without ohmic losses.* Absorption sets a strong limitation, with the appearance of a backbending in the dispersion curves of surface plasmons, preventing the excitation of high in-plane k's.

Here, since $\varepsilon_{||}\varepsilon_{\perp} < 0$, we have an extreme anisotropy and the isofrequency surface expands into an open hyperboloid (having two manifolds; see Fig. 18.7). Such a phenomenon requires the material to behave like a metal in two directions of polarization and like a dielectric (insulator) in the third one. This does not readily occur in nature at optical frequencies but can be achieved using artificial nanostructured EM media, as shown by the results of the static-limit homogenization of composite materials in Chapter 15.

The most important property of such media, called *hyperbolic media*, is related to the behaviour of waves with large wavevector magnitudes. In vacuum, these waves are evanescent and decay exponentially in at least one direction. However, in hyperbolic media, the open shape of the isofrequency surface allows for propagating waves with arbitrarily large wavevectors.

For instance, if k_x or k_y becomes increasingly large, e.g. much larger than ω/c, we find that $k_z^2 > 0$ in Eq. (18.8), since $\varepsilon_{||}\varepsilon_{\perp} < 0$, implying that a mode may propagate with infinitely large transverse momentum. Thus, there are no evanescent waves in such a medium in the ideal case when loss is neglected.

It is considerably easier to produce hyperbolic structures than it is to produce isotropic metamaterials that possess both negative permittivity and negative permeability, as the only essential criterion for hyperbolic structures is that the motion of free electrons be constrained in one or two spatial directions. Perhaps the simplest hyperbolic structure consists in a stack of alternating layers of dielectric and metallic materials, with $\varepsilon_{||} = \langle \varepsilon(z) \rangle$ and $\varepsilon_{\perp} = \langle 1/\varepsilon(z) \rangle^{-1}$ behaving as a metal and a dielectric, respectively (see Chapter 15). Due to their easy fabrication, they triggered many studies on radiative decay rate enhancement, heat transport, enhanced absorption, and superresolution imaging in pursuit of the perfect lens.

Figure 18.8 shows the first demonstration of near-field imaging with a very elementary hyperbolic metamaterial, an insulator–silver–insulator stack with a single 35-nm-thick silver layer. In the experiment layout shown in Fig. 18.8a, patterns formed

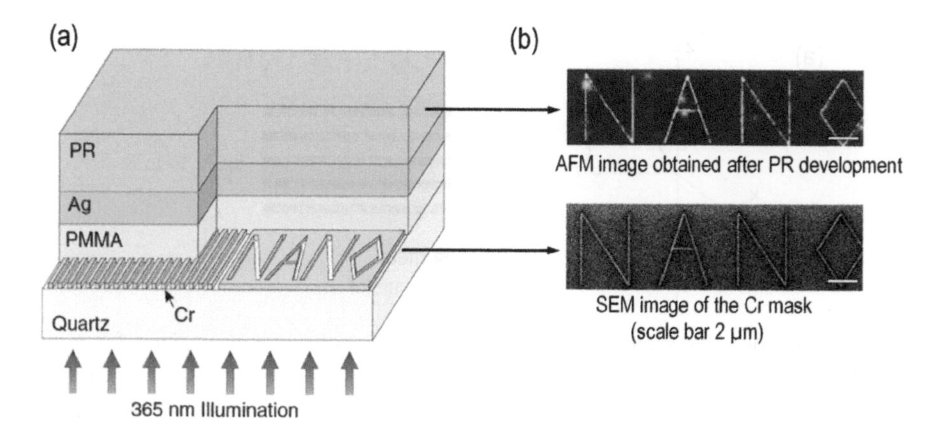

Fig. 18.8 Superlensing with hyperbolic metamaterials (in the present case, simply an insulator–metal–insulator stack). (a) Optical experiment; (b) sharp image formed by the 'superlens' of an arbitrary object 'NANO' etched in a chromium (Cr) mask, as revealed by the atomic force microscopy scan of the photoresist (PR) after development. The image is much sharper than one taken without the Ag layer. After ref. (Fang et al., 2005).

of narrow slits opened into a chromium (Cr) mask are covered by the stack, with an intermediate 40-nm-thick polymethyl metacrylate (PMMA) spacer. The mask is illuminated from the rear side at $\lambda = 365\,\text{nm}$. The image of the object is recorded using photoresist (PR) on the top of the silver superlens. The pattern inscribed in the PR after development is inspected with atomic force microscopy (AFM). Fig. 18.8b shows that the pattern 'NANO' is reproduced with a high fidelity (at this very short imaging distance, it is still reproduced without the Ag layer but much less sharply). The authors of this observation claim a 90 nm ($\lambda/4$) resolution.

Most of the applications of hyperbolic media may be understood in terms of Fermi's Golden Rule, which stipulates that the emission rate into a particular photonic mode is proportional to the local density of modes (LDOS). For a scatterer or an emitter placed in the vicinity of a hyperbolic medium, e.g. the chromium mask in the previous experiment, light will preferentially excite the metamaterial. The mode that plays a fundamental role is the dominant Bloch mode of the metamaterial, the supermode formed by the coupling of the gap-plasmon modes (see Fig. 18.6) of every dielectric layer in the stack. For the superlensing experiment, the dominant mode is the fundamental mode of the insulator–silver–insulator stack. This mode has properties very similar to those of gap modes. It is symmetric like in Fig. 18.6 and has 'no' cutoff and high-$k_{||}$ values as the metal thickness decreases (see details in Chapter 13).

However, we should emphasize that, because of ohmic losses in the metal, the high-$k_{||}$ modes that are excited at a fixed frequency by the scatterer 'NANO' placed in the near-field are highly damped in the hyperbolic metamaterial. What actually happens is that, when considering loss in Eq. (18.8) with realistic complex values for $\varepsilon_{||}$ and $\varepsilon_{\perp}$ (taking into account absorption), k_z becomes complex too for a fixed $k_{||} \equiv (k_x, k_y)$. Actually, it is an easy exercise to consider Eq. (18.8), set a small imaginary part in $\varepsilon_{||}$, and calculate k_z to evidence that it becomes complex with an imaginary part that grows with $k_{||}$. Thus, the most attractive property of hyperbolic metamaterial is hampered by ohmic losses (Khurgin, 2015), and the idealized transfer of high-$k_{||}$ waves, as suggested by the hyperboloid dispersion diagram, is not observed in reality.

18.4 Dielectric metasurfaces

Metasurfaces are often considered as the 2D versions of 3D metamaterials, but they are not. By metasurface, one generally means any nanostructured surface capable of manipulating light into the subwavelength regime to achieve a macroscopic function. Metasurfaces might be conceived around the concept of homogenization (see chapters 3 and 15), but most often they rather exploit unusual guidance or resonance effects in nanostructured high-index materials. These effects can be achieved either with metallic nanostructures that support surface plasmons or with semiconductor nanostructures that can be electrically doped and gated to realize active subwavelength devices. In general, dielectric nanostructures offer reduced dissipative losses compared to metallic nanostructures at optical frequencies, and will be documented in more detail hereafter.

With the developments in nanoscale technologies, optical metasurfaces have received increasing attention in the 2010s, with the aim of implementing various functions by controlling the amplitude, phase, polarization, orbital momentum, absorption, reflectance, and emissivity of light at high spatial resolution (Genevet et al., 2017). This had led scientists and engineers to revisit various traditional applications in optics with a different point of view.

It is presently difficult to anticipate the genuine added value of this booming field or to exhaustively review it. Thus the following discussion is restricted to tutorial examples, emphasizing the basic physical effects that are exploited rather than the objectives or means for which they have been conceived. We also try to establish clear links with early works. The present section mostly focuses on dielectric metasurfaces, starting with homogenization concepts, while section 18.5 studies metasurfaces for manipulating surface plasmons on a single interface.

18.4.1 Phase plates

We start with a basic example directly inspired by homogenization ideas on the form birefringence of artificial dielectrics. For high-refractive-index materials, the birefringence is much larger than that of naturally available ones. Hence, in polarization phase-plate settings, binary subwavelength gratings composed of tiny dielectric wires may provide phase retardation that is substantial even for grating thicknesses smaller than the wavelength. This can be easily proven from the averaging discussed in Chapter 15 with the static-limit formulations: for a two-media lamellar grating with relative permittivities ε_1 and ε_2 and filling factor f, one gets ordinary and extraordinary refractive indices given by

$$n_o = \left((1-f)\varepsilon_1 + f\varepsilon_2\right)^{1/2}, \tag{18.9}$$

$$n_e = \left((1-f)/\varepsilon_1 + f/\varepsilon_2\right)^{-1/2}. \tag{18.10}$$

Thus a phase plate producing a phase birefringence φ can be implemented with a grating depth $h = \lambda \, \frac{\varphi}{2\pi} \, \frac{1}{n_o - n_e}$. Quarter-wave plates ($\varphi = \frac{\pi}{2}$) can be realized with gratings consisting of air grooves etched into a high-index material ($n = 3.5$). The grating thickness is small enough and the plates can be manufactured at low cost, e.g. by embossing a resist and transferring the pattern to the high-index material.

The use of high-index semiconductor materials leads to large chromatic dispersions for both n_o and n_e. By wisely engineering the dispersion of the composite indices n_e and n_o, however, it is possible for the phase retardation of a given composite slab of thickness d, $\Delta\varphi = \frac{2\pi d}{\lambda}\left(n_o(\lambda) - n_e(\lambda)\right)$, to remain constant over a broad spectral interval. For instance, by etching a binary subwavelength grating into a silicon substrate, the quarter-wave phase components exhibiting a phase retardation of $\pi/4$ with a few per cent accuracy have been manufactured for operation from 3.5 to 5.0 μm (Deguzman and Nordin, 2001).

18.4.2 Antireflection coatings

The critical feature of an optical surface that gives rise to unwanted Fresnel reflections is the sudden transition from one optical medium to another. If the transition is made more gradual and extended over at least a significant fraction of a wavelength, the reflection (or impedance mismatch) can be significantly reduced. This has been achieved in glass by appropriately corroding the surface with acid. Matter is leached out of the glass, creating nanometric pores on a scale much finer than the wavelength, and the open structure left in the region of the surface becomes gradually more dense as one penetrates into the glass. So the gradual corrosion translates into a gradual change of refractive index at the open surface of glass. The reflection of the surface is reduced to a level as low as that achieved with very complex multilayer antireflection coatings.

An analogue of the leached-glass antireflection surface is to be found on the eye of the night-flying moth. The cornea is covered with a fine regular hexagonal array of 'egg box' protuberances that have a period of about 200 nm with a similar depth, and a cross section that is approximately sinusoidal. This natural corrugation was discovered by Bernhard in 1967 (Bernhard, 1967) who postulated that it resulted in a gradual change of refractive index that reduced the reflection over a wide spectral and angular bandwidth and significantly improved the moth's camouflage.

The problem of surface reflection is particularly acute for semiconductors, on account of their very large index of refraction in the visible and near-infrared ranges. The reflection losses at normal incidence are up to $\simeq$40% for Si, Ge, or GaAs. Reducing them is of interest when maximizing absorption for solar-cell applications and other optoelectronic devices. Fig. 18.9a shows the reflectivity of a silicon substrate with high-aspect-ratio pyramid arrays. The reflectivity under illumination at normal incidence does not exceed 4% over four octaves starting from 200 nm, a level as low as that achieved with very complex multilayer antireflection coatings.

For structured films with periods only slightly smaller than the wavelength, accurate design should not only consider reflection due to effective refractive index mismatch, like in graded-index films, but also the reflection due to the mode profile mismatch, which is often the dominant effect (Fig. 18.9b).

Apart from their optical performance, structures of this type offer the additional advantages that there are no problems with adhesion or with diffusion of one material into another. Since they are monolithic and introduce no foreign materials, they also

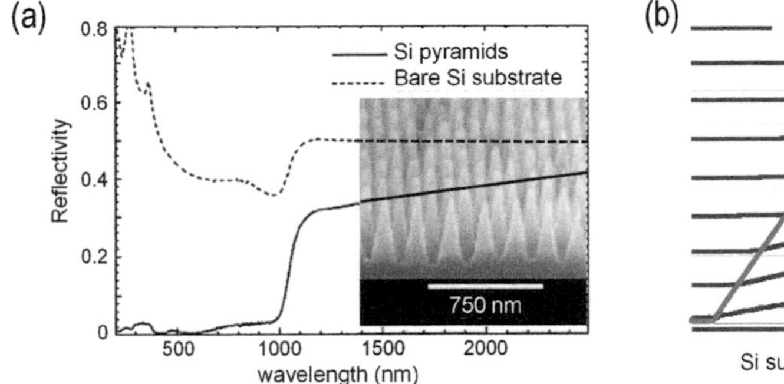

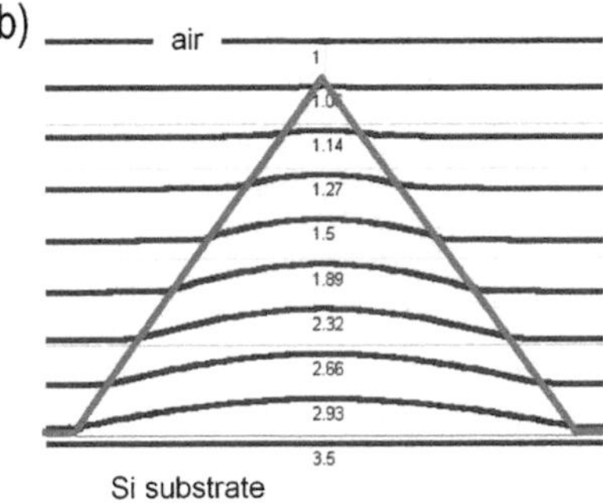

Fig. 18.9 (a) Antireflection behaviour of a square array of 350-nm-high pyramids etched into silicon (solid curve) compared to the bare silicon surface reflectivity (dashed curve). The inset shows a scanning electron micrograph of the antireflection surface. Note that the abrupt reflectivity increase at $\lambda = 1100\,\mathrm{nm}$ is not due to the pyramids. For incident beams with energies above 1.1 eV, silicon becomes transparent and the beam is reflected on the rear interface of the silicon substrate. (b) Progressive Bloch-mode profile transformation (blue curves, with corresponding effective index values indicated) in pyramid arrays computed with the Rigorous Coupled Wave Analysis (RCWA), from plane wave in air to plane wave in silicon. Fig. (a) is reprinted with permission from ref. (Kanamori et al., 1999) ©(1999) Optical Society of America.

tend to be more stable and robust than multilayer dielectric materials, particularly when used with high-power lasers and high-fluence pulses.

18.4.3 Generalized Kerker conditions for metasurfaces

Nanophotonic structures composed of dielectric resonators arrayed on a surface offer a large variety of functionalities (Kuznetsov et al., 2016). At first approximation, it is easier to consider that the resonators operate independently and that the properties of the array are driven by the Mie resonances (or quasinormal modes) of the individual meta-atoms. For subwavelength dielectric particles, a single multipole usually dominates and determines the magnetic or electric nature of the mode (see section 12.2.1).

For meta-atoms with a size comparable to the wavelength in the material $\lambda/\sqrt{\epsilon_d}$, each Mie resonance is a combination of different multipoles including both magnetic and electric contributions. In addition, different Mie modes may overlap spectrally. The combination of different multipolar contents can be used for engineering the amplitude and directionality of the scattered fields. A special case is realized when the electric and magnetic resonances spectrally overlap to produce orthogonal MD and ED excitations of equal strength with synchronized phases which fulfil the so-called Kerker condition. We may view the metasurface as supporting both electric and fictitious magnetic surface current densities that are excited by the incident wave. Because both currents produce reflected waves that are exactly out of phase with each other, their sum can cancel out entirely if the total induced effective magnetic current exactly balances the total induced electric current (Moreau et al., 2012), leading to unidirectional scattering. These metasurfaces are often referred to as Huygens metasurfaces.

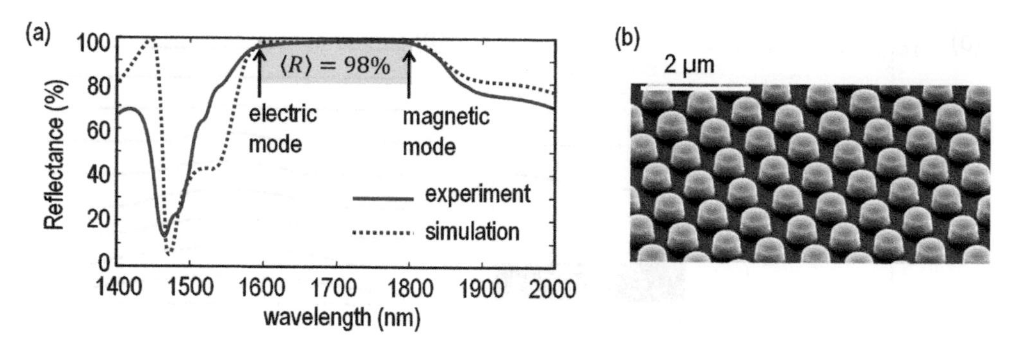

Fig. 18.10 Example of a high-performance ($R \gtrsim 98\%$) broadband dielectric mirror based on generalized Kerker conditions. (a) Reflectance spectrum; (b) scanning electron microscope tilted view (45°) of the metasurface consisting of an array of Si cylinders. Adapted with permission from ref. (Moitra et al., 2015) ©(2015) American Chemical Society.

Similar to the case of perfect transmission, it is easy to realize broadband perfect reflection (Collin, 2014; Liu and Kivshar, 2018). Transmission elimination is induced by a significant forward scattering that interferes destructively with the incident wave. The most widely studied case is the resonant excitation of ED and/or MD modes, but a similar effect can be achieved with the higher-order multipoles. Standard implementations at telecom wavelengths involve arrays of silicon nanowires which reflect light for one polarization of the incident light (Taillaert et al., 2002; Mateus et al., 2004) or 2D arrays of silicon nanodiscs (Liu and Kivshar, 2018). Optimized all-dielectric metasurfaces in reflection easily surpass 95% reflection. One simple example is shown in Fig. 18.10, where the incident wave can be fully reflected by a lattice of silicon nanodiscs. Electric-like (with a null tangential E-field component) or magnetic-like (with a null tangential H-field component) modes can be implemented by controlling the phase symmetries of multipoles of different natures (Liu, 2017).

The presentation of Kerker-like effects from the perspective of an individual nanoparticle is quite simplistic. In reality, meta-atoms that combine several resonances strongly interact when they are arrayed. It is not just near-field coupling: long-distance radiative coupling also plays a critical role. Therefore, the Kerker condition implemented in periodic metasurfaces should be thought of as a collective resonance resulting from this interaction, which can be referred to as 'generalized' Kerker conditions for metasurfaces (Liu and Kivshar, 2018). The immediate consequence, already encountered in section 18.2.6 for negative-index metamaterials, is spatial dispersion, implying that the generalization Kerker conditions hold only for a specific incidence angle. For instance, even a minute tilt of the incident light in the experiment of Fig. 18.10 may result in sharp and deep dips in the transmittance spectra, a phenomenon that is well explained with a Bloch-mode analysis (Gigli et al., 2021). Note that these dips are not encountered in classical dielectric Bragg reflectors (see Chapter 15).

Resonant dielectric periodic metasurfaces have been studied for plenty of applications in the decade from 2010 to 2020 (Kuznetsov et al., 2016). Broadband absorption by simultaneously eliminating both reflection and transmission is one example. Structural colour is another. Other applications targeting large field enhancements include biosensing (see Chapter 20), non-linear harmonic generation, and enhanced coupling to excitons in 2D materials (Liu and Kivshar, 2018).

This section has focused on dielectric meta-atoms, but very similar discussions and conclusions hold for metallic meta-atoms. In general, owing to their higher permittivity, metallic meta-atoms with a prescribed multipolar content are smaller than dielectric meta-atoms and offer less spatial dispersion. But in return, they display ohmic losses (Khurgin, 2015).

18.4.4 Metalenses and metagratings

For antireflection, index gradients are implemented perpendicular to the substrate. They may also be manufactured parallel to the substrate. We then enter the field of the so-called 'subwavelength flat optics' that shapes the phase of an incident wave in free space through subwavelength structures, a concept that is parent to that of holograms.

It is interesting to first recall the principle of operation of conventional diffractive optical components with 'saw-tooth' profiles, such as the blazed grating depicted in Fig. 18.11a or Fresnel lenses, which implement a (wrapped) 0–2π-phase difference across their surface by a $\lambda_0/(n-1)$ excursion of the diffractive surface thickness. Nowadays, these elements are manufactured at high volume and low cost by large-area replication technologies (embossing, moulding, or casting), with minimal light loss in durable materials. However, they suffer from at least three main limitations:

- The discontinuities of the wrapped phase introduce a shadow that wastes light by exciting spurious orders. The effect is stringent for saw-tooths with short periods, to the extent that the saw-tooth approach is essentially useless for high-numerical-aperture (NA) lenses (see Fig. 18.11a and the blue curve in Fig. 18.11b).
- The wrapping relies on a nominal design wavelength λ_0 for which the 2π phase excursion is achieved, implying that the imprinted phase varies with the wavelength of the incident light and thus the efficiency η decreases as the illumination wavelength λ departs from λ_0: $\eta = \text{sinc}^2(1-\lambda_0/\lambda)$ for large periods (see Exercise 18.5).
- Not only does the efficiency of a diffractive lens vary with the wavelength; so does its focal length: $f(\lambda_2)\lambda_2 = f(\lambda_1)\lambda_1$. Actually, f decreases as the wavelength increases, in sharp contrast with refractive lenses based on common glasses, for which the refractive index (resp. focal length) decreases (resp. increases) with increasing wavelength. Note that this limitation is turned into a natural advantage for manufacturing achromatic hybrid doublets that reduce the weight and size of optical systems (Stone and George, 1988).

Subwavelength flat optics has challenged these three fundamental limitations, but not always with full success. Two main physical effects have been exploited.

The first approach relies on guidance along metallic microgrooves or high-index dielectric nanopillars, which implement variable and local wavefront phase differences. Metagratings relying on guidance were first achieved by arraying microgrooves for operation at thermal infrared wavelengths (Haidner et al., 1993). At visible frequencies, the metal approach is not efficient owing to ohmic losses, and guidance along nanopillars is preferable. Such an approach provides high diffraction efficiencies, $\approx$80% routinely.

Figure 18.11b summarizes the main results obtained in the late 1990s for a series of gratings operating in the visible with increasing deviation angles (i.e. decreasing periods) and using TiO_2 nanopillars. A remarkable property is that the efficiency drop of saw-tooth gratings at large deviation angles (solid blue curve) no longer occurs, giving room for large deviation angles or, equivalently, high-NA metalenses (Lee et al., 2002). This is explained by the fact that every nanopillar individually behaves as a nanowaveguide that confines light and guides it vertically throughout the metasurface, thus removing the obliqueness involved in the shadow effect. The waveguiding operation also guarantees that the imprinted phase is resilient to changes of the illumination direction, implying that the same structure remains nearly optimal for various incidences of the illumination, effectively focusing off-axis plane waves or collimating diverging beams.

The guidance approach attracted considerable attention in the mid 2010s, and new functionalities that cannot be achieved with alternative approaches (e.g. refractive or gradient-index optics) have been demonstrated. For instance, complete control of the artificial birefringence at the level of every individual meta-atom has been implemented with nanopillars with elliptical cross sections, enabling metasurfaces that offer independent responses for orthogonal polarizations (Arbabi et al., 2015; Lin et al., 2014). Metalenses with a fixed focal length for operation over a broad spectral range have also been reported (Khorasaninejad et al., 2017; Shrestha et al., 2018; Wang et al., 2021).

The guidance approach is intrinsically a non-resonant mechanism. In the second approach, nanoresonators instead of nanowaveguides are used to imprint the phase excursion with falter elements. As light excites a non-degenerate resonator, the scattered field is always proportional to $1/(\omega - \tilde{\omega}_R)$, $\tilde{\omega}_R$ being its complex resonance frequency (see Chapter 7). Then, by progressively varying the shape of a series of subwavelength resonators etched on a surface, one may implement a gradual resonance frequency shift and imprint, through the phase of $\omega - \tilde{\omega}_R$, a controllable phase shift on the scattered wave.

Phase control with nanoresonators was first implemented with a complicated design based on tiny V-shaped plasmonic nanoantennas (Yu et al., 2011) with very low efficiencies and then further developed with dielectric nanoresonators to avoid ohmic losses in the metal. With single resonances, the phase shift is intrinsically limited from zero to π. To achieve the required 2π phase excursion for transmiting metasurfaces, the idea is to combine two resonances (Pfeiffer and Grbic, 2013). Both resonances can be excited simultaneously within simple nanoresonator geometries, such as silicon nanodiscs for operation at optical frequencies (Staude et al., 2013). Additionally, since the two dipole radiations interfere, the radiation pattern in the backward direction can be strongly diminished (Kerker effect), providing enhanced transmission. This approach, already encountered in section 18.4.3, has been popularized under the name of Huygens' metasurfaces.

However, the efficiencies ($\simeq$40–60%) of these metasurfaces are significantly lower than those obtained with the guidance approach. This is due to the difficulty of designing resonant meta-atoms offering a prescribed phase and amplitude relation between *only two* dipole moments. Artificial magnetism requires resonators with a size that is typically equal to the wavelength in the particle (Merlin, 2009). Therefore, dielectric meta-atoms are multipolar in nature and, even if only the two desired resonant

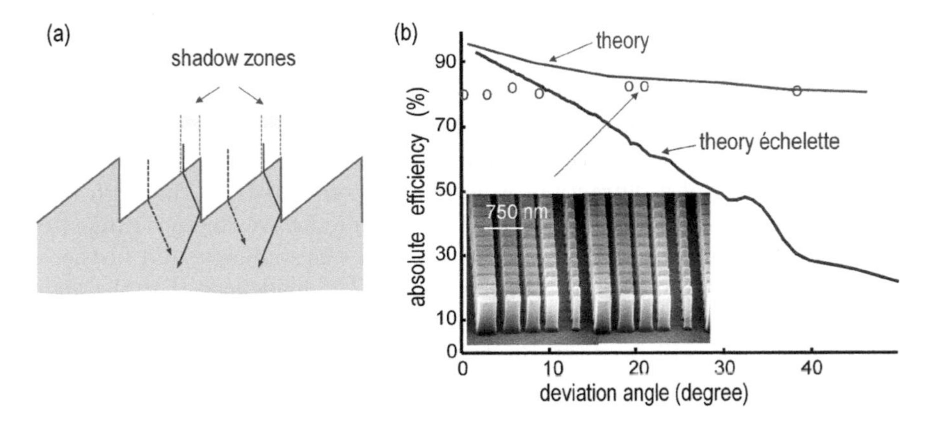

Fig. 18.11 Metagratings operating under the guidance approach with efficiencies larger than those of saw-tooth gratings. (a) Ray-tracing picture of the shadowing effect in blazed saw-tooth gratings. The effect illustrated for normally incident beams worsens at oblique incidence. See Exercise 18.4 for an illumination from the back side. (b) First-order transmission efficiencies of metagratings as a function of their deviation angle in air (the grating period is varying) for a normally incident and unpolarized plane wave at $\lambda = 633$nm. Circles: experimental results. Grey curve: theoretical results. Thick blue curve: theoretical efficiency of saw-tooth gratings in glass, illustrating the detrimental impact of the shadow on the efficiency as the grating period decreases. The lower inset is an electron micrograph of a section of a TiO_2 metagrating operating at a 20° deviation angle. Adapted from ref. (Lee et al., 2002).

dipoles are excited by the driving field owing to symmetry, for instance, long-range mutual interaction between the meta-atoms excites high-order multipoles and phase control is not implemented locally at the individual meta-atom level but at a larger scale involving several meta-atoms (Gigli et al., 2021). Inevitable collective behaviours, which are avoided in the guidance approach, then occur. We are facing a fundamental limitation, already encountered in the design of negative-index optical materials.

18.5 Plasmonic metasurfaces

As we saw in Chapter 13, bare metallo-dielectric surfaces support surface-plasmon modes. The latter can be launched and manipulated by drilling subwavelength indentations in the surface, e.g. isolated slits, grooves, or dimples, arrays of parallel or radial grooves, or hole arrays. These decorated surfaces—we may call them plasmonic metasurfaces—give birth to many phenomena and applications over a broad spectral range, from visible to THz and microwaves. Their optical properties are impacted not only by geometrical parameters, e.g. the indentation features and their density, but also by the EM interaction between neighbouring indentations of the surface.

18.5.1 Wood's anomaly

Robert W. Wood was the first to evidence surface-plasmon resonances at optical frequency when he was studying the spectrum of a continuous light source reflected by a metallic surface in which a periodic array of tiny grooves was etched away: 'I was astounded to find that under certain conditions, the drop from maximum illumination to minimum, a drop certainly of from 10 to 1, occurred within a range of wavelengths not greater than the distance between the sodium lines' (Wood, 1902).

The abrupt change in such a narrow spectral range immediately drew considerable attention. By considering the metal as perfectly conducting, in 1907, Lord Rayleigh proposed the first explanation of the anomalies, suggesting that the abrupt change was due to the passing-off of a diffraction order (now often called the Rayleigh anomaly) and that it occurred at wavelengths for which a scattered wave emerges tangentially to the grating surface. However, this first interpretation was simplistic and further experimental study of Wood's anomalies by J. Strong in 1936 evidenced that the anomalies occur at a wavelength systematically larger than that predicted by the grating equation.

To explain the red-shift from the Rayleigh grating-equation condition, U. Fano introduced a microscopic model in his seminal article published in 1941 (Fano, 1941). Fano's model is much less mathematically involved than the theoretical work by Lord Rayleigh. It rather relies on a Huygens-type intuitive interpretation and, importantly, it suggests that a surface mode with a parallel momentum greater than the free-space momentum be involved in the energy transport between adjacent grooves. It is retrospectively interesting and amazing to see how the surface mode is introduced in Fano's model. Fano first considered the parallel propagation constants of modes of a glass plate sandwiched between metal and vacuum and asks himself if there is any mode left when the glass-layer thickness vanishes. By solving the bi-interface problem analytically, like we did in Chapter 6, he showed that one and only one bound mode (the surface-plasmon mode) exists in the limit of vanishingly small glass thicknesses for TM polarization, with a complex propagation constant whose real part is always slightly larger than the modulus k_0 of the wavevector in a vacuum. Indeed, Fano's surface mode is nothing other than the surface plasmon of the flat interface, introduced 16 years afterwards by R. H. Ritchie in another context (see Chapter 13).

Fano therefore made the ansatz that Wood's anomaly originates from a collective resonance of the decorated surface (see Fig. 18.12) in which part of the wave scattered

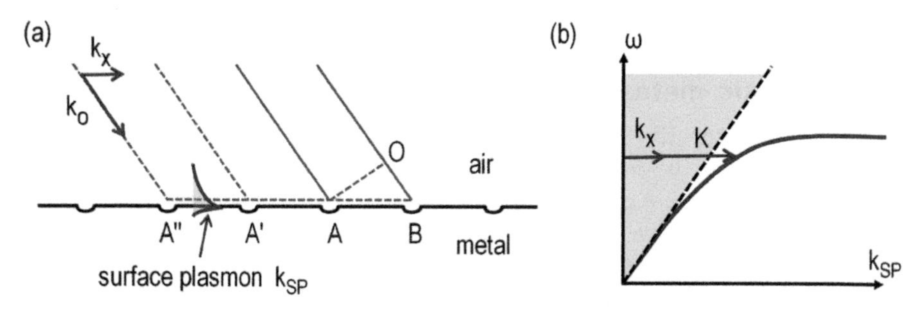

Fig. 18.12 Wood's anomaly. (a) The part of the incident wave that is scattered by groove A launches a surface plasmon that travels on the metal surface. If the plasmon reaches the neighbouring groove B in phase with the incident wave and with waves scattered by grooves A′, A″, a resonance builds up. Note the net difference between Fano's explanation (a resonance) and Rayleigh's one (the passing-off of a diffraction order). Adapted from ref. (Fano, 1941). (b) Phase-matching condition of Eq. (18.11). The dotted line corresponds to the air light line, $k = \omega/c$. k_x is the parallel component of the incident wavevector and $K = 2\pi/a$ corresponds to the momentum needed to couple to the surface-plasmon mode below the light line. (Note that, for simplicity, the backbending for large k's is not represented).

by groove A excites the bound-plasmon mode, then travels along the surface between the grooves with a phase velocity smaller than the vacuum one c, and gives a resonance whenever it reaches the neighbouring groove B in phase with the incident wave. Denoting the complex propagation constant of the surface plasmon by $k_{\text{SP}} = \frac{\omega}{c}\left(\frac{\varepsilon_d \varepsilon_m(\omega)}{\varepsilon_d + \varepsilon_m(\omega)}\right)^{1/2}$ (see Chapter 13) and assuming that the grooves are infinitely small, the microscopic interpretation by Fano leads to the following phase-matching condition for exciting the resonance,

$$\text{Re}(k_{\text{SP}}) \simeq k_{inc,x} + \frac{2\pi}{a}, \tag{18.11}$$

where the real part of the propagation constant is matched to the parallel wavevector $k_{inc,x}$ of the incident plane wave through a wavevector $2\pi/a$ of the 1D reciprocal lattice associated with the grating period a. This *approximate* surface-plasmon resonant excitation is completely equivalent to the resonant excitation of waveguide gratings detailed in section 17.3.2.

It is noticeable that Eq.(18.11) is qualitative only. Actually it is the parallel wavevector of the surface mode of the periodically corrugated interface that should enter the phase-matching condition. The wavevector can be computed only with the help of advanced numerical tools. Exercise 18.6 proposes a simplified analysis, which relies on a microscopic model that effectively clarifies the quantity to be considered in the phase-matching condition. The big success of introducing a surface-plasmon mode to convincingly explain the experimental red-shift was a milestone result, and since Fano's work, our understanding of Wood's anomalies has been intimately linked to the resonant excitation of surface plasmons. This view is simplistic. Section 18.6 will give a more comprehensive view; it explains why the red-shift is also observed in the near- and mid-infrared without recourse to any surface plasmons.

18.5.2 Wire-grid polarizers and slit-array filters

The wire-grid polarizer was probably the earliest device to exploit the form birefringence of subwavelength metallic gratings. It was used by Hertz to test the properties of the newly discovered radiowave in the late nineteenth century. It consists of a fine grid of parallel metal wires with subwavelength spacing. To simply understand why the grid polarizes light, let us insert into Eqs (18.9) and (18.10) a large negative value for the relative permittivity $\varepsilon_2 = \varepsilon_m \ll -1$ of the metal and assuming $\varepsilon_1 = 1$. Then, due to the fact that $|\varepsilon_m| \gg \varepsilon_1$, one gets

$$n_o = (f\varepsilon_m)^{1/2}, \tag{18.12}$$

$$n_e = ((1-f)/\varepsilon_1)^{-1/2}. \tag{18.13}$$

Taking $f = 0.5$, for instance, it appears that waves polarized along the wires see a good metal with a large and negative permittivity $n_o^2 \simeq \frac{\varepsilon_m}{2}$, whereas waves polarized perpendicular to the wires see a low-index material with a refractive index $\simeq 1.4$, since $n_e^2 = \left(\frac{1}{2}\right)^{-1}$. There are various ways of producing wire-grid polarizers. For radiowaves and microwaves up to the far-infrared, it is possible to wind a wire around a suitable frame and produce a free-standing grid. For shorter wavelengths in the near-infrared, obtaining a device of substantial area requires a series of very fine parallel wires on a transparent substrate. The shortest wavelength at which a grid works as a polarizer is dictated by its period, and progress in nanolithography has led to wire grids being manufactured in the visible (Tamada et al., 1997) and even in the ultraviolet, where dichroic materials providing a preferential absorption of light in particular polarization directions are rather missing.

In reality, the principle of operation of wire-grid polarizers is substantially different from the naïve homogenization picture. To be approximately valid, Eqs (18.12) and (18.13) require that the dominant Bloch modes resemble plane waves. For dielectric materials, such as polymers or even semiconductors, this is approximately valid for a wavelength-to-period ratio λ/a equal or greater than two or three (see section 15.4.2). For metallo-dielectric composites, it is additionally required that Bloch modes penetrate the metal almost uniformly.

Another scale then plays a key role: the skin depth, which turns out to be $\simeq 25\,\text{nm}$ for noble metal in the visible and near-infrared. The metal-wire cross sections are much larger, even for wire-grid polarizers operating in the visible, so that the genuine Bloch-mode homogenization regime is not reached in practice.

So for most wire grids, the EM field is essentially null in the metal and light is effectively funnelled into the slits between the wires. Fig. 18.13c shows the transmission

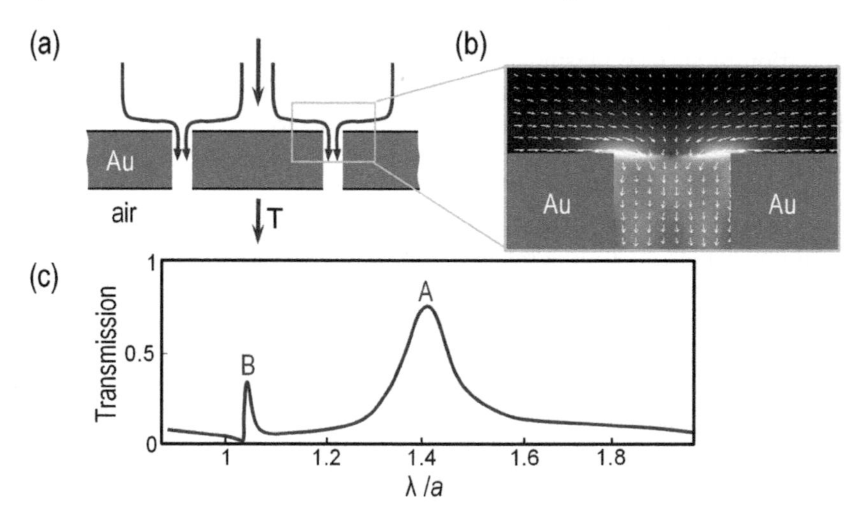

Fig. 18.13 Principle of operation of subwavelength wire gratings. (a) Sketch of the geometry. The grating is composed of a periodic array of tiny slits in a metal film; (b) funnelling effect at resonance (point A in (c)) illustrated by the distribution of the EM field's intensity (in grey) at the entrance of an illuminated slit. The white arrows indicate the direction and norm of the Poynting vector; (c) transmission spectrum for a TM-polarized normally incident plane wave. The computation is performed for gold wires of width 3μm and height 4μm. The slit width is 0.5μm. Resonance A does not shift as the angle of incidence is varied, while resonance B does shift.

spectrum of a gold metal plate perforated with tiny subwavelength slits (0.5μm) in the infrared (period $a = 3.5$μm) for TM polarization (magnetic field parallel to wires). The field map in Fig. 18.13b confirms that the electromagnetic field is indeed null in the metal ridges that are much larger than the skin depth. It also shows that the field inside the slits is much larger than the incident field, therein implying that the incident light is efficiently funnelled into the slit and gives rise to a transmission much larger than the normalized aperture.

What happens is that the incident light excites the fundamental mode of the slit (the gap-plasmon mode with a normalized propagation constant close to unity; see Chapter 13 and Fig. 18.5). The mode bounces back and forth between the upper and lower interfaces, just like in a Fabry–Perot resonance (Adams and Botten, 1979). Therefore, every slit behaves as a resonator with a cross section larger than the slit aperture, on the order of one wavelength at resonance (in 1D, the cross-section dimension is a length). It is not surprising, then, that an array with a subwavelength slit-to-slit distance almost fully transmits light at resonance. This is exactly what is happening at point A of Fig. 18.13c. Additionally, since the slits are weakly coupled at point A, we further understand why the resonance wavelength at point A does not disperse when the incident plane wave is tilted or when the positions of the slits are randomly and independently shifted.

For TE polarization, the fundamental slit mode is cut off for slit widths smaller than half the wavelength, so that the dominant Bloch mode is evanescent. The incident light is exponentially damped (as predicted by Eq. (18.12) but with the wrong attenuation), and light is essentially reflected. In general, large wavelength-to-period ratios and thick grating depths warrant a large contrast between TE and TM transmissions, making the wire grid a better polarizer. However, the desirable TM transmission is degraded owing to ohmic losses in the metal, especially in the visible and ultraviolet regions of the spectrum.

The other feature seen in the spectrum is resonance peak B. Unlike peak A, this peak B does evolve (it 'disperses') as the angle of incidence of the incident plane wave is varied. What is happening is the following: part of the incident energy that impinges on the grating is coupled to the slit, but not all of it, leaving a significant part not reflected but coupled to the surface, where it propagates up to the adjacent slits and couples to them. This also happens at point A. However, when this light that is recoupled from the surface is in phase with the light that is directly coupled from the incident wave (this happens only at point B when the grating equation is approximately satisfied, $k_{||} \pm \frac{2\pi}{a} = \pm k_0$), the slits becomes coherently coupled via the field scattered by the interface. This coupling has been attributed extensively to surface plasmons in the literature since Fano's seminal work (Fano, 1941) on Wood's anomaly, but surface plasmons only play a casual role, which is not even predominant in the visible and negligible at infrared frequencies, like in the example of Fig. 18.13. More is said in section 18.6 on surface waves and surface modes on metal surfaces.

The asymmetric resonance response of peak B has interesting consequences. The response lineshape, which is quite distinctively different from that of conventional symmetric resonance curves, such as at peak A, is known as a Fano lineshape. Originally introduced by U. Fano in the context of quantum systems interacting with a continuum of states, Fano-like resonances have been found in plasmonic nanoparticles, photonic crystals, and EM metamaterials. Their steep dispersion profiles (see the blue side of the resonance peak A) promise applications in sensors, lasing, switching, and non-linear devices; see ref. (Luk'yanchuk et al., 2010) for a comprehensive review.

18.5.3 Hole arrays and the extraordinary transmission

In the 1990s, progress in nanotechnologies made it possible to drill complex patterns into metal surfaces with high precision to elaborate 'plasmonic metasurfaces' that harness plasmons for implementing compact photonic devices. The topic saw a great revival following the report of the extraordinary optical transmission through arrays of subwavelength holes milled in an opaque metal film, i.e. the observation in the transmission spectrum of such an array of a set of peaks with a substantial efficiency per hole, much larger than the efficiency of one isolated hole, in particular when the holes are so small that all propagation modes are cut off and only evanescent fields are present. This report (Ebbesen et al., 1998) generated considerable interest, stimulated much fundamental research, and promoted subwavelength apertures in metal films as a building block of new optical devices (Maier et al., 2001; Atwater and Polman, 2010; Stockman, 2011). As we saw in the previous section and in Chapter 13, periodic metallic structures can convert light into surface plasmons by providing the necessary momentum conservation, and the extraordinary optical transmission can be thus seen as yet another manifestation of Wood's anomaly (Liu and Lalanne, 2008).

An emblematic example of plasmonic metasurfaces is illustrated in Fig. 18.14a, where the optical properties of a single hole are modified by carving a series of shallow subwavelength radial grooves in the surrounding surface. Such modifications give rise to much higher transmission than single holes at selected wavelengths. The grooves act as an antenna and couple the incident light into surface plasmons (see Fig. 18.14c). As a consequence, the EM fields at the surface become intense in the aperture, resulting in very high transmission efficiencies at well-defined wavelengths (Fig. 18.14b). Here the

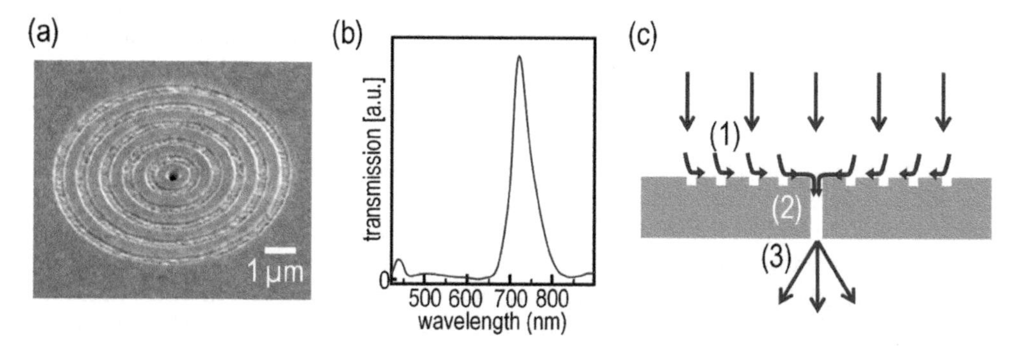

Fig. 18.14 Single hole dressed with annular periodic grooves. (a) Micrograph. Fabrication relies on focused ion beam (hole diameter 300nm, period 650nm); (b) Transmission spectrum. (c) The transmission process can be divided into three steps: (1) coupling of light to surface waves on the incident surface, (2) transmission through the hole to the second surface, and then (3) re-emission from the second surface. Adapted from (Lezec et al., 2002).

resonant wavelength is mainly determined by the periodicity of the grooves according to Eq. (18.11). Absolute quantification is difficult, but compared to an isolated single hole of identical dimensions, the transmission enhancement can be an order of magnitude larger when the phase-matching condition is satisfied.

If the output surface surrounding the aperture is also decorated, a narrow beam can be generated, having a divergence of a couple of degrees. This is because the light emerging from the hole couples to the pseudo-periodic structure of the exit surface, which, in turn, scatters the surface waves into freely propagating light. Such double-sided structures implement many functionalities. They may focus light, like a lens, just below the output surface. If grooves with different periods are on either side of the film next to a slit, they may act like two separate independent gratings connected by a pinhole, making the direction of the output beam independent of the input beam direction. The antenna can also be engineered such that the light directly coupled to the tiny central aperture and the surface waves coupled in the aperture mutually cancel, so that no light goes through. This device may act as a dark-field sensor, for which a tiny perturbation of the decorated surface by a foreign object unbalances the exact cancellation and allows light transmission (Zheng et al., 2010).

18.5.4 Broadband absorbers

The pyramid-array presented in section 18.4.2 prevents reflection by tapering the incoming light into the substrate. However, if the pyramid material absorbs light and the pyramid height is large enough, all the incident light may be absorbed before reaching the substrate, making the array an excellent candidate for broadband absorption.

Alternative approaches exist using gap-plasmon resonances of MIM waveguides. If the insulator thickness d is small, the effective index of the gap plasmon increases ($n_{\text{eff}} \propto d^{-1}$; see section 13.5.2) and Fabry–Perot resonance can be enabled at a subwavelength scale. Fig. 18.15 depicts three configurations that rely on this physics and provide remarkable absorption. In Fig. 18.15a, a metal substrate is etched with tiny grooves with different widths and depths. To achieve a broadband absorption, we may optimise the dimensions of each groove, so that the grooves resonate for a discrete set of wavelengths, e.g. at blue, green, and red wavelengths. Additionally, if critical coupling (see Chapter 7) is achieved for every individual resonator and the resonators combined together have negligible mutual influences on each other, perfect absorption can be implemented over a broad spectral range and a large angular acceptance.

Similar ideas have also been validated with ultra-thin superabsorbers composed of MIM resonators with tiny transverse dimensions lying horizontally on a metal substrate, pointing out the intriguing possibility of achieving near unity and broadband optical absorption in a deep-subwavelength layer. As suggested in Fig. 18.15b, by controlling the transverse dimension of the metal patch, the resonance wavelength can be monitored. Such surfaces are known as 'frequency selective surfaces' (Munk, 2000), or 'reflect and transmit arrays' in the microwave community.

In Fig. 18.15c, another way of embracing multiple resonances working at distinct frequencies within the same unit cell is depicted, based on the concept of the collective effect of multiple distinct oscillators. The resonators mixed within each unit cell can work on the same principle as in (a) and (b). Again, if the dielectric layers are much thinner than the wavelength, the MIM gap plasmons have large effective indices and Fabry–Perot resonances hold for unit cells that are smaller than the wavelength.

Other cheaper nanotechnology approaches exist to change the properties of almost any metal, rendering them literally black, e.g. illumination by femtosecond laser pulses that force the metal surface to form nanopits (Vorobyev and Guo, 2008).

18.6 Surface waves and surface modes on metal surfaces

Transmission through hole arrays has been characterized in great detail both theoretically and experimentally. When the anomalously high transmission was first observed, it was attributed to the resonant excitation of surface plasmons '*à* la Fano'. However, subsequent calculations and experimentations showed that subwavelength holes drilled in a *perfect* conductor gave rise to similar anomalous transmission, even though the free surface of a perfect conductor has no surface modes, and questioned the actual origin of the effect.

18.6.1 Fields scattered by tiny apertures drilled on metallic surface

To understand this apparent contradiction, we need to consider the properties of the field scattered by tiny apertures drilled on metallic surfaces. This field has been of long-standing interest in electromagnetism, since well before the development of plasmonic metasurfaces. In the 1900s, radiowave technology prompted theoretical studies to explain why very long-distance transmission could be achieved with such waves above the earth (trans-ocean transmission was achieved in 1907 by G. Marconi). The genuine reason is radiowave guiding by the ionosphere layers, but in the early twentieth century, the explanation was thought to be due to the nature of the surface waves launched on the conductive earth by the emitting antennas acting as EDs.

Sommerfeld was the first to determine the complete EM field radiated by an oscillating dipole at the interface between two semi-infinite half-spaces (Sommerfeld, 1926). He verified that his complicated solution, far away from the antenna on the surface, can be decomposed as the sum of a 'direct contribution' known as the Norton wave (Collin,

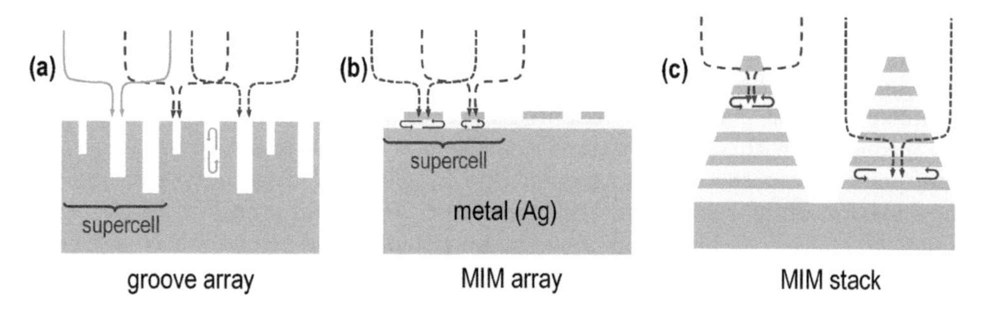

Fig. 18.15 Different metasurface layout for broadband absorbance with MIM gap-plasmon nanoresonators. (a) Grooves. (b) MIM patches. (c) Vertically arrayed MIMs. The interested reader may find quantitative discussions and much more information in ref. (Cui et al., 2014), for instance.

2004; Baños, 1966) and a so-called bound Zenneck mode, the analogue of surface plasmons for metals at optical frequencies. He also verified that the direct contribution decays algebraically as $1/r^2$ (note that a spherical wave emitted in free space follows a slower decay, as $1/r$) and is dominant at an asymptotically long distance from the dipole, since the bound mode decays exponentially.

In plasmonic metasurfaces, subwavelength indentations illuminated with a driving field essentially behave as ED or MD sources. Thus, from a mathematical point of view, the solution of the plasmonic problem is identical to that of the radiowave emission problem. Only the perspectives are different (García de Abajo, 2007). Plasmonics is rather concerned by short-distance (rather than long-distance) EM interactions, since the distance between two nearby apertures rarely exceeds a few wavelengths.

Two geometries are of general interest: metasurfaces with 1D subwavelength indentations, e.g. ridges, grooves, or slits, or those with 0D subwavelength indentations, e.g. holes or nanodiscs. For 1D indentations and TE-polarized illumination, the induced electric line source J_y is polarized parallel to the indentation y-axis. Because of the field-continuity relation, the electric field is nearly null at the interface around the line source, and the indentations of the metasurface are scattering light independently. For TM polarization, two electric line sources, one, J_z, being polarized perpendicular to the interface (along the z-axis), and the other, J_x, parallel to the interface (along the x-axis), are induced.

Figure 18.16a, b shows the magnetic field emitted at $\lambda = 1\,\mu\text{m}$ by a line source, J_z, polarized vertically and placed on an air–gold interface. The emission by a line source J_x would give nearly identical results, except for a scaling factor on the magnitude of the emission. Away from the surface, the radiated field behaves as a cylindrical Bessel wave with a $1/\sqrt{x}$ decay, just as if the source was radiating into free space. Close to the surface ($z < \lambda$), the radiated field is more complex. Following Sommerfeld, it is customary to decompose it as a sum of a bound plasmon with an exponential damping in the x-direction (thick blue dashed curve in Fig. 18.16b) and a direct wave (thin blue solid curve). The latter takes two asymptotic forms. It initially decays as a cylindrical wave ($\sim 1/x^{1/2}$) at small distances from the source, before decaying at a faster rate ($\sim 1/x^{3/2}$) at large propagation distances. As expected, the direct wave that could be viewed as being akin to the Norton wave is the dominant contribution at long distances, where the field vanishes.

More important for plasmonic metasurfaces is the fact that it is also dominant at short distances $x < \lambda$, implying that the interaction between nanoslits or nanogrooves at subwavelength separation distances on a decorated surface is mainly driven by the direct fields, *and not by the surface-plasmon field.* This conclusion holds for noble metals in the near-infrared and at longer wavelengths, but is not valid at smaller wavelengths in the visible, as shown in Fig. 18.16c. As one progressively moves from the visible to the infrared, the direct-wave contribution at short distances is unchanged, whereas the initial plasmon contribution rapidly drops as the metal permittivity increases, $H_{y,\text{SP}}(x/\lambda \to 0, z = 0) \simeq |\varepsilon_m|^{-1/2}$, due to increasing delocalization of the plasmon field in the dielectric medium. Thus, at visible wavelengths ($\lambda = 0.63\,\mu$m), the plasmon contribution dominates even at very short distances, the plasmon and direct-wave contributions being actually equal for the crossing-over distance $x_c \simeq \lambda/6$ (x_c is

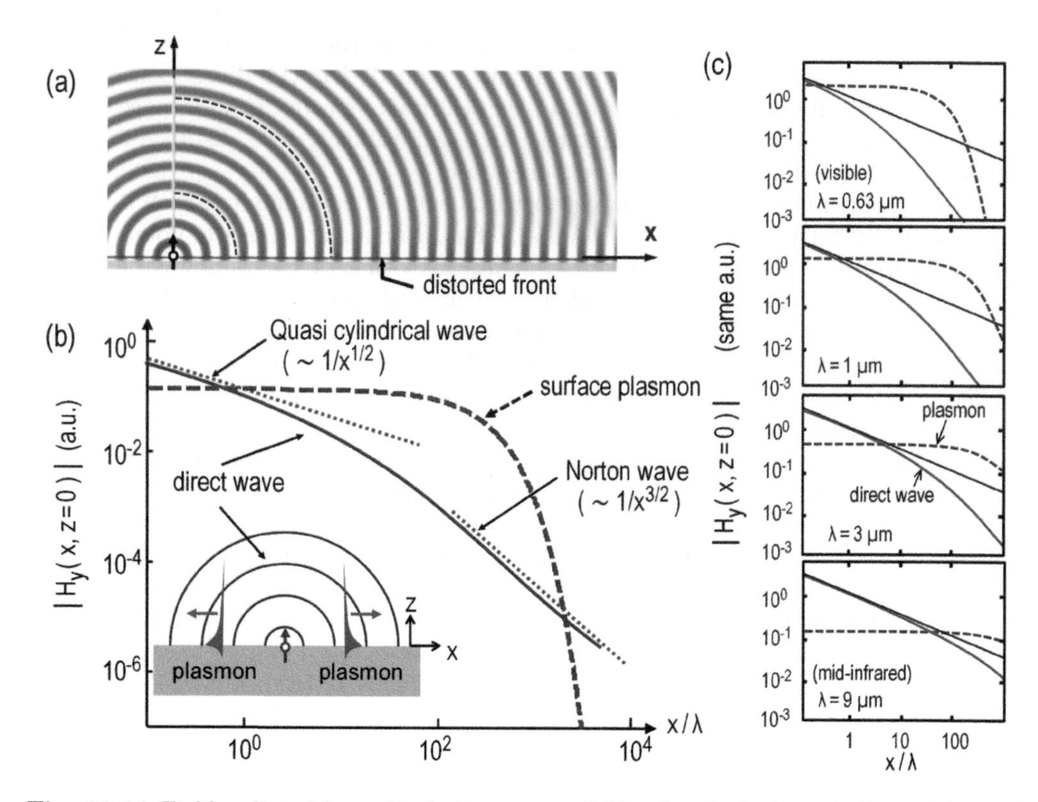

Fig. 18.16 Field radiated by a dipole line source $J_z\delta(x, z)$ polarized vertically and located just above an air–Au interface ($\varepsilon_{\mathrm{Au}} = -46.8 + 3.5i$ at $\lambda = 1\mu\mathrm{m}$). (a) Away from the surface, the radiated magnetic field $H_y(x, z)$ behaves as a cylindrical Bessel wave shown with the dashed semi-rings. Close to the surface ($z < \lambda$), the field pattern is more complex; it is composed of a bound surface plasmon and a 'direct-wave' contribution that is a remnant of the cylindrical Bessel wave; the wavefronts are distorted with respect to semi-rings. (b) The total field $H_y(x, z = 0)$ on the surface is composed of two contributions: an exponentially damped surface plasmon (thick blue dashed curve) and a power-law-damped direct wave (thin blue solid curve). (c) Evolution as a function of the wavelength. It can be shown that the initial crossing-over distance x_c, for which the direct-wave and surface-plasmon contributions are equal, increases with metal permittivity modulus according to $x_c \simeq \lambda|\varepsilon_m|/\left(2\pi\varepsilon_d^{3/2}\right)$. Adapted from ref. (Lalanne and Hugonin, 2006).

explicit in Fig. 18.16). At $\lambda = 9\,\mu\mathrm{m}$, the direct wave is predominant up to distances as large as $100\,\lambda$.

The field radiated by a 0D oscillating point dipole can be similarly decomposed into a plasmon contribution and a direct-wave contribution. For dipoles oriented perpendicular to the interface, the in-plane component of the radiated field is radially polarized and is independent of the azimuthal angle θ. For in-plane dipoles, there is an azimuthal dependence. Fig. 18.17 shows the radial electrical fields generated by an in-plane point

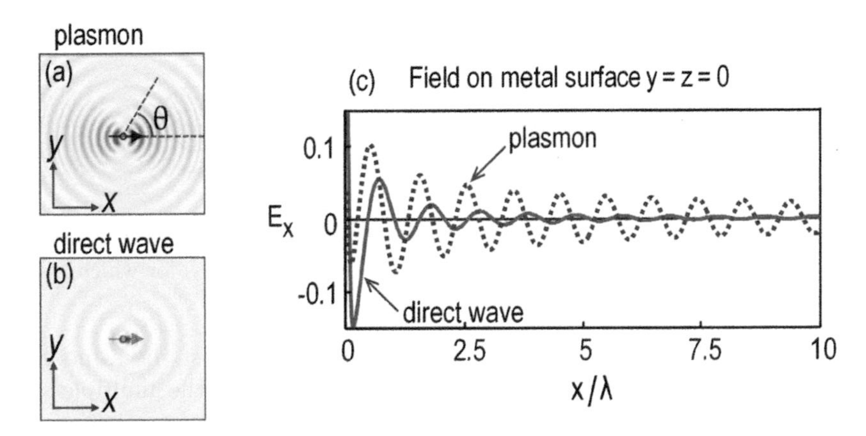

Fig. 18.17 Radial electric field radiated at $\lambda = 800\,\text{nm}$ along an Au–air surface by a dipole point source polarized along the x-axis, thus parallel to the surface. Surface maps are shown for (a) the plasmon field and (b) the field of the direct wave, with the same grey-level map. As the metal conductivity increases, the perfect-conductor limit is reached and the quasispherical wave (in air) becomes a spherical wave with a $r^{-1}\exp(i\,k_0\,r)$ dependence. (c) Radial field (here $E_r \equiv E_x$) plotted as a function of x for $y = 0$. The blue dashed curve corresponds to the plasmon field and the bluish solid curve to the direct wave.

dipole. The plasmon field on the surface is proportional to $r^{-1/2}\ \exp(iK_{\text{SP}}r)$ with r the radial distance; its electric vector is mainly perpendicular to the surface with a small in-plane component. Along any direction (different from $\theta = \pi/2$), the in-plane plasmon field tends to be radially polarized, as $E_\theta/E_r \simeq \tan\theta$, where the subscripts θ and r are used to denote the orthoradial and radial components of the fields. The direct wave initially varies as a quasispherical wave with an $r^{-1}\ \exp(ik_0r)$ spatial dependence at small distances from the dipole, but then, at longer distances, its electric-field amplitude decays faster, as $r^{-2}\exp(ik_0r)$, like the Norton radiowave. Its electric field mainly points along the direction perpendicular to the interface like the surface-plasmon field, and its in-plane components also satisfy $E_\theta/E_r \simeq \tan(\theta)$.

The dual-wave picture (direct wave and surface plasmon) explains well why many properties of metasurfaces (the extraordinary optical transmission, for instance; see ref. (Lalanne et al., 2008)), observed at optical frequencies, are 'faithfully reproduced' at longer wavelengths without plasmons by scaling the geometrical parameters. The main reason is that the surface-plasmon field on the surface has much in common with the direct-wave field. For instance, the plasmon propagation constant $\text{Re}[k_{\text{SP}}]$ is close to the free-space propagation constant k_0 of the direct wave. Thus, constructive or destructive interferences of plasmons in optics or direct waves at longer wavelengths occur for nearly the same inter-indentation separation distances. In addition, the field intensity of the surface plasmons launched in the visible or near-infrared (see the dashed curve of the two upper panels in Fig. 18.16c) at a typical inter-indentation separation distance of, say, 1λ is only slightly larger than the field intensity launched on the surface by the direct wave at any wavelength (solid curves in all panels). Thus, indentations

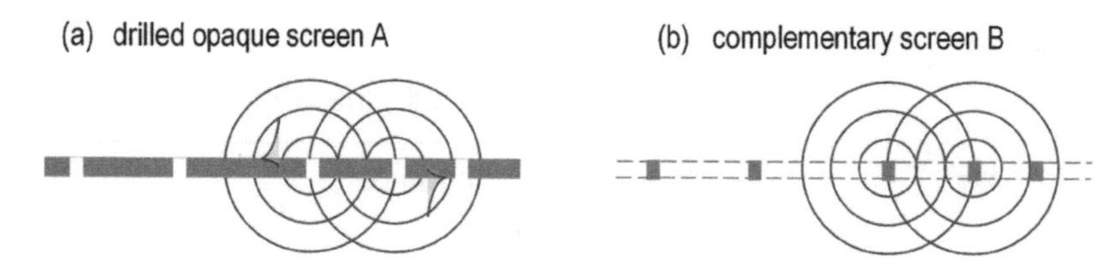

Fig. 18.18 The Babinet principle for complementary opaque films applies to plasmonic metasurfaces over a broad range of frequencies, except at visible frequency, for which the plasmon contribution cannot be neglected.

excite their adjacent wavelength-distant neighbours with almost the same force, and this holds at any wavelength. It is therefore noticeable that the multiple-scattering processes that are responsible for the properties of metasurfaces are largely identical at all wavelengths, except for a reinforcement due to plasmons at visible frequency.

18.6.2 The legitimacy of the Babinet principle for metallic metasurfaces

Babinet's principle (Born, 1999) is most often considered in optics, but it is also true for other forms of EM radiation and is, in fact, a general theorem of diffraction in wave mechanics. It states that the diffraction pattern from an opaque body is identical to that from the complementary screen, except for the overall forward-specular-beam intensity. For instance, the fields scattered by a hole in a perfect-metal thin screen or a perfect-metal thin disc are identical. A test of Babinet's principle that is strongly related to the extraordinary transmission at optical frequencies can be found in the seminal work by Ulrich (Ulrich, 1967) on the relation between hole arrays in thin metal screens and their complementary screens for applications of frequency-selective surfaces in the far-infrared. Although initially formulated for perfect metals and tested at long wavelengths, Babinet's principle is often used in plasmonics without further justification—see the related discussion in ref. (García de Abajo, 2007).

Assume an original diffracting body A, composed of apertures (e.g. slits) in a thin opaque film (see Fig. 18.18a), and its complement, i.e., a body B that, for a given region in space, is opaque where A is transparent and transparent where A is opaque (hence at infinity). Further assume that the apertures are tiny enough to be able to scatter like oscillating dipoles. For long wavelengths, according to what has just been said, each aperture scatters a direct wave, which is strictly a cylindrical Bessel wave and which is exactly the wave scattered by each individual wire of the complementary screen (the 2D Green functions are strictly identical). Therefore, the radiation pattern of A must be the same as that of B (see Fig. 18.18b). At shorter wavelengths, the wavefront cross sections of the Bessel waves start to be distorted with respect to circles, but essentially the radiation patterns remain similar. In fact, the radiation patterns of the bodies A and B start to differ only at visible frequencies; the direct waves generated by the slits or the wires still have similar properties, but surface plasmons are launched in body A, whereas they are absent in body B. Bodies A and B may have significantly different optical properties.

18.6.3 Spoof surface-plasmon waves in a perfect metal

In the visible and near-infrared domains, the conductivity of noble metals is low. Surface-plasmon modes have a short penetration depth into metals and a relatively short penetration depth into dielectrics. In the microwave regime, metals have a very large complex refractive index, $n + ik$, with $n \sim k \sim 10^3$ (see Chapter 3), and behave almost as perfect conductors. For these long wavelengths, waveguides such as optical fibres are not used because of compactness, and guiding is mostly performed by narrow dielectric slabs sandwiched between two parallel metal plates separated by a distance smaller than half a wavelength, i.e. by MIM waveguides similar to those studied for optical frequencies in Chapter 13 but with much weaker loss for the former than the latter.

If guiding is possible with metal sandwiches over long distances, then it means that the mode penetration in the metal is small. This also implies that surface modes at metallic interfaces do not exist in millimetre-wave integrated circuits or, more precisely, that the surface-plasmon mode extends over so many wavelengths into the dielectric that it is very nearly a plane wave; it is not a confined surface mode. In particular, this precludes the usefulness of applying to the microwave regime near-field concepts that were introduced for the visible domain.

An artificial way of synthesizing surface modes on metals at microwave frequencies is to etch the surface with regularly spaced subwavelength holes (Ulrich and Tacke, 1973; see Fig. 18.19). The idea uses the fact that the subwavelength holes in the metal are hollow metallic waveguides and therefore have a cutoff frequency. Below this, no propagating modes are allowed and any incident fields fall off exponentially with distance down the hole. Thus, below their cutoff frequency, the holes behave as reservoirs of evanescent fields and the penetration depth into the metal can be controlled by varying the depth and size of the holes. At the hole aperture, the evanescent fields radiate in much the same way as oscillating dipoles and generate direct waves that propagate on the flat surface. Like the pure plasmonic model of Exercise 18.6, a multiple scattering process involving the evanescent fields and the direct-wave fields takes place to create a surface mode that propagates over considerable distances at the structured interface (Pendry et al., 2004).

If the period is on a scale much smaller than the wavelength of probing radiation, it is possible to describe the structured surfaces by an effective dielectric function (Pendry et al., 2004) of the plasma form with an effective plasma frequency $\omega_{p,\text{eff}}$, and the surface wave is a truly bound mode with a dispersion relation very similar to that of surface plasmons at optical frequency. Notably, at low frequencies the surface mode approaches the light line asymptotically, and the fields associated with the mode expand into the vacuum. At large $k_{||}$, the frequency of the mode approaches $\omega_{p,\text{eff}}$. Fig. 18.19b shows a sketch of the dispersion relation of the bound mode. The latter is often called the 'spoof plasmon', to emphasize its artificial plasmon character. Interestingly, the plasma frequency of the effective dielectric function is entirely determined by the geometry of the surface and may therefore be chosen from anywhere within the microwave spectrum. If the period is progressively extended without changing the hole diameters, holes always act as evanescent-field reservoirs, but the artificial media approach ceases to be

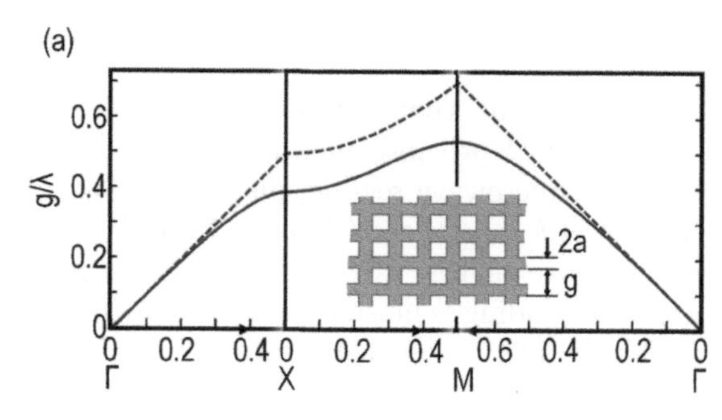

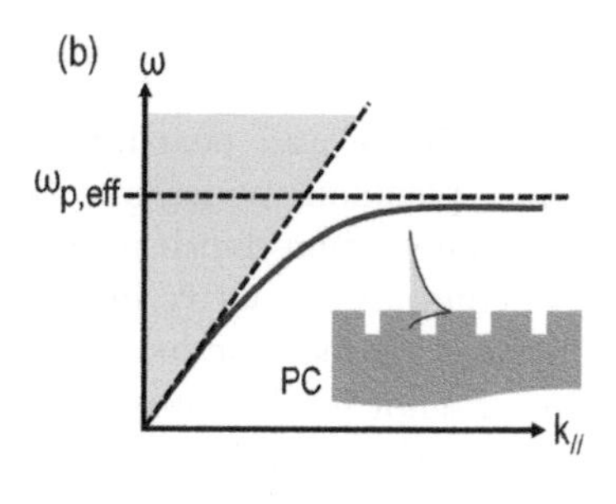

Fig. 18.19 Spoof surface plasmon at microwave frequencies on nearly perfect metals (PC). (a) Dispersion relation along the three main sections through the first Brillouin zone (see Chapter 17) of a copper film of thickness 7μm perforated by a regular mesh of square holes with a period $g = 127\mu\text{m}$ (inset). The dashed curve gives the light cone. (b) Sketch of the plasmon-like dispersion relation of spoof plasmons on PC metals. Fig. (a) is reprinted with permission from ref. (Ulrich and Tacke, 1973) ©(1973) American Institute of Physics.

valid. Still, the picture of an array of evanescent-field reservoirs in the metal connected by the direct wave on the surface continues to be valid, and the spoof plasmon becomes a leaky surface wave with diffracted beams that can be excited by a probe field, just like the resonant Bloch modes of Chapter 17.

Bound states can be synthesized with metal surfaces perforated by arrays of sub-wavelength holes or grooves of finite depth, surfaces covered with SRRs, and with many materials from bulk metals to ultra-thin flexible copper films; they have been extensively studied and have proven inspiring for many applications.

Exercises

Exercise 18.1 Equivalence between material properties and coordinate transforms: perfect lensing. Although the perfect lens is more a myth than a reality, it is an interesting concept. The purpose of this exercise is to try to understand the principle of operation of the perfect lens and the equivalence between material properties and coordinate transforms.

We consider a monochromatic EM wave that is invariant in the y-direction, with a polarization such that the only non-zero field components are E_x, E_z, and H_y. The wave satisfies the source-free Maxwell equations

$$\partial_z E_x - \partial_x E_z = i\omega\mu_{yy}H_y, \quad \partial_z H_y = i\omega\varepsilon_{xx}E_x, \quad \partial_x H_y = -i\omega\varepsilon_{zz}E_z, \tag{18.14}$$

where all off-diagonal tensor components of the permittivity and permeability are assumed to be null. We introduce the coordinate change $z' = \alpha z$.

(1) Show that

$$\partial_{z'} E_x - \partial_x \frac{E_z}{\alpha} = i\omega\frac{\mu_{yy}}{\alpha}H_y, \quad \partial_{z'} H_y = i\omega\frac{\varepsilon_{xx}}{\alpha}E_x, \quad \partial_x H_y = -i\omega(\alpha\varepsilon_{zz})\frac{E_z}{\alpha}, \tag{18.15}$$

implying that, if $(E_x,\ E_z,\ H_y)$ is a field distribution solution of the Maxwell equations in a uniform material with permittivity $\varepsilon_{xx}, \varepsilon_{zz}$ and permeability μ_{yy}, then $(E_x,\ \frac{E_z}{\alpha},\ H_y)$ is a field distribution solution of Maxwell equations in a uniform material with permittivity $\frac{\varepsilon_{xx}}{\alpha}, \alpha\varepsilon_{zz}$ and permeability $\frac{\mu_{yy}}{\alpha}$.

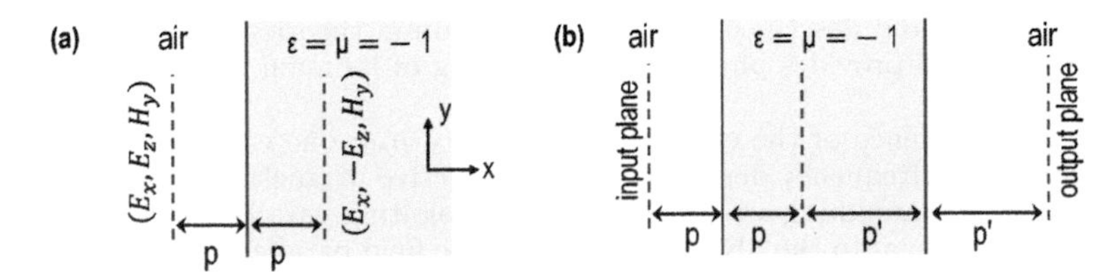

Fig. 18.20 Perfect imaging with a perfect lens. (a) Interface between air and a uniform medium with $\varepsilon = \mu = -1$. (b) Double interface.

(2) In Fig. 18.20a, consider that the field in a plane located at a distance p on the left side of the interface is E_x, E_z, H_y. Potentially, the field contains progressive plane waves and evanescent ones with a parallel momentum larger than $\frac{\omega}{c}\sqrt{\varepsilon\mu}$ (with a transverse momentum $k_\perp$ such that $k_\perp^2 + k_\parallel^2 = \varepsilon\mu(\frac{\omega}{c})^2$). Upon propagation towards the interface, progressive waves acquire a phase difference and evanescent waves are damped. We are in air, and we know that propagation acts as a low-pass filter. By applying $\alpha = -1$ in the system of Eqs (18.15), show that the field at a separation distance p on the right side of the interface is exactly E_x, $-E_z$, H_y.

(3) Finally, consider Fig. 18.20b with the two interfaces and argue that the field of the input plane is *perfectly* reconstructed in the output plane.

Exercise 18.2 Effective permeability of arrays of SRRs. It is well known from elementary electromagnetics that an MD source can be realized by a circulating current on a closed loop, such as a closed SRR (CSRR), which leads to a magnetic moment, m, with magnitude given by the product of the current and the area of the CSRR and direction perpendicular to the plane of the SRR. Therefore, a CSRR behaves as an inductor, storing magnetic energy $U = \int_0^B m\, dB = LI^2/2$, where L is the self-inductance of the loop, B is the magnetic field ($B = \mu\mu_0 H$), I is the current in the loop, and μ_0 is the free-space magnetic permeability. If a CSRR is combined with a capacitor with capacitance C (see Fig. 18.3), then one obtains an LC circuit with a resonance frequency $\omega_{LC} = 1/\sqrt{LC}$. Such a capacitor can be realized by making a cut in the ring, leading to an actual SRR. Thus, the SRR acts like an EM resonator, producing resonant circular currents at ω_{LC} leading to resonant magnetization, i.e. resonant effective permeability.

(1) With reference to Fig. 18.3, assuming some dielectric lossless medium of relative constant ε_c to replace air, explain what approximation is made to obtain the capacitance of the gap as $C = \varepsilon_c\varepsilon_0 wt/d$.

(2) Using the solenoid formula with a loop line density $1/t$, deduce the self-inductance L of a loop of size ℓ (comment on the approximation as well).

(3) Deduce that the resonant frequency is $\omega_{LC} = c\ell^{-1}(w\varepsilon_c)^{-1/2}d^{1/2}$, with c the light velocity in vacuum.

(4) Show that the corresponding wavelength λ_{LC} is proportional to ℓ, the SRR size, but that the ratio λ_{LC}/ℓ can be very large with appropriate w, d values. This is a fundamental difference between left-handed materials (LHMs) and the other important class of EM metamaterials, which includes photonic crystals, where the frequencies of operation correspond to wavelengths of the same order of magnitude as the unit cell size. The fact that in LHMs $\lambda_{LC} \gg \ell$ implies that a LHM can very accurately be considered as a homogeneous

effective medium and described using effective medium theories, which greatly simplifies its description and provides physical understanding of its main features.

(5) Frequency dependence of the effective permeability μ_{eff}: one can easily obtain a simple expression for the frequency dependence of the effective magnetic permeability $\mu_{\text{eff}}(\omega)$ for a lattice of SRRs, assuming an incident wave propagating parallel to the SRR plane with magnetic field normal to the SRR plane and electric field parallel to the uncut sides of the SRRs. Under these conditions, and according to the Kirchhoff loop rule, the self-induced voltage in the inductance L, U_L plus the voltage drop across the capacitance C, U_C equals the voltage induced by the external magnetic flux variation $U_L + U_C = U_{\text{ind}} = d\Phi/dt$ with $\Phi = \mu_0 \ell^2 H$ the external flux.

Assuming a time-harmonic dependence $H = H_0 \exp(-i\omega t)$ and using $U_L = dI/dt$ and $U_C = C^{-1} \int I\, dt$, show that the current has a resonant behaviour $I_0 \propto \omega^2/[\omega^2 - (1/LC)]$.

(6) Deduce from the above the expressions for a single SRR magnetic moment $m = I\ell^2$ and the magnetization $M = (N_{LC}/V)m$, defining the SRR density by $N_{LC}/V = 1/a^3$ if they stand in a cubic lattice arrangement of side a. Deduce the magnetic permeability in the form $\mu_{\text{eff}}(\omega) = 1 - F\omega^2/[\omega^2 - \omega_{LC}^2]$. Verify that $F = \ell^2 t/a^3$.
Apart from the ω^2 factor in the numerator, this represents a Lorentz oscillator resonance for a magnetic atom. Here, we have lumped the various parameters into the dimensionless quantity F that must clearly not exceed unity: $F < 1$. As often done, all the losses and the scattering mechanisms could be lumped into a damping factor, $i\omega\Gamma_M$, inserted in the resonant denominator.

(7) Show that the domain of negative $\mu_{\text{eff}}(\omega)$ is given by the interval $[\omega_{LC}, \omega_{LC}(1 - F)^{-1/2}]$.

Exercise 18.3 Effective permittivity of metallic post arrays. For determining the effective permittivity of a lattice of metallic posts (see Fig. 18.21), we shall use a very simple model that gives a good approximation. The derivation is based on the relationship between current and electric field with a square unit cell of size a. Note that a is the distance between the posts as shown in Fig. 18.21, and it is also approximately equal to the length of the posts within every unit cell. Let us first find the current in a single post of length a and radius r, for an incident electric field E parallel to the post axis. According to Ohm's law and neglecting skin-depth effects, the free electrons of the metal move and the induced current I is then equal to $I = Ea/Z_w$, where Z_w is the impedance of the wire, $Z_w = R + iL\omega$. The expressions for the resistance and inductance are classical (Collin, 2004):

$$R = a/(\pi r^2 \sigma) \quad \text{and} \quad L = (\mu_0/2\pi)[\ln(2a/r) - 3/4], \tag{18.16}$$

σ being the metal conductivity.

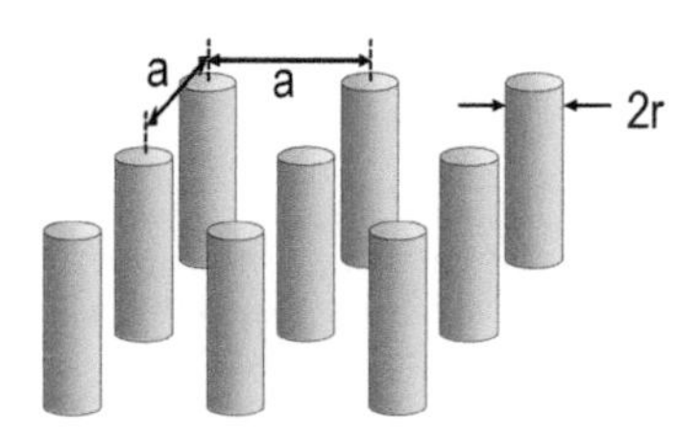

Fig. 18.21 Periodic array of metallic posts.

(1) Average the current density J_{eff} flowing along the posts: since every unit cell contains a single wire, write the current *density* $\mathbf{J}_{\text{eff}}$ as a function of $\mathbf{E}$, Z_w, and a.

(2) The classical derivation of the permittivity of metals is achieved by adding the $\nabla \times \mathbf{H}$ contributions of bound and free electrons. Writing the contribution to $\nabla \times \mathbf{H}$ of bound electrons as $i\omega\varepsilon_r\varepsilon_0\mathbf{E}$ with $\varepsilon_r = 1$ in vacuum, show that one obtains for an array of posts in free space

$$\varepsilon_{\text{eff}} = 1 - i/(\omega Z_w a \varepsilon_0). \tag{18.17}$$

Transform this expression using $Z_w = R + iL\omega$.

(3) It is important to relate this expression to the classical expression of the dielectric constant for a metal described by an electron gas with the Drude model. To this end, cast the dielectric constant in the form

$$\varepsilon_{\text{eff}} = 1 - \omega_p^2/(\omega^2 + i\gamma(\omega)\omega). \tag{18.18}$$

Two time scales are included in the model. On one hand, the plasma frequency is the mode frequency of the charge-density oscillation given by $\omega_p^2 = Ne^2/\varepsilon_0 m$ with the classical equations of motion of the electrons including the friction term ($\lambda_p = 227\,\text{nm}$ for gold and $324\,\text{nm}$ for silver). The plasma frequency lies in the near-ultraviolet for most metals. On the other hand, a second time scale appears in this formula, namely the relaxation time $\tau = 1/\gamma$. This relaxation time that describes the relaxation processes for excited electrons is on the order of 10 fs for noble metals for DC. The Drude model is quite accurate for silver and gold in the visible and near-infrared. In the ultraviolet, additional terms taking into account interband transitions are needed.

(4) Show that formally identifying in Eq. (18.18) the terms that were deduced from Eq. (18.17) leads to the definition of an effective plasma frequency and an effective relaxation time

$$\omega_p^2 = (a\varepsilon_0 L)^{-1} \quad \text{and} \quad \gamma(\omega) = R/L. \tag{18.19}$$

As an example, let us take $a = 6\,\text{mm}$, $r = 0.03\,\text{mm}$ (small fill fraction in metals), and $\sigma = 10^7\,\text{S/m}$ for copper, as used in Smith's experiments. The resulting effective plasma frequency can be calculated to be 8.7 GHz, well in the microwave region, whereas the plasma frequency of noble bulk metals or copper is usually a few $10^{15}\,\text{s}^{-1}$. Since $\omega \gg \gamma(\omega)$, we further neglect absorption and the effective dielectric constant can be written

$$\varepsilon_{\text{eff}} = 1 - (a\varepsilon_0 L)^{-1}/\omega^2. \tag{18.20}$$

The important fact is that the post array behaves at microwave frequencies like a metal does at optical frequencies: the effective dielectric constant is almost a negative real number, whereas it is usually almost a pure imaginary number for bulk metals at microwave frequencies. For frequencies below the effective plasma frequency of Eq. (18.19), the post lattice reflects EM waves, while it transmits them at larger frequencies.

Exercise 18.4 Shadow effect: Illumination from the substrate. Referring to Fig. 18.11a, use ray tracing to study a saw-tooth grating illuminated from the substrate side. Show that, although the rays are all refracted in the first-order direction, a shadow-like effect is observed in the transmitted beam. Explain why this effect feeds all diffracted orders.

Exercise 18.5 Efficiency of saw-tooth diffractive elements. The exercise is inspired from the MIT Technical Report 854 (MIT, 1989) by G. J. Swanson. We consider an *arbitrary* refractive element with phase $\Phi(x, y)$ made of a material with refractive index n immersed in a background with refractive index n_0. Material dispersion is neglected. The corresponding diffractive optical element (DOE), designed for operation at a wavelength λ_0 in the order m, is obtained by removing blocks of height $H = m\lambda_0/(n - n_0)$. This means that, compared to the thickness $t_{ref}(x, y)$ of the refractive element, the thickness $t(x, y)$ of the DOE is much smaller: $t(x, y) = t_{ref}(x, y)$ modulo H. Thus, the DOE, a Fresnel lens or a grating like in Fig. 18.11a, has a saw-tooth profile.

We assume that the phase is slowly varying at the scale of the wavelength, so that the scalar theory of diffraction may be applied with confidence. The DOE has a transmittance $\exp(i\phi_0)$ at wavelength λ_0 with $\phi_0(x, y) = \frac{2\pi}{\lambda_0}(n - n_0)t(x, y)$.

(1) Plot ϕ_0 vs Φ. The *key point* is to note that ϕ_0 is a periodic function of Φ.

(2) At λ, the DOE transmittance is $\exp(i\phi(x, y))$. Express ϕ vs ϕ_0, λ, and λ_0.

(3) Note that the transmittance of the DOE at wavelength λ is a periodic function of ϕ_0 with period $2m\pi$. Deduce that the DOE transmittance at any wavelength λ can be expanded in a Fourier series of the form $\exp(i\phi) = \sum_p c_p \exp(i\frac{p}{m}\phi_0)$. Show that $c_p = \text{sinc}(p - m\frac{\lambda_0}{\lambda})$.

(4) Calculate the diffraction efficiency η at wavelength λ in order p of a DOE designed to provide an arbitrary phase λ_0 at wavelength λ_0 in order m.

(5) Application: Consider a DOE lens for $m = 1$. How does that lens diffract light at wavelength λ? What are its diffracted orders?

Exercise 18.6 Microscopic model of surface modes on metallic metasurfaces. Our objective is to build a simple model (à la Fano (Fano, 1941)) that explains the formation of surface modes on metallic metasurfaces decorated by periodic arrays of subwavelength indentations. The present model is inspired by ref. (Liu and Lalanne, 2008). It retains the main ingredients of Fano's model: (i) it relies on the fact that surface modes are built from the accumulation of in-phase scattering events by the indentations; and (ii) it conveniently considers the restrictive assumption that the EM interaction between adjacent indentations is provided only by surface plasmons (see the last question for the analysis of the direct-wave contribution). However, with the present model we go one step further and give a correction to the classical surface-plasmon resonant excitation of Eq. (18.11). In particular, we show that the surface-plasmon resonant excitation occurs when the wavelength is slightly larger than that predicted by Eq. (18.11). In addition to providing a deep understanding of surface-plasmon modes, the exercise gives the opportunity to learn more on the difference between normal ($\tilde{k}(\omega)$) and quasinormal ($\tilde{\omega}(k)$) modes on periodic metal surfaces. For simplicity, we consider a 2D metal surface decorated by a periodic array of indentations (grooves or ridges) with arbitrary shapes. The period is denoted by a. Let us first introduce the reflection and transmission coefficients, ρ and τ, of a surface plasmon that is reflected or transmitted by a single indentation (see Fig. 18.22a). We assume that the indentations are symmetric, so that the reflection of the left-going surface plasmon is also ρ. Let us also introduce the modal coefficients of right- and left-going surface plasmons, A_n and B_n, which are excited in between the indentations in the array (see Fig. 18.22b).

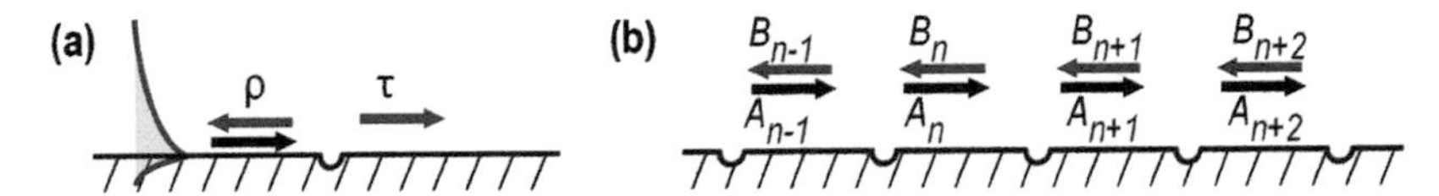

Fig. 18.22 Microscopic model of Wood's anomalies. (a) Elementary scattering coefficients for surface plasmons scattered by a subwavelength indentation. (b) Multiple scattering of surface plasmons on a periodic array of identical indentations. A_n and B_n denote the coefficients of the right- and left-going surface plasmons excited in between indentations $n-1$ and n in the array.

(1) Show that the modal coefficients A_n and B_n satisfy the following conditions:

$$A_{n+1} = \tau A_n \exp(iK_{\mathrm{SP}}a) + \rho B_{n+1} \exp(iK_{\mathrm{SP}}a), \tag{18.21}$$

$$B_{n+1} = \tau B_{n+2} \exp(iK_{\mathrm{SP}}a) + \rho A_{n+1} \exp(iK_{\mathrm{SP}}a), \tag{18.22}$$

with $K_{\mathrm{SP}}(\omega) = \frac{\omega}{c}(\frac{\varepsilon_d \varepsilon_m(\omega)}{\varepsilon_d+\varepsilon_m(\omega)})^{1/2}$ the surface-plasmon propagation constant. A surface mode (guided-like or quasinormal, see section 13.4.1) is found as a source-free solution of the Maxwell equations, and thus can be found by finding the self-consistent solution of the previous system of equations.

(2) We now introduce the Bloch–Floquet condition, $A_{n+1} = A_n \exp(ik_x a)$ and $B_{n+1} = B_n \exp(ik_x a)$, which fixes a Bloch wavevector k_x. Show that the system of Eqs (18.21) and (18.22) admits a solution if

$$[1-\tau \exp(-ik_x a)\exp(iK_{\mathrm{SP}}a)][1-\tau \exp(ik_x a)\exp(iK_{\mathrm{SP}}a)] - \rho^2 \exp(2iK_{\mathrm{SP}}a) = 0. \tag{18.23}$$

(3) We further assume that $|\rho| \ll 1$, i.e. that the indentations weakly scatter the incident plasmons, and neglect the ρ^2 term. Show that Eq. (18.23) admits a solution for (k_x, ω) pairs that verify

$$\tau(\omega)\exp(\pm ik_x a)\exp(iK_{\mathrm{SP}}(\omega)a) = 1. \tag{18.24}$$

(4) In a first solution, we fix ω real and compute the complex propagation constant of the normal mode. What are the three physical phenomena that make k_x complex (rather than real)? Calculate k_x and show that $\mathrm{Re}(k_x) \neq \mathrm{Re}(K_{\mathrm{SP}})$ and that the damping rate of the surface mode is slightly larger than that of the surface plasmon ($\mathrm{Im}(k_x) > \mathrm{Im}(K_{\mathrm{SP}})$). For better accuracy, what is the quantity that should be involved in Eq. (18.11) in place of $\mathrm{Re}(K_{\mathrm{SP}})$?

(5) In a second solution, we consider quasinormal surface-plasmon modes, fix k_x real in the first Brillouin zone, and calculate the complex frequency $\tilde{\omega}$. What are the two physical phenomena that make ω complex rather than real? The transcendental Eq. (18.24) can be easily solved if neglecting metal absorption, $\mathrm{Im}(\varepsilon_m) = 0$. With this approximation, show that

$\mathrm{Re}(K_{\mathrm{SP}}) = k_x - \frac{\arg(\tau)}{a} + m\frac{2\pi}{a}$, with m an integer, and show that

$\frac{\omega''}{c} = a^{-1}\ln(|\tau|)(\frac{\varepsilon_d+\varepsilon_m(\omega')}{\varepsilon_d\varepsilon_m(\omega')})^{1/2}$, where $\tilde{\omega} = \omega' + i\omega''$.

In the absence of metal loss, what is the sole reason for the finite mode lifetime? Verify that the sign of ω'' is correct.

(6) At normal incidence, $k_x = 0$, the two terms written in brackets in Eq. (18.23) become equal. Show that the system of Eqs (18.21) and (18.22) admits two solutions, $1 = (\rho \pm \tau)\exp(iK_{\mathrm{SP}}a)$, at frequencies such that

$$\mathrm{Re}(K_{\mathrm{SP}}) = \frac{\arg(\rho \pm \tau)}{a} + m\frac{2\pi}{a}, \tag{18.25}$$

for which the + sign corresponds to dark modes and the − sign, to leaky modes.

(7) Many optical phenomena in subwavelength metallic surfaces, which are observed with metallic nanostructures at visible frequencies, can be ‘reproduced’ at longer wavelengths by scaling the geometrical parameters. To realize why, imagine a microscopic model for long wavelengths in which the interaction between neighbouring indentations is mediated not by surface plasmons but by direct waves. For gratings illuminated with a wavelength comparable to the period a, what is the critical period for which the grating properties are governed by surface-plasmon interactions?

Part VI

Confined photons: Nanoantennas, microcavities, and optoelectronic devices

19 Controlling light–matter interaction at the nanoscale with cavities and nanoantennas

19.1 Introduction

In this chapter, we consider light–matter interaction processes such as emission, absorption, elastic or inelastic scattering, and fluorescence by a quantum object such as an atom, a molecule, or a quantum dot. The idea of controlling these processes with a resonator (a cavity, as in Chapter 7, or a resonant nanoantenna) is to use an intermediate structure which is coupled to both plane waves in vacuum and the emitters. It is possible to control spontaneous emission, namely to modify some emission features such as lifetime, emission angular pattern, or polarization. Possible applications include the design of light sources, from light-emitting diodes (LEDs) to single photon sources. It is also possible to control absorption, namely to increase the absorption cross section or to modify the absorption spectrum. Applications include detector design, photovoltaic energy conversion, molecular detection, and visual-aspect control.

Engineering the coupling of waves to subwavelength objects is something familiar to everybody. Antennas are widely used in radiowaves to efficiently couple waves propagating in open space to electrical lines, which are subsequently coupled to detectors or emitters. The wires are subwavelength in the radiowave regime, so that antennas confine electromagnetic (EM) waves well below the diffraction limit, something that cannot be achieved with lenses and mirrors. Cavities are very familiar in acoustics. Musical instruments such as guitars or pianos have sources (the strings) which need cavities to be efficiently coupled to acoustic plane waves in the air. The challenge is to bring these concepts down to the nanoscale, so that light emission by elementary emitters such as quantum dots (QDs) or molecules can be controlled.

In order to understand the operating principle, let us take a look at a guitar. A string hardly emits acoustic waves without a cavity. A guitar string is attached to the bridge, which is a piece of wood attached to the guitar cavity. At the bridge, the string vibration efficiently excites the guitar cavity. Hence, the string energy is transferred to the cavity, which in turn can efficiently radiate acoustic waves.

Let us now consider a two-level system radiating in vacuum. It is possible to modify its emission by coupling it to an antenna. In what follows, we use the terms antenna or resonator for any structure sustaining a mode, such as a metallic antenna, a dielectric cavity, or a waveguide. In this broad sense, a guitar is an antenna. An important task is to make sure that a two-level system transfers its energy to the antenna. This requires

Introduction to Nanophotonics. Henri Benisty, Jean-Jacques Greffet, and Philippe Lalanne, Oxford University Press.
 DOI: 10.1093/oso/9780198786139.003.0019

a large coupling term $\mathbf{d} \cdot \mathbf{E}_{ant}$, where $\mathbf{d}$ is the transition dipole moment and $\mathbf{E}_{ant}$ is the electric field of the resonator mode. Hence, the two-level system should be located at a hot spot of the antenna mode which plays the role of the bridge in the guitar. It is useful to introduce the branching ratio β describing the fraction of power transferred by a two-level system to the antenna. By introducing the energy decay rate γ_0 in vacuum and the decay rate $\gamma = \gamma_0 + \gamma_{ant}$ in the presence of the antenna, the fraction of energy coupled to the antenna is given by γ_{ant}/γ. The second part of the control of emission is the coupling between the antenna and the surrounding medium. In the case of the guitar, a large vibrating surface can radiate efficiently. Similarly, antennas and cavities need to be much larger than a two-level system to radiate efficiently. By designing their structure, the angular and polarization dependence of the emission can be controlled.

In this chapter, we use a unified approach to discuss any type of resonator through the concept of the Green tensor introduced in Chapter 12. While this common vision is now widely shared, the connection between cavity quantum electrodynamics, near-field optics and radiowave antennas was not trivial in the 1990s. Section 19.2 reviews different models for dealing with spontaneous emission with the aim of clarifying the connection between classical and quantum points of view and pointing out what is classical and what is quantum. We then review how to modify the local density of states (LDOS) using dielectrics, interfaces, cavities, and antennas, with particular attention to the quenching process in the case of lossy antennas. We also discuss the regime of Rabi oscillations where the energy of an initially excited atom is periodically exchanged between the macroscopic resonator and the atom (Haroche and Raimond, 2006). After this analysis of spontaneous emission, we discuss different light–matter interaction processes such as Raman scattering and fluorescence in both the linear and saturation regimes. Finally, we discuss how to optimize absorption and introduce the concept of impedance for an optical nanoantenna. Reviews on optical antennas can be found in refs (Bharadwaj et al., 2009; Biagioni et al., 2012; Agio and Alu, 2013); see also a pioneering paper by D. Pohl (Pohl, 1999).

19.2 Spontaneous emission

The goal of this section is to discuss how the presence of a resonator modifies the spontaneous emission of a two-level system characterized by an electric dipolar transition. We will compare the classical antenna point of view and the quantum point of view for radiation emission. We will specifically discuss what is quantum and what is classical in the spontaneous emission process. We start by reviewing different points of view to analyse spontaneous emission in vacuum before analysing the role of an antenna.

19.2.1 Decay rate of a classical oscillator

We start by establishing the connection between the power radiated in vacuum and the decay rate of the energy of the oscillator. We model the emitter using a harmonic oscillator with position $x(t) = x_0(t)\cos(\omega_0 t)$, where $x_0(t)$ is a slowly varying amplitude, velocity $v(t) = \dot{x}(t)$, and electric dipole moment $-ex(t)$. The oscillator has a resonant frequency ω_0 and a mass m. The instantaneous energy is given by

$$U = \frac{1}{2}mv^2(t) + \frac{1}{2}m\omega_0^2 x^2(t). \tag{19.1}$$

The amplitude $x_0(t)$ is a slowly decaying function of time due to radiative losses. Assuming a slow amplitude decay $\dot{x_0} \ll \omega_0 x_0$, we can approximate the energy by

$$U = \frac{1}{2} m \omega_0^2 x_0^2(t). \tag{19.2}$$

We can write an energy-conservation equation where P_R stands for the Larmor formula of the power radiated by an electric dipole (see Eq. (2.48)):

$$\frac{\mathrm{d}U}{\mathrm{d}t} = -P_R(t) = -\frac{\omega_0^4 e^2 x_0^2(t)}{12\pi\varepsilon_0 c^3} = -\gamma_{cl} U, \tag{19.3}$$

where the classical decay rate γ_{cl} is given by

$$\gamma_{cl} = \frac{e^2 \omega_0^2}{6\pi\varepsilon_0 c^3 m}. \tag{19.4}$$

As expected, the larger the dipole moment, the larger the radiated power and the decay rate.

19.2.2 Quantum spontaneous-emission decay rate in vacuum

The full quantum calculation of the decay rate in vacuum gives a very different result (Loudon, 2000) with a different frequency dependence:

$$\gamma_0 = \frac{|d_{12}|^2 \omega^3}{3\pi\varepsilon_0 c^3 \hbar}, \tag{19.5}$$

where d_{12} is the matrix element of the dipole operator and $2\pi\hbar$ is Planck's constant. The presence of $\hbar$ stresses the quantum nature of this result. It can be shown using the Wigner–Weisskopf method that the decay is exponential (Novotny and Hecht, 2012). The decay rate of this exponential can be found using the Fermi Golden Rule (Loudon, 2000):

$$\gamma_0 = \frac{2\pi}{\hbar^2} |\hat{W}_{if}|^2 g(\omega), \tag{19.6}$$

where $\hat{W}_{if} = \langle f | \hat{d} \cdot \hat{E} | i \rangle$ is the matrix element of the interaction Hamiltonian between the initial and the final states, and $g(\omega)\mathrm{d}\omega$ is the number of EM states in the frequency interval $\mathrm{d}\omega$ with the electric field parallel to the dipole moment. Let us stress that an EM state is nothing but a mode of Maxwell equations (e.g. a mode in a cavity or waveguide). In vacuum, the states or modes are plane waves characterized by a frequency and a wavevector. The projected density of states (DOS) is given by[1]

$$g(\omega) = V \frac{\omega^2}{3\pi^2 c^3}. \tag{19.7}$$

[1] In k-space, the volume corresponding to EM modes with frequency in the interval $[\omega, \omega + \mathrm{d}\omega]$ is $4\pi k^2 \mathrm{d}k$, where $k = \omega/c$. Using the periodic boundary conditions for a cubic volume with side L, it is shown that the volume associated with a single mode is given by $(2\pi/L)^3 = 8\pi^3/V$. A factor 2 needs to be included to account for the two polarizations for each plane wave. Finally, a factor 1/3 is included to account for the fact that a linear electric dipole oriented along the z-axis couples only to the z-component of the electric field. This is called the projected DOS.

Using the quantized EM field with amplitude $\sqrt{\hbar\omega/2\varepsilon_0 V}$, it is found that the interaction Hamiltonian matrix element $\hat{W}_{if}$ is given by:

$$|\hat{W}_{if}|^2 = \frac{|d_{12}|^2\ \hbar\omega}{2\varepsilon_0 V}. \tag{19.8}$$

Upon inserting Eqs (19.7) and (19.8) into Eq. (19.6), we obtain the radiative decay rate in vacuum, Eq. (19.5). Although the quantum result is very different from the classical one, it is interesting to note that it is possible to recover the quantum result by simply replacing two terms in the classical calculation by their quantum counterpart. The first term is the classical energy $U = \frac{1}{2}m\omega_0^2 x_0^2$, which can be replaced by $\hbar\omega$, the second term is the classical dipole moment ex_0, which can be replaced by $2d_{12}$.

19.2.3 Classical radiation in vacuum: Radiation reaction work

In this section, we establish a connection between the power radiated and the power exchanged between the dipole and the radiation reaction field, namely the field radiated by a source on itself. We start by considering the power radiated by a classical dipole in vacuum. We consider a dipole source inside a volume V enclosed in a surface A (see Fig. 19.1a, b). The integral form of the energy conservation Eq. (1.59) yields:

$$\int_V \frac{\partial u}{\partial t}\,\mathrm{d}^3\mathbf{r} = -\int_V \mathbf{j}(\mathbf{r})\cdot\mathbf{E}(\mathbf{r})\,\mathrm{d}^3\mathbf{r} - \int_A \mathbf{S}\cdot\mathrm{d}\mathbf{A}, \tag{19.9}$$

where u is the EM energy density, $\mathbf{j}$ is the current density, and $\mathbf{E}$ is the electric field. We now consider a monochromatic point-like dipole source with dipole moment

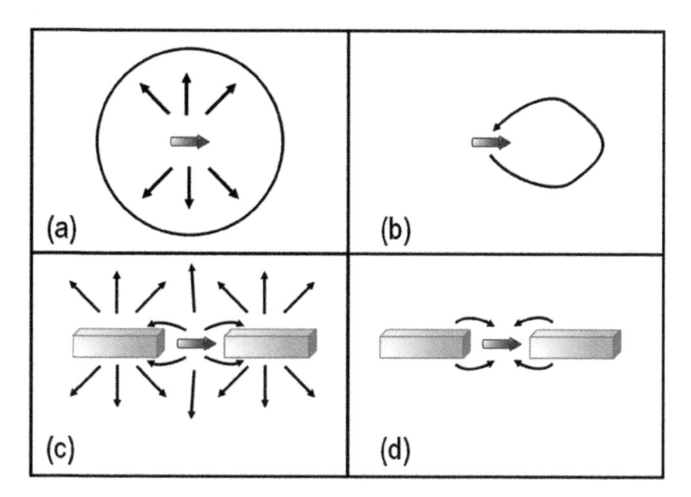

Fig. 19.1 Illustration of the energy conservation: the work done by the radiation reaction (b and d) is equal to the power radiated (a and c). (a) A dipole radiating in vacuum. The black lines symbolically represent the lines of the Poynting vector. (b) The red line symbolically represents the field produced by the dipole acting on itself. (c) The black lines represent the flow of energy radiated directly or through the antenna. (d) the red line symbolically represents the radiation reaction produced by the antenna on the dipole.

$\mathbf{p}$ in stationary regime such that $\mathbf{j}(\mathbf{r}) = -i\omega\mathbf{p}\delta(\mathbf{r}-\mathbf{r}')$. We can now compute the time-averaged energy conservation relation using complex amplitudes:

$$\frac{1}{2}\mathrm{Re}\left[i\omega\mathbf{p}\cdot\mathbf{E}^*\right] = \int_A \langle\mathbf{S}\rangle\cdot\mathrm{d}\mathbf{A} = P_R, \tag{19.10}$$

where we have introduced the radiated power P_R. The term on the left involves the electromagnetic field generated by the dipole at its own position, namely the radiation reaction. The electric field generated by the dipole source can be cast in the form $\mathbf{E}(\mathbf{r}) = \mu_0\omega^2 \overleftrightarrow{G}^{(0)}(\mathbf{r},\mathbf{r}')\mathbf{p}$, where we have introduced the vacuum Green tensor (Van Bladel, 1991; Novotny and Hecht, 2012):

$$\overleftrightarrow{G}^{(0)}(\mathbf{r},\mathbf{r}') = PV[\overleftrightarrow{I} + \frac{1}{k_0^2}\nabla\nabla]\frac{\exp[ik_0R]}{4\pi R} - \frac{\overleftrightarrow{I}}{3k_0^2}\delta(\mathbf{r}-\mathbf{r}'), \tag{19.11}$$

and where PV stands for the principal value of the integral and $k_0 = \omega/c$. Using Eq. (19.10), the power generated by a dipole $\mathbf{p} = p\mathbf{e}_z$ can be cast in the form

$$P_R^{(0)} = \frac{\mu_0\omega^3}{2}\mathrm{Im}\left[G_{zz}^{(0)}(\mathbf{r},\mathbf{r})\right]|\mathbf{p}_z|^2. \tag{19.12}$$

Although the real part of $G_{zz}^{(0)}$ diverges, its imaginary part is well defined and takes the value $\omega/(6\pi c)$, so that we recover the usual Larmor formula for the power radiated by a dipole:

$$P_R^{(0)} = \frac{\mu_0\omega^3}{2}\frac{\omega}{6\pi c}|\mathbf{p}_z|^2 = \frac{1}{4\pi\varepsilon_0}\frac{\omega^4|\mathbf{p}|^2}{3c^3}. \tag{19.13}$$

19.2.4 The Lorentz oscillator model accounting for radiation reaction

In the previous section, we showed that the emitted power can be viewed as the work done by the radiation reaction. In what follows, we adopt an alternative point of view to illustrate the role of the radiation reaction. Here, we write a dynamic equation for the dipole moment in vacuum. Let us consider for the sake of simplicity that the dipole consists of a charge $-e$ oscillating along the axis $0z$. We assume that the forces due to the atom are modelled by a parabolic potential $m\omega_0^{*2}z^2/2$. Furthermore, there is an electric field produced by the charge itself (namely the radiation reaction), given by $\mu_0\omega^2 G_{zz}^{(0)}(\mathbf{r},\mathbf{r})(ez)$. Newton's equation yields:

$$-m\omega^2 z = -m\omega_0^{*2}z + e\mu_0\omega^2 G_{zz}^{(0)}(\mathbf{r},\mathbf{r})(ez). \tag{19.14}$$

It is seen that the real part of the Green tensor provides a correction term of the resonance frequency. The eigenfrequency is a solution of the equation

$$\omega^2 - \omega_0^2 + i\frac{e^2\mu_0\omega^2}{m}\mathrm{Im}[G_{zz}^{(0)}] = 0, \tag{19.15}$$

where we have introduced the notation ω_0^2 corresponding to the observed resonance frequency, which accounts for the radiation reaction. It is seen that accounting for the radiation reaction introduces both a frequency shift and an imaginary part describing the radiation damping. The frequency shift is the Lamb shift (Busch et al., 2000).[2] The eigenfrequency can be approximated by

$$\omega \approx \omega_0 - i\frac{\gamma_{cl}}{2}, \tag{19.16}$$

where we recover the classical decay rate,

$$\gamma_{cl} = \frac{e^2\mu_0\omega_0}{m}\mathrm{Im}[G^{(0)}_{zz}] = \frac{e^2\omega_0^2}{6\pi m\varepsilon_0 c^3}. \tag{19.17}$$

19.2.5 Classical radiation in the presence of a resonator

In this subsection, we consider the power radiated by a classical dipole source with dipole moment $\mathbf{p}$ *in the presence of a resonator*. What is called a resonator here can be any object made of any arbitrary material. It could be a cavity, a nanoantenna as shown in Fig. 19.1, or any complex medium. The dipole and the resonator are located in a volume V. The dipole drives a current density $\mathbf{j}_{ant}(\mathbf{r}')$ in the resonator which leads to both radiation and a material dissipation rate $\mathbf{j}_{ant}(\mathbf{r}')\cdot\mathbf{E}(\mathbf{r}')$, where $\mathbf{E}(\mathbf{r}')$ is the electric field induced in the resonator. We can repeat the analysis of the previous section and find a spectral shift and a decay rate given by the Green tensor G_{zz} instead of $G^{(0)}_{zz}$. A quantum calculation of these two effects can be found in ref. (Wylie and Sipe, 1984).

We can also write the energy conservation in the form

$$\frac{1}{2}\mathrm{Re}\left[i\omega\mathbf{p}\cdot\mathbf{E}^*\right] = \int_V \mathbf{j}_{ant}(\mathbf{r})\cdot\mathbf{E}(\mathbf{r})\,\mathrm{d}^3\mathbf{r} + \int_A \mathbf{S}\cdot\mathrm{d}\mathbf{A}. \tag{19.18}$$

This equation shows that the power delivered by the dipole is equal to the sum of the radiated power P_R and a non-radiative power dissipated in the resonator by losses P_{NR}. It is thus seen that the energy conservation can now be written in the form

$$\frac{\mu_0\omega^3}{2}\mathrm{Im}[G_{zz}(\mathbf{r},\mathbf{r})]\,|\mathbf{p}|^2 = P_{NR} + P_R. \tag{19.19}$$

It is important to note that the Green tensor is no longer the vacuum Green tensor. It provides the field produced by the dipole in the presence of the resonator:

$$\mathbf{E}(\mathbf{r}) = \mu_0\omega^2 \overleftrightarrow{G}(\mathbf{r},\mathbf{r}')\mathbf{p}. \tag{19.20}$$

In particular, the radiation reaction includes a term that is the field radiated by the resonator back onto the dipole source. Hence, the only formal modification to the previous classical calculation is to replace $G^{(0)}$ by G. The decay rate is thus proportional to $\mathrm{Im}[G_{zz}]$. The decay rate γ_{ant} in the presence of the antenna is given by:

[2] When using a point-like model for the electron, it is seen that the real part of the Green tensor diverges. We can ignore this divergence, which is a consequence of the point-like assumption, and hereafter use the frequency in vacuum. It is of interest to realize that modifying the environment with cavities and resonators leads to a modification of the Green tensor and therefore a further shift in the frequency which can be observed.

$$\frac{\gamma_{ant}}{\gamma_0} = \frac{\text{Im}[G_{zz}]}{\text{Im}[G_{zz}^{(0)}]} = \frac{P_R + P_{NR}}{P_R^{(0)}}. \tag{19.21}$$

Although the spontaneous-emission decay rate depends on $\hbar$ and on the quantum electric-dipole-moment matrix element, the spontaneous-emission decay rate γ_{ant} in an arbitrary environment is simply the vacuum decay rate γ_0 in vacuum multiplied by a correction factor to account for the modification of the radiation reaction. This is simply given by the ratio of the imaginary part of the Green tensor or the ratio of the power dissipated by a classical dipole in a given environment $P_R + P_{NR}$ and in vacuum $P_R^{(0)}$.

Using the quantum point of view, we can rephrase this statement by invoking the modification of the DOS. However, our previous discussion based on the radiation reaction clearly pointed out that the decay rate depends on the exact position of the emitter. In terms of density of modes, this requires us to use the concept of *local* density of states (LDOS), $\rho(\mathbf{r}, \omega)$, already introduced in Chapter 12. Let us remind the reader of the origin of the spatial dependence of the DOS using the example of the spatial structure of a mode in a gap between two mirrors forming a cavity. The modes have nodes and antinodes. The amplitude of a mode excited by a dipole is given by Eq. (12.72) and is proportional to the scalar product of the dipole moment and the mode field. It is thus maximum at an antinode and null at a node.

Finally, we stress that the decay rate can be increased either by the increase of the radiated power or by the additional decay channels due to losses in the antenna, as shown in Eq. (19.13). The calculation of the LDOS based on the imaginary part of the Green tensor takes into account the non-radiative decay or, in other words, the coupling to non-radiative modes. These modes can be neglected for microwave or dielectric cavities but play a key role for plasmonic nanoantennas, as will be discussed later.

19.2.6 When quantum physics meets classical physics

We now compare the classical and quantum approaches. The connection between the two points of view can be established in a simple manner. In vacuum, we have:

$$\frac{\omega^2}{\pi^2 c^3} = \frac{6\omega}{\pi c^2}\text{Im}[G_{zz}^{(0)}(\mathbf{r}, \mathbf{r}')], \tag{19.22}$$

where the term on the left-hand side was introduced as a density of EM states and the term on the right was introduced as the radiation reaction. If we now consider a more general situation than vacuum by introducing resonators, the radiation reaction is still well defined. It turns out that it can be shown (Joulain et al., 2003) that the LDOS $\rho(\mathbf{r}, \omega)$ is proportional to the imaginary part of the trace of the Green tensor:

$$\rho(\mathbf{r}, \omega) = \frac{2\omega}{\pi c^2}\text{Im}[\text{tr}\, \overleftrightarrow{G}(\mathbf{r}, \mathbf{r})]. \tag{19.23}$$

This formula is a general formula for the LDOS in the presence of lossy objects (Joulain et al., 2003).[3] When dealing with spontaneous emission, we restrict the LDOS to the

[3] The formula is valid in a non-lossy medium close to lossy media but is not valid inside a lossy local medium. The formula predicts an unphysical divergence of the DOS. The origin of the divergence

axis parallel to the dipole moment to obtain the projected LDOS. Inserting this form in the spontaneous emission decay rate given in Eq. (19.6), we obtain:

$$\gamma_R = \frac{2\pi}{\hbar^2}|\hat{W}_{if}|^2 \frac{2\omega}{\pi c^2}\text{Im}[G_{zz}(\mathbf{r},\mathbf{r})]. \tag{19.24}$$

It is seen that the structure of the decay rate is given by the product of a quantum term $(\frac{2\pi}{\hbar^2}|\hat{W}_{if}|^2)$ and the LDOS, which is a classical quantity. Upon comparing the decay rate with antenna and in vacuum, again we recover the simple result:

$$\gamma_{ant} = \gamma_0 \frac{\text{Im}[G_{zz}]}{\text{Im}[G_{zz}^{(0)}]}. \tag{19.25}$$

To summarize, the classical point of view identifies the power radiated with the power due to the radiation reaction, which is proportional to $\text{Im}[G_{zz}]$. The quantum point of view shows that the decay rate is proportional to the DOS, which is proportional to $\text{Im}[G_{zz}]$. At this point, we emphasize that both the classical and the quantum approaches show that the spontaneous decay rate is proportional to $\text{Im}[G_{zz}]$. The key remark is that modifying the environment of an emitter amounts to modifying the LDOS and the radiation reaction, which are two sides of the same coin. It follows that a classical calculation of the modification of the DOS allows the modification of the spontaneous decay rate to be predicted.

19.3 Controlling the spontaneous decay rate by modifying the environment

We saw in the previous section that the lifetime of an emitter can be modified by changing the environment. In this section, we discuss different types of environment: modification of the bulk refractive index, presence of an interface, the role of losses, presence of a cavity, and presence of an antenna.

19.3.1 Modification of the spontaneous decay rate in a dielectric

We first consider the modification of the decay rate γ_R of an emitter located in a homogeneous transparent dielectric with a real refractive index n, as compared to the decay rate in vacuum γ_0. The first obvious consequence of the presence of a refractive index is that light speed is changed from c to $v = c/n$, so that the DOS is multiplied by n^3 according to Eq. (19.7). However, this does not mean that the spontaneous emission rate has been multiplied by the same factor. Indeed, the vacuum permittivity ε_0 is replaced by $\varepsilon_0\varepsilon_r = n^2\varepsilon_0$. Finally, it is seen from Eq. (19.5) that the decay rate in a dielectric is given by $n\gamma_0$.

We now use the radiation reaction point of view to address the same issue. We have seen in Eq. (19.12) that the power radiated by a classical dipole with a fixed dipole moment is proportional to the imaginary part of the Green tensor. In a homogeneous

is a consequence of the absence of cutoff frequencies for wavevectors when using a local model of the permittivity. This is similar to the divergence of the number of phonon modes in a crystal which is avoided by introducing the Debye cutoff frequency.

dielectric, the imaginary part of the Green tensor is given by $\frac{n\omega}{6\pi c}$. Hence, we see from Eq. (19.12) that the power radiated by a classical dipole in a lossless dielectric with refractive index n is multiplied by a factor n.

19.3.2 Modification of the spontaneous decay rate close to an interface

We now discuss the behaviour of the decay rate in the vicinity of an interface separating two homogeneous isotropic materials. This effect was first observed experimentally by Drexhage (Drexhage et al., 1968) for europium ions located above a silver mirror. The distance was varied by inserting an adjustable number of organic monolayers. A seminal contribution by Chance, Prock, and Silbey (Chance et al., 1974) showed that this result could be explained using a classical dipole model. The reason for the success of this classical model was just discussed in the previous sections: radiation reaction and DOS are classical concepts. Here, we will discuss the case of an interface between two dielectrics. We start by displaying the result in Fig. 19.2 for a dielectric interface. This question was studied extensively in ref. (Lukosz and Kunz, 1977).

It is seen that the decay rate oscillates as the distance to the interface changes. This behaviour can be understood in terms of interferences. As the radiated field is the sum of the field emitted upwards and the field emitted downwards and reflected on the substrate, it is seen that the total field radiated depends on the distance to the interface. An alternative and equivalent interpretation is to consider the structure of the modes in the presence of an interface. The modes in vacuum are plane waves. In the presence of interfaces, the modes are the sum of an incident and a reflected wave on one side of the interface and a transmitted wave on the other side. The interference between the incident and reflected waves produces an oscillating field with a period that depends on the z-component of the wavevector. By adding all possible angles of incidence, the visibility is smeared out except close to the interface, where the path

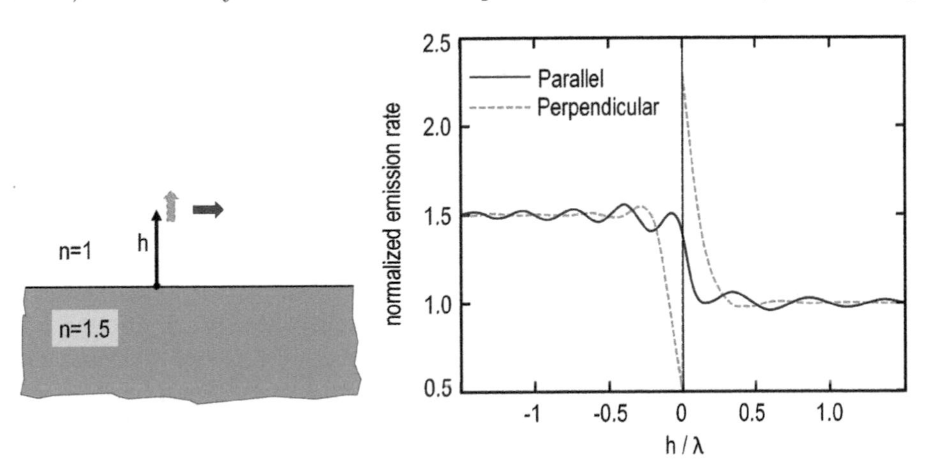

Fig. 19.2 Decay rate close to an interface. The decay rate normalized by its value in vacuum has an asymptotic value of 1.5 in the dielectric away from the interface. Note the very different behaviour depending on the dipole orientation. For both dipole orientations, the presence of the interface is responsible for oscillations.

difference is null. Indeed, at the interface, the boundary condition imposes the phase of the interference pattern. For instance, for a perfect conductor, the field has to be zero on the interface whatever the angle of incidence, so that all the interference patterns have a node at the interface.

We observe in Fig. 19.2 that the decay rate close to a dielectric interface presents a clear maximum in the vacuum for a dipole normal to the interface. This variation is observed close to the interface. It seems that the emitters in vacuum benefit from the larger LDOS of the dielectric material, while the LDOS in the dielectric decays close to the interface. The decay of the LDOS in the dielectric can be analysed as a consequence of the interferences between an incident wave and the reflected wave. In total internal reflection, the z-component of the reflected field changes sign, so that the LDOS decreases as compared to the bulk in dielectric. We also observe that the LDOS increases in vacuum. This is due to the modes existing in the dielectric and totally reflected at the interface. These modes have an evanescent tail which extends in vacuum, providing an additional contribution to the LDOS in vacuum. This mechanism also explains why the enhancement is confined close to the interface at a distance roughly given by $\lambda/2\pi n$.

To understand why the LDOS is larger in vacuum than in the dielectric, we note that the normal component of $\mathbf{D}$ is continuous across the interface, so that the normal electric field is larger by a factor ε in vacuum. The ratio of LDOS is proportional to the ratio of intensity and is thus given by $\varepsilon^2 = n^4$. In the example shown in Fig. 19.2, a factor 5.06 is predicted by this simple argument in good agreement with the exact numerical calculation. By contrast, for parallel polarization, the electric field is continuous, so that no discontinuity of the LDOS is expected.

We now consider the case of an interface between a metal and vacuum. The result is shown in Fig. 19.3a in the vacuum above the interface. We observe again the oscillatory behaviour due to interferences. It is seen that for the field polarized perpendicular to the interface, the LDOS diverges when the distance to the interface goes to zero. This large LDOS in the near field is associated with the excitation of non-radiative modes in the metal. These non-radiative modes decay into heat. We discuss this topic in the following section.

19.3.3 Radiation quenching

One of the major limitations of nanoantennas is the competition between losses and radiation. The goal of this section is to provide a physical picture of the mechanisms at work in this process. We start with an analysis in the framework of classical electrodynamics. We consider a monochromatic electric dipole located above a metallic interface at a distance d. The electric field produced by the electric dipole at a distance r varies, as $1/r^3$ when $r \ll \lambda$. If a subwavelength electric dipole is brought at distances smaller than λ from an interface-separating vacuum from a lossy material, it induces large current densities, which results in large losses. This is illustrated in Fig. 19.4. These losses may become much larger than the power radiated, so that the radiative efficiency drops to zero. This phenomenon is called quenching and has been known for many years. Although the previous discussion is made in the framework of classical electrodynamics, it is valid for quantum emitters such as atoms or QDs (Ford and

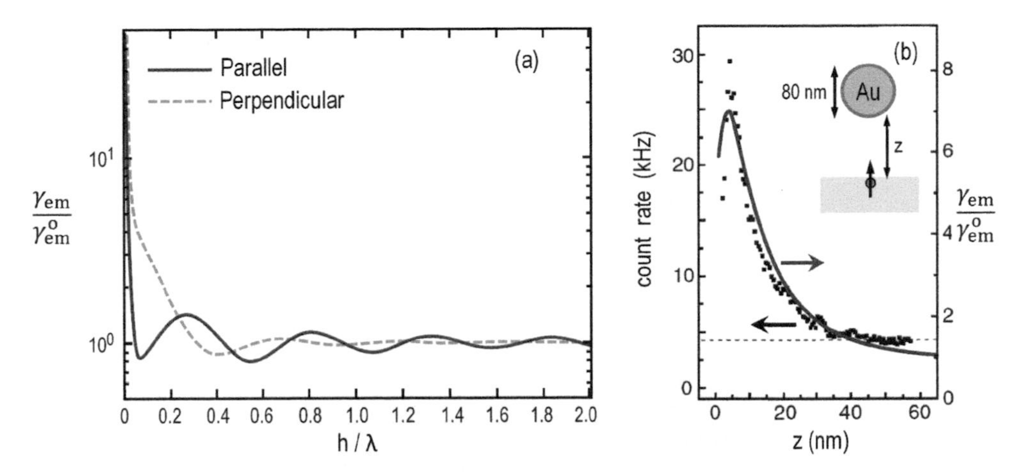

Fig. 19.3 Influence of a metal on the spontaneous emission. (a) Theoretical decay rate γ_{em} of an electric dipole normalized by the decay rate $\gamma^{\circ}_{\text{em}}$ in vacuum. h is the distance in μm to a gold film. It is seen that the decay rate increases at small distances due to the non-radiative energy transfer to the surface. (b) Fluorescence signal of a molecule at a distance d from a gold nanosphere. It is seen that the signal decays for a distance smaller than 5 nm, indicating non-radiative energy transfer to the metal nanoparticle. The fluorescence is quenched. Fig. (b) is reprinted with permission from ref. (Anger et al., 2006). ©2006 American Physical Society.

Weber, 1984; Chance et al., 1974). Non-radiative losses in the metal are due to interband absorption, electron–hole excitation, and electron-scattering losses. The trade-off between quenching and fluorescence is well illustrated in Fig. 19.3b, where the fluorescence enhancement of a molecule as a function of the distance to a gold nanosphere is plotted. This result suggests that fluorophores should be at distances over 5 nm from the surface to prevent quenching. To understand this phenomenon, we consider the emission by a vertical dipole located above a silver–vacuum interface. We first analyse the decay rate using the imaginary part of the Green tensor, which accounts for the LDOS. To this aim, we simply derive the asymptotic behaviour of the LDOS for small distances (Joulain et al., 2003):

$$\frac{\text{Im}[G_{zz}]}{\text{Im}[G^{(0)}_{zz}]} = \frac{\text{Im}(\varepsilon)}{|\varepsilon + 1|^2} \frac{3}{4(k_0 d)^3}. \tag{19.26}$$

This result shows clearly that the divergence of the LDOS is proportional to $\text{Im}(\varepsilon)$ and therefore associated with losses in the metal. From this point of view, the large decay is due to the coupling of the dipole to the large number of non-radiative modes existing in the metal. The physical origin of these non-radiative additional states in a metal is the presence of electrons. More generally, the increased DOS close to matter is due to the presence of electronic excitations which are coupled to the EM field. An alternative formulation of the physical mechanism is based on the classical description of losses. Fig. 19.4 displays the map of the square modulus of the electric field generated by a dipole in the metal. It is seen that the near-field component of the electric field

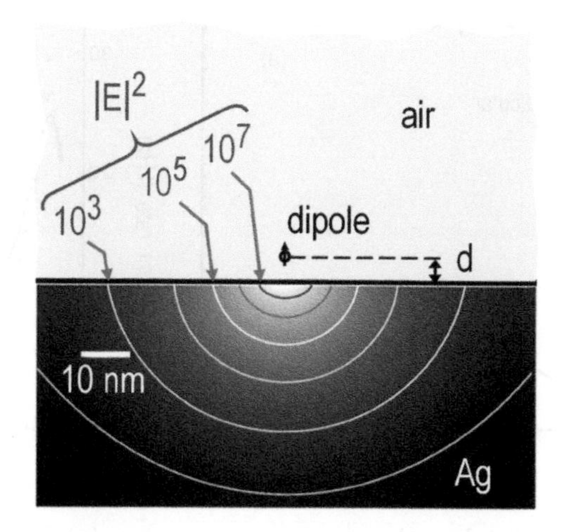

Fig. 19.4 Map of the field intensity in a silver substrate illuminated by a vertical dipole at 632.8nm located at a distance $d = 5$nm above the interface. The Joule effect is induced in the metal.

generates large Joule losses. The $1/d^3$ dependence can be understood as follows: the near-field decays as $1/r^3$, so that Joule losses vary as $1/r^6$ and after integration over the metal, the total losses vary as $1/d^3$.

So far, we have outlined that the total energy provided by a dipole above a vacuum–metal interface is given by the LDOS and illustrated the role of losses in the near-field region. However, part of the power is radiated in propagating waves and surface plasmons. It is of interest to analyse how the total power P_{tot} transferred from the dipole to its environment is split into radiation, plasmon excitation, and losses in the metal. In order to compare the power transferred into the different terms, we proceed as follows. The total power P_{tot} transferred from the dipole to its environment can be derived from the *exact* form of the LDOS. This is similar to the discussion of the extinction by a small particle in Chapter 14. The losses of a small particle are given by the static limit of the polarizability, but retardation needs to be included to account for radiative losses. The power transferred to the plasmon P_{sp} can be derived by computing the energy transfer between the dipole and the plasmon field using the exact form of the plasmon field at the origin. The total radiated power P_R can be derived by integrating the flux of the Poynting vector across a plane at constant z. The quantity $P_{tot} - P_{sp} - P_R$ is the rest of the power dissipated in the metal by the non-radiative modes. By normalizing these powers to the power emitted by a dipole in vacuum, we obtain the contributions to the total decay rate, $\gamma_{tot} = P_{tot}/P^{(0)}, \gamma_{sp} = P_{sp}/P^{(0)}, \gamma_R = P_R/P^{(0)}$. These three quantities are represented in Fig. 19.5.

We start looking at large distances. In that case, the dipole is slightly perturbed by the mirror and its decay rate is close to its value in vacuum. It is seen that the radiative decay rate γ_R accounts for most of the total decay rate. The dipole is too far away from the interface to be able to excite the plasmon and the other evanescent modes. We remind that the coupling to a mode involves an overlap integral of the mode and

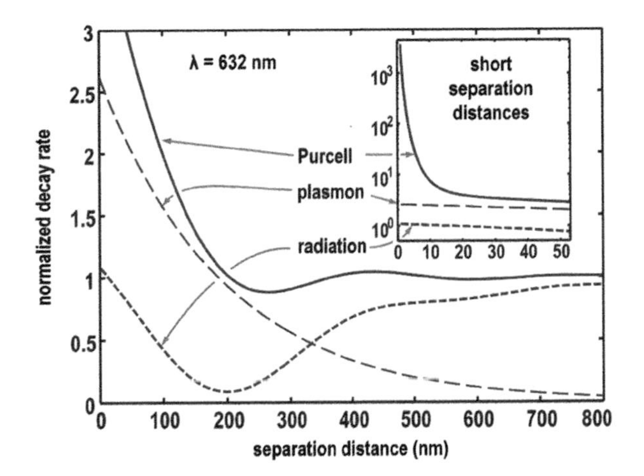

Fig. 19.5 Contributions of the surface plasmon and radiation to the total decay rate (denoted Purcell). The distances in the inset are in nanometers.

the source. Evanescent modes with a large wavevector k have a typical decay length given by $1/k$ as discussed in section 10.1.2, so that the number of evanescent modes that can be excited increases as the distance d decreases. This discussion in terms of modes is equivalent to the previous discussion in terms of Joule losses. It corresponds to the increase by orders of magnitude of the decay rate at distances smaller than 10 nm. It is seen in the inset that this increase does not correspond to an increase of the plasmon excitation or the radiated fields. In this regime, the emission by a dipole will be quenched.

Between the $d < 10$ nm and the $d > 600$ nm regimes, there are two regimes. Between 200 nm and 600 nm, the decay rate oscillates around one. This regime has been discussed in section 19.3.2. When the distance d becomes smaller than 200 nm, the dipole can excite the tail of the surface plasmon in vacuum and plasmon excitation gradually becomes the leading decay mechanism until 10 nm, where the number of evanescent waves becomes larger.

Note that this discussion is for silver at 632.8 nm, an energy much smaller than the energy of the asymptote of the dispersion relation of the surface plasmon corresponding to $\mathrm{Re}(\varepsilon) = -1$. For that particular frequency, the near-field decay rate is further enhanced by the excitation of the high k surface plasmons, as can be seen from the presence of the resonant term $\frac{1}{|\varepsilon+1|}$ in Eq. (19.26). Finally, we note that for distances smaller than 1 nm, non-local effects need to be taken into account (Ford and Weber, 1984).

In summary, the decay rate at a distance smaller than 8 nm from a sphere or a plane is dominated by quenching, thus preventing their use in enhancing spontaneous emission. Fortunately, it is possible to design nanostructures to support highly localized plasmonic modes, producing a very large local density at some given points. If this contribution can be made larger than the contribution of the non-radiative modes, quenching can be avoided. This is the case for dimers of nanospheres, nanocones, and patch antennas, to name a few antennas studied in the literature with large decay rate and radiation efficiencies above 50%. Further discussions can be found

(Faggiani et al., 2015; Marquier and Greffet, 2013; Koenderink, 2017). In the next section, we introduce the idea that a single mode can produce a very large *density* of states per unit volume and a given frequency interval if it is properly confined in space.

Before closing this discussion, we stress that it is not possible to write an energy-conservation equation such as $P_{tot} = P_{sp} + P_R + P_{NR}$. While this is possible for the fields

$$\mathbf{E}(\mathbf{r}) = \mathbf{E}_{sp} + \mathbf{E}_R + \mathbf{E}_{NR}, \tag{19.27}$$

it is seen that the Joule losses in the metal cannot be expressed as the integral over the metal volume of $(\omega \mathrm{Im}(\varepsilon_r)\varepsilon_0/2)[|\mathbf{E}_{sp}|^2 + |\mathbf{E}_R|^2 + |\mathbf{E}_{NR}|^2]$. It is clear that interference terms such as $(\omega \mathrm{Im}(\varepsilon_r)\varepsilon_0)\mathrm{Re}[\mathbf{E}_{sp}^* \cdot (\mathbf{E}_R + \mathbf{E}_{NR})]$, which may be either positive or negative, appear, so that the losses are not the sum of each field component. This is another consequence of the non-linearity of absorption discussed in Chapter 3.

19.3.4 Controlling $\mathrm{Im}[G_{zz}]$ with a cavity: LDOS and the Purcell factor

The first proposal to control spontaneous emission by modifying the environment was introduced in the context of nuclear magnetic resonance by Purcell in the microwave regime (Purcell, 1946). We now give a back-of-the-envelope derivation of the modification of the spontaneous emission rate given by a multiplicative factor F_P called the Purcell factor. Let us assume that there is an emitter in a single-mode cavity with a mode volume V and a decay rate κ. Although the cavity has a single mode, the presence of radiative and non-radiative losses of the cavity introduces a broadening of the cavity spectrum. If the cavity spectral width κ is much larger than the two-level width γ_R, the cavity can be considered to be a quasi-continuum to which the two-level system is coupled, so that the Fermi Golden Rule can be used. We further assume that the resonance is Lorentzian, so that the spectral density is given by the normalized function

$$g(\omega) = \frac{1}{2\pi} \frac{\kappa}{(\omega - \omega_0)^2 + \kappa^2/4}. \tag{19.28}$$

At resonance, the spectral density is $g(\omega_0) = 2/\pi\kappa$. The LDOS is now found as follows. We have one mode in a volume V with a spectral density $g(\omega_0)$. We also assume that the mode is uniform in the cavity for the sake of simplicity. It follows that the local density of EM states is given by $g(\omega_0)/V$. Eq. (19.21) shows that the acceleration factor of the decay rate, also called the Purcell factor, F_P, γ_{cav}/γ_R, is given by the ratio of LDOS in the cavity and in vacuum Eq. (19.7):

$$F_P = \frac{\gamma_{cav}}{\gamma_R} = \frac{2}{\pi\kappa V} \frac{3\pi^2 c^3}{\omega^2 n^3} = \frac{3Q}{4\pi^2} \frac{\lambda^3}{V}, \tag{19.29}$$

where $2\pi/\lambda = n\omega/c$ and the cavity quality factor (Q-factor) is defined by $Q = \frac{\omega}{\kappa}$. This is the form originally introduced by Purcell. It is seen that this factor measures the confinement of the mode in space (λ^3/V) and in the frequency domain (Q). The LDOS can thus be increased by confining the mode spectrally (large Q-factor) and spatially (small mode volume). For a dielectric microcavity, the Q-factor can be higher than 10^5. On the other hand, the ratio λ^3/V is typically on the order of 10 or less. Conversely, plasmonic resonators provide large spatial confinements with modest Q-factors.

19.3.5 Controlling $\mathrm{Im}[G_{zz}]$ with a nanoantenna

The key difference between a microcavity and a nanoantenna is that the antenna usually has a Q-factor on the order of 10 and a mode volume that is significantly subwavelength. In other words, the electric field is extremely confined in space, while the spectrum is poorly confined in the frequency domain. This makes nanoantennas very interesting for broadband emitters. The bandwidth of a radio antenna is of course an important figure of merit. It is desirable to use the same antenna while scanning different transmission channels with different carrier frequencies.

It is usually difficult to estimate a mode volume associated with the antenna. The field confinement is mostly associated with evanescent waves, whereas the field of a microcavity is associated with stationary waves in a cavity. A powerful technique has been introduced recently by Sauvan et al. (Sauvan et al., 2013) in the framework of quasinormal modes. It allows the interaction between spectrally overlapping modes to be modelled at the cost of introducing a complex volume.

On the other hand, it is possible to derive the Green tensor and thus compute the LDOS without using Eq. (19.29), so that introducing a mode volume is not necessary inasmuch as the Green tensor can be computed directly. An elementary derivation of a mode volume is made for a toy model in the case of a metallic sphere modelled by a dipole in ref. (Carminati et al., 2006).

19.3.6 Controlling $\mathrm{Im}[G_{zz}]$ with a cavity: The cavity quantum electrodynamics (CQED) point of view

In this section, we adopt the notations of the CQED community to describe the modification of spontaneous emission by the cavity. We aim to establish a link between different notations and points of view used by different communities when dealing with the same concepts. Losses are responsible for a broadening, so that, if $\kappa \gg \gamma_R$, the cavity can be modelled as a continuum of states and the Fermi Golden Rule can be used. The DOS is given by $g(\omega_0) = 2/\pi\kappa$. In CQED notations, the interaction energy $|\hat{W}_{if}|$ is denoted by $\hbar g$. It follows from Eq. (19.6) that

$$\gamma_{cav} = \frac{2\pi}{\hbar^2}(\hbar g)^2 \frac{2}{\pi\kappa} = \frac{4g^2}{\kappa}. \tag{19.30}$$

The Purcell factor in the cavity is thus given by:

$$F_P = \frac{\gamma_{cav}}{\gamma_R} = \frac{4g^2}{\gamma_R \kappa}. \tag{19.31}$$

With these notations, the Purcell factor appears as a dimensionless quantity. It is worth mentioning that the quantity $g^2/(\kappa\gamma_R)$ is often called cooperativity or single-photon cooperativity and denoted by C_1. This name was introduced (Bonifacio and Lugiato, 1978) in the context of non-linear transmission by a Fabry–Perot cavity filled with N molecules. The dimensionless parameter that describes the collective (or co-operative) interaction between the N molecules and the cavity is $C_N = Ng^2/(\kappa\gamma_R)$. Obviously, with a single emitter, there are no cooperative or collective effects, but the comparison shows that $F_P = 4C_1$, hence the name cooperativity often used. However, this number is very useful for characterizing the interaction between the cavity and the emitter.

19.3.7 Remark on the definition of the Purcell factor

Purcell introduced the spontaneous emission acceleration factor in the context of a closed microwave cavity. When considering an open cavity such as a Fabry–Perot cavity, the radiative decay of an emitter can be done either by coupling to the cavity with a decay rate γ_{cav} or by coupling to the vacuum modes with a decay rate γ_R. It follows that there are two possible definitions for what is called the 'Purcell factor'. As both of them are used depending on the authors, we clarify the two points of view. By taking the point of view that the Purcell factor measures the acceleration of the decay rate, it is then defined as the ratio of the LDOS:

$$F_{P1} = \frac{\gamma_R + \gamma_{cav}}{\gamma_R} = \frac{\mathrm{Im}(G_{zz})}{\mathrm{Im}(G^{(0)}_{zz})} = 1 + \frac{\mathrm{Im}(S_{zz})}{\mathrm{Im}(G^{(0)}_{zz})}. \tag{19.32}$$

The second definition takes the point of view that the Purcell factor measures the coupling to the cavity mode only. It is thus given by:

$$F_{P2} = \frac{\gamma_{cav}}{\gamma_R} = \frac{\mathrm{Im}(S_{zz})}{\mathrm{Im}(G^{(0)}_{zz})}. \tag{19.33}$$

This second point of view is used in CQED.

19.3.8 The Rabi oscillation regime

In this section, we give a brief introduction to the strong coupling regime in the context of a single emitter coupled to a resonator. This section is also an introduction to the terminology and notations used in cavity quantum electrodynamics. The concept of weak and strong coupling for two classical coupled oscillators has been introduced in section 7.8.4. The reader is referred to ref. (Barnes et al., 2003) for an extensive discussion in the context of light–matter interaction. Let us consider a single atom in the cavity depicted in Fig. 19.6. Here, we ask the question: will a photon trapped in the cavity be absorbed by the atom or leave the cavity without being absorbed? To answer that question, we simply use the Beer–Lambert law: the transmission through a scattering medium with thickness L is given by $\exp(-L/l_{abs})$. The absorption length is given by $l_{abs} = 1/\rho\sigma$, where ρ is the number of atoms per unit volume and σ is the absorption cross section. A photon in the cavity spends on average a time $\tau = 1/\kappa$ in the cavity, so that it travels over a distance $L = (c/n)(1/\kappa)$. The corresponding transmission is $\exp(-L/l_{abs})$. The distance normalized by the absorption length $L/l_{abs} = c/(n\kappa l_{abs})$ is called optical thickness and characterizes the absorption probability. With one atom in the cavity with volume V, we have $\rho = 1/V$, and the absorption cross section at resonance is given by $\sigma = 3\lambda^2/2\pi$.[4] The optical thickness is thus given by

[4] Here a vocabulary note may be useful. In the context of atoms, absorption means that the atom is excited. The absorbed energy will be subsequently radiated by spontaneous emission. In the case of a molecule with a more complex energy spectrum, there will be fluorescence and the emission frequency may be smaller than the exciting frequency. In the case of a particle, absorption means that the EM incident energy will be transformed into heat in the particle.

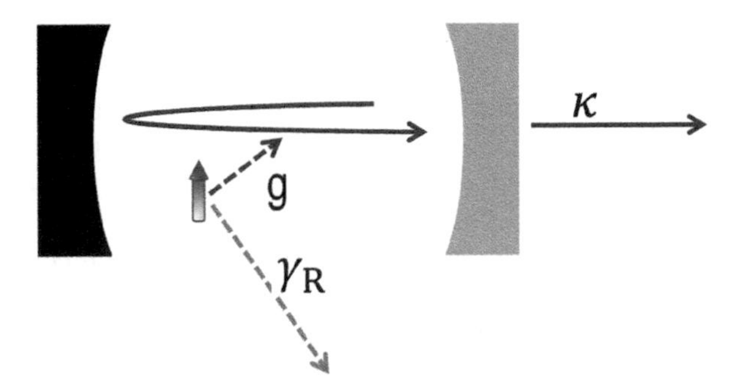

Fig. 19.6 Increasing the spontaneous emission rate with a cavity. The spontaneous emission decay rate is denoted by γ_R, the coupling rate between the two-level system and the cavity is denoted by g. The cavity decay rate is denoted by κ; it is due to the transmission of the partially reflecting mirror for a Fabry–Perot cavity. The Purcell regime corresponds to the situation where the coupling to the cavity rate $4g^2/\kappa$ is faster than the coupling to the vacuum γ_R. In the bad-cavity regime, $2g \ll \kappa$, the cavity energy is released. Note that the same picture applies to a nanoantenna.

$$\frac{c}{n\kappa}\rho\sigma = \frac{c}{n\kappa}\frac{1}{V}\frac{3\lambda^2}{2\pi} = \frac{c}{n\omega}\left(\frac{\omega}{\kappa}\right)\frac{1}{V}\frac{3\lambda^2}{2\pi} = \frac{3Q}{4\pi^2}\frac{\lambda^3}{V} = F_P = 4C_1. \tag{19.34}$$

It is seen that $4C_1$ appears to be the optical thickness of the cavity with one two-level system at resonance. With this argument, it is seen that a cooperativity C_1 larger than one means that a photon in the cavity will be absorbed by the atom before escaping the cavity. This suggests that the energy can be periodically exchanged between the atom and the cavity, a phenomenon called Rabi oscillation. It was shown in section 7.8.4 that the Rabi oscillation frequency Ω_R is given by $2g$. Hence, a qualitative argument suggests that the oscillations are observed if their period is smaller than the residence time in the cavity $1/\kappa$. Two regimes are particularly interesting. In the regime

$$\gamma_R \ll g \ll \kappa, \quad \text{and} \ \frac{g^2}{\kappa\gamma_R} \gg 1, \tag{19.35}$$

the emitter couples to the cavity with a good efficiency $\beta = \gamma_{cav}/[\gamma_{cav} + \gamma_R]$. Once the cavity is excited, the photon is released without being reabsorbed, as the cavity decay time is short: $g \ll \kappa$. This is the so-called bad-cavity limit. The corresponding regime is often called the weak-coupling regime or Purcell regime. The emission decay rate is increased and the photon is efficiently collected by the cavity. Antennas operate in this regime. Conversely, if the cavity has a large Q-factor ω/κ such that

$$\gamma_R \ll \kappa \ll g, \tag{19.36}$$

the photon is reabsorbed before leaving the cavity. The system oscillates between a photonic state and an atomic state: it is the Rabi oscillation regime. In this regime, the eigenstates of the system are a linear superposition of the excited emitter state and the excited cavity state (photon). The spectrum is characterized by a Rabi splitting given by $\Omega_R = 2g$. This splitting is visible if $\Omega_R = 2g$ is larger than the peak linewidth given by $(\gamma + \kappa)/2$ (see section 7.8.4).

The terminology strong coupling is often used in this context, albeit with different meanings. For some authors, the regime of strong coupling is the regime of Rabi oscillation $\gamma_R \ll \kappa \ll g$, whereas the regime of weak coupling is the Purcell or bad cavity regime. This choice is consistent with the discussion of Chapter 7 illustrated in Fig. 7.11. Other authors use the term 'strong coupling' if the cooperativity is large ($g^2 \gg \kappa\gamma_R$), i.e. for both the case of Rabi oscillation and the bad-cavity regime with large Purcell factor. Here, the term strong coupling is used to indicate that the cavity significantly modifies the dynamics of the emitter.

Finally, we conclude by noting that the key difference between cavities and antennas is the Q-factor. The usual situation for antennas is to operate in the bad-cavity limit. The small Q-factor is compensated by a very small mode volume. The observation of the Rabi oscillation regime with plasmonic nanoantennas is extremely challenging because the cavity decay rate is extremely fast due to metallic losses. On the other hand, it is possible to reduce the mode volume to extremely small values and therefore to enhance the coupling parameter g. Experiments have been reported with spectral-scattering signatures showing the Rabi splitting (Chikkaraddy et al., 2016; Santhosh et al., 2016).

19.4 Enhanced Raman scattering

We now address the question of the control of the Raman-scattering process using a nanoantenna (Esteban et al., 2009). Before discussing the phenomena, we need to define the setup. We consider a confocal microscope setup as depicted in Fig. 19.7 with two pinholes in front of the source and the detector. The light coming from the source S is collimated and focused on the sample by an objective. The antenna is located at the focus of the objective. The molecules are thus illuminated and scatter the light, which is collected by the same objective and then reflected by the dichroic mirror towards the

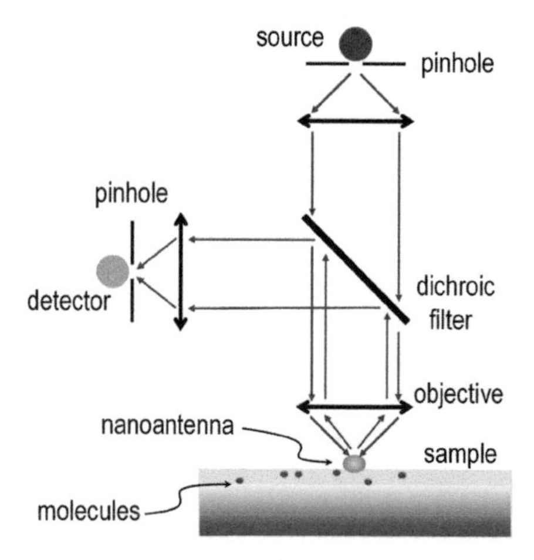

Fig. 19.7 Confocal setup for Raman scattering. The antenna is represented by an amber nanosphere, the fluorophores are represented by red dots.

detector. Note that the detector is placed behind a pinhole in a confocal setup, so that its position is the symmetric of the source position. We are interested in the analysis of the light scattered at frequency ω_2 by the molecules illuminated with a laser light at frequency ω_1. To analyse the system, we consider that the incident light is produced by a polarized point-like source with dipole moment $\mathbf{p}_S$ located at the pinhole position $\mathbf{r}_S$. It follows that the field at $\mathbf{r}$, close to the focus of the objective, can be cast in the form

$$\mathbf{E}(\mathbf{r}) = \mu_0\omega^2\mathbf{G}(\mathbf{r},\mathbf{r}_S,\omega_1)\mathbf{p}_S. \tag{19.37}$$

In the presence of the antenna, the field is modified and we denote it by

$$\mathbf{E}_{ant}(\mathbf{r}) = \mu_0\omega^2\mathbf{G}_{ant}(\mathbf{r},\mathbf{r}_S,\omega_1)\mathbf{p}_S. \tag{19.38}$$

The effect of the antenna is to enhance the electric-field amplitude by a factor denoted by $K(\omega_1)$:

$$K(\omega_1) = \frac{|\mathbf{G}_{ant}(\mathbf{r},\mathbf{r}_S,\omega_1)\mathbf{p}_S|}{|\mathbf{G}(\mathbf{r},\mathbf{r}_S,\omega_1)\mathbf{p}_S|}. \tag{19.39}$$

We stress that this field enhancement differs from the LDOS enhancement (see also Eq. (14.25)). The former deals with the value of the field in a specific illumination condition, whereas the latter is an intrinsic quantity. The former is related to $|\mathbf{G}_{ant}(\mathbf{r},\mathbf{r}_S,\omega_1)\mathbf{p}_S|$, whereas the latter is proportional to $\mathrm{Im}[G(\mathbf{r},\mathbf{r},\omega_1)]$. The field produced at the detector position by the induced dipole at frequency ω_2 due to the Raman shift is proportional to the Green tensor $\mathbf{G}_{ant}(\mathbf{r}_D,\mathbf{r},\omega_2)$. By noting that D and S are symmetrical, we can replace the Green tensor by $\mathbf{G}_{ant}(\mathbf{r}_S,\mathbf{r},\omega_2)$. We now make use of the reciprocity of the Green tensor $G_{nm}(\mathbf{r},\mathbf{r}',\omega) = G_{mn}(\mathbf{r}',\mathbf{r},\omega)$.[5] It follows that $G_{nm,ant}(\mathbf{r}_S,\mathbf{r},\omega_2) = G_{mn,ant}(\mathbf{r},\mathbf{r}_S,\omega_2)$. According to Eq. (19.39), the latter is simply proportional to the amplitude enhancement at frequency ω_2. This result demonstrates that the antenna serves to increase the amplitude of both the field illuminating the dipole and the field radiated by the dipole towards the detector. The reciprocity theorem shows that these are two sides of the same coin. Thus, we find that the field amplitude at the detector is multiplied by $K(\omega_1)K(\omega_2)$ and the intensity is multiplied by the square of this quantity. Finally, if we assume that the antenna and the microscope do not change between the two frequencies, we conclude that the presence of the antenna amounts to multiplying the Raman signal by a factor of $K^4(\omega_1)$. It is worth emphasizing that the field enhancement can be two orders of magnitude, so that the signal enhancement can be eight orders of magnitude.

19.5 Fluorescence in the stationary regime

We now study the effect of an antenna on fluorescence (Esteban et al., 2009). It cannot be summarized in a few words. It depends on the illumination: pulsed or stationary, intensity in the linear or in the saturated regime. It also depends on the nature of the

[5] This equality follows from the equality $\mathbf{p}_1 \cdot \mathbf{E}_2 = \mathbf{p}_2 \cdot \mathbf{E}_1$, derived from Maxwell equations where the electric field $\mathbf{E}_i$ is produced by the dipole $\mathbf{p}_i$ in an arbitrary environment with the only restriction being that the materials have a symmetric permittivity tensor, $\varepsilon_{ij} = \varepsilon_{ji}$. Inserting $\mathbf{E}_n(\mathbf{r}_1) = \mu_0\omega^2 G_{nm}(\mathbf{r}_1,\mathbf{r}_2,\omega)p_{2m}$, we find that reciprocity implies $G_{nm}(\mathbf{r}_2,\mathbf{r}_1,\omega) = G_{mn}(\mathbf{r}_1,\mathbf{r}_2,\omega)$.

emitter: the effect on a two-level system or a dye molecule is not the same. The effect of the antenna may depend on the incident intensity, as there may be saturation effects. The aim of this section is to give a simple analysis of the dependence of the fluorescence signal on the parameters of the antenna.

19.5.1 Fluorescence signal

In the absence of an antenna, the fluorescence signal is proportional to the radiative rate of the molecule and the number of molecules N_2 in the excited state. We denote by γ_R the radiative rate of the molecule without antenna and by Γ_R the radiative rate in the presence of the antenna. The signals with and without antenna S_{ant} and S_0 are thus given by

$$S_{ant} = C f_{coll}^{ant} \Gamma_R N_2 \;\; ; \;\; S_0 = C f_{coll} \, \gamma_R N_2, \tag{19.40}$$

where C is a constant that accounts for the detector response and f_{coll} is a number smaller than one that accounts for the fraction of collected radiation (collection efficiency). We now need to write population equations to find the number N_2 of molecules in the excited state.

19.5.2 Population equations

Let us begin by introducing a simplified level structure for the fluorophore. The energy scale is depicted in Fig. 19.8. The fundamental level is denoted by zero. A laser pump at frequency ω_p illuminates the system and excites the fluorophore to level 3. A fast non-radiative decay process then populates level 2. The fluorescent radiative decay takes place between level 2 and level 1. Finally, level 1 has a fast non-radiative relaxation towards level 0. If we consider that the non-radiative decay of level 3 and 1 are very fast, it is possible to consider that their populations are negligible. The molecule has a radiative decay rate γ_R and a non-radiative decay rate γ_{NR}, leading to an intrinsic quantum yield η_{int}:

$$\eta_{int} = \frac{\gamma_R}{\gamma_R + \gamma_{NR}}. \tag{19.41}$$

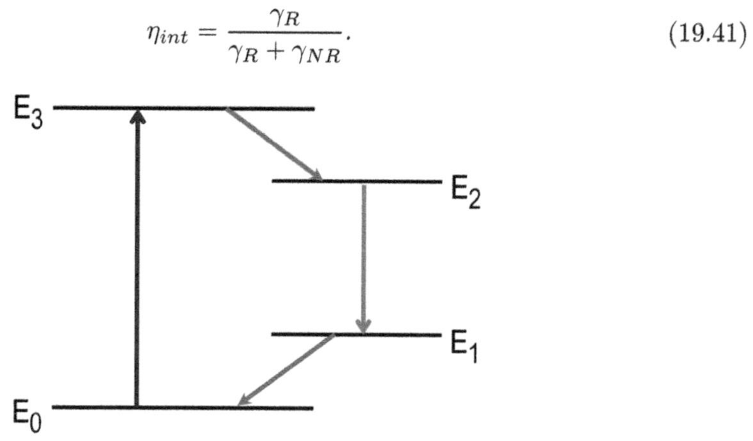

Fig. 19.8 Scheme of the fluorophore energy levels.

With obvious notations, we can then write that the total number of fluorophores N satisfies:

$$N = N_0 + N_2. \tag{19.42}$$

In the stationary regime, the rate of variation of N_0 is given by

$$\frac{dN_0}{dt} = \Gamma_{tot} N_2 + B\, W_f N_2 - B'\, W_p N_0 = 0, \tag{19.43}$$

where W_f is the power density at the fluorescence frequency, $W_p = K^2 W_{inc}$ is the power density at the pump frequency W_{inc} enhanced by K^2 by the antenna, and $\Gamma_{tot} = \Gamma_R + \Gamma_{NR} + \gamma_{NR}$ is the total decay rate of the molecule, accounting, respectively, for the radiative decay in the presence of the antenna, the non-radiative decay in the presence of the antenna, and the internal non-radiative decay rate. B and B' are the Einstein coefficients for stimulated emission. The solution yields

$$N_0 = N_2 \frac{\Gamma_{tot} + BW_f}{B'W_p}. \tag{19.44}$$

Combining Eqs (19.42) and (19.44) yields

$$N_0 = N \frac{\Gamma_{tot} + BW_f}{\Gamma_{tot} + BW_f + B'W_p} \;; \quad N_2 = N \frac{B'W_p}{\Gamma_{tot} + BW_f + B'W_p}. \tag{19.45}$$

19.5.3 Saturation regime

Here, we consider the case of large pumping intensity $B'\, W_p \gg \Gamma_{tot} + BW_f$, so that $N_2 \approx N$. In this so-called saturation regime, all molecules are excited. The signal is then given by

$$S_{ant}^{(sat)} = C\, f_{coll}^{ant}\, \Gamma_R N. \tag{19.46}$$

In the saturation regime, the signal is limited by the spontaneous emission rate Γ_R. In simple words, the pumping efficiency is so large that all the molecules are excited. On average, an excited molecule needs a time $1/\Gamma_R$ to decay radiatively, and this does not depend on the incident intensity. Without an antenna, the saturated signal S_0 is proportional to γ_R, so that the effect of the antenna is given by

$$S_{ant}^{(sat)} = \frac{\Gamma_R}{\gamma_R} \frac{f_{coll}^{ant}}{f_{coll}} S_0^{(sat)}. \tag{19.47}$$

In this regime, the Purcell effect can increase the signal, but the incident field enhancement does not play any role.

19.5.4 Linear regime

We now consider the low-pumping regime, where we can approximate the population of the excited state by

$$N_2 = N \frac{B'W_p}{\Gamma_{tot}}, \tag{19.48}$$

so that the population is proportional to the incident intensity. This is the linear regime. The fluorescence signal is then given by:

$$S_{ant}^{(lin)} = C\, f_{coll}^{ant} \Gamma_R \frac{B' W_p}{\Gamma_R + \Gamma_{NR} + \gamma_{NR}}. \tag{19.49}$$

In the case $\Gamma_R + \Gamma_{NR} \gg \gamma_{NR}$, the result can be approximated by

$$S_{ant}^{(lin)} \approx C\, f_{coll}^{ant} \Gamma_R \frac{B' W_p}{\Gamma_R + \Gamma_{NR}} = C\, f_{coll}^{ant} \eta_{ant} B' K^2 W_{inc}. \tag{19.50}$$

Without an antenna, the signal is

$$S_0^{(lin)} = C\, f_{coll} \eta_{int} B' W_{inc}, \tag{19.51}$$

so that the antenna signal is given by

$$S_{ant}^{(lin)} = \frac{\eta_{ant}}{\eta_{int}} \frac{f_{coll}^{ant}}{f_{coll}} K^2 S_0^{(lin)}. \tag{19.52}$$

The effect of the antenna is therefore multiple. The intensity is multiplied by the local intensity-enhancement factor K^2. The antenna can concentrate the emission pattern into the detection solid angle of the receiving system, thereby increasing f_{coll}. Finally, the quantum yield of the molecule η_{int} is replaced by the radiative efficiency of the antenna η_{ant} if $\Gamma_R + \Gamma_{NR} \gg \gamma_{NR}$. It is seen that the antennas are of particular interest for molecules with low intrinsic quantum yield (Khurgin et al., 2008). The physical reason is simple: a low quantum yield means that the non-radiative decay channels are faster than the radiative decay channel ($\gamma_{NR} \gg \gamma_R$), so that light cannot be emitted efficiently. If the antenna introduces a new decay channel faster than the non-radiative decay channel ($\Gamma_R > \gamma_{NR}$), the molecule energy can be funnelled to the antenna. The intrinsic quantum efficiency is then replaced by the antenna radiation efficiency.

19.6 Controlling the fluorescence spectrum of broad spectrum emitters

The previous discussion considered the case of emitters with a single fluorescence transition between level 2 and level 1. The effect of the antenna is markedly different (Ringler et al., 2008) if the emission has a broad spectrum, as it is the case of dye molecules or QDs, as discussed in Fig. 9.8. We consider the linear regime. The roles of the intensity enhancement and the collection factor remain the same. By contrast, the quantum yield and the emission spectrum are modified.

To analyse this behaviour, we introduce $f_0(\omega)$, the integral-normalized spectrum of emission by the naked emitter. It depends on the spectral distribution of states and on their occupation given by a canonical distribution at temperature T. Then, the probability density $\Pi(\omega)$ of decay through emission at frequency ω in the presence of an antenna is given by

$$\Pi(\omega) = f_0(\omega) \frac{\Gamma_R(\omega)}{\Gamma_{tot}}, \tag{19.53}$$

where $\Gamma_{tot} = \int d\omega f_0(\omega)[\Gamma_R(\omega) + \Gamma_{NR}(\omega)] + \gamma_{NR}$. It is seen that the emission spectrum is now proportional to the product $f_0(\omega)\Gamma_R(\omega)$, where $\Gamma_R(\omega)$ is the radiative decay rate in the presence of the antenna. For an antenna with a spectrum narrower than the emitter spectrum, the emission spectrum can thus be significantly modified. If a large Q-factor antenna largely detuned from the emitter is used, the emitter emission spectrum may be suppressed and an emission peak at the antenna frequency may appear on the tail of the naked-emitter spectrum.

Regarding the enhancement of the integrated signal, Eq. (19.52) is still valid, provided that the antenna quantum efficiency is computed using

$$\eta_{ant} = \frac{\int d\omega f_0(\omega)\Gamma_R(\omega)}{\Gamma_{tot}}. \tag{19.54}$$

19.7 Controlling angular pattern and polarization emission

We consider an emitter radiating in vacuum and we aim to control its emission angular pattern and polarization. The basic mechanism is to couple the emitter to a single mode of a resonator, be it a cavity or a nanoantenna. Since the cavities and nanoantenna modes have radiative losses, the power transferred from the emitter to the resonator is then radiated by the resonator. It follows that the angular pattern and polarization properties are determined by the design of the cavity or resonator. This qualitative discussion is supported by inspection of Eqs (12.71) and (12.72), which show that the radiated field can be expanded over the modes of the system with a mode amplitude proportional to the scalar product of the dipole source and the mode electric field. The latter is nothing but the coupling term g introduced in section 19.3.6.

Examples of resonators well suited to controlling the angular emission of a single emitter are Yagi–Uda antennas (Curto *et al.*, 2010) and semiconductor micropillars (Dousse et al., 2008; Senellart et al., 2017). In order to efficiently control the emission, most of the power of the emitter should be coupled to the mode of the resonator. Following laser physics, the figure of merit is given by the fraction β of power released by an emitter which is coupled to the cavity mode.

In order to design bright light sources, it is necessary to couple the antenna to a large number of emitters. This can be done using Fabry–Perot cavities or metasurfaces supporting extended modes, so that they can couple to a layer of emitters. By properly engineering the radiative leakage of the metasurface modes, it is possible to control the angular emission pattern. We will expand this discussion in Chapter 21 using a statistical model of light emission.

19.8 Plasmonic antenna design rules for light emission

The purpose of this section is to briefly analyse the rules for an efficient plasmonic nanoantenna operating in the Purcell regime. More specifically, we address the issue of an efficient coupling between a plane wave and a quantum emitter while preserving a good radiation efficiency. We are thus interested in discussing the trade-off between local field enhancement and quenching. On one hand, the coupling to an antenna mode is increased by reducing the mode volume. Typical examples are a dimer of nanorods or a bow-tie antenna. The field is increased as the gap between the two parts of the antenna is reduced. However, as the gap is reduced below 10 nm, an emitter placed at the centre is at a distance of 5 nm to the metal surfaces, so that quenching is expected to start playing a negative role. Hence, enhancing the LDOS by reducing gaps to values below 10 nm has been considered to be plagued by the onset of non-radiative losses. Finally, we note that when the distance becomes smaller than 1 nm, a tunnel effect becomes possible and considerably modifies the enhancement (Savage et al., 2012; Esteban et al., 2012; Scholl et al., 2013).

19.8.1 Radiative and non-radiative modes

A key issue is the competition between radiative and non-radiative modes. As discussed in section 19.3.3, the LDOS can be four or five orders of magnitude larger than the LDOS in vacuum. This is usually mostly due to the very large numbers of non-radiative modes in metallic structures. However, it has also been shown that the LDOS due to a single mode can be very large, provided that the mode volume is very small and its Q-factor is good. The key to a 50% efficiency is simply to ensure that the antenna or cavity mode provides a contribution to the LDOS that matches the contribution of the non-radiative modes. This can be achieved by two means: (i) using a hot spot to take advantage of the light confinement; (ii) using a plasmonic resonance to further enhance the mode field. The light confinement can be obtained by engineering the antenna. For instance, a tip produces a singular behaviour of the field and takes advantage of the so-called lightning rod effect, a dimer of nanospheres produces a highly localized field between the spheres, etc. By combining light confinement and plasmonic resonance, it is possible to design a hot spot characterized by a LDOS which can be five orders of magnitude larger than in vacuum.

19.8.2 Radiative efficiency of the antenna mode

Generating a hot spot is by no means sufficient to obtain a good antenna. A hot spot merely ensures that the emitter will excite the antenna mode. A second issue is the radiative efficiency of the antenna mode. Indeed, the antenna mode can relax either by emitting radiation or through losses in the material. It is thus necessary to reduce non-radiative losses and to increase the radiated power. Non-radiative losses can be reduced by designing metallo-dielectric antennas. Radiated power can be increased by properly designing the size of the antenna and its structure. This requires a minimum size. As a rule of thumb, it is useful to keep in mind that the absorption cross section of a metallic sphere is equal to its scattering cross section for a diameter on the order of 40 nm. For smaller radii, absorption dominates, while radiative losses (scattering) dominate for larger radii. In summary, an efficient nanoantenna needs to be large to radiate efficiently and to have some localized hot spots to couple efficiently to the emitter. This is the case of all the recently proposed nanoantennas, such as nanocones, nanocubes, and nanopatches.

19.9 Nanoantenna impedance: Optimizing absorption

In this section, we introduce the concept of impedance for nanoantennas. An extended discussion can be found in ref. (Greffet et al., 2010; Marquier and Greffet, 2013). The motivation for the definition of the concept of impedance is based on the fact that there are concepts such as impedance matching and radiation resistance which are widely used in antennas and can be useful in optics. Of particular importance is the idea of impedance matching: how can we optimize the transfer of the power collected by the antenna to a load? This is a key issue for engineering detectors and for photovoltaics applications (Atwater and Polman, 2010).

We note that this issue can be addressed without using the concept of impedance. By computing the Green tensor of the structure, the field is known and, therefore, any quantity of interest can be computed. Yet, it is of practical interest to reformulate the multiple-scattering equations of optics using the vocabulary of electrical circuits. When addressing the question of the impedance of a cavity, a nanoantenna, or an atom or molecule, the first difficulty is the definition of the feed points. In electrical engineering, there is a line connecting the antenna to the source or detector. It is possible to define the line impedance. The line is connected to the antenna at two feed points, and it is possible to define the voltage difference between these two points and the current intensity in the line. If we now consider an atom radiating light (the source) in proximity to a metallic nanoparticle (the antenna), it is not possible to define the line or the feed points.

In this section, we first introduce an impedance for the antenna and for a two-level system. We then apply this concept to discuss the maximal power absorbed by a nanoparticle close to an antenna. This is, for example, the goal of an antenna designed to funnel energy into a subwavelength detector. Indeed, enhancing the absorption by a nanoparticle requires two conditions to be fulfilled: (i) a lot of power must be collected with an antenna with large absorption cross section; and (ii) this power must be transferred efficiently from the antenna to the absorber. Impedance matching ensures that the second condition is fulfilled in radiowaves. Our goal is to identify this concept in the context of optical nanoantennas.

19.9.1 Nanoantenna impedance

Since impedance is the concept that allows the power radiated by an antenna to be discussed, let us try to introduce the impedance heuristically by analysing the power emitted by a dipolar antenna. We start from the power P_0 delivered by the dipole to the optical field. The time-averaged value is given by $P_0 = \langle -\frac{d\mathbf{p}}{dt} \cdot \mathbf{E} \rangle = \frac{1}{2}\mathrm{Re}(i\omega \mathbf{p} \cdot \mathbf{E}^*)$. There is a clear similarity between the structure of this equation and the familiar form of the electrical power P dissipated in a load $P = \frac{1}{2}\mathrm{Re}(IU^*)$. This suggests a linear relation should be introduced between the dipole moment and the field. This is analogous to the relation $U = Z_L I$, where Z_L is the impedance of the load in an electrical circuit. Such a relation can be written using the Green tensor,[6] which yields the field radiated at $\mathbf{r}_0$ by a dipole located at $\mathbf{r}_0$:

$$\mathbf{E}(\mathbf{r}_0) = \overleftrightarrow{\mathbf{G}}(\mathbf{r}_0, \mathbf{r}_0, \omega) \cdot \mathbf{p} = \frac{\overleftrightarrow{\mathbf{G}}(\mathbf{r}_0, \mathbf{r}_0, \omega)}{-i\omega} \cdot [-i\omega \mathbf{p}]. \tag{19.55}$$

Fig. 19.9 shows the dipole near a nanoantenna. The dipole is assumed to be oriented along the z-axis, so that $\mathbf{p} \cdot \overleftrightarrow{\mathbf{G}}^*(\mathbf{r}_0, \mathbf{r}_0, \omega)\mathbf{p}^* = |\mathbf{p}|^2 G^*_{zz}(\mathbf{r}_0, \mathbf{r}_0, \omega)$. The energy transferred

[6] Note that the definition of the Green tensor used here follows ref. (Greffet et al., 2010) and differs by a factor $\mu_0\omega^2$ from the definition used earlier in the chapter.

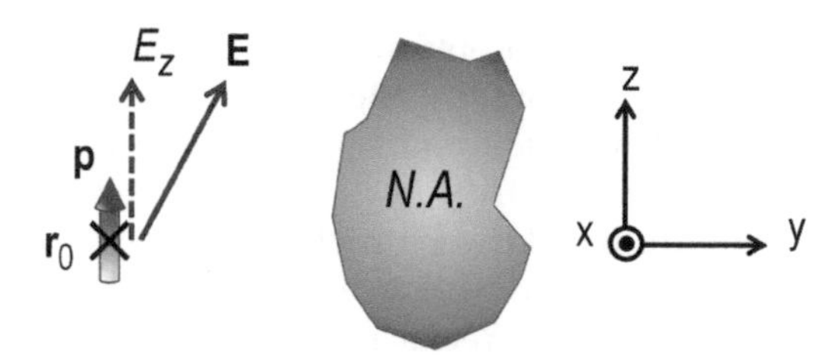

Fig. 19.9 Sketch of the system. N.A. stands for Nanoantenna. A dipole is located at $\mathbf{r}_0$ near the nanoantenna and the dipole moment is oriented along the z-axis. E_z is the z-component of the electric field $\mathbf{E}$ at $\mathbf{r}_0$.

by the dipole to the field can be written in the form $P_o = \frac{1}{2}\text{Im}\left(\frac{G_{zz}(\mathbf{r}_0,\mathbf{r}_0,\omega)}{\omega}\right)|\omega\mathbf{p}|^2$. This has the structure of the power $\frac{1}{2}\text{Re}(Z_L)|I|^2$ dissipated in a load Z_L. A comparison between the two forms of the power delivered by a source suggests the following identification:

$$
\begin{aligned}
I &\leftrightarrow -i\omega p_z, \\
U &\leftrightarrow -E_z(\mathbf{r}_0), \\
Z &\leftrightarrow \frac{-iG_{zz}(\mathbf{r}_0,\mathbf{r}_0,\omega)}{\omega},
\end{aligned}
\tag{19.56}
$$

where we have $U = ZI$.[7] As for lumped elements, losses are given by the real part R of the impedance $Z = R + iY$. The resistive part of the impedance is thus $\text{Im}(G_{zz})/\omega$ and accounts for both radiative and non-radiative losses. It is important to emphasize that this impedance has the dimension of $\Omega\,.\text{m}^{-2}$. It is thus an impedance per unit area. The difference stems from the fact that we use $-i\omega p_z$ (in Am) instead of I (in A) and E_z (in V/m) instead of U (in V). We shall therefore use the term *specific* impedance. We note that this approach is similar to the so-called emf method introduced by Brillouin (Brillouin, 1922) to derive the input impedance in the case of a wire antenna.

19.9.2 Two-level system impedance

Defining an internal specific impedance for the source amounts to finding a linear relation between the induced dipole moment $\mathbf{p}^{\text{ind}}$ and the electric field. Far from the saturation regime, such a linear relation is simply given by the polarizability of a two-level system: $\mathbf{p}^{\text{ind}} = \alpha\varepsilon_0\mathbf{E}$. We can thus write $\mathbf{E} = Z_\text{S}(-i\omega\mathbf{p}^{\text{ind}})$, where we have introduced a scalar specific impedance for the source:

$$
Z_\text{S} = \frac{i}{\omega\alpha\varepsilon_0}. \tag{19.57}
$$

Let us consider a two-level system for the sake of illustration. The density matrix formalism allows the particular form of the polarizability for such a system to be derived. The transition frequency ω_0 is defined by $\hbar\omega_0 = E_2 - E_1$ (from a classical point of

[7]Note that the minus signs are due to the relations $\mathbf{j} = \frac{\partial p}{\partial t} = -i\omega\mathbf{p}$ and $\mathbf{E} = -\nabla\phi$, where ϕ is a scalar potential.

view, ω_0 is the bare frequency of the dipole decoupled from all fields), where the atomic Hamiltonian eigenvalues are denoted by E_1 and E_2, respectively. The polarizability of the system can be written in the form (Loudon, 2000)

$$\alpha(\omega) = \frac{\alpha_0}{\omega_0^2 - \omega^2 - i\gamma_0\omega}, \tag{19.58}$$

where α_0 can be written, using the oscillator strength f, as $\alpha_0\varepsilon_0 = (e^2/m)f$ or $\alpha_0\varepsilon_0 = 2d_{12}^2\omega_0/\hbar$, with $d_{12} = |p|$ the dipole moment of the transition. The term γ_0 accounts for the broadening of the resonance. It consists of several contributions. First of all, the radiative emission is characterized by a contribution γ_R. It corresponds to a population decay of level 2. Coupling to the environment can also provide inelastic interactions, leading to a population decay of the excited state. It corresponds to a non-radiative decay and is included in the term $\gamma_{NR} = 1/T_1$. Finally, elastic collisions produce a dephasing of the wave function without modifying the population of the excited state. This contribution to the resonance broadening is called dephasing and is characterized by the contribution $\gamma_2^* = 2/T_2^*$, so that we have $\gamma_0 = \gamma_R + \gamma_{NR} + \gamma_2^*$. Finally, the impedance can be cast in the form

$$Z_S = \frac{\gamma_0}{\alpha_0\varepsilon_0} + i\frac{\omega_0^2}{\alpha_0\varepsilon_0}\frac{1}{\omega} - i\omega\frac{1}{\alpha_0\varepsilon_0} \tag{19.59}$$

$$= R - \frac{1}{iC\omega} - iL\omega. \tag{19.60}$$

This impedance has the structure of the impedance of a RLC series circuit, where $R = \gamma_0/\alpha_0\varepsilon_0$, $L = 1/\alpha_0\varepsilon_0$, and $C = \alpha_0\varepsilon_0/\omega_0^2$. Let us consider the resistance here. It is proportional to the sum $\gamma_R + \gamma_{NR} + \gamma_2^*$. The first term corresponds to spontaneous emission in vacuum. The corresponding resistance is denoted by R_0. The other terms account for non-radiative interactions with the environment. From the impedance point of view, it is easier to consider the radiative resistance as an external load and to write the impedance of the source using only $\gamma = \gamma_{NR} + \gamma_2^*$ instead of γ_0. We will now write $R_S = \gamma/\alpha_0\varepsilon_0$, L_S, and C_S, the lumped elements of the source.

19.9.3 Conjugate impedance-matching condition

We consider here the absorption of energy coming from an external source. In electricity, it is well known that a conjugate impedance-matching condition must be fulfilled to optimize absorption in a load impedance. Here, we will apply the concept of impedance to analyse the absorption by a nanoparticle or an atom when located in a given environment. This analysis will provide a guideline to design a structure that enhances the absorption by a nanoparticle or a two-level system.

The nanoparticle (or the atom) has a specific impedance Z_1 and the environment acts as a load Z_2. The system is illuminated by an external field $\mathbf{E}_{\text{ext}}$. The amplitude of the dipole moment of the system, illuminated by a field that is the sum of the incident field and the field scattered by the environment, is given by $p_z = \alpha\varepsilon_0[E_{z,\text{ext}} + G_{zz}p_z]$, where α is the nanoparticle (or the atom) polarizability and G_{zz} is the Green tensor that accounts only for the environment. This equation can be reformulated as

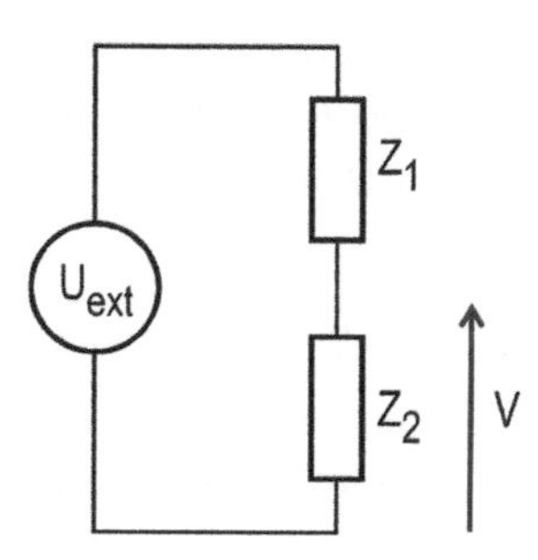

Fig. 19.10 Equivalent circuit of a nanoparticle in a given environment. An external field is represented by a source voltage. The voltage difference for the load Z_2 is V.

$E_{z,\text{ext}} = \left[\frac{i}{\omega\alpha\varepsilon_0} + \frac{-iG_{zz}}{\omega}\right](-i\omega p_z) = [Z_1 + Z_2](-i\omega p_z)$. We now sketch the equivalent circuit in Fig. 19.10.

The illuminating field is represented by a voltage U_{ext} applied to Z_1 and Z_2 in series. The voltage difference for the load is V. We recognize a voltage divider, so that $V = Z_2/(Z_2+Z_1)U_{\text{ext}}$. It is now a simple matter of deriving the condition for maximum power dissipated in the load. Let us write $Z_1 = R_1+iY_1$ and $Z_2 = R_2+iY_2$. The power dissipated in an environment such as a nanoantenna, due to both radiative and ohmic losses, is given by

$$P_0 = \frac{1}{2}\frac{R_2|U_{\text{ext}}|^2}{(R_1+R_2)^2+(Y_1+Y_2)^2}, \tag{19.61}$$

where R_1 and R_2 are always positive, accounting, respectively, for the losses of the source and the LDOS at the position of the source. Y_1 and Y_2 can be either positive or negative, depending on the frequency. The maximum power dissipated by the nanoantenna is obtained when $Y_1+Y_2 = 0$ and $R_1 = R_2$. This condition is the usual conjugate impedance-matching condition $Z_1 = Z_2^*$ that ensures an optimized power transfer between a source and a load. This condition is nothing but the critical-coupling condition already discussed in the context of coupled oscillators in Chapter 7 and when optimizing the absorbed power by a nanoparticle in Chapter 14.

19.9.4 Multiple scattering and impedance in series

In this short section, we point out that there is a strict equivalence between two familiar equations: the multiple scattering between two particles and the addition of two impedances in series. We have seen in Chapter 12 that an equation of the form $E = E_0 + KE$ where K is a linear operator can be solved iteratively. In the context of multiple scattering, this leads to the single, double, etc. scattering terms. Using the definition of the impedances in Eq. (19.56), we have

$$\begin{aligned} p_z &= \alpha\varepsilon_0[E_{z,\text{ext}} + G_{zz}p_z], \\ I &= -\frac{U_{ext}}{Z} - \frac{Z_S}{Z}I. \end{aligned} \tag{19.62}$$

The second equation, which is merely a reformulation of $-U_{ext} = (Z+Z_S)I$, appears to have the structure of a multiple-scattering equation. We can thus analyse a circuit in terms of multiple scattering or analyse an emitter coupled to an antenna in terms of impedances in series.

For a simple circuit with two impedances in series, it is obvious that the intensity circulating across the two lumped elements depends on the sum of the two impedances. It is usually less intuitive that the dipole moment of a particle depends on the impedance of the environment because in many situations it is only a correction term. In nanophotonics, this term can be very large.

When discussing the optimization of absorption, this remark becomes critical. The existence of a maximum in the power absorbed by a load in an electrical circuit is easy to understand. On the one hand, for low values of the impedance Z_2 of the load, the losses increase as R_2 increases. On the other hand, if R_2 becomes too large, the intensity in the circuit will decrease, leading to a decay of the absorbed power. This obvious behaviour is less intuitive when dealing with an emitter or a nanoparticle coupled to a cavity or an antenna. The usual wisdom is that the antenna should generate a large local field and the reaction of the absorber on the antenna is ignored. It is enlightening to exhibit the equivalence of the two visions and to realize that the presence of the absorber may modify the system (the total impedance) and therefore reduce the losses. In summary, when designing an antenna to enhance the absorption by an atom or a detector, it is necessary to account for the interaction between the absorber and the antenna: this is the impedance-matching condition.

19.A Complement: Fluorescence in the impulse regime

In this complement, we derive a model for the fluorescence signal when using a periodic pulse excitation. Considering an ensemble of N emitters, we first derive the number of emitters in the excited state as a function of time, $N_2(t)$. We consider periodic pulses with a repetition rate R of small duration δt and irradiance ϕ_{inc}. After the pulse, the emitters in the ground state are excited with a probability Π. We denote by $N_2(0^-)$ the number of excited emitters before the pulse. The number of excited emitters after the pulse is thus $N_2(0^+) = N_2(0^-) + \Pi[N - N_2(0^-)]$. This population decays with a rate Γ and the number of excited emitters at the end of a cycle is $N_2(t = 1/R) = N_2(0^+)\exp(-\Gamma/R)$. Because the excitation is periodic, we have $N_2(t = 1/R) = N_2(0^-)$. We consequently derive the number of excited emitters as a function of time:

$$N_2(t) = N_2(0^+)\exp(-\Gamma t) \ ; \ N_2(0^+) = N\frac{\Pi}{1-(1-\Pi)e^{-\Gamma/R}}. \tag{19.63}$$

We can now use the population to derive the emitted intensity. The radiative decay rate of the emitter is Γ_R and the collection efficiency of the system is f_{coll}. During an excitation cycle, the signal consisting of the mean intensity collected (in photons s^{-1}) per pulse is

$$S = R\int_{0^+}^{1/R} \Gamma_R \, f_{coll} \, N_2(t)\mathrm{d}t, \tag{19.64}$$

which yields:

$$S = \frac{\Gamma_R}{\Gamma} R \, f_{coll} \, N\frac{\Pi}{1-(1-\Pi)e^{-\Gamma/R}}(1 - e^{-\Gamma/R}). \tag{19.65}$$

Possible saturation effects are included in the dependence of the excitation probability Π on the pump irradiance. We introduce this effect using rate equations for populations of an ensemble of emitters. The excitation is assumed to be proportional to the product of the absorption cross section σ_{abs}, the pump irradiance ϕ_{inc}, and the population of emitters in the ground state. In this simple model, we have assumed that internal relaxation processes after excitation are faster than the radiative decay. During the excitation (duration δt), the population equation is:

$$\frac{\mathrm{d}N_2(t)}{\mathrm{d}t} = \frac{\sigma_{abs}\phi_{inc}}{\hbar\omega_{inc}}[N - N_2(t)] - \Gamma_{tot}N_2(t). \tag{19.66}$$

We solve this equation using the initial condition $N_2(0) = 0$. The transition probability Π for an emitter initially in its ground state can then be cast in the form

$$\Pi(\phi_{inc}) = \frac{N_2(\delta t)}{N} = \frac{1}{1 + \frac{\Gamma\hbar\omega_{inc}}{\sigma_{abs}\phi_{inc}}}(1 - e^{-\frac{\delta t\sigma_{abs}}{\hbar\omega_{inc}}\phi_{inc}}e^{-\Gamma\delta t}). \tag{19.67}$$

These results give a model for the fluorescence collected per pulse, regardless of the relative values of the repetition rate R and the decay rate Γ, and for any intensity regime. We now consider the particular case of a low-excitation regime $\phi_{inc} \ll \frac{\Gamma\hbar\omega_{inc}}{\sigma_{abs}}$, so that $\Pi \ll 1$ and a low repetition rate $\Gamma/R \ll 1$. We get the following form of the fluorescence signal:

$$S = \frac{\Gamma_R}{\Gamma}\, R\, f_{coll}\, N\, \frac{\sigma_{abs}\phi_{inc}\delta t}{\hbar\omega_{inc}}. \tag{19.68}$$

This result shows that under these conditions, the fluorescence signal per pulse has a similar structure as in the stationary regime. It is proportional to the antenna radiative efficiency, intensity enhancement, and collection efficiency. However, there is an additional factor that introduces a difference. The total fluorescence signal is proportional to the repetition rate R. As we have made the approximation that the pulse repetition rate R satisfies $\Gamma/R \gg 1$, the signal is bounded by Γ. Hence, in the pulse regime, the signal can be increased by increasing the decay rate.

Exercises

Exercise 19.1 Link between classical and quantum spontaneous-emission decay rate. The goal of this exercise is to compare the classical spontaneous emission rate with its quantum form. Starting from Eq. (19.3), replace the classical dipole moment ex_0 with the matrix element d_{12} and replace the classical form of the energy of the oscillator U with $\hbar\omega$. Compare the result with Eq. (19.5).

Exercise 19.2 Purcell factor and coherence volume. In Exercise 12.7, we introduced the coherence volume V_{coh} defined as the volume corresponding to a single state in vacuum with a given spectral bandwidth. Show that the Purcell factor can be cast in the form of the ratio of two volumes:

$$F_P = \frac{V_{coh}}{V}. \tag{19.69}$$

The Purcell factor can be large using either cavities with volumes on the order of λ^3 and Q-factors that are typically in the range $10^3 - 10^6$ or plasmonic antennas with nanoscale volumes and Q-factors on the order of 10. Compare V_{coh} and V for both cases.

Exercise 19.3 Enhancing fluorescence. We consider a nanoantenna used to enhance fluorescence of a four-level molecule. The nanoantenna provides an enhancement K of the field locally illuminating the molecule. The radiative decay rate without antenna is γ_R and the molecule intrinsic quantum yield is denoted by η_{int}. We consider two antennas. First, a plasmonic antenna with an enhancement $K = 10$ at the pumping frequency and a radiative efficiency $\eta_{ant} = 0.2$. Second, an antenna with an enhancement $K = 2$ at the pumping frequency and a radiative efficiency $\eta_{ant} = 0.5$. These antennas are used to enhance the fluorescence of two different molecules. The only difference is the intrinsic quantum yield η_{int}, which is 1 for molecule A and 10^{-2} for molecule B. Assuming that all the fluorescence light is collected, what is the enhancement in the linear regime for the four cases? Does it depend on the Purcell effect?

Exercise 19.4 Rabi oscillation of a molecule with a nanoantenna. In this exercise, we aim to derive a condition for observing the Rabi oscillation regime for a molecule coupled to a plasmonic cavity. A necessary condition is $2g > \kappa$. We estimate $\hbar g = dE$ using

$$E = \sqrt{\frac{\hbar\omega}{2\varepsilon_0\varepsilon_r V}},$$

and we introduce the oscillator strength f using $d^2 = f\frac{\hbar e^2}{2m\omega}$. Show that

$$\frac{2g}{\kappa} = Q\sqrt{f\frac{\lambda^3}{V}}\sqrt{\frac{r_e}{\pi\varepsilon_r\lambda}},$$

where $r_e = e^2/(4\pi\varepsilon_0 mc^2)$ is the classical electron radius. Show that for $\lambda = 600$ nm, a Q-factor on the order of 10, a volume of 30 nm^3 and a permittivity of 2, the strong coupling condition is satisfied.

20

From nanophotonics to devices

20.1 Passive, active, and emitting devices

In this chapter we shall exploit the understanding gained from previous ones to get insight into devices and functions that are deemed long-lasting in twenty-first-century photonics. We leave aside the case of device-related metasurfaces addressed in Chapter 18. Their use indeed enhances *extended* devices (lenses, etc.), whereas we focus on devices having at least one wave dimension already confined in some way (by a guide, plasmon, etc.) and thus more directly related to the nanoworld. We also refer the reader to existing literature such as refs (Atwater and Polman, 2010; Mokkapati and Catchpole, 2012) for nanophotonic and plasmonic applications to the broad field of photovoltaics. Within these provisions, we choose to address passive and active devices separately, the latter kind meaning the presence of photon emission.

We start by treating passive devices with emphasis on index sensing in section 20.2. The general rationale is to detect very weak amounts of matter diluted in some fluid or the presence of specific molecules. Thanks to various chemical approaches, the matter or the molecules of interest can be immobilized on a surface. The attainment of sufficient selectivity depends on both thermodynamic factors (chemical affinity: think of complementary DNA strands) and kinetic ones (speed and nature of molecular motion: drift or diffusion). The typical system to be probed is a nanometric layer: in the so-called label-free approach (no small fluorescent 'label' molecules are grafted to the macromolecules of interest), the issue is to detect this layer in spite of its weak index contrast against surrounding fluid: say, $\Delta n \sim 0.1$ for a full layer, proportionally less for a 'partially occupied' layer described by a lower effective contrast. So a tight quantitative accuracy is always welcome. We believe it is also a helpful entry point to the whole topic of index engineering within nanodevices. We will scan through dielectric and plasmonic solutions and, in this latter case, through devices exploiting planar or localized forms of plasmonic structures (see Chapters 13 and 14).

We shall only give brief hints regarding the large realm of passive integrated optics (IO) and the possible benefits from added nanostructures in section 20.3. We will look at two kinds of devices that have been workhorses of IO since 1990: Mach–Zehnder (MZ) modulators for electro-optic processing and spectral multiplexing/demultiplexing devices ('mux/demux', essentially spectrometers) such as the arrayed waveguide grating (AWG).

As for the active devices treated in section 20.4, it is logical to further distinguish two categories: those based on stimulated emission—lasers, and, in particular, laser diodes—and those based on spontaneous emission, mainly light-emitting diodes

Introduction to Nanophotonics. Henri Benisty, Jean-Jacques Greffet, and Philippe Lalanne, Oxford University Press.
 DOI: 10.1093/oso/9780198786139.003.0020

(LEDs). The nanoantennas discussed in Chapter 19 could be among the next-generation devices but are not discussed here.

20.2 Index sensing with confined waves

20.2.1 Sensing of the cladding of a symmetric dielectric waveguide

The classical instrument for measuring the refractive indices of bulk fluids is the Abbe refractometer, which is based on the dependence of the limit angle of total internal reflection on the index of a drop of the fluid, and whose index sensitivity goes to the fifth digit. However, as evoked above, the indices of interest are those of layers near solid interfaces where large molecules transported by fluids—notably biological molecules such as DNA, proteins, antibodies, and antigens—can be immobilized. Additionally, a microfluidic environment is welcome to dynamically control single assays or manage multiple parallel assays. None of these features are supported by the Abbe refractometer. The waveguide-based solutions are more suited to this aim (this change of the very principle of the device for a given need is at variance with the case of metasurfaces, which aims to replace *existing* devices).

We use the case of the dielectric waveguide to present the issues at stake for index sensing. We only consider sensing the real part of the dielectric constant, although absorption-monitoring avenues would not make the discussion fundamentally different. We start with a simple symmetric planar waveguide, Fig. 20.1a, using the same notations as in Figs 6.2 and 6.5 except the thickness denoted W: the core index is n_2; the cladding index is n_1; the fundamental TE-polarized guided-mode profile at wavelength λ or frequency ω has the form $E(x,z,t) = F(x)\exp(ik_{||}z - \omega t)$, with $k_{||}$ the propagation constant, related to the effective index by

$$k_{||} = n_{\text{eff}}(\omega/c) = n_{\text{eff}}k_0 \tag{20.1}$$

and $F(x)$, a continuous, derivable, even function for TE polarization. We assume that the guide is part of an overall device able to sense n_{eff}: for instance, an interference setup able to sense an optical path difference $\delta = n_{\text{eff}} \times \ell$ such as a MZ interferometer, possibly in integrated versions (see section 20.3). Periodic waveguide devices can also serve this purpose (see section 20.2.5).

We now assume that the index is perturbed in the top cladding (see ref. (Yariv and Yeh, 1984) for a classical waveguide perturbation theory), with some spatial profile $P(x) \leq 1$ reaching $P(x_{\max}) = 1$ at a peak perturbation position $x_{\max}$ in the cladding:

$$n_{\text{clad}} = n_{\text{clad}} + P(x)\,\Delta n \quad \text{with} \quad P(x_{\max}) = 1 \text{ and } P(x) \leq 1. \tag{20.2}$$

For a uniform thin layer of thickness d lying on top of the core, we write, for instance, $P(x) = \text{rect}\,([x + d/2 - W/2]/d)$ with rect the usual ‘hat’ function with unity width and unity integral. Examples are provided in Fig. 20.1a with various scenarios (i)–(iv) for discussion below. The sensitivity of n_{eff} to this *cladding index modification* can be seen as the basic parameter of the system. It can be defined as:

$$S_{\text{clad}} = \frac{\partial n_{\text{eff}}}{\partial \Delta n}. \tag{20.3}$$

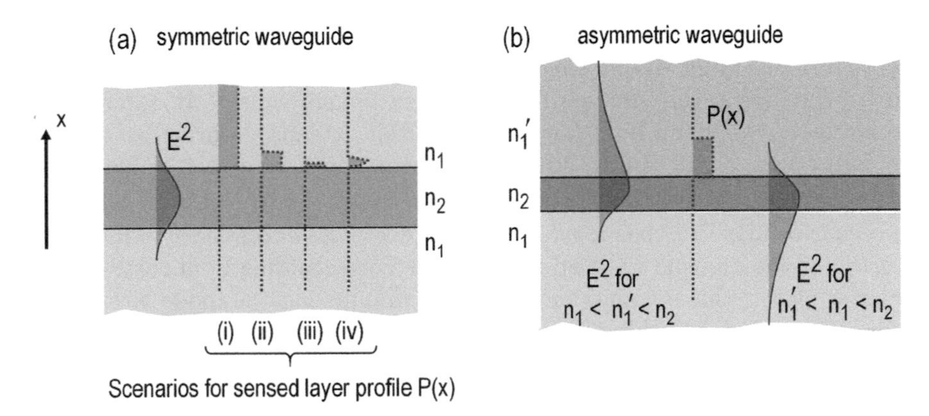

Fig. 20.1 (a) Basic symmetric dielectric waveguide with core and cladding indices (n_1 and n_2) indicated. Four scenarios are proposed for the profile $P(x)$ of the index variation: (i) semi-infinite; (ii) finite of the same order as typical decay length; (iii) very thin; (iv) peaked $P(x)$ profile. (b) Asymmetric version with large overlap $\Gamma_{\text{clad,top}}$ either with the top cladding (left side, case $n_1 < n'_1 < n_2$) or the bottom cladding, often the device substrate (right side, case $n'_1 < n_1 < n_2$). $P(x)$ is pictured as relatively large in comparison to the guide width W.

From a perturbation analysis, since n_{eff}^2 is the eigenvalue of the wave equation, its variations are first order in the perturbation (of the potential $V(x)$ in quantum mechanics), while those of the field (the wavefunction in quantum mechanics, the mode profile $E(x)$ here) are of second order. The exact (and more general) result is that the effective index variation is given by

$$\delta n_{\text{eff}} = \frac{\int \Delta n \left(P(x) \times \text{power flux at position } x\right)\, dx}{\text{Total power flux integrated over } x}. \tag{20.4}$$

According to this approach, since the power flux of a TE mode is directly proportional to $|E(x)|^2$ (based on the $E_y\, H_x$ product), the factor ruling S_{clad} is the so-called (partial) confinement factor Γ (see Exercise 6.4 and Eq. (8.43), here it is the fraction of $|E|^2$ weighted by the function $P(x)$:

$$\Gamma_{\text{Sensing}} = \frac{\int_{x>\frac{W}{2}} P(x)|E(x)|^2\, dx}{\int_{-\infty}^{+\infty} |E(x)|^2\, dx}. \tag{20.5}$$

Thus it suffices that $P(x)$ is integrable with the $|E(x)|^2$ kernel, a soft condition, given its decay in the cladding. Furthermore, Eq. (20.4) grants $S_{\text{clad}} < 1$. And if a single cladding is considered, it is intuitively true that the stricter limit $S_{\text{clad}} < 0.5$ holds. The various scenarios of Fig. 20.1a give rise to different optimal conditions.

- For scenario (i), $P(x)$ is the whole top cladding (a bulk fluid, for instance). The optimum profile is the 'deconfined' guided mode, with vanishing index contrast $n_2 - n_1 \to 0$ or vanishing core thickness $W \to 0$. The guided mode then reduces to its two similar tails, so that $n_{\text{eff}} \to n_{\text{clad}}$ and $S_{\text{clad}} \to 0.5$ can be expected.

- For scenario (ii), the finite size d of $P(x)$ entails qualitatively that the guide mode characteristic extent (decay length) should not exceed this size. While minimization of the field in the core is still welcome, we cannot squeeze W too much unless we increase $n_2 - n_1$ to limit how much the tail extends. In practice, high indices are limited ($n_1 \lesssim 2.0$–2.4 in the visible, for instance), so a limit value is met and $S_{\text{clad}} \sim 0.25$ could decently represent a good device.
- In scenario (iii), $P(x)$ has a vanishing width d. The aim is now to maximize the overlap by maximizing the field value at the core-cladding boundary. For a given index contrast, this amounts intuitively to minimizing the mode size. The exact value of the squared field at this boundary will then be some large fraction of the maximum (say 50–70%), but Γ_{Sensing} now depends more critically on the ratio of layer size d to the mode size than on this particular value.
- In scenario (iv), $P(x)$ is no longer a piecewise square function. However, as long as it is a peaked function with characteristic size smaller than the smaller decay length allowed by the design (thus dictated by λ, n_{eff} and $n_1 - n_2$, but typically above 70 nm with materials that sustain waveguiding in the visible or near-infrared), the trends of scenarios (ii) and (iii) can still be seen to hold qualitatively.

20.2.2 Sensing with an asymmetric dielectric waveguide

Turning to Fig. 20.1b, we study an asymmetric waveguide, with the top cladding having index $n_1' \neq n_1$. We can reason on scenario (ii) without loss of generality, as it tends to scenarios (i) and (iii) naturally. In principle, we would like the left-side situation: extending the field of the perturbed side to maximize Γ_{Sensing} and adjusting the thickness of the core index to fit the tail size to the perturbation profile.

Insofar as fluids are concerned, however, their indices often fall in a low range. Aqueous solutions in the visible or near-infrared have indices in the range $n = 1.33$–1.45 and become very viscous beyond (glycerol is a common addition to get variable indices), causing problematic microfluidic management. Some solvents and oils reach indices in the range $n = 1.48$–1.53, but not much above this in common uses of index sensing. So the index n_1' of our problem is rather low.

The difficulty is then to find at least an index n_1 that would be much lower than n_1' in order to get the left-side situation. For water, this is almost impossible with bulk material. Two possible solutions are (i) commercial fluoropolymers that can achieve very low cladding indices, close to 1.33 or so. Speciality products (e.g. Cytop®) are available that produce low-loss homogeneous materials, further enabling the deposition of the core layer on them, not a simple task for very thin cores as desired; (ii) nanoporous layers, for instance nanoporous silica, for which an effective medium approach (Maxwell–Garnett or Bruggeman, see Chapter 15) predicts indices below 1.33 for air fraction above ~40%. The difficulty of the latter approach is to limit the unavoidable light scattering induced by the spatial index fluctuations. Their characteristic scale is much larger in these porous realizations (say, 1–15 nm) than those of their non-porous counterparts (bulk cladding such as silica, or thick resists, whose small local-index fluctuations dictate the equivalent of Rayleigh scattering). This can considerably reduce the useful length.

Thus, the customary situation on, say, a glass substrate with a standard waveguiding layer (deposited silicon nitride or an oxide such as tantalate or titania), is much more often the one pictured on the right-hand side of Fig. 20.1b, with the larger field tail unfortunately on the substrate side. Obviously, this keeps S_{clad} to a very low value—say, $S_{\mathrm{clad}} \lesssim 0.1$—even for thick-layer sensing.

Based on this relatively frustrating condition, more variants are investigated to counteract this low sensitivity. The impediment depends very much on the overall device, however. For an interferometric device, for instance, even if a thin layer is used and the bad asymmetric situation is retained, it is instructive to understand what can be obtained with S_{clad} as low as, say, $3.3\,10^{-3}$. If the guide length used is around $\ell = 1\,\mathrm{cm}$ (the volume sensed is still much smaller than that of an Abbe refractometer and still relevant to microfluidics), then the optical path difference (OPD) takes the form $\mathrm{OPD} = \ell n_{\mathrm{eff}} \times 3.3\,10^{-3}\Delta n \sim (5\,10^{-5}\,\mathrm{m} \;\; \times \Delta n)$. If we consider an interference device (say, an MZ) with $\lambda/100$ sensitivity and a wavelength of 500 nm, an OPD of 5 nm can be detected. This still corresponds from the above expression to $\Delta n = 10^{-4}$. This is at the low end of acceptable performance, indeed, but still honourable. Conversely, when investigations are made to track $\Delta n < 10^{-5}$, several instrumental issues are met, such as the need for very tight temperature control.

In an actual design, several other parameters would be taken into account. For instance, if the guide core is very thin in the central testing section, it might be desired to have more common and larger guide cores elsewhere, to apply known coupling recipes (for integrated sources or for 'end fire' of the waveguide by external sources). Then the feasibility of a so-called taper section, whereby the guide thickness W is changed, should be considered. An adiabatic change is needed to avoid spilling light out of the taper, with a slow enough variation.

Such considerations will be helpful to understand, by contrast, the success met in our next topic: the use of surface-plasmon resonance.

20.2.3 Surface plasmon polariton sensing

Surface plasmon. The surface plasmon polariton (SPP) has been described in various parts of this book (see chapters 13 and 14). Here, we refer to Fig. 20.2a, where the typical transverse-magnetic (TM)-polarized SPP can be recognized. The field sketched here is the simpler horizontal field component E_x (free of discontinuity, unlike the z component, which must follow the change of sign between ε_{fluid} and ε_{metal}, Fig. 13.3).

The similarity to the dielectric waveguide is quite valid: the propagation constant of such a plasmon typically corresponds to an effective index n_{eff}, which is larger by a few per cent than the top medium index, $n \equiv \sqrt{\varepsilon_{fluid}}$. Hence, the tail decays with a typical scale of 100–200 nm for visible light. This is still good for sensing a typical 2–10 nm-thick biomolecular layer, as shown in the zoom of Fig. 20.2b. The small confinement factor of such a thin layer still gives a proper idea of the index dependence, but the perturbation analysis should be rewritten properly for this electromagnetic configuration, with (x, z) electric-field components instead of y for transverse-electric (TE) ones.

Assessing the effective index of such a plasmon polariton is the topic of surface-plasmon resonance (SPR). It is preferably performed through the Kretschmann geometry, Fig. 20.2b, (or Fig. 13.7b) using a tunnel effect through the thin metal layer

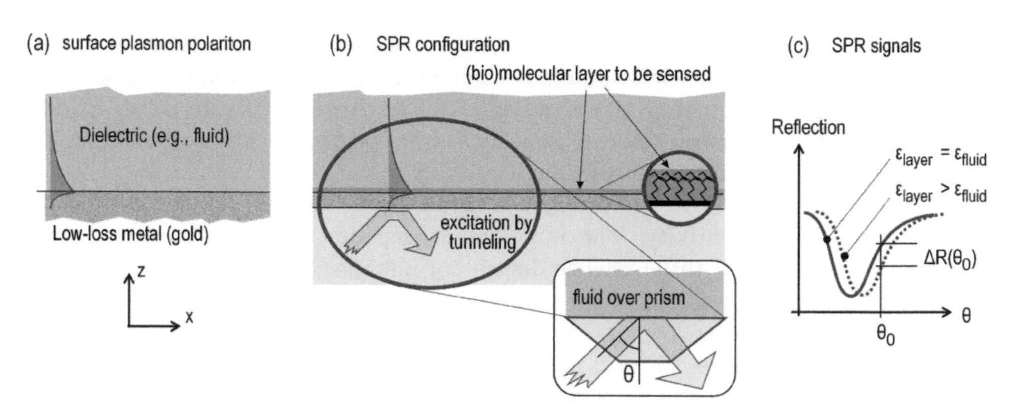

Fig. 20.2 (a) Surface plasmon polariton field profile at a metal–dielectric interface (in the form $|E_z|$). (b) Excitation of this mode by the bottom of a metal layer of adequate thickness. The top right zoom shows the biomolecular layer being sensed. The bottom-right inset shows the optical beam path with the prism coupling technique. (c) Principle of SPR: the resonant reflection dip shifts because the layer index differs from the displaced fluid one. At a wavelength λ_0 with large $dR(\theta)/d\theta$ slope, the reflectivity modulation $\Delta R(\theta_0)$ is largest.

(often gold), and a bottom medium (a prism) of high enough index n_{prism}. When the incident wave has an angle of incidence θ_{prism} such that $k_{||} = \text{Re}(n_{\text{eff}})k_0 = n_{prism} \sin\theta_{prism} \simeq k_{||,\text{plasmon}}$, the energy is coupled to the surface mode, where it resides for a few tens of wavelengths, instead of being reflected. This thus causes a dip in the reflection, Fig. 20.2c, with a typical angular width of a few degrees (which is dictated by the imaginary part of $k_{||}$). One then uses a given angle and wavelength to measure the largest possible reflection modulation ΔR by using one of the two large slopes available. The final choice is also a compromise between linearity, sensitivity, and range of analyte concentration to be sensed.

Localized plasmon. A metallic object of finite extent, in particular a subwavelength particle, can host a localized plasmon; this gives rise to a strong resonance for good plasmonic metals (gold, silver): see Chapter 14. Fig. 20.3a illustrates the resonant field of a spherical nanoparticle for excitation along the vertical axis. The field is most intense close to the particle; for the smallest particles, it is dominated by the electrostatic component of the field. Thus the degree of field confinement is dictated by the particle size, neglecting extra phenomena such as skin depth.

Transduction takes place here through the dependence of the resonance frequency on the surrounding medium index. Hence, as illustrated in Fig. 20.3b, measurement of either an individual particle or an ensemble of monodisperse particles (all of nearly the same size and dilute enough to avoid interactions between them) shows how much the resonant spectrum shifts. This shift can then be used to track the index surrounding the particles, much as we discussed for SPR, i.e. with assumptions on the profile of the matter to be sensed (molecular film, etc.). The more subtle aspect of developing this strategy is taking into account which particle shapes offer the largest sensitivity to the surrounding index. The dipolar excitation that we have sketched here is only

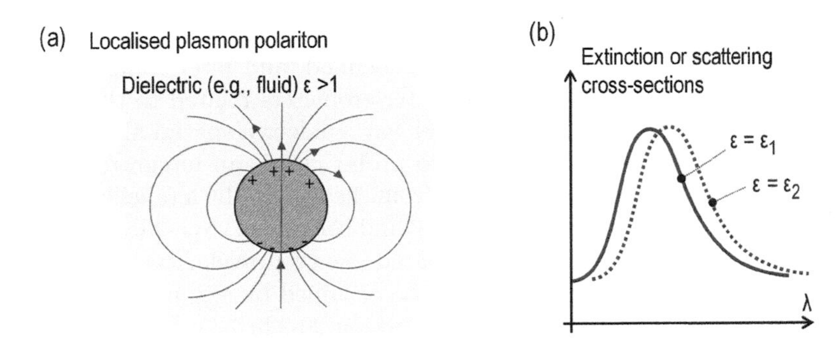

Fig. 20.3 (a) Sketch of the dipole mode of a plasmonic (metallic) nanoparticle, typically gold or silver. (b) Resonant scattering response of a single nanoparticle and its shift under the influence of an external index change.

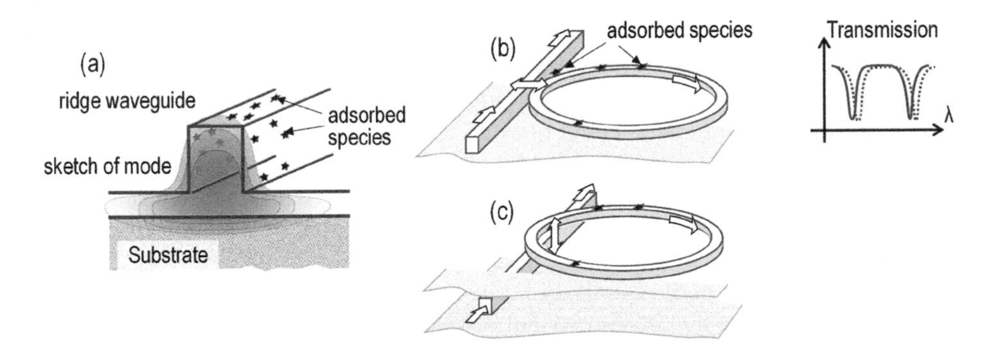

Fig. 20.4 Channel waveguide: (a) Ridge; (b) Straight bus guide and ring coupled in the same plane; (c) Vertical bus-ring makes coupling immunity to outer elements.

a canonical example; elongated shapes or star shapes with many tips offer sensitive 'hot spots'. The gap between two close nanoparticles offers a particularly sensitive 'hot spot', if it can be properly exploited.

20.2.4 Channel geometry

IO mostly makes use of channel waveguiding. A preferred geometry is a ridge fabricated in a guiding (core) layer by partially etching away said layer around the ribbon (see Fig. 20.4a). The mode field is more complex than in a planar guide, with six non-zero field components if symmetries are lost, as in this example.

But often a single component of E dominates; hence, the knowledge from scalar field and planar guide is still valuable, with Γ simply redefined as an integral over x and y. Still, two features differ in interesting respects: (i) there is more freedom to design a mode distribution that favourably samples the fluid; (ii) chemically, it may be difficult to get to a controlled situation on the top and side surfaces of the ridge, which have undergone different processes and have different hydrodynamic fluid coupling. A nanostructured coating may act to restore a more uniform situation.

For propagation in a single-mode channel, the phase of the mode $\varphi(\ell) = k\ell = k_0 n_{\text{eff}}\ell$ along a length ℓ (curvilinear possibly) is well defined, and interferences do not suffer from the 'localization' issues of free-space interferometers related to the source extent and variable paths. In other words, a channel waveguide is a spatial filter, acting all the way along. A preferred non-resonant way to probe $\varphi(\ell)$ is an integrated MZ. Assume one MZ arm without change (e.g. protected from the analyte by a repellent coating) and the other arm engineered for high sensitivity and selectivity. Specifically, a high *group* effective index boosts sensitivity (the associated low group velocity v_g also means that light spends more time for probing). It can be obtained by segmenting the waveguide in a subwavelength manner, as described in section 20.2.5.

Channel waveguides may also be formed into resonant sensors, such as the two variants of the ring geometry in Fig. 20.4b,c. Field enhancement in the resonator enhances the signal (see Chapter 7). In Fig. 20.4b, a straight bus waveguide skirts the ring. Transmission dips manifest the coupling, as shown. Critical coupling is targeted (see the asymmetric Fabry–Perot still in Chapter 7) associated with highly contrasted dips. These dips then shift due to adsorbed-species index changes on the ring walls; the sensitivity is mainly dictated by the local slope and thus the quality factor Q of the ring (the ring quasinormal mode's Q, since we have an open system).

The coupling itself may be affected in the tiny coupling area if, for instance, species are easily trapped there. The variant of Fig. 20.4c with the bus under the ring, embedded in the substrate, as shown in Fig. 20.4c, circumvents this issue: then only the top ring surface plays the requested role, and no change in coupling occurs under actual test conditions.

Another geometry of high sensitivity is the so-called slotted waveguide geometry. If a channel waveguide has a thin, largely subwavelength slot in its middle (say of width <80 nm in practice), then, due to the boundary condition of the x field component normal to the slot interfaces for quasi-transverse-electric polarization, a large value of the mode field inside the slot is achieved, and hence a particular sensitivity to species that arrive at the interior. This is a weaker form of the field reinforcement in the so-called metal–insulator–metal (MIM) structures seen for metamaterials and plasmons in chapters 18 and 13.

20.2.5 Periodicity along the propagation axis

- **Subwavelength grating.** We start this discussion of periodicity in sensing with the subwavelength channel geometry: in Fig. 20.5a, we show a channel waveguide which has been patterned into small dashes at a subwavelength scale of period a. This increases the sensing surface, with structures that are more open than the usual pores. But what about the way the field now senses the surrounding volume? In a subwavelength regime, we are in the first photonic band. Upon approaching the first gap, as seen in section 16.2, the mode tends to concentrate in the dielectric 'dashes'. Fortunately, even though this counteracts negatively (especially in the slow-light regime), the presence of the weakened field in the whole space between the dashes still increases the sensitivity in comparison to the unstructured guide one.

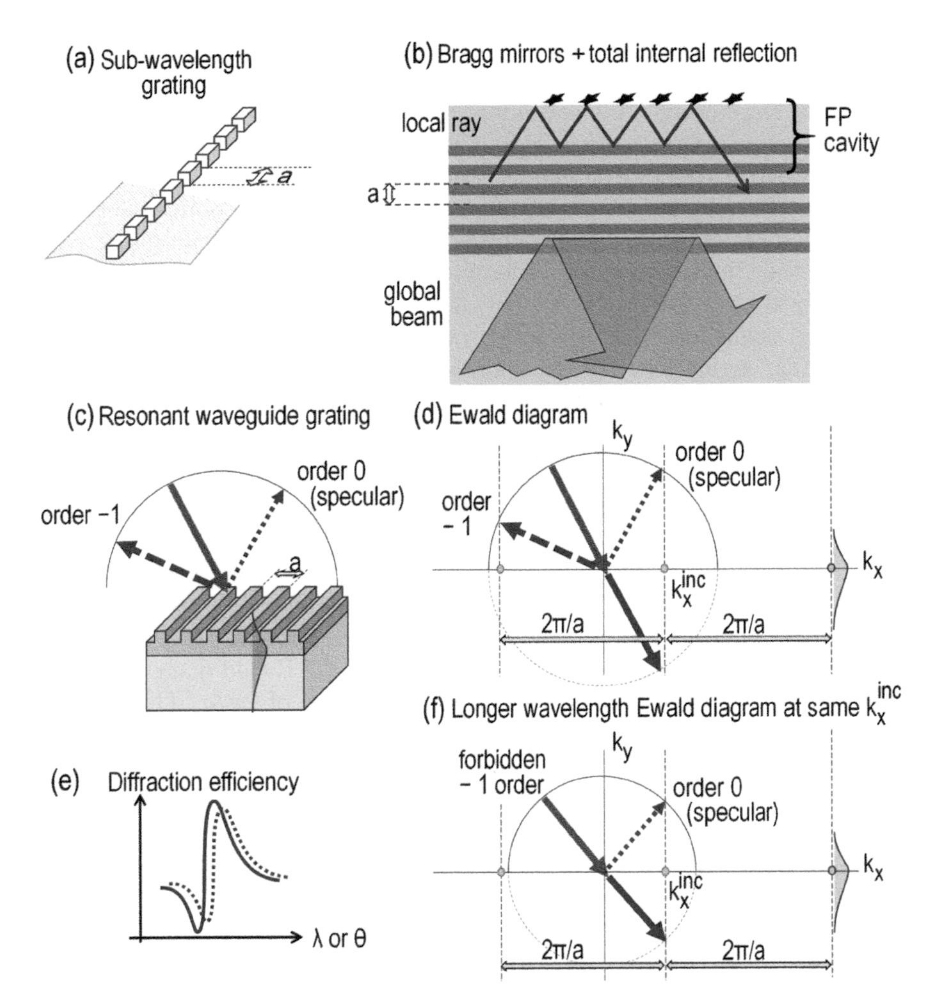

Fig. 20.5 Uses of periodicity: (a) subwavelength waveguide; (b) dielectric Fabry–Perot stack using a bottom Bragg mirror to enhance surface sensitivity, preferably used at total internal reflection condition; (c) resonant waveguide grating geometry with impinging beam, diffracted orders, and quasi-guided mode; (d) corresponding Ewald diagram at resonance: the guided mode wavevector differs by a multiple of $2\pi/a$ from the projection k_x^{inc} of the incident one; (e) typical Fano shape of the resonant diffraction efficiency vs wavelength λ or angle θ, and influence of index changes; (f) same as (d) but for a lower frequency, with only a specular order allowed on the top side.

- **Bragg mirrors.** We have dealt with truly guided waves, but such waves, are accessed only at the ends of the guide (the light is *end-fired*). Operation with a usual optical beam impinging through a prism, much as in SPR, is achieved in Fig. 20.5b: if the incidence is beyond the total internal reflection (TIR) limit at the sensed top interface, it is possible to strongly localize light there (Sinibaldi et al., 2013). This can be seen as an asymmetric Fabry–Perot cavity sandwiched between a

Bragg mirror and a mirror due to TIR. The near-perfect top mirror is the sensor surface. Only evanescent waves probe the fluid, experiencing the Goos–Hänchen phase shift (Exercise 5.6). The coupling to the guided mode is resonant essentially as in SPR, but with a much narrower linewidth, owing to the low-loss nature of such a dielectric system: coupling is done through an evanescent Bloch mode. For a really perfect asymmetric Fabry–Perot, as discussed in Complement 7.B, there are, instead of intensity dips and peaks, only abrupt phase jumps at resonance (those of the Gires–Tournois interferometer, popular in non-linear optics), which are awkward to probe in the context of sensors. However, there are always some losses, and operating at a well-chosen point not too far from critical coupling (in other words, engineering the quasinormal mode), a sizeable and very narrow dip appears as in the SPR technique.

20.2.6 Resonant waveguide grating (RWG) sensing

Now periodicity is added in a waveguide (hence an RWG). Therefore, the mode $k_{||}(\omega)$ can be resonantly excited if the matching condition $k_{x,\text{inc}} = k_{||}(\omega) \pm \frac{2\pi}{a}$ is satisfied. Waves are incoming not along the periodic direction, but, as shown in Fig. 20.5c, from above or below, as we saw about periodic waveguides in Chapter 17, when the situation in the dispersion diagram $\omega(k_{||})$ was above the so-called light line (see Fig. 17.9). Coupling of incident light to a guided mode is pictured in the Ewald diagram (wavevector diagram) of Fig. 20.5d with shifts $\Delta k_{||}$ integer multiples of $2\pi/a$. The exact picture that should be used is that of a quasinormal mode that we excite resonantly if we hit the right k-matching condition. The response of an RWG is due to the mixing of the direct specular wave (if we work in reflection) and a wave that comes, in naïve terms, from resonantly stored energy in the guided mode scattered by the grating. This results in the so-called Fano lineshape, as illustrated in Fig. 20.5e, and as is often met in nanophotonics. Of course, it is the presence of an abrupt slope (here shown in the specular order) that makes the system highly sensitive to the refractive index at its surface.

Sensitivity is dictated by the quasinormal-mode dispersion, mainly the guided-mode dispersion with the $2p\pi/a$ translation (integer p). This may cause, for instance, a strong change in angle for a small change in frequency, notably larger than the same change in an ordinary non-resonant grating. What such an RWG dictates is a link between wavelength and angle, pinned by the underlying dispersion of the waveguide mode. The simplest situation is in principle that of Fig. 20.5f, associated with a longer wavelength–period ratio than in Fig. 20.5d: there is only one specular and transmitted order in the light cone, but, nevertheless, the distribution of light among them shall vary sharply at resonance.

20.2.7 Extensions

For all the resonant devices we have seen, including the RWG, it is possible to also detect absorption instead of index change. The leading trend is of course, that more absorption provides more damping of the resonances. But the actual signal change is not trivial and depends notably on whether one works above or below the critical coupling condition: a faint resonant-reflection peak can become larger when adding absorption to the system if this absorption brings it closer to the critical point. We also note as another interesting extension that 'tensor' optical quantities can be sensed as well, such as birefringence or gyrotropy (rotatory power of fluids with chiral molecules, optical activity, or Faraday rotation). The intrinsically birefringent nature of all integrated optics waveguides is not an impediment. One can measure, for example, TE and TM resonances of a ring or an RWG, and compare the shifts. It can help with probing the reconfiguration of molecules (such as proteins or collagen). Such conformational changes may cause negligible 3D-averaged sensed index changes but should induce clearer birefringence or dichroism changes.

A more fundamental extension can be seen in inelastic photonic processes such as Raman scattering or fluorescence. Waveguides or resonators can be efficient on the excitation side or on the emission side, or even both. Some of the overlap factors invoked above will therefore acquire a double role (excitation and emission), and a large variety of enhancement configurations can be envisioned, but they would bring us much beyond the scope of this section.

20.3 Selected IO: passive and electro-optic devices

In this short account, we exemplify, thanks to a selection among the decades-old teachings of IO (well rendered in books such as ref. (März, 1994)), how a clever combination of spatial and modal engineering was exploited to build up useful devices. Since the desirable trend of miniaturization in IO generally involves large index contrasts, ray tracing is rather useless and modal engineering much needed. These games of modes may serve as an inspiration in nanophotonics, notably as regards tolerance to fabrication imperfections. The four devices of Fig. 20.6 are specifically selected as heuristic triggers along this line.

The MZ interferometer, Fig. 20.6a, requires a pair of so-called 1×2 beam splitters. In each arm, the phase can be controlled by an electro-optic change of the material index (core, cladding, or both). The electro-optic material that proved best to operate in the near-infrared at modulation frequencies well above 10 GHz was lithium niobate $LiNbO_3$. Unlike semiconductors, it lends itself very poorly to wavelength-scale confinement or to being neatly etched. Guidance is instead ensured by ion diffusion through an adequate mask. The diffused ions induce tiny index changes $\delta n \sim 10^{-2}$.

As for the beam-splitter function, two basic options are shown in Fig. 20.6b: the first is the Y-junction, but this often suffers from the problem that the very tip of the Y, exposed to the antinode of the incoming mode, must be perfect to within ~10 nm, a very hard task with optical lithography; the second is the directional coupler, whereby evanescent transfer couples energy into a side guide (as could be modelled by spatial coupled-mode theory; see Chapter 7). In this latter case, it is the control of evanescent transfer that implies extremely accurate fabrication (somehow as in the microring devices seen above). We mentioned above a more tolerant vertical coupling geometry, Fig. 20.4c, which obviously adds its own demands in terms of stacking and burying the waveguides. Hence, a preferred splitter that does not meet these difficulties is the so-called multimode interferometer (MMI), which is simply a piece of broad multimode waveguide, Fig. 20.6c, for which a physical explanation based on the series formed by the modes' phase differences is given in Exercise 20.2. Note that some of the

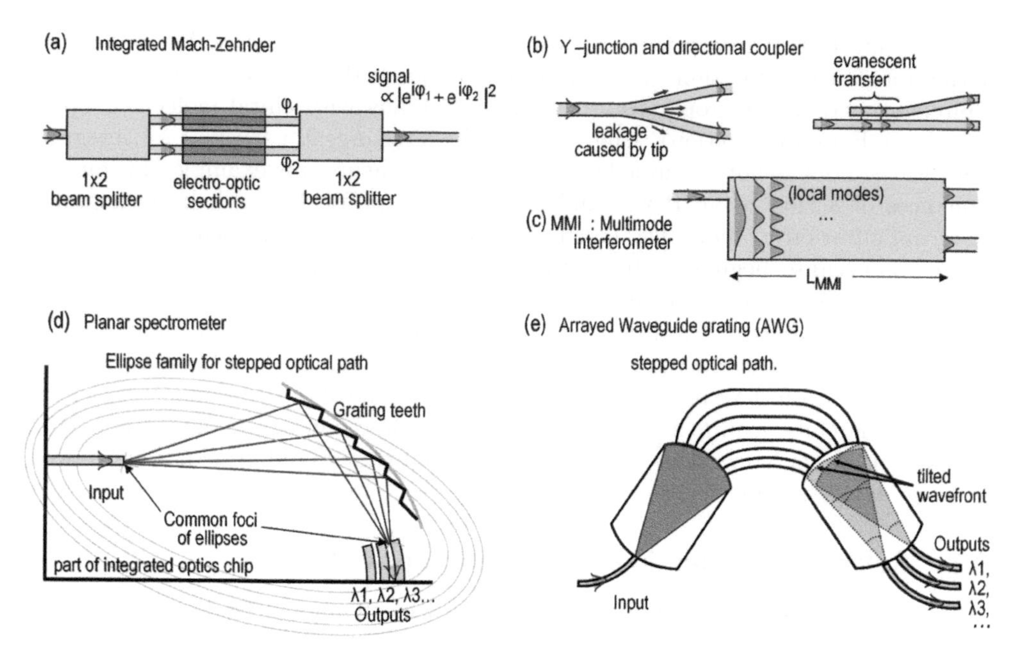

Fig. 20.6 (a) Mach-Zehnder modulator; (b) classical beam splitters: Y-junction and directional coupler; (c) multimode interferometer; (d) planar spectrometer: the design uses the constant optical path between ellipse foci, hence teeth facets are along ellipses, and ellipses whose focus-to-focus optical paths are stepped by a multiple of the central wavelength are chosen to reconstruct a converging wave at the second focus; (e) arrayed waveguide grating: optical paths between different waveguides connecting the splitting and recombining areas are also stepped by a multiple of the central wavelength.

aberration limitations of MMIs (imperfect phase difference series) can be tackled with clever nanophotonic solutions (Maese-Novo et al., 2013).

We sketch two kinds of spectrometer in Fig. 20.6d, e, with the intent of showing an evolution from classical ray design to modal design: the first, the planar spectrometer, Fig. 20.6d, is much like any bulk grating spectrometer: we refer here to the Rowland design, where a curved grating merges the diffractive and focusing functions (not the more common Czerny–Turner design with flat grating). What integrated optics easily allows is a custom phase portrait of the grating. Here, the requested constant optical paths for various directions (stigmatism) exploit the property of ellipses between their two foci. To also capture the grating function, the grating is built on a family of ellipses whose focus-to-focus optical paths differ by $\delta = m\lambda_0$ (m integer). This causes reconstruction of reflected light into a wavefront converging at the second focus for wavelength λ_0. Neighbouring wavelengths exit with a tilted front so as to focus on nearby exits, which are purposely implemented there. The good design of grating teeth that behave well under tilted incidences can be seen as a nanophotonic topic related, for instance, to the engineering of Bloch-mode reflection in cleverly structured systems.

In Fig. 20.6e, we show the integrated optics solution known as the AWG, invented by M. Smit in 1986 and also known as the 'Phasar' (Smit and Van Dam, 1996). It is very popular to mux (multiplex) and demux (de-multiplex) signals of different wavelengths, a key function in all the fibre-optical networks commonly deployed in the telecom industry. It is reminiscent of both the prism and the grating. Again, after a wavefront has been formed in the entry splitter or star, delays are imposed by adequately stepping the different waveguide optical lengths. At the design wavelength, a reciprocal focusing process takes place in the exit star, while at neighbouring wavelengths tilted wavefronts are formed, focusing the intensity at neighbouring exits. We remark that optimizing the robustness of design against fabrication imperfection, and also against temperature variations, for instance, can be seen as an 'anti-sensing problem', reciprocal of the sensing one discussed earlier, whereby *minimal* influence of the cladding index is sought.

20.4 Active devices and their multiple efficiency definitions

20.4.1 Stimulation, transparency, gain, and threshold

A description of light emission regimes through the use of Einstein coefficients was discussed in section 8.3 on semiconductors. Note that this description is equally valid for other emitters (e.g. organic dyes, film with quantum dots, etc., see section 19.5).

In all these systems, an electrical or optical pump feeds excited states. For low relative occupation of excited states, the regime of spontaneous emission prevails. For stronger occupation, stimulated emission acquires a larger role. Macroscopically, as pumping increases, we turn the bulk material optical absorption into optical gain for a given $1 \rightarrow 2$ transition of photon energy $E_{12} = \hbar\omega_{12}$, crossing a transparency regime at some point.

Above the transparency condition, optical waves are amplified. If there is a feedback, notably if amplification takes place in a resonator, we have the laser physics seen in section 8.7, whereby we have the threshold for lasing action. The usual vocabulary is then that optical modes are amplified (for a general modal discussion of open but gain-free systems, see quasinormal modes in Chapter 7). In classical laser language, the threshold is the point at which the gain factor $\exp(gL_{cav})$ can balance the mirror losses $r_1 r_2$ (and possibly other extinction losses caused not by the bulk medium but, for instance, by scattering along its walls if it is a guide).

Above this threshold, once we have the laser effect, a large number of photons build up into a few optical modes. Indeed the 'winner takes all' metaphor would apply in the absence of spontaneous emission, and all the energy would end up in a single mode. Due to spontaneous emission, subsequently amplified if geometry allows, other modes are also excited, leading to a broad multi-line spectrum. These lines are individually broadened by the random spontaneous-emission-induced phase diffusion, according to the textbook Schawlow–Townes physics. Nevertheless, what generally remains of the 'winner takes all' is still a huge channelling of photon energy into a few well-defined modes. This is the crucial point that makes laser sources much more power-efficient than spontaneous ones. The price to pay is the more complex laser device, with its feedback, and the onset of a large coherence of the generated field.

We are going to provide a metric for these effects. We only mention in passing the so-called superradiant sources. They have no feedback, and hence no strong modal structure. Still, geometry allowing, a large photon population is generated by the sole amplified spontaneous emission, with a much weaker degree of coherence than lasers.

20.4.2 Efficiency of lasers and β factor

To address the efficiency issue, we must see where generated photons are going. If the first two photons generated were kind enough to readily go into the same mode, stimulated emission would start with almost no threshold (a rate of two pump electrons/ns corresponds to a current flow of $I = 0.32\,\text{nA}$). The reason for a larger threshold is the huge number of other modes that receive the first emitted photons with about the same probability as the future lasing mode. The mode-counting argument for semiconductor lasers amounts to determining this number of modes as the ratio of the cavity volume to an elementary modal volume of $(\lambda/2n)^3$ (this can be found by adapting section 8.2.1 to the 3D or 2D photon case, $E \propto k$ and $s = 1$ in Eq. (8.9)). With a typical $300\,\mu\text{m}$ length in a laser-diode cavity, such a ratio quickly exceeds 1000.

The concept of the β-factor embodies this issue. It is the fraction of photons sent into the future lasing mode when still in spontaneous regime. In the simplified view of N-modes that are identically fed by emitted photons, we would just get $\beta = \frac{1}{N}$. Hence, wavelength-scale nanophotonic devices, featuring low N, are good for larger β factors. There is some more actual complexity, as even inside a simple resonator, all modes are not channelling the same amount of emission. In few-mode systems, the exact spatial positions of emitters matter: the photon probability of presence scales like $|E(x, y, z)|^2$, being zero at nodes and maximal at antinodes of the mode, so a local β could be defined, peaking at some value β_{max} at antinodes. For large systems (e.g. 300-μm-long edge-emitting laser diodes), all antinodes are similar, and the mode of interest is essentially a stationary wave made from two interfering plane waves, e.g. $E^2 \propto \cos^2(kx)$. In such a simple case, the overall β-factor would quickly average to $\frac{\beta_{\text{max}}}{2}$ if we dealt with electron–hole pairs uniformly distributed over the x-degree of freedom, typically one of the extended dimensions of a quantum well (QW) within a laser-diode stack.

When a log-log plot of the power output P_{out} of a laser vs the pump power P_{pump} or the pump current is made, two parallel asymptotes separated by essentially the β factor appear. At low power, $P_{out} \simeq \eta_{QE}\beta P_{pump}$ with η_{QE} an internal quantum efficiency, whereas at large power, ideally, $P_{out} \simeq \eta_{QE} P_{pump}$, since a large photon number shall channel almost all photons into the lasing mode: the 'winner takes all' argument.

20.4.3 Efficiency of LEDs

For LEDs, we work only in the spontaneous emission regime. To get sizeable power, there are generally many modes, although one particular direction may be quantized. For instance, we could deal with a QW emitting inside a wavelength-size Fabry–Perot vertical resonator of large horizontal extent.

From a narrow engineering perspective, there are only two macroscopic efficiencies, the most macroscopic being the *wall-plug efficiency*:

$$\eta_{\text{wall-plug}} = \frac{\text{Light output power}}{\text{Electrical input power}}. \tag{20.6}$$

For the specific use of lighting, $\eta_{\text{wall-plug}}$ is turned into lumen/Watt (W), where lumens (lm) account for the human-eye sensitivity, peaking at 555 nm; the record for white LEDs in the lab is around 300 lm/W. Note that the largest monochromatic efficiency is 683 lm/W for the absolute limit $\eta_{\text{wall-plug}} = 1$. But to produce white light, the admixture of less efficient wavelengths on the red and blue sides of the 555 nm green peak must reduce the peak value to around 400 lm/W (depending on the required white cue 'quality': the color rendering index, or CRI in technical terms), so the best devices at 300 lm/W are not very far from the absolute limit $\eta_{\text{wall-plug}} \simeq 1$.

The other efficiency is the *external quantum efficiency* (EQE): it is the number of emitted photons (effectively going outside the device) per injected electron. In other words, *quanta* are counted, not watts. For monochromatic emission with photon energy $\hbar\omega_{\text{emi}}$ (the typical spectral spread of 'monochromatic' LED emission is ~5–10% for photons $\hbar\omega_{\text{emi}}$ of 1 to 3 eV), the EQE reads:

$$\eta_{EQE} = \frac{\text{Light output power (in W)}}{\text{Electrical current (in A)}} \times \frac{1}{(\hbar\omega_{\text{emi}} \text{ (in eV)})}. \tag{20.7}$$

The product appearing in the denominator is the power when each injected electron is plainly converted into the same quantum, $\hbar\omega_{\text{emi}}$. The reader can see the reciprocity with detector/photodiode sensitivities, documented in A/W. Digging deeper into LED physics, the EQE is the product of two factors:

$$\eta_{EQE} = \eta_{QE} \times \eta_{\text{extraction}}. \tag{20.8}$$

The first term η_{QE}, is the quantum efficiency which describes the probability of an electron actually being converted into an *internal* photon (rather than heat), also discussed in section 19.6. The second term, $\eta_{\text{extraction}}$, known as the *extraction efficiency*, describes the probability of such photons making it to the outside of the device. This could be restricted to a limited solid angle outside—say, a cone—if light escaping from the chip edges is considered useless.

Extraction is a strong hurdle to overcome in order to exploit spontaneous emission from solids. For randomly emitted photons in a macroscopic cube of index n, it can be shown that the scaling of $\eta_{\text{extraction}}$ from a single face is $\frac{1}{4n^2}$ (Exercise 20.3), which is only 2.5% for the index $n \sim 3.2$ of the GaAlAs alloy family. Even for the smaller index $n \sim 2.5$ of nitrides (GaN), the extraction from a single face is only $\eta_{\text{extraction}} \sim 4\%$.

Hence, in LED physics, strategies based on nanophotonics attempt to repel photons from useless non-extracted modes to further channel them towards modes that do get out. Furthermore, increased $\eta_{\text{extraction}}$ must happen without degrading η_{QE}. For instance, the metal in the candidate plasmonic structures of Chapter 13, or in nanoantennas of Chapter 19, may induce too many non-radiative recombination channels, as has been studied since the 1980s (Ford and Weber, 1984).

20.5 Stimulated emission: Laser diodes

20.5.1 Edge-emitting lasers

The laser diode is an iconic optoelectronic component. Its edge-emitting version is a reference device with which we deal here. We have already substantiated how electrons and holes are confined to produce a gain medium in Chapter 8. The crucial concept for relating bulk gain and modal gain is again the confinement factor Γ, now for the active layer. We discuss in the following the *photonic* improvements to these devices (as opposed to electronic issues, doping, material quality, etc.). To fix the ideas, one can think of the basic interaction as that of one or several stacked QWs with a guided mode with reasonable confinement (see Fig. 20.1a).

20.5.2 The Fabry–Perot laser

The basis, sketched in Fig. 20.7a, is the Fabry–Perot laser diode, whereby a waveguide is delimited by two facets of the chip that serve as mirrors and thus form the Fabry–Perot cavity. The waveguide is vertically monomode; hence, its thickness is on the order of $\sim \lambda/2n \sim 300\,\text{nm}$ for typical near-infrared situations. Laterally, the guide is generally broader. There are a variety of lateral confinement strategies: they are quite technologically involved, as they must ensure compatibility with current injection (the top electrode is suggested atop a so-called *buffer layer* or current-spreading layer, the bottom one is far away vertically, the chip being traversed by current lines). A rather broad waveguide can be taken as a generic result—say, $2\,\mu\text{m}$ wide—the risk of multimode operation being mitigated by (i) a weak lateral index contrast, and (ii) the gain lateral profile, which best fits the fundamental mode.

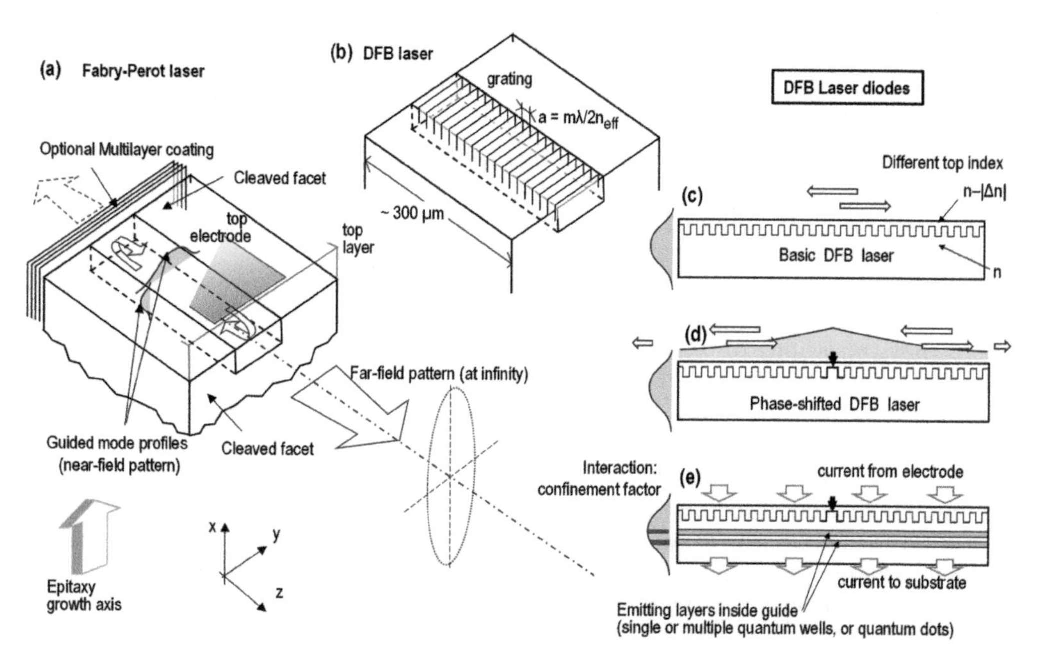

Fig. 20.7 (a) Edge-emitting laser diode with modal structure and far-field pattern; (b) distributed feedback (DFB) laser; (c) side view of DFB laser; (d) phase-shifted DFB laser with the phase-shift defect (black arrow) forming a cavity; (e) same with sketch of the embedded active layers and current path, from p-doped side (top) to n-doped side (bottom).

As for any Fabry–Perot cavity, the reflectivities can be varied. The starting point for facet reflectivity r is the semiconductor–air interface, with $r^2 \sim 30\%$ Fresnel coefficients for $n \sim 3.2$. A highly reflective multilayer coating can be deposited on the rear facet to privilege light output on the front facet. This also diminishes the threshold gain, g_{th}, according to the basic threshold condition $\exp(g_{\text{th}} L_{cav}) r_1 r_2 = 1$.

The output beam shape (say in air) comes from the diffraction of the near-field pattern, i.e. the field distribution $E(x, y)$ at the facet. The far field is the Fourier transform of this near field, in agreement with the plane-wave expansion of Chapter 14. It is thus of an elliptical shape, diverging more in the vertical direction (narrowest mode extent) than horizontally. This beam shape (which is not astigmatism: the two foci in horizontal and vertical planes still nearly coincide) causes a penalty for fibre coupling with modes of circular symmetry. It requires anamorphic prism-based optics which are awkward to implement. Efforts to change the mode shape can be based on nanophotonics. Adding a more complex architecture than the multilayer coating or defining some constrictions onto the waveguide path close to the targeted end can help shape the output mode.

What is the trouble with the simple Fabry–Perot laser diode? The fact that it supports multiple longitudinal modes. This is due to the broad gain profile (see Chapter 8), typically spreading over a 1–2% relative spectral width, $\Delta\omega/\omega$. In comparison, for a cavity length of $L_{\text{cav}} = 300\,\mu\text{m}$, the round-trip optical path is $\ell_{\text{round-trip}} = 2nL_{\text{cav}} \sim 2000\,\mu\text{m}$. Hence for the near-infrared, $\lambda \sim 1\,\mu\text{m}$, the relative free spectral range (FSR) is $\Delta\omega/\omega = (\lambda/\ell_{\text{round-trip}}) \sim 1/2000 = 0.05\%$. Hence there are a few dozen modes close to the top of the gain curve. In other words, the FSR is below 1 nm, while the gain exceeds a 20 nm width.

Hence, either lasing is multimode and thus poorly coherent, or, by some non-linear mechanism impossible to predict at the fabrication stage, a single mode occurs. Even so, the stability of this mode at variable current or against ageing cannot be granted. Photonic wavelength selection mechanisms intend to overcome these drawbacks.

20.5.3 The uniform distributed-feedback (DFB) laser

For photons with a specific wavelength to be favoured, their feedback must be increased inside the cavity. This can be done in a distributed manner, hence the name DFB, by inserting a grating of period a as shown in Fig. 20.7b, c. From the knowledge we have of the periodic waveguide, Chapter 17, it is known that the Bragg reflection stopbands (1D photonic gap) are dictated by the condition $a = m\frac{\lambda}{2n_{\text{eff}}}$ $(m = 1, 2, \ldots)$.

The grating itself, as shown in Fig. 20.7c, is generally first formed by etching the grown layers to obtain a pattern of adequate depth and aspect ratio. Next, the grating is embedded in a material of slightly lower index, typically with $\Delta n \sim -0.1$.

The width of the bandgaps is related to the strength of the relevant Fourier component of the dielectric constant (see Chapter 16). In the IO community, this is known as the grating coupling constant κ, and directly enters the coupled-mode equations in spatial version, as hinted in Chapter 7. Since we deal with guided waves of known profile (forward or backward propagating), assuming uniformity along the horizontal lateral direction y, the situation can be made similar to a bulk plane wave when seen from the top. What matters—for instance, for the first gap $(m = 1)$—is essentially the normalized integral $\int E^2(x)\Delta\varepsilon_1(x)dx/\int E^2(x)dx$, where $\Delta\varepsilon_1(x)$ is the first Fourier component along z (associated with period a) of the dielectric profile $\varepsilon(z,x)$ at height x.

A practical limit here is to avoid etching inside the active layers (see Fig. 20.7e) with sufficient margin, as this can induce non-radiative recombinations. Hence, the grating layer can only occupy a modest fraction of the vertical profile $E(x)$. This is not so problematic, though: the strength of the Fourier component for the suggested $|\Delta n| \sim 0.1$ value and a confinement factor of about 10% would typically result in a gap width on the order of 1 nm ($\Delta\omega/\omega = 0.1\%$), which is generally narrow enough to select a single mode among those that are still more or less imposed by the cavity facets (it is customary for DFB lasers to apply antireflection coatings on them, though).

The laser spectrum can now be guessed: the gap means that no propagating mode may exist in the frequency range. Stimulated emission from *propagative* modes then selects the two slowest modes in the curved regions of the dispersion branches: there is one such mode on each side of the stopband. Nevertheless, we have not won the battle of single-mode lasing. We can have two modes (frequently), and one mode by chance (facet cleavage), but even so, we do not know on which side of the stopband it is, resulting in a 1–2 nm uncertainty. Sampling among such lasers one by one is then undertaken to fulfil the requirements of telecom networks. For integrated chips with multiple lasers, no sampling can be done: *all* lasers should lase at their designed wavelength in a deterministic manner. We will examine in the following how a grating 'defect' in the so-called phase-shifted DFB laser diodes addresses this demanding performance.

20.5.4 The phase-shifted DFB laser

We are going to see how to favour safe single-mode DFB laser operation at the Bragg wavelength, and not at the stopband edges, as occurred in the uniform DFB case, by introducing a defect in the middle of the grating arrangement. We must leave the vision of a specific Bloch mode extended over the whole grating. But we can still consider Bloch modes for each half of the grating and wonder how we 'connect' them. We shall answer using the phase consistently, because the buildup of a resonant mode amounts to a round-trip phase being a multiple of 2π. Along this line of thought, the uniform grating is a useful basis to exploit, because at the exact Bragg wavelength λ_B, the fact that we are at the stopband centre means an optimal antiresonance condition. Thus, if we cut the infinite grating of Fig. 20.7b, c into two fictitious halves at one of its symmetry planes, the reflected phase φ_R of each half must *maximally prevent* the buildup of the mode. This occurs for $\varphi_R = \pi/2$: the round-trip phase $\varphi_{\text{round-trip}}$ seen at the symmetry plane is then $\varphi_{\text{round-trip}} = 2\varphi_R = \pi$ and causes destructive interference.

Conversely, if we now *add an appropriate extra section* between the two halves, $\varphi_{\text{round-trip}}$ can increase to $2\varphi_R + \pi = 2\pi$: we have gained the possibility of oscillating (and thus lasing) at the gap centre. This is the device sketched in Fig. 20.7d with a half-period ($\lambda/4n_{\text{eff}}$) spacer added in the middle of the grating (black arrow): one of the teeth or grooves has size a instead of $a/2$. It has become customary to call this laser a *quarter-wave* phase-shifted DFB laser or PSDFB laser.

From the point of view of photonic crystals, such an inserted longer section is nothing but a crystal defect. And it is well known that the mode sustained by a defect decays in the cladding if the frequency lies inside a stopband. This is also the case here. It can be shown either by full calculations or, more elegantly, by the coupled-mode approach that the Bloch mode profile envelope along z, $E(z)$ has the convex-hat shape of Fig. 20.7d. The coupling strength mentioned above, κ, often given in cm^{-1} by practitioners, is actually the quantity that defines the characteristic decay length of the mode field according to $L_{\text{decay}}\kappa = 1$.

Let us put some typical numbers on this. The strength of gratings was seen to give $\Delta\omega/\omega \sim 0.1\%$. It also intuitively means that the wave samples a thousand periods to sufficiently build up the feedback. Hence, since for telecom lasers at 1550 nm the period is $a = \frac{\lambda}{2n_{\text{eff}}} \sim 230\,\text{nm}$ (for first-order feedback, i.e. $m = 1$), this means a decay length of $230\,\mu\text{m}$ and a $460\,\mu\text{m}$ chip length of the laser diode in Fig. 20.7d. Note that facets are often antireflection-coated to suppress their perturbative role.

Overall, modal management of a PSDFB laser diode is shown by the two schemes of Fig. 20.7d, e: (d) shows the longitudinal envelope of the Bloch modes peaking at the grating defect; (e) reminds the reader that the diode works by injecting current along the vertical axis into active QWs (as in Fig. 8.14a), and that these QWs require a large overlap with the guided mode. They should also lie far enough from the grating layer to avoid the local damage induced upon etching the grating.

20.5.5 Vertical laser diodes: VCSELs

We now examine vertical cavity surface emitting lasers (VCSELs). A substantial difficulty with edge-emitting laser diodes is the need to cleave them to get the device and the light out, hence huge packaging, mounting, and aligning costs. Around 1990, it became possible to make lasers with vertical beams confined between two Bragg mirrors (often called distributed Bragg reflectors, or DBRs); see Fig. 20.8. Obviously, these lasers can emit prior to any cleavage and can be clustered on chips so as to form dense arrays of tens of lasers, providing as many information channels for very-high-aggregated-rate telecom transceivers. Epitaxial growth of high-quality $\text{Ga}_{1-x}\text{Al}_x\text{As}$ alloys, alternating between very different Al contents x and reaching a substantial refractive index contrast ($\Delta n \sim 0.5\Delta x$ for the GaAlAs family), made it possible to produce highly reflective Bragg mirrors suffering only minute absorption and scattering losses. For instance, attaining $R \simeq 99\%$ requires $\sim$20 layer pairs.

The reason why such highly reflective mirrors are needed is as follows. The gain layer in a VCSEL cannot be very thick. Remember that pumping a QW to transparency entails a current density of $J \sim 100\,\text{A/cm}^2$ (see Chapter 8). For multiple QWs (QWs), the optical gains are now clearly adding up in series, vertically. But note that the *currents* that feed them are still in parallel and add up. Currents are arranged much as they were in edge-emitting lasers, with electron and hole flows of opposite direction (thus with corresponding currents of the same downward direction) coexisting between the individual QWs. Hence, to remain with reasonable current densities, e.g. $J < 1000\,\text{A/cm}^2$, no more than a dozen QWs can be stacked. This amounts to having about 200 nm of bulk active layer if we compare to the bulk, because the wave does not travel *along* the QWs, as occurred inside edge-emitting lasers: it now travels *across* them. It has been seen that in order to work with weak mirrors (facets with Fresnel reflection $r^2 \sim 30\,\%$, as seen above), a cavity length $>100\,\mu\text{m}$ is needed because bulk

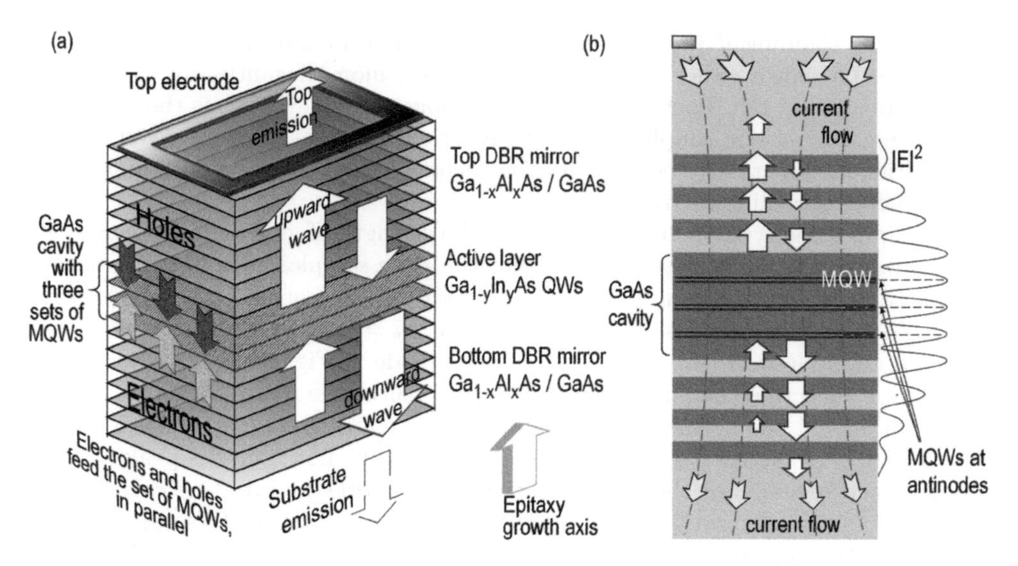

Fig. 20.8 (a) Overview of a vertical cavity surface emitting laser, lasing between top and bottom DBR mirrors; (b) profile of the resonant mode with several antinodes, Bragg mirrors, and quantum wells organized as three sets of multi-quantum-wells (MQWs). Current lines are also indicated. Note that the MQWs are, electrically speaking, fed in parallel, as suggested by the electron and hole flows in (a); these flows coexist in the GaAs central region.

gain is typically limited to 100–500 cm^{-1}. With an effective active medium length L_{active} as short as $0.2\,\mu\text{m}$, the gain per pass across the QW stack is at best 0.01 ($\exp(gL_{\text{active}}) = 1.01$). To lase, each mirror should spoil less than 1% of incoming photons, leading to the $r^2 \lesssim 99\,\%$ reflectivity lower bound.

Working with many QWs embedded in the central GaAs cavity is thus recommended. The best compromise is currently to use $\text{Ga}_{1-x}\text{In}_x\text{As}$ QWs emitting around 960 nm, where index contrast is large and injection is efficient (thus not compromising QE). Nevertheless, the standard for VCSEL transceivers has become the less optimal 850 nm one, using basic GaAs QWs sandwiched in between GaAlAs Bragg mirrors. This is because strained In-based QWs were not mastered when stabilizing this initial standard, which addresses mostly short-haul optical links, given fibre absorption limits.

One physical aspect of VCSEL strongly differs from edge-emitting lasers. The standing wave now has nodes and antinodes between the Bragg mirrors (see Fig. 20.8b). This stronger local intensity means a stronger stimulated emission: QWs must be preferably located at field *antinodes.* It is not possible, however, to locate, say, 10 QWs at the same antinode: with a pitch of 20–5 nm between QWs to account for electron and hole barriers, 250 nm would be needed: more than the distance between two antinodes (the Bragg period is $\lambda/2n \simeq 130\,\text{nm}$ at $\lambda \simeq 850\,\text{nm}$). Hence, it is preferable to slightly expand the cavity thickness to accommodate two to four antinodes and to locate about three QWs per antinode. Overall, beyond the technical demand of mastered growth of long non-trivial layer sequences, this is worth noticing as a case where light–matter interaction has to be carefully engineered with an accuracy of 10–20 nm.

A last point concerns the current traversing the system. We only mention the electronic issue of balancing the recombination currents among all 10 or so QWs—in other words, the issue of vertical injection uniformity, a demanding one in terms of modelling or probing. As for horizontal uniformity, it is actually not desired. The current is fed from a hollow ring on the top, inducing larger densities on the side. But more local-field intensity and more current density is needed in the device centre to fit the in-plane mode profile that dictates local stimulated emission rates. To get this uneven distribution, diaphragms for current are defined in the form of inserted insulating rings in the upper Bragg mirror that also act as focusing lenses, assisting the mode field concentration inside the VCSEL pillar rather than at its circular boundary. Nanophotonic improvements may further be implemented on the top surface to manage in-plane polarization issues such as polarization degeneracies, etc.

20.6 Other surface-emitting lasers

In this section, we describe lasers whose cavity extends in the wafer plane but features an added photonic structure to extract light above or below it, and provide other interesting glimpses into modal control. Considering in a first step systems along a single direction, the band structure $\omega = \omega(k)$ of a periodic guide like those of Chapter 17 has the generic features of 1D photonic crystals underlined in Fig. 20.9a: neglecting various other dispersions, the dispersion relation $\omega(k)$ without periodicity ('empty lattice') just follows $\omega = kc/n_{\text{eff}}$. With a grating, $\omega(k)$ is modified by gaps induced by Fourier harmonics of the grating modulation, as seen for DFB lasers. The gap for a first-order ($m = 1$) DFB laser at $k = \frac{\pi}{a}$, well outside the light cones of top and bottom media, cannot provide *surface* emission.

The gap of interest for possibly radiating outside is the second gap (with period $a \simeq \lambda/n_{\text{eff}}$; i.e. $m = 2$ with respect to the equation of DFB lasers in section 20.5.3). When looking at the unfolded and folded versions of $\omega(k)$, a point just below this gap corresponds to the guided mode for the unfolded k location indicated by [ii], $k = k^{(+1)}$. However, targeting now the point closer to the $k = 0$ region (the Γ point) indicated by [i], $k = k^{(0)}$, it corresponds to a plane wave radiating vertically. Such considerations are closely related to those about RWG (see section 20.2.6).

A main difference, though, is that the source is inside, not outside. Another point is that the coherence of the lasing wave is caused by the grating feedback at second order (the coupled guided modes differ by $\Delta k = \frac{4\pi}{a}$), while the very same grating provides coupling to the air at first order (with a small k-value), differing by $\frac{2\pi}{a}$ from the guided wave. From the Floquet–Bloch (and quasinormal-mode) viewpoint, these are not independent waves, only various harmonics of the same mode.

The bottom scheme associated with [i] shows the waves going to air and substrate at a given frequency. There are four waves because $\pm k^{(0)}$ correspond to the same frequency, with an ω-dependent angle $n(\omega/c)\sin(\pm\theta) = \pm k^{(0)}$ (much as in RWG). There is a degenerate limiting case, $k^{(0)} = 0$, whereby the four waves merge into two waves. The finite system size blurs the k-resolution, though, so the outside beam is never a plane wave; rather, it is a diffraction-limited beam, coherent over the whole

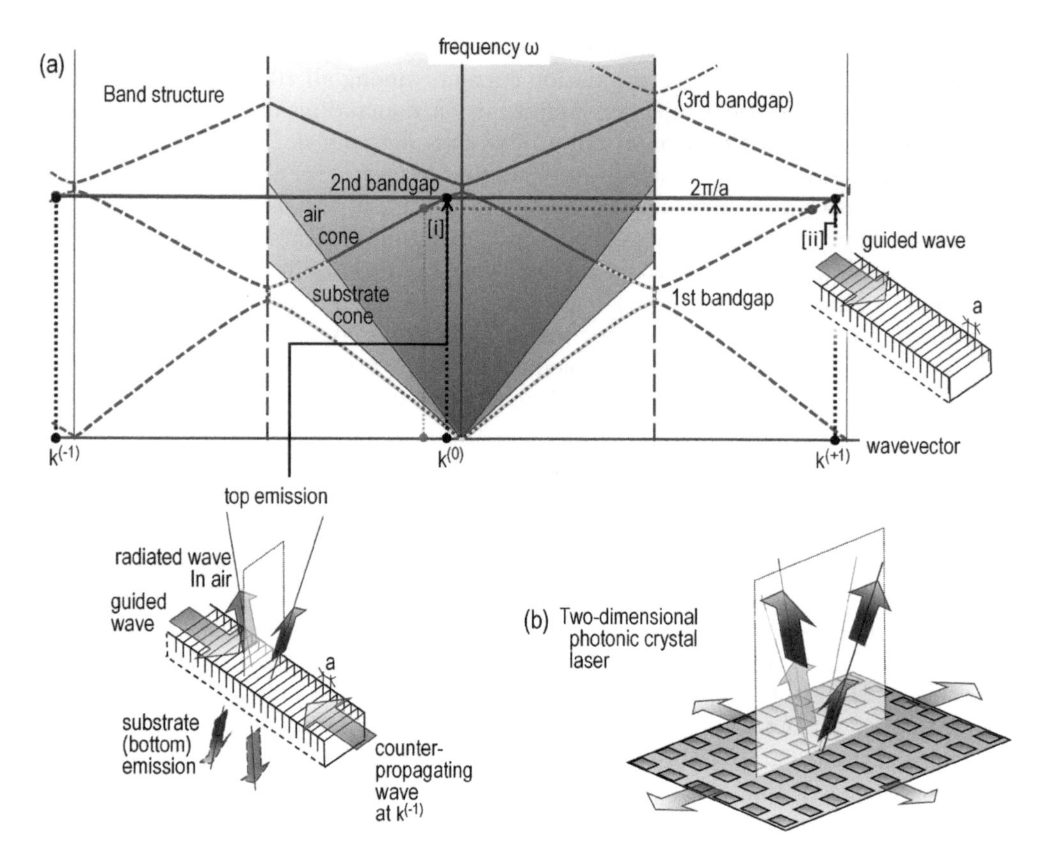

Fig. 20.9 (a) 1D band structure and relation to the surface-emitting laser; (b) 2D surface-emitting laser that can also be seen as a laser with feedback through a photonic crystal.

device length. Thus, for a few thousand periods, it has a divergence well below 1 mrad: much less than other types of lasers seen above (e.g. VCSELs or edge-emitting).

All these results can be generalized for a 2D periodic extracting grating, which can be considered here as a 2D photonic crystal (see Fig. 20.9b). The system can then lase coherently in the plane in an adequately pumped area, with several guided beams/spatial harmonics that obey the same Floquet–Bloch rule: they are related by one of the reciprocal lattice wavevectors $\mathbf{K}$ of the photonic crystal. Here, for the sketched square lattice, a given $\mathbf{k}^{(0)}$ harmonic is linked to at least four other harmonics using $\pm\frac{2\pi}{a}$ shifts along x and y. It is also of interest here to approach the degenerate case, and to get an outside beam that is diffraction-limited by the whole device size (at the supposedly single lasing wavelength). Recent realizations of such lasers do offer such impressive beams, which are unusual for these very flat 'coin-shaped' laser sources (Noda et al., 2017).

Note that the situation of gratings that both extract out of plane and provide in-plane feedback was known early in laser optoelectronics, when the appeal of DFB structures had to circumvent the technical challenge of carving gratings for first-order

DFBs with periods below 250 nm. Operating at the second order led to easier processes, with the exposure of resist masks becoming more feasible by holographic techniques. Playing with the Fourier components through the mark–space ratio (also known as the fill factor), it is not possible to suppress the first gap, but feedback at the second bandgap can still be favoured.

20.7 Spontaneous emission devices

20.7.1 LEDs: Main categories

In LEDs, electron–hole pairs are generated, and their recombination produces internal photons yet to be extracted. This picture is valid for standard inorganic LEDs based on semiconductors. It is also valid for organic LEDs (OLEDs), whereby the electronic band edges are to be replaced by so-called LUMO (lowest unoccupied molecular orbital, akin to a conduction band) and HOMO (highest occupied molecular orbital, akin to a valence band). The much lower conductivity of organic compounds translates into a much lower limit to the maximum areal current density J_{max} of the devices. This results in a much lower irradiance (also called brightness), the spectrally integrated power per unit area and unit solid angle (W/sr/m^2).

Specifically for lighting purposes, units of lumen, candelas, etc., take into account the human-eye sensitivity in the frame of Commission Internationale de l'Eclairage (CIE) norms. All in all, for use as a source in an apparatus such as a video projector, a very bright source is needed: OLEDs are poor in this respect. On the contrary, for very extended sources, say above 5–10 cm in size, direct tiling by semiconductor chips is very expensive, whereas an OLED solution is comparatively much cheaper.

The issue of extraction holds for both LEDs and OLEDs, but often differs in detail, even among LEDs. We distinguish three categories below, to make main trends more apparent.

(1) **Classical semiconductor LEDs.** Historical LEDs were made using the GaAlAs family, based on the availability of GaAs substrates (LEDs must be produced by billions!). A sketch of a chip structure is shown in Fig. 20.10a, alongside a packaged version in an epoxy shape such as a gently focusing dome in Fig. 20.10b. In the (AlGaIn)As family, a range of wavelengths from near-infrared to deep red can be emitted (660–1000 nm). The orange and yellow-green ranges have been addressed for a long time by the GaP family, which does not lend itself to heterojunctions: specific material defects, e.g. based on nitrogen, are at work in these old LEDs (however, the lattice-matched GaAlInP alloys emit up to the orange-red range: ~600 nm). Thus, modal control through growth of dedicated layers was mostly done using the (AlGaIn)As family.

(2) **Nitride LEDs.** The prospects for efficient blue emitters were poor until the 1990s. Only SiC, a crystal probed by Russian pioneer Losev in around 1920, gave a faint broadband blue glow. The II-VI family, based on ZnSe substrates, sparked some hopes, but was unable to produce reliable blue-green laser diodes. Then, around 1990, came the finding by Japanese teams that gallium nitride (GaN, whose gap is in the near-ultraviolet, $\lambda \simeq 393$ nm) could be grown and doped in both types,

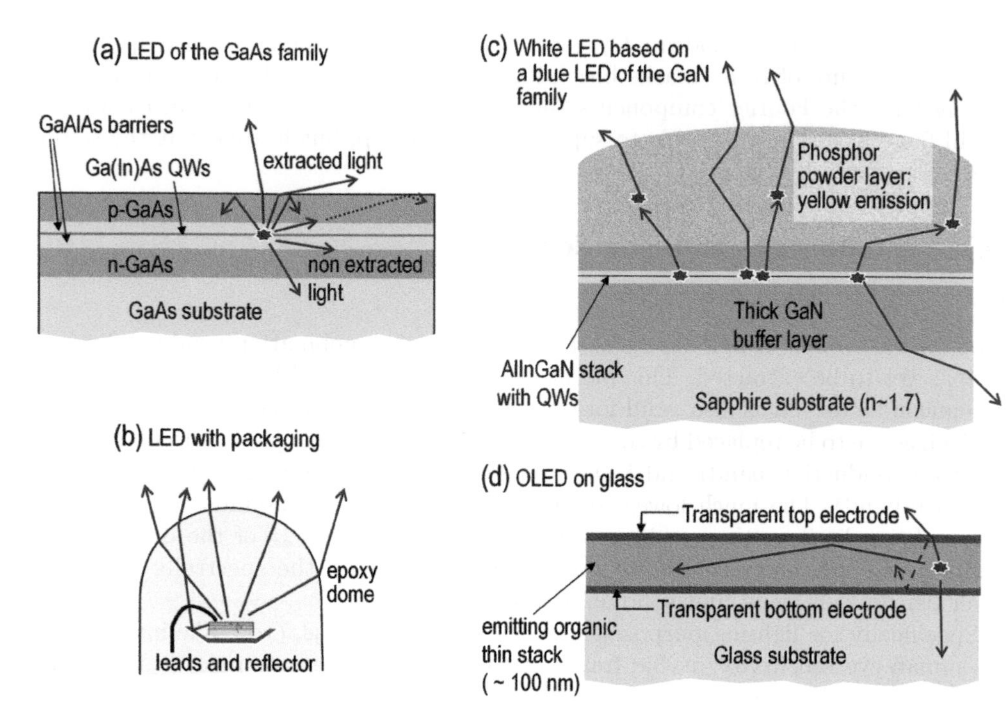

Fig. 20.10 Three families of light-emitting diodes: (a) based on gallium arsenide (infrared to red); (b) packaging with reflector and epoxy dome; (c) based on gallium nitride GaN (blue-green) with yellow phosphor on top to get white light; (d) OLED with thin conductive layers serving as transparent contacts.

especially *p*-type. They were awarded no less than the Nobel Prize in 2014. GaN (or, more precisely, the blue-emitting GaInN), appeared to be an extremely efficient blue emitter, in spite of a huge density of defects due to its highly mismatched growth on sapphire (crystalline Al_2O_3) substrates.

Altogether, GaAs and GaN families do cover blue and red-orange ranges, but leave an 'efficiency gap' in the yellow-green range (500–550 nm): just that of maximum eye sensitivity. The current solution to get white light, which triggered a revolution in lighting and energy saving, is to spread a relatively thick layer of fluorescent phosphors onto a blue LED (see Fig. 20.10c). A large fraction of blue emission (but not all of it) is absorbed by the phosphor, a powder based on cerium-doped garnet. It emits brightly in a broad yellow band. The combination (blue + yellow) is seen as white, but the bumpy spectrum may alter the vision of colour nuances. Affordable improvements are sought, e.g. through nanophotonic effects.

(3) **OLEDs.** Organic LEDs capitalize on the availability of emitting species with bright yellow-green emission, just in the 'efficiency gap' of LEDs. As a rule, an OLED stack is much thinner than a semiconductor one. Unlike LEDs, it is not contacted through side or top tree-like electrodes, but through two thin transparent conductive layers (Fig. 20.10d), e.g. conductive oxides such as the so-called ITO (indium tin oxide), which is also used for the ubiquitous touch-sensitive screens.

20.7.2 Extraction in LEDs

Nanophotonic approaches can be helpful to increase $\eta_{\text{extraction}}$. What is the starting point? A GaAs-based LED is essentially a cube with an index around $n = 3.5$. The extraction from a cube face being $\eta_{\text{extraction}} = \frac{1}{4n^2}$, it is as low as 2.0%, thus leaving large room for improvement. In the GaN LEDs, however, $n \lesssim 2.5$, and furthermore, the sapphire substrate, mentioned above, features $n \sim 1.7$. An interest of larger extraction is also that any extra radiation obviously limits heating, given that LEDs commonly and universally fade upon heating (see Chapter 8: if one reasons at constant current and electron and hole densities, more non-radiative channels are available in the higher-energy areas of the bands, i.e. those that must be more occupied by Fermi–Dirac statistics at higher temperatures). This is unlike classical tungsten-filament bulbs (hot black-body emitters in the 2000–3000 K range) with peak emission classically in the near-infrared, which blue shifts favourably upon heating. So, for LEDs, enhanced extraction gives a positive second-order feedback on efficiency. Finally, in OLEDs, the core thin sheet is 20–40 nm thick, and features an index $n = 1.6$–1.9. Thus extraction might seem less of an issue. However, the basic starting point is still $\eta_{\text{extraction}} \lesssim 50\%$, and often ~30%. On the remaining 70%, a large part goes into a guided mode or a mode with some plasmonic component if the contact layer has enough electron density to induce a metallic dielectric constant (think of the Drude model). Methods that extract a guided mode or avoid feeding such modes are certainly quite interesting for application to OLEDs.

20.7.3 Enhancing extraction efficiency by planar means

What is the impact of a vertical microcavity on light emission? If an emitter of planar geometry (e.g. a QW) is embedded in a Fabry–Perot (micro-)cavity, its emission feeds the various Fabry–Perot modes (or quasinormal modes). Assume the cavity is a slab of index n and width W, and the Fabry–Perot mirrors are ideal ($R + T = 1$). We assume that the emission is the same in each mode because the density of photon states is the same for all modes (see Exercise 20.3).

Thus, for a cavity supporting five modes at the central emission wavelength λ_0 (cavity order $m_c = 5$), each mode channels 20% of the emitter power (see Fig. 20.11a). The modes have quantized vertical $k_z \equiv m\pi/W$ with $m = 1, \ldots, 5$, as pictured in the wavevector diagram (bottom left). Their in-plane momentum is $k_{||} = (n^2k_0^2 - k_z^2)^{1/2} = (n^2k_0^2 - (m\pi/W)^2)^{1/2}$. Only the highest-order mode, $m = 5$, has a chance, if well tuned, of ensuring a momentum $k_{||} < k_0$ inside the air light cone (steep oblique line in wavevector diagram). All four other modes are outside the light cone. Still, this scheme achieves $\eta_{\text{extraction}} = 20\%$ for $n = 3.5$ instead of $\frac{1}{4n^2} = 2.0\%$. If the mirrors are asymmetric ($1 - R_{\text{bottom}} \ll 1 - R_{\text{top}}$), then light gets out by the top face only. We have thus gained a factor of 10 over the bulk trend by means of a microcavity. The cavity order chosen here $m_c = 5$ is realistic, even if the mirrors are distributed ones (GaAlAs Bragg mirrors, with a reflected phase behaviour similar to a remote mirror, causing $m_c = 2.5$ for each of them; see Fig. 16.5 and Eq. (16.19) on penetration depth).

The general rule is to compare the fraction of light in each mode $1/m_c$ (with a perfect bottom mirror) to the bulk fraction at one face, doubled by an incoherent mirror effect

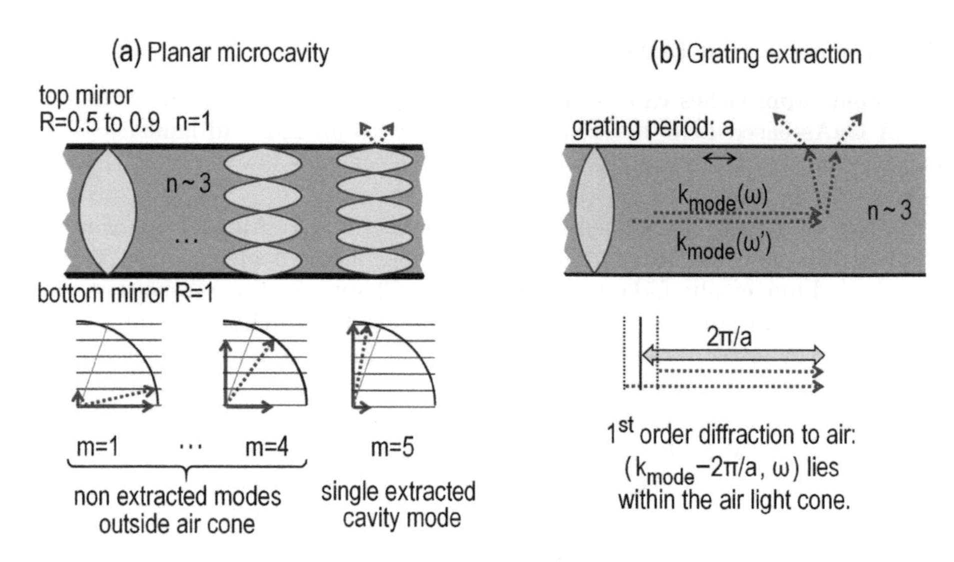

Fig. 20.11 Extraction enhancement principles for light-emitting diodes: (a) based on a planar microcavity, simple case with five modes whose field patterns in the cavity are shown with corresponding wavevectors below; (b) based on grating extraction.

(e.g. due to a metal contact at the chip bottom); hence, $\frac{2}{4n^2} = \frac{1}{2n^2}$(Benisty et al., 1998): if the equivalent order m_c is such that

$$m_c < 2n^2, \tag{20.9}$$

then extraction enhancement results from the cavity.

20.7.4 Enhancing extraction efficiency by micropillar structures

Using microcavities in III-V semiconductors as a basis, a simple and successful option is to confine light in the plane by etching a micropillar structure, typically requiring steep near-vertical sidewalls of several μm depths. Such pillars are akin to vertical fibres with a very high index contrast. They prevent light emission sideways and, if of adequate size, efficiently discretize lateral degrees of freedom towards a 3D microcavity. If emitters such as quantum dots (QDs) are used, then they are reasonably immune to etching damage and can be easily photopumped through, e.g., the III-V Bragg mirrors as in VCSELs. These monolithic structures have been analysed in detail since the late 1990s, as they embodied the first demonstrations of sizeable Purcell effects of monolithic systems (Gérard et al., 1998). They can also feature high β factors of spontaneous emission in the lasing mode and even reach the strong coupling regime at low temperature (Dousse et al., 2008). Transforming them into electrically tunable bright sources of single photons for use in quantum optics is a main application direction (Senellart et al., 2017). Tunability aims here at aligning the InAs QD narrow emission line with the microcavity resonance frequency and can be based on the Stark effect.

20.7.5 Enhancing extraction efficiency by in-plane structures

Can in-plane structures such as gratings assist extraction in otherwise planar LEDs? Reasoning on a semiconductor membrane held in air may clarify the issue (see Fig. 20.11b), even though current injection is awkward. There are in this case two clear-cut fates for generated photons: either they are extracted to the air through the interfaces, or they feed guided modes along the membrane. So the extraction issue can be restated as: 'can guided modes be scattered to the air'? It was shown in sections 20.2.6 and 20.6 and Chapter 17 that a guided mode becomes a leaky quasi-guided mode once its (folded) dispersion enters the light cone of the target medium.

For a given lattice, this means having a high enough frequency or, conversely, for a given frequency, a large enough lattice pitch, to be able to fold the guided mode dispersion early enough for the targeted frequency to correspond to a dispersion point folded in the air (or substrate) light cone.

Not all gratings of sufficiently large pitch extract guided modes well in actual structures, though. Firstly, not all modes are guided modes. Some leak at oblique incidence without being either guided or extracted. They should be minimized. Since they often exist only in the lower half-space, they correspond by reciprocity to light bouncing at the top interface with total internal reflection. It is then likely that a field *node* exists in the field pattern. Then its location might be a good height to locate a QW, as it would not emit in this useless direction.

Another point is that the guided modes do not have an infinite life: they have their own extinction length, due to the combination of reabsorption and scattering by items other than the grating. Extracting them means competing with this extinction so as to get photons out before their energy dissipates. This entails that the grating features should overlap all relevant modes strongly enough. This can be an issue, especially in GaN LEDs, which usually exploit thick buffer layers and for which the few fundamental guided modes, which are well located to collect a substantial fraction of the emitted power (they have antinodes at the QW height), may be buried at quite some depth under the surface. In OLEDs, the difficulty rather comes from the compatibility of the grating structure with the thin sophisticated sandwich.

Overall, the grating strategy is a natural one. It has paid off in most cases. It does not demand that the active part of the stack be engineered, only the top part. The extraction pattern can be borrowed from remarks in section 20.6: gratings induce an angular-spectral dependence on extracted light which will possibly lead to annoying chromatic effects for LEDs. Then the regular, ordered nature of the grating must be challenged. Given the isotropically distributed guided light inside the LED active plane, there is no actual need for the grating to have a crystalline order. What matters most is to scatter light out with the proper local periodicity. Hence, the various orientationally disordered geometries offered by modern physics can be employed: quasi-crystals, non-periodic tilings, superperiodic tilings, fractal rules, multiperiod gratings, etc.

Finally, a topic that is still open to discussion is the comparison between a dielectric grating obtained by etching or by addition of materials, and a more metallic grating which is liable to absorb but also possibly better at strongly scattering a guided mode, hence better at competing with its extinction length.

To conclude, in spite of the simpler perception generally associated with LEDs in comparison to lasers, optimizing photonic nanostructures for extraction from LEDs brings us to a rich spectrum of concepts.

Exercises

Exercise 20.1 Asymmetric waveguide for sensing.

(1.) Based on the asymmetric profile of Fig. 20.1b, consider a core $n_2 = 2.2$ (i.e. oxides such as Ta, Ti, or Hf-based ones), a bottom cladding $n_1 = 1$ (air) and a top cladding of aqueous solution, $n_1' = 1.35$. Try to find the thickness of the core from the limit case $n_{\text{eff}} \simeq n_1'$ (apart from the systematic wavelength scaling, so no specific wavelength is proposed). The profile in the core is then a cosine with an extremum at the higher-index-cladding interface (the top). Discuss how a similarity can be drawn with a symmetric case with double core thickness and same effective index. From the figures of Chapter 6, locate the symmetric case vs the case of minimal profile width.

(2.) For a glass or silica bottom cladding ($n_1 = 1.5$), is it desirable that n_{eff} approaches n_1 (from above)? What range of effective index is liable to mitigate the adverse effect of this high-index bottom cladding? What is the corresponding core thickness?

(3.) What could be of interest in drilling deep nanohole arrays in the case of a glass or silica bottom cladding (see chapters 15, 16, and 17)?

(4.) Alternatively, nanoporous silica with $n \sim 1.22$ can be used as the bottom cladding. Using Chapter 15 on composites, what is the air fraction needed to achieve this (use $n = 1.48$ for bulk silica)? What would be the issue if $n \sim 1.05$ was desired?

Exercise 20.2 Multimode interferometer (MMI), Fig.20.6c. We can use the slab notations to treat the MMI as a 2D geometry, even though it is also confined in the direction normal to the figure. The field created at an input can be seen as an exciting field, $|\Psi_{exc}\rangle \equiv \{\mathsf{E}_{exc}(x), \mathsf{H}_{exc}(x)\}$. It can only excite forward-propagating modes $|\Psi_m\rangle$, which form a complete set and carry away all the exciting power (no reflection). To know how much each guided mode takes away, we write the field in the form $\sum_m a_m |\Psi_m\rangle$ and equate it to the field at the input. The coefficients a_m can be calculated from their projections $a_m \equiv \langle \Psi_m | \Psi_{exc} \rangle_{\mathrm{P}}$.

- Discuss the validity of the approximate profile form of the perfect-conductor boundary $\mathsf{E}_m(x) \simeq d^{-1/2} \sin(\frac{m\pi x}{d})$ for such modes: see Complement 6.D (note that neglect of radiated modes is built-in in this picture).

- Using the Helmholtz equation and the sine second derivative inducing a factor $-(\frac{\pi}{d})^2$, show that

$$\beta_m \simeq k_0 n_{core} - m^2 (\pi/d)^2/(2k_0 n_{core}). \tag{20.10}$$

- Deduce that the propagation constants of all the lower-order modes (those with the largest β and n_{eff}) are spaced by multiples of $\delta\beta = (\beta_2 - \beta_1)/3$.

- Denote ϕ_L by $-L(\pi/d)^2/(2k_0 n_{core})$. Analyse the phases $\varphi_{m,L} = \beta_m L$ taken by the different modes m upon a single pass L ($L = L_{MMI}$ of Fig. 20.6c). Find length L such that ϕ_L is a multiple of 2π and show that the mode phases $\varphi_{m,L}$ also differ by such multiples. What must the corresponding field contributions at $z = L$ (within the overall phase shift $k_0 n_{core} L$)

then be, when compared to those at $z = 0$? This imaging property of the entrance field $|\Psi_{exc}\rangle$ at the plane $z = L$, although remarkable, has a limit in terms of resolution. Where does this limit come from in terms of transverse momentum?

- Carry out a similar analysis for different distances L such that the phase differences $L\delta\beta \equiv L(m^2 - m'^2)(\pi/d)^2/(2k_0 n_{core})$ are now multiples of $\pi, 2\pi/3$, or $\pi/2$ etc., instead of multiples of 2π. Discuss the formation and spatial arrangement of multiple images based on the values of $(m^2 - m'^2)$ for the modes.

- Relationship with Talbot diffraction of gratings: in the ray picture, the rays from the initial inputs are multiply reflected by the sidewalls. This creates multiple families of images that replicate the initial source in the $z = 0$ plane. Explain why this is similar to the situation of the Talbot imaging effect in the Fresnel diffraction regime of a periodic grating illuminated by a plane wave.

Exercise 20.3 Sensing with a ring and Fano interferences. Study the device of Fig. 20.12. Path 2 of a Mach-Zehnder (MZ) interferometer has a resonant ring attached to it, while path 1 is a reference arm. The resonant ring is entirely similar to a so-called Gires–Tournois interferometer with 100% reflectivity, although this may seem non-intuitive: treat the ring
(in → out) response as the (incident → reflected) channel of a particular Fabry–Perot with incidence onto a top mirror R and a perfect 100% reflecting bottom mirror; hence, light can only be reflected, never transmitted. You should find that the phase response (the phase of the reflection for the Fabry–Perot, but the phase of the transmission along the arm for the ring, because the ring-arm coupler plays exactly the same role as mirror R) makes abrupt jumps when the resonance condition is satisfied inside the resonator, while the modulus of the reflection is always unity.

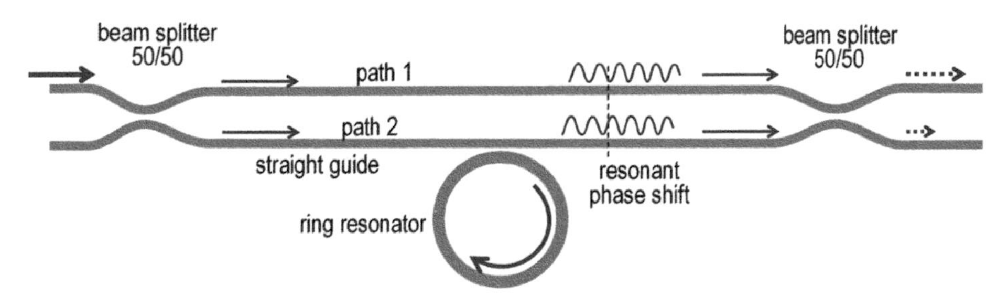

Fig. 20.12 Mach–Zehnder interferometer with a resonant ring on one arm, providing a Fano response at resonance. The ring can be seen as a Gires–Tournois interferometer, i.e. a Fabry–Perot with a perfect mirror, whose resonant character is manifested by resonant phase-shifts rather than transmitted intensity resonances.

Insert this phase response as appropriate to describe the two-wave interference of the MZ interferometer. Depending on the top-arm slowly evolving phase (you can assign it an arbitrary evolution with an arm length of the same order as the ring perimeter), the overall MZ output offers Fano resonances, which can be sharper than the Airy function of a Fabry–Perot with the same quality factor would be. With a computer model or an analytical approach, quantify this interferometric gain.

Part VII

Fluctuational electrodynamics

21 Fluctuational electrodynamics

21.1 Introduction

Black-body radiation is usually viewed as a purely incoherent phenomenon resulting from uncorrelated spontaneous emission events. It is usually described in the framework of radiometry, which is a phenomenological model. Fluctuational electrodynamics revisits this field in the framework of Maxwell equations. Lorentz already tried to establish a link between black-body radiation and the electromagnetic (EM) framework at the beginning of the twentieth century (Lorentz, 1916). He computed the 'radiant heat' emitted by a metal as radiation emitted by electrons randomly moving due to thermal excitation. This involves a major conceptual step: recognizing that sources are no longer described by a deterministic function but rather by a statistical random process which can be analysed using statistical physics. At that time, electrons were described classically and the connection to temperature was done through the energy equipartition theorem $\langle mv_x^2/2\rangle = k_B T/2$, where k_B is Boltzmann constant, so that it was impossible to recover Planck's function. A major step forward has been the derivation of the quantum form of the fluctuation–dissipation theorem in 1951 (Callen and Welton, 1951) which allowed the correct form of the power spectral density of the current density fluctuations to be derived (see Complement 1.A for a brief summary of spectral analysis of random processes). Soon after, Rytov established the foundations of fluctuational electrodynamics. The basic idea is simple and similar to the Langevin model of Brownian motion. In the Langevin model, a random force is inserted into Newton's equations to generate the random movement of the particles. In fluctuational electrodynamics, a random current density is inserted into the Maxwell equations to generate the random fields. A review of earlier work can be found in English in ref. (Rytov et al., 1989). A major result of this formalism is that it provides the form of thermal radiation in the near field, thereby predicting effects which cannot be explained in the framework of radiometry. A remarkable effect is the large increase of electromagnetic energy close to emitters due to evanescent waves (Shchegrov et al., 2000). This type of model is not limited to thermal radiation. It has also been developed to explore the noise of optoelectronic devices (Henry and Kazarinov, 1996) and to model light emission by electrically pumped semiconductors (Henry and Kazarinov, 1996) and semiconductors in resonators or metasurfaces (Greffet et al., 2018). The model also captures the quantum fluctuations and can be used to derive forces due to quantum fluctuations between two surfaces (Casimir force) and between a surface and an atom (Casimir–Polder force) (Casimir, 1948; Lamoreaux, 1997; Intravaia and Lambrecht, 2005; Decca et al., 2003; Henkel et al., 2004).

Introduction to Nanophotonics. Henri Benisty, Jean-Jacques Greffet, and Philippe Lalanne, Oxford University Press.
 DOI: 10.1093/oso/9780198786139.003.0021

Initially, fluctuational electrodynamics was mostly developed in the context of radiowaves. It was realized in the 1980s that thermal radiation in the near field is particularly interesting in the infrared because surface phonon polaritons can be excited thermally (Zhizhin et al., 1982).

The purpose of this chapter is to revisit black-body radiation and show how it can be tailored both in the near and far fields. In the rest of the chapter, we briefly introduce the basic ideas of fluctuational electrodynamics. We then present some unexpected properties of thermal radiation in the near field. Section 21.2 is devoted to tailoring thermal emission in the far field: it is possible to design incandescent emitters which are directional and/or quasi-monochromatic. Section 21.3 deals with radiative heat transfer at the nanoscale. The complement provides some background on spatial coherence.

21.1.1 Thermal emission as an antenna phenomenon

Let us briefly summarize the fluctuational electrodynamics framework developed by Rytov (Rytov et al., 1989). Random current densities $\mathbf{j}^{\mathrm{f}}(\mathbf{r}',\omega)$ accounting for the thermal motion of charges (e.g. electrons in metals, ions in iono-covalent crystals) are introduced to generate black-body radiation. Computing the field thermally emitted by a body amounts to summing the contributions of all the volume elements in the material. The fluctuating electric field emitted by any body in local thermodynamic equilibrium is given by (Eq. (12.12))

$$\mathbf{E}^{\mathrm{f}}(\mathbf{r},\omega) = \mathrm{i}\omega\mu_0 \int_V \mathrm{d}\mathbf{r}' \overleftrightarrow{G}(\mathbf{r},\mathbf{r}',\omega)\cdot\mathbf{j}^{\mathrm{f}}(\mathbf{r}',\omega), \tag{21.1}$$

where the integral is taken over the volume V, which contains the fluctuating source currents; $\overleftrightarrow{G}$ is the electric Green tensor. Note, in particular, that each volume element can be characterized by a fluctuating dipole $\mathbf{p}_{\mathrm{f}}$ such that $-i\omega\mathbf{p}_{\mathrm{f}} = \mathbf{j}^{\mathrm{f}}(\mathbf{r}',\omega)\mathrm{d}\mathbf{r}'$. It follows that thermal radiation can be reduced to the emission of a set of random time-dependent dipoles located in the volume of the emitter.

We now discuss a few consequences of the statistical properties of the sources. The mean value of the random currents is zero, so that the mean value of the radiated fields is also zero. However, the power carried by the field fluctuations is given by the Poynting vector, which is a quadratic quantity that is non-zero on average.

We assume that the process is stationary. This introduces a difficulty, as the Fourier transforms of the fields and currents are not defined in the sense of function, as they are not square-integrable. As summarized in Complement 1.A, the spectral analysis of a stationary random process is well defined using the power spectral density. However, it is possible and convenient to formally use the Fourier transform of the fields and sources when computing quadratic quantities. They can be related to the power spectral density $W_{j_n j_m}(\mathbf{r},\mathbf{r}',\omega)$ using:

$$\langle j_n^{\mathrm{f}}(\mathbf{r},\omega) j_m^{\mathrm{f}}(\mathbf{r}',\omega')\rangle = 2\pi\delta(\omega+\omega')W_{j_n j_m}(\mathbf{r},\mathbf{r}',\omega), \tag{21.2}$$

where the delta function $2\pi\delta(\omega+\omega')$ accounts for the stationarity of the system. Note that $j_m^f(t)$ is real, so that $j_m^f(\mathbf{r}',-\omega) = j_m^{f*}(\mathbf{r}',\omega)$. The power spectral density of the current-density fluctuation is given at thermodynamic equilibrium by the fluctuation–dissipation theorem (Landau et al., 1980),

$$W_{j_n j_m}(\mathbf{r}, \mathbf{r}', \omega) = \omega \varepsilon_0 i \big[\varepsilon^*_{mn}(\omega) - \varepsilon_{nm}(\omega)\big] \Theta(\omega, T) \delta(\mathbf{r} - \mathbf{r}'), \tag{21.3}$$

where $i^2 = -1$, $j^{\mathrm{f}}_n(\mathbf{r}, \omega)$ is a spatial component of the fluctuating current density at the frequency ω. The subscripts n or m stand for the x, y, or z component of the vector. $\varepsilon_{nm}(\omega)$ is the relative permittivity tensor of the emitter, and the function

$$\Theta(\omega, T) = \frac{\hbar\omega}{2} + \frac{\hbar\omega}{\mathrm{e}^{\hbar\omega/(k_B T)} - 1} \tag{21.4}$$

is the mean energy of a harmonic oscillator with frequency ω in thermodynamic equilibrium with a thermostat at temperature T. k_B is Boltzmann's constant and $2\pi\hbar$ is Planck's constant.[1] In Eq. (21.3), the delta function $\delta(\mathbf{r} - \mathbf{r}')$ shows up because we have neglected spatial dispersion.

21.1.2 Novel effects at the nanoscale

We already pointed out that a key advantage of fluctuational electrodynamics as compared to radiometry is the possibility of modelling electromagnetic fields in the near field. In the infrared and optical regimes, another key advantage is to be able to predict the role of surface phonon polaritons. Accounting for the role of surface polaritons led to the prediction of unexpected effects, which will be presented in this section: spatial coherence of black-body radiation (Carminati and Greffet, 1999), temporal coherence of black-body radiation (Shchegrov et al., 2000), and enhanced radiative heat transfer at the nanoscale due to the contribution of surface phonon polaritons (Mulet et al., 2001; Mulet et al., 2002).

Black-body radiation in the near field. It has been predicted (Rytov et al., 1989) that the spectral energy density per unit volume $u(z, \omega)$ defined by

$$u(z) = \varepsilon_0 \frac{\mathbf{E}^2(\mathbf{r}, t)}{2} + \frac{\mathbf{B}^2(\mathbf{r}, t)}{2\mu_0} = \int_0^\infty \frac{\mathrm{d}\omega}{2\pi} u(z, \omega) \tag{21.5}$$

can be increased by several orders of magnitude in the vicinity of an interface-separating vacuum from a homogeneous and isotropic material with dielectric constant $\varepsilon(\omega)$. This work was primarily focused on radiowaves. It was later predicted (Shchegrov et al., 2000) that if the interface supports a surface phonon polariton in the infrared, the density is further increased at a frequency such that $\mathrm{Re}[\varepsilon(\omega)] + 1 = 0$, so that the spectrum becomes quasi-monochromatic in the near field. In other words, thermal radiation becomes temporally coherent at the nanoscale (see Complement 1.A). The coherence time is given by the inverse of the line width, which is essentially the surface phonon polariton lifetime. This result is illustrated in Fig. 21.1, where the spectrum of the energy density above an interface between vacuum and SiC at 300 K is shown at different distances. Only the emitted energy is shown here. In other words, the material is at a temperature of 300 K, while the environment is at 0 K. These spectra are normalized by the maximum value u_0 of the black-body spectrum in vacuum. Note that

[1] The quantum form of the fluctuation–dissipation theorem deals with non-commuting operators. The resulting power spectral density is not an even function, in contrast with the classical case. Here, we have used a symmetrized form of the correlation function, which restores the symmetry.

the upper spectrum (Fig. 21.1a) is similar to the black-body spectrum except in a small spectral window where the material does not emit. This is due to the low emissivity and therefore large reflectivity of the material in the so-called Reststrahlen region, where the refractive index is close to being a pure imaginary number. The scale on Fig. 21.1c shows that the energy density is increased by four orders of magnitude at a distance of 100 nm from the interface. These two predictions (large increase of the energy density in the near field and a black-body spectrum which is quasi-monochromatic) have been confirmed experimentally by two groups (Jones and Raschke, 2012; Babuty et al., 2013).

An asymptotic analysis of the energy density normalized by the energy density in vacuum is given by the formula (Joulain et al., 2003)

$$u_{tot} = \frac{1}{4} \frac{Im[\varepsilon(\omega)]}{|\varepsilon(\omega)+1|^2} \frac{1}{(k_0 z)^3} \Theta(\omega, T), \tag{21.6}$$

where z is the distance to the interface, $k_0 = \omega/c$. The $1/(k_0 z)^3$ dependence is the signature of the near-field effect. The energy density is given by the product of the

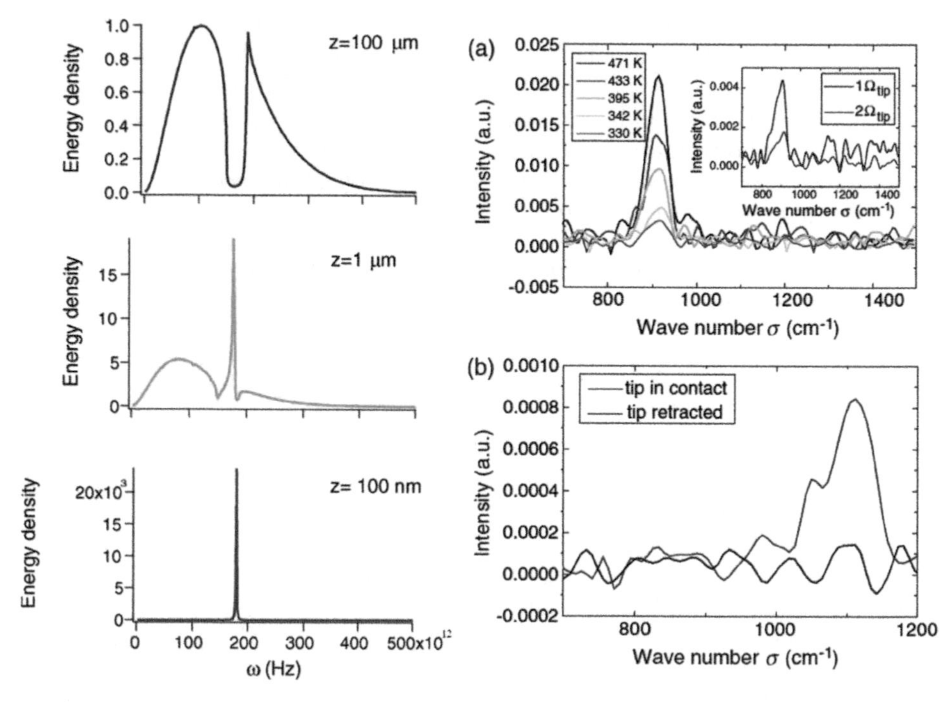

Fig. 21.1 Spectrum of the energy density $u(z, \omega)$ produced by a half-space at 300 K in the near field normalized by the maximum value u_0 of the black-body spectrum in vacuum. (a–c) It is seen that as the distance to the interface decreases, the intensity increases by several orders of magnitude and the spectrum becomes quasi-monochromatic. (d) Experimental measurements of the spectrum scattered by a tip in contact with an SiC surface for different temperatures. (e) When the tip is retracted, the signal vanishes, confirming that it is a near-field effect. See ref. (Babuty et al., 2013) for further details on the experimental setup.

mean energy of a harmonic oscillator by the local density of states (LDOS) above an interface introduced in Eq. (12.105).

A qualitative discussion will explain the origin of the large energy density and the quasi-monochromatic spectrum. The physical origin of the infrared emission is the radiation by electric dipoles due to the Si–C link. As the atoms have different electronegativities, they carry partial charges. Hence, the mechanical vibration of the link induced by thermal motion results in the excitation of an electric dipole. Such a dipole has a spectral resonance due to the vibrational resonance in the infrared. We now discuss the origin of the enhancement of the energy density in the near field. The electric field produced by the molecular dipoles has a leading contribution decaying as $1/r^3$, where r is the distance between the dipole and the observation point. As discussed in Section 10.2, this term becomes dominant for distances smaller than $1/k_0$. The corresponding energy density decays as $1/r^6$. After integration over a half-space, a $1/z^3$ dependence is recovered. These intense fields exist in the near field and do not propagate, so that they cannot be observed in the far field. Nevertheless, they can be detected using near-field microscopy.

We can now repeat this qualitative discussion by thinking in terms of angular spectrum and using evanescent waves instead of looking at the field in direct space. We note that the presence of the denominator $|\varepsilon(\omega)+1|$ in Eq. (21.6) is the signature of the surface phonon polariton resonance. The peak in the spectrum corresponds to the frequency ω_{sp}, satisfying $\text{Re}[\varepsilon(\omega_{sp})] = -1$. Here, we explain the spectral peak as a consequence of the large LDOS due to the surface wave at a particular frequency which is the horizontal asymptote of the dispersion relation of a surface phonon polariton (see also Complement 12.D.1). In the dispersion-relation picture, the large intensity close to the surface is understood from the fact that the surface mode provides a contribution to the field at a distance z with an amplitude proportional to $\exp(i\gamma z) \approx \exp(-kz)$. Hence, the number of modes contributing to the energy increases as z decreases.

Coherent thermal emission. We have just discussed the role of surface waves in the modification of the thermal radiation close to the surface. An unexpected consequence of the existence of surface waves is the development of spatial coherence along the interface. Indeed, a random source excites surface waves that can propagate along the interface with a decay length δ_x, which is typically tens or hundreds of wavelengths (see section 13.4.2). It follows that the field thermally emitted is correlated at points $\mathbf{r}_1$ and $\mathbf{r}_2$ if the same surface wave contributes to the field at these two points. This is the case, provided that both points are close to the surface and separated by a distance smaller than the decay length δ_x of the surface wave. More generally, the existence of surface waves provides a means of tailoring thermal emission: it is possible to engineer an interface in order to control both the emission direction and the emission spectrum. This will be the subject of section 21.2.

Radiative heat transfer at the nanoscale. We have seen that the electromagnetic energy density is significantly enhanced close to the interface. Let us now consider a nanoparticle which is brought close to the surface. Such a particle will be illuminated by electromagnetic fields with an intensity that increases as $1/z^3$. The absorbed power is therefore significantly increased. By reciprocity, the power transferred from the particle

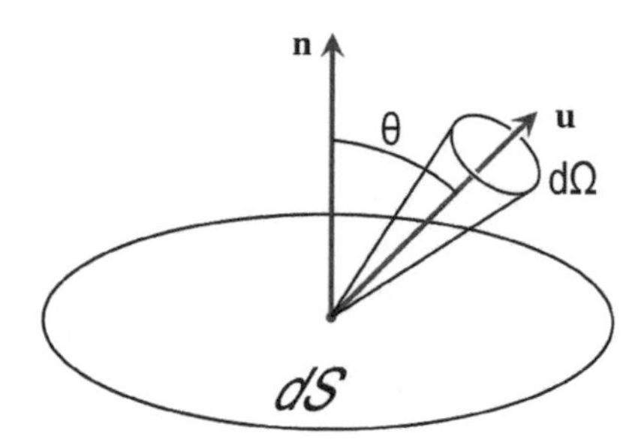

Fig. 21.2 Sketch of the notations used to describe the field radiated by an area $\mathrm{d}S$ in the solid angle $\mathrm{d}\Omega$.

to the material below the interface is also increased. It follows that the heat transfer due to radiation in the near field is expected to be increased by orders of magnitude. This will be the topic of section 21.3.

21.2 Harnessing black-body radiation with metasurfaces

21.2.1 Thermal emission in the framework of radiometry

In radiometry, the power $\mathrm{d}Q^e$ emitted by an elementary opaque surface $\mathrm{d}S$ at temperature T in an elementary frequency range $\mathrm{d}\omega$ around the circular frequency ω in a solid angle $\mathrm{d}\Omega$ around a direction $\mathbf{u}$ making an angle θ with the normal to the surface (see Fig. 21.2) is given by

$$\mathrm{d}Q^e(\omega,\theta) = I^e(\omega,T)\mathrm{d}S\cos\theta\mathrm{d}\omega\mathrm{d}\Omega, \tag{21.7}$$

where $I^e(\omega,T)$ is the specific intensity (or spectral radiance) of the emitted radiation. A body cannot radiate more than a black-body at the same temperature. Indeed, at thermodynamic equilibrium, the intensity leaving the body has to be the specific intensity at thermodynamic equilibrium. It is also the sum of the emitted and reflected intensities, which are both positive. It follows that the emitted specific intensity is a fraction of the equilibrium specific intensity which can be cast in the form

$$I^e(\omega,T) = E(\omega,\theta)I_{BB}(\omega,T), \tag{21.8}$$

where $E(\omega,\theta)$ is the emissivity (a real positive number smaller than one) and $I_{BB}(\omega,T)$ is the specific intensity at thermodynamic equilibrium (also known as black-body specific intensity) given by

$$I_{BB}(\omega,T) = \frac{c}{4\pi}\times\frac{\omega^2}{\pi^2c^3}\times\frac{\hbar\omega}{\exp(\hbar\omega/k_BT)-1}. \tag{21.9}$$

In the above equation, the first term is the ratio between the specific intensity and the energy per unit volume, the second term is the density of electromagnetic modes in vacuum, and the third is the mean energy per mode.

21.2.2 Radiative flux

We now compute the radiative flux, which is the rate of energy exchange by the body through radiation. The body gains energy by absorption and loses energy by emission. The corresponding absorbed power $\mathrm{d}Q^{abs}$ in an elementary frequency range $\mathrm{d}\omega$ around the circular frequency ω in a solid angle $\mathrm{d}\Omega$ around a direction $-\mathbf{u}$ making an angle θ with the normal to the surface is given by

$$\mathrm{d}Q^{abs}(\omega,\theta) = A(\omega,\theta) I^{inc}(\omega) \mathrm{d}S \cos\theta \mathrm{d}\omega \mathrm{d}\Omega, \tag{21.10}$$

where $A(\omega,\theta)$ is the absorptivity of the material. The elementary radiative flux is thus given by

$$\mathrm{d}Q^{R}(\omega,\theta) = \mathrm{d}Q^{e}(\omega,\theta) - \mathrm{d}Q^{abs}(\omega,\theta). \tag{21.11}$$

21.2.3 Kirchhoff law in a nutshell

In this section, we discuss the physical meaning of the emissivity and the different ways of engineering this quantity. It characterizes the ability of a material to produce thermal radiation in a given angle and polarization. In radiometry, it is introduced as a phenomenological quantity. Using energy-conservation arguments for an opaque body (no transmission), it can be shown that it is related to the interface reflectivity (Rytov et al., 1989) by a simple relation,

$$E(\omega,\theta) = 1 - R(\omega,\theta) = A(\omega,\theta), \tag{21.12}$$

where $R(\omega,\theta)$ is the Fresnel reflection factor and $A(\omega,\theta)$ is the absorptivity. This equality between emissivity and absorptivity is known as the Kirchhoff law (Kirchhoff, 1860).

We now discuss the physical origin of this relation and show that both emissivity and absorptivity are related to the interface transmission factor. We first consider the case of a beam incident on the interface coming from vacuum (medium 1), as shown in Fig. 21.3a. The incident energy is either reflected or transmitted by the interface with a transmission factor T_{12}. We note that the energy transmitted by the interface is absorbed in medium 2 as it propagates. As medium 2 is considered to be a half-space (in practice, thicker than the decay length), all the energy transmitted by the interface is finally absorbed. Thus, we conclude that the absorptivity is nothing but the interface transmission $A = T_{12}$. We now consider the emission by the lower half-space in local thermodynamic equilibrium with temperature T (see Fig. 21.3b). While the medium is opaque, it contains charges (ions, electrons) which are randomly moving in the system and therefore it radiates. The corresponding radiation is in local thermodynamic equilibrium with other excitations such as electrons and phonons. It is thus characterized by a black-body radiation, although the medium is opaque. Obviously, radiation cannot propagate over long distances, as the medium is lossy. There is a typical decay length given by the skin depth. Nevertheless, photons emitted within a distance from the interface smaller than the skin depth can be transmitted through the interface. Once these photons are transmitted by the interface, they are considered to be emitted. Hence, we conclude that the emission is proportional to the transmission factor T_{21}. Now it can be shown, using the reciprocity theorem, that the transmission

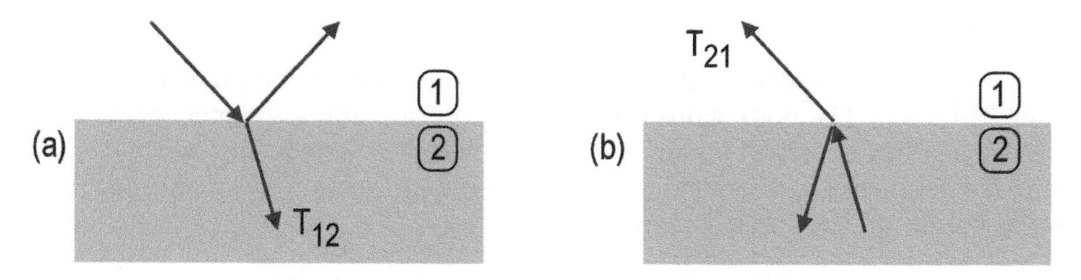

Fig. 21.3 The Kirchhoff law. (a) An incident beam is either reflected or transmitted by the interface. Transmitted light is subsequently absorbed in an opaque medium. (b) The light emitted by the hot body is transmitted by the interface. The equality between absorption and emission follows from the equality $T_{12} = T_{21}$.

factor of the interface satisfies the equality $T_{12} = T_{21}$. This equality is the physical origin of the Kirchhoff law: both absorption and emission are, in fact, nothing but transmission by an interface. With this physical picture in mind, it now appears that engineering thermal emission amounts to tailoring the interface transmission factor. This can be achieved using coatings, multilayers, and metasurfaces. For instance, using a grating to resonantly excite surface waves allows these properties to be modified for very well-defined frequencies and angles.

21.2.4 The Kirchhoff law for subwavelength bodies

The Kirchhoff law in its original form is useful inasmuch as the concepts of reflectivity and absorptivity of an interface are meaningful. Thus, the absorption and emission by a body are proportional to its area and the specific intensity. We discussed the limit of these concepts in section 10.4.2 and showed that they become meaningless at a subwavelength scale, so that the fluxes cannot be computed using Eq. (21.7). However, the absorption cross section introduced in Chapter 12 can be defined for an object with arbitrary size, including subwavelength objects. It turns out that the Kirchhoff law is still valid using the concepts of absorption cross section (Rytov 1989, Greffet 2018). In other words, if we consider an arbitrary body at temperature T with absorption cross section $\sigma_{abs}(\mathbf{u}, \omega, l)$ at frequency ω, polarization $l = s, p$, and direction $\mathbf{u}$, the emitted power $\mathrm{d}Q^e$ in the solid angle $\mathrm{d}\Omega$ around direction $-\mathbf{u}$, frequency ω, and polarization l is given by

$$\mathrm{d}Q^e = \sigma_{abs}(\mathbf{u}, \omega, l)\frac{I_{BB}(\omega, T)}{2}\mathrm{d}\Omega. \tag{21.13}$$

As discussed in Chapter 12, the cross section can be viewed as the effective area of a perfect absorber at normal incidence. The factor 1/2 accounts for the fact that we consider a single polarization.

We now apply this formula to a body much larger than the wavelength with area A, whose surface makes an angle θ with the direction $\mathbf{u}$ and is perfectly absorbing for polarizations and any angle. Its absorption cross section is thus proportional to the projected area $A\cos(\theta)$. We find an emitted power given by the familiar formula $I_{BB}A\cos(\theta)\mathrm{d}\Omega$. The derivation of Eq. (21.13) can be found in ref. (Rytov et al., 1989).

An extension of the Kirchhoff law to deal with electroluminescence and photoluminescence is given in ref. (Greffet et al., 2018).

21.2.5 Radiative transfer between subwavelength bodies

Let us now consider the radiative flux between two subwavelength bodies at temperatures T_1 and T_2 with absorption cross sections $\sigma_{1,abs}$ and $\sigma_{2,abs}$ separated by a distance d_{12} such that evanescent waves have no contributions. The power emitted by body 2 into a solid angle $\mathrm{d}\Omega$ subtended by an area $\mathrm{d}S$ at a distance d_{12} and in polarization l is given by

$$\mathrm{d}Q^e = \sigma_{2,abs}(\mathbf{u},\omega,l)\frac{I_{BB}(\omega,T_2)}{2}\mathrm{d}\Omega = \sigma_{2,abs}(\mathbf{u},\omega,l)\frac{I_{BB}(\omega,T_2)}{2}\frac{\mathrm{d}S}{d_{12}^2}, \tag{21.14}$$

so that the incident power per unit area on body 1 emitted by body 2 in polarization l is $\sigma_{2,abs}(\mathbf{u},\omega,l)I_{BB}(\omega,T_2)/2d_{12}^2$. Multiplying by the absorption cross section of body 1 yields

$$\mathrm{d}Q^{1,abs} = \sigma_{1,abs}(-\mathbf{u},\omega,l)\frac{\sigma_{2,abs}(\mathbf{u},\omega,l)I_{BB}(\omega,T_2)}{2d_{12}^2}. \tag{21.15}$$

The elementary radiative transfer exchanged between body 1 and body 2 in polarization l is thus simply given by

$$\mathrm{d}Q^R(\omega,l) = \mathrm{d}Q^{1,e} - \mathrm{d}Q^{1,abs} = \frac{\sigma_{1,abs}(-\mathbf{u},\omega,l)\sigma_{2,abs}(\mathbf{u},\omega,l)}{d_{12}^2}\frac{I_{BB}(\omega,T_1) - I_{BB}(\omega,T_2)}{2}. \tag{21.16}$$

Integration over all frequencies and summing over the two polarizations yields the radiative flux. Note that we have neglected energy multiply scattered between both bodies.

21.2.6 Spatially coherent thermal emission in the far field

We saw in section 21.2.3 the origin of the Kirchhoff law. It follows from this discussion that the spectral and angular dependence of emission can be controlled by shaping the transmittivity. Remarkably, this provides a means of controlling the spatial coherence of the source, as discussed in Complement 21.A. Briefly, the angular emission pattern is the Fourier transform of the field correlation function in a planar source. Hence, designing a spatially coherent thermal source amounts to designing an interface with a narrow angular transmission factor or a narrow angular absorptivity peak.

The first example of such a spatially coherent incandescent source was discussed in (Greffet et al., 2002; Marquier et al., 2004) and took advantage of the angular absorption peak due to the resonant excitation of a surface wave mediated by a grating (see Fig. 21.4). Let us emphasize that the coherence is due to not the grating but the surface wave. The role of the grating is merely to couple the surface wave to a propagating wave.

The combination of a grating and a surface wave is not the only means of controlling the direction of emission. Different schemes have been proposed to produce partially coherent thermal sources. They are based on a filtering of the emission pattern in order to reduce the angular width of the emission pattern. A more conventional type of filter can be designed using multilayer systems.

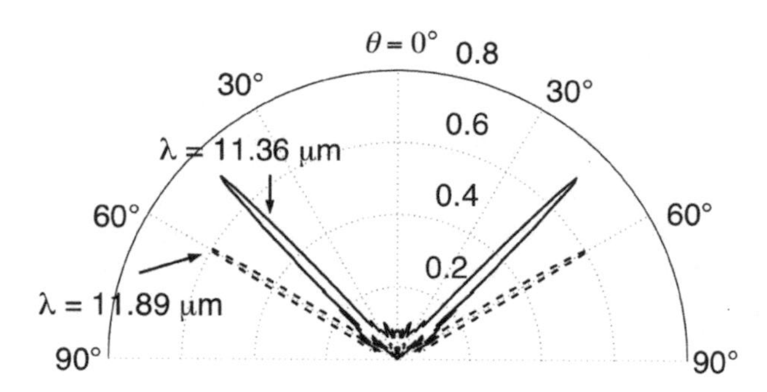

Fig. 21.4 Experimental angular emission of an SiC grating at two different wavelengths. The angular pattern has the characteristic shape of an antenna. It demonstrates the spatial coherence of the thermal source. Measurements are taken at 800 K. From ref. (Marquier et al., 2004).

21.2.7 Controlling the emission spectrum: Temporal coherence

In the previous section, we discussed the connection between directivity and spatial coherence of the field emitted by an incandescent source. In this section, we discuss how to control the emission spectrum of an incandescent source. The formal connection between the spectrum and the temporal coherence of the emitted field is discussed in Complement 1.A.

Different strategies have been implemented to design thermal sources with a controlled spectrum in the infrared. A quasi-monochromatic source can be useful for gas detection. It is also useful for thermophotovoltaics in order to convert energy from a heat source into a monochromatic radiation matching the gap of a photovoltaic cell. A simple solution is to add a filter on top of the emitting body (Rephaeli and Fan, 2009). Such a filter can be made with a multilayer deposited on the emitter surface. The key difference with placing a filter between a usual thermal source and the detector is the significant reduction of energy emitted by the source. In other words, the same amount of radiation is emitted in the useful spectral window, while no radiation is emitted in the rest of the spectrum, so that the efficiency of the source can be increased by one or two orders of magnitude. This is an essential property for many applications.

A different approach consists in starting from a material transparent in the infrared such as a semiconductor and adding a quantum well (QW) to produce losses at a well-controlled frequency. The drawback of this approach is the very large number of QWs required, given that a single QW has an absorption cross section on the order of 0.01. The advantage is a very low background emissivity and therefore a good efficiency. By designing a photonic-crystal structure, it is also possible to control the emission direction (De Zoysa et al., 2012).

Finally, let us mention a class of sources using metasurfaces made with metal–insulator–metal resonators explored by many authors (Puscasu and Schaich, 2008; Liu et al., 2011; Bouchon et al., 2012; Costantini et al., 2015). An example is given in Fig. 21.5. These structures are based on the design of a resonator taking advantage of the plasmonic mode in the gap. The design of the size and shape of the resonator allows the resonance frequency to be tuned. These resonators are deposited on a metallic

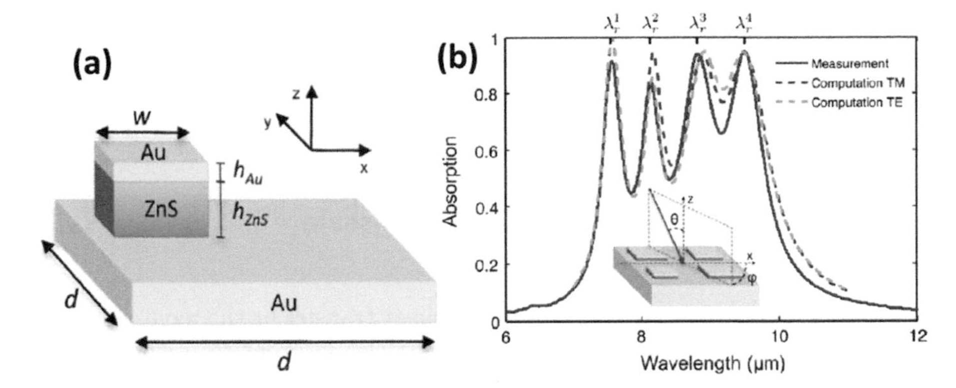

Fig. 21.5 Controlling the absorption spectrum with metal–insulator–metal resonators. (a) An example of a patch antenna. The height h_{ZnS} controls the phase velocity of the mode, which is much smaller than c, so that the resonator widths (w) are much smaller than half a wavelength in vacuum. (b) An example of a periodic array of four different patches producing four different absorption peaks. From ref. (Bouchon et al., 2012).

substrate with low absorption. By properly designing a periodic array of these resonators, it is possible to reach the so-called critical-coupling condition corresponding to a total absorption. Furthermore, the geometrical size of these resonators is much smaller than their absorption cross section. It is thus possible to pack different resonators in a cell of the periodic array, thereby designing surfaces with total absorption at different frequencies. An important issue for these applications is to find materials which may support resonances and also sustain high temperatures. Platinum is a very robust material that can be deposited on silicon nitride (Wojszvzyk et al., 2021). A new class of plasmonic materials has been put forward (Naik et al., 2013). When the period is much smaller than the wavelength, these metasurfaces are quasi-isotropic. However, it is possible to take advantage of the periodicity to modulate the absorptivity/emissivity as a function of the angle (Costantini et al., 2015).

21.2.8 Modulation of thermal emission at high frequency

In this section, we discuss fast modulation of the intensity of an incandescent source. The usual incandescent sources are hot membranes heated using the Joule effect. They can be modulated by turning the power on and off. The modulation frequency is limited by the membrane cooling due to heat diffusion, resulting in a typical maximum frequency on the order of a few tens of Hz. However, if nanostructures on a cold substrate are heated, they may cool down rapidly. Modulation up to 10 MHz has been demonstrated (Wojszvzyk et al., 2021).

A completely different approach has been proposed in ref. (Vassant et al., 2013). The idea is to keep the body at high temperature and modulate its emissivity, so that thermal diffusion time is no longer a limiting factor. Here, the limitations come from the physics of the phenomenon used to modulate the emissivity. Noda's group proposed to apply a voltage to a set of QWs in a semiconductor. Here, the absorption is due to the intersubband absorption in the QWs. By depleting the electron population, the absorption due to electrons can be modified, resulting in a strong modulation of the thermal emission. Using this technique, a time modulation of 600 kHz has been reported in ref. (Inoue et al., 2014), with a peak emissivity close to 1 and a very low background emission over a large spectrum.

21.3 Radiative heat transfer at the nanoscale

21.3.1 Introduction

In this section, we give an overview of radiative heat transfer at the nanoscale. We will qualitatively discuss the case of two spherical particles in order to provide a physical picture of the origin of the enhanced radiative heat transfer at the nanoscale.

When addressing the issue of radiative heat transfer between two particles, the physical mechanism is usually considered to be the emission of photons by one particle and their absorption by the second particle as in section 21.2.5. Let us note that if photons are exchanged by the particles, some energy is exchanged as well as some momentum, leading to a force. As far as forces are concerned, this so-called recoil force is known to be much smaller than the van der Waals force due to near-field (i.e. dipole–dipole) interaction between particles. The same is true for energy transfer. The near-field contribution (dipole–dipole interaction) is much larger than the far-field (photon-exchange) contribution. In the case of molecules, this non-radiative energy transfer is known as Förster energy transfer. A similar mechanism exists between lossy nanoparticles. The physical picture is as follows: the fluctuating currents in a particle P_1 can be described by a random dipole if the particle is much smaller than the wavelength and if the interparticle distance r is larger than a few diameters. This dipole induces currents in the particle P_2. These currents are dissipated through losses. Since the field varies as $1/r^3$ in the near field, where r is the distance between the particles, the power exchanged varies as $1/r^6$. Furthermore, if the particles support a polariton resonance, the heat flux is resonantly enhanced at a frequency corresponding to the pole of the polarizability (see Chapter 14).

A similar behaviour is expected for the heat transfer between plane surfaces. Random currents generate electromagnetic fields in the vicinity of the interface between a material and vacuum. At distances d such that $d \ll \lambda/2\pi$, the near-field contributions provide the leading contribution.

21.3.2 Heat transfer between two particles

We now calculate the heat transfer between two spherical nanoparticles with radius a held at different temperatures and separated in vacuum. This is a simple system where the calculations can be performed analytically in order to reveal the physical mechanism of heat transfer at the nanoscale (Volokitin and Persson, 2007; Domingues, 2005). Let us consider two nanoparticles whose dielectric constants are ϵ_1 and ϵ_2 and whose temperatures are T_1 and T_2. We first calculate the power dissipated in particle 2 by the electromagnetic field induced by particle 1 using the dipolar approximation

$$P_{1\to 2}(\omega) = \varepsilon_0 \frac{\omega}{2} \mathrm{Im}(\alpha_2) |\mathbf{E}_{inc}(\mathbf{r}_2, \omega)|^2, \tag{21.17}$$

where $\mathbf{r}_2$ denotes the position of the particle 2 and α_2 is the polarizability of a sphere of radius a:

$$\alpha_2 = 4\pi a^3 \frac{\epsilon_2 - 1}{\epsilon_2 + 2}. \tag{21.18}$$

The field incident on particle 2 created by the thermal fluctuating dipole of particle 1 located at $\mathbf{r}_1$ at temperature T_1 is given by

$$\mathbf{E}_{inc}(\mathbf{r}_2, \omega) = \mu_0 \omega^2 \overleftrightarrow{\mathbf{G}}(\mathbf{r}_2, \mathbf{r}_1, \omega) \cdot \mathbf{p}, \tag{21.19}$$

where $\overleftrightarrow{\mathbf{G}}(\mathbf{r}_2, \mathbf{r}_1, \omega)$ is the vacuum Green tensor,

$$\overleftrightarrow{\mathbf{G}}(\mathbf{r}_2, \mathbf{r}_1, \omega) = [\overleftrightarrow{I} + \frac{1}{k_0^2}\nabla\nabla] \frac{\exp[i\omega R/c]}{4\pi R}, \tag{21.20}$$

where $R = |\mathbf{r}_2 - \mathbf{r}_1|$. To proceed, we need the correlation function of the dipole given by the fluctuation–dissipation theorem (Landau et al., 1980). In the near field, we keep only the leading terms decaying as $1/R^3$, which is equivalent to using the form of the field produced by an electrostatic dipole. We finally obtain the heat exchange between two spherical nanoparticles at temperatures T_1 and T_2:

$$P_{1\leftrightarrow 2} = \frac{3}{4\pi^3} \frac{\mathrm{Im}[\alpha_1(\omega)]\mathrm{Im}[\alpha_2(\omega)]}{|\mathbf{r}_2 - \mathbf{r}_1|^6} \left[\Theta(\omega, T_1) - \Theta(\omega, T_2)\right]. \tag{21.21}$$

Let us note the $1/r^6$ spatial dependence of the heat transfer. This dependence is typical of the dipole–dipole interaction and reminiscent of Förster energy transfer for molecules. We find that nanoparticles follow a similar behaviour with a resonance when the dielectric constant approaches -2. The particle resonances appear in the visible part of the spectrum for metals and in the infrared for polar materials.

21.3.3 Heat transfer between two planes

We now focus on the heat transfer between two semi-infinite half-spaces separated by a vacuum gap and whose temperatures T_1 and T_2 are uniform. The main change that occurs in the near-field regime is the fact that evanescent waves can contribute to the heat transfer through tunnelling. We summarize the results in the next sections. Detailed derivations can be found in refs (Mulet et al., 2002; Polder and Van Hove, 1971), for instance. In the phenomenological theory, the radiative flux is given by

$$q = \int_0^{2\pi} \cos\theta d\Omega \int_0^{\infty} d\omega \frac{\varepsilon'_{1\omega}\varepsilon'_{2\omega}}{1 - \rho'_{1\omega}\rho'_{2\omega}} \times \left[I_{BB}(\omega, T_1) - I_{BB}(\omega, T_2)\right], \tag{21.22}$$

where the $\varepsilon'_{i\omega}$ are the directional monochromatic emissivities and the $\rho'_{i\omega}$ are the directional monochromatic reflectivities. The denominator $1-\rho'_{1\omega}\rho'_{2\omega}$ accounts for incoherent multiple reflections in the gap.

In order to account for near-field effects, a full theory was derived by Polder and Van Hove (Polder and Van Hove, 1971) in the framework of fluctuational electrodynamics.

The flux can be written as the sum of two terms: $q = q^{prop} + q^{evan}$. The first term, q^{prop}, is the contribution of propagating waves:

$$q^{prop} = \sum_{j=s,p} \int d\omega\, d\Omega\, \cos\theta \left[\frac{(1 - |r_{31}^j|^2)(1 - |r_{31}^j|^2)}{|1 - r_{31}^j r_{32}^j e^{2i\gamma_3 d}|^2} \right] \frac{[I_{BB}(\omega, T_1) - I_{BB}(\omega, T_2)]}{2}. \tag{21.23}$$

Let us note that $1-|r_{31}^{s,p}|^2$ and $1-|r_{32}^{s,p}|^2$ are the transmission energy coefficients between media 1 and 3 and 2 and 3 for the s- or p-polarization. The black-body intensity is divided by a factor 2 to account for the two polarizations. These coefficients can be identified as a polarized emissivity in the same way as defined for a single interface. Let us remark that this expression for the contribution of the propagating waves to the radiative flux between two semi-infinite media is very close to the form obtained using radiometry. Only the denominators are different, because interferences are not taken into account in the phenomenological model. Nevertheless, if one considers a frequency interval small in comparison with the frequency but larger than c/d, the variation of $e^{i\gamma_3 d}$ with ω is much faster than the Fresnel reflection coefficient variations. The integration over this interval yields an average value of $|1 - r_{31}^j r_{32}^j e^{2i\gamma_3 d}|^2$ which is $1 - |r_{31}^j|^2 |r_{32}^j|^2$. Matching the reflectivity with the Fresnel reflection energy coefficient, one can then identify this expression for the radiative flux with Eq. (21.22). In Fig. 21.6, we represent the heat-transfer coefficient h^R computed at $T = 300$ K. It is seen that for a distance d larger than $10\,\mu$m, which is the peak wavelength at this temperature, the heat transfer does not depend on the distance, as expected from radiometry.

The second term, q^{evan}, is the contribution of the evanescent waves with $k_{\|} > \omega/c$:

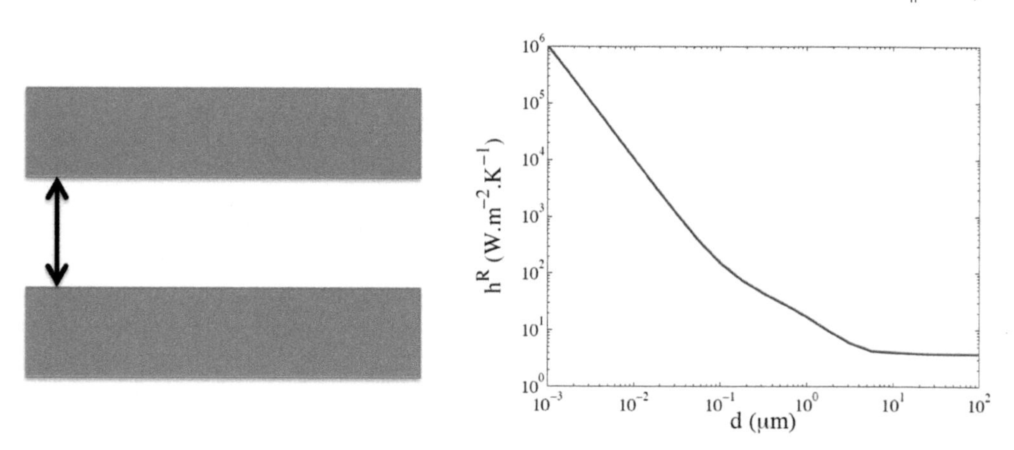

Fig. 21.6 Radiative heat transfer between two half-spaces. Left: Two semi-infinite half-spaces (media 1 and 2) at temperatures T_1 and T_2 separated by a gap with width d. Right: Heat transfer coefficient $h^R = \lim_{T_1 \to T_2} q/(T_1 - T_2)$ as a function of d for two half-spaces of SiC at 300 K.

$$q^{evan} = \sum_{j=s,p} \int d\omega \int_{\omega/c}^{\infty} \frac{\mathrm{d}^2\mathbf{k}_{\|}}{k_0^2} \left[\frac{\mathrm{Im}(r_{31}^j)\mathrm{Im}(r_{32}^j)e^{-2\mathrm{Im}(\gamma_3)d}}{|1 - r_{31}^j r_{32}^j e^{2i\gamma_3 d}|^2} \right] \frac{[I_{BB}(\omega, T_1) - I_{BB}(\omega, T_2)]}{2},$$
$$= \sum_{j=s,p} \int \frac{d\omega}{2\pi} \int_{\omega/c}^{\infty} \frac{\mathrm{d}^2\mathbf{k}_{\|}}{4\pi^2} \left[\frac{\mathrm{Im}(r_{31}^j)\mathrm{Im}(r_{32}^j)e^{-2\mathrm{Im}(\gamma_3)d}}{|1 - r_{31}^j r_{32}^j e^{2i\gamma_3 d}|^2} \right] \frac{[\Theta(\omega, T_1) - \Theta(\omega, T_2)]}{2},$$
(21.24)

where $k_0 = \omega/c$ and $\gamma_3 = [\varepsilon_3 k_0^2 - k_{\|}^2]^{1/2}$. Contrary to the single-interface case, the contribution of evanescent waves does not vanish because of the existence of both upward and downward evanescent waves in the space between the two media. In the short-distance regime, the contribution of evanescent waves dominates and yields a $1/d^2$ contribution (see Fig. 21.6). To understand this behaviour, we now make a rough estimate of the flux due to evanescent waves. The key idea is that the number of electromagnetic modes contributing to the heat transfer is increased as compared to vacuum. We start by noting that an electromagnetic mode is characterized by a frequency ω and a wavevector $(k_x, k_y, [\omega^2/c^2 - k_x^2 - k_y^2]^{1/2})$. Each electromagnetic mode has a mean energy given by $\Theta(\omega, T)$. The number of modes can be estimated using the periodic boundary conditions, so that $(k_x = n_x 2\pi/L, k_y = n_y 2\pi/L)$ where n_x, n_y are integers. Each dot in Fig. 21.7 represents a mode, so that two neighbouring dots are separated by $2\pi/L$. Hence, the area corresponding to a single mode in the plane (k_x, k_y) is $4\pi^2/L^2$.

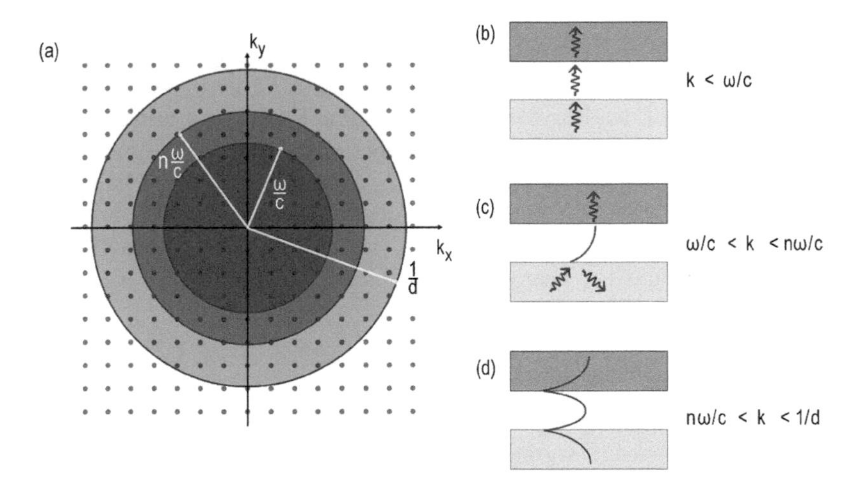

Fig. 21.7 Representation of the electromagnetic modes with frequency ω in a gap (width d) between two half-spaces which both have a complex refractive index with real part denoted by n. (a) Distribution of modes in the plane (k_x, k_y); (b) heat transfer due to propagating transmitted modes; (c) heat transfer due to partial frustration of total internal reflection by tunnelling. Here, the number of modes is increased by a factor n^2. (d) Heat transfer due to tunnelling of evanescent modes. With this picture, it is seen that the flux increases because the number of contributing modes increases as $1/d^2$. Adapted from ref. (Biehs et al., 2010).

In the plane (k_x, k_y), the propagating modes lie in a disc with radius $\omega/c = 2\pi/\lambda$. It follows that the number of modes per unit area is given by

$$\pi \left[\frac{4\pi^2}{\lambda^2}\right] \frac{L^2}{4\pi^2} = \pi \frac{L^2}{\lambda^2}.$$

We now estimate the number of modes contributing to the heat transfer in the near field, i.e. when the gap width d is smaller than $\lambda/2\pi$. A mode with large wavevector decays as $\exp(-|k_z|z)$ where $|k_z| \approx k_\parallel$ where $k_\parallel = \sqrt{k_x^2 + k_y^2}$. It follows that the amplitude of this mode at a distance d is approximated by $\exp(-k_\parallel d)$. It is seen that $1/d$ appears to be a cutoff spatial frequency. For a gap with width d, modes with wavevectors smaller than $1/d$ will contribute significantly to the heat transfer. The corresponding number of modes N is given by

$$N \approx \pi \left[\frac{1}{d^2}\right] \frac{L^2}{4\pi^2} = \frac{L^2}{4\pi d^2}.$$

This behaviour is illustrated in Fig. 21.7. In summary, the flux increases in the near field because the number of modes contributing to the heat flux increases as $1/d^2$. To fully justify this simple argument, it is necessary to show that each mode has the same contribution to the flux. We will show later that indeed, each mode with a given wavevector k provides, after integration over all frequencies ω, a contribution to the flux which is the product of a universal quantum of thermal conductance and a transmission factor.

This transmission factor is given by the term in brackets in Eq. (21.24). It can be checked that this term is always smaller than one. It can also be seen that the denominator may cancel, leading to a resonant behaviour. It turns out that searching for poles of a transmission factor yields the dispersion relation of the modes of the gap. If surface waves such as surface phonon polaritons are supported by the interfaces, the gap modes are a linear combination of these surface modes coupled through the gap. In other words, the transmission factor takes significant values when the gap modes are excited. It follows that the heat transfer is mostly dominated by a few frequencies when surface modes are present.

21.3.4 Experimental results

The effect of enhanced radiative heat transfer at the nanoscale was first observed experimentally by Hargreaves (Hargreaves, 1969) with metals. Measurements are difficult to achieve with metals because the transmission factors are low. A new generation of experiments has been reported after it was realized that the fluxes are an order of magnitude larger for dielectrics than for metals due to the contribution of surface phonon polaritons (Mulet et al., 2001; Mulet et al., 2002). The first generation of experiments (Shen et al., 2009; Rousseau et al., 2006) was conducted by measuring the heat exchanged between a sphere and a plane surface. A sphere is used for practical reasons in order to avoid the difficulty of keeping two plane surfaces parallel with a gap on the order of a few tens of nanometres. The theory is in agreement with the measurements, as seen in Fig. 21.8.

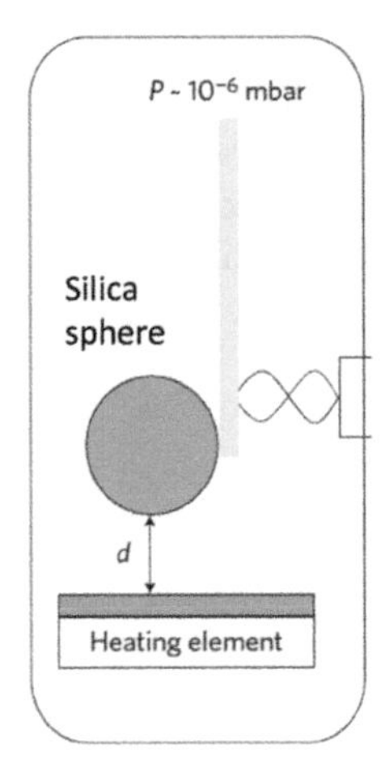

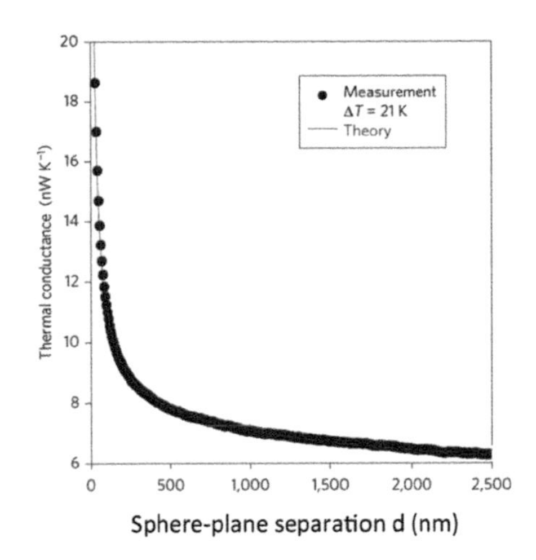

Fig. 21.8 Heat flux exchanged between a sphere and a plane surface as a function of the distance d. For a small temperature difference ΔT, the heat flux can be written as $\Phi = h^R \Delta T$, where h^R is the thermal conductance. Left: Setup, a silica sphere with a radius of 40 µm is at a distance d from the heated surface. The sphere is glued on a SiN cantilever coated with gold. As the heat flux changes, the cantilever temperature changes, producing a bending due to the bimorph effect. The cantilever displacement is detected optically. Right: Comparison between theory and experiment. From ref. (Rousseau et al., 2006).

21.3.5 A mesoscopic approach to radiative heat transfer: Quantum of thermal conductance

In this section, we introduce the concept of quantum of thermal conductance associated with each spatial mode $\exp(i\mathbf{k}_\parallel \cdot \mathbf{r})$. We start from Eq. (21.24). In order to identify a linear conductance for each mode $(j, \mathbf{k}_\parallel)$, where j stands for the polarization, it is necessary to linearize, assuming a weak temperature difference, $\Theta(\omega, T_1) - \Theta(\omega, T_2) = \frac{\partial \Theta}{\partial T}(T_1 - T_2)$. Assuming that a mode with given wavevector $k_\parallel$ has a transmission factor equal to 1, it is possible to find the thermal conductance g_0 associated with this mode (Pendry, 1983),

$$\int_0^\infty \frac{\mathrm{d}\omega}{2\pi} \left[\Theta(\omega, T_1) - \Theta(\omega, T_2)\right] \approx g_0 [T_1 - T_2], \tag{21.25}$$

where

$$g_0 = \int_0^\infty \frac{\mathrm{d}\omega}{2\pi} \frac{\partial \Theta}{\partial T} = \frac{\pi^2 k_B^2 T}{3h} \tag{21.26}$$

is the quantum of thermal conductance. We now analyse the linearized form of the flux, keeping the angular dependence of the transmission factor:

$$q = \sum_{j=\{\mathrm{s,p}\}} \int \frac{\mathrm{d}^2\mathbf{k}_\parallel}{(2\pi)^2} \int_0^\infty \frac{\mathrm{d}\omega}{2\pi} \mathcal{T}_j(\omega, \mathbf{k}_\parallel, d) \frac{\partial \Theta}{\partial T} \left[T_1 - T_2\right]. \tag{21.27}$$

This equation shows that we can attribute to each electromagnetic mode $(j, \mathbf{k}_{\parallel})$ a contribution to the conductance which is bounded by the quantum of thermal conductance. It appears very clearly in this context that the enhancement of the flux in the near field is due to the increase of the number of modes $N(\mathbf{k}_{\parallel}, d)$. Let us establish a rigorous form of the flux in the mesoscopic framework. To this end, we start with Eq. (21.27). Following ref. (Biehs et al., 2010), we can introduce a mean transmission factor weighted by $\frac{\partial\Theta}{\partial T}$ and averaged over frequencies:

$$\overline{\mathcal{T}}_j = \frac{\int_0^\infty \mathrm{d}x\, f(x) \mathcal{T}_j(x, \mathbf{k}_{\parallel}; d)}{\int_0^\infty \mathrm{d}x\, f(x)} \tag{21.28}$$

with $f(x) = x^2 \mathrm{e}^x/(\mathrm{e}^x - 1)^2$ and $x = \hbar\omega/(k_\mathrm{B}T)$. This mean transmission factor is always smaller than 1. Using this quantity, we obtain a Landauer-like expression for the heat flux as derived in ref. (Biehs et al., 2010):

$$q = g_0 \Big(\sum_{j=\mathrm{s,p}} \int \frac{\mathrm{d}^2\mathbf{k}_{\parallel}}{(2\pi)^2} \overline{\mathcal{T}}_j \Big) [T_1 - T_2]. \tag{21.29}$$

This equation shows that the thermal conductance is a sum over all the modes $(j, \mathbf{k}_{\parallel})$. Each mode has a conductance which is given by the quantum of thermal conductance g_0 weighted by the transmission factor $\overline{\mathcal{T}}_j$, averaged over frequencies. The transmission factors associated with evanescent waves are exponentially decaying with the distance d, except when surface waves produce resonances. It follows from the linearized expression of the heat flux in Eq. (21.29) that two different strategies can be followed to enhance the heat flux between two media. The first one consists in increasing the number of modes $(j, \mathbf{k}_{\parallel})$, while the second one consists in increasing their average transmission factor.

21.3.6 The Stefan–Boltzmann constant revisited

Let us now focus on the maximum radiative flux in the far field. At large distances (i.e. in the far field), only propagating waves are involved. We see from Eq. (21.23) that the maximum heat flux corresponds to a situation where $\mathcal{T} = 1$ for $k_{\parallel} < \omega/c$. This transmission factor corresponds to perfectly absorbing media also called black-bodies. For a semi-infinite body, this situation is realized when the Fresnel reflection factors are exactly zero for both polarizations. The heat flux can then be computed from Eq. (21.23), yielding

$$q_\mathrm{BB} = \int_0^\infty [\Theta(\omega, T_1) - \Theta(\omega, T_2)] \left(\frac{\omega^2}{c^3\pi^2} \right) \frac{c}{4} = \sigma_\mathrm{BB}(T_1^4 - T_2^4), \tag{21.30}$$

which is the Stefan–Boltzmann law for the heat flux between two black-bodies with the Stefan constant $\sigma_\mathrm{BB} = 2\pi^5 k_B^4/15h^3c^2 = 5.67\,10^{-8}\,\mathrm{Wm^{-2}K^{-4}}$.

This well-known equation can be cast in a different form in the spirit of a Landauer approach. By linearizing $(T_1^4 - T_2^4)$ in the form $4T^3(T_1 - T_2)$ with $T = (T_1 + T_2)/2$, we find

$$q_{\mathrm{BB}} = g_0 \frac{2\pi}{5\lambda_T^2}(T_1 - T_2). \tag{21.31}$$

In other words, in the linearized regime, the usual Stefan–Boltzmann law is the quantum of thermal conductance g_0 times the number of modes per unit area roughly given by $\frac{1}{\lambda_T^2}$.

21.A Complement: Spatial coherence

In this complement, we analyse the spatial correlation of the electromagnetic field produced by random sources. The existence of spatial correlation is known as spatial coherence in optics. More precisely, we will discuss the connection between the angular far-field behaviour of a source and the spatial coherence in the source plane. In simple terms, a spatially partially coherent source is a source that radiates a field which has a narrow angular aperture at a given wavelength. The typical examples of coherent sources are lasers and antennas. These sources have an emission with well-defined angular lobes. These sources also have fields that are highly coherent over the width of the source. In what follows, we will show that a narrow angular-emission lobe is a signature of the spatial coherence of the field in the plane of the source. We will use the same tools as seen in Complement 1.A and replace time by spatial coordinate and frequency by wavevector component.

We start by introducing an analogue of the Wiener–Khinchin theorem (WKT) to analyse the spatial coherence of the field. For a translationally invariant random field, the Fourier transform of the field does not converge in the sense of a function. Yet, one can define a field $\mathrm{E}_A(\mathbf{r}_\parallel, \omega)$ equal to the random field in a square of area A and null outside. We can now define the Fourier transform of this field in the plane $z = 0$ as:

$$\mathrm{E}_A(\mathbf{r}_\parallel, \omega) = \int \frac{\mathrm{d}^2\mathbf{k}_\parallel}{4\pi^2}\, \mathrm{E}_A(\mathbf{k}_\parallel, \omega) \exp(i\mathbf{k}_\parallel \cdot \mathbf{r}_\parallel). \tag{21.32}$$

It can be shown (Goodman, 2015) that the WKT yields a relation between the spatial correlation function and the power spectral density of the field:

$$\left\langle E(\mathbf{r}_\parallel, \omega) E^*(\mathbf{r}_\parallel + \mathbf{r}'_\parallel, \omega) \right\rangle = \int \frac{d^2\mathbf{k}_\parallel}{4\pi^2} \lim_{A\to\infty} \frac{1}{A} \left\langle E_A(\mathbf{k}_\parallel, \omega) E_A^*(\mathbf{k}_\parallel, \omega) \right\rangle \exp(i\mathbf{k}_\parallel \cdot \mathbf{r}'_\parallel). \tag{21.33}$$

This relation implies that if the power spectral density has a bandwidth smaller than $2\pi/\lambda$, the coherence length is larger than λ. The second step is to show that the bandwidth of the power spectral density $\lim_{A\to\infty} \left\langle \mathrm{E}_A(\mathbf{k}_\parallel, \omega) \mathrm{E}_A^*(\mathbf{k}_\parallel, \omega) \right\rangle / A$ is given by the emission pattern in the far field. According to Eq. (4.14), the field propagating in vacuum can be written everywhere as a function of the Fourier transform of the field in the source plane

$$\mathrm{E}_A(\mathbf{r}, \omega) = \int \frac{\mathrm{d}^2\mathbf{k}_\parallel}{4\pi^2}\, \mathrm{E}_A(\mathbf{k}, \omega) \exp(i\mathbf{k}_\parallel \cdot \mathbf{r}_\parallel + i\gamma z), \tag{21.34}$$

where $\mathbf{k} = (\mathbf{k}_\parallel, \gamma)$ and γ is given by $\gamma^2 = k_0^2 - \mathbf{k}_\parallel^2$. This field can be evaluated asymptotically in the far field using the stationary-phase approximation (Mandel and Wolf, 1995). It can be cast in the form

$$\mathrm{E}_A(\mathbf{r},\omega) = \frac{K\exp(ikr)}{r}\,\mathrm{E}_A\left(\mathbf{k}_\parallel = \frac{2\pi}{\lambda}\hat{\mathbf{r}}_\parallel,\omega\right), \tag{21.35}$$

where K is a constant. The power $\mathrm{d}P$ flowing through an element of surface $\mathrm{d}S = r^2\mathrm{d}\Omega$ is given by the flux of the Poynting vector. In the far field, the Poynting vector has a plane-wave structure locally, so that its time-averaged amplitude is given by $\varepsilon_0 c|\mathrm{E}|^2/2$:

$$\mathrm{d}P = \frac{\varepsilon_0 c}{2}|K|^2\left|\mathrm{E}_A\left(\mathbf{k}_\parallel = \frac{2\pi}{\lambda}\hat{\mathbf{r}}_\parallel,\omega\right)\right|^2\mathrm{d}\Omega, \tag{21.36}$$

where $\hat{\mathbf{r}}$ is the unit vector $\mathbf{r}/|\mathbf{r}|$. This relation completes the discussion of the link between the directivity of the emitted field and the coherence of the field in the plane of the source. It follows from the properties of a Fourier transform that a directional source with a narrow K-spectrum implies a large correlation length. In other words, the angular width is a direct estimate of the inverse of the correlation length.

Exercises

Exercise 21.1 Tailoring spectral absorption with an array of plasmonic nanoantennas. Fig. 21.5b shows that a periodic array of four different plasmonic nanoantennas (resonators) tuned to four different wavelengths can produce an almost total absorption at each frequency. We consider for simplicity a periodic array of square patches with a width of 2 μm and a period of 5 μm. Thinking of absorption in terms of ray optics and assuming that each resonator is totally absorbing what should the maximum absorption be?

Geometrical optics is not valid at a subwavelength scale, and we have to use the cross-section concept (see Complement 14.A.1). Assuming that the small resonator behaves as a resonant electric dipole, what is the expected extinction cross section on resonance? We assume that the resonator resonates at approximately 10 μm. Comparing the extinction cross section with d^2, where d is the period of the periodic array of resonators, estimate the maximum period for observing total absorption.

Exercise 21.2 Estimation of the coherence length. Adapting the discussion of section 2.4.2, show that the angular width of an emission peak is given by the ratio of the wavelength by the coherence length. From Fig. 21.4, it can be measured that the angular width is on the order of 3°. Derive an estimate of the coherence length of the source.

Exercise 21.3 Quantum of thermal conductance. Derive the form of the quantum of thermal conductance given by Eq. (21.26). Note that $\int_0^\infty \frac{u\,du}{e^u-1} = \frac{\pi^2}{6}$.

Exercise 21.4 Radiative heat transfer along a metallic wire. We consider a metallic wire that can support guided waves on the microwave regime with a non-dispersive velocity c. We assume that this wire connects two thermostats at temperature T_1 and T_2. In analogy with Eq. (21.9), write the form of the (1D) specific intensity:

$$I_{BB,\omega}(T) = \frac{1}{2\pi}\times\frac{\hbar\omega}{\exp(\hbar\omega/k_BT)-1}.$$

Show that for the 1D problem, $I_{BB} = uc$, where u is the spectral energy per unit volume and $1/(2\pi c)$ is the density of states per unit length. Prove that the flux is given by Eq. (21.25).

By introducing the dimensionless variable $x = \hbar\omega/k_B T$, show that the temperature dependence of the black-body radiation along the wire is not T^4 but T^2.

Exercise 21.5 Spatial dependence of the radiative heat flux in the near field. Use simple arguments to derive the power-law dependence on the distance d of the heat flux exchanged between two nanoparticles ($1/d^6$), between a nanoparticle and a surface ($1/d^3$), and between two surfaces ($1/d^2$).

References

Adams, J. L. and Botten, L. C. (1979). 'Double gratings and their applications as Fabry-Perot interferometers'. *J. Opt.*, **10**(3), 109–17.

Agarwal, G. S. (1975). 'Quantum electrodynamics in the presence of dielectrics and conductors. iii. Relations among one-photon transition probabilities in stationary and nonstationary fields, density of states, the field-correlation functions, and surface-dependent response functions'. *Phys. Rev. A*, **11**, 253–64.

Agio, M. and Alu, A., Editors (2013). *Optical Antennas.* Cambridge University Press.

Agranovich, V. M. and Gartstein, Y. N. (2006). 'Spatial dispersion and negative refraction of light'. *Physics-Uspekhi*, **49**(10), 1029.

Agranovich, V. M. and Ginzburg, V. L. (1984). *Crystal Optics with Spatial Dispersion, and Excitons.* Springer, Berlin.

Agranovich, V. M. and Mills, D. L., Editors (1982). *Surface Polaritons Electromagnetic Waves at Surfaces and Interfaces.* Volume 1, Modern Problems in Condensed Matter Sciences. North-Holland.

Akahane, Y., Asano, T., and Song, B. S. et al. (2003). 'High-Q photonic nanocavity in a two-dimensional photonic crystal'. *Nature*, **425**, 944–7.

Allen, L. and Eberly, J. H. (1987). *Optical Resonance and Two-level Atoms.* Dover, New York.

Allen, P. B. (1971). 'Electron-phonon effects in the infrared properties of metals'. *Phys. Rev. B*, **3**, 305–20.

Anger, P., Bharadwaj, P., and Novotny, L. (2006). 'Enhancement and quenching of single-molecule fluorescence'. *Phys. Rev. Lett.*, **96**, 113002.

Arakawa, Y. and Sakaki, H. (1982). 'Multidimensional quantum well laser and temperature dependence of its threshold current'. *Appl. Phys. Lett.*, **40**(11), 939–41.

Arbabi, A., Horie, Y., Ball, A. J., Bagheri, M., and Faraon, A. (2015). 'Subwavelength-thick lenses with high numerical apertures and large efficiency based on high-contrast transmitarrays'. *Nat. Commun.*, **6**, 7069–75.

Archambault, A., Teperik, T. V., Marquier, F., and Greffet, J.-J. (2009). 'Surface plasmon Fourier optics'. *Phys. Rev. B*, **79**, 195414.

Arfken, G. B., Weber, H. J., and Harris, F. E. (2013). *Mathematical Methods for Physicists: A Comprehensive Guide.* Elsevier Science.

Arhab, S., Soriano, G., Ruan, Y., Maire, G., Talneau, A., Sentenac, D., Chaumet, P. C., Belkebir, K., and Giovannini, H. (2013). 'Nanometric resolution with far-field optical profilometry'. *Phys. Rev. Lett.*, **111**, 053902.

Ash, E. A. and Nicholls, G. (1972). 'Super-resolution aperture scanning microscope'. *Nature*, **237**(5357), 510–12.

Ashcroft, N. W. and Mermin, N. D. (1976). *Solid State Physics.* Holt, Rinehart and Winston, New York.

Aspnes, D. E. (1982). 'Local-field effects and effective-medium theory: A microscopic perspective'. *Am. J. Phys.*, **50**(8), 704–9.

Atwater, H. A. and Polman, A. (2010). 'Plasmonics for improved photovoltaic devices'. *Nat. Mater.*, **9**, 205–13.

Authier, A. (2001). *Dynamical Theory of X-ray Diffraction.* IUCR Crystallographic Symposia Series. Oxford University Press, Oxford.

Babuty, A., Joulain, K., Chapuis, P.-O., Greffet, J.-J., and De Wilde, Y. (2013). 'Blackbody spectrum revisited in the near field'. *Phys. Rev. Lett.*, **110**, 146103.

Balanis, C. A. (2016). *Antenna Theory: Analysis and Design.* Wiley, New York.

Band, Y. B. and Heller, D. F. (1988). 'Relationship between the absorption and emission of light in multilevel systems'. *Phys. Rev. A*, **38**, 1885.

Barnes, W. L., Dereux, A., and Ebbesen, T. W. (2003). 'Surface plasmon subwavelength optics'. *Nature*, **424**, 824–30.

Bastard, G. (1992). *Wave Mechanics Applied to Semiconductor Heterostructures.* Les Editions de Physique, Paris.

Baños, A. (1966). *Dipole Radiation in the Presence of a Conducting Half-space.* Pergamon, New York.

Belkebir, K., Chaumet, P. C., and Sentenac, A. (2006). 'Influence of multiple scattering on three-dimensional imaging with optical diffraction tomography'. *J. Opt. Soc. Am. A*, **23**(3), 586–95.

Benisty, H. (1995). 'Reduced electron-phonon relaxation rates in quantum-box systems: Theoretical analysis'. *Phys. Rev. B*, **51**, 13281–93.

Benisty, H., De Neve, H., and Weisbuch, C. (1998). 'Impact of planar microcavity effects on light extraction-Part I: Basic concepts and analytical trends'. *IEEE J. Quantum Electron.*, **34**(9), 1612–31.

Benisty, H., Sotomayor-Torrès, C. M., and Weisbuch, C. (1991). 'Intrinsic mechanism for the poor luminescence properties of quantum-box systems'. *Phys. Rev. B*, **44**, 10945–8.

Bernhard, C. G. (1967). 'Structural and functional adaptation in a visual system'. *Endeavour*, **26**, 79–84.

Berini, P. (2009). 'Long-range surface plasmon polaritons'. *Advances in Optics and Photonics*, **1**(3), 484–588.

Berry, M. V. and Popescu, S. (2006). 'Evolution of quantum superoscillations and optical superresolution without evanescent waves'. *J. Phys. A: Math. Gen.*, **39**(22), 6965–77.

Betzig, E. (1995). 'Proposed method for molecular optical imaging'. *Opt. Lett.*, **20**(3), 237–9.

Betzig, E., Patterson, G. H., Sougrat, R., Lindwasser, O. W., Olenych, S., Bonifacino, J. S., Davidson, M. W., Lippincott-Schwartz, J., and Hess, H. F. (2006). 'Imaging intracellular fluorescent proteins at nanometer resolution'. *Science*, **313**(5793), 1642–5.

Bharadwaj, P., Deutsch, B., and L., Novotny (2009). 'Optical antennas'. *Adv. Opt. Photon.*, **1**(3), 438–83.

Biagioni, P., Huang, J.-S., and Hecht, B. (2012). 'Nanoantennas for visible and infrared radiation'. *Rep. Prog. Phys.*, **75**(2), 024402.

Biehs, S.-A., Rousseau, E., and Greffet, J.-J. (2010). 'Mesoscopic description of radiative heat transfer at the nanoscale'. *Phys. Rev. Lett.*, **105**, 234301.

Bisi, O., Ossicini, S., and Pavesi, L. (2000). 'Porous silicon: A quantum sponge structure for silicon based optoelectronics'. *Surf. Sci. Rep.*, **38**(1), 1–126.

Boardman, A. (1982). *Electromagnetic Surface Modes.* Wiley, Chichester.

Bohren, C. and Huffman, R. (1983). *Absorption and Scattering of Light by Small Particles.* Wiley, New York.

Boltasseva, A. and Atwater, H. A. (2011). 'Low-loss plasmonic metamaterials'. *Science*, **331**(6015), 290–1.

Bonifacio, R. and Lugiato, L. A. (1978). 'Optical bistability and cooperative effects in resonance fluorescence'. *Phys. Rev. A*, **18**, 1129–44.

Born, M. and Wolf, E. (1999). *Principles of Optics: Electromagnetic Theory of Propagation, Interference and Diffraction of Light* (7th edn). Cambridge University Press.

Bouchon, P., Koechlin, C., Pardo, F., Haïdar, R., and Pelouard, J.-L. (2012). 'Wideband omnidirectional infrared absorber with a patchwork of plasmonic nanoantennas'. *Opt. Lett.*, **37**(6), 1038–40.

Bougot-Robin, K., Pang, C., Pommarède, X., Itawi, A., Talneau, A., Hugonin, J. P., and Benisty, H. (2014*a*). 'Embedded effective-index-material in oxide-free hybrid silicon photonics characterized by prism deviation'. *J. Lightwave Technol.*, **32**(19), 3283–9.

Bougot-Robin, K., Talneau, A., and Benisty, H. (2014*b*). 'Low-index nanopatterned barrier for hybrid oxide-free III-V silicon conductive bonding'. *Opt. Express*, **22**(19), 23333–8.

Bozhevolnyi, S. I. and Khurgin, J. B. (2016). 'Fundamental limitations in spontaneous emission rate of single-photon sources'. *Optica*, **3**(12), 1418–21.

Bozhevolnyi, S. I., Volkov, V. S., Devaux, E., Laluet, J.-Y., and Ebbesen, T. W. (2006). 'Channel plasmon subwavelength waveguide components including interferometers and ring resonators'. *Nature*, **440**, 508–11.

Brandes, T. and Kettemmann, S., Editors (2003). *Anderson Localization and Its Ramification.* Springer, Berlin.

Brandt, O., Tapfer, L., Ploog, K., Bierwolf, R., Hohenstein, M., Phillipp, F., Lage, H., and Heberle, A. (1991). 'InAs quantum dots in a single-crystal GaAs matrix'. *Phys. Rev. B*, **44**, 8043–53.

Brekhovskikh, L. M. (1960). *Waves in Layered Media.* Volume 6, Applied Mathematics and Mechanics. Academic Press, New York.

Brillouin, L. (1922). 'Sur l'origine de la résistance de rayonnement'. *Radioélectricité*, **3**, 147.

Brillouin, L. (1960). *Wave Propagation and Group Velocity.* Volume 8, Pure and Applied Physics. Academic Press.

Busch, K., Vats, N., John, S., and Sanders, B. C. (2000). 'Radiating dipoles in photonic crystals'. *Phys. Rev. E*, **62**, 4251–60.

Callen, H. B. and Welton, T. A. (1951). 'Irreversibility and generalized noise'. *Phys. Rev.*, **83**, 34–40.

Carminati, R. and Greffet, J.-J. (1999). 'Near-field effects in spatial coherence of thermal sources'. *Phys. Rev. Lett.*, **82**, 1660–3.

Carminati, R., Greffet, J.-J., Henkel, C., and Vigoureux, J. M. (2006). 'Radiative and non-radiative decay of a single molecule close to a metallic nanoparticle'. *Opt. Commun.*, **261**(2), 368–75.

Carminati, R., Sáenz, J. J., Greffet, J.-J., and Nieto-Vesperinas, M. (2000). 'Reciprocity, unitarity, and time-reversal symmetry of the S matrix of fields containing evanescent components'. *Phys. Rev. A*, **62**, 012712.

Casimir, H. B. G. (1948). 'On the attraction between two perfectly conducting plates'. *Proc. Kon. Ned. Akad. Wet.*, **51**, 793.

Chance, R. R., Prock, A., and Silbey, R. (1974). 'Lifetime of an emitting molecule near a partially reflecting surface'. *J. Chem. Phys.*, **60**, 2744–8.

Cheben, P., Halir, R., Schmid, J. H., Atwater, H. A., and Smith, D. R. (2018). 'Subwavelength integrated photonics'. *Nature*, **560**, 565–72.

Chen, F.-C. and Chew, W. C. (1998). 'Experimental verification of super resolution in nonlinear inverse scattering'. *Appl. Phys. Lett.*, **72**(23), 3080–2.

Chikkaraddy, R., de Nijs, B., Benz, F., Barrow, S. J., Scherman, O. A., Rosta, E., Demetriadou, A., Fox, P., Hess, O., and Baumberg, J. J. (2016). 'Single-molecule strong coupling at room temperature in plasmonic nanocavities'. *Nature*, **535**, 127–30.

Chutinan, A. and Noda, S. (2000). 'Waveguides and waveguide bends in two-dimensional photonic crystal slabs'. *Phys. Rev. B*, **62**, 4488–92.

Cohen-Tannoudji, C., Diu, B., and Laloë, F. (1991). *Quantum Mechanics, Volume 1.* Wiley.

Collin, R. E. (1960). *Field Theory of Guided Waves.* McGraw-Hill.

Collin, R. E. (2004). 'Hertzian dipole radiating over a lossy earth or sea: some early and late 20th-century controversies'. *IEEE Antennas Propag. Mag.*, **46**(2), 64–79.

Collin, S. (2014). 'Nanostructure arrays in free-space: Optical properties and applications'. *Rep. Prog. Phys.*, **77**(12), 126402.

Colom, R., Devilez, A., Bonod, N., and Stout, B. (2016). 'Optimal interactions of light with magnetic and electric resonant particles'. *Phys. Rev. B*, **93**, 045427.

Costantini, D., Lefebvre, A., Coutrot, A.-L., Moldovan-Doyen, I., Hugonin, J.-P., Boutami, S., Marquier, F., Benisty, H., and Greffet, J.-J. (2015). 'Plasmonic metasurface for directional and frequency-selective thermal emission'. *Phys. Rev. Applied*, **4**, 014023.

Crawford, F. S. (1968). *Waves.* McGraw-Hill.

Cui, Y., He, Y., Jin, Y., Ding, F., Yang, L., Ye, Y., Zhong, S., Lin, Y., and He, S. (2014). 'Plasmonic and metamaterial structures as electromagnetic absorbers'. *Laser Photonics Rev.*, **8**(4), 495–520.

Curto, Alberto G., Volpe, Giorgio, Taminiau, Tim H., Kreuzer, Mark P., Quidant, Romain, and van Hulst, Niek F. (2010). Unidirectional emission of a quantum dot coupled to a nanoantenna. *Science*, **329**(5994), 930–933.

David, A., Benisty, H., and Weisbuch, C. (2012). 'Photonic crystal light-emitting sources'. *Rep. Prog. Phys.*, **75**(12), 126501.

Dawson, P., de Fornel, F., and Goudonnet, J.-P. (1994). 'Imaging of surface plasmon propagation and edge interaction using a photon scanning tunneling microscope'. *Phys. Rev. Lett.*, **72**, 2927–30.

De Zoysa, M., Asano, T., Mochizuki, K., Oskooi, A., Inoue, T., and Noda, S. (2012). 'Conversion of broadband to narrowband thermal emission through energy recycling'. *Nat. Photonics*, **6**(8), 535–9.

Decca, R. S., López, D., Fischbach, E., and Krause, D. E. (2003). 'Measurement of the Casimir force between dissimilar metals'. *Phys. Rev. Lett.*, **91**, 050402.

Deguzman, P. C. and Nordin, G. P. (2001). 'Stacked subwavelength gratings as circular polarization filters'. *Appl. Opt.*, **40**(31), 5731–7.

Del Fatti, N. (1999). *Dynamique Electronique Femtoseconde dans les Systèmes Métalliques Massifs et Confinés*. PhD thesis, Ecole Polytechnique.

Ding, Y. and Magnusson, R. (2004). 'Use of nondegenerate resonant leaky modes to fashion diverse optical spectra'. *Opt. Express*, **12**(9), 1885–91.

Dousse, A., Lanco, L., Suffczyński, J., Semenova, E., Miard, A., Lemaître, A., Sagnes, I., Roblin, C., Bloch, J., and Senellart, P. (2008). 'Controlled light-matter coupling for a single quantum dot embedded in a pillar microcavity using far-field optical lithography'. *Phys. Rev. Lett.*, **101**, 267404.

Drexhage, K. H., Kuhn, H., and Schäfer, F. P. (1968). 'Variation of the fluorescence decay time of a molecule in front of a mirror'. *Berichte der Bunsengesellschaft für physikalische Chemie*, **72**(2), 329.

Dubrovskii, V. G., Cirlin, G. E. Cirlin, and Ustinov, V. M. (2009). 'Semiconductor nanowhiskers: Synthesis, properties, and applications'. *Semiconductors*, **43**, 1539–84.

Eagen, C. F., Weber, W. H., McCarthy, S. L., and Terhune, R. W. (1980). 'Time-dependent decay of surface-plasmon-coupled molecular fluorescence'. *Chem. Phys. Lett.*, **75**(2), 274–277.

Ebbesen, T. W., Lezec, H. J., Ghaemi, H. F., Thio, T., and Wolff, P. A. (1998). 'Extraordinary optical transmission through sub-wavelength hole arrays'. *Nature*, **391**, 667–9.

Esteban, R., Borisov, A., Nordlander, P., and Aizpurua, J. (2012). 'Bridging quantum and classical plasmonics with a quantum-corrected model'. *Nat. Commun.*, **3**, 825.

Esteban, R., Laroche, M., and Greffet, J.-J. (2009). 'Influence of metallic nanoparticles on upconversion processes'. *J. Appl. Phys.*, **105**(3), 033107.

Etchegoin, P. G., Le Ru, E. C., and Meyer, M. (2006). 'An analytic model for the optical properties of gold'. *J. Chem. Phys.*, **125**(16), 164705.

Faggiani, R., Yang, J., and Lalanne, P. (2015). 'Quenching, plasmonic, and radiative decays in nanogap emitting devices'. *ACS Photonics*, **2**(12), 1739–44.

Faist, J., Capasso, F., Sivco, D. L., Sirtori, C., Hutchinson, A. L., and Cho, A. Y. (1994). 'Quantum cascade laser'. *Science*, **264**(5158), 553–6.

Fang, N., Lee, H., Sun, C., and Zhang, X. (2005). 'Sub-diffraction-limited optical imaging with a silver superlens'. *Science*, **308**(5721), 534–7.

Fano, U. (1941). 'The theory of anomalous diffraction gratings and of quasi-stationary waves on metallic surfaces (Sommerfeld's waves)'. *J. Opt. Soc. Am.*, **31**(3), 213–22.

Fehrembach, A.-L., Talneau, A., Boyko, O., Lemarchand, F., and Sentenac, A. (2007). 'Experimental demonstration of a narrowband, angular tolerant, polarization independent, doubly periodic resonant grating filter'. *Opt. Lett.*, **32**(15), 2269–71.

Feibelman, P. J. (1982). 'Surface electromagnetic fields'. *Prog. Surf. Sci.*, **12**(4), 287–407.

Felsen, L. and Marcuvitz, N. (1996). *Radiation and Scattering of Waves.* Oxford University Press, Oxford.

Ford, G. W. and Weber, W. H. (1984). 'Electromagnetic interactions of molecules with metal surfaces'. *Phys. Rep.*, **113**(4), 195–287.

Gall, D. (2016). 'Electron mean free path in elemental metals'. *J. Appl. Phys.*, **119**(8), 085101.

Garcia, N. and Nieto-Vesperinas, M. (2002). 'Left-handed materials do not make a perfect lens'. *Phys. Rev. Lett.*, **88**, 207403.

García de Abajo, F. J. (2007). 'Colloquium: Light scattering by particle and hole arrays'. *Rev. Mod. Phys.*, **79**, 1267–90.

Geim, A. K. (2009). 'Graphene: Status and prospects'. *Science*, **324**(5934), 1530–4.

Geim, A. K. and Novoselov, K. S. (2007). 'The rise of graphene'. *Nat. Mater.*, **6**, 183–91.

Genet, C., Lambrecht, A., and Reynaud, S. (2000). Temperature dependence of the Casimir effect between metallic mirrors. *Phys. Rev. A*, **62**, 012110.

Genevet, P., Capasso, F., Aieta, F., Khorasaninejad, M., and Devlin, R. (2017). 'Recent advances in planar optics: From plasmonic to dielectric metasurfaces'. *Optica*, **4**(1), 139–52.

Gigli, C., Li, Q., Chavel, P., Leo, G., Brongersma, M. L., and Lalanne, P. (2021). 'Fundamental limitations of Huygens' Metasurfaces for optical beam shaping'. *Laser Photonics Rev.*, **15**, 2000448.

Goldstein, L., Glas, F., Marzin, J.-Y., Charasse, M.-N., and Le Roux, G. (1985). 'Growth by molecular beam epitaxy and characterization of InAs/GaAs strained-layer superlattices'. *Appl. Phys. Lett.*, **47**(10), 1099–101.

Goodman, J. W. (2005). *Introduction to Fourier Optics.* McGraw-Hill Physical and Quantum Electronics Series. W. H. Freeman.

Goodman, J. W. (2015). *Statistical Optics* (2nd edn). Wiley Series in Pure and Applied Optics. Wiley, New York.

Gramotnev, D. K. and Bozhevolnyi, S. I. (2010). 'Plasmonics beyond the diffraction limit'. *Nat. Photonics*, **4**, 83–91.

Greffet, J.-J., Bouchon, P., Brucoli, G., and Marquier, F. (2018). 'Light emission by nonequilibrium bodies: Local Kirchhoff law'. *Phys. Rev. X*, **8**, 021008.

Greffet, J.-J., Carminati, R., Joulain, K., Mulet, J.-P., Mainguy, S., and Chen, Y. (2002). 'Coherent emission of light by thermal sources'. *Nature*, **416**(6876), 61–4.

Greffet, J.-J., Laroche, M., and Marquier, F. (2010). 'Impedance of a nanoantenna and a single quantum emitter'. *Phys. Rev. Lett.*, **105**, 117701.

Grynberg, G., Aspect, A., Fabre, C., and Cohen-Tannoudji, C. (2010). *Introduction to Quantum Optics: From the Semi-classical Approach to Quantized Light.* Cambridge University Press.

Grüning, U., Lehmann, V., Ottow, S., and Busch, K. (1996). 'Macroporous silicon with a complete two-dimensional photonic band gap centered at 5 μm'. *Appl. Phys. Lett.*, **68**(6), 747–9.

Gustafsson, M. G. L. (2005). 'Nonlinear structured-illumination microscopy: Wide-field fluorescence imaging with theoretically unlimited resolution'. *Proc. Natl. Acad. Sci.*, **102**(37), 13081–6.

Guyot-Sionnest, P. (2008). 'Colloidal quantum dots'. *Comptes Rendus Physique*, **9**(8), 777–87. Recent Advances in Quantum Dot Physics/Nouveaux développements dans la physique des boîtes quantiques.

Gérard, J. M., Sermage, B., Gayral, B., Legrand, B., Costard, E., and Thierry-Mieg, V. (1998). 'Enhanced spontaneous emission by quantum boxes in a monolithic optical microcavity'. *Phys. Rev. Lett.*, **81**, 1110–3.

Hache, F., Ricard, D., and Flytzanis, C. (1986). 'Optical nonlinearities of small metal particles: Surface-mediated resonance and quantum size effects'. *J. Opt. Soc. Am. B*, **3**(12), 1647–55.

Haidner, H., Kipfer, P., Sheridan, J. T., Schwider, J., Streibl, N., Collischon, M., Hutfless, J., and März, M. (1993). 'Diffraction grating with rectangular grooves exceeding 80% diffraction efficiency'. *Infrared Physics*, **34**(5), 467–75.

Halevi, P. (1995). 'Hydrodynamic model for the degenerate free-electron gas: Generalization to arbitrary frequencies'. *Phys. Rev. B*, **51**, 7497–9.

Halir, R., Ortega-Moñux, A., Benedikovic, D., Mashanovich, G. Z., Wangüemert-Pérez, J. G., Schmid, J. H., Molina-Fernández, I., and Cheben, P. (2018). 'Subwavelength-grating metamaterial structures for silicon photonic devices'. *Proc. IEEE*, **106**, 2144–57.

Hao, F. and Nordlander, P. (2007). 'Efficient dielectric function for FDTD simulation of the optical properties of silver and gold nanoparticles'. *Chem. Phys. Lett.*, **446**(1), 115–18.

Hargreaves, C. M. (1969). 'Anomalous radiative transfer between closely-spaced bodies'. *Phys. Lett. A*, **30**(9), 491–2.

Haroche, S. and Raimond, J.-M.(2006). *Exploring the Quantum* (1st edn). Oxford University Press, Oxford.

Haus, H. A. (1984). *Waves and Fields in Optoelectronics.* Solid State Physical Electronics. Prentice-Hall.

Hauss, H. A. and Melcher, J. R. (1989). *Electromagnetic Fields and Energy* (1st edn). Prentice-Hall, Englewood Cliffs.

Hell, S. W. (2015). 'Nanoscopy with focused light (nobel lecture)'. *Angewandte Chemie International Edition*, **54**(28), 8054–66.

Hell, S. W. and Wichmann, J. (1994). 'Breaking the diffraction resolution limit by stimulated emission: stimulated-emission-depletion fluorescence microscopy'. *Opt. Lett.*, **19**(11), 780–2.

Henkel, C., Joulain, K., Carminati, R., and Greffet, J.-J. (2000). 'Spatial coherence of thermal near fields'. *Opt. Commun.*, **186**(1), 57–67.

Henkel, C., Joulain, K., Mulet, J.-P., and Greffet, J.-J. (2004). 'Coupled surface polaritons and the Casimir force'. *Phys. Rev. A*, **69**, 023808.

Henry, C. H. (1986). 'Phase noise in semiconductor lasers'. *IEEE. J. Lightwave Technol.*, **4**(3), 298–311.

Henry, C. H. and Kazarinov, R. F. (1996). 'Quantum noise in photonics'. *Rev. Mod. Phys.*, **68**, 801–53.

Ho, K. M., Chan, C. T., and Soukoulis, C. M. (1990). 'Existence of a photonic gap in periodic dielectric structures'. *Phys. Rev. Lett.*, **65**, 3152–5.

Hopfield, J. J. (1958). 'Theory of the contribution of excitons to the complex dielectric constant of crystals'. *Phys. Rev.*, **112**, 1555–67.

Hutley, M. C. and Maystre, D. (1976). 'The total absorption of light by a diffraction grating'. *Opt. Commun.*, **19**(3), 431–6.

Inoue, T., De Zoysa, M., Asano, T., and Noda, S. (2014). 'Realization of dynamic thermal emission control'. *Nat. Mater.*, **13**(10), 928–31.

Intravaia, F. and Lambrecht, A. (2005). 'Surface plasmon modes and the Casimir energy'. *Phys. Rev. Lett.*, **94**, 110404.

Jackson, J. D. (1975). *Classical Electrodynamics.* Wiley.

Javaux, L., Mahler, B, Dubertret, B., Shabaev, A., Rodina, A. V., Efros, Al. L., Yakovlev, D. R., Liu, F., Bayer, M., Camps, G., Biadala, L., Buil, S., Quelin, X., and Hermier, J. P. (2013). 'Thermal activation of non-radiative auger recombination in charged colloidal nanocrystals.' *Nat. Nanotechnol.*, **8**, 206.

Jean, P., Gervais, A., LaRochelle, S., and Shi, W. (2020). 'Slow light in subwavelength grating waveguides.' *IEEE J. Sel. Top. Quantum Electron.*, **26**, 8200108.

Jennewein, S., Brossard, L., Sortais, Y. R. P., Browaeys, A., Cheinet, P., Robert, J., and Pillet, P. (2018). 'Coherent scattering of near-resonant light by a dense, microscopic cloud of cold two-level atoms: Experiment versus theory'. *Phys. Rev. A*, **97**, 053816.

Joannopoulos, J. D., Johnson, S. G., Winn, J. N., and Meade, R. D. (2008). *Photonic Crystals: Molding the Flow of Light (Second Edition).* Princeton University Press, Princeton.

John, S. (1987). 'Strong localization of photons in certain disordered dielectric superlattices'. *Phys. Rev. Lett.*, **58**, 2486–9.

Johnson, P. B. and Christy, R. W. (1972). 'Optical constants of the noble metals'. *Phys. Rev. B*, **6**, 4370–9.

Johnson, S. G., Fan, S., Villeneuve, P. R., Joannopoulos, J. D., and Kolodziejski, L. A. (1999). 'Guided modes in photonic crystal slabs'. *Phys. Rev. B*, **60**, 5751–8.

Jones, A. C. and Raschke, M. B. (2012). 'Thermal infrared near-field spectroscopy'. *Nano Lett.*, **12**(3), 1475–81.

Jones, D. S. (1989). *Acoustic and Electromagnetic Waves.* Oxford Science Publications. Clarendon Press.

Joulain, K., Carminati, R., Mulet, J.-P., and Greffet, J.-J. (2003). 'Definition and measurement of the local density of electromagnetic states close to an interface'. *Phys. Rev. B*, **68**, 245405.

Joulain, K., Mulet, J.-P., Marquier, F., Carminati, R., and Greffet, J.-J. (2005). 'Surface electromagnetic waves thermally excited: Radiative heat transfer, coherence properties and Casimir forces revisited in the near field'. *Surf. Sci. Rep.*, **57**(3), 59–112.

Jourdan, G., Lambrecht, A., Comin, F., and Chevrier, J. (2009). 'Quantitative non-contact dynamic Casimir force measurements'. *EPL (Europhysics Letters)*, **85**(3), 31001.

Kanamori, Y., Sasaki, M., and Hane, K. (1999). 'Broadband antireflection gratings fabricated upon silicon substrates'. *Opt. Lett.*, **24**(20), 1422–4.

Kaveh, M. and Wiser, N. (1984). 'Electron-electron scattering in conducting materials'. *Adv. Phys.*, **33**(4), 257–372.

Kazarinov, R. and Henry, C. H. (1985). 'Second-order distributed feedback lasers with mode selection provided by first-order radiation losses'. *IEEE J. Quantum Electron.*, **21**(2), 144–50.

Khorasaninejad, M., Shi, Z., Zhu, A. Y., Chen, W. T., Sanjeev, V., Zaidi, A., and Capasso, F. (2017). 'Achromatic metalens over 60 nm bandwidth in the visible and metalens with reverse chromatic dispersion'. *Nano Lett.*, **17**(3), 1819–24.

Khurgin, J. B. (2010). 'Slow light in various media: A Tutorial'. *Adv. Opt. Photon.*, **2**(3), 287–318.

Khurgin, J. B. (2015). 'How to deal with the loss in plasmonics and metamaterials'. *Nat. Nanotechnol.*, **10**(3), 2–6.

Khurgin, J. B., Sun, G., and Soref, R. A. (2008). 'Electroluminescence efficiency enhancement using metal nanoparticles'. *Applied Physics Letters*, **93**(2), 021120.

Kihm, H. W., Lee, K. G., Kim, D. S., Kang, J. H., and Park, Q.-H. (2008). 'Control of surface plasmon generation efficiency by slit-width tuning'. *Appl. Phys. Lett.*, **92**(5), 051115.

King, R. W. P., Owens, M., and Wu, T. T. (1992). *Lateral Electromagnetic Waves*. Springer.

Kittel, C. (1966). *Quantum Theory of Solids*. Wiley.

Kittel, C. (2004). *Introduction to Solid State Physics*. Wiley.

Koenderink, A. F. (2017). 'Single-photon nanoantennas'. *ACS Photonics*, **4**(4), 710–22.

Kogelnik, H. (1969). 'Coupled wave theory for thick hologram gratings'. *Bell Syst. Tech. J.*, **48**(9), 2909–47.

Krauss, T. F. (2007). 'Slow light in photonic crystal waveguides'. *J. Phys. D: Appl. Phy.*, **40**(9), 2666–70.

Kreibig, U. and Genzel, L. (1985). 'Optical absorption of small metallic particles'. *Surf. Sci.*, **156**, 678–700.

Kuznetsov, A. I., Miroshnichenko, A. E., Brongersma, M. L., Kivshar, Y. S., and Luk'yanchuk, B. (2016). 'Optically resonant dielectric nanostructures'. *Science*, **354**(6314), aag2472.

Lalanne, P., Coudert, S., Duchateau, G., Dilhaire, S., and Vynck, K. (2019). 'Structural slow waves: Parallels between photonic crystals and plasmonic waveguides'. *ACS Photonics*, **6**(1), 4–17.

Lalanne, P. and Hugonin, J.-P. (2006). 'Interaction between optical nano-objects at metallo-dielectric interfaces'. *Nature*, **2**(6), 551–6.

Lalanne, P., Hugonin, J.-P., and Chavel, P. (2006). 'Optical properties of deep lamellar gratings: A coupled Bloch-mode insight'. *J. Lightwave Technol.*, **24**(6), 2442–9.

Lalanne, P., Hugonin, J. P., and Rodier, J.-C. (2005). 'Theory of surface plasmon generation at nanoslit apertures'. *Phys. Rev. Lett.*, **95**, 263902.

Lalanne, P., Sauvan, C., and Hugonin, J.-P. (2008). 'Photon confinement in photonic crystal nanocavities'. *Laser Photonics Rev.*, **2**, 514–26.

Lalanne, P., Yan, W., Vynck, K., Sauvan, C., and Hugonin, J.-P. (2018). 'Light interaction with photonic and plasmonic resonances'. *Laser Photonics Rev.*, **12**(5), 1700113.

Lamoreaux, S. K. (1997). 'Demonstration of the Casimir force in the 0.6 to 6μm range'. *Phys. Rev. Lett.*, **78**, 5–8.

Landau, L. D., Lifshitz, E. M., and Pitaevskii, L. P. (1980). *Statistical Physics Part I* (1 edn). Volume 8, Course of Theoretical Physics. Pergamon, Amsterdam.

Landau, L. D., Lifshitz, E. M., and Pitaevskii, L. P. (1984). *Electrodynamics of Continuous Media* (2nd edn). Volume 8, Course of Theoretical Physics. Pergamon, Amsterdam.

Laroche, T. and Girard, C. (2006). 'Near-field optical properties of single plasmonic nanowires'. *Appl. Phys. Lett.*, **89**(23), 233119.

Le Feber, B., Rotenberg, N., Beggs, D. M., and Kuipers, L. (2013). 'Simultaneous measurement of nanoscale electric and magnetic optical fields'. *Nature Photonics*, **8**, 43–6.

Lecamp, G., Hugonin, J.-P., and Lalanne, P. (2007). 'Theoretical and computational concepts for periodic optical waveguides'. *Opt. Express*, **15**(18), 11042–60.

Lee, M. S. L., Lalanne, P., Rodier, J.-C., P., Chavel, Cambril, E., and Chen, Y. (2002). 'Imaging with blazed-binary diffractive elements'. *J. Opt. A: Pure Appl. Opt.*, **4**(5), S119–S124.

Leonhardt, U. (2006). 'Notes on conformal invisibility devices'. *N. J. Phys.*, **8**, 118.

Lezec, H. J., Degiron, A., Devaux, E., Linke, R. A., Martin-Moreno, L., Garcia-Vidal, F. J., and Ebbesen, T. W. (2002). 'Beaming light from a subwavelength aperture'. *Science*, **297**(5582), 820–2.

Li, K., Stockman, M. I., and Bergman, D. J. (2003). 'Self-similar chain of metal nanospheres as an efficient nanolens'. *Phys. Rev. Lett.*, **91**, 227402.

Lin, D., Fan, P., Hasman, E., and Brongersma, M. L. (2014). 'Dielectric gradient metasurface optical elements'. *Science*, **345**(6194), 298–302.

Liu, W. (2017). 'Generalized magnetic mirrors'. *Phys. Rev. Lett.*, **119**, 123902.

Liu, W. and Kivshar, Y. S. (2018). 'Generalized kerker effects in nanophotonics and meta-optics'. *Opt. Express*, **26**(10), 13085–13105.

Liu H. T. and Lalanne, P. (2008). 'Microscopic theory of the extraordinary optical transmission'. *Nature*, **452**(7188), 728–731.

Liu, X., Tyler, T., Starr, T., Starr, A. F., Jokerst, N. M., and Padilla, W. J. (2011). 'Taming the blackbody with infrared metamaterials as selective thermal emitters'. *Phys. Rev. Lett.*, **107**, 045901.

Lorentz, H. A. (1916). *The Theory of Electrons and Its Applications to the Phenomena of Light and Radiant Heat.* B. G. Teubner, Leipzig.

Loudon, R. (2000). *The Quantum Theory of Light.* Oxford University Press, Oxford.

Lukosz, W. and Kunz, R. E. (1977). 'Fluorescence lifetime of magnetic and electric dipoles near a dielectric interface'. *Opt. Commun.*, **20**(2), 195–9.

Luk'yanchuk, B., Zheludev, N. I., Maier, S. A., Halas, N. J., Nordlander, P., Giessen, H., and Chong, C. T. (2010). 'The Fano resonance in plasmonic nanostructures and metamaterials'. *Nat. Materials*, **9**, pages 707–15.

Luo, X., Qiu, T., Lu, W., and Ni, Z. (2013). 'Plasmons in graphene: Recent progress and applications'. *Materials Science and Engineering: R: Reports*, **74**(11), 351–76.

Maese-Novo, A., Halir, R., Romero-García, S., Pérez-Galacho, D., Zavargo-Peche, L., Ortega-Moñux, A., Molina-Fernández, I., Wangüemert-Pérez, J. G., and Cheben, P. (2013). 'Wavelength independent multimode interference coupler'. *Opt. Express*, **21**(6), 7033–40.

Mahler, B., Spinicelli, P., Buil, S., Quelin, X., Hermier, J.-P., and Dubertret, B. (2008). 'Towards non-blinking colloidal quantum dots'. *Nat. Mater.*, **7**, 659.

Maier, S. A., Brongersma, M. L., Kik, P. G., Meltzer, S., Requicha, A. A. G., and Atwater, H. A. (2001). 'Plasmonics: A route to nanoscale optical devices'. *Adv. Mater.*, **13**(19), 1501–5.

Mandel, L. and Wolf, E. (1995). *Optical Coherence and Quantum Optics.* Cambridge University Press, Cambridge.

Marquier, F. and Greffet, J.-J. (2013). 'Impedance of a nanoantenna'. In *Optical Antennas* (ed. A. Mario and A. Alù), pp. 26–45. Cambridge University Press, Cambridge.

Marquier, F., Joulain, K., Mulet, J.-P., Carminati, R., Greffet, J.-J., and Chen, Y. (2004). 'Coherent spontaneous emission of light by thermal sources'. *Phys. Rev. B*, **69**, 155412.

März, R. (1994). *Integrated Optics Design and Modelling.* Artech House Inc.

Marzin, J.-Y., Gérard, J.-M., Izraël, A., Barrier, D., and Bastard, G. (1994). 'Photoluminescence of single InAs quantum dots obtained by self-organized growth on GaAs'. *Phys. Rev. Lett.*, **73**, 716–9.

Mateus, C. F. R., Huang, M. C. Y., Chen, L., J., Chang-Hasnain C., and Suzuki, Y. (2004). 'Broad-band mirror (1.12-1.62 μm) using a subwavelength grating'. *IEEE Phot. Technol. Lett.*, **16**, 1676–8.

Merlin, R. (2009). 'Metamaterials and the Landau–Lifshitz permeability argument: Large permittivity begets high-frequency magnetism'. *Proc. Nat. Acad. Sci.*, **106**(6), 1693–8.

Milton, G. W. (2002). *The Theory of Composites.* Cambridge Monographs on Applied and Computational Mathematics. Cambridge University Press, Cambridge.

Milton, G. W., Nicorovici, N.-A. P., and McPhedran, R. C. (2007). 'Opaque perfect lenses'. *Physica B: Condensed Matter*, **394**(2), 171–5.

Moerner, W. E. and Kador, L. (1989). 'Optical detection and spectroscopy of single molecules in a solid'. *Phys. Rev. Lett.*, **62**, 2535–8.

Moitra, P., Slovick, B. A., Li, W., Kravchencko, I. I., Briggs, D. P., Krishnamurthy, S., and Valentine, J. (2015). 'Large-scale all-dielectric metamaterial perfect reflectors'. *ACS Photonics*, **2**(6), 692–8.

Mokkapati, S. and Catchpole, K. R. (2012). 'Nanophotonic light trapping in solar cells'. *J. Appl. Phys.*, **112**(10), 101101.

Moneron, G., Dubois, D., and Boccara, A. C. (2005). 'Three-dimensional multiple nanometric local probes microscopy'. In *Three-Dimensional and Multidimensional Microscopy: Image Acquisition and Processing XII* (ed. J.-A. Conchello, C. J. Cogswell, and T. Wilson), Volume 5701, pp. 112–15. International Society for Optics and Photonics: SPIE.

Moreau, A., Ciraci, C., Mock, J. J., Hill, R. T., Wang, Q., Wiley, B. J., Chilkoti, A., and Smith, D. R. (2012). 'Controlled-reflectance surfaces with film-coupled colloidal nanoantennas'. *Nature*, **492**, 86–9.

Mudry, E., Belkebir, K., Girard, J., Savatier, J., Le Moal, E., Nicoletti, C., Allain, M., and Sentenac, A. (2012). 'Structured illumination microscopy using unknown speckle patterns'. *Nat. Photonics*, **6**, 312–15.

Mulet, J.-P., Joulain, K., Carminati, R., and Greffet, J.-J. (2001). 'Nanoscale radiative heat transfer between a small particle and a plane surface'. *Appl. Phys. Lett.*, **78**(19), 2931–3.

Mulet, J.-P., Joulain, K., Carminati, R., and Greffet, J.-J. (2002). 'Enhanced radiative heat transfer at nanometric distances'. *Microscale Thermophysical Engineering*, **6**(3), 209–222.

Munk, B. A. (2000). *Frequency Selective Surfaces: Theory and Design* (1st edn). Wiley.

Naik, G. V., Shalaev, V. M., and Boltasseva, A. (2013). 'Alternative plasmonic materials: Beyond gold and silver'. *Adv. Mater.*, **25**(24), 3264–94.

Nelayah, J., Kociak, M., Stéphan, O., Garcìa de Abajo, F. J., Tencé, M., Henrard, L., Taverna, D., Pastoriza-Santos, I., Liz-Marzàn, L. M., and Colliex, C. (2007). 'Mapping surface plasmons on a single metallic nanoparticle'. *Nat. Physics*, **3**, 348–53.

Nicolet, A., Remacle, J.-F., Meys, B., Genon, A., and Legros, W. (1994). 'Transformation methods in computational electromagnetism'. *J. Appl. Phys.*, **75**, 6036–8.

Noda, S., Kitamura, K., Okino, T., Yasuda, D., and Tanaka, Y. (2017). 'Photonic-crystal surface-emitting lasers: Review and introduction of modulated-photonic crystals'. *IEEE J. Sel. Top. Quantum Electron.*, **23**(6), 1–7.

Notomi, M. (2010). 'Manipulating light with strongly modulated photonic crystals'. *Rep. Prog. Physics*, **73**(9), 096501.

Notomi, M., Yamada, K., Shinya, A., Takahashi, J., Takahashi, C., and Yokohama, I. (2001). 'Extremely large group-velocity dispersion of line-defect waveguides in photonic crystal slabs'. *Phys. Rev. Lett.*, **87**, 253902.

Novotny, L. (2007*a*). 'The history of near-field optics'. In *Progress in Optics* (ed. E. Wolf), Volume 50, Progress in Optics, pp. 137–84. Elsevier.

Novotny, L. (2007*b*). 'Effective wavelength scaling for optical antennas'. *Phys. Rev. Lett.*, **98**, 266802.

Novotny, L. and Hecht, B. (2012). *Principles of Nano-Optics* (2nd edn). Cambridge University Press, Cambridge.

Ordal, M. A., Bell, R. J., Alexander, R. W., Long, L. L., and Querry, M. R. (1985). 'Optical properties of fourteen metals in the infrared and far infrared: Al, Co, Cu, Au, Fe, Pb, Mo, Ni, Pd, Pt, Ag, Ti, V, and W'. *Appl. Opt.*, **24**(24), 4493–9.

Orrit, M. and Bernard, J. (1990). 'Single pentacene molecules detected by fluorescence excitation in a p-terphenyl crystal'. *Phys. Rev. Lett.*, **65**, 2716–19.

Padjen, R., Gérard, J.-M., and Marzin, J.-Y. (1994). 'Analysis of the filling pattern dependence of the photonic bandgap for two-dimensional systems'. *J. Mod. Optics*, **41**(2), 295–310.

Palik, E. D. (1997). *Handbook of Optical Constants of Solids.* Volume 4, Handbook of Optical Constants of Solids. Academic Press, Boston, MA.

Pendry, J. B. (1983). 'Quantum limits to the flow of information and entropy'. *J. Phys. A: Math. Gen.*, **16**(10), 2161–71.

Pendry, J. B. (2000). 'Negative refraction makes a perfect lens'. *Phys. Rev. Lett.*, **85**, 3966–9.

Pendry, J. B., Martín-Moreno, L., and Garcia-Vidal, F. J. (2004). 'Mimicking surface plasmons with structured surfaces'. *Science*, **305**(5685), 847–8.

Peng, S. T., Tamir, T., and Bertoni, H. L. (1975). 'Theory of periodic dielectric waveguides'. *IEEE Trans. Microw. Theory Tech.*, **23**, 123–33.

Petit, R., Editor (1980). *Electromagnetic Theory of Gratings.* Topics in Current Physics, 22. Springer, New York.

Pfeiffer, C. and Grbic, A. (2013). 'Metamaterial Huygens' surfaces: Tailoring wave fronts with reflectionless sheets'. *Phys. Rev. Lett.*, **110**, 197401.

Pietroy, D., Tishchenko, A. V., Flury, M., and Parriaux, O. (2007). 'Bridging pole and coupled wave formalisms for grating waveguide resonance analysis and design synthesis'. *Opt. Express*, **15**(15), 9831–42.

Pines, D. (1964). *Elementary Excitations in Solids: Lectures on Phonons, Electrons and Plasmons.* W. A. Benjamin, London.

Pitarke, J. M., Silkin, V. M., Chulkov, E. V., and Echenique, P. M. (2006). 'Theory of surface plasmons and surface-plasmon polaritons'. *Rep. Prog. Phys.*, **70**(1), 1–87.

Plihal, M. and Maradudin, A. A. (1991). 'Photonic band structure of two-dimensional systems: The triangular lattice'. *Phys. Rev. B*, **44**, 8565–71.

Pohl, D.W. (1999). 'Near-Field Optics Principles and Applications'. In *Near field optics seen as an antenna problem* (ed. Zhu, X. Z. and Ohtsu, M.), pp. 9–21. World Scientific: Singapore

Pohl, D. W., Denk, W., and Lanz, M. (1984). 'Optical stethoscopy: Image recording with resolution $\lambda/20$'. *Appl. Phys. Lett.*, **44**(7), 651–3.

Polder, D. and Van Hove, M. (1971). 'Theory of radiative heat transfer between closely spaced bodies'. *Phys. Rev. B*, **4**, 3303–14.

Porto, J. A., Carminati, R., and Greffet, J.-J. (2000). 'Theory of electromagnetic field imaging and spectroscopy in scanning near-field optical microscopy'. *J. Appl. Phys.*, **88**(8), 4845–50.

Powell, C. J. and Swan, J. B. (1959). 'Origin of the characteristic electron energy losses in aluminum'. *Phys. Rev.*, **115**, 869–75.

Prodan, E., Radloff, C., Halas, N. J., and Nordlander, P. (2003). 'A hybridization model for the plasmon response of complex nanostructures'. *Science*, **419**, 302.

Purcell, E. M. (1946). 'Spontaneous emission probabilities at radio frequencies'. *Phys. Rev.*, **69**, 681.

Puscasu, I. and Schaich, W. L. (2008). 'Narrow-band, tunable infrared emission from arrays of microstrip patches'. *Appl. Phys. Lett.*, **92**(23), 233102.

QDLASER website https://www.qdlaser.com/en/technology.html#modal2, last checked 31.12.2021.

Rabouw, F. T., van der Bok, J. C., Spinicelli, P., Mahler, B., Nasilowski, M., Pedetti, S., Dubertret, B., and Vanmaekelbergh, D. (2016). 'Temporary charge carrier separation dominates the photoluminescence decay dynamics of colloidal CdSe nanoplatelets'. *Nano Lett.*, **16**(3), 2047–53.

Raether, H. (1988). *Surface Plasmons.* Springer, Berlin.

Rephaeli, E. and Fan, S. (2009). 'Absorber and emitter for solar thermo-photovoltaic systems to achieve efficiency exceeding the shockley-queisser limit'. *Opt. Express*, **17**(17), 15145–59.

Reza, A, Dignam, M. M., and Hughes, S. (2008). 'Can light be stopped in realistic metamaterials'. *Nature*, **455**(7216), E10–E11.

Ringler, M., Schwemer, A., Wunderlich, M., Nichtl, A., Kürzinger, K., Klar, T. A., and Feldmann, J. (2008). 'Shaping emission spectra of fluorescent molecules with single plasmonic nanoresonators'. *Phys. Rev. Lett.*, **100**, 203002.

Roelkens G., et al. (2010). 'Bridging the gap between nanophotonic waveguide circuits and single mode optical fibers using diffractive grating structures'. *J. Nanosci. Nanotechnol.*, **10**, 1551–62.

Röhrich, R. and Koenderink, A. F. (2021). 'Double moire localized plasmon structured illumination microscopy'. *Nanophotonics*, **10**(3), 1107–21.

Rosenblatt, D., Sharon, A., and Friesem, A. A. (1997). 'Resonant grating waveguide structures'. *IEEE J. Quantum Electron.*, **33**(11), 2038–59.

Rousseau, E., Siria, A., Jourdan, G., Volz, S., Comin, F., Chevrier, J., and Greffet, J.-J. (2006). 'Radiative heat transfer at the nanoscale'. *Nat. Photonics*, **3**(9), 514–17.

Rust, M. J, Bates, M., and Zhuang, X. (2006). 'Sub-diffraction-limit imaging by stochastic optical reconstruction microscopy (STORM)'. *Nature Methods*, **3**(10), 793–6.

Rytov, S. M., Kravtsov, Y. A., and Tatarskii, V. I. (1989). *Principles of statistical radiophysics. 3. Elements of Random Fields.* Springer, New York.

Santhosh, K., Bitton, O., Chuntonov, L., and Haran, G. (2016). 'Vacuum rabi splitting in a plasmonic cavity at the single quantum emitter limit'. *Nat. Commun.*, **7**, 11823.

Sauvan, C., Hugonin, J. P., Maksymov, I. S., and Lalanne, P. (2013). 'Theory of the spontaneous optical emission of nanosize photonic and plasmon resonators'. *Phys. Rev. Lett.*, **110**, 237401.

Sauvan, C., Wu, T., Zarouf, R., Muljarov, E.A. and Lalanne, P. (2022). 'Normalization, orthogonality, and completeness of quasinormal modes of open systems: the case of electromagnetism'. *Opt. Express,* **30**(5), 6846–85.

Savage, K. J., Hawkeye, M. M., Esteban, R., Borisov, A. G., Aizpurua, J., and Baumberg, J. J. (2012). 'Revealing the quantum regime in tunneling plasmonics'. *Nature*, **491**, 574–7.

Scholl, J. A., García-Etxarri, A., Koh, A. L., and Dionne, J. A. (2013). 'Observation of quantum tunneling between two plasmonic nanoparticles'. *Nano Lett.*, **13**(2), 564–9.

Schot, S. H. (1992). 'Eighty years of Sommerfeld's radiation condition'. *Historia Mathematica*, **19**(4), 385–401.

Schuller, J. A., Barnard, E. S., Cai, W., Jun, Y. C., White, J. S., and Brongersma, M. L. (2010). 'Plasmonics for extreme light concentration and manipulation'. *Nat. Mater.*, **9**(3), 193–204.

Scullion, M. G., Di Falco, A., Krauss, T. F. (2011) Slotted photonic crystal cavities with integrated microfluidics for biosensing applications. Biosensors and Bioelectronics **27**(1), 101–105.

Senellart, P., Solomon, G., and White, A. (2017). 'High-performance semiconductor quantum-dot single-photon sources'. *Nat. Nanotech.*, **12**(11), 1026–39.

Shapiro, M. A., Shvets, G., Sirigiri, J. R., and Temkin, R. J. (2006). 'Spatial dispersion in metamaterials with negative dielectric permittivity and its effect on surface waves'. *Opt. Lett.*, **31**(13), 2051–3.

Shchegrov, A. V., Joulain, K., Carminati, R., and Greffet, J.-J. (2000). 'Near-field spectral effects due to electromagnetic surface excitations'. *Phys. Rev. Lett.*, **85**, 1548–51.

Shelby, R. A., Smith, D. R., and Schultz, S. (2001). 'Experimental verification of a negative index of refraction'. *Science*, **292**(5514), 77–9.

Shen, S., Narayanaswamy, A., and Chen, G. (2009). 'Surface phonon polaritons mediated energy transfer between nanoscale gaps'. *Nano Lett.*, **9**(8), 2909–13.

Shrestha, S., Overvig, A. C., Lu, M., Stein, A., and Yu, N. (2018). 'Broadband achromatic dielectric metalenses'. *Light: Science & Applications*, **7**, 85.

Siegman, A. E. (1986). *Lasers.* University Science Books, Sausalito.

Sinibaldi, A., Rizzo, R., Figliozzi, G., Descrovi, E., Danz, N., Munzert, P., Anopchenko, A., and Michelotti, F. (2013). 'A full ellipsometric approach to optical sensing with Bloch surface waves on photonic crystals'. *Opt. Express*, **21**(20), 23331–44.

Sipe, J. E. (1987). 'New Green-function formalism for surface optics'. *J. Opt. Soc. Am. B*, **4**(4), 481–9.

Smit, M. K. and Van Dam, C. (1996). 'PHASAR-based WDM-devices: Principles, design and applications'. *IEEE J. Sel. Top. Quantum Electron.*, **2**(2), 236–50.

Smith, D. R., Padilla, W. J., Vier, D. C., Nemat-Nasser, S. C., and Schultz, S. (2000). 'Composite medium with simultaneously negative permeability and permittivity'. *Phys. Rev. Lett.*, **84**, 4184–7.

Smith, J. B. and Ehrenreich, H. (1982). 'Frequency dependence of the optical relaxation time in metals'. *Phys. Rev. B*, **25**, 923–30.

Snyder, A. W. and Love, J. D. (1983). *Optical Waveguide Theory* (2nd edn). Springer, New York.

Solymar, L. and Shamonina, E. (2009). *Waves in Metamaterials.* Oxford University Press, Oxford.

Sommerfeld, A. N. (1926). 'Propagation of waves in wireless telegraphy'. *Ann. Phys. (Leipzig)*, **81**, 1135–53.

Song, B. S., Noda, S., and Akahane, Y. (2005). 'Ultra-high Q photonic double-heterostructure nanocavity'. *Nat. Mater.*, **4**, 207–10.

Staelin, D. H., Morgenthaler, A. W., and Kong, J. A. (1994). *Electromagnetic Waves.* Prentice Hall International, Englewood Cliffs, New York.

Staude, I., Miroshnichenko, A. E., Decker, M., Fofang, N. T., Liu, S., Gonzales, E., Dominguez, J., Luk, T. S., Neshev, D. N., Brener, I., and Kivshar, Y. (2013). 'Tailoring directional scattering through magnetic and electric resonances in subwavelength silicon nanodisks'. *ACS Nano*, **7**(9), 7824–32.

Stockman, M. I. (2011). 'Nanoplasmonics: The physics behind the applications'. *Physics Today*, **64**, 39–44.

Stone, T. and George, N. (1988). 'Hybrid diffractive-refractive lenses and achromats'. *Appl. Opt.*, **27**(14), 2960–71.

Sze, S. M. (2008). *Semiconductor Devices: Physics and Technology.* Wiley, New York.

Taflove, A., Oskooi, A., and Johnson, S. G. (2013). *Electromagnetic Wave Source Conditions in 'Advances in FDTD Computational Electrodynamics: Photonics and Nanotechnology'.* Artech House Antennas and Propagation Library. Artech House.

Tai, C. (1994). *Dyadic Green Functions in Electromagnetic Theory.* IEEE Press Publication Series. IEEE Press.

Taillaert, D., Bogaerts, W., Bienstman, P., Krauss, T. F., Van Daele, P., Moerman, I., Verstuyft, S., De Mesel, K., and Baets, R. (2002). 'An out-of-plane grating coupler for efficient butt-coupling between compact planar waveguides and single-mode fibers'. *IEEE J. Quantum Electron.*, **38**, 949–55.

Takahara, J., Yamagishi, S., Taki, H., Morimoto, A., and Kobayashi, T. (1997). 'Guiding of a one-dimensional optical beam with nanometer diameter'. *Opt. Lett.*, **22**(7), 475–7.

Tamada, H., Doumuki, T., Yamaguchi, T., and Matsumoto, S. (1997). 'Al wire-grid polarizer using the s-polarization resonance effect at the 0.8-μm-wavelength band'. *Opt. Lett.*, **22**, 419–21.

Teperik, T. V., Nordlander, P., Aizpurua, J., and Borisov, A. G. (2013). 'Quantum effects and non locality in strongly coupled plasmonic nanowire dimers'. *Opt. Express*, **21**, 27306.

Tessier, M. D., Javaux, C., Maksimovic, I., Loriette, V., and Dubertret, B. (2012). Spectroscopy of single CdSe nanoplatelets. *ACS Nano*, **6**(8), 6751–8.

Le Thomas, N., Zhang, H., Jágerská, J., Zabelin, V., and Houdré, R. (2009). Light transport regimes in slow light photonic crystal waveguides, *Phys. Rev. B*, **80**, 125–332.

Ulrich, R. (1967). 'Far-infrared properties of metallic mesh and its complementary structure'. *Infrared Physics*, **7**(1), 37–55.

Ulrich, R. and Tacke, M. (1973). 'Submillimeter waveguiding on periodic metal structure'. *Appl. Phys. Lett.*, **22**(5), 251–3.

Valentine, J., Zhang, S., Zentgraf, T., Ulin-Avila, E., Genov, D., Bartal, G., and Zhang, X. (2008). 'hree-dimensional optical metamaterial with a negative refractive index'. *Nature*, **455**, 376–9.

Van Bladel, J. (1991). *Singular Electromagnetic Fields and Sources.* Wiley-IEEE Press, New York.

van de Linde, S., Löschberger, A., Klein, T., Heidbreder, M., Wolter, S., Heilemann, M., and Sauer, M. (2011). 'Direct stochastic optical reconstruction microscopy with standard fluorescent probes'. *Nature Protocols*, **6**, 991–1009.

Vassallo, C. (1991). *Optical Waveguide Concepts.* Elsevier, Amsterdam.

Vassant, S., Moldovan Doyen, I., Marquier, F., Pardo, F., Gennser, U., Cavanna, A., Pelouard, J. L., and Greffet, J.-J. (2013). 'Electrical Modulation of Emissivity'. *Appl. Phys. Lett.*, **102**(8), 081125.

Veselago, V. G. (1968). 'The electrodynamics of substances with simultaneously negative values of ε and μ'. *Soviet. Phys. Usp*, **10**, 509–14.

Vial, A., Grimault, A.-S., Macías, D., Barchiesi, D., and de la Chapelle, M. L. (2005). 'Improved analytical fit of gold dispersion: Application to the modeling of extinction spectra with a finite-difference time-domain method'. *Phys. Rev. B*, **71**, 085416.

Villeneuve, P. R. and Piché, M. (1994). 'Photonic bandgaps in periodic dielectric structures'. *Progress in Quantum Electronics*, **18**(2), 153–200.

Vincent, P. and Nevière, M. (1979). 'Corrugated dielectric waveguides: A numerical study of the second-order stop bands'. *Appl. Phys.*, **20**, 345–51.

Voisin, C., Del Fatti, N., Christofilos, D., and Vallée, F. (2001). 'Ultrafast electron dynamics and optical nonlinearities in metal nanoparticles'. *J. Phys. Chem. B.*, **105**(12), 2264–80.

Volokitin, A. I. and Persson, B. N. J. (2007). 'Near-field radiative heat transfer and noncontact friction'. *Rev. Mod. Phys.*, **79**, 1291–329.

Vorobyev, A. Y. and Guo, C. (2008). 'Colorizing metals with femtosecond laser pulses'. *Appl. Phys. Lett.*, **92**(4), 041914.

Walford, J. N., Porto, J.-A., Carminati, R., and Greffet, J.-J. (2002). 'Theory of near-field magneto-optical imaging'. *J. Opt. Soc. Am. A*, **19**(3), 572–83.

Wang, Y., Chen, Q., Yang, W., Ji, Z., Jin, L., Ma, X., Song, Q., Boltasseva, A., Han, J., Shalaev, V. M., and Xiao, S. (2021). 'High-efficiency broadband achromatic metalens for near-IR biological imaging window'. *Nat Commun*, **12**, 55–60.

Wang, F. and Shen, Y. R. (2006). 'General properties of local plasmons in metal nanostructures'. *Phys. Rev. Lett.*, **97**, 206806.

Wang, J. J. H. (1990). 'Generalised moment methods in electromagnetics'. *IEE Proceedings H (Microwaves, Antennas and Propagation)*, **137**, 127–32.

Wang, S. S. and Magnusson, R. (1994). 'Design of waveguide-grating filters with symmetrical line shapes and low sidebands'. *Opt. Lett.*, **19**(12), 919–21.

Weisbuch, C., Nishioka, M., Ishikawa, A., and Arakawa, Y. (1992). 'Observation of the coupled exciton-photon mode splitting in a semiconductor quantum microcavity'. *Phys. Rev. Lett.*, **69**, 3314–17.

Wijnands, F., Pendry, J. B., Garcia-Vidal, F. J., Bell, P. M., Roberts, P. J., and Martiń-Moreno, L. (1997). 'Green's functions for Maxwell's equations: Application to spontaneous emission'. *Opt. Quantum Electron.*, **29**(2), 199–216.

Wojszvzyk, L., Nguyen, A., Coutrot, A.-L., Zhang, C., Vest, B., and Greffet, J. J. (2021). 'An incandescent metasurface for quasimonochromatic polarized mid-wave infrared emission modulated beyond 10 MHz'. *Nat. Commun.*, **12**, 1492.

Wood, R. W. (1902). 'On a remarkable case of uneven distribution of light in a diffraction grating spectrum'. *The London, Edinburgh, and Dublin Philosophical Magazine and Journal of Science*, **4**(21), 396–402.

Wu, T., Gurioli, M., and Lalanne, P. (2021). 'Nanoscale light confinement: the q's and v's'. *ACS Photonics*, **8**(6), 1522–38.

Wylie, J. M. and Sipe, J. E. (1984). 'Quantum electrodynamics near an interface'. *Phys. Rev. A*, **30**, 1185–93.

Yablonovitch, E. (1987). 'Inhibited spontaneous emission in solid-state physics and electronics'. *Phys. Rev. Lett.*, **58**, 2059–62.

Yaghjian, A. D. (1980). 'Electric dyadic Green's functions in the source region'. *Proc. IEEE*, **68**(2), 248–63.

Yariv, A. (1989). *Quantum Electronics* (3rd edn). Wiley, New York.

Yariv, A. (1990). *Optical Electronics.* Oxford Series in Electrical and Computer Engineering. Oxford University Press, Oxford.

Yariv, A. and Yeh, P. (1984). *Optical Waves in Crystals: Propagation and Control of Laser Radiation.* Wiley, New York.

Yeh, P. (1988). *Optical Waves in Layered Media.* Wiley, New York.

Yokoyama, H. and Ujihara, K. (1995). *Spontaneous Emission and Laser Oscillation in Microcavities.* Laser & Optical Science & Technology. Taylor & Francis.

Yu, N., Genevet, P., Kats, M. A., Aieta, F., Tetienne, J.-P., Capasso, F., and Gaburro, Z. (2011). 'Light propagation with phase discontinuities: Generalized laws of reflection and refraction'. *Science*, **334**(6054), 333–7.

Zangwill, A. (2013). *Modern Electrodynamics.* Cambridge University Press, Cambridge.

Zayats, A. V., Smolyaninov, I. I., and Maradudin, A. A. (2005). 'Nano-optics of surface plasmon polaritons'. *Phys. Rep.*, **408**(3), 131–314.

Zheng, G., Cui, X., and Yang, C. (2010). 'Surface-wave-enabled darkfield aperture for background suppression during weak signal detection'. *Proc. Natl. Acad. Sci.*, **107**(20) 9043–8.

Zhizhin, G. N., Vinogradov, E. A., Moskalova, M. A., and Yakovlev, V. A. (1982). 'Applications of surface polaritons for vibrational spectroscopic studies of thin and very thin films'. *Appl. Spectroscop. Rev.*, **18**(2), 171–263.

Ziman, J. M. (1972). *Principles of the Theory of Solids* (2nd edn). Cambridge University Press, Cambridge.

Zory, P. S. (1993). *Quantum Well Lasers.* Academic Press, Boston, MA.

Index